New Symmetry Principles in Quantum Field Theory

NATO ASI Series

Advanced Science Institutes Series

A series presenting the results of activities sponsored by the NATO Science Committee, which aims at the dissemination of advanced scientific and technological knowledge, with a view to strengthening links between scientific communities.

The series is published by an international board of publishers in conjunction with the NATO Scientific Affairs Division

A	**Life Sciences**	Plenum Publishing Corporation
B	**Physics**	New York and London
C	**Mathematical and Physical Sciences**	Kluwer Academic Publishers
D	**Behavioral and Social Sciences**	Dordrecht, Boston, and London
E	**Applied Sciences**	
F	**Computer and Systems Sciences**	Springer-Verlag
G	**Ecological Sciences**	Berlin, Heidelberg, New York, London,
H	**Cell Biology**	Paris, Tokyo, Hong Kong, and Barcelona
I	**Global Environmental Change**	

Recent Volumes in this Series

Volume 290—Phase Transitions in Liquid Crystals
edited by S. Martellucci and A. N. Chester

Volume 291—Proton Transfer in Hydrogen-Bonded Systems
edited by T. Bountis

Volume 292—Microscopic Simulations of Complex Hydrodynamic Phenomena
edited by Michel Mareschal and Brad Lee Holian

Volume 293—Methods in Computational Molecular Physics
edited by Stephen Wilson and Geerd H. F. Diercksen

Volume 294—Single Charge Tunneling: Coulomb Blockade Phenomena in Nanostructures
edited by Hermann Grabert and Michel H. Devoret

Volume 295—New Symmetry Principles in Quantum Field Theory
edited by J. Fröhlich, G.'t Hooft, A. Jaffe, G. Mack, P. K. Mitter, and R. Stora

Volume 296—Recombination of Atomic Ions
edited by W. G. Graham, W. Fritsch, Y. Hahn, and J. A. Tanis

Volume 297—Ordered and Turbulent Patterns in Taylor-Couette Flow
edited by C. David Andereck and F. Hayot

Series B: Physics

New Symmetry Principles in Quantum Field Theory

Edited by

J. Fröhlich

ETH-Zürich
Zürich, Switzerland

G. 't Hooft

Institute for Theoretical Physics
Utrecht, The Netherlands

A. Jaffe

Harvard University
Cambridge, Massachusetts

G. Mack

II Institut für Theoretische Physik
Hamburg, Germany

P. K. Mitter

University of Paris VI
Paris, France

and

R. Stora

Laboratory of Particle Physics, Annecy
Annecy-le-Vieux, France

Springer Science+Business Media, LLC

Proceedings of a NATO Advanced Study Institute on
New Symmetry Principles in Quantum Field Theory,
held July 16–27, 1991,
in Cargèse, France

NATO-PCO-DATA BASE

The electronic index to the NATO ASI Series provides full bibliographical references (with key-words and/or abstracts) to more than 30,000 contributions from international scientists published in all sections of the NATO ASI Series. Access to the NATO-PCO-DATA BASE is possible in two ways:

—via online FILE 128 (NATO-PCO-DATA BASE) hosted by ESRIN, Via Galileo Galilei, I-00044 Frascati, Italy.

—via CD-ROM "NATO-PCO-DATA BASE" with user-friendly retrieval software in English, French, and German (© WTV GmbH and DATAWARE Technologies, Inc. 1989)

The CD-ROM can be ordered through any member of the Board of Publishers or through NATO-PCO, Overijse, Belgium.

Library of Congress Cataloging-in-Publication Data

New symmetry principles in quantum field theory / edited by J. Fröhlich ... [et. al].
p. cm. -- (Nato ASI series. Series B, Physics ; v. 295)
"Published in cooperation with NATO Scientific Affairs Division."
"Proceedings of a NATO Advanced Study Institute on New Symmetry Principles in Quantum Field Theory, held July 16-27, 1991, in Cargèse, France"--Copr. p.
Includes bibliographical references and index.
ISBN 978-1-4613-6538-9 ISBN 978-1-4615-3472-3 (eBook)
DOI 10.1007/978-1-4615-3472-3
1. Quantum field theory--Congresses. 2. Symmetry (Physics)--Congresses. I. Fröhlich, Jürg. II. North Atlantic Treaty Organization. Scientific Affairs Division. III. Series.
QC174.45.A1N477 1992
530.1'43--dc20 92-18441
CIP

Additional material to this book can be downloaded from http://extra.springer.com.

ISBN 978-1-4613-6538-9

Originally published by Plenum Press in 1992

PREFACE

Soon after the discovery of quantum mechanics, group theoretical methods were used extensively in order to exploit rotational symmetry and classify atomic spectra. And until recently it was thought that symmetries in quantum mechanics should be groups. But it is not so. There are more general algebras, equipped with suitable structure, which admit a perfectly conventional interpretation as a symmetry of a quantum mechanical system. In any case, a "trivial representation" of the algebra is defined, and a tensor product of representations. But in contrast with groups, this tensor product needs to be neither commutative nor associative. Quantum groups are special cases, in which associativity is preserved.

The exploitation of such "Quantum Symmetries" was a central theme at the Advanced Study Institute.

Introductory lectures were presented to familiarize the participants with the algebras which can appear as symmetries and with their properties. Some models of local field theories were discussed in detail which have some such symmetries, in particular conformal field theories and their perturbations. Lattice models provide many examples of quantum theories with quantum symmetries. They were also covered at the school. Finally, the symmetries which are the cause of the solubility of integrable models are also quantum symmetries of this kind. Some such models and their nonlocal conserved currents were discussed.

There are several different approaches by which local field theory models with quantum symmetry can be treated. One can start from canonical quantization. One can also start from a lattice approximation and take the continuum limit. Finally there is the algebraic field theory approach. All three approaches were covered at the school.

Quantum symmetries in low dimensional local field theories come together with braid group statistics. The occurrence of braid group statistics is experimentally established through the phenomena observed in the fractional quantum Hall effect. Braid group statistics and the quantum Hall effect were discussed in lectures at the school.

There is another "new symmetry", which is new only in the quantum physics, but old in classical physics. This is the general covariance in the theory of gravity. The study of matrix models and of topological field theories has offered new insights into the problems of quantum gravity. This was discussed in detail in lectures at this school. There are also rich interrelations with integrable systems and with models of conformal field theories which are now used to describe black holes. A special set of lectures was devoted to black hole physics.

The ultimate solution to the quantum gravity problem may well turn out to be quantum space time, and noncommutative differential geometry. Lectures on non-commutative differential geometry and a geometric version of the standard model of elementary particle physics were presented at the school.

In addition to the main lectures there was a limited number of seminars presented by participants of the school.

The Editors

March, 1992

CONTENTS

LECTURERS

Quantum Symmetries in 2D Massive Field Theories 1
D. Bernard

On Knot and Manifold Invariants 37
R. Bott

The Metric Aspect of Noncommutative Geometry 53
A. Connes and J. Lott

Intersection Theory, Integrable Hierarchies and
Topological Field Theory 95
R. Dijkgraaf

Quantum Symmetry in Conformal Field Theory
by Hamiltonian Methods 159
L.D. Faddeev

Observables, Superselection Sectors and Gauge Groups 177
C. Fredenhagen

Incompressible Quantum Fluids, Gauge-Invariance,
and Current Algebra 195
J. Frölich and U. Studer

Non-Compact WZW Conformal Field Theories 247
K. Gawedzki

S-Matrix Theory for Black Holes 275
G. 't Hooft

Non-Commutative Geometry and Mathematical Physics 295
A. Jaffe

Whitham Theory For Integrable Systems and
Topological Quantum Field Theories 309
I. Krichever

Quantum Symmetry in Quantum Theory 329
G. Mack and V. Schomerus

Quantum Groups in Lattice Models 355
V. Pasquier

Lagrangian Conformal Models 381
R. Stora

Semi-Classical Liouville Theory, Complex Geometry of Moduli Spaces, and Uniformization of Riemann Surfaces 383
L. Takhtajan

SEMINAR SPEAKERS

Integrability Properties of the Collective String Field Theory 407
J. Avan

BRST Analysis of Physical States for 2D (Super) Gravity Coupled to (Super) Conformal Matter 413
P. Bouwknegt, J. McCarthy, and K. Pilch

Correlation Functions of Local Operators in 2D Gravity Coupled to Minimal Matter 423
Vl. Dotsenko

W-Algebras and Langlands-Drinfeld Correspondence 433
E. Frenkel

Non-Tannakian Categories in Quantum Field Theory 449
T. Kerler

Extra States in $c < 1$ String Theory 483
S. Mukhi

W(sl(n)): Existence, Cartan Basis and Infinite Abelian Subalgebras 493
M. Niedermaier

Aspects of Quantizing Lorentz Symmetry 505
A. Schirrmacher

Fock Space Representations of $A_1^{(1)}$ and Topological Representations of $U_q(sl_2)$ 513
G. Felder and C. Wieczerkowski

Index 523

QUANTUM SYMMETRIES IN 2D MASSIVE FIELD THEORIES

Denis Bernard

Service de Physique Théorique de Saclay *
F-91191 Gif-sur-Yvette, France

ABSTRACT

We review various aspects of (infinite) quantum group symmetries in 2D massive quantum field theories. We discuss how these symmetries can be used to exactly solve the integrable models. A possible way for generalizing to three dimensions is shortly described.

INTRODUCTION

Symmetries of the S-matrices of 4D quantum field theories are subject to the severe constraints of the Coleman-Mandula theorem [1]. In general, these possible symmetries do not allow for non-perturbative solutions of the theories. In two dimensions, this theorem breaks down and there is room for richer symmetries. The aim of the lecture is to describe new symmetries of 2D massive QFT, known as quantum group symmetries. This paper is mainly a review of published papers completed by few remarks and comments.

There at least two motivations for studying quantum symmetries in 2D quantum field theories:

(i) The study of the possible symmetries of 2D QFT. The quantum group symmetries we will analyse in these lectures are characterized by the fact that, unlike standard Lie algebra symmetries, they do not act additively, and by the fact that they are generated by non-local currents having in general non-integer Lorentz spins. They thus provide non-Abelian extension of the Lorentz group.

---

* Laboratoire de la Direction des sciences de la matière du Commissariat à l'énergie atomique.

New Symmetry Principles in Quantum Field Theory, Edited by
J. Fröhlich et al., Plenum Press, New York, 1992

(ii) An algebraic formulation of 2D QFT and exact solutions. Our (desesperate ?) goal is to formulate an algebraic approach to 2D QFT, based on their symmetries (local and non-local), which could offer a way to solve the integrable two-dimensional quantum field theories from symmetry data, in analogy with the approach used in conformal field theories [2].

It is of some interest to compare the approach which as been used recently in CFT and in 2D integrable models. a) Rational conformal field theories are invariant under chiral vertex operator algebras, which could be local, e.g. the Virasoro or affine algebras, or non-local, e.g. the parafermionic algebras. The conformal field theories are reformulated as representation theories of the chiral algebras: Hilbert spaces of the CFT's are direct sums of representations of the chiral algebras; conformal primary fields are intertwiners for the chiral algebras, etc... The chiral algebras are non-abelian and this is, for a large part, at the origin of the exact solvabality of the CFT's. Being completely integrable, the CFT's also possess infinitely many local integrals of motion in involution. However, almost none of these integrals of motion are actually used to solve the conformal models. (It is even difficult to express them in terms the generators of the chiral algebras.) b) In contrast, a way of studying integrable models [3] consists in extracting local integrals of motion which are in involution. These integrals of motion thus form an abelian algebra. There existence ensures the integrability of the theory and the factorization of the S-matrix. In general, they do not provide enough informations for solving the theory, e.g. for determining the S-matrix.

Thus, almost none of the techniques used in one of these fields is used in the other. However, besides these local integrals of motion, the 2D integrable models and the conformal models also possess non-local integrals of motion. These non-local conserved charges are the generators of non-abelian algebras known as the quantum symmetry algebras of the models. It is hoped that these new symmetry algebras will allow us to define a framework which could be apply simultaneously to the conformal field theories and to the massive integrable models.

To characterize the massive quantum field theories uniquely by their symmetry algebras requires :
(I) that the asymptotic particles form multiplets of the symmetry algebras and that the invariance for the S-matrix determine it uniquely. As we will see in the course of this lecture, because quantum group symmetries in 2D do not commute with the Lorentz group, they provide algebraic relations on the S-matrices.
(II) that all the fields of the QFT can be gathered into field multiplets transforming covariantly under the symmetry algebras and that the intertwining properties for

the field multiplets determine them uniquely (in analogy with the minimal assumption in rational conformal field theories). This amounts to demand that the Ward identities have unique solutions. As we will describe, the components of the field multiplets, the descendents and the highest vector fields, are related through the Ward identities, once more in complete analogy with conformal field theories.

The symmetry algebras having these requirements could be called complete symmetry algebras. For a given model there could be more than one complete symmetry algebra. The problem of solving the integrable massive models reduce to the problem of finding a complete symmetry algebra. As we already said, the local integrals of motion which are involution do not form (in general) a complete symmetry algebra. The quantum symmetry we are going to describe provide in general more informations than these abelian integrals of motion. To our knownledge, it is not known if they form or not a complete symmetry algebra. However, it is tempting to conjecture that the algebra generated by the local integrals of motion together with the generators of the quantum symmetry form a complete symmetry algebra.

An example: $\widehat{sl_q(2)}$ -symmetry in the Sine-Gordon models. The quantum sine-Gordon theory is described by the Euclidean action

$$S = \frac{1}{4\pi} \int d^2 z \left[\partial_z \Phi \partial_{\bar{z}} \Phi + \lambda : \cos\left(\widehat{\beta}\Phi\right) : \right] . \tag{1}$$

The parameter $\widehat{\beta}$ is a coupling constant; it is related to the conventially normalized coupling by $\widehat{\beta} = \beta/\sqrt{4\pi}$. For $\widehat{\beta} \leq \sqrt{2}$ the action can be renormalized by normal-ordering the $\cos\left(\widehat{\beta}\Phi\right)$ interaction and absorbing the infinities into λ; the coupling constant $\widehat{\beta}$ is thereby unrenormalized [4]. The sine-Gordon theory has a well known topological current: $J^\mu(x,t) = \frac{\widehat{\beta}}{2\pi}\, \epsilon^{\mu\nu}\partial_\nu \Phi(x,t)$ where $\epsilon^{\mu\nu} = -\epsilon^{\nu\mu}$. The topological charge is:

$$\mathcal{T} = \frac{\widehat{\beta}}{2\pi} \int_{-\infty}^{+\infty} dx\ \partial_x \Phi = \frac{\widehat{\beta}}{2\pi} \Big(\Phi(x=\infty) - \Phi(x=-\infty) \Big). \tag{2}$$

The topological solitons that correspond to single particles in the quantum theory are described classically by field configurations with $\mathcal{T} = \pm 1$. These solitons are kinks that connect two neighboring vacua in the $\cos(\widehat{\beta}\Phi)$ potential.

The sine-Gordon model possesses infinitely many local integrals of motion with odd Lorentz spins, we denote them by $\mathcal{J}_n$. Besides those, the sine-Gordon model also admits four non-local conserved currents [5]:

$$\partial_\mu\ J^\pm_\mu(x,t) = \partial_\mu \overline{J}^\pm_\mu(x,t) = 0. \tag{3}$$

The Lorentz spin s of the currents $J_\mu^\pm$ $\left(\overline{J}_\mu^\pm\right)$ are $s = \frac{2}{\beta^2}$ $\left(-\frac{2}{\beta^2}\right)$. ¿From these conserved currents we define four conserved charges, $Q_\pm$ and $\overline{Q}_\pm$, respectively associated to the currents $J_\mu^\pm(x,t)$ and $\overline{J}_\mu^\pm(x,t)$. The Lorentz spins of the conserved charges are :

$$\text{spin}\,(Q_\pm) \;=\; -\text{spin}\,(\overline{Q}_\pm) \;=\; \frac{2-\widehat{\beta}^2}{\widehat{\beta}^2} = \frac{8\pi-\beta^2}{\beta^2}. \tag{4}$$

The conserved currents whose exact expressions are given in ref. [5] are Mandelstam like vertex operators [6] and are thus non-local.

The algebra of the non-local charges is :

$$Q_\pm\,\overline{Q}_\pm - q^2\,\overline{Q}_\pm\,Q_\pm \;=\; 0 \tag{5a}$$

$$Q_\pm\,\overline{Q}_\mp - q^{-2}\,\overline{Q}_\mp\,Q_\pm \;=\; a\Big(1-q^{\pm 2\mathcal{T}}\Big) \tag{5b}$$

$$\Big[\,\mathcal{T}\,,\,Q_\pm\,\Big] \;=\; \pm 2\,Q_\pm \tag{5c}$$

$$\Big[\,\mathcal{T}\,,\,\overline{Q}_\pm\,\Big] \;=\; \pm 2\,\overline{Q}_\pm, \tag{5d}$$

where $q = \exp(-2\pi i/\widehat{\beta}^2)$. and a some constant. The algebra (5) is a known infinite dimensional algebra, namely the q-deformation of the $sl(2)$ affine Kac-Moody algebra, denoted $\widehat{sl_q(2)}$, with zero center [7] [8]. Only the Serre relations for $\widehat{sl_q(2)}$ are missing in (5).

The non-local charges (3) provide relevent information; for example, the S-matrix of the Sine-Gordon solitons [9] can be deduced from this $\widehat{sl_q(2)}$ symmetry plus its unitary and crossing symmetry property. However, they probably do not form a complete symmetry algebra because the local integrals of motion do not seem to be generated by them. To prove the conjecture that the local conserved charges $\mathcal{J}_n$ and these non-local charges generate a complete symmetry algebra for the sine-Gordon models will be very illuminating.

1. QUANTUM SYMMETRIES IN 2D LATTICE FIELD THEORY

We consider vertex models, i.e. models of two-dimensional statistical mechanics in which the discrete spin variables live on the midpoints of the links of a square lattice, and the Boltzmann weights are associated to the vertices of the lattice [10]. The Boltzmann weight of a given vertex depends on the spin variables $\sigma_1, \ldots, \sigma_4$ at the four sites surrounding the vertex, and is denoted by $R_{\sigma_1\sigma_2}^{\sigma_3\sigma_4}$. It is useful to view R as an operator $V \otimes V \to V \otimes V$, where V is the vector space spanned by a set of basis vectors e_σ labeled by the possible values of the spin variable:

$$Re_{\sigma_1} \otimes e_{\sigma_2} = R_{\sigma_1\sigma_2}^{\sigma_3\sigma_4}\; e_{\sigma_3} \otimes e_{\sigma_4}. \tag{1.1}$$

Partition function and correlation functions are defined as follows. Consider the system in a finite square box of size $N \times N$. Then the square lattice $\Lambda = \mathbb{Z}^2/N^2\mathbb{Z}^2$ contains points with integer coordinates (x,t) called space and time. The spin variables live on the lattice Λ' of points of the form $(x+\frac{1}{2},t)$ and $(x,t+\frac{1}{2})$ with x, t integers modulo N. For each $i \in \Lambda'$ introduce a copy V_i of the space V. For $(x,t) \in \Lambda$ define $R(x,t)$ be the matrix R mapping $V_{(x-\frac{1}{2},t)} \otimes V_{(x,t-\frac{1}{2})}$ to $V_{(x+\frac{1}{2},t)} \otimes V_{(x,t+\frac{1}{2})}$. The matrix of Boltzmann weights,

$$B = \bigotimes_{(x,t)\in\Lambda} R(x,t), \tag{1.2}$$

is an operator from $\bigotimes_{i\in\Lambda'} V_i$ to $\bigotimes_{j\in\Lambda'} V_j$.

The partition function is $Z_N = \operatorname{tr} B$.

For any operator $\mathcal{O} \in \operatorname{End}V$ define the insertion $\mathcal{O}(j)$ of $\mathcal{O}$ at the point $j \in \Lambda'$ to be the operator $1 \otimes \cdots \otimes 1 \otimes \mathcal{O} \otimes 1 \otimes \cdots \otimes 1$ acting on V_j in the tensor product $\bigotimes_{i\in\Lambda'} V_i$. The correlation functions of operator insertions are defined as

$$\langle \mathcal{O}_1(j_1) \cdots \mathcal{O}_n(j_n)\rangle_N = \frac{1}{Z_N} \operatorname{tr}\Big(\prod_{k=1}^{n} \mathcal{O}_k(j_k)\, B \Big). \tag{1.3}$$

Classical examples are $\mathcal{O}e_\sigma = \sigma e_\sigma$ and $\mathcal{O}e_\sigma = \delta_{\sigma\overline{\sigma}} e_\sigma$ for some $\overline{\sigma}$. Their correlation functions are the usual spin correlation functions and the joint probabilities that the spin σ_{j_k} assume given values.

An alternative formalism is the transfer matrix formulation. In this formalism, one assigns to each $x = 1, \cdots, N$ a copy V_x of V. The transfer matrix T is:

$$T = \operatorname{tr}_0\Big(R_{0N} \cdots R_{02}\, R_{01} \Big) \tag{1.4}$$

with R_{nm} the matrix R acting on V_n and V_m in $V_0 \otimes V_1 \otimes \cdots \otimes V_N$, and the trace is over V_0.

For $\mathcal{O} \in \operatorname{End}V$ and $x \in \{1, \cdots, N\}$ define $\mathcal{O}(x)$ as $\mathcal{O}$ acting on V_x in $V_1 \otimes \cdots \otimes V_N$ and $\mathcal{O}(x-\frac{1}{2})$ as $\mathcal{O}(x-\frac{1}{2}) = T^{-1}\operatorname{tr}_0\Big(R_{0N} \cdots R_{0x}\, \mathcal{O}_0\, R_{0,x-1} \cdots R_{01}\Big)$. Heisenberg fields $\mathcal{O}(j)$, $j \in \Lambda'$ are defined as:

$$\begin{aligned} \mathcal{O}(x-\frac{1}{2},t) &= T^{-t}\, \mathcal{O}(x-\frac{1}{2})\, T^t \\ \mathcal{O}(x,t-\frac{1}{2}) &= T^{-t}\, \mathcal{O}(x)\, T^t \end{aligned} \tag{1.5}$$

The partition function in the transfer matrix formalism is $Z = \operatorname{tr}T^N$, and if the time coordinates of $j_1, \cdots, j_n$ are ordered (with smaller times on the right of larger times) the correlation function $\langle \mathcal{O}_1(j_1) \cdots \mathcal{O}_n(j_n)\rangle_N$ defined above coincides with

$$\langle \mathcal{O}_1(j_1) \cdots \mathcal{O}_n(j_n)\rangle_N = \frac{1}{Z_N} \operatorname{tr}\Big(\mathcal{O}_1(j_1) \cdots \mathcal{O}_n(j_n)\, T^N \Big) \tag{1.6}$$

We have defined point-like operator insertions in two different formalisms. It is sometimes useful to define operator insertions associated to some finite set of neighboring points in Λ' as linear combination of products of point-like insertions at the points of the set. This is the lattice analogue of the operator product expansion of field theory.

1a) Quantum symmetries and conserved currents.

(i) Lie algebra symmetry. Suppose that the Boltzmann weights are invariant under some Lie algebra $\mathcal{G}$. This means that V carries a representation of $\mathcal{G}$ and for each generator T_a of $\mathcal{G}$ in that representation we have

$$R(T_a \otimes 1 + 1 \otimes T_a) = (T_a \otimes 1 + 1 \otimes T_a)R \tag{1.7}$$

i.e. R is an intertwiner. It is useful to represent this equation graphically. If T_a is represented by a little cross, we have

(1.8)

Introduce now a local current $J^\mu(x,t;X)$, linear in $X \in \mathcal{G}$, for each vertex (x,t) of the lattice. The components $J^t(x,t;X)$, $J^x(x,t;X)$ are defined by the insertion of the matrix $X = \sum_a X_a T_a$ at the site $(x, t-\frac{1}{2})$ or the site $(x-\frac{1}{2}, t)$, respectively. Graphically,

$J^t(x,t;X) =$ 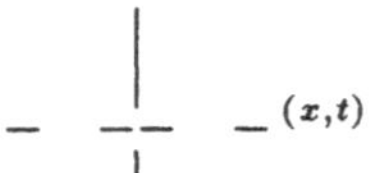

(1.9a)

$J^x(x,t;X) =$ 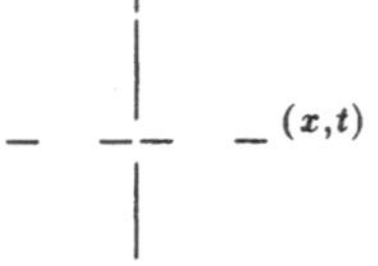

(1.9b)

Then, in any correlation function (with no insertion of other fields at the sites surrounding (x,t)), (1.7) reads

$$J^t(x,t+1;X) - J^t(x,t;X) + J^x(x+1,t;X) - J^x(x,t;X) = 0 \tag{1.10}$$

which is the lattice version of the continuity equation $\partial_\mu J^\mu = 0$. As in the continuum, this equation implies the conservation of the charge $Q(X) = \sum_x J^t(x,t;X)$.

As it is obvious from the pictures (1.9), the operators $J^\mu(x;X)$ are local operators: they satisfy equal-time commutation relations:

$$J^\mu(x;X)\; J^\nu(y;Y) \;=\; J^\nu(y;Y)\; J^\mu(x;X) \quad ; \quad \forall\, x \neq y \tag{1.11}$$

for all $X, Y \in \mathcal{G}$.

(ii) Quantum invariance. We now generalize [11] the preceding construction to an invariance under a Hopf algebra. Recall that a Hopf algebra A is an algebra with unit 1 and associative product $m : A \otimes A \rightarrow A$, equipped with a coproduct $\Delta : A \rightarrow A \otimes A$, a counit $\epsilon : A \rightarrow \mathbb{C}$, and an antipode $S : A \rightarrow A$ so that: (i) Δ, ϵ are algebra homomorphisms, S is an algebra antihomomorphism; (ii) $(1 \otimes \Delta)\Delta(X) = (\Delta \otimes 1)\Delta(X)$; (iii) $(1 \otimes \epsilon)\Delta(X) = (\epsilon \otimes 1)\Delta(X) = X$; (iv) $m(1 \otimes S)\Delta(X) = m(S \otimes 1)\Delta(X) = \epsilon(X)1$, for all $X \in A$. The Hopf algebra A we will consider are those generated by elements T_a, Θ_a^b, $\widehat{\Theta}_a^b$ with, among the relations, $\Theta_a^c \widehat{\Theta}_c^b = \widehat{\Theta}_a^c \Theta_c^b = \delta_a^b$. We also assume that the comultiplication in A is defined by:

$$
\begin{aligned}
\Delta(T_a) &= T_a \otimes 1 + \Theta_a^b \otimes T_b \\
\Delta(\Theta_a^b) &= \Theta_a^c \otimes \Theta_c^b \\
\Delta(\widehat{\Theta}_b^a) &= \widehat{\Theta}_b^c \otimes \widehat{\Theta}_c^a
\end{aligned}
\tag{1.12}
$$

The definition of the counit and the antipode in A are found from the Hopf algebra axioms. The motivation for introducing these algebras will be given latter. Lie superalgebras provide an example of such algebras.

The correct generalization of the invariance eq. (1.7) is $R\,\Delta(X) = \sigma \circ \Delta(X)\,R$, with $\sigma X \otimes Y = Y \otimes X$. Explicitly,

$$R\,(T_a \otimes 1 + \Theta_a^b \otimes T_b) = (1 \otimes T_a + T_b \otimes \Theta_a^b)\,R \tag{1.13a}$$

$$R\,\Theta_a^c \otimes \Theta_c^b = \Theta_c^b \otimes \Theta_a^c\,R \tag{1.13b}$$

These equations have a graphical interpretation. The generators T_a, Θ_a^b, $\widehat{\Theta}_a^b$ are conveniently represented in terms of crosses and oriented wavy lines:

$$T_a = {}^a \times \quad ; \quad \Theta_a^b = {}^a \qquad {}^b \quad ; \quad \widehat{\Theta}_a^b = {}^b \qquad {}^a$$

The graphical representation of $(1.13a)$ is then

with the convention that where pieces of wavy lines join an implicit contraction of indices is understood. The currents $J^\mu(x,t;X)$, $X = \sum_a X_a T_a$, are then constructed as for parafermionic currents, namely with a disorder line (the wavy line) attached:

$$J^t_a(x,t) = J^t(x,t;T_a) =$$ 

(1.14a)

$$J^x_a(x,t) = J^x(x,t;T_a) =$$ 

(1.14b)

The disorder line ends at some specified point on the boundary of the lattice. The identity (1.13b) implies that the disorder line may be deformed (away from insertions of observables) without changing the value of correlation functions. It behaves as the holonomy of a flat connection, just as for ordinary disorder fields [12]. Equation (1.13a) implies the continuity equation (1.10) for non-local currents.

In the operator formalism, the time component of the current is an operator (in the Schrödinger picture) acting on the finite volume Hilbert space $V \otimes \cdots \otimes V$ (N factors) as

$$J^t_a(x) = J^t(x;T_a) = \Theta^{a_1}_a \otimes \Theta^{a_2}_{a_1} \otimes \cdots \otimes \Theta^{a_{x-1}}_{a_{x-2}} \otimes T_{a_{x-1}} \otimes 1 \otimes \cdots \otimes 1. \qquad (1.15)$$

The space component has a more cumbersome operator representation which we omit.

(iii) The braiding relations. By construction the currents (1.14) are non-local. They satisfy braided equal-time commutation relations. These braiding relations arise due to the topological obstructions that one encounters when trying to move the wavy string attached to the currents through a point on which a field is located. In order to write simple closed formula for the braiding relations we now assume that we have completed the set of generators, T_a, Θ^b_a, $\widehat{\Theta}^b_a$ such that they close under the adjoint action. This implies that there exists a c-number matrix R^{bd}_{ac} such that:

$$\begin{aligned} \Theta^n_a \, T_c \, \widehat{\Theta}^b_n &= R^{bd}_{ac} \, T_d \\ R^{ab}_{nm} \, \Theta^n_c \, \Theta^m_d &= \Theta^b_m \, \Theta^a_n \, R^{nm}_{cd} \end{aligned} \qquad (1.16)$$

Then, a simple computation shows that:

$$J^\mu_a(x) \, J^\nu_b(y) = R^{cd}_{ab} \, J^\nu_d(y) J^\mu_c(x), \qquad \text{for } x > y \qquad (1.17)$$

(iv) Global symmetry algebra. The algebra A acts on the Hilbert space $V^{\otimes N}$ by the coproduct Δ_N, defined recursively by $\Delta_2 = \Delta$, $\Delta_{n+1} = \Delta_n(1 \otimes \Delta)$. For generators, we have the formulae

$$
\begin{aligned}
\Delta_N(\Theta_a^b) &= \Theta_a^{a_1} \otimes \Theta_{a_1}^{a_2} \otimes \cdots \otimes \Theta_{a_{N-1}}^b \\
\Delta_N(\widehat{\Theta}_a^b) &= \widehat{\Theta}_{a_1}^b \otimes \widehat{\Theta}_{a_2}^{a_1} \otimes \cdots \otimes \widehat{\Theta}_a^{a_{N-1}} \\
\Delta_N(T_a) &= \sum_{x=1}^{N} \Theta_a^{a_1} \otimes \Theta_{a_1}^{a_2} \otimes \cdots \otimes \Theta_{a_{x-2}}^{a_{x-1}} \otimes T_{a_{x-1}} \otimes 1 \otimes \cdots \otimes 1.
\end{aligned}
\tag{1.18}
$$

Comparing with (1.15), we see that $\Delta_N(T_a)$ is the charge corresponding to the current $J_a^\mu(x)$:

$$
\Delta_N(T_a) = \sum_{x=1}^{N} J_a^t(x). \tag{1.19}
$$

The global charges $\Delta_N(\Theta_a^b)$ can be interpreted as topological charges.

The global charges $\Delta_N(T_a)$ and $\Delta_N(\Theta_a^b)$, satisfy the same algebra as the original generators T_a, Θ_a^b (because the comultiplication Δ is an homorphism from A to $A \otimes A$). If we assume, as we did in the previous section, that the generators T_a, Θ_a^b are closed under the adjoint action, then there also exist c-numbers f_{bc}^a such that :

$$
T_a T_b - R_{ab}^{cd}\, T_d T_c \;=\; f_{ab}^c\, T_c. \tag{1.20}
$$

These are generalized braided Lie commutation relations.

Remark: Because the currents are non-local, the local conservation laws for the currents do not systematically imply those of the global charges. The conservation laws for the charges can be broken by boundary terms which depends on the sector on which the charges are acting.

1b) Fields multiplets

The symmetry algebra A acts also on the field operators. In the operator formalism, the fields are operators $\mathcal{O} \in \mathrm{End}(V^{\otimes^N})$. Any element $X \in A$ of an Hopf algebra A acts an operator $\mathcal{O}$ by

$$
Q_X\ \mathcal{O} = \sum_i\ X_i\ \mathcal{O}\ S(X^i) \tag{1.21}
$$

with $\Delta(X) = \sum_i X_i \otimes X^i$ and S the antipode in A. In our case, for the generators T_a and Θ_a^b, this becomes:

$$
\begin{aligned}
Q_a \mathcal{O} &= \Delta_N(T_a)\ \mathcal{O}\ - (Q_a^b \mathcal{O})\ \Delta_N(T_b) \\
Q_a^b \mathcal{O} &= \Delta_N(\Theta_a^c)\ \mathcal{O}\ \Delta_N(\widehat{\Theta}_c^b)
\end{aligned}
\tag{1.22}
$$

The multiplets of fields are collections of fields transforming in a representation of the symmetry algebra A. More precisely, let V_Λ be a representation space for A. A field multiplet at x is an operator $\Phi^\Lambda(x;v)$ acting on $V^{\otimes N}$ depending linearly on a vector v in V_Λ, with transformation property

$$Q_X \Phi^\Lambda(x;v) = \Phi^\Lambda(x;Xv). \tag{1.23}$$

It is clear that, in general, the fields $\Phi^\Lambda(x;v)$ are necessarily non-local. However, in a given multiplet there can be local fields.

Field multiplets can be constructed from the following data: Representation spaces W, W' and operators ϕ and Ω, $\phi : V_\Lambda \otimes W \to W'$ and $\Omega : V_\Lambda \otimes V \to V_\Lambda \otimes V$, with (twisted) intertwining properties

$$\begin{aligned} \phi\Delta(X) &= X\phi \\ \Omega\Delta(X) &= \sigma \circ \Delta(X)\Omega, \end{aligned} \tag{1.24}$$

for all $X \in A$. The spaces W, W' are in the simplest case equal to V, or may be tensor products $V^{\otimes n}$, $V^{\otimes n'}$. In terms of a basis $\{e_i\}$ of V_Λ, with $\phi(e_i \otimes w) = \phi_i w$, $\Omega(e_i \otimes v) = e_j \otimes \Omega_i^j v$, $Xe_i = e_j X_i^j$, the field multiplets are operators acting on $V^{\otimes N}$ defined by,

$$\Phi^\Lambda(x;e_i) \equiv \Phi_i^\Lambda(x) = \Omega_i^{i_1} \otimes \Omega_{i_1}^{i_2} \otimes \cdots \otimes \Omega_{i_{x-2}}^{i_{x-1}} \otimes \phi_{i_{x-1}} \otimes 1 \otimes \cdots \otimes 1, \tag{1.25}$$

with ϕ_{i_x} acting on the tensor product of the xth to the $(x-1+n)$th factor *. Graphically this is represented as

$\Phi_i^\Lambda(x) \equiv$ 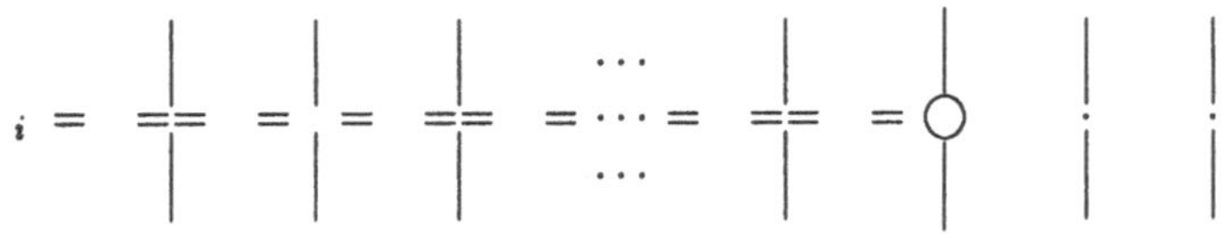

where the circle represents an insertion of ϕ and each crossing of the double line with a single line represents an Ω.

The transformation property (1.23) follows from (1.24) and Hopf algebra properties. The action of the algebra of fields is by definition (1.22):

$$\begin{aligned} Q_a^b \Phi^\Lambda(x;v) &= \Theta_a^{a_1} \otimes \Theta_{a_1}^{a_2} \otimes \cdots \otimes \Theta_{a_{N-1}}^{c} \; \Phi^\Lambda(x;v) \; \widehat{\Theta}_{b_1}^{b} \otimes \widehat{\Theta}_{b_2}^{b_1} \otimes \cdots \otimes \widehat{\Theta}_{c}^{b_{N-1}}, \\ Q_a \Phi^\Lambda(x;v) &= \sum_{y=1}^{N} \Big(J_a^t(y) \; \Phi^\Lambda(x;v) - Q_a^b \Phi^\Lambda(x;v) \; J_b^t(y) \Big). \end{aligned} \tag{1.26}$$

---

* If $n \neq n'$ the insertion of a field produces a deformation of the lattice. It is understood that we consider correlation functions where the total n is equal to the total n' so that at infinity the lattice is the regular square lattice.

Because the terms with $y > x + n - 1$ cancel in the sum, the action of the charges Q_a can written as a contour integral on the lattice. Indicating graphically the summation by an integration contour on the dual lattice, we have:

$Q_a\Phi_i^\Lambda(x) \equiv$ 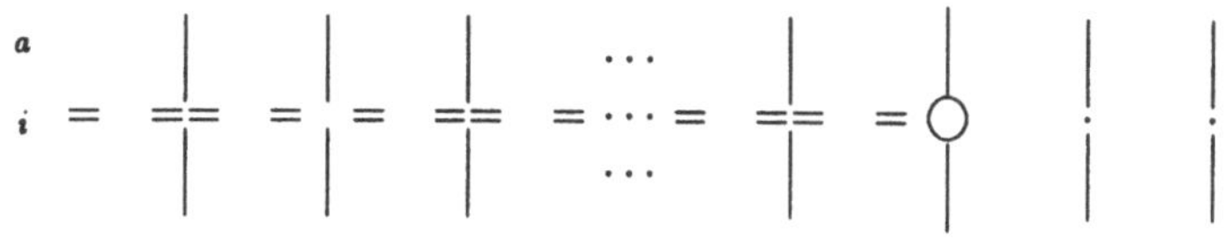 (1.27)

The integration contour is surrounding the fields. The intertwining properties (1.24) of the microscopic data ϕ and Ω imply that the field multiplets defined in (1.25) transform covariantly:

$$\begin{aligned} Q_a^b\ \Phi_i^\Lambda(x) &= \Theta_{ai}^{bj}\Phi_j^\Lambda(x) \\ Q_a\ \Phi_i^\Lambda(x) &= T_{ai}^{j}\Phi_j^\Lambda(x) \end{aligned} \tag{1.28}$$

Remark: Because the field multiplets are non-local they satisfy equal-time braiding relations. But the braiding relations between the field multiplets are constrained by the quantum invariance. Let V_Λ and $V_{\Lambda'}$ be two representation spaces of A with basis $e_i \in V_\Lambda$ and $e'_\alpha \in V_{\Lambda'}$. Let $\Phi_i(x) \equiv \Phi(x; e_i)$ and $\Phi'_\alpha \equiv \Phi'(y; e'_\alpha)$ be two field multiplets. Denote by $\mathcal{R} : V_{\Lambda'} \otimes V_\Lambda \to V_{\Lambda'} \otimes V_\Lambda$, with $\mathcal{R}(e'_\alpha \otimes e_i) = \mathcal{R}_{\alpha i}^{\beta j}\ e'_\beta \otimes e_j$, the braiding matrix of these field multiplets:

$$\Phi_i(x)\Phi'_\alpha(y) = \mathcal{R}_{\alpha i}^{\beta j}\ \Phi'_\beta(y)\Phi_j(x) \qquad \text{for } x > y \tag{1.29}$$

By quantum invariance,

$$\mathcal{R}\ \Delta(X) = \sigma \circ \Delta(X)\ \mathcal{R} \quad , \quad \forall X \in A \tag{1.30}$$

Thus, the braiding matrices intertwines the quantum algebra. Examples will be given in the following sections.

1c) Examples. (i) Yangian invariance. The Yangians are deformations of loop algebras which have been introduced by Drinfel'd [7]. They are related to rational solutions of the quantum Yang-Baxter equation.

Let us first recall what are the Yangians. Let $\mathcal{G}$ be a simple Lie algebra with structure constants f_{abc} in an orthonormalized basis. The Yangian, denoted $Y(\mathcal{G})$, is the associative algebra with unity generated by the elements t_a and $\mathcal{T}_a$, $a = 1, \cdots, \dim\mathcal{G}$, satisfying the relations:

$$\begin{aligned} \Big[\, t_a \,,\, t_b \,\Big] &= f_{abc}\, t_c \\ \Big[\, t_a \,,\, \mathcal{T}_b \,\Big] &= f_{abc}\, \mathcal{T}_c \\ \Big[\mathcal{T}_a, \Big[\mathcal{T}_b, t_c\Big]\Big] - \Big[t_a, \Big[\mathcal{T}_b, \mathcal{T}_c\Big]\Big] &= A_{abc}^{lmn}\, \Big\{t_l, t_m, t_n\Big\} \end{aligned} \tag{1.31}$$

with $A^{def}_{abc} = \frac{1}{24} f_{adk} f_{bel} f_{cfm} f^{klm}$ and $\{x_1, x_2, x_3\} = \sum_{i \neq j \neq k} x_i x_j x_k$. In particular, the elements t_a generate the Lie algebra $\mathcal{G}$ and the elements $\mathcal{T}_a$ are $\mathcal{G}$ - intertwiners taking values in the adjoint representation of $\mathcal{G}$. (for $\mathcal{G} = SU(2)$ one must add another Serre-like relation.) The Yangians $Y(\mathcal{G})$ are Hopf algebras with comultiplication Δ, counit ϵ and antipode S defined by:

$$\Delta\, t_a = t_a \otimes 1 + 1 \otimes t_a \; ; \qquad (1.32a)$$

$$\epsilon(t_a) = 0 \quad ; \quad S(t_a) = -t_a$$

$$\Delta\, \mathcal{T}_a = \mathcal{T}_a \otimes 1 + 1 \otimes \mathcal{T}_a - \frac{1}{2} f_{abc}\, t_b \otimes t_c \; ; \qquad (1.32b)$$

$$\epsilon(\mathcal{T}_a) = 0 \quad ; \quad S(\mathcal{T}_a) = -\mathcal{T}_a - \frac{C_{\mathrm{Ad}}}{4} t_a$$

with C_{Ad} the Casimir in the adjoint representation of $\mathcal{G}$: $f_{abc} f_{bcd} = C_{\mathrm{Ad}}\, \delta_{ad}$.

The Yangian invariant R - matrices are those which satisfy the intertwining relation $R\,\Delta(Y) = \sigma\circ\Delta(Y)\, R$ for all $Y \in Y(\mathcal{G})$. We suppose that the vertex models we are considering in this section are defined from $Y(\mathcal{G})$ - invariant Boltzmann weights. The non-local conserved currents we will describe in this section are the lattice analogues of those hidden in 2D massive current algebras [13] [14], see section 4.

From the defining relations of $Y(\mathcal{G})$, it is obvious that the Yangians $Y(\mathcal{G})$ possess the properties we need in order to apply our formalism. We therefore can define Yangian currents. For simplicity, we just define the currents associated to the generators t_a and $\mathcal{T}_a$; we denote them $J^\mu_a(x,t)$ and $\mathcal{J}^\mu_a(x,t)$, respectively. By applying the general construction, we deduce that the conserved currents $J^\mu_a(x,t)$ are local (as it should be); they are defined by local insertions of the matrices t_a representing the Lie algebra $\mathcal{G}$. The conserved currents $\mathcal{J}^\mu_a(x,t)$ are non-local; in the operator formalism we have:

$$\mathcal{J}^\mu_a(x,t) = \mathcal{T}^\mu_a(x,t) + \frac{1}{2} f_{abc}\, J^\mu_b(x,t)\, \phi_c(x,t) \qquad (1.33)$$

with

$$\phi_c(x,t) = \sum_{y<x} J^t_c(y,t) \qquad (1.34)$$

In eq. (1.33) and (1.34), the notations $J^\mu_a(x,t)$ and $\mathcal{T}^\mu_a(x,t)$ refer to insertions of the matrices t_a or $\mathcal{T}_a$ on the link oriented in the direction μ and ending at the point (x,t). Notice the similarity between the expression of these lattice non-local currents and of their continuous partners [13] [14] and section 4b.

The braiding relations for the currents $J_a^\mu(x,t)$ and $\mathcal{J}_a^\mu(y,t)$ follows from our previous discussion:

$$\begin{aligned}
J_a^\mu(x,t)\ J_b^\nu(y,t) &= J_b^\nu(y,t)\ J_a^\mu(x,t) & ;\ \forall\, x \neq y \\
\mathcal{J}_a^\mu(x,t)\ J_b^\nu(y,t) &= J_b^\nu(y,t)\ \mathcal{J}_a^\mu(x,t) & ;\ \text{for } x < y \\
\mathcal{J}_a^\mu(x,t)\ J_b^\nu(y,t) &= J_b^\nu(y,t)\ \mathcal{J}_a^\mu(x,t) & \\
&\quad - \frac{1}{2} f_{anm}\Big(f_{nbk} J_k^\nu(y,t)\Big)\ J_m^\mu(x,t) & ;\ \text{for } x > y
\end{aligned} \tag{1.35}$$

(ii) Quantum universal enveloping algebras. For any (affine) Kac-Moody algebra $\mathcal{G}$ with Cartan matrix a_{ij}, $0 \leq i,j \leq r$ and any complex number $q \neq 0$, Drinfel'd [7] and Jimbo [8]define an universal quantum enveloping algebra (QUEA) $U_q(\mathcal{G})$. Let d_i be positive integers such that the matrix $d_i a_{ij}$ is symmetric, and let $q_i = q^{d_i}$. The algebra $A = U_q(\mathcal{G})$ has generators E_i^+, E_i^-, K_i^2, K_i^{-2}, $0 \leq i \leq r$, and relations

$$\begin{aligned}
K_i^2 K_j^{\pm 2} &= K_j^{\pm 2} K_i^2, \\
K_i^2 K_i^{-2} &= K_i^{-2} K_i^2 = 1, \\
K_i^2 E_j^\pm &= q_i^{\pm a_{ij}} E_j^\pm K_i^2, \\
E_i^+ E_j^- - q_i^{-a_{ij}} E_j^- E_i^+ &= \delta_{ij}(K_i^4 - 1),
\end{aligned} \tag{1.36}$$

plus Chevalley-Serre relations to be written below. The Hopf algebra structure is defined by the coproduct

$$\begin{aligned}
\Delta(K_i^{\pm 2}) &= K_i^{\pm 2} \otimes K_i^{\pm 2}, \\
\Delta(E_i^\pm) &= E_i^\pm \otimes 1 + K_i^2 \otimes E_i^\pm,
\end{aligned} \tag{1.37}$$

counit $\epsilon(E_i^\pm) = 0$, $\epsilon(K_i^{\pm 2}) = 1$, and antipode $S(E_i^\pm) = -K_i^{-2} E_i^\pm$, $S(K_i^{\pm 2}) = K_i^{\mp 2}$. The adjoint representation is then defined as usual, eq. (1.21), and the Chevalley-Serre relations are

$$\mathrm{Ad}_{E_i^\pm}^{1-a_{ij}} E_j^\pm = 0. \tag{1.38}$$

We see that this is a very simple example of the Hopf algebras described in the introduction: the generators T_a are $E_i^\pm$ and Θ_a^b is diagonal with entries K_i^2.

Statistical models with QUEA symmetry are defined by trigonometric solutions of the Yang-Baxter equation, the simplest case being the six-vertex model [10].

The simple currents are defined by insertions of $E_i^\pm$ with disorder lines given by insertions of K_i^2. In the operator formalism, the time components of the currents are

$$J_i^{t\pm}(x) = K_i^2 \otimes \cdots \otimes K_i^2 \otimes E_i^\pm \otimes 1 \otimes \cdots \otimes 1. \tag{1.39}$$

The corresponding charges are the generators $\Delta_N(E_i^{\pm})$ acting on the whole space $V^{\otimes N}$. The braiding relations are, for $x > y$:

$$\begin{aligned} J_i^{\mu\pm}(x)J_j^{\nu\pm}(y) &= q_i^{\pm a_{ij}} J_j^{\nu\pm}(y)J_i^{\mu\pm}(x), \\ J_i^{\mu\pm}(x)J_j^{\nu\mp}(y) &= q_i^{\mp a_{ij}} J_j^{\nu\mp}(y)J_i^{\mu\pm}(x). \end{aligned} \tag{1.40}$$

These relations are the same as the braiding relations of chiral vertex operators of a free massless field ϕ taking value in the Cartan subalgebra of $\mathcal{G}$ with canonical momentum π. This suggests the continuum limit identification

$$J_j^{t\pm} \sim \exp\left(i\beta\alpha_j\left(\pm\phi(x) + \int_{-\infty}^{x} \pi(y)dy\right)\right), \tag{1.41}$$

with $q = e^{i\beta^2}$, α_j the simple roots, with inner product $\alpha_i\alpha_j = d_i a_{ij}$. The space component in the continuum limit is $J_j^{x\pm} = \mp i J_j^{t\pm}$.

2. CLASSICAL ORIGIN OF QUANTUM SYMMETRIES: DRESSING TRANSFORMATIONS

The dressing transformations form the (hidden) symmetry groups of solitons equations. Dressing transformations were first introduced by V. Zakharav and A. Shabat [15] and futher developped by the Kyoto group in their Tau-function appraoch to soliton equations [16]. Their Poisson structure was disantangled by M. Semenov-Tian-Shansky [17]. The author's understanding of these transformations emerged from a joint work with O. Babelon [18].

2a) What are the dressing transformations?

(i) Equations of motion and Lax connexions. Suppose that the equations of motion of a set of fields ϕ are described by a set of non-linear differential equations. Suppose moreover that these equations admit a Lax representation. This means that there exists a field dependent connexion, called the Lax connexion, $\mathcal{D}_\mu$,

$$\mathcal{D}_\mu = \partial_\mu - A_\mu[\phi],$$

such that the equations of motion are equivalent to the zero curvature condition for $\mathcal{D}_\mu$,

$$\Big[\, \mathcal{D}_\mu \,,\, \mathcal{D}_\nu \,\Big] = 0 \tag{2.1}$$

The Lax connexion takes value in some Lie algebra $\mathcal{G}$ with Lie group G.

Notice that, thanks to the zero-curvature condition, the Lax connexion is a pure gauge; i.e. there exists a G-valued function $\Psi(x,t)$ such that:

$$\Big(\, \partial_\mu - A_\mu \,\Big)\Psi = 0 \qquad \text{or} \qquad A_\mu = \Big(\partial_\mu \Psi\Big)\, \Psi^{-1} \tag{2.2}$$

The function $\Psi(x,t)$ is defined up to a right multiplication by a space-time independent group element. This freedom is fixed by imposing a normalization condition on Ψ; e.g. $\Psi(x_0) = 1$ for some point x_0.

(ii) Construction of the dressing transformations. The dressing transformations are non-local gauge transformations acting on the Lax connexion $A_\mu \rightarrow A_\mu^g$ and leaving its form invariant. They thus induce a transformation of the field variables $\phi \rightarrow \phi^g$ mapping a solution of the equations of motion into another.

They are constructed as follows. First let us study the set of gauge transformations mapping the Lax connexion A_μ on a given connexion A_μ^g. Suppose that there exist two G valued functions, Θ_+^g and Θ_-^g, such that:

$$A_\mu^g = \left(\partial_\mu \Theta_\pm^g\right) \Theta_\pm^{g\ -1} + \Theta_\pm^g \ A_\mu \ \Theta_\pm^{g\ -1} \tag{2.3}$$

Since A_μ is a pure gauge, $A_\mu = (\partial_\mu \Psi) \Psi^{-1}$, A_μ^g is also a pure gauge, $A_\mu^g = \partial_\mu \left(\Theta_\pm^g \Psi\right) \left(\Theta_\pm^g \Psi\right)^{-1}$. This implies that $\left(\Theta_+^g \Psi\right)$ and $\left(\Theta_-^g \Psi\right)$ differ by a right multiplication by a space-time independent group element which we denote by g. Equivalently:

$$\Theta_-^{g\ -1} \ \Theta_+^g = \Psi \ g \ \Psi^{-1} \tag{2.4}$$

The main idea underlaying the dressing tranformations is to consider eq. (2.4) as a factorization problem; i.e. we look for two subgroups $B_\pm \subset G$ such that any element $h \in G$ admits a unique decomposition $h = h_-^{-1} \ h_+$ with $h_\pm \in B_\pm$. The requirement that $\Theta_\pm$ belongs to $B_\pm$ then specify them uniquely from eq. (2.4). The subgroups $B_\pm$ are found by demanding that the transformations (2.3) preserve the form of the Lax connexion.

The factorization problem in G,

$$g = g_-^{-1} \ g_+ \quad \text{with} \quad g_\pm \in B_\pm \tag{2.5}$$

is a called an algebraic Riemann Hilbert problem (by analogy with the classical Riemann Hilbert problem). For the dressing transformations to be well-defined this decomposition as to be unique.

The gauge transformation (2.3) induces a transformation of the group valued function Ψ: $\Psi \rightarrow \Psi^g$. Decompose the group element $g \in G$ as $g = g_-^{-1} \ g_+$ with $g_\pm \in B_\pm$, in the way specified by the algebraic factorization problem discussed above, then,

$$\Psi^g = \left(\Psi g \Psi^{-1}\right)_+ \ \Psi \ g_+^{-1} = \left(\Psi g \Psi^{-1}\right)_- \ \Psi \ g_-^{-1} \tag{2.6}$$

The transformation (2.6) is well defined on the phase space because it preserves the normalization condition $\Psi(x_0) = 1$.

(iii) The composition law for the dressing transformations. It is not the compostion law in G [17] [19]. Let $g,\ h \in G$ with decomposition, $g = g_-^{-1}\ g_+$ and $h = h_-^{-1}\ h_+$, their composition law in the dressing group is:

$$(h_+, h_-) \bullet (g_+, g_-) \ = \ (h_+ g_+, h_- g_-) \tag{2.7}$$

In particular, the plus and minus components commute. We denote by G_R the new group equipped with this multiplication law. The group law (2.7) can be derived by using the dressing of Ψ. First, we dress $\Psi \to \Psi^g$ by the $g = g_-^{-1} g_+$ according to eq. (2.6). Then, we dress $\Psi^g \to (\Psi^g)^h$ by $h = h_-^{-1} h_+$:

$$(\Psi^g)^h \ = \Theta^h_\pm\ \Psi^g\ h_\pm^{-1} \quad \text{with} \quad \Theta^h_\pm = \left(\Psi^g\ h\ \Psi^{g\ -1}\right)_\pm \tag{2.8}$$

Using the definition (2.6) of Ψ^g, the factorization of $\Psi^g h \Psi^{g\ -1}$ can be written as follows:

$$\Theta^{h\ -1}_- \Theta^h_+ = \Psi^g h \Psi^{g\ -1} = \Theta^g_- \Psi (h_- g_-)^{-1} (h_+ g_+) \Psi^{-1} \Theta^g_+$$

This implies that $(\Theta^h_- \Theta^g_-)^{-1} (\Theta^h_+ \Theta^g_+) = \Psi (h_- g_-)^{-1} (h_+ g_+) \Psi^{-1}$, or equivalently:

$$\left(\Psi (h_- g_-)^{-1} (h_+ g_+) \Psi^{-1}\right)_\pm \ = \ \left(\Psi^g h \Psi^{g\ -1}\right)_\pm \left(\Psi g \Psi^{-1}\right)_\pm \ = \ \Theta^h_\pm\ \Theta^g_\pm$$

This proves eq. (2.7).

For infinitesimal transformations, $g \simeq 1 + X$, $X \in \mathcal{G}$, and $g_\pm \simeq 1 + X_\pm$ with $X = X_+ - X_-$, the dressing transformations are:

$$\delta_X\ \Psi \ = Y_\pm\ \Psi - \Psi X_\pm \qquad \text{with} \qquad Y_\pm = \left(\Psi X \Psi^{-1}\right)_\pm . \tag{2.9}$$

The composition law is, for $X, Z \in \mathcal{G}$:

$$\Big[\, X\ ,\ Z\, \Big]_R = \Big[\, X_+\ ,\ Z_+\, \Big] - \Big[\, X_-\ ,\ Z_-\, \Big] \tag{2.10}$$

This defines a new Lie algebra $\mathcal{G}_R$ which is the Lie algebra of G_R.

2b) Few of their properties

(i) *They are non-local.* This is obvious from their definition as Ψ is non-local. The dressing transformations can be used to construction solutions of the soliton equations having non-trivial topological numbers from solutions with trivial topological numbers:

$$\phi(x) \ \to \ \phi^g(x) \quad ; \quad \forall g \in G_R; \tag{2.11}$$

In particular, by dressing local conserved currents, the dressing transformations provide a way to construct non-local conserved currents :

$$J_\mu(x,t) \to J^g_\mu(x,t) \qquad ; \quad \forall g \in G_R. \tag{2.12}$$

In the quantum theories, these non-local currents are turned into the generators of the quantum group symmetries.

(ii) *They induce Lie Poisson actions.* The dressing transformations induce an action of the group G_R on the space of solutions of the classical equations of motion, i.e. on the phase space. This action is (in general) compatible with the Poisson structure; more precisely, it is a Lie Poisson action. It means that the Poisson brackets transform covariantly if the group G_R is equipped with a non-trivial Poisson structure. This Poisson structure, which, by construction, is compatible with the multiplication in G_R, turns the group G_R into a Lie Poisson group.

Let us be a more precise. Denote by $\mathcal{P}$ the phase space and by $\{,\}_{\mathcal{P}}$ the Poisson bracket on it. The dressing transformations define an action of G_R on the function over the phase space: $f^g(x) = f(g^{-1}\cdot x)$ for $f \in \mathrm{Funct}(\mathcal{P})$ and $x \in \mathcal{P}$. Suppose that the group G_R of is equipped with a Poisson bracket which we denote by $\{,\}_{G_R}$. The statement that the dressing transformation are Lie Poisson action is equivalent to the covariance of the Poisson brackets:

$$\{f_1, f_2\}^g_{\mathcal{P}} = \{f_1^g, f_2^g\}_{\mathcal{P}\times G_R} \qquad \forall\, f_1, f_2 \in \mathrm{Funct}(\mathcal{P})\ ;\ \forall\, g \in G_R \tag{2.13}$$

The Poisson bracket on $\mathcal{P} \times G_R$ is the product Poisson structure.

(iii) *A standard example.* As is well known, a Lie group G can be equipped with the following Poisson bracket (Sklyanin's Poisson bracket) [20]:

$$\Big\{\ \Psi(x) \otimes,\ \Psi(x)\ \Big\} = \Big[\ r^\epsilon\ ,\ \Psi(x) \otimes \Psi(x)\ \Big] \tag{2.14}$$

with matrices r^ϵ, $\epsilon = \pm$, solutions of the classical Yang-Baxter equation. By G-invariance, in eq. (2.14) we can choose any of two solutions r^+ or r^- of the classical Yang-Baxter equation provided that their difference is the tensor Casimir $C = r^+ - r^-$. A direct computation [17] [21] shows that the Sklyanin's Poisson brackets are covariant under the transformation (2.6), $\Psi \to \Psi^g$, only if there are non-trivial Poisson brackets among the g's but vanishing Poisson brackets between the g's and the fields Ψ. The Poisson brackets in G_R are:

$$\Big\{\ g_+ \ \otimes,\ g_+\ \Big\} = \Big[\ r^\pm\ ,\ g_+ \otimes g_+\ \Big] \tag{2.15a}$$
$$\Big\{\ g_- \ \otimes,\ g_-\ \Big\} = \Big[\ r^\pm\ ,\ g_- \otimes g_-\ \Big] \tag{2.15b}$$
$$\Big\{\ g_- \ \otimes,\ g_+\ \Big\} = \Big[\ r^-\ ,\ g_- \otimes g_+\ \Big] \tag{2.15c}$$
$$\Big\{\ g_+ \ \otimes,\ g_-\ \Big\} = \Big[\ r^+\ ,\ g_+ \otimes g_-\ \Big] \tag{2.15d}$$

For $g = g_-^{-1} g_+$, the Poisson brackets are the Semenov-Tian-Shansky brackets:

$$\begin{aligned}\left\{ g \otimes, g \right\} &= (g \otimes 1)r^+(1 \otimes g) + (1 \otimes g)r^-(g \otimes 1) \\ &\quad -(g \otimes g)r^{\pm} - r^{\mp}(g \otimes g)\end{aligned} \tag{2.16}$$

It is easy to check that the multiplication in G_R (not in G) is a Poisson mapping for the Poisson structure defined in eq. (2.16), or in eq. (2.15). Therefore G_R is a Poisson Lie group and the actions (2.6) are Lie Poisson actions.

3. AN EXAMPLE OF DRESSING TRANSFORMATIONS: CURRENT ALGEBRAS

We illustrate the general construction explained in the previous section on the example of the classical current algebras. We essentially follow [22]. Dressing transformations in the Toda theories were traited in [18].

3a) The equations of motion

The field variables are one-forms, denoted by $J_\mu(x)$, valued in a semi-simple Lie algebra $\mathcal{G}$: $J_\mu(x) = \sum_a J_\mu^a(x)t^a$ where t^a, $a = 1, \cdots, \dim\mathcal{G}$, form a basis of $\mathcal{G}$ [1]. By definition the equations of motion impose to $J_\mu(x)$ to be a curl-free conserved current:

$$\begin{aligned}&\partial_\mu J_\mu^a(x) = 0 \\ &\partial_\mu J_\nu^a(x) - \partial_\nu J_\mu^a(x) + f^{abc} J_\mu^b(x) J_\nu^c(x) = 0\end{aligned} \tag{3.1}$$

The equations of motion (3.1) admit a Lax representation: they are equivalent to the zero curvature condition, $[\mathcal{D}_\mu(\lambda), \mathcal{D}_\nu(\lambda)] = 0$, for the connexion $\mathcal{D}_\mu(\lambda)$,

$$\mathcal{D}_\mu(\lambda) = \partial_\mu + \frac{\lambda^2}{\lambda^2 - 1} J_\mu(x) + \frac{\lambda}{\lambda^2 - 1} \epsilon_{\mu\nu} J_\nu(x) \tag{3.2}$$

The Lax connexion is an element of the loop algebra $\widehat{\mathcal{G}} = \mathcal{G} \otimes C[\lambda, \lambda^{-1}]$.

Remark 1: The linear problem $(\partial_\mu - A_\mu(x))\,\Psi(x) = 0$ associated to the Lax representation (3.2) is equivalent to the following ($\epsilon_{\mu\nu}\epsilon^{\nu\sigma} = \delta_\mu^\sigma$):

$$\left(\partial_\mu - \lambda\epsilon_{\mu\nu}\partial_\nu - \lambda\epsilon_{\mu\nu}J_\nu\right) \Psi(x) = 0 \tag{3.3}$$

Remark 2: In the light-cone components of the Lax connexion are ($\epsilon_\pm^\pm = \pm 1$):

$$A_\pm = -\frac{\lambda}{\lambda \mp 1} J_\pm \tag{3.4}$$

---

[1] We suppose the t^a orthonormalized. We use the convention: $\left[t^a, t^b\right] = f^{abc}t^c$ where f^{abc} denote the structure constants of $\mathcal{G}$.

The Lax connexion is therefore completely characterized by the following two conditions: i) $A_\pm$ have a simple pole at $\lambda = \pm 1$ (we then set $\mathrm{Res}_{\lambda=\pm 1}\, A_\pm = \mp J_\pm$) and ii) $A_\pm(\lambda = 0) = 0$. Therefore, for a gauge transformation to be a symmetry it only has to preserve these two conditions.

Remark 3: The gauge condition $A_\pm(\lambda = 0) = 0$ implies that $\Psi(\lambda = 0)$ is space-time independent. In the following we fix the gauge on Ψ by setting $\Psi(\lambda = 0) = 1$. Moreover we have:

$$J_\nu(x) = \partial_\lambda \left(\epsilon_{\nu\mu}\, A_\mu(x)\right)_{\lambda=0} = \partial_\lambda \left(\epsilon_{\nu\mu}\,(\partial_\mu \Psi)\,\Psi^{-1}\right)_{\lambda=0}. \tag{3.5}$$

3b) Dressing transformations and the Riemann Hilbert problem

First, because the Lax connexion takes value in the loop algebra, we have to define the factorization problem (2.5) in the loop group. It can be formulated as follows: Let Γ be a contour around the origin $\lambda = 0$. Denote by Γ_- (Γ_+) the exterior (interior) domain of Γ. We choose Γ such that the points $\lambda = \pm 1$ belong to Γ_-. The factorization problem consists in factorizing any regular element of the loop group, $G(\lambda)$, $\lambda \in \Gamma$, into the product of two λ-dependent group elements $G_\pm(\lambda)$ respectively analytic on $\Gamma_\pm$:

$$G(\lambda) \;=\; G_-^{-1}(\lambda)\; G_+(\lambda) \qquad ; \qquad \lambda \in \Gamma \tag{3.6}$$

This is the Riemann-Hilbert factorization problem. It is known that it admits a unique solution up to a left multiplication, $G_\pm \to M G_\pm$, by a λ-independent group element M. The definition of the Riemann-Hilbert factorization is cooked up such that the transformations we will now define are symmetries of the equations of motion of the classical current algebras.

To dress the Lax connexion (3.2), we follow the general procedure:

(i) We pick up an element $G(\lambda)$ of the loop group. We fix the gauge in the Riemann-Hilbert factorization by imposing $G_+(\lambda = 0) = 1$.

(ii) We define $\Theta^G(\lambda) = \Psi\; G(\lambda)\Psi^{-1}$ and factorize it according to the Riemann-Hilbert problem:

$$\Theta^G(\lambda) = \Psi\; G(\lambda)\Psi^{-1} = \Theta^G_-(\lambda)^{-1}\; \Theta^G_+(\lambda). \tag{3.7}$$

We impose the gauge condition $\Theta^G_+(\lambda = 0) = 1$. The solution to eq. (3.7) is then unique.

(iii) We define the dressed Lax connexion by:

$$A^G_\mu \;=\; (\partial_\mu \Theta^G_\pm)\;\Theta^{G\,-1}_\pm \;+\; \Theta^G_\pm\, A_\mu\, \Theta^{G\,-1}_\pm. \tag{3.8}$$

Because we can implement the dressing either using Θ_+ or using Θ_-, it easy to check that the dressed connexion A_μ^G possesses the same poles with the same orders than the original connexion A_μ. The gauge conditions we choose for the Riemann-Hilbert factorization also ensure that A_μ^G satisfy the same gauge condition as A_μ. Therefore, the dressing transformation $A_\mu \to A_\mu^G$, preserving the structure of the Lax connexion, induce a symmetry of the equations of motion. The dressed currents $J_\mu^G(x)$ are defined via the eq. (3.5) with A_μ^G instead of A_μ:

$$J_\mu^G(x) \;=\; J_\mu(x) \;+\; \epsilon_{\mu\nu}\partial_\nu\Big(\partial_\lambda\Theta_+^G\Big)_{\lambda=0} \tag{3.9}$$

(iv) For infinitesimal transformations, $G(\lambda) = 1 + X(\lambda) + \cdots$, where $X(\lambda) \in \widehat{\mathcal{G}}$, $X(\lambda) = X_+(\lambda) - X_-(\lambda)$ with $X_\pm(\lambda)$ analytic in $\Gamma_\pm$, the dressings are:

$$\begin{aligned} \delta_X\, A_\mu \;&=\; \partial_\mu\, Y_\pm + \Big[Y_\pm, A_\mu\Big] \\ \delta_X\, \Psi \;&=\; Y_\pm\, \Psi \;-\; \Psi\, X_\pm \end{aligned} \tag{3.10}$$

with $Y(\lambda) = \left(\Psi X(\lambda)\Psi^{-1}\right)(\lambda) = Y_+(\lambda) - Y_-(\lambda)$. In particular for the current:

$$\delta_X\, J_\mu \;=\; \epsilon_{\mu\nu}\partial_\nu\Big(\partial_\lambda Y_+\Big)_{\lambda=0} \tag{3.11}$$

3c) Non-local conserved currents

The problem consists now in solving the Riemann-Hilbert factorization, eq. (3.6). The main point is that we will find differential equations which solve this problem recursively. In the following we restrict ourselves to the dressing of the current $J_\mu(x)$.

(i) *Projection on* $\widehat{\mathcal{G}}_+$. Recall that by definition, eq. (3.6), $\widehat{\mathcal{G}}_+$ is the algebra of $\mathcal{G}$-vector fields regular at the origin $\lambda = 0$. Therefore, if $Y(\lambda)$ is an element of the loop algebra $\widehat{\mathcal{G}}$, its projection $Y_+(\lambda)$ on $\widehat{\mathcal{G}}_+$ is:

$$Y_+(\lambda) \;=\; \oint_\Gamma \frac{dz}{2i\pi}\,\frac{Y(z)}{z-\lambda} \qquad ; \qquad \lambda \in \Gamma_+ \tag{3.12}$$

(ii) *The dressing transformations act on* J_μ *by non-local gauge transformations.* The dressing of J_μ is defined in eq. (3.9), or its infinitesimal form (3.11). To compute it we use the explicit expression of $Y(\lambda)$ and the projector (3.12):

$$\begin{aligned} \partial_\mu\Big(\partial_\lambda Y_+\Big)_{\lambda=0} \;&=\; \oint \frac{dz}{2i\pi z^2}\Big[\left(\partial_\mu\Psi(z)\right)\Psi^{-1}(z)\;,\;\Psi(z)X(z)\Psi^{-1}(z)\Big] \\ &=\; \epsilon_{\mu\nu}\Big(\;\partial_\nu Z_+ \;+\; [J_\nu, Z_+]\;\Big) \end{aligned} \tag{3.13}$$

with:

$$Z_+ = \oint \frac{dz}{2i\pi z}\Big(\Psi(z)X(z)\Psi^{-1}(z)\Big). \tag{3.14}$$

To derive the last equation we used the linear problem in the form (3.3). The variation of J_μ is therefore:

$$\delta_X\, J_\mu = \partial_\mu Z_+ + [J_\mu, Z_+]. \tag{3.15}$$

(iii) *Recursion relation for* Z_+. The last step consists in solving for Z_+ by recursion. Let $X \in \widehat{\mathcal{G}}$ be $X^n(\lambda) = v\lambda^{-n}$ with $n = 0, 1, \cdots$ and $v \in \mathcal{G}$. Denote by Z^n the corresponding solution to eq. (3.14):

$$Z^n = \oint \frac{dz}{2i\pi z}\Big(\Psi(z)v\Psi^{-1}(z)\Big)z^{-n}. \tag{3.16}$$

Then using once more the differential equation (3.3), we have:

$$\partial_\mu Z^{n+1} = \epsilon_{\mu\nu}\Big(\, \partial_\nu Z^n + [J_\nu, Z^n]\,\Big). \tag{3.17}$$

As advertised, this solves recursively the Riemann-Hilbert problem. The dressed currents, $\delta^n_v J_\mu$, are recursively defined by eqs. (3.15) and (3.17). This recursive construction is equivalent to the construction of ref. [23]. The conservation law for the dressed currents $\delta^n_v J_\mu$ can be checked directly.

(iv) *The two first conserved currents.* The first ones are the local currents $J^a_\mu(x)$ since, for $X = v \in \mathcal{G}$,

$$\delta^0_v\, J_\mu(x) = \Big[J_\mu(x), v\Big]. \tag{3.18}$$

For $X = v\lambda^{-1}$, $v \in \mathcal{G}$, we have $\partial_\mu Z^1 = \epsilon_{\mu\nu}\,[J_\nu, v]$, or equivalently,

$$Z^1(x) = \Big[\Phi(x), v\Big] \quad \text{with} \quad \Phi(x) = \int_{C_x} \star J \tag{3.19}$$

where C_x is a curve ending at the point x. The dressing of J_μ is:

$$\delta^1_v J_\mu(x) = \epsilon_{\mu\nu}\Big[J_\nu(x), v\Big] + \Big[J_\mu(x), \Big[\Phi(x), v\Big]\Big]. \tag{3.20}$$

In particular, projecting on the adjoint representation, we find the following non-local conserved currents:

$$\begin{aligned} &f^{abc}\delta^1_{t^b}\, J^c_\mu(x) \propto {J^{(1)}}^a_\mu(x) \\ &{J^{(1)}}^a_\mu(x) = \epsilon_{\mu\nu} J^a_\nu(x) + \frac{1}{2} f^{abc}\, J^b_\mu(x)\, \Phi^c(x) \end{aligned} \tag{3.21}$$

4. QUANTIFICATION: YANGIANS IN MASSIVE CURRENT ALGEBRAS

We use the example of the massive current algebras in order to describe how non-local conserved currents can be defined in a non-perturbative way and to illustrate few of their properties, (e.g. how they act on the states, on the fields, etc...). But the approach is more general, see e.g. ref. [5].

The currents ${J^{(1)}}^a_\mu(x)$ are the currents we want to quantize. There are different ways to specify the quantum theory, e.g. by defining it on the lattice, or as perturbation of its U.V. fixed point, etc... Here we use an alternative approach: we look for the conditions that we have to impose on the operator algebra in order to be able to define the quantum non-local conserved currents. Therefore, we are interested in a quantum models satisfying the following hypothesis:

(a) There exist quantum local conserved currents, $J^a_\mu(x)$, taken values in the Lie algebra $\mathcal{G}$:

$$\partial_\mu \; J^a_\mu(x) \;=\; 0. \tag{4.1}$$

Furtheremore, because the currents J^a_μ have to be one-forms, we impose that they have scaling dimensions one.

(b) The currents $J^a_\mu(x)$ satisfy the quantum version of the equations of motion (3.1); i.e. the quantum currents are curl-free:

$$\partial_\mu J^a_\nu(x) - \partial_\nu J^a_\mu(x) + \; f^{abc} \; : J^b_\mu(x) \; J^c_\nu(x) : = 0 \tag{4.2}$$

where the double dots denote an appropriate regularization of $f^{abc} J^b_\mu(x) J^c_\nu(x)$, e.g. by a point splitting. This hypothesis imposes constraints on the operator product expansion (OPE) of the currents.

(c) The only fields taking values in the adjoint representation of $\mathcal{G}$ and having scaling dimensions zero, one or two are either J^a_μ or $\partial_\nu J^a_\mu$. This fixes the OPE $f^{abc} J^b_\mu(x) J^c_\nu(0)$ up to the order $\mathcal{O}(|x|^{1-0})$:

$$f^{abc} \; J^b_\mu(x) J^c_\nu(0) = \mathcal{C}^\rho_{\mu\nu}(x) \; J^a_\rho(0) + \mathcal{D}^{\sigma\rho}_{\mu\nu}(x) \Big(\partial_\sigma J^a_\rho(0) \Big) + \mathcal{O}(|x|^{1-0}) \tag{4.3}$$

The quantum currents $J^a_\mu(x)$ satisfying these three hypthesis generate what could be called a massive current algebra.

4a) OPE's in massive quantum current algebras

We now show that these hypothesis ensure that the currents satisfy the commutation relations of a Kac-Moody algebra but also that they satisfy the following OPE's [2]:

---

[2] We used the following space-time conventions $x^\nu \equiv (x^0 = t, x^1 = x)$; $x^\pm = x \pm t$; and $ds^2 = \eta_{\mu\nu} dx^\mu dx^\nu = dt^2 - dx^2$.

$$J^b_\pm(x)J^c_\pm(0) = -\frac{k\delta^{ab}}{8i\pi}\frac{1}{(x^\pm)^2} - \frac{f^{abc}}{2i\pi}\frac{J^c_\pm(0)}{x^\pm} + \mathcal{O}(|x|^{-0}) \tag{4.4a}$$

$$\frac{1}{2}f^{abc}\left(J^b_+(x)J^c_-(0) - J^b_-(x)J^c_+(0)\right) \tag{4.4b}$$
$$= \frac{C_{Adj}}{8i\pi}\log\left(M^2x^+x^-\right)\left(\partial_+J^a_-(0) - \partial_-J^a_+(0)\right) + \mathcal{O}(|x|^{1-0})$$

C_{Adj} is the Casimir of $\mathcal{G}$ in the adjoint representation and M is the mass scale. The product $J^a_\pm(x)J^c_\mp(0)$ is logarithmically divergent. We solve for the OPE (4.3) using our hypothesis. The proof goes in few steps:

(i) First, locality, PT-invariance and Lorentz covariance determine the general tensor form of $\mathcal{C}^\rho_{\mu\nu}(x)$ and $\mathcal{D}^{\sigma\rho}_{\mu\nu}(x)$. Notice also that the conservation law for J^a_μ allows us to choose $\mathcal{D}^{\sigma\rho}_{\mu\nu}(x)$ to be traceless: $\eta_{\sigma\rho}\mathcal{D}^{\sigma\rho}_{\mu\nu}(x) = 0$. Therefore, under the conditions (a) to (c), the generators $J^a_\mu(x)$ of a massive current algebra satisfy the following OPE's [13] :

$$\begin{aligned} f^{abc}\, J^b_\mu(x)J^c_\nu(0) \;=\; & \\ \Big(C_1\, x^2\eta_{\mu\nu}x^\rho + C_2\, x^2\left(x_\mu\delta^\rho_\nu + x_\nu\delta^\rho_\mu\right) + C_3\, x_\mu x_\nu x^\rho\Big) & \left(J^a_\rho(0) + \frac{1}{2}x^\sigma\partial_\sigma J^a_\rho(0)\right) \\ +\Big(D_1\, x^\rho\left(x_\mu\delta^\sigma_\nu - x_\nu\delta^\sigma_\mu\right) + D_2\, x^\sigma\left(x_\mu\delta^\rho_\nu - x_\nu\delta^\rho_\mu\right)\Big) & \left(\partial_\sigma J^a_\rho(0)\right) \\ + \mathcal{O}\left(|x|^{1-0}\right) & \end{aligned} \tag{4.5}$$

The coefficients C_i and D_i only depend on x^2. Furthermore, the conservation law for the currents implies the following differential equations for the functions C_i and D_i [13] :

$$x^2\frac{d}{dx^2}C_2 = -\frac{1}{2}\left(C_1 + 5C_2\right) \tag{4.6a}$$
$$x^2\frac{d}{dx^2}\left(C_1 + C_2 + C_3\right) = -\left(C_1 + C_2 + 2C_3\right) \tag{4.6b}$$

and

$$x^2\frac{d}{dx^2}D_1 = -D_1 - \frac{x^2}{4}C_1 \tag{4.7a}$$
$$x^2\frac{d}{dx^2}D_2 = -D_2 - \frac{x^2}{4}C_2 \tag{4.7b}$$
$$x^2\frac{d}{dx^2}\left(D_1 + D_2\right) = \frac{x^2}{4}C_3 \tag{4.7c}$$

(ii) The differential equations (4.6) and (4.7) do not specify uniquely the unknown coefficients $C_i(x^2)$ and $D_i(x^2)$. But we can use the hypothesis on the scaling dimension of the currents to fixe the leading behaviour of the functions $C_i(x^2)$:

$$C_i(x^2) = \frac{\alpha_i}{(x^2)^2} + \mathcal{O}(|x|^{-3-0}) \quad ; \quad i = 1,\ 2,\ 3, \tag{4.8}$$

with α_i some constants. We assume that there is no leading logarithmic corrections.

Solving the differential equations (4.6) and (4.7), we find:

$$C_1(x^2) = -\frac{\alpha}{(x^2)^2} + \mathcal{O}(|x|^{-3-0}) \tag{4.9a}$$

$$C_2(x^2) = \frac{\alpha}{(x^2)^2} + \mathcal{O}(|x|^{-3-0}) \tag{4.9b}$$

$$C_3(x^2) = -\frac{\gamma\alpha}{(x^2)^2} + \mathcal{O}(|x|^{-3-0}) \tag{4.9c}$$

$$D_k(x^2) = \frac{-\alpha_k}{4x^2}\log\left(-M_k^2\, x^2\right) + \mathcal{O}(|x|^{-1-0}) \quad ; \quad k=1,2 \tag{4.9d}$$

The constants M_k are related to the mass scale and $\gamma = 2\log(M_2/M_1)$. The constant α depends on the normalization of the currents: we fix the normalization such that $\alpha = -\frac{C_{adj}}{2i\pi}$

(iii) We finally impose the zero curvature condition. From eqs. (4.5) and (4.9), we have:

$$\begin{aligned}\epsilon_{\mu\nu}\left[f^{abc}J^b_\mu(x)J^c_\nu(0) \; + \; Z(-x^2)\; \left(\partial_\mu J^a_\nu(0) - \partial_\nu J^a_\mu(0)\right)\right] \\ = -\frac{\alpha\gamma}{4x^2}\frac{x^\mu}{2}\left(x^\rho\epsilon^{\mu\sigma} + x^\sigma\epsilon^{\mu\rho}\right)\left(\partial_\sigma J^a_\rho(0) + \partial_\rho J^a_\sigma(0)\right)\end{aligned} \tag{4.10}$$

with $Z(-x^2) = \frac{\alpha}{4}\log(-M_1M_2x^2)$.

The curl-free equation (4.2) is then an immediat consequence of (4.10) if γ vanishes. The normal order in (4.2) is defined in such a way to cancel the logarithmic divergence in $f^{abc}J^b_\mu J^c_\nu$. Therefore, the curl-free equation fixes the two mass scale to be equal

$$\gamma \;=\; 2\log(M_2/M_1) \;=\; 0$$

The same conclusion could have been reached by imposing the chiral splitting of the leading terms of the OPE (4.5)[3]; the approach based on the chiral splitting assumption was described in ref. [14]. **(iv)** *Commutation relations of the currents.* Finally, other current OPE's can be deduced using the same techniques. In particular we have:

$$\begin{aligned}J^a_\mu(x)J^c_\nu(0) \;=\; & -\frac{k\delta^{ab}}{2i\pi}\frac{1}{(x^2)^2}\left(x_\mu x_\nu - \frac{1}{2}x^2\eta_{\mu\nu}\right) \\ & -\frac{f^{abc}}{2i\pi}\frac{1}{x^2}\left(x_\mu\delta^\rho_\nu + x_\nu\delta^\rho_\mu - x^2\eta_{\mu\nu}x^\rho\right)J^a_\rho(0) \;+\; \mathcal{O}(|x|^{-0}).\end{aligned} \tag{4.11}$$

---

[3] The case $\gamma \neq 0$ is also quite interesting; it probably corresponds to the 2D $O(n)$ models. In particular, in this case the leading terms of the OPE of the currents do not satisfy the chiral splitting. In other words the chiral components of the currents, J^a_- and J^a_+, are mixed in the leading terms of the OPE.

These OPE's reduce to eq. (4.4a). The products of the quantum operators are defined by:

$$J^a_\mu(x,t)J^b_\nu(y,t) = \lim_{\epsilon\to 0^+} J^a_\mu(x,t+i\epsilon)J^b_\nu(y,t). \tag{4.12}$$

Therefore, using $\lim_{\epsilon\to 0^+} \frac{i\epsilon}{x^2+\epsilon^2} = i\pi\delta(x)$, the OPE's (4.11) implies:

$$\begin{aligned}
\left[J^a_t(x)\ ,\ J^b_x(0)\right] &= f^{abc}J^c_x(0)\delta(x) - \frac{k}{2}\delta^{ab}\delta'(x)\\
\left[J^a_t(x)\ ,\ J^b_t(0)\right] &= f^{abc}J^c_t(0)\delta(x)\\
\left[J^a_x(x)\ ,\ J^b_x(0)\right] &= f^{abc}J^c_t(0)\delta(x)
\end{aligned} \tag{4.13}$$

They are the commutation relations of a current algebra: the light cone component $J^a_\pm$ satisfy the commutation relations of the affine Kac-Moody algebra $\mathcal{G}^{(1)}$.

Remark 1: Two hidden consequences of the definition of the massive current algebras we choose are: i) their ultra-violet limits are WZW models with $\mathcal{G}^{(1)}\otimes\mathcal{G}^{(1)}$ symmetry; and ii) they describe perturbations of affine Kac-Moody algebras by the perturbing fields $\Phi_{\text{pert.}}(x) = \sum_a J^a_\mu(x)\ J^a_\mu(x)$.

Remark 2: The massive current algebras are characterized by the level K of the affine Kac-Moody algebras. However the OPE's (4.4) and the curl-free equation (4.2) are model independent in the sense that they do not depend on the level.

Remark 3: Because the WZW models are the U.V. fixed point of the massive chiral algebras, they are also $Y(\mathcal{G})$ invariant. Actually, The WZW models are $Y(\mathcal{G})\otimes Y(\mathcal{G})$ invariant (at least classically [24]). They are also $U_q(\mathcal{G})\times U_q(\mathcal{G})$ invariant. The perturbing field $J^a_\mu J^a_\mu$ breaks these symmetries down to the diagonal $Y(\mathcal{G})$ symmetry times a fractional supersymmetry. It could be interesting to solve the WZW models from their non-local symmetries. This will provide a test of the idea we are trying to develop for the massive integrable models.

4b) The quantum non-local conserved currents

(i) Their definition. Having proved that the quantum conserved currents satisfy the quantum form (4.2) of the equations of motion (3.1), it is now easy to defined the quantum conserved currents $J^{(1)}(x,t)$. We define them by a point splitting regularization ($\delta > 0$):

$$\begin{aligned}
{J^{(1)}}^a_\mu(x,t) &= \lim_{\delta\to 0^+} {J^{(1)}}^a_\mu(x,t|\delta)\\
{J^{(1)}}^a_\mu(x,t|\delta) &= Z(\delta)\epsilon_{\mu\nu}J^a_\nu(x,t) + \frac{1}{2}f^{abc}\ J^b_\mu(x,t)\phi^c(x-\delta,t)
\end{aligned} \tag{4.14}$$

where $\phi^c(x,t)$, which satisfies $d\phi^c = \star J^c$, is defined by: $\phi^c(x,t) = \int_{C_x} \star J^c$ The contour of integration C_x is a curve from $-\infty$ to x.

The renormalization constant $Z(\delta)$ is fixed by requiring that ${J^{(1)}}^a_\mu(x,t)$ are finite and conserved. First it is easily seen from eq. (4.4b) that ${J^{(1)}}^a_\mu(x,t)$ is finite whenever $Z(\delta) = \frac{\alpha}{2}\log(\delta) + \text{constant}$. The constant is fixed by demanding the conservation law for ${J^{(1)}}^a_\mu$. (The other subleading terms in $Z(\delta)$ are meaningless.) Using eq. (2.4) we deduce,

$$\partial_\mu {J^{(1)}}^a_\mu(x,t|\delta) = \frac{1}{2}\epsilon_{\mu\nu}\Big[Z(\delta)\left(\partial_\mu J^a_\nu - \partial_\nu J^a_\mu\right)(x,t) + f^{abc}\, J^b_\mu(x,t) J^c_\nu(x-\delta,t)\Big] \quad (4.15)$$

Therefore, from eq. (4.4b) or (4.10), we learn that $\partial_\mu {J^{(1)}}^a_\mu(x,t|\delta)$ vanishes when $\delta \to 0$ if $Z(\delta) = \frac{\alpha}{2}\log(M\delta) + \mathcal{O}(\delta^{1-0})$.

(ii) Non-locality: the braiding relations. The non-local character of the currents $J^{(1)}(x,t)$ is encoded in their braiding relations, the equal time commutation relations. The latter are described as follows: Let $\Phi(y,t)$ be a quantum field local with respect to the currents $J^a_\mu(x,t)$. Then it satisfies the following equal-time braiding relations [14]:

$$\begin{aligned} {J^{(1)}}^a_\mu(x,t)\Phi(y,t) &= \Phi(y,t){J^{(1)}}^a_\mu(x,t) \quad ; \qquad \text{for } x < y \qquad (4.16a)\\ {J^{(1)}}^a_\mu(x,t)\Phi(y,t) &= \Phi(y,t){J^{(1)}}^a_\mu(x,t) - \frac{1}{2} f^{abc}\, Q^b_0\Big(\Phi(y,t)\Big) J^c_\mu(x,t) \\ &\qquad ; \quad \text{for } x > y \qquad (4.16b) \end{aligned}$$

where Q^b_0 are the global charges associated with the local conserved current J^b_μ. They are the same as on the lattice, eq. (1.35).

The proof of the braiding relations (4.16) is the same as the proof of the braiding relations for disorder fields. It only relies on the way to deform the contour $\mathcal{C}_x$ entering in the definition of the currents $J^{(1)}(x,t)$ The relative positions of the contours $\mathcal{C}_x$ depend if $J^{(1)}(x,t)$ acts first or second: if $J^{(1)}(x,t)$ acts first (second) the contour is slightly under (above) the equal-time slice $t = \text{cst}$, we denote them $\mathcal{C}^-_x$ ($\mathcal{C}^+_x$). (Remember that product of operators are defined by time ordering.) The relation (4.16a) follows because, in this case, there is no topological obstruction for moving the contour from the configuration $\mathcal{C}^+_x$ to the configuration $\mathcal{C}^-_x$. In the case of the relation (4.16b), these is an obstruction for moving the contour $\mathcal{C}^+_x$ onto the contour $\mathcal{C}^-_x$. This implies non-trivial braiding relations. All the non-locality of the currents ${J^{(1)}}^a_\mu(x,t)$ is concentrated in the fields $\phi^c(x,t)$, eq. (2.4). For $x > y$ the exchange relation between $\phi^c(x,t)$ and $\Phi(y,t)$ is:

$$\begin{aligned} \phi^c(x,t)\Phi(y,t) &= \int_{z\in\mathcal{C}^+_x} \star J^c(z)\Phi(y,t) \\ &= \int_{z\in\gamma(y)} \star J^c(z)\Phi(y,t) + \int_{z\in\mathcal{C}^-_x} \star J^c(z)\Phi(y,t) \qquad (4.17)\\ &= Q^c_0\big(\Phi(y,t)\big) + \Phi(y,t)\ \phi^c(x,t) \end{aligned}$$

The contour $\gamma(y)$ is a small contour surrounding the point y. Plugging back eq. (4.17) into the definition of the non-local current ${J^{(1)}}^a_\mu$ proves the braiding relations (4.16b).

4c) The non-local conserved charges and their algebra

Given conserved currents the associated charges are defined by integrating their dual forms along some curves. The charges depend weakly on the contours of integration because the dual forms are closed. The global conserved charges acting on the states of the physical Hilbert space are defined by choosing the domain of integration to be an equal-time slice. Namely for a current $\mathcal{J}_\mu(x,t)$:

$$Q = \int_{t=cst} dx\ \mathcal{J}_t(x,t) \tag{4.18}$$

We denote by Q_0^a and Q_1^a the global charges associated to the currents $J_\mu^a(x)$ and ${J^{(1)}}^a_\mu(x)$.

The (non-local) conserved charges generate a non-abelian extension of the two-dimensional Lorentz algebra. In two dimensions the Poincaré algebra which is generated by the momentum operators P_μ and the Lorentz boosts L is abelian. The momentum operators P_μ are the global charges associated with the conserved stress-tensor $T_{\mu\nu}(x)$: $\partial_\mu T_{\mu\nu}(x) = 0$ The Lorentz boost L is the global charge associated with the conserved boost current:

$$L_\mu(x) = \frac{1}{2}\epsilon^{\rho\sigma}\Big(x_\rho T_{\mu\sigma}(x) - x_\sigma T_{\mu\rho}(x)\Big) \tag{4.19}$$

The (non-local) charges satisfy the following algebraic relations:

$$\begin{aligned} \Big[\, Q_0^a \,,\, Q_0^b \,\Big] &= f^{abc} Q_0^c \quad ; \quad \Big[\, Q_0^a \,,\, Q_1^b \,\Big] = f^{abc} Q_1^c \\ \Big[\, L \,,\, Q_0^a \,\Big] &= 0 \quad ; \quad \Big[\, L \,,\, Q_1^a \,\Big] = -\frac{C_{Adj}}{4i\pi}\, Q_0^a \end{aligned} \tag{4.20}$$

The relations (4.20) are part of the defining relations of the semi-direct product of the Yangians $Y(\mathcal{G})$ by the Poincaré algebra. Only the Serre relations are missing. (They are more difficult to prove because they involve commutation relations between the non-local charges.) Moreover, as we will soon show, the comultiplications are those in $Y(\mathcal{G})$.

The three first relations are easily proved. The last relation is more interesting and can be proved in geometrical way. It consists in imposing a Lorentz boost $\mathcal{R}_{2\pi}$ of angle $(i2\pi)$ to the non-local currents ${J^{(1)}}^a_\mu(x,t)$. It is a rotation of (2π) in the Euclidian plane. Because the currents ${J^{(1)}}^a_\mu(x,t)$ are non-local this transformation does not act trivially on them: the string $\mathcal{C}_x$ winds around the point x. By decomposing this winding contour into the sum of a contour from $-\infty$ to x plus a small contour surrounding x we obtain:

$$\mathcal{R}_{2\pi}\, {J^{(1)}}^a_\mu(x,t)\, \mathcal{R}_{2\pi}^{-1} \;=\; {J^{(1)}}^a_\mu(x,t) - \frac{1}{2} f^{abc}\, Q_0^c\Big(\, J^b_\mu(x,t)\,\Big) \tag{4.21}$$

Integrating the time-component of eq. (4.21) over an equal-time slice gives

$$\mathcal{R}_{2\pi}\, Q_1^a\, \mathcal{R}_{2\pi}^{-1} \;=\; Q_1^a - \frac{1}{2}\, C_{Adj}\, Q_0^a \tag{4.22}$$

in agreement with the relations (4.20) because $\mathcal{R}_{2\pi} = \exp(i2\pi L)$.

4d) Action on the asymptotic states and the S-matrices. Non-perturbative results on the S-matrices can be deduced by looking at the action of the quantum charges on the asymptotic states. The constraints on the S-matrices we obtain arise by requiring that they commute with the non-local charges. These commutation relations imply algebraic equations which are nothing but the exchange relations for the quantum symmetry algebra (the Yangians $Y(\mathcal{G})$ in the case of massive current algebras). In general, these algebraic equations implies non-trivial constraints on the S-matrices which are sometimes enough to determine them.

Example: the SO(N) Gross-Neveu models. The $SO(N)$ Gross-Neveu models are equivalent to the $SO(N)$ massive current algebras at level $K = 1$. In the $SO(N)$ Gross-Neveu models the fundamental asymptotic particles are Majorana fermions taking values in the vector representation of $SO(N)$. In the SO(N) Gross-Neveu models, the $Y(SO(N))$ charges acting on the asymptotic fermions are given by:

$$Q_0^{kl} \;=\; T^{kl} \tag{4.23a}$$

$$Q_1^{kl} \;=\; -\,\frac{\theta\,(N-2)}{i\pi}\,\left(T^{kl}\right) \tag{4.23b}$$

$$\Delta Q_1^{kl} = Q_1^{kl}\otimes 1 + 1\otimes Q_1^{kl} - \sum_n \left(T^{kn}\otimes T^{nl} - T^{ln}\otimes T^{nk}\right) \tag{4.23c}$$

where the T^{kl}'s form the vector representation $\sqcup$ of $SO(N)$: $\left(T^{kl}\right)^{mn} = \delta^{km}\delta^{ln} - \delta^{lm}\delta^{kn}$. The charges Q_0^a and Q_1^a defined in eq. (4.23) satisfy the algebra (4.20); on-shell the boost operator L acts as $\frac{\partial}{\partial\theta}$. They define an irreducible representation of the $SO(N)$- Yangians in the vector representation of $SO(N)$. Eq. (4.23c) is the comultiplication in $Y((SO(N))$.

Denote by $S(\theta_{12})$, $\theta_{12} = \theta_1 - \theta_2$, the S-matrix of the two-fermion scattering. $S(\theta)$ acts from $\sqcup\otimes\sqcup$ into itself. As an $SO(N)$ representation the tensor product $\sqcup\otimes\sqcup$ decomposes into $(\,\sqcap\!\!\!\sqcup + \;\;\perp\;\; + \bullet)$. We denote by P_-, P_+ and P_0 the respective projectors. By $SO(N)$-invariance, $S(\theta)$ decomposes on these projectors:

$$S(\theta) \;=\; \sigma_+(\theta)P_+ + \sigma_-(\theta)P_- + \sigma_0(\theta)P_0 \tag{4.24}$$

where $\sigma_n(\theta)$ are scattering amplitudes. The non-local charges Q_1^{kl} are conserved and therefore they commute with the S-matrix. For the two-fermion scattering, the $Y(SO(N))$ exchange relations imply the following algebraic relations bewteen the scattering amplitudes:

$$\frac{\sigma_-(\theta)}{\sigma_+(\theta)} = \frac{\theta(N-2)+i2\pi}{\theta(N-2)-i2\pi} \quad ; \quad \frac{\sigma_0(\theta)}{\sigma_-(\theta)} = \frac{\theta+i\pi}{\theta-i\pi} \tag{4.25}$$

Eq. (4.25) determine $S(\theta)$ up to an overall function which could be fixed by closing the bootstrap program [9].

4e) Action on the fields and the field multiplets

(i) The definition of the action. The definitions of charges acting on the states and on the fields differ by the choice of the contour along which the conserved current is integrated. The charges acting on a field $\Phi(y)$ located at a point y are defined by choosing the contour of integration $\gamma(y)$ from $-\infty$ to $-\infty$ but surrounding the point y:

$$Q_k^a\Big(\Phi(y)\Big) = \int_{z\in\gamma(y)} dz_\mu \epsilon^{\nu\mu}\; J^{(k)}{}^a_\nu(z)\Phi(y) \tag{4.26}$$

Compare with the lattice definition (1.27).

For the currents $J_\mu^a(x)$ and the charges Q_0^a deforming the contour $\gamma(y)$ proves that

$$Q_0^a\Big(\Phi(y)\Big) = Q_0^a\;\Phi(y) - \Phi(y)\;Q_0^a \tag{4.27}$$

When the currents and the field $\Phi(y)$ are not respectively local the situation is more subtle. The contour $\gamma(y)$ can no more be closed and the action of the charges on the field is no more a pure commutator. For the non-local conserved currents $J^{(1)}(x)$ the relation between the global charges (4.18) acting on the states and the charges (4.26) acting on the fields is the following:

$$Q_1^a\Big(\Phi(y)\Big) = Q_1^a\;\Phi(y) - \Phi(y)\;Q_1^a + \frac{1}{2}f^{abc}\;Q_0^b\Big(\Phi(y)\Big)\;Q_0^c \tag{4.28}$$

The proof of eq. (4.28) consists in decomposing the contour of integration $\gamma(y)$ into the difference of two contours γ_+ and γ_- which are respectively above and under the point y, and in using the braiding relation (4.16b) when the current $J^b(x)$ is on γ_-.

(ii) The comultiplications. We now derive the comutiplication from the braiding relations. The comultiplications just encode how the charges act on a product of fields, say $\Phi_1(y_1)\Phi_2(y_2)\cdots$. We denote them by Δ. In the case of the charges Q_0^a

and for fields $\Phi_n(y_n)$ which are local with respect to the currents $J^a_\mu(x)$ all the contours can be deformed without troubles and we have:

$$\begin{aligned} Q^a_0\Big(\Phi_1(y_1)\Phi_2(y_2)\Big) &= Q^a_0\Big(\Phi_1(y_1)\Big)\Phi_2(y_2) + \Phi_1(y_1)Q^a_0\Big(\Phi_2(y_2)\Big) \\ \Delta Q^a_0 &= Q^a_0 \otimes 1 + 1 \otimes Q^a_0 \end{aligned} \tag{4.29}$$

It is the standard Lie algebra comultiplication as it should be.

In the case of the non-local charges Q^a_1 the standard comultiplication is deformed due to the non-trivial braiding relations between the non-local currents and the fields. Let $\Phi_n(y_n)$ be quantum fields local with respect to the currents $J^a_\mu(x)$. Then we have the following comultiplication for the non-local conserved charges Q^a_1:

$$\begin{aligned} Q^a_1\Big(\Phi_1(y_1)\Phi_2(y_2)\Big) &= Q^a_1\Big(\Phi_1(y_1)\Big)\,\Phi_2(y_2) + \Phi_1(y_1)\,Q^a_1\Big(\Phi_2(y_2)\Big) \\ &\quad - \frac{1}{2}f^{abc}\,Q^b_0\Big(\Phi_1(y_1)\Big)Q^c_0\Big(\Phi_2(y_2)\Big) \end{aligned} \tag{4.30a}$$

$$\Delta Q^a_1 = Q^a_1 \otimes 1 + 1 \otimes Q^a_1 - \frac{1}{2}f^{abc}\,Q^b_0 \otimes Q^c_0 \tag{4.30b}$$

Eqs. (4.29) and (4.30) are the comultiplication in $Y(\mathcal{G})$. Equation (4.30a) can be proved by decomposing the contour γ_{12} used in defining the action of Q^a_1 on the product $\Phi_1(y_1)\Phi_2(y_2)$. The contour γ_{12} is surrounding the two points y_1 and y_2. It decomposes into the sum of two contours γ_1 and γ_2 surrounding y_1 and y_2 respectively. But on the contour γ_2 we have to use the braiding relations (4.16) in order to pass the string $\mathcal{C}_z$ through the point y_1. Eq. (4.30) can also be proved starting from the graded commutators (4.28).

(iii) The field multiplets. To any (local) field $\Phi(x,t)$ is associated a multiplet which is constructed by acting on the field with as many charges as possible:

$$Q^{A_1} \cdots Q^{A_P}\ \Phi(x,t) \tag{4.31}$$

with, in the case $Y(\mathcal{G})$ symmetry, $Q^A = Q^a_0$, Q^a_1 or any element of the algebra generated by them. By construction, the fields (4.31) form a field multiplet in the sense of eq. (1.22). In general the field multiplets are infinite dimensional.

The main property of the field multiplets resides, (assuming the knowledge of the action on the asymptotic states), in the fact that if the field $\Phi(x,t)$ is known, then all its descendents, $Q^{A_1} \cdots Q^{A_P}\ \Phi(x,t)$ are also known. In other words, the descendents are completely determined by the data of the fields $\Phi(x,t)$ and of the values of the charges on the asymptotic states.

This property follows from the Ward identities expressing the quantum invariance:

$$\Delta^{(M)}\left(Q^A\right)\ \langle\Phi_1(x_1)\cdots\Phi_M(x_M)\rangle = 0 \tag{4.32}$$

where $\Delta^{(M)}$ the Mth comultiplication with $\Delta Q^A = Q^A \otimes 1 + \Theta^A_B \otimes Q^B$. Here we have assumed that the vaccuum is quantum group invariant: $Q^A|0\rangle = 0$, $\langle 0|Q^A = 0$. The identity (4.32) can be formulated on the form factors The form factors are the matrix elements of the fields between asymptotic states. By crossing symmetry, only matrix elements between the vacuum and the asymptotic particles are relevent. Let us denote by $Z^\alpha(\theta)$ the asymptotic particles with rapidity θ; they form a representation W of the quantum symmetry algebra. The form factors of the fields $\Phi_i(x,t)$ are defined by:

$$\mathcal{F}_i^{\alpha_1,\cdots\alpha_M}(\theta_1,\cdots,\theta_M) = \langle 0|\Phi_i^\Lambda(0)|Z^{\alpha_1}(\theta_1)\cdots Z^{\alpha_M}(\theta_M)\rangle. \tag{4.33}$$

On the form factors, the Ward identities (4.32) become:

$$\begin{aligned}
&\langle 0|\Big(Q^A\Phi_i(x)\Big)|Z^{\alpha_1}(\theta_1),\cdots,Z^{\alpha_M}(\theta_M)\rangle \\
&\quad = -\langle 0|\Big(\Theta^A_B\Phi_i(x)\Big)\Big(\Delta^{(M)}Q^B|Z^{\alpha_1}(\theta_1)\cdots Z^{\alpha_M}(\theta_M)\rangle\Big) \\
&\quad = \langle 0|\Phi^\Lambda(0)\Big(\Delta^{(M)}s(Q^A)|Z^{\alpha_1}(\theta_1)\cdots Z^{\alpha_M}(\theta_M)\rangle\Big)
\end{aligned} \tag{4.34}$$

with s the antipode. Eqs. (4.34) give the form factors of the field $Q^A(\Phi_i(x,t))$ in terms of those of the fields $\Theta^A_B\Phi_i(x,t)$ and of the action of the charges Q^A on the asymptotic particles $Z^\alpha(\theta)$.

Example: action on the stress-tensor. In massive current algebras, the stress-tensor and the current are in the same $Y(\mathcal{G})$ - multiplets. We have:

$$Q^a_1(T_{\mu\nu}) \propto \epsilon_{\mu\rho}\partial_\rho J^a_\nu + \epsilon_{\nu\rho}\partial_\rho J^a_\mu. \tag{4.35}$$

This relation was proved in [25] using form factor technique, it can aslo be deduced from the hypothesis we made for defining the massive current algebras.

Remark 1: The Ward identity (4.34) can written for any element in the enveloping algebra. Choosing a particular element associated to the square of the antipode leading to the so-called deformed KZ equations for the form factors [26].

Remark 2: Assuming, as in conformal field theory, that the (complete) symmetry algebra possess free field vertex representations, the form factors will also admit free field representations. The Zamoldchikov creation operators $Z^\alpha(\theta)$ as well as the field operators $\Phi^\Lambda(x)$ will be represented as quantum vertex operators in analogy with the vertex operator representations of the quantum affine algebras. This is suggested by the explicit formula for the form factors found by Smirnov [27]. Their generic forms are as follows:

$$\begin{aligned}
\mathcal{F}(\theta_1,\cdots,\theta_M) = &\int \prod_k d\mu(\alpha_k)\ P(\alpha_1,\cdots,\alpha_k|\theta_1,\cdots,\theta_M) \\
&\times \prod_{k<l} G_1(\alpha_k-\alpha_l)\prod_{i<j} G_2(\theta_i-\theta_j)\prod_{k,j} G_3(\alpha_k-\theta_j)
\end{aligned} \tag{4.36}$$

with $d\mu(\alpha)$ some integration measure, the functions G_n are some models dependent kernels and $P(\alpha_k|\theta_i)$ are polynomials. Formula (4.36) suggest the following vertex operator representations:

$$\mathcal{F}(\theta_1, \cdots, \theta_M) = \int \prod_k d\mu(\alpha_k) \langle \mathcal{O} \prod_k V(\alpha_k) \prod_i Z(\theta_i) \rangle \tag{4.37}$$

where the $V(\alpha_k)$'s are "screening" operators, the $Z(\theta)$'s are vertex operator representations of the Zamolodchikov operators and $\mathcal{O}$ an operator representing the fields. The kernel between these operators can be deduced from the formula (4.36).
Remark 3: The braiding relations between the quantum field multiplets are determined by the quantum symmetries: the braiding matrices intertwine the quantum symmetries, cf e.g. eqs. (1.29) and (1.30). Moreover the braiding relations are scaled invariant; i.e. they are renormalization group invariant. This is obvious from their definitions but this also follows from the topological origin of the braiding relations. The braiding relations just reflect the monodromy of the field multiplet correlation functions. Therefore, the renormalization group induces isomonodromy deformations [28]. The connection between isomonodromy deformations and quantum group symmetries could provide another starting point for determining the correlation functions in massive two dimensional quantum field theories.

5. CONCLUSIONS

Few open problems: Quantum symmetries have been used with some success to study integrable perturbations of conformal field theories. Some examples are [29]: the $\Phi_{(1,3)}$ and the $\Phi_{(1,2)}$, $\Phi_{(2,1)}$ perturbations of the minimal conformal models, the $G_K \otimes G_L/G_{K+L}$ cosets models, the Z_N parafermionic models, and the fractional supersymmetric models, etc... Most of the results obtained in these papers concern the S-matrices of these massive models. The main open problem is the derivation of the off-shell properties of the models, (the correlation functions and the form factors), from their quantum symmetries. As we mentioned in the introduction, this requires checking if the quantum symmetries form a complete symmetry algebra or not. Other few technical problems have been formulated in the previous sections, most of them as remarks. In particular, the connection between isomonodromy deformations and quantum group symmetries could open a new way of solving for the correlation functions.

3D generalizations? Let us discuss how these constructions could possibly be generalized to three dimensions. In two dimensions, quantum group symmetries require non-local currents: the non-locality, which is reflected in the equal-time commutation relations, imply the non-trivial comultiplications. The 2D non-local currents are fields localized on points but with a "string" attached to them. The currents are generically express as products of disorder fields (the "wavy string" in the lattice description) by spin fields (which are local fields). This is analogue to the definition of the 2D parafermions.

In any dimensions, to have more general symmetry than supersymmetry we need fields with non-trivial equal-time commutation relations. In three dimensions, this requires to consider fields localized on curves (with a sheet attached to them). Once again, examples are provided by disorder and parafermionic fields. The latters can be described as follows: consider a group G invariant spin lattice model in three dimensions. The disorder fields $\mu_g(C)$, $g \in G$, are defined by splitting all the spin variables σ which leave on a surface Σ_C bounded by C: $\sigma \to g\sigma$. By G-invariance, $\mu_g(C)$ depend weakly on Σ_C. The 3D parafermions $\Psi_g(C;x)$ are defined as product of disorder fields $\mu_g(C)$ by spin fields $\sigma(x)$:

$$\Psi_g(C;x) \;=\; \mu_g(C)\,\sigma(x) \qquad ; \qquad x \in C. \tag{5.1}$$

They satisfy non-trivial commutation relations analogous to the two-dimensional case. Anions are particular examples of this construction, with the curve C a small two-dimensional cone extending to the spacial infinity [30]. Generalizing eq. (5.1) by considering product of spin fields all along the curve C gives the parafermionic string which has been considered in the 3D Ising model [31].

Thus, if quantum group symmetry exists in three dimensions, it is a theory of quantum fields localized on curves, i.e. a theory of quantum loops. A (formal) example is given by Polyakov's string representation of gauge theories in three dimensions [32]. Let $W(C)$ be the Wilson loops:

$$W(C) \;=\; P\exp\left(\oint_C A\right) \tag{5.2}$$

where A is the Yang-Mills connection. Define the functional current $\mathcal{J}_\mu(C;x)$ by:

$$\mathcal{J}_\mu(C;x) \;=\; W(C)^{-1}\,\frac{\delta}{\delta x_\mu} W(C) \tag{5.3}$$

This functional current is conserved and curl-free:

$$\begin{aligned}
&\frac{\delta}{\delta x_\mu}\mathcal{J}_\mu(C;x) = 0\\
&\frac{\delta}{\delta x_\mu}\mathcal{J}_\nu(C;x') - \frac{\delta}{\delta x'_\nu}\mathcal{J}_\mu(C;x) + \Big[\mathcal{J}_\mu(C;x)\;,\;\mathcal{J}_\nu(C;x')\Big] = 0\\
&t_\mu \mathcal{J}_\mu(C;x) = 0
\end{aligned} \tag{5.4}$$

with t_μ the vector tangent to the curve C at the point x. Formally the following non-local current, $\mathcal{J}_\mu^{(1)}(C;x)$, localized on the curve C is functionally conserved:

$$\mathcal{J}_\mu^{(1)}(C;x) = \epsilon_{\mu\nu\rho} t_\nu \mathcal{J}_\rho(C;x) + \frac{1}{2}\Big[\, \mathcal{J}_\mu(C;x)\ ,\ \mathcal{P}(C;x)\,\Big] \tag{5.5}$$

with $\delta\mathcal{P}(C;x)/\delta x_\mu = \epsilon_{\mu\nu\rho} t_\nu \mathcal{J}_\rho(C;x)$. The analogy with the 2D current algebras is appealing, cf. eq. (3.21). It seems to indicate the possibility of having generalized non-local symmetry in 3D gauge theories. Unfortunately, the equations of motion (5.4) are not very rigorous; only the discretized lattice version has been proved and, up to our knowledge, no concrete results on the quantum continuous case has never been proved. However, to construct generalized quantum group symmetry in three dimensions remains a very attractive challenge.

Acknowledgements: It is a pleasure to thank my collaborators, Olivier Babelon, Giovanni Felder and André Leclair. I thank the organizers of the 91 Cargese school gor giving me the opportunity to present this lecture in a very pleasant atmosphere.

References

[1] S. Coleman and J. Mandula, Phys. Rev. 159 (1967) 1251

[2] A. Belavin, A. Polyakov and A. Zamolodchikov, Nucl. Phys. B241 (1984) 333

[3] See e.g. O. Babelon and C.-M. Viallet, *Integrable models, Yang-Baxter equation and quantum groups*, LPTHE-Paris preprint (1989), to appear (?) as a book; or L.D. Faddeev and L. Takhtadjan, *Hamiltonian methods in theory of solitons*, Springer Verlag (1987)

[4] S. Coleman, Phys. Rev. D11 (1975) 2088

[5] D. Bernard and A. Leclair, *"Quantum Group Symmetries and Non-Local Currents in 2D QFT"*, to appear in Comm. Math. Phys.

[6] S. Mandelstam, Phys. Rev. D11 (1976) 3026

[7] V. Drinfel'd, Sov. Math. Dokl. 32 (1985) 254; Sov. Math. Dokl. 36 (1988) 212

[8] M. Jimbo. Lett. Math. Phys. 10 (1985) 63; Lett. Math. Phys. 11 (1986) 247

[9] A. Zamolodchikov and Al. Zamolodchikov, Annals Phys. 120 (1979) 253

[10] Cf e.g. R.J. Baxter, *Exactly solved models in statistical mechanics*, Academic Press 1982

[11] D. Bernard and G. Felder, *Quantum group symmetries in 2D lattice quantum field theory*, to appear in Nucl. Phys. B

[12] L.P. Kadanoff and H. Ceva, Phys. Rev. B3 (1971) 3918;
E. Fradkin and L.P. Kadanoff, Nucl. Phys. B170 (1980) 1;
J. Fröhlich and P. A. Marchetti, Comm. Math. Phys. 112 (1987) 343

[13] M. Lüscher, Nucl. Phys. B135 (1978) 1

[14] D. Bernard, Comm. Math. Phys. 137 (1991) 191

[15] V. Zakharov and A. Shabat, Funct. Anal. 13 (1979) 166

[16] E. Date, M. Jimbo, M. Kashiwara and T. Miwa, Proc. Japan Acad. 57A (1981) 342; ibid. 57A (1981) 387; J. Phys. Soc. Japan 50 (1981) 3806; Physica 4D (1982) 343; Publ. RIMS 18 (1982) 1111; J. Phys. Soc. Japan 50 (1981) 3813

[17] M. Semenov-Tian-Shansky, Funct. Anal. Appl. 17 (1983) 259; Publ. RIMS 21 (1983) 1237

[18] O. Babelon and D. Bernard, Phys. Lett. 260B (1991) 81; *"Symmetries of the Heisenberg models"*, in preparation

[19] J. Avan and M. Bellon, Phys. Lett. B213 (198) 459

[20] E. Sklyanin, *"On the complete integrability of the Landau-Lifshitz equation"*, preprint LOMI E-3-79 (1980) Leningrad

[21] O. Babelon, unpublished

[22] K. Uneo and Y. Nakamura, Phys. Lett. 117B (1982) 208

[23] E. Brezin et al, Phys. Lett. 82B (1979) 442

[24] M. Abdalla Phys. Lett. 152B (1985) 215

[25] A. Leclair and F. Smirnov, *Infinite quantum symmetries of fields in massive quantum field theories*, to appear in J. Mod. Phys. A

[26] I. Frenkel and N. Reshetikhin, to appear; F. Smirnov, preprint RIMS-772 (1991)

[27] F. Smirnov, *Form factors in completly integrable models of quantum field theory*, to be published in World Scientific

[28] See e.g. the review by M. Jimbo, Proc. of Symposia in Pure Math. 49 (1989) 379

[29] A. Zamolodchikov, Adv. Studies Pure math. 19 (1989) 641; N. Reshetikhin and F. Smirnov, Comm. Math. Phys. 131 (1990) 157; D. Bernard and A. Leclair, Nucl. Phys. B340 (1990) 721; F. Smirnov, Int. J. Mod. Phys. A6 (1991) 1253; D. Bernard and A. Leclair, Phys. Lett. B247 (1990) 309; C. Ahn, D. Bernard and A. Leclair, Nucl. Phys. B 346 (1990) 409; H. de Vega and V. Fateev, preprint LPTHE-90-36; V. Fateev, Int. J. Mod. Phys. A6 (1991) 2109; etc.

[30] F. Wilczek, Phys. Rev. Lett. 48 (1982) 1144; Y.S. Wu, Phys. Rev. Lett. 52 (1984) 2103; J. Frölich and P. Marchetti, Comm. Math. Phys. 121 (1989) 121

[31] Vl. Dotsenko and A. Polyakov, Adv. Studies Pure Math. 16 (1988) 171

[32] A. Polyakov, Phys. Lett. 82B (1979) 247

ON KNOT AND MANIFOLD INVARIANTS

Raoul Bott[1]

Department of Mathematics, Harvard University
Cambridge, Massachusetts, USA

LECTURE 1

In these two lectures I would like to discuss the new direction in which topology has moved largely through the impetus of physics-inspired ideas. But to start us off I thought I would, in this first lecture, present an "old" invariant of knots albeit in a "new" guise. Then we will compare it to its "old" manifestation.

Let me, therefore, teach you L. Kaufman's "state algorithm" for the "Conway polynomial" $\nabla_K(t)$ of the trefoil knot [4]. This polynomial is a slight refinement of the Alexander polynomial $\Delta_K(t)$, and the two are related by

$$\Delta_K(z) = \nabla_K(\sqrt{z} - 1/\sqrt{z}).$$

A planar projection of K takes the form

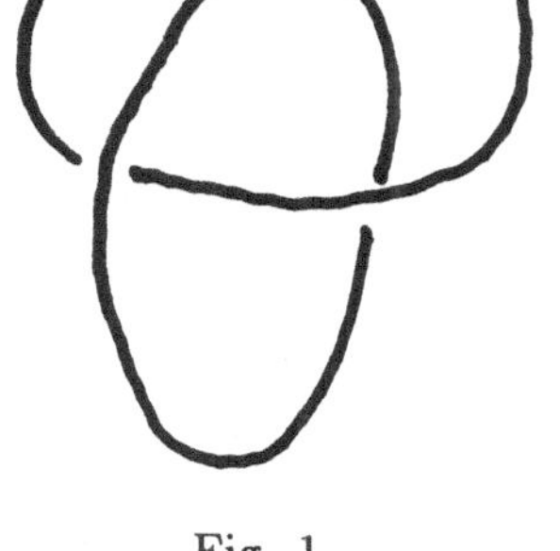

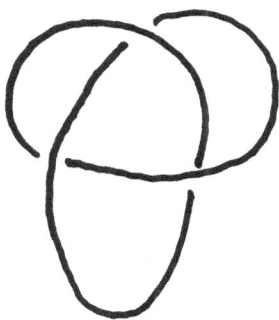

Fig. 1

and the first move towards Kaufman's statemodel is to choose two adjacent regions and to mark them with *'s:

[1]Supported in part by NSF Grant No. DMS-98 07995

New Symmetry Principles in Quantum Field Theory, Edited by
J. Fröhlich et al., Plenum Press, New York, 1992

Fig. 2

This corresponds to a base-point choice.

A state of this "pointed figure" is now by definition a marker "→" placed in each remaining region, pointing to one of its vertexes – *but with the proviso that each vertex receives only one pointer.*

Thus in our situation there are three possible states:

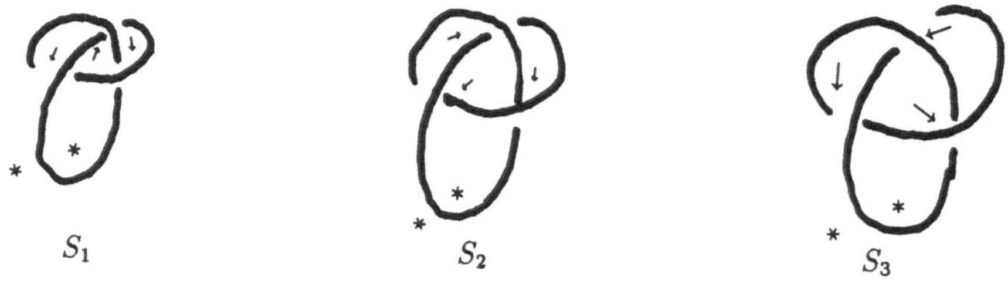

Fig. 3

Next associate a weight with each state as follows. The overpasses can be rotated to fit one of two types:

Fig. 4

and one marks these alternatives according to the principle:

Fig. 5

Thus these "vertex weights" can be marked on our K to yield:

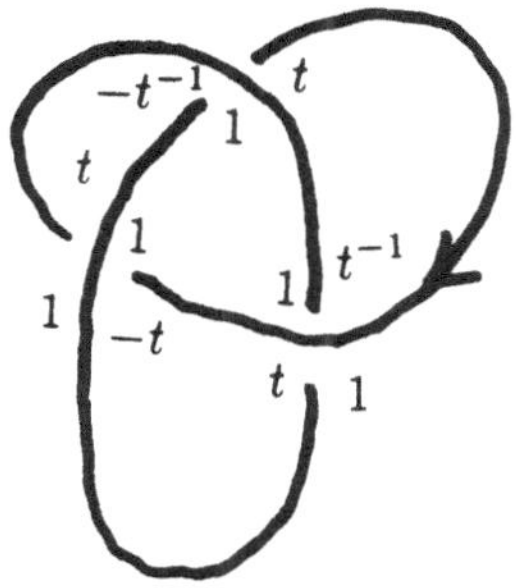

Fig. 6

Now, then, Kaufman counts each state of K by multiplying the "weights" to which the markers of that state point.E.g.,

$$\langle K|S_1\rangle = -1, \quad \langle K, S_2\rangle = tk^{-2}, \quad \langle K|S_3\rangle = t^2.$$

Finally, he sums there monomials, to arrive at a polynomial, in t which turns out to reproduce the Conway polynomial ∇_K of K. Thus:

$$\nabla_K = \sum \langle K|S\rangle = -1 + t^{-2} + t^2$$

for our trefoil. Finally, if one rewrites $\nabla_K(t)$ in terms of the variable $z = \sqrt{t} - \frac{1}{\sqrt{t}}$, then one obtains a representative of the classical Alexander polynomial $\Delta_K(z)$. Hence in our case:

$$\Delta_K(z) = z^2 + 1.$$

Classically the Alexander polynomial is only defined up to a unit in the ring of Laurent polynomials, i.e., a multiplicative factor of $\pm t^n$. Hence the Conway polynomial removes this ambiguity and is therefore a refinement.

The great virtue of these polynomials is of course that they are invariants of the knot – i.e., they are independent of the projective chosen to compute them. From the combinatorial point of view this means that they are invariant under the "Reidemeister movers" which mediate between such different projections.

These "moves" are as follows:

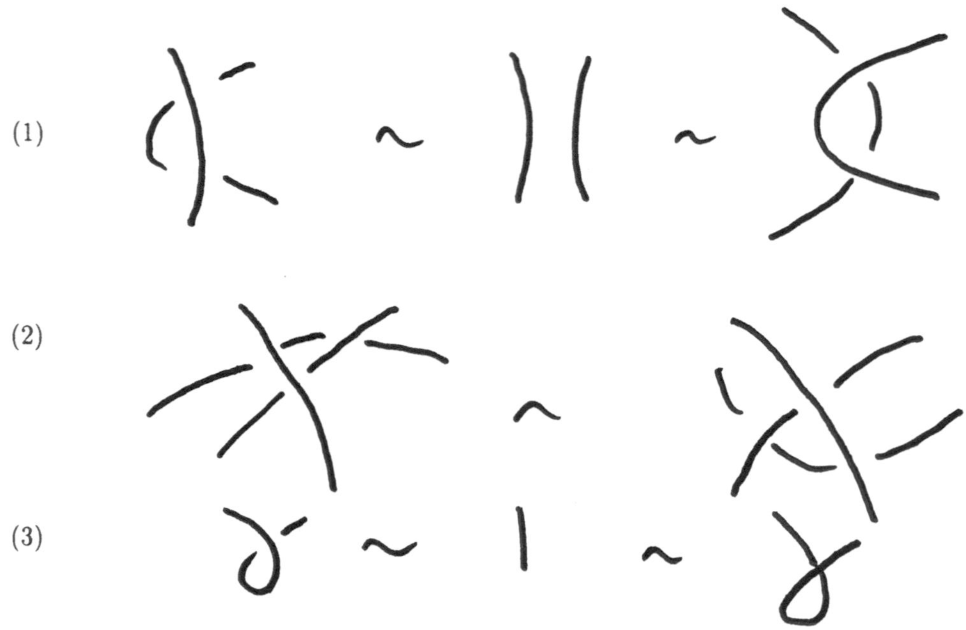

Fig. 7

and it is intuitively clear that they do not alter a knot. Conversely, Reidemeister showed that any two projections of isotopic knots can be connected by a sequence of such moves.

Now, simple as this algorithm is, it has the disadvantage of all statistical models: The computational complexity growth exponentially in the number of crossings, and so even from a completely combinatorial point of view one would like to "understand" this algorithm and find a more tractable form for it. In short, the question is: For what is this construction a recipe? The answer – already known to Alexander – is very beautiful, but to explain it properly I have to recall a little bit of "covering-space" theory.

The coverings of a reasonable space, X, e.g., a manifold finite complex, etc., are controlled by its fundamental group $\pi_1(X)$, in the sense that the distinct coverings correspond to the conjugacy classes of subgroups of π_1. In fact, if $\widetilde{X} \xrightarrow{\pi} X$ is a "covering", then the image of the fundamental group of $\pi_1(\widetilde{X})$ in $\pi_1(X)$ determines $\widetilde{X}$ and if this image is a normal subgroup, then the quotient group $\pi_1(X)/\pi_1(\widetilde{X})$ *acts on* $\widetilde{X}$ as "covering transformations", i.e., by permuting the leaves. For example, if

$X = S^1 \vee S^2 =$ 

Fig. 8

Then $\pi_1(X) = \pi_1(S^1) = \mathbf{Z}$ and corresponding to the trivial subgroup of $\mathbf{Z}$ we have the *universal* cover of X which looks like an infinite string of regularly spaced balloons:

Fig. 9

The covering transformations are just the translations. We, therefore, see clearly that although $H_2(\widetilde{X})$ is infinite dimensional, it is generated by just one element as a $\pi_1(X)$ module. Precisely, $H_2(\widetilde{X})$ is a free module of rank one over $\mathbb{Z}\pi_1(X)$, the group ring of $\pi_1(X)$, which can be thought of as the ring of Laurent series with integer coefficients in an indeterminate T.

$$\mathbb{Z}\pi_1(X) = \mathbb{Z}[T, T^{-1}]$$

With this warm-up understood, consider the complement of our knot K in $\mathbf{R}^3 \cup \infty = S^3$.

$$X = S^3 - K.$$

Because K is connected, $H_1(X) = \mathbb{Z}$, the generator being a small circle looping K once. But $H_1 = \pi_1/[\pi_1, \pi_1]$, so that if $\widetilde{X}$ is the covering of X associated to the *commutator subgroup of* π_1, then $\mathbb{Z} = H_1$ acts on $\widetilde{X}$ via covering transformations! In short, $H_1(\widetilde{X})$ is naturally a $\mathbb{Z}[T, T^{-1}]$ module, but this time it turns out that $H_1(\widetilde{X}) \otimes \mathbf{R}$ is a *finite dimensional* vector space! (We say that $H_1(\widetilde{X})$ is a torsion-module for $\mathbb{Z}[T, T^{-1}]$.) It follows that multiplication by T is given by a finite dimensional matrix A, and its characteristic polynomial

$$\det(T - A)$$

annihilates all elements of $H_1(\widetilde{X})$. Setting

$$A_K(T) = \det(T - A)$$

brings us to the "old" and most conceptual definition of the Alexander polynomial of K.

Now as physicists your natural question will be whether all this theory has brought us anything of computational value and to convince you that it has, let me push on and describe the nature of the answer in terms of the "Seifert Surface" of the knot K.

Because $\mathbf{R}^3$ is intractible, K is the boundary of a 2-chain which in fact can be taken to be a smooth oriented 2-manifold with a small disc removed, smoothly imbedded in $\mathbf{R}^3$. For our K the following picture illustrates this fact.

Fig. 10

$K = \partial S$ $\qquad$ S -- the shaded region

S being a surface of genus g.

One now creates $\widetilde{X}$, the covering of $X = S^3 - K$ we are after, by first "blowing X up" along the submanifold $Y = S - K$. You may think of the resulting space W as obtained from X by cutting along Y, or more technically as replacing $Y \subset X$ by its "normal ray bundle".

The resulting space is then a manifold W whose boundary consists of two copies of Y, which after a suitable orientation is chosen, we denote by Y_+ and Y_-.

$$\partial W = Y_+ - Y_- .$$

In terms of W, $\widetilde{X}$ is created by taking $\mathbf{Z}$ copies of W and glueing them end to end with Y_+ of the nth copy attached to Y_- of the $(n+1)$st. Put differently, this construction yields a Meyer-Vittoris picture of $\widetilde{X}$ of the type

$$0 \longleftarrow \widetilde{X} \longleftarrow \coprod_{n \in \mathbf{Z}} W_n \leftleftarrows \coprod_{n \in \mathbf{Z}} Y_n \longleftarrow 0$$

with the deck-transformation acting by translation. This finally leads to a description of $H_1(\widetilde{X})$ as a quotient module of two free $\mathbf{Z}[T, T^{-1}]$ modules:

$$0 \longleftarrow H_1(\widetilde{X}) \longleftarrow H_1(W)[T_1 T^{-1}] \xleftarrow{\alpha} H_1(Y)[T, T^{-1}] \longleftarrow 0.$$

It should also be intuitively clear now that in this description the Meyer-Vietoris map α is given by:

$$\alpha = \iota^+ - T\iota^-$$

where ι^+ and ι^- map $H_1(Y)$ into $H_1(W)$ and correspond to the act of pushing a cycle on Y into W "upward" or "downward". For instance, in Fig. 11, I have drawn the surface of an anchor-ring in $\mathbf{R}^3$ with a small disc at the top removed:

In this instance, a basis for $H_1(Y)$ is given by α and β. If we take the + normal to Y as indicated, then we see that $\iota^+\alpha$ is nontrivial in $H_1(W)$, whereas $\iota^-\alpha$ is homologous to zero. On the other hand, $\iota^+\beta$ is homologous to zero and $\iota^-\beta$ is non trivial. In fact,

$$\iota^+ - \iota^- : H_1(Y) \longrightarrow H_1(W)$$

is now clearly an isomorphism. This phenomenon is true in general and explicitly describes "Alexander duality" in this situation.

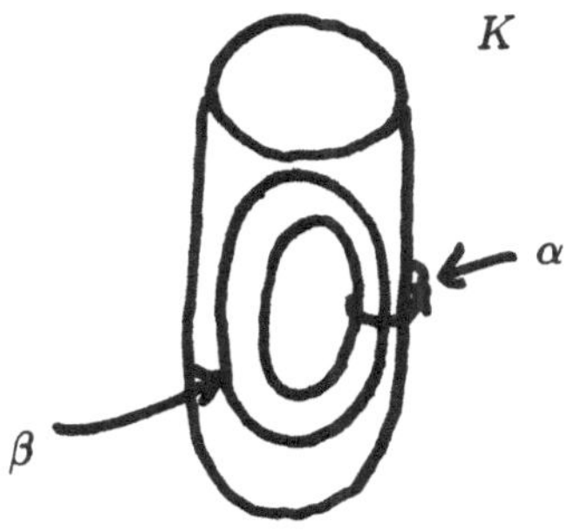

Fig. 11

You may think of this picture as a rather "inefficient" Seifert surface for "unknot" K.

The upshot is that if one uses the isomorphism

$$\iota_+ - \iota_- : H_1(Y) \simeq H_1(W)$$

to identify these two vector spaces we can write our "resolution of $H_1(\widetilde{X})$" in the form:

$$0 \longleftarrow H_1(\widetilde{X}) \longleftarrow H_1(Y)[T,T^{-1}] \xleftarrow{\tilde{\alpha}} H_1(Y)[T,T^{-1}] \longleftarrow 0$$

with

$$\tilde{\alpha} : \Gamma_+ - T\Gamma_-$$

where Γ_+ and Γ_- are endomorphisms of $H_1(Y) = H_1(S)$ inducing ι_+ and ι_-:

$$\iota_+ = (\iota_+ - \iota_-)\Gamma_+ \qquad \iota_- = (\iota_+ - \iota_-)\Gamma_-$$

so that $\Gamma_+ - \Gamma_- = 1$. Now suppose that Γ_- is an isomorphism. Then inverting by Γ_- we see that in $H_1(\widetilde{X})$, T acts by the matrix $A = \Gamma_+\Gamma_-^{-1}$. Hence $\det(T - A)$ yields our Alexander invariant.

More generally, it is not hard to see that we can reduce the general case to this one by taking $\tilde{\alpha}$ to act from the Laurent polynomials on $H_1(Y)/(\ker\Gamma_+ \oplus \ker\Gamma_-)$ to those on $H_1(Y)/\Gamma_- \ker\Gamma_+ \oplus \Gamma_+ \ker\Gamma_-$. QED

In this geometric setting the Alexander polynomial is the characteristic polynomial of the matrix $A = \Gamma_+\Gamma_-^{-1}$ which arises naturally in studying the homology inclusions of a spanning surface S of the knot K into the blow-up along S of its complement. All this and more really goes back to Stiefel and is beautifully explained in Kaufman's book [K] where one can also find Stiefel's formulae for the homology of the n-fold branched covers of S^3 along K. These are essentially obtained from the present discussion by dividing $\widetilde{X}$ by T^n, and so can also be determined in terms of Γ_+ and Γ_-.

To summarize then: Whereas the complexity of the statistical definition growth exponentially in the # of vertexes, the "topological" definition has the complexity of the characteristic polynomial of a $g \times g'$ matrix, with $g' \leq$ genus of a Seifert surface. To a topologist there is, of course, no question which of these two definitions is the "proper one". After all, the second one depends squarely on the well known mainstays of algebraic topology, homotopy and homology.

LECTURE 2

Let me start this second lecture by now computing the Jones polynomial of our trefoil K for you, again using an algorithm due to L. Kaufman [4]. His construction goes in two steps. First he constructs his "bracket polynomial" $\langle\ \rangle$, which is defined on knot diagrams but is only invariant under the first two Reidemeister moves. Note that these two do not violate "regularity" (i.e., smoothly turning tangents) of the immersions depicted, whereas the third one does. It is, therefore, prudent to take care of these moves in different ways. For his $\langle\ \rangle$ Kaufman first constructs a polynomial in the letters A, B, and d, for knot diagrams, using the following "Skein Rules":

(1)
$$\begin{cases} \langle \ \diagdown \ \rangle = A\langle \ = \ \rangle + B\langle \ )(\ \rangle \\ \langle \ \diagup \ \rangle = B\langle \ = \ \rangle + A\langle \ )(\ \rangle \end{cases}$$

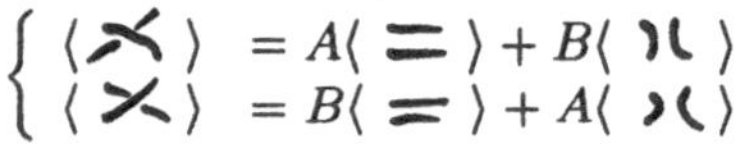

(2)
$$\langle 0 \amalg K \rangle = d\langle K \rangle$$

(3)
$$\langle 0 \rangle = 1$$

These are rules explaining how $\langle\ \rangle$ changes under alterations of a diagram at a particular crossing and they clearly amount to an algorithm which associates to every way of removing all crossings in K a monomial in A, B, and d. In fact, the following recipe for computing $\langle K \rangle$ is valid.

Now define a state "S" of K to be the assignment of a transversal marker (line segment) to every crossing, as indicated below for our trefoil

$S:$

Fig. 12

One next names the regions abutting each crossing, as A and B regions, the A regions being those swept out by the *overpassing* curve as it is rotated *counter-clockwise* in the plane. In our example this gives:

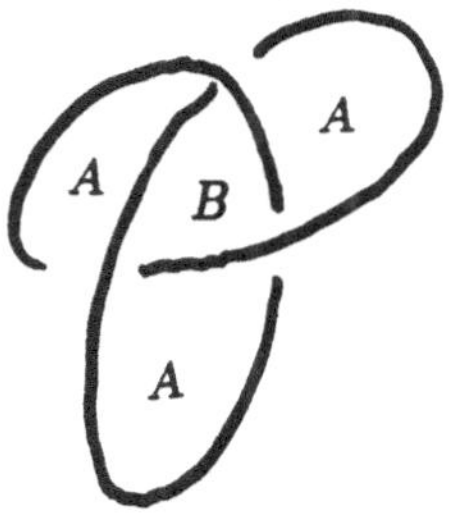

Fig. 13

With this understood it is clear that every state S determines a monomial in A and B, just counting the number of A-markers and B-markers. For our illustration one obtains B^3, and in general this monomial is denoted by $\langle K|S\rangle$. Kauffman now uses the markers as barriers which uniquely describe a "dismantling" of the knot into a collection of disjoint circles. In our case this yields

Fig. 14

The number of components of this collection of circles is notated $|S|$. With this understanding, the polynomial $\langle K\rangle$ is defined by

$$\langle K\rangle = \sum_S \langle K|S\rangle d^{|S|-1}.$$

In our example, this gives:

$$\langle K\rangle = A^3d + B^3d^2 + 3AB^2d + 3A^2B$$

corresponding to the remaining diagrams.

Fig. 15

By specializing A, B and d, Kauffman next observes that his $\langle K \rangle$ can be made to obey the *regular* Reidemeister moves. The specialization is given by:

$$A = B^{-1} \qquad \text{and} \qquad d = -A^2 - A^{-2}.$$

Under the remaining move this $\langle K \rangle$ is not invariant – but picks up a multiplicative factor $(-A)^{w(K)}$ where $w(K)$ is a numerical invariant of *oriented* links under regular moves, which goes by the lovely name of the writhe of K. It is computed by counting the crossings with a sign ϵ determined by the rule:

$$\epsilon(\searrow) = +1 \qquad \epsilon(\nearrow) = -1.$$

The marvelous upshot is that

$$f(K) = (-A^3)^{w(K)} \langle K \rangle$$

is an invariant of oriented links, which if A is replaced by $t^{-1/4}$ precisely yields the Vaughan Jones polynomial! By the way, in our example, I get

$$\langle K \rangle = A^{-7} - A^5 - A^{-3} \qquad w(K) = 3,$$

$$\begin{aligned} f(K) &= -A^{-9}(A^{-7} - A^5 - A^{-3} \\ &= -A^{-16} + A^{-4} + A^{-12}, \end{aligned}$$

which shows that $\langle K \rangle$ is not invariant under reflection as reflection corresponds to the transformation $A \longmapsto A^{-1}$, as the Alexander polynomial was.

For this fine new invariant $f(K)$ the question: *For what is $f(K)$ an algorithm*, can seemingly not be reduced to the flesh and bones of algebraic topology. Rather, here the answers proposed so far seems to lead to compact Lie groups and Riemann surfaces and conformal field theory.

But before I try to comment on this, let me take a minute to "remind" you of some combinatorial invariants which I devised long ago, in 1950, just after having learned about the Euler characteristic. These invariants were published in 1952 [B], and have slumbered peacefully in the literature since then. The word "reminding" is therefore meant here in its Platonic sense. However, as you will see, they fit beautifully into a state model framework. One starts with a finite n-cell complex X. A state S of X is now defined as simply a *finite subset* of the n-cells of X and X_S denotes the closed subcomplex of X obtained by deleting precisely the interior of the n-cells contained in S. We define:

$$\left\{\begin{array}{ll} \alpha(X) & = \dim H^n(X_S) \\ \\ \beta(X) & = \dim H^{n-1}(X_S) \\ \\ |S| & = \text{cardinality of } S, \end{array}\right\}$$

a numerical invariant for M, and with this understood the polynomials:

$$R(X;\lambda) = \sum_S (-1)^{|S|} \lambda^{\alpha(X_S)}$$

and

$$S(X;\lambda) = \sum (\sqrt{-1})^{|S|+\beta(X_S)} \lambda^{\alpha(X_S)}$$

are then seen to be combinatorial invariants of X.

The proof is very elementary: one checks invariance under "elementary" cell subdivision and then the multiplicative magic of state models combines with Meyer Vietoris to yield invariance. From this perspective the Kauffman $\langle\quad\rangle$ is seen to combine this sort of ingredient with the additional data furnished by "resolving" all double points in "all possible ways". I would therefore conjecture that there are many diffeomorphism invariants, relating say to higher knottedness phenomena, constructed by statistical means along these lines, and that our children will have interesting times explaining their "true" significance.

Evidence for this complexity is also to be found in the combinatorial invariants of compact three manifolds recently described by Turaev and Viro [T,V].

They start with an elaborate but finite system of colors I and nonzero weights $\omega_i \in \mathrm{C}$ for each color $i \in I$, and an auxiliary complex number $\omega \in \mathrm{C}^*$. They further postulate a notion of "admissible triples" in $I^3 = I \times I \times I$. The state model of a triangulated 3-manifold M is then built on the concept of "admissible colorings" φ of the edges of M. These correspond to colorings which assign to every boundary of a 2-simplex an admissible triplet of colors. The "initial data" $\{I, \omega_1, \omega; \text{adm } I\}$ is next enhanced by defining an admissible 6-tuple $\{i,j,k;\ell,m,n\} \in I^6$ to be one which arises from an admissible coloring of a tetrahedron T with i, j, k the colors of the boundary of a 2-simplex of T and ℓ, m, n the colors corresponding to the opposite edges of i, j, k. Finally, to such 6-tuples they postulate "weights" written

$$\begin{vmatrix} i & j & k \\ \ell & m & n \end{vmatrix},$$

also in $\mathbf{C}^*$.

These data are then subjected to an elaborate set of numerical constraints and symmetry conditions such as

$$\begin{vmatrix} i & j & k \\ \ell & m & n \end{vmatrix} = \begin{vmatrix} j & i & k \\ m & \ell & n \end{vmatrix}, \text{etc.}$$

(induced by the symmetry group of T) and

$$\sum w_j^2 w_{j_4}^2 \begin{vmatrix} j_2 & j_1 & j \\ j_3 & j_5 & j_4 \end{vmatrix} \begin{vmatrix} j_3 & j_1 & j_6 \\ j_2 & j_3 & j \end{vmatrix} = \delta_{j_4 j_6}, \quad \text{etc.}$$

I will return to the rationale for these identities in a moment, but first let me explain how they finally lead to a numerical invariant for M. Indeed, for any admissible φ we may now compute M_φ the "weight of φ" by setting:

$$M_\varphi = \prod_{\text{vertexes}} \omega^{-2} \prod_{\text{edges}} \omega_{\varphi(e)} \cdot \prod_{\text{tetrahedra}} T_\varphi,$$

where $T_\varphi = \begin{vmatrix} i & j & k \\ \ell & m & n \end{vmatrix}$ is the weight assigned to T by the $6j$-symbols of its coloring.

Finally: form the sum

$$|M| = \sum_\varphi M_\varphi \qquad \text{(admissible coverings } \varphi\text{).}$$

This is their invariant!

Where does this elaborate result come from? I only have time for a very brief indication. What most probably is happening here is that we are encountering a set of numerical constraints which one can associate with a very interesting extension of the group concept. Recall that every group G (say, finite) has as its dual the category of $R(G)$ finite dim representations of G over $\mathbf{C}$, and that G can be recovered from this category via the duality theory of Chevalley-Tanaka-Krein-Deligne. Now the simplest invariant of $R(G)$ is its character ring – or equivalently, its K-theory, or character table. However, $R(G)$ has more subtle invariants which are implicitly needed for the recovery of G. In particular, every choice of irreducibles, say $(A, B, C, \cdots)$ in $R(G)$ carries with it a definite set of "$6j$-symbols", and from this point of view some of the identities postulated by Turaev and Viro are the identities known to hold for these $6j$-symbols for quantum $SU(2)$ at various levels.

It is interesting that the $6j$-symbols have been used by physicists for a long time, but I know of no previous instance of their occurrence in the mathematical literature. Here is what they deal with.

Once a complete set of irreducibles in $R(G)$ is selected, we have decomposition formulae of the type:

$$A \otimes B \cong \sum_X W^X_{A,B} \otimes X$$

the sum being taken over our complete set of irreducibles $\{X\}$, and the $W^X_{AB} = \mathrm{Hom}^G(X, A \otimes B)$ being vector spaces over $\mathbf{C}$. Now, suppose these isomorphisms have been chosen. Then the associativity of the tensor product leads to the following isomorphisms:

$$\begin{aligned} A \otimes (B \otimes C) &\cong \sum_X A \otimes W^X_{BC} \otimes X \cong \sum_{XY} W^X_{BC} \otimes W^Y_{AX} \otimes Y \\ (A \otimes B) \otimes C &\cong \sum_Z W^Z_{AB} \otimes Z \otimes C \cong \sum_{ZY} W^Z_{AB} \otimes W^Y_{ZC} \otimes Y. \end{aligned}$$

As the direct sums are canonical here, the isomorphisms between the coefficients of a fixed Y, are given by homomorphisms

$$W^X_{BC} \otimes W^Y_{AX} \longrightarrow W^Z_{AB} \otimes W^Y_{ZC}$$

which are clearly indexed by 6 irreducibles, namely $\{A\ B\ C\ X\ Y\ Z\}$. In the $SU(2)$-case the W^X_{BC} are 0 or one-dimensional, and these arrows can then be converted into numbers once a suitable sign convention is adopted, and these are the classical $6j$ symbols $\begin{vmatrix} i & j & k \\ \ell & m & n \end{vmatrix}$ with the entries 1/2 integers indicating the spin of the corresponding irreducibles. The squares of these numbers are canonically determined as the traces of the corresponding arrows. (Note that the upper and lower letters have opposite variance so that each arrow has a natural "trace".)

In any case it is these symbols – however, in their q-deformations – that give rise to interesting invariants, and whose properties were undoubtedly the inspiration for the Turaev-Viro "initial data".

In a course given by D. Kazhdan last year, a different framework for these matters was proposed, which I find conceptually very appealing.

The first step is to extend the notion of $R(G)$ to a broader one – namely for that of a *Rigid-Balanced-Category.* Grosso modo this is a category very much like $R(G)$ in the sense that it has a $*$ operation and associative tensor product, $A \otimes B$, but where the commutativity isomorphism

$$S_{AB} : A \otimes B \longrightarrow B \otimes A$$

need not have square one.

Instead, one demands that (1) S satisfy braiding relations, and (2) that every object A in τ have a balancing automorphism $\tau_A \subset \mathrm{Aut}(A)$ with respect to which S^2

obeys a cocycle condition:

$$S_{B\otimes A} \circ S_{A\otimes B} = \tau_{A\otimes B} \cdot \tau_A^{-1} \otimes \tau_B^{-1}.$$

The braiding relations imposed upon S are of the following type.

$$\begin{array}{ll} S_{A\cdot B} \otimes 1 & \downarrow \\ 1 \otimes S_{A\cdot C} & \downarrow \\ S_{BC} \otimes 1 & \downarrow \\ 1 \otimes S_{BA}^{-1} & \downarrow \\ S_{C\otimes A}^{-1} \otimes 1 & \downarrow \\ 1 \otimes S_{CB}^{-1} & \downarrow \end{array}$$

Fig. 16

with the composition of these arrows = 1 and together these two properties allow one to develop a "scattering theory" for links colored by objects in X. In particular, a diagram

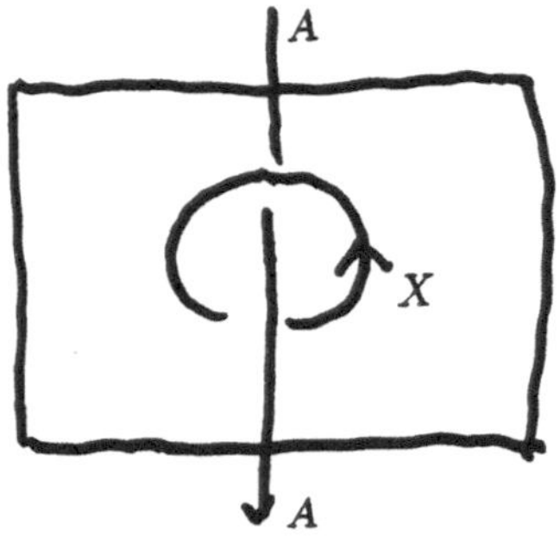

Fig. 17

determines an arrow $A \longrightarrow A$ describing the "scattering effect" of X on A, and if τ contains a linear combination of objects $X = \sum \omega_j A_i$, such that "scattering of A by X" equals τ_A for all objects in τ, then such an X determines a 3-manifold invariant. From this perspective, then, the w_i of Turaev and Viro seem to correspond to the components of such an X, in a suitable balanced category.

Presumably the conceptual flow-chart of the new invariants can therefore also be taken to be: given a compact Lie group, a Riemannian surface, and a level, the conformal field theory of this situation determines a balanced category, and if this category contains a representing element X of τ, then these data produce new potential 3-manifold invariants!

It is difficult to give precise references for these categorical concepts as there are many very similar axiom systems in the literature at present. In abstract algebra and category theory the names Drinfeld, Deligne, A. Joyal, R. Street, Joyce, Freyd,

Yetter, see [F;Y] should be mentioned, whereas in the arena of actual computations of 3-manifold invariants there are the accounts of Turaev, of Reshetikhin, of Kohno, of Likorish, of Walker, of Whitten, and of Jones - just to get you started! It is all a little bewildering and an article which traced a path between them would be very welcome.

But it seems to me that on the whole three very clear questions now arise:

(1) Are these "statistical" invariants of manifolds confined to low dimension?

(2) Are there concrete examples in which these "new" invariants are not just "statistical means" of Reidemeister torsion and other more "standard" topological invariants?

(3) Can one find solutions of the Turaev-Viro initial data, which are quite independent of the "holomorphic examples" furnished by the quantum groups, or balanced categories?

To explain this last question let me close with a word about the most geometric and for obvious reasons - for me - the most appealing point of view towards these 3 manifold invariants. That is, that they are all Morse-theoretic in nature! In the final analysis they can be traced to the fact that if f is a Morse function on a Riemann 3-fold, then the level sets $f^{-1}(c)$ with non-critical c, are oriented surfaces with an induced holomorphic structure.

One can therefore hope to make holomorphic constructions on these level sets - which are locally constant away from the critical levels and whose behavior at the critical points can be analyzed. Witten's beautiful infinite dimensional statistical interpretation of the Jones polynomial can certainly be cast in this language. Many other presently known computations can be placed in this context, and from this perspective the privileged nature of 3 dimensions is clear. In short, I am frankly torn between my instinct that new statistical invariants of diffeomorphism types of immersion should exist, and on the other hand, that in the final analysis these are all low-dimensional "fundamental group" phenomena.

BIBLIOGRAPHY

1. R. Bott, Two new combinatorial invariants for polyhedra, Portugalia Math., Vol. 11, Fasc. 1, pp. 35-40.

2. P.J. Freyd and D.H. Yetter, Braided compact closed categories with applications to low-dimensional topology, Advances in Math. 77 (1989), 156-182.

3. V.F.R. Jones, Index of subfunctors, Invent. Math. 72 (1983), 1-25.

4. L. Kaufman, On knots, Annals of Math. Studies # 115, Princeton University Press, 1987.

5. T. Kohno, Topological invariants of 3-manifolds using representations of mapping class groups I. (To appear)

6. W.B.R. Lickorish, Three-manifolds and the Temperley-Lieb algebra. To appear.

7. N. Reshetikhin and V.G. Turaev, Ribbon graphs and their invariants derived from quantum groups, Comm. Math. Phys. 198.

8. V.G. Turaev and O.V. Viro, State sum invariants of 3-manifolds and quantum $6j$ symbols (1990). Preprint.

9. Kevin Walker, On Witten's 3-manifold invariants (1990). Preprint, MSRI.

10. E. Witten, Quantum field theory and the Jones' polynomial, Comm. Math. Phys. 121 (1989), 356-399.

THE METRIC ASPECT OF NONCOMMUTATIVE GEOMETRY

Alain Connes[1] and John Lott [2*]

[1]IHES, Bures-sur-Yvette, France
[2]Department of Mathematics, University of Michigan-Ann Arbor, USA

Most of the previous work on "noncommutative geometry" could more accurately be labeled as noncommutative differential topology, in that it deals with the homology of differential forms on noncommutative spaces (cyclic homology) and vector bundles on noncommutative spaces (K-theory) [Co1]. However, the essence of geometry has to do with the metric properties of spaces.

In this paper we shall begin to investigate metric properties of noncommutative spaces. First, we shall show that the metric data of differential geometry can be reformulated in operator theoretic terms. This will be done using a familiar differential operator, the Dirac operator D. We shall see how the metric structure on a Riemannian manifold, namely the geodesic distance

(1) $d(p,q)$ = Infimum of the length of paths γ from p to q,

can be recovered from the selfadjoint operator D, acting in the Hilbert space $\mathfrak{h}$ of L^2 spinors, together with the representation in $\mathfrak{h}$ of the algebra of functions on the manifold.

Our data of a noncommutative metric space will consist of a triple $(\mathcal{A}, \mathfrak{h}, D)$ of a Hilbert space $\mathfrak{h}$, an involutive algebra $\mathcal{A}$ of operators on $\mathfrak{h}$ and a selfadjoint unbounded operator D on $\mathfrak{h}$. This new notion of a metric space includes as examples the following list of spaces, besides Riemannian manifolds:

α) Finite spaces.
β) Spaces of non integral Hausdorff dimension.
γ) Group rings of discrete subgroups of Lie groups.
δ) Configuration spaces in supersymmetric quantum field theory.
ε) "Quantum" tori.

Our task will be to show that this new class of spaces still deserves the name of geometry. For this, we shall replace the tools of the differential and integral calculus by operator theoretic tools. We shall develop a differential calculus on noncommutative spaces which, on a Riemannian manifold, reproduces the calculus of differential forms. The main tool of integration will be a nonstandard trace on operators, the Dixmier trace. It will allow us to develop the analogue of the Yang Mills action. (As an indication that we have the correct mathematical notion of the noncommutative Yang Mills action, we shall give a very general lower bound for the action in terms of a topological quantity.)

---

*Supported in part by the Humboldt Foundation and NSF Grant DMS-9101920.

New Symmetry Principles in Quantum Field Theory, Edited by
J. Fröhlich et al., Plenum Press, New York, 1992

The main example of a space to which all these considerations will be applied is an extended Euclidean space time. We shall give a geometric interpretation of the experimentally confirmed (at least at present energies) model of particle physics, namely the standard model. Our geometric interpretation of this model is as a pure gauge theory, but on a space time $E' = E \times F$, the product of ordinary Euclidean space time by a finite space. The geometry of the finite space is specified by a pair (h, D) as above, where h is finite dimensional and the self adjoint operator D encodes the fermion masses and the Kobayashi-Maskawa mixing parameters of the standard model.

Our analysis is limited to the classical level, and does not adress at the moment the questions related to renormalization. Nevertheless our geometric interpretation of the standard model gives an indication that the phenomenology of particle physics is saying something about the small-scale geometric structure of space time.

Most of the results of this paper appeared in [CL], and will appear as a chapter in [Co2].

1. RIEMANNIAN MANIFOLDS AND THE DIRAC OPERATOR

Let X be a compact Riemannian spin manifold and $D = \partial_X$ the corresponding Dirac operator (cf [LM]). Thus D is an unbounded self adjoint operator acting on the Hilbert space h of L^2 spinors on the manifold X.

To fix notation, we can write $D = \Sigma \gamma^j \nabla_j$, where the matrices $\{\gamma^j\}$ are skew-Hermitian. Then if f is a function on X, also regarded as a multiplication operator on h, we have $[D, f] = \Sigma \gamma^j \partial_j f$, an operator of Clifford multiplication which we shall denote by $\gamma(df)$.

We shall give 4 formulae below which show how to reconstruct the *metric space* (X, d) with d being the geodesic distance, the *volume measure* dv on X, the space of *gauge potentials* and finally the *Yang-Mills* action functional, from the following purely operator theoretic data :

$(\mathcal{A}, h, D)$,

where D is the Dirac operator on the Hilbert space h and $\mathcal{A}$ is the abelian algebra of continuous functions on X .

By Gelfand's theorem, we can recover the compact topological space X from $\mathcal{A}$. Namely, a point p of X gives a $*$ homomorphism $\rho : \mathcal{A} \to \mathbb{C}$ by setting $\rho(a) = a(p)$ for all $a \in \mathcal{A}$. And conversely, any such homomorphism ρ is given by evaluation at some point p, so X can be identified with the space of all such homomorphisms.
This is somewhat qualitative information, and we now come to the first interesting formula, giving us a natural distance function.

Formula 1
For any pair of points $p, q \in X$, their geodesic distance is given by:

(2) $$d(p,q) = \mathrm{Sup}\,\{\, |a(p) - a(q)| \,;\, a \in \mathcal{A} \,,\, \| [D,a] \| \leqslant 1 \,\}$$

The proof is straightforward, but is worth going through. The operator $[D,a]$, which is bounded iff a is Lipschitz, is given by the Clifford multiplication $\gamma(da)$ by the differential of a. This differential is a section of the cotangent bundle T^*X of X and one has :

$$\| [D,a] \| = \text{Essential Sup } \| da \| = \text{Lipschitz norm of } a$$

So in (2), we are imposing that |grad a| is everywhere ≤ 1, and then upon expressing a(p) - a(q) as a line integral, it follows that the right hand side of (2) is ≤ the geodesic distance d(p,q). But fixing the point p and considering the function

$$a(q) = d(q,p),$$

one checks that a is Lipschitz with constant one, so that $\| [D,a] \| \leq 1$ and one gets the desired equality. Note that (2) is in essence dual to the original formula (1) in that instead of involving arcs, namely copies of $\mathbb{R}$ inside the manifold X, it involves *functions*, i.e. maps from X to $\mathbb{R}$ (or $\mathbb{C}$).

This is an essential point for us, since in the case of discrete spaces or of noncommutative spaces X there may be no interesting arcs in X, but nevertheless there are plenty of *functions*, namely the elements $a \in \mathcal{A}$ of the defining algebra. We note rightaway that the right hand side of (2) makes sense in that general context, and technically defines a distance on the space of *states* of the C* algebra A:

$$d(\varphi, \psi) = \operatorname{Sup} \{ | \varphi(a) - \psi(a) | ; \| [D,a] \| \leq 1 \}.$$

We have now recovered from our original data $(\mathcal{A}, h, D)$ the metric space (X,d), with d being the geodesic distance. We still need tools of Riemannian geometry which are not immediately implied by the metric structure, the first being the measure given by the volume form :

$$f \mapsto \int_X f \, dv,$$

where in local coordinates, one has :

$$dv = (\det (g_{\mu\nu}))^{1/2} | dx^1 \wedge \dots \wedge dx^n |$$

This takes us to our second formula, which is nothing but a restatement of the H. Weyl theorem about the asymptotic behaviour of eigenvalues of elliptic differential operators. It does, however, involve a new tool, *the Dixmier trace* Tr_ω which, unlike asymptotic expansions, will make sense in full generality, and will be the correct operator theoretic replacement for integration.

Formula 2
For any $f \in \mathcal{A}$ one has :

(3) $$\int_X f \, dv = \text{const.}(d) \ \mathrm{Tr}_\omega (f | D |^{-d}), \qquad d = \dim X.$$

For the detailed definition and properties of the Dixmier trace Tr_ω we refer to [Di, Co2]. For the time being we can interpret the right hand side as the limit of the sequence :

$$\frac{1}{\operatorname{Log} N} \sum_0^N \lambda_j$$

where the λ_j 's are the eigenvalues of the compact operator $f | D |^{-d}$, or equivalently, as the residue of the function

$$\zeta(s) = \operatorname{Trace} (f | D |^{-sd}) \qquad \operatorname{Re} s > 1$$

at the point s = 1.

For us, the crucial fact is that the Dixmier trace is defined for operators on *any* Hilbert space, and that properties of the integral $\int_X f \, dv$ become corollaries of the general properties of the Dixmier trace :

A) Positivity : $\mathrm{Tr}_\omega(T) \geq 0$ if T is a *positive* operator.

B) Finiteness : $\mathrm{Tr}_\omega(T) < \infty$ if the characteristic values of T satisfy $\sum_0^N \mu_n(T) = O(\mathrm{Log}\, N)$.

C) Covariance : $\mathrm{Tr}_\omega(U\,T\,U^*) = \mathrm{Tr}_\omega(T)$ for any unitary U.

D) Vanishing : $\mathrm{Tr}_\omega(T)$ vanishes if T has finite trace in the usual sense.

Property D is the counterpart of locality in our framework. It shows that the Dixmier trace of an operator is unaffected by a finite rank perturbation.

One can see more clearly the locality of the Dixmier trace for a special class of operators, namely the pseudodifferential operators of order -d, acting on a vector bundle over X. If P is such an operator then, up to an overall constant which we shall neglect, the Dixmier trace is the integral, over the cosphere bundle of X, of the local trace of the symbol of P. Equation (3) is a special case of this last statement.

We shall now focus on defining the noncommutative Yang-Mills action. Before doing so, it may be worth making some general comments. First, the usual Yang-Mills action involves zeroth order information about the metric, i.e. no derivatives. We do not address the question of noncommutative *Riemannian* geometry, in the sense of Riemannian curvature, in this paper. Second, we shall shortly define a differential algebra on a noncommutative space, and use this to do gauge theory on the noncommutative space. If one is only interested in the differential calculus of gauge theory, it is enough to have a differential *Lie* algebra, such as, for example, the graded Lie algebra of g-valued differential forms for some Lie algebra g. (One still needs an inner product on the curvature forms, for which we shall use the Dixmier trace.) This fact was used to construct physical models in [CES]. However, one then loses the notion of an underlying space, which corresponds to an algebra. Given an algebra, one obtains a Lie algebra by putting $[x,y] = xy - yx$, but not every Lie algebra arises in this way. In this paper we base everything on *algebras*, which, as we shall see, corresponds to a generalization of electrodynamics. However, we cannot say for *a priori* reasons that either approach is physically right or wrong.

Let us then clearly state our aim; it is to recover the Yang Mills functional making use only of the following data :

Definition 1

A K cycle (h, D) *over an involutive algebra* $\mathcal{A}$ *is given by a representation of* $\mathcal{A}$ *on a Hilbert space* h *and a (possibly unbounded) selfadjoint operator* D *such that* $(1 + D^2)^{-1}$ *is compact and* $[D,a]$ *is bounded for all* $a \in \mathcal{A}$. *We shall say that a K cycle is even if, in addition, there is a self-adjoint operator* Γ *on* h, *the* $\mathbb{Z}/2$ *grading-operator, such that* $\Gamma^2 = 1$, $\Gamma D + D\Gamma = 0$ *and* $\Gamma a = a\Gamma$ *for all* $a \in \mathcal{A}$.

If the eigenvalues λ_n of $|D|$ are of the order of $n^{1/d}$ as $n \to \infty$, we say that the K cycle is (d, ∞) summable. On the algebra of smooth functions on a compact Riemannian spin manifold, the Dirac operator determines a K cycle which is (d,∞) summable where $d = \dim X$. If X is even-dimensional then the K cycle is even, with Γ being the chirality operator. We shall call this the Dirac K cycle. (The term K cycle comes from K-homology theory.)

The value of the following construction is that it will also apply when the $*$ algebra $\mathcal{A}$ is not commutative, or when D is no longer the Dirac operator. The reader can have in mind both the Riemannian case and the slightly more involved case where the algebra $\mathcal{A}$ is the $*$ algebra of matrix-valued functions on a Riemannian manifold, just in order to have in mind that the usual notion of exterior product does not make sense in the latter case.

We shall begin by the notion of a connection on the trivial bundle, i.e. the case of "electromagnetism", and define the vector potentials and Yang Mills action. We shall then treat the case of arbitrary hermitian bundles.

We want to define k-forms over $\mathcal{A}$ as operators on $\mathfrak{h}$ of the form

$$\omega = \Sigma\ a_0^j\,[D, a_1^j\,] \ldots [D, a_k^j\,]$$

where the a_i^j are elements of $\mathcal{A}$, represented as operators on $\mathfrak{h}$. This idea arises because although the operator D fails to be invariant under the representation on $\mathfrak{h}$ of the unitary group $\mathcal{U}$ of $\mathcal{A}$:

$$\mathcal{U} = \{ u \in \mathcal{A} \,;\, u^* u = uu^* = 1 \}$$

the following equality shows that the failure of invariance is governed by a 1-form in the above sense :

$$u\,D\,u^* = D + \omega_u\,, \quad \omega_u = u\,[D, u^*].$$

Note that ω_u is self adjoint as an operator on $\mathfrak{h}$ and it is thus natural to adopt the following definition :

Definition 2

A vector potential V *is a self adjoint element of the space of 1-forms :* $\Sigma\ a_0^j\ [D\ , a_1^j\ ]\ ,\ a_k^j \in \mathcal{A}$.

One can immediately check that in the basic example of the Dirac operator on a spin Riemannian manifold X, a vector potential in the above sense is exactly given by an imaginary 1-form v on X, the corresponding operator on spinors being $V = \gamma(v)$.

The action of the unitary group $\mathcal{U}$ on vector potentials is such that it replaces the operator $D + V$ by $u(D + V)\,u^*$. It is thus given by the algebraic formula :

$$\gamma_u(V) = u[D, u^*] + u\,V\,u^* \qquad u \in \mathcal{U}$$

Note that it is *not true* in general that $u\,V\,u^* = V$, as happens in the case of Riemannian manifolds.

We now just need to define the curvature or field strength θ for a vector potential, and use the analogue of equation (3) above to integrate the square of θ. So

$$YM(V) = Tr_\omega(\theta^2\,|D|^{-d})$$

should give us the Yang Mills action.

The formula for θ should be of the form :

$$\theta = dV + V^2$$

and the only difficulty is to define properly the "differential" dV of a vector potential, as an operator on $\mathfrak{h}$. The naive formula is :

$$\text{If } V = \Sigma\ a_0^j\,[D, a_1^j\,] \text{ then } dV = \Sigma\,[D, a_0^j\,]\,[D, a_1^j\,].$$

Before we point out what the difficulty is, let us check that if we replace V by $\gamma_u(V)$,

$$\gamma_u(V) = u\,[D, u^*\,] + \Sigma\ u\,a_0^j\,[D, a_1^j\,]\,u^*$$

then the curvature transforms in a covariant way :

$$d(\gamma_u(V)) + \gamma_u(V)^2 = u(dV + V^2)\, u^*$$

This computation is instructive so we shall do it in detail. First, to write $\gamma_u(V)$ in the same form as V, we use :

$$[D, a_1^j]\, u^* = [D, a_1^j u^*] - a_1^j [D, u^*]$$

Thus

$$\gamma_u(V) = u[D, u^*] + \Sigma\; u a_0^j\; [D, a_1^j u^*] - \Sigma\, u a_0^j a_1^j [D, u^*]$$

and one has :

$$d\gamma_u(V) = [D,u]\, [D,u^*] + \Sigma\, [D, u a_0^j\,]\, [D, a_1^j u^*] \;-\; \Sigma\, [D, u a_0^j a_1^j\,]\, [D, u^*].$$

Now we can see that the following operators on $\mathfrak{h}$ are equal :

α) $\quad d\,\gamma_u(V) + \gamma_u(V)^2$

β) $\quad u(dV + V^2)\, u^*.$

Indeed, the operator α) is equal to :

$$d\gamma_u(V) + (u[D,u^*] + u\, V\, u^*)^2 =$$

$$d\,\gamma_u(V) + u[D, u^*]\, u[D,u^*] + u[D,u^*]\, u\, V\, u^* + u\, V\, u^*\, u[D,u^*] + u\, V^2\, u^* =$$

$$d\gamma_u(V) - [D,u]\, [D,u^*] - [D,u]\, Vu^* + u\, V[D,u^*] + u\, V^2\, u^* =$$

$$\Sigma\, [D, u a_0^j\,]\, [D, a_1^j u^*] - \Sigma\, [D, u\, a_0^j a_1^j\,]\, [D,u^*] - [D,u]\, Vu^* + uV\, [D,u^*] +$$

$$u\, V^2 u^* =$$

$$u\,(d\,V)\, u^* + u\, V^2\, u^*,$$

where the last equality follows from :

$$\Sigma\, [D,u]\, a_0^j\, [D, a_1^j u^*] - \Sigma\, [D, u]\, a_0^j a_1^j\, [D, u^*] = [D,u]\, Vu^*,$$

$$\Sigma\, u\, [D, a_0^j]\, [D, a_1^j u^*] - \Sigma\, u\, [D, a_0^j\,]\, a_1^j\, [D, u^*] = u\, d\, V\, u^*,$$

$$\Sigma\, u\, a_0^j\, [D, a_1^j\,]\, [D, u^*] = u\, V\, [D, u^*].$$

However, there is a big difficulty that we overlooked, namely that the same vector potential V might be written in several ways as $V = \Sigma\;\; a_0^j\, [D, a_1^j\,]$, and so the definition of dV as $\Sigma\, [D, a_0^j\,]\, [D, a_1^j\,]$ is ambiguous.

To understand the nature of the problem, let us introduce some algebraic notation. We let $\Omega^*(\mathcal{A})$ be the universal differential graded algebra over $\mathcal{A}$. It is a formal object equal to $\mathcal{A}$ in degree 0 and generated by symbols $da,\ a \in \mathcal{A}$, of degree 1 with the following relations :

α) $\quad d(ab) = (da)\, b + a\, db \qquad \forall\, a, b \in \mathcal{A}$

β) $\quad d1 = 0.$

The involution $*$ of A extends uniquely to an involution on $\Omega^*(\mathcal{A})$ by the rule :

γ) $\quad (da)^* = -\, da^*$

The differential d on $\Omega^*(\mathcal{A})$ is defined *unambiguously* by

$$d(a^0 da^1 \dots da^n) = da^0\, da^1 \dots da^n \quad , \quad \text{for all} \quad a^j \in \mathcal{A}$$

and it satisfies :

δ) $d^2\omega = 0$ for all $\omega \in \Omega^*(\mathcal{A})$

ε) $d(\omega_1\, \omega_2) = (d\omega_1)\, \omega_2 + (-1)^{\deg \omega_1}\, \omega_1\, d\omega_2$ for all $\omega_j \in \Omega^*(\mathcal{A})$.

We will abbreviate $\Omega^*(\mathcal{A})$ by Ω^*, and $\Omega^k(\mathcal{A})$ by Ω^k.

Proposition 3

1) *The following equality defines an involutive representation* π *of the algebra* Ω^* *on* $\mathfrak{h}$ *:*

$$\pi(a^0\, da^1 \dots da^n) = a^0\, [D, a^1] \dots [D, a^n] \qquad \text{for all} \quad a^j \in \mathcal{A}.$$

2) ***Let*** $J_0 = \operatorname{Ker} \pi \subset \Omega^*$ ***be the two sided ideal of*** Ω^* ***given by*** $J_0^{(k)} = \{\, \omega \in \Omega^k \ , \ \pi(\omega) = 0 \,\}$. ***Then*** $J = J_0 + dJ_0$ ***is a two sided ideal of*** Ω^*, ***invariant under*** *d.*

The first statement is easy to see. Using it, we can define the Yang-Mills action unambiguously for any self-adjoint element of Ω^1. Let us discuss 2). By construction, J_0 is a two sided ideal, but it is not in general a ***differential*** ideal i.e. if $\omega \in \Omega^k$ and $\pi(\omega) = 0$ one does not have in general $\pi(d\omega) = 0$. This is exactly the reason why the above definition of $\Sigma\, [D, a_j^0]\, [D, a_j^1]$ as the differential of the 1-form $\Sigma\ a_j^0\, [D, a_j^1]$ was ambiguous.

Let us show, however, that $J = J_0 + dJ_0$ is still a two sided ideal. Since we have $d^2 = 0$, it is obvious that J is then a differential ideal. Let $\omega \in J^{(k)}$ be a homogenous element of J. Then ω can be written in the form $\omega = \omega_1 + d\omega_2$, where $\omega_1 \in J_0 \cap \Omega^k$, $\omega_2 \in J_0 \cap \Omega^{k-1}$. Choose $\omega' \in \Omega^{k'}$ and let us show that $\omega\omega' \in J^{(k+k')}$. One has

$$\omega\omega' = \omega_1\omega' + (d\omega_2)\, \omega' = \omega_1\, \omega' + d(\omega_2\, \omega') - (-1)^{k-1}\, \omega_2\, d\omega' =$$
$$(\omega_1\omega' + (-1)^k\, \omega_2\, d\omega') + d(\omega_2\, \omega').$$

But the first term belongs to $J_0 \cap \Omega^{k+k'}$, and $\omega_2\, \omega' \in J_0 \cap \Omega^{k+k'-1}$. Similarly, one shows that $\omega'\, \omega$ is in $J^{(k+k')}$. QED

Now using proposition 3, we can introduce the following graded differential algebra :

$$\Omega_D^* = \Omega^*/J.$$

(Quotienting out by J has the effect of eliminating the spurious fields of [CL], and makes the comparison easier with other treatments, such as [CES].)

Let us look at Ω_D^0 , Ω_D^1 and Ω_D^2 .

We have $J \cap \Omega^0 = J_0 \cap \Omega^0 = \{0\}$ if we assume, as we shall, that $\mathcal{A}$ is embedded in $\mathcal{L}(h)$, the bounded operators on h. Thus $\Omega_D^0 = \mathcal{A}$. Next,

$$J \cap \Omega^1 = J_0 \cap \Omega^1 + d(J_0 \cap \Omega^0) = J_0 \cap \Omega^1.$$

Thus Ω_D^1 is the quotient of Ω^1 by the kernel of π, and so it is exactly the space $\pi(\Omega^1)$ of operators ω of the form :

$$\omega = \Sigma\, a_j^0\, [D, a_j^1] \quad ; \quad a_j^k \in \mathcal{A}.$$

Now $J \cap \Omega^2 = J_0 \cap \Omega^2 + d(J_0 \cap \Omega^1)$ and the representation π gives us an isomorphism :

(4) $$\Omega_D^2 \simeq \pi(\Omega^2) / \pi\,(d\,(J_0 \cap \Omega^1)).$$

More precisely, this means that we can view an element ω of Ω_D^2 as a class of elements ρ of the form :

$$\rho = \Sigma\, a_j^0\, [D, a_j^1]\, [D, a_j^2] \quad ; \quad a_j^k \in \mathcal{A}$$

modulo the subspace of elements of the form :

$$\rho_0 = \Sigma\, [D, b_j^0]\, [D, b_j^1] \quad ; \quad b_j^k \in \mathcal{A} \quad , \quad \Sigma\, b_j^0\, [D, b_j^1] = 0.$$

It is clear now that because we work *modulo the subspace* $\pi(d(J_0 \cap \Omega^1))$, the question of the ambiguity in the definition of $d\omega$ for $\omega \in \pi(\Omega^1)$ no longer arises.

Note that equality (4) makes sense for all k :

(5) $$\Omega_D^k \simeq \pi(\Omega^k) / \pi(d(J_0 \cap \Omega^{k-1}))$$

and allows us to define an inner product on Ω_D^k : for each k, let h_k be the Hilbert space completion of $\pi(\Omega^k)$ with respect to the inner product

$$< T_1 , T_2 >_k = \mathrm{Tr}_\omega (T_2^* T_1 \,|D|^{-d}) \ , \ \forall\ T_j \in \pi\,(\Omega^k).$$

Let P be the orthogonal projection from h_k onto the orthogonal of the subspace $\pi(d(J_0 \cap \Omega^{k-1}))$. By construction, for $\omega_j \in \pi(\Omega^k)$, the inner product $< P\omega_1 , P\omega_2 >$ only depends upon the classes of the ω_j in Ω_D^k. We let Λ^k be the Hilbert space completion of Ω_D^k for this inner product; it is of course equal to Ph_k.

Proposition 4

1) *The actions of* $\mathcal{A}$ *on* Λ^k *by left and right multiplication define commuting unitary representations of* $\mathcal{A}$ *on* Λ^k.

2) *The functional* $YM(V) = < dV + V^2 , dV + V^2 >$ *is positive, quartic and invariant under gauge transformations,*

$$\gamma_u(V) = udu^* + uVu^* \qquad \forall\ u \in \mathcal{U}(\mathcal{A})$$

3) *The functional* $I(\alpha) = \mathrm{Tr}_\omega(\theta^2\,|D|^{-d})$, $\theta = \pi(d\alpha + \alpha^2)$ *is positive, quartic and gauge invariant on* $\{\alpha \in \Omega^1 : \alpha = \alpha^*\}$.

4) *One has* $YM(V) = \mathrm{Inf}\{ I(\alpha);\ \pi(\alpha) = V \}$

Let us say a few words about the easy proof. Since $\pi(d(J_0 \cap \Omega^{k-1})) \subset \pi(\Omega^k)$ is invariant under left and right multiplication by $\mathcal{A}$, and since the left and right actions of $\mathcal{A}$ on h_k are unitary, it follows that $P(a\xi b) = aP(\xi)b$ for all $a, b \in \mathcal{A}$ and $\xi \in h_k$. So 1) follows. For 2), one just notes that by the above calculation, with dV unambiguous now, $\theta = dV + V^2$ is covariant under gauge transformations. For 3), one uses again the above calculation to show that $d\alpha + \alpha^2$ transforms covariantly under gauge transformations. Finally, to see 4), note that if $\pi(\alpha) = V$ then $dV + V^2$, as an element of Λ^2, is equal to $P(\pi(d\alpha + \alpha^2))$. Thus $I(\alpha) \geqslant YM(V)$. To see that the infimum is attained, fix V and take any α such that $\pi(\alpha) = V$. We have that $\pi(d\alpha + \alpha^2) - (dV + V^2)$ is an element of $\pi(d(J_0 \cap \Omega^1))$, say $\pi(d\sigma)$. Put $\beta = \alpha - \sigma$. Then $\pi(\beta) = V$ and $\pi(d\beta + \beta^2) = dV + V^2$, so $I(\beta) = YM(V)$. QED

Stated in simpler terms the meaning of proposition 4 is that the ambiguity that we met above in the definition of the Yang-Mills action can be resolved by taking the infimum over all possibilities. The obtained action is nevertheless quartic by 4.2.

Example 1: We shall now check that in the case of Riemannian manifolds, with the Dirac K cycle, the graded differential algebra Ω_D^* is the same as the de Rham algebra of differential forms on X, with its usual prehilbert space structure. We now specialize to the Riemannian case, where $\mathcal{A}$ is the algebra of functions (with some regularity) on an even-dimensional compact spin manifold X and $D = \partial_X$ is the Dirac operator on the Hilbert space $L^2(X,S)$ of spinors. We let C be the bundle over X whose fiber at each $p \in X$ is the complexified Clifford algebra $\mathrm{Cliff}_{\mathbb{C}}(T_p^*(X))$ of the cotangent space at $p \in X$. Any measurable bounded section ρ of C defines a bounded operator $\gamma(\rho)$ on $h = L^2(X,S)$. For any $f_0, \ldots, f_n \in \mathcal{A}$, $f_0 df_1 \ldots df_n$ is an element of $\Omega^*(A)$ and one has $\pi(f_0 df_1 \ldots df_n) = f_0 df_1 \cdot \ldots \cdot df_n$, where on the right hand side, the usual differential df is considered as a section $[D, f]$ of C, and $\cdot$ denotes the product in C.

For each $p \in X$, the Clifford algebra C_p has a filtration by $\{C_p^{(k)}\}$, where $C_p^{(k)}$ is the subspace spanned by products of $\leq k$ elements of $T_p^*(X)$. The associated graded algebra $\{C_p^{(k)} / C_p^{(k-1)}\}$ is isomorphic to the complexified exterior algebra $\Lambda_{\mathbb{C}}(T_p^*(X))$ and we shall let $\sigma_k : C^{(k)} \to \Lambda_{\mathbb{C}}^k(T^*)$ be the quotient map.

Using the inner product on C given by the trace in the spinor representation, one can also identify $\Lambda_{\mathbb{C}}^k$ with the orthogonal complement of $C^{(k-1)}$ in $C^{(k)}$, or equivalently, if we let C^k be the subspace of $C^{(k)}$ of elements of the same parity as k then

$$\Lambda_{\mathbb{C}}^k = C^k \ominus C^{k-2}$$

Lemma 5

Let (h, D) *be the Dirac K cycle on the algebra* $\mathcal{A}$ *of functions on* X. *For* $k \in \mathbb{N}$, *a pair* T_1, T_2 *of operators on* h *is of the form* $T_1 = \pi(\omega)$, $T_2 = \pi(d\omega)$ *for some* $\omega \in \Omega^k$ *iff there exist sections* ϱ_1, ϱ_2 *of* C^k *and* C^{k+1} *such that*

$$T_j = \gamma(\varrho_j),\ j = 1, 2, \quad d\sigma_k(\varrho_1) = \sigma_{k+1}(\varrho_2).$$

Here $\sigma_k(\varrho_1)$ is an ordinary k-form on X and d is its ordinary differential. Note that for $k > d = \dim X$, one has $\sigma_k(\varrho) = 0$. We shall omit the proof of Lemma 5.

We can now easily determine the graded differential algebra Ω_D^*. First, let us identify $\pi(\Omega^k)$ with the space of sections of $C^{(k)}$. Then lemma 5 shows that :

$$\pi(d(J_0 \cap \Omega^{k-1})) = \operatorname{Ker} \sigma_k.$$

(If ϱ is a section of C^k with $\sigma_k(\varrho) = 0$ then the pair $\varrho_1 = 0$, $\varrho_2 = \varrho$ in C^{k-1} and C^k fulfills the condition of lemma 5, and so $\varrho = \pi(d\omega)$ for some ω with $\pi(\omega) = 0$.)

Thus σ_k is an isomorphism : $\Omega_D^k \simeq$ Sections of $\wedge_{\mathbb{C}}^k(T^*)$, which again by Lemma 5 commutes with the differential. We can then state :

Formula 3

The map $a^0\, da^1 \ldots da^n \to a^0\, da^1 \wedge \ldots \wedge da^n$ *from* Ω^* *to* $\wedge_{\mathbb{C}}^k(X)$ *extends to an isomorphism of the differential graded algebra* Ω_D^* *with the de Rham algebra of differential forms on* X. *Under this isomorphism, the inner product on* Ω_D^k *is the Riemannian inner product of k-forms :*

$$< \omega, \omega' > = \int_X \omega \wedge * \omega'.$$

The last equality follows from the computation of the Dixmier trace for the operator on $h = L^2(X,S)$ associated to a section ϱ of the bundle C of Clifford algebras:

$$\operatorname{Tr}_\omega (\varrho\, |D|^{-d}) = \int_X \operatorname{trace}(\varrho(p))\, dv(p).$$

As an immediate corollary of formula 3 we get :

$$YM(V) = \int_X \| dV \|^2\, dv$$

for any vector potential V on X. **End of Example 1.**

Let us now consider the generalized fermionic action. A fermion field will simply be an element of the Hilbert space h. The operator $D + V$ is self-adjoint.

Definition 6

The fermionic action is $<\psi, (D + V)\psi>$, *for* $\psi \in h$ *and* $V \in \Omega_D^1$.

By construction, the fermionic action is gauge invariant in that for any $u \in \mathcal{U}(\mathcal{A})$, it is invariant under the transformations

$$\psi \to u\psi, \quad V \to \gamma_u(V).$$

Example 2: *Massless chiral electrodynamics.*

We can call the following the action for a generalized "massless chiral electrodynamics":

$$L(V, \psi) = g^{-2}\, YM(V) + \langle \psi, (D + V)\, \psi \rangle \qquad \text{for } \psi \in h \text{ and } V \in \Omega_D^1.$$

The reason that the name is appropriate is that if we use the Dirac K cycle on a 4-dimensional spin manifold, then we obtain exactly the usual Euclidean action of massless electrodynamics with one fermion field, and with g being the physical coupling constant. We should note that one cannot really write a Euclidean action for chiral fermions, and the ψ field above is a 4-component spinor field. The reason to call it "chiral" electrodynamics is that one can Wick-rotate to Minkowski space, and then impose $\Gamma\psi = \psi$.

Similarly, to obtain the action of massless electrodynamics with N_+ right-handed fermions and N_- left-handed fermions, we would take the Hilbert space to be $h = h_+ \oplus h_-$, where h_+ consists of N_+ copies of $L^2(X,S)$ and h_- consists of N_- copies. With the grading operator given by $\Gamma = (\gamma^5 \otimes I_{N+}) \oplus (-\gamma^5 \otimes I_{N-})$, we can then write out $L(V,\psi)$, Wick-rotate and impose $\Gamma\psi = \psi$.

End of Example 2.

Let us now extend the definition of the Yang Mills action to connections on arbitrary hermitian vector bundles.

First of all, we need to express in algebraic terms, i.e. using only the involutive algebra $\mathcal{A} = C(X)$, the notion of a hermitian vector bundle over X. A vector bundle E is entirely characterized by the vector space $\mathcal{E}$ of its sections. Furthermore, this vector space has an $\mathcal{A}$-action, which we shall take to be on the right. In other words, $\mathcal{E}$ is a right $\mathcal{A}$-module. The local triviality of E and the finite dimensionality of its fibers translate algebraically to saying that there is an $\mathcal{E}'$ such that $\mathcal{E} \oplus \mathcal{E}'$ is $\mathcal{A}^N$ for some finite N, or in more fancy terms, that $\mathcal{E}$ is a finite projective module over $\mathcal{A}$.

The Hermitian structure on E, i.e. the inner product $(\xi, \eta)_p$ on each fiber E_p, allows us to construct a sesquilinear map :

$$(\ ,\ ) : \mathcal{E} \times \mathcal{E} \to \mathcal{A}$$

given by $(\xi, \eta)(p) = (\xi(p), \eta(p))_p$.

This map (,) satisfies the following conditions :

1) $(\xi a, \eta b) = a^* (\xi, \eta)\, b$ for all $\xi, \eta \in \mathcal{E}$, $a, b \in \mathcal{A}$

2) $(\xi, \xi) \geqslant 0$ for all $\xi \in \mathcal{E}$

3) $\mathcal{E}$ is self dual for (,).

Thus the hermitian vector bundles over X correspond to the hermitian finite projective modules over $\mathcal{A}$ in the following sense :

Definition 7

Let $\mathcal{A}$ be an algebra with an involution $*$ and a unit. Then a hermitian structure on a finite projective module $\mathcal{E}$ over $\mathcal{A}$ is given by a sesquilinear map $(\ ,\ ) : \mathcal{E} \times \mathcal{E} \to \mathcal{A}$ satisfying 1, 2 and 3 .

One can show that all hermitian structures on a given finite projective module $\mathcal{E}$ over $\mathcal{A}$ can be obtained as follows: one writes $\mathcal{E}$ as $e\mathcal{A}^N$ for appropriate e and N, where e, an NxN matrix with entries in $\mathcal{A}$, is selfadjoint and satisfies $e^2 = e$. One then restricts to $\mathcal{E}$ the hermitian structure on $\mathcal{A}^N$ given by :

$$(\xi, \eta) = \Sigma\, \xi_i^* \eta_i \in \mathcal{A} \qquad \text{for all } \xi = (\xi_i)\ ,\ \eta = (\eta_i).$$

The algebra $\mathrm{End}_{\mathcal{A}}(\mathcal{E})$ of endomorphisms of $\mathcal{E}$, that is, linear maps T from $\mathcal{E}$ to $\mathcal{E}$ which commute with the $\mathcal{A}$-action, has a natural involution, given by :

$$(T^*\xi, \eta) = (\xi, T\eta) \quad \text{for all } \xi, \eta \in \mathcal{E}.$$

With this involution, $\mathrm{End}_{\mathcal{A}}(\mathcal{E})$ is isomorphic to the algebra of matrices $eM_N(\mathcal{A})e$.

As before, we let (h, D) be a K cycle over $\mathcal{A}$.

Definition 8

The Hilbert space of "gauged spinors" is $\mathcal{E} \otimes_{\mathcal{A}} h$. Its inner product is given by

$$<\xi_1 \otimes \eta_1, \xi_2 \otimes \eta_2> = <\eta_1, (\xi_1, \xi_2)\eta_2>.$$

If h has a grading operator Γ, it extends to a grading operator on $\mathcal{E} \otimes_{\mathcal{A}} h$.

Definition 9

Let $\mathcal{E}$ be a hermitian finite projective module over $\mathcal{A}$. Then a connection on $\mathcal{E}$ is given by a linear map $\nabla: \mathcal{E} \to \mathcal{E} \otimes_{\mathcal{A}} \Omega_D^1$ such that

$$\nabla(\xi a) = (\nabla\xi)a + \xi \otimes da \quad \text{for all } \xi \in \mathcal{E}, a \in \mathcal{A}.$$

A connection ∇ is compatible (with the metric) iff :

$$<\xi, \nabla\eta> - <\nabla\xi, \eta> = d<\xi, \eta> \quad \text{for all } \xi, \eta \in \mathcal{E}.$$

Both sides of the last equation lie in Ω_D^1. (In computations, one should remember that $(da)^* = -da^*$ for all $a \in \mathcal{A}$, and if $\nabla\xi = \Sigma\, \xi_i \otimes \omega_i$ with $\omega_i \in \Omega_D^1$ then $<\nabla\xi, \eta> = \Sigma\, \omega_i^* <\xi_i, \eta>$.)

Such connections always exist, for with $\mathcal{E}$ expressed as $e\mathcal{A}^N$, one may take ∇ to be :

$$\nabla_0\xi = e\, d\xi.$$

Two connections ∇ and ∇' on $\mathcal{E}$ differ by an element of $\mathrm{Hom}_{\mathcal{A}}(\mathcal{E}, \mathcal{E} \otimes_{\mathcal{A}} \Omega_D^1)$. Any compatible connection can be written as $\nabla\xi = e\, d\xi + \rho\, \xi$, where ρ is a self-adjoint NxN matrix of 1-forms satisfying $e\rho = \rho e = \rho$.

As in proposition 4 we shall now give two equivalent definitions of the action functional $YM(\nabla)$ on the affine space $C(\mathcal{E})$ of compatible connections.

The group $\mathcal{U}(\mathcal{E})$ of unitary automorphisms of $\mathcal{E}$,

$$\mathcal{U}(\mathcal{E}) = \{u \in \mathrm{End}_{\mathcal{A}}(\mathcal{E})\ ;\ uu^* = u^*u = 1\}$$

will be the unitary gauge group, in that it acts by gauge-transformations (explicitly, $\gamma_u(\nabla) = u\nabla u^*$) on the space $C(\mathcal{E})$. To define the curvature θ of a connection ∇, we first need to define the covariant derivative of a vector-valued form. Put

$$\mathcal{E}' = \mathcal{E} \otimes_{\mathcal{A}} \Omega_D^*,$$

the space of vector-valued forms. Extend ∇ to a unique linear map from $\mathcal{E}'$ to $\mathcal{E}'$, which we shall also denote by ∇, by

$$\nabla(\xi \otimes \omega) = (\nabla\xi)\omega + \xi \otimes d\omega \qquad \text{for all } \xi \in \mathcal{E},\ \omega \in \Omega_D^*.$$

One finds that this linear map ∇ satisfies :

$$\nabla(\eta\omega) = (\nabla\eta)\,\omega + (-1)^{\deg\eta}\,\eta\, d\omega$$

for any homogeneous $\eta \in \mathcal{E}'$ and $\omega \in \Omega_D^*$.

It follows that $\nabla^2(\eta\omega) = (\nabla^2\eta)\,\omega$, i.e. ∇^2 is an endomorphism of the right Ω_D^* module $\mathcal{E}'$. It is determined by its restriction to $\mathcal{E}$, which we shall denote by θ :

$$\theta \in \mathrm{Hom}_{\mathcal{A}}(\mathcal{E}, \mathcal{E} \otimes_{\mathcal{A}} \Omega_D^2).$$

If we use the representation $\mathcal{E} = e\mathcal{A}^N$ then θ will be a self-adjoint NxN matrix of 2-forms such that $e\theta = \theta e = \theta$.

Next, using the inner product on Ω_D^2 and the hermitian structure on $\mathcal{E}$, one gets a natural inner product on

$$\mathrm{Hom}_{\mathcal{A}}(\mathcal{E}, \mathcal{E} \otimes_{\mathcal{A}} \Omega_D^2).$$

Using this we define

Definition 10 $\quad YM(\nabla) = \langle \theta, \theta \rangle.$

By construction, this action is gauge invariant, positive and quartic.

Formula 4
Let X be a Riemannian spin manifold with its Dirac K cycle (h, D). Then the notion of connection (Def. 9) is the usual one, and one has :

$$YM(\nabla) = \int_M \|\theta\|^2\, dv,$$

where θ is the usual curvature of ∇.

This follows immediately from Formula 3.

Thus we recover in this case the usual Yang Mills action. For the fermionic action, we define the gauged Dirac operator on $\mathcal{E} \otimes_{\mathcal{A}} h$ by

$$D_\nabla(\xi \otimes \eta) = \xi \otimes D\eta + (\nabla\xi)\,\eta \qquad \text{for all } \xi \in \mathcal{E} \text{ and } \eta \in h.$$

Definition 11
The fermionic action is $\langle\psi, D_\nabla \psi\rangle$, for $\psi \in \mathcal{E} \otimes_{\mathcal{A}} h$ and ∇ a compatible connection.

Example 3: *U(N) gauge theory with chiral fermions.*

Given an even $(4,\infty)$-summable K cycle, take $\mathcal{E}$ to be $\mathcal{A}^N$. Then $\mathcal{E} \otimes_{\mathcal{A}} h$ is simply N copies of h. Let ∇ be a compatible connection on $\mathcal{E}$. Consider the action

$$\mathcal{L}(\nabla, \psi) = g^{-2}\, YM(\nabla) + \langle\psi, D_\nabla \psi\rangle \qquad \text{for } \psi \in \mathcal{E} \otimes_{\mathcal{A}} h.$$

Its group of gauge invariances consists of the unitary NxN matrices over $\mathcal{A}$.

In the case of the Dirac K cycle on a 4-dimensional spin manifold, we obtain the usual Euclidean action of a U(N)-gauge theory with one N-tuple of fermion fields in the fundamental representation of U(N). **End of Example 3.**

We shall now mention the easy adaptation of proposition 4.4 to the general case. First of all, any compatible connection in the sense of Definition 9 is the composition with π of a *universal compatible connection,* i.e. a linear map

$$\nabla : \mathcal{E} \to \mathcal{E} \otimes_{\mathcal{A}} \Omega^1$$

fulfilling exactly the conditions of Definition 9.

To see the surjectivity of the map :

$$\pi : CC(\mathcal{E}) \to C(\mathcal{E})$$

(where $CC(\mathcal{E})$ is the space of universal compatible connections), it is enough to note that the special "Grassmannian" connection ∇_0 is of this form, and that π is a surjection of Ω^1 onto Ω_D^1. Next, a universal compatible connection extends uniquely to a linear map :

$$\nabla : \mathcal{E} \otimes_{\mathcal{A}} \Omega^* \to \mathcal{E} \otimes_{\mathcal{A}} \Omega^*$$

such that

$$\nabla (\eta\omega) = (\nabla\eta)\, \omega + (-1)^{\deg(\eta)}\eta\, d\omega$$

for any homogeneous $\eta \in \mathcal{E} \otimes_{\mathcal{A}} \Omega^*$ and $\omega \in \Omega^*$.

The curvature $\theta = \nabla^2$ is then an endomorphism of the module $\mathcal{E}' = \mathcal{E} \otimes_{\mathcal{A}} \Omega^*$ over Ω^*, and $\pi(\theta)$ makes sense as a bounded operator on the Hilbert space $\mathcal{E} \otimes_{\mathcal{A}} h$. Then the analogue of the action I of proposition 4 is given by :

$$I(\nabla) = \mathrm{Tr}_\omega (\pi(\theta)^2 | D_\nabla |^{-d})$$

One proves in the same way as before that for a given compatible connection $\nabla \in C(\mathcal{E})$, one has :

$$YM(\nabla) = \mathrm{Inf}\,\{ I(\nabla_1) \,;\, \pi(\nabla_1) = \nabla\}.$$

Let us briefly state the inequality between the Yang-Mills action and a topological quantity. For the background notions and notations, we refer to [Co1]. Suppose that we have an even $(4,\infty)$-summable K cycle. Define a Hochschild cocycle by

$$\Phi(a^0\, da^1 \dots da^4) = \mathrm{Tr}_\omega (\Gamma a^0 [D,a^1] \dots [D, a^4] D^{-4}) \quad \text{for all } a^j \in \mathcal{A}.$$

Let B be the operator :

$$B : H^4(\mathcal{A}, \mathcal{A}^*) \to HC^3(\mathcal{A}).$$

Then one can show that $B\Phi = 0$. We shall, however, need :

Hypothesis 12 $B\Phi = 0$ as a cochain.

(In the case of the Dirac operator on a 4-dimensional manifold one has :

$$\Phi(f^0, .., f^4) = \int f^0\, df^1 \wedge df^2 \wedge \dots \wedge df^4,$$

which satisfies the hypothesis.)

Since $B_0\Phi$ is already cyclic :

$$B_0\Phi(a^0, a^1, a^2, a^3) = \Phi(1, a^0, a^1, a^2, a^3) = \mathrm{Tr}_\omega (\Gamma [D, a^0] \dots [D, a^3] D^{-4})$$

the condition $B\Phi = 0$ means that in fact, $B_0\Phi = 0$. This, together with $b\Phi = 0$, implies that Φ is a cyclic cocycle.

Theorem 13
For any hermitian finite projective module $\mathcal{E}$ *over* $\mathcal{A}$ *one has :*

$$YM(\nabla) \geq < [\mathcal{E}] , \Phi > \qquad \text{for all } \nabla \in C(\mathcal{E})$$

The right hand side is the pairing between K theory and cyclic cohomology. In the case of the Dirac operator on a compact 4-dimensional spin manifold X, one recovers the usual lower bound for the Yang-Mills action in terms of the topological charge of the vector bundle.

2. PRODUCT OF CONTINUUM BY DISCRETE AND THE SYMMETRY BREAKING MECHANISM

We have shown how to extend the notions of gauge potentials and Yang Mills action to finitely summable K cycles (h, D) over an algebra $\mathcal{A}$, and we have also defined the fermion action.

In this section we shall give two examples of computations of these actions:
a) The case of a discrete 2 pt space.
b) The case of a product of a 4 dimensional manifold by case a).

We first need a brief discussion of product spaces. Suppose that we have two triples :

$$(\mathcal{A}_1, h_1, D_1) \quad , \quad (\mathcal{A}_2, h_2, D_2).$$

We assume that one of them is *even* , i.e. we are given a $\mathbb{Z}/2$ grading, say Γ_1, on h_1. We define the product to be the triple $(\mathcal{A}, h, D)$:

$$\mathcal{A} = \mathcal{A}_1 \otimes \mathcal{A}_2 \quad , \quad h = h_1 \otimes h_2 \; ,$$

$$D = D_1 \otimes 1 + \Gamma_1 \otimes D_2 \, .$$

One can check that if our two triples are Dirac K cycles coming from two Riemannian manifolds, then the product K-cycle corresponds to the Dirac K cycle of the product manifold. If we have finite hermitian projective modules $\mathcal{E}_j$ over the $\mathcal{A}_j$, then $\mathcal{E}_1 \otimes \mathcal{E}_2$ is a finite hermitian projective module over $\mathcal{A}$.

Next, the formula $D^2 = D_1^2 \otimes 1 + 1 \otimes D_2^2$, which follows from the anticommutation of D_1 with Γ_1, shows that dimensions add up, that is, if the D_j are (p_j , ∞) summable then D is $(p_1 + p_2 , \infty)$ summable. Moreover, one can show that

$$\mathrm{Tr}_\omega((T_1 \otimes T_2) \, |D|^{-(p_1+p_2)}) = \mathrm{Tr}_\omega (T_1 |D_1|^{-p_1}) \; \mathrm{Tr}_\omega (T_2 |D_2|^{-p_2}),$$

$$\text{for all } T_j \in \mathcal{L}(h_j).$$

More precisely, this is true provided that $p_j \geqslant 1$, but in the case of interest (Example 5), we have $p_1 = 4$ and $p_2 = 0$. The corresponding formula turns out to be :

$$\mathrm{Tr}_\omega((T_1 \otimes T_2) \, |D|^{-p}) = \mathrm{Tr}_\omega (T_1 |D_1|^{-p}) \; \mathrm{Tr}(T_2)$$

Thus in the 0-dimensional case, we should replace the Dixmier trace by the ordinary trace, and the Yang-Mills action $YM(\nabla)$ is just given by $\mathrm{Tr}\,(\theta^2)$.

Example 4
The space we are dealing with has *two points* a and b . Thus the algebra $\mathcal{A}$ is just $\mathbb{C} \oplus \mathbb{C}$, the direct sum of two copies of $\mathbb{C}$. An element $f \in \mathcal{A}$ is given by two complex numbers $f(a)$, $f(b) \in \mathbb{C}$. Let (h , D , Γ) be a 0-dimensional K cycle over $\mathcal{A}$. Then h is *finite dimensional* and the representation of $\mathcal{A}$ on h corresponds to a decomposition of h as a direct sum $h = h_a \oplus h_b$, with the action of $\mathcal{A}$ given by :

$$f \in \mathcal{A} \rightarrow \begin{matrix} f(a) & 0 \\ 0 & f(b) \end{matrix}$$

If we write D as a 2 x 2 matrix in this decomposition :

$$D = \begin{matrix} D_{aa} & D_{ab} \\ D_{ba} & D_{bb} \end{matrix}$$

we can ignore the diagonal elements since they commute with the action of $\mathcal{A}$. We shall thus take D to be of the form :

$$D = \begin{matrix} 0 & D_{ab} \\ D_{ba} & 0 \end{matrix}$$

where $D_{ba} = D_{ab}^*$ and D_{ab} is a linear map from h_b to h_a. We shall denote by M this linear map. As will become clear, a "standard" geometry on our 2-point space corresponds to M = 0. Thus, although our algebra in this example is *commutative*, our more general notion of geometry allows us to consider nonstandard geometries on this "commutative" space.

We shall take for Γ the $\mathbb{Z}/2$ grading given by the matrix

$$\Gamma = \begin{matrix} 1 & 0 \\ 0 & -1 \end{matrix}$$

So the geometry of our 2-point space is given by :

$$\mathcal{A} = \mathbb{C} \oplus \mathbb{C} \ , \quad h = h_a \oplus h_b \ , \quad D = \begin{matrix} 0 & M \\ M^* & 0 \end{matrix} \ , \ \Gamma = \begin{matrix} 1 & 0 \\ 0 & -1 \end{matrix}$$

Let us first compute the metric on the space F = {a , b}, using Formula 1. Given $f \in \mathcal{A}$, one has :

$$[D\ ,f] = \begin{matrix} 0 & M \\ M^* & 0 \end{matrix} \ , \ \begin{matrix} f(a) & 0 \\ 0 & f(b) \end{matrix}$$

$$= \begin{matrix} 0 & M(f(b)-f(a)) \\ -M^*(f(b)-f(a)) & 0 \end{matrix} = (f(b)-f(a)) \begin{matrix} 0 & M \\ -M^* & 0 \end{matrix}$$

Thus the norm of this commutator is $|\,f(b) - f(a)\,|\ \|M\|$, where $\|M\|$ is the largest characteristic value of M. Hence :

$$d(a, b) = \mathrm{Sup}\ \{\,|\,f(a) - f(b)\,|\ ,\ \|[D,f]\| \leqslant 1\,\} = 1/\|M\|$$

Let us now determine the space of gauge potentials, the curvature and the action in two cases :

α) $\mathcal{E} = \mathcal{A}$ (i.e. the trivial bundle over F).

First, let e be the idempotent $e \in \mathcal{A}$ given by $e(a) = 1$ and $e(b) = 0$. For notational simplicity, let us write e' for the idempotent $1 - e$. Then $\mathcal{A}$ is spanned by e and e', and

$de = - de'$. Thus the space Ω^1 of universal 1-forms over $\mathcal{A}$ is a 2 dimensional space, with the basis $\{e\, de, e'\, de\}$. So every element of Ω^1 can be written in the form $\delta_a\, e\, de + \delta_b\, e' de$ for some complex numbers δ_a and δ_b. We shall denote this 1-form by the pair (δ_a, δ_b). Using the identities

$$e\, de = (de)\, e', \quad e'\, de = de\,(e),$$

one finds that the action of $\mathcal{A}$ on Ω^1 is given by

$$f\,(\delta_a, \delta_b) = (f(a)\,\delta_a, f(b)\,\delta_b); \qquad (\delta_a, \delta_b)\, f = (\,\delta_a\, f(b), \delta_b\, f(a)).$$

We see that although the algebra $\mathcal{A}$ is commutative, the algebra Ω^* is *not* graded-commutative.

The differential $d : \mathcal{A} \to \Omega^1$ is essentially a *finite difference* operator:

$$df = (\Delta f,\ \Delta f) \quad , \quad \Delta f = f(a) - f(b).$$

One computes:

$$\pi((\delta_a, \delta_b)) = \begin{pmatrix} 0 & -\delta_a M \\ \delta_b M^* & 0 \end{pmatrix} \in \mathcal{L}(h).$$

So provided $M \neq 0$, the representation $\pi : \Omega^* \to \mathcal{L}(h)$ is 1-1 on Ω^1, and $\Omega^1 = \Omega^1_D$.

Next, let us find what Ω^2_D is. Recall that $\Omega^2_D = \pi(\Omega^2)/\pi(d(J_0 \cap \Omega^1))$. Now any element of Ω^2 can be written as $h_a\, e\, de\, de + h_b\, e'\, de\, de$, which we shall denote by the pair of complex numbers (h_a, h_b). One computes:

$$\pi(h_a, h_b) = \begin{pmatrix} -h_a MM^* & 0 \\ 0 & -h_b M^*M \end{pmatrix} \in \mathcal{L}(h).$$

So if $M \neq 0$ then the representation $\pi : \Omega^* \to \mathcal{L}(h)$ is 1-1 on Ω^2. And since π is 1-1 on Ω^1, $J_0 \cap \Omega^1 = 0$. Thus $\Omega^2_D = \Omega^2$. The differential

$d : \Omega^1 \to \Omega^2$ is given by $d(\delta_a, \delta_b) = (\delta_a - \delta_b)\, de\, de = (\delta_a - \delta_b, \delta_a - \delta_b)$.

And the multiplication

$\Omega^1 \times \Omega^1 \to \Omega^2$ is given by $(\delta_a, \delta_b)\,(\delta_a', \delta_b') = (\delta_a\, \delta_b', \delta_b\, \delta_a')$.

We now have enough of the differential calculus on the 2-point space to compute the gauge theory. A vector potential is given by a self adjoint element of Ω^1_D, or in our case, by $V = (-\Phi, \Phi^*)$, with Φ a complex number. Its curvature is :

$$\theta = dV + V^2 = -(\,|\Phi + 1|^2 - 1,\ |\Phi + 1|^2 - 1\,).$$

This gives the following formula for the Yang Mills action :

$$YM(V) = 2(\,|\Phi + 1|^2 - 1\,)^2 \quad \mathrm{Trace}\,((M^*M)^2\,).$$

The action of the gauge group $\mathcal{U} = U(1) \times U(1)$ on the space of vector potentials, i.e. on Φ, is given by $\gamma_u(V) = u\, du^* + u\, V\, u^*$, which, with $u = u_a\, e + u_b\, e'$, gives:

$$\gamma_u(V) = (1 - u_a u_b^* (\Phi + 1), -1 + u_b u_a^* (\Phi^* + 1)).$$

On the variable $\Phi + 1$, this just means multiplication by $u_a u_b^*$.

Thus in this very simple case, our action $YM(V)$ reproduces the symmetry-breaking Higgs potential. It has nonunique minima, given by $|\Phi + 1| = 1$, which are acted upon nontrivially by the gauge group.

The fermionic action is given by $< \psi , (D + V) \psi >$, where the operator $D + V$ is equal to :

$$\begin{pmatrix} 0 & M \\ M^* & 0 \end{pmatrix} + \begin{pmatrix} 0 & \Phi M \\ \Phi^* M^* & 0 \end{pmatrix} = \begin{pmatrix} 0 & (1+\Phi)M \\ (1+\Phi)^* M^* & 0 \end{pmatrix}$$

Writing ψ as $(\psi_+, \psi_-)^T$, we see that the fermionic action is

$$< \psi , (D + V)\psi > = \psi_+^* (1+\Phi) M \psi_- + (\text{complex conjugate}),$$

which is a sort of a primitive Yukawa coupling. Let us note that for the "standard" geometry with M = 0, the two points are infinitely far apart, and the Yang-Mills action vanishes.

β) Let us take for $\mathcal{E}$ the non trivial bundle over $F = \{a, b\}$ with fibers of dimension n_a and n_b on a and b respectively. (This does not affect the differential calculus that we worked out before.) The bundle is nontrivial iff $n_a \neq n_b$ and we shall consider the simple case of $n_a = 1$ and $n_b = 2$. The finite projective module $\mathcal{E}$ is of the form :

$$\mathcal{E} = f \mathcal{A}^2$$

where the idempotent $f \in M_2(\mathcal{A})$ is given in terms of the notation of α by the formula

$$f = \begin{pmatrix} (1,1) & 0 \\ 0 & (0,1) \end{pmatrix} = \begin{pmatrix} 1 & 0 \\ 0 & e' \end{pmatrix}$$

To the idempotent f corresponds a particular compatible connection on $\mathcal{E}$, given by $\nabla_0 \xi = fd\xi$ with obvious notations. An arbitrary compatible connection on $\mathcal{E}$ has the form :

$$\nabla \xi = \nabla_0 \xi + \rho \xi$$

where ρ is a self adjoint element of $M_2(\Omega_D^1)$ such that $f\rho = \rho f = \rho$. If we write ρ as a matrix,

$$\rho = \begin{pmatrix} \rho_{11} & \rho_{12} \\ \rho_{21} & \rho_{22} \end{pmatrix},$$

these conditions are:

(6) $$e'\rho_{21} = \rho_{21} \ , \ e'\rho_{22} = \rho_{22} = \rho_{22} e' \ , \ \rho_{12} e' = \rho_{12}.$$

Thus we get :

$$\rho_{11} = -\Phi_1 \, ede + \Phi_1^* \, e'de \ , \ \rho_{21} = \Phi_2^* \, e'de \ , \ \rho_{12} = -\Phi_2 \, ede \ , \ \rho_{22} = 0,$$

where Φ_1 and Φ_2 are arbitrary complex numbers.

The curvature θ is given by $\theta = fdfdf + f(d\rho)f + \rho^2$

$$= \begin{pmatrix} 0 & 0 \\ 0 & e'dede \end{pmatrix} + \begin{pmatrix} d\rho_{11} & (d\rho_{12})e \\ e'd\rho_{21} & 0 \end{pmatrix} + \begin{pmatrix} \rho_{11}\rho_{11} + \rho_{12}\rho_{21} & \rho_{11}\rho_{12} \\ \rho_{21}\rho_{11} & \rho_{21}\rho_{12} \end{pmatrix}$$

Explicitly, the components of θ are:

$$\theta_{11} = (1 - |\Phi_1 + 1|^2 - |\Phi_2|^2)\, edede + (1 - |\Phi_1 + 1|^2)\, e'dede,$$

$$\theta_{12} = -\Phi_2(\Phi_1 + 1)^*\, e'dede,$$

$$\theta_{21} = -\Phi_2^*(\Phi_1 + 1)\, e'dede,$$

$$\theta_{22} = (1 - |\Phi_2|^2)\, e'dede.$$

An easy calculation then gives the action $YM(\nabla)$ in terms of the variables Φ_1 and Φ_2 :

$$YM(\nabla) = (1 + 2(1 - (|\Phi_1 + 1|^2 + |\Phi_2|^2))^2) \quad Tr((M^*M)^2).$$

It is by construction invariant under the gauge group U(1) x U(2). We see that the minimum of $YM(\nabla)$ is strictly positive, and so the bundle $\mathcal{E}$ does not admit any compatible connection with vanishing curvature. We also see, after the fact, that there is nothing special about ∇_0 ; any connection with $|\Phi_1 + 1|^2 + |\Phi_2|^2 = 1$ also minimizes the Yang-Mills action.

To write the fermionic action, note that $\mathcal{E} \otimes_{\mathcal{A}} h$ is $f\mathcal{A}^2 \otimes_{\mathcal{A}} h = f h^2$. Let us write a typical element of $\mathcal{E} \otimes_{\mathcal{A}} h$ as $\psi = ((e_R, e_L), (0, \nu_L))^T$. Then the fermionic action is

$$< \psi, D_\nabla \psi > = < \psi, (D \otimes 1_2 + \rho)\psi > = e_R^*(1 + \Phi_1) M e_L + e_R^* \Phi_2 M \nu_L$$
$$+ \text{(complex conjugate)},$$

which is like the leptonic Yukawa coupling. **End of Example 4.**

Example 5. 4 dim. Riemannian manifold x (2-point space)

To fix notations, we let X be a compact Riemannian spin 4 manifold, $\mathcal{A}_1$ the algebra of functions on X and (h_1, D_1, Γ_1) the Dirac K-cycle on $\mathcal{A}_1$. We shall let the triple $(\mathcal{A}_2, h_2, D_2)$ to be as in example 4 above, i.e.

$$\mathcal{A}_2 = \mathbb{C} \oplus \mathbb{C}, \quad h_2 = h_{2,a} \oplus h_{2,b}, \quad D_2 = \begin{pmatrix} 0 & M \\ M^* & 0 \end{pmatrix}$$

We put $\mathcal{A} = \mathcal{A}_1 \otimes \mathcal{A}_2$, $h = h_1 \otimes h_2$, $D = D_1 \otimes 1 + \Gamma_1 \otimes D_2$.

The algebra $\mathcal{A}$ is commutative. It is the algebra of complex valued functions on the space Y = X x F, which is the union of two copies of the manifold X: $Y = X_a \cup X_b$.

Let us first compute the metric on Y associated to the K-cycle (h, D) :

$$d(p,q) = \sup_{f \in \mathcal{A}} \{ |f(p) - f(q)| \,;\, \| [D, f] \| \leqslant 1 \}.$$

Every $f \in \mathcal{A}$ is a pair (f_a, f_b) of functions on X. Also, to the decomposition of h_2 as :

$$h_2 = h_{2,a} \oplus h_{2,b}$$

corresponds a decomposition $h = h_a \oplus h_b$, on which the action of $f = (f_a , f_b) \in \mathcal{A}$ is diagonal :

$$f \to \begin{pmatrix} f_a & 0 \\ 0 & f_b \end{pmatrix} \in \mathcal{L}(h).$$

In this decomposition, the operator D becomes :

$$D = \begin{pmatrix} \partial_X \otimes 1 & \gamma_5 \otimes M \\ \gamma_5 \otimes M^* & \partial_X \otimes 1 \end{pmatrix}$$

where ∂_X is the Dirac operator on X and γ_5 the $\mathbb{Z}/2$ grading of its spinor bundle.

This gives us the following formula for the "differential" $[D,f]$ of a function f :

$$[D,f] = \begin{pmatrix} \gamma(df_a) \otimes 1 & (f_b - f_a)\gamma_5 \otimes M \\ (f_a - f_b)\gamma_5 \otimes M^* & \gamma(df_b) \otimes 1 \end{pmatrix}$$

Thus the differential $[D,f]$ contains three parts :

α) The usual differential df_a of the restriction of f to the copy X_a of X.

β) The usual differential df_b of the restriction of f to the copy X_b of X.

γ) The finite difference $\Delta f = f(p_a) - f(p_b)$ where p_a and p_b are the points of X_a and X_b above a given point p of X.

One can then show:

Proposition 14

1) The restriction of the metric d on $X_a \cup X_b$ to each copy (X_a or X_b) of X is the Riemannian geodesic distance of X.

2) For each point $p = p_a$ of X_a, the distance $d(p_a , X_b)$ equals $\|M\|^{-1}$ and is attained at the unique point p_b.

Let us now pass to the computation of the space Ω_D^1 of 1-forms on Y. As a 1-form is a sum of terms of the form $\pi(f_0\, df_1)$, the above computation of $[D,f] = \pi(df)$ shows that an element α of Ω_D^1 is given by :

α) A complex 1-form ω_a on X_a

β) A complex 1-form ω_b on X_b

γ) A pair of complex valued functions δ_a , δ_b on X.

The corresponding operator on h is given by :

$$\begin{pmatrix} \gamma(\omega_a) \otimes 1 & -\delta_a \gamma_5 \otimes M \\ \delta_b \gamma_5 \otimes M^* & \gamma(\omega_b) \otimes 1 \end{pmatrix}$$

The action of $\mathcal{A}$ on Ω_D^1 is given, with obvious notations, by :

$$(f_a, f_b)(\omega_a, \omega_b, \delta_a, \delta_b) = (f_a\omega_a, f_b\omega_b, f_a\delta_a, f_b\delta_b)$$

$$(\omega_a, \omega_b, \delta_a, \delta_b)(f_a, f_b) = (f_a\omega_a, f_b\omega_b, f_b\delta_a, f_a\delta_b)$$

The involution on Ω_D^1 is given by $(\omega_a, \omega_b, \delta_a, \delta_b)^* = -(\omega_a^*, \omega_b^*, \delta_b^*, \delta_a^*)$.

The differential $f : \mathcal{A} \to \Omega_D^1$ is given by:

$$f = (f_a, f_b) \to (df_a, df_b, f_a - f_b, f_a - f_b) \in \Omega_D^1.$$

When we write things in terms only of X, we can view Ω_D^1 as a 10 dimensional bundle over X, given by two copies of the complexified cotangent bundle and a trivial 2 dimensional bundle, so that over a point p in X, the fiber consists of

$$T_p^*(X)_{\mathbb{C}} \oplus T_p^*(X)_{\mathbb{C}} \oplus \mathbb{C} \oplus \mathbb{C}$$

But we have to keep in mind the nontrivial multiplication structure in the last two terms.

As in the case of the Dirac operator on Riemannian manifolds (Lemma 6), let us compute the pairs of operators of the form $\pi(\rho) = T_1$, $\pi(d\rho) = T_2$ for $\rho \in \Omega^1(\mathcal{A})$.

Given $\rho = \Sigma\, f_j\, dg_j \in \Omega^1(\mathcal{A})$, with f_j, $g_j \in \mathcal{A}$, one has :

$$\pi(\rho) = \begin{matrix} \gamma(\omega_a) \otimes 1 & -\delta_a \gamma_5 \otimes M \\ \delta_b \gamma_5 \otimes M^* & \gamma(\omega_b) \otimes 1 \end{matrix}$$

with $\omega_a = \Sigma\, f_{ja}\, dg_{ja}$, $\omega_b = \Sigma\, f_{jb}\, dg_{jb}$ and :

$$\delta_a = \Sigma\, f_{ja}(g_{ja} - g_{jb}), \quad \delta_b = \Sigma\, f_{jb}(g_{ja} - g_{jb}).$$

One has $\pi(d\rho) = \Sigma\, \pi(df_j)\, \pi(dg_j)$, which gives the 2 x 2 matrix :

$$\pi(d\rho) = \begin{matrix} \gamma(\xi_a) \otimes 1 + (\delta_b - \delta_a) \otimes MM^* & -\gamma(\eta_a)\gamma_5 \otimes M \\ \gamma(\eta_b)\gamma_5 \otimes M^* & \gamma(\xi_b) \otimes 1 + (\delta_b - \delta_a) \otimes M^*M \end{matrix}$$

where $\xi_a = \Sigma\, df_{ja} \cdot dg_{ja}$ and $\xi_b = \Sigma\, df_{jb} \cdot dg_{jb}$ are sections of the Clifford algebra bundle C^2 over X , while

$$\eta_a = \Sigma\,((f_{jb} - f_{ja})\, dg_{jb} - (g_{jb} - g_{ja})\, df_{ja}) \quad \text{and}$$

$$\eta_b = \Sigma\,((g_{ja} - g_{jb})\, df_{jb} - (f_{ja} - f_{jb})\, dg_{ja}).$$

Using the equalities :

$$d\delta_a = \Sigma\, f_{ja}(dg_{ja} - dg_{jb}) + (g_{ja} - g_{jb})\, df_{ja}$$

$$d\delta_b = \Sigma\, f_{jb}(dg_{ja} - dg_{jb}) + (g_{ja} - g_{jb})\, df_{jb}$$

$$\omega_a = \Sigma\, f_{ja}\, dg_{ja} \quad , \quad \omega_b = \Sigma\, f_{jb}\, dg_{jb}$$

we can rewrite η_a and η_b as follows :

$$\eta_a = \omega_b - \omega_a + d\delta_a$$

$$\eta_b = \omega_b - \omega_a + d\delta_b.$$

Thus knowing T_1 fixes δ_a, δ_b, η_a and η_b. As in the Riemannian case (Lemma 5), the sections ξ_a, ξ_b of C^2 are arbitrary except for $\sigma_2(\xi_a) = d\omega_a$ and $\sigma_2(\xi_b) = d\omega_b$.

This shows that the subspace $\pi(d(J_0 \cap \Omega^1))$ of $\pi(\Omega^2)$ is the space of 2 x 2 matrices of operators of the form :

$$T = \begin{matrix} \gamma(\xi_a) \otimes 1 & 0 \\ 0 & \gamma(\xi_b) \otimes 1 \end{matrix}$$

where ξ_a and ξ_b are sections of C^0, i.e. are just arbitrary scalar valued functions on X.

A general element of $\pi(\Omega^2)$ is a 2 x 2 matrix of operators of the form :

$$T = \begin{matrix} \gamma(\alpha_a) \otimes 1 - h_a \otimes MM^* & -\gamma(\beta_a)\gamma_5 \otimes M \\ \gamma(\beta_b)\gamma_5 \otimes M^* & \gamma(\alpha_b) \otimes 1 - h_b \otimes M^*M \end{matrix}$$

where α_a and α_b are arbitrary sections of C^2, h_a and h_b are arbitrary functions on X and β_a and β_b are arbitrary sections of C^1 (i.e. 1-forms). We thus get :

Lemma 15

Assume that M^*M is not a scalar multiple of the identity matrix. Then an element of Ω_D^2 is given by

1) a pair of complex 2-forms σ_a, σ_b on X
2) a pair of complex 1-forms β_a, β_b on X
3) a pair of complex functions h_a, h_b on X

The hypothesis $M^*M \neq$ const. Id enters because otherwise the functions h_a and h_b are eliminated when we quotient out by $\pi(d(J_0 \cap \Omega^1))$.

Using the above computation of $\pi(d\rho)$ we can compute the differential $d\omega$ of an element $\omega = (\omega_a, \omega_b, \delta_a, \delta_b)$ of Ω_D^1. We get :

1) $\sigma_a = d\omega_a$, $\sigma_b = d\omega_b$

2) $\beta_a = \omega_b - \omega_a + d\delta_a$, $\beta_b = \omega_b - \omega_a + d\delta_b$

3) $h_a = \delta_a - \delta_b$, $h_b = \delta_a - \delta_b$.

So we see that the differential $d\omega \in \Omega_D^2$ involves both the differential terms $d\omega_a$, $d\omega_b$, $d\delta_a$ and $d\delta_b$, and the finite difference terms $\omega_a - \omega_b$ and $\delta_a - \delta_b$, but in combinations imposed by $d(df) = 0$.

Next, let us compute the product $\omega\,\omega' \in \Omega_D^2$ of elements $\omega = (\omega_a, \omega_b, \delta_a, \delta_b)$ and $\omega' = (\omega'_a, \omega'_b, \delta'_a, \delta'_b)$ of Ω_D^1. We get :

1) $\sigma_a = \omega_a \wedge \omega'_a$ $\quad$ $\sigma_b = \omega_b \wedge \omega'_b$

2) $\beta_a = -\delta_a\,\omega'_b + \delta'_a\,\omega_a$ $\quad$ $\beta_b = -\delta_b\,\omega'_a + \delta'_b\,\omega_b$

3) $h_a = \delta_a\,\delta'_b$ $\quad$ $h_b = \delta_b\,\delta'_a$.

Comparing these formulae with those of Example 4, one can summarize the results by saying that the differential algebra on Y is simply the *graded tensor product* of the differential algebras on X and F.

The next step is to determine the inner product on the space Ω_D^2 of 2-forms given in section 1 . By definition we take the orthogonal of $\pi(d(J_0 \cap \Omega^1))$ in $\pi(\Omega^2)$, gifted with the inner product $< T_1 , T_2 > = \mathrm{Tr}_\omega (T_1^* T_2 |D|^{-4})$.

An easy calculation then gives :

Lemma 16

Let $P(M^*M)$ *be the orthogonal projection of the matrix* M^*M *on the scalar matrices* const. Id. *Then the square norm of an element* $(\sigma_a , \sigma_b , \beta_a , \beta_b , h_a , h_b)$ *of* Ω_D^2 *is given by*

$$\int_X (N_a \| \sigma_a \|^2 + N_b \| \sigma_b \|^2) \, dv + \mathrm{tr}\,(M^*M) \int_X (\| \beta_a \|^2 + \| \beta_b \|^2) \, dv +$$
$$\mathrm{tr}\,((M^*M - P(M^*M))^2) \int_X (\| h_a \|^2 + \| h_b \|^2) \, dv$$

where $N_a = \dim\, h_a$, $N_b = \dim\, h_b$.

We are now ready to compute the action $YM(\nabla)$. For $\mathcal{E}$, we shall take the space of sections of the hermitian vector bundle E on $Y = X_a \cup X_b$ which has fiber $\mathbb{C}$ on the copy X_a of X, and fiber $\mathbb{C}^2$ on the copy X_b of X. In other words we consider the product of Example 2 and Example 4 β. We can say immediately that the gauge group $\mathcal{U} = \mathrm{End}_{\mathcal{A}} (\mathcal{E})$ of our gauge theory is the unitary gauge group of the vector bundle E over $Y = X_a \cup X_b$, or equivalently the group :

$$\mathcal{U} = \mathrm{Map}\,(X , U(1) \times U(2)).$$

As in Example 4 β, we can write $\mathcal{E}$ as $f\mathcal{A}^2$, where $f \in M_2(\mathcal{A})$ is the idempotent

$$f = \begin{matrix} 1 & 0 \\ 0 & e' \end{matrix} \quad \text{and} \quad e' = (0, 1) \in \mathcal{A} .$$

Then a compatible connection ∇ has the form

$$\nabla \xi = f d\xi + \rho \xi \quad \in \quad \mathcal{E} \otimes_{\mathcal{A}} \Omega_D^1 \quad , \quad \text{for all } \xi \in \mathcal{E},$$

where ρ is a self adjoint element of $M_2(\Omega_D^1)$ which satisfies the conditions (6). Using the description of an element ω of Ω_D^1 as a quadruple $(\omega_a, \omega_b, \delta_a, \delta_b)$, we find that the entries of the 2 x 2 matrix ρ have the form:

$$\rho_{11} = (\omega^a_{11} , \omega^b_{11} , -\Phi_1 , \Phi_1^*)$$
$$\rho_{12} = (0 , \omega^b_{12} , -\Phi_2 , 0)$$
$$\rho_{21} = (0 , \omega^b_{21} , 0 , \Phi_2^*)$$
$$\rho_{22} = (0 , \omega^b_{22} , 0 , 0),$$

where ω^a is a u(1)-valued 1-form on X, ω^b is a u(2)-valued 1-form on X and (Φ_1 , Φ_2) is a pair of complex-valued functions on X. In other words, a compatible connection on $\mathcal{E}$ consists of

α) A u(1)- connection ∇_a on the restriction of E to X_a
β) A u(2)- connection ∇_b on the restriction of E to X_b
γ) A linear map (Φ_1 , Φ_2) from E over X_b, to E over X_a.

The action of the gauge group on ∇_a and ∇_b is the obvious one, and the action on (Φ_1, Φ_2) is given by composition.

Next, the curvature θ is the following element of $f M_2(\Omega_D^2) f$:

$$\theta = f\, df\, df + f\, d\rho\, f + \rho^2,$$

which is easily computed using the above formulae for

$$d : \Omega_D^1 \to \Omega_D^2 \quad \text{and} \quad \wedge : \Omega_D^1 \times \Omega_D^1 \to \Omega_D^2.$$

Let us write an element of Ω_D^2 as a sextuple $(\sigma_a, \sigma_b, \beta_a, \beta_b, h_a, h_b)$. Then one finds that the components of θ are:

$$\theta_{11} = (F^a_{11}, F^b_{11}, -D(1+\Phi_1), D(1+\Phi_1)^*, 1 - |\Phi_1 + 1|^2 - |\Phi_2|^2, 1 - |\Phi_1 + 1|^2)$$

$$\theta_{12} = (0, F^b_{12}, -D\Phi_2, 0, 0, -\Phi_2(\Phi_1 + 1)^*)$$

$$\theta_{21} = (0, F^b_{21}, 0, D\Phi_2, 0, -\Phi_2^*(\Phi_1 + 1))$$

$$\theta_{22} = (0, F^b_{22}, 0, 0, 0, 1 - |\Phi_2|^2).$$

Here F^a and F^b are the curvatures of ω^a and ω^b respectively, and

$$D(1+\Phi_1, \Phi_2) = d(1+\Phi_1, \Phi_2) - (1+\Phi_1, \Phi_2)\begin{pmatrix} \omega^b_{11} - \omega^a_{11} & \omega^b_{12} \\ \omega^b_{21} & \omega^b_{22} - \omega^a_{11} \end{pmatrix}$$

(Note that the calculation of the (h_a, h_b)'s is exactly the same as in Example 4 β.)

Applying Lemma 16 gives that the Yang-Mills action is the integral of a Lagrangian density $\mathcal{L}_B$ over X, with

$$\mathcal{L}_B = c_1 \|F^a\|^2 + c_2 \|F^b\|^2 + c_3 \operatorname{Tr}(M^*M) \|D(1+\Phi_1, \Phi_2)\|^2 +$$
$$c_4 \operatorname{Tr}((M^*M - P(M^*M))^2)\,(1 + 2(1 - (|\Phi_1 + 1|^2 + |\Phi_2|^2))^2).$$

The c_i's are various positive constants; we shall come back to their meaning in the next section.

The fermionic action is even easier to compute. Note that $\mathcal{E} \otimes_{\mathcal{A}} h$ is $f h^2$. Let us write a typical element of $\mathcal{E} \otimes_{\mathcal{A}} h$ as $\psi = ((e_R, e_L), (0, \nu_L))^T$. Then the fermionic action is

$$<\psi, D_\nabla \psi> = <\psi, (D \otimes 1_2 + \rho)\psi>,$$

which is the integral of a Lagrangian density $\mathcal{L}_F$ over X, with

$$\mathcal{L}_F = e_R^* \partial_a e_R + (e_L, \nu_L)^* \partial_b (e_L, \nu_L) +$$
$$[e_R^* (1+\Phi_1) M \gamma_5 e_L + e_R^* \Phi_2 M \gamma_5 \nu_L + (\text{complex conjugate})].$$

Here ∂_a is the Dirac operator on X, when coupled to the u(1)-gauge field ω^a, and similarly ∂_b is the Dirac operator on X, when coupled to the u(2)-gauge field ω^b.

It should now be clear that the total Lagrangian density $\mathcal{L}_B + \mathcal{L}_F$ of our noncommutative gauge theory is almost the same as that of the Glashow-Weinberg-Salam (GWS) model of leptons [GWS]. In fact, there are only two differences. First, the global gauge group of the GWS model is not U(1) x U(2), but U(1) x SU(2). In order to reduce our gauge group, we impose the

Ad Hoc Condition : $tr(\omega^a) = \omega^b$.

We shall give a less ad hoc formulation of this condition in Section 4. The second difference is that we need for the fermions to be chiral. To achieve this, we simply Wick-rotate to Minkowski space and impose the condition $\Gamma\psi = \psi$.

End of Example 5.

3. BIMODULES

In the discussion so far, we have had a single algebra $\mathcal{A}$ acting on the Hilbert space $\mathfrak{h}$. In fact, it turns out to be natural to extend this to having two algebras $\mathcal{A}$ and $\mathcal{B}$ acting on $\mathfrak{h}$, whose actions commute. We can express this by saying that $\mathcal{A}$ acts on $\mathfrak{h}$ on the left, and $\mathcal{B}$ acts on $\mathfrak{h}$ on the right. Alternatively, we can say that $\mathcal{A} \otimes \mathcal{B}$ acts on $\mathfrak{h}$ on the left. Given the first description, we get the second description by putting

$$(a \otimes b)\eta = a\eta b \quad \text{for all } a \in \mathcal{A},\ b \in \mathcal{B} \text{ and } \eta \in \mathfrak{h}.$$

This situation of having two algebras acting arises when one want to extend Poincaré duality to an algebraic setting. We shall not need the details of this, for which we refer to [CS,Ka,Co2], and will only broadly state the ideas . Recall that if X is a closed oriented manifold then Poincaré duality gives an isomorphism between the cohomology and homology of X. Similarly, if X is a spin^c manifold then X is K-oriented and there is an isomorphism between the K-theory $K^*(X)$ and K-homology $K_*(X)$ of X.

Let us consider what the analogous statement would be for general algebras. On the level of K-groups, it would be an isomorphism between $K_*(\mathcal{A})$ and $K^*(\mathcal{B})$. (In the special case that X is a spin^c manifold, we can take both $\mathcal{A}$ and $\mathcal{B}$ to be C(X).) Of course, one needs additional structure to have such an isomorphism. The essential piece of information needed is a K cycle $(\mathcal{A} \otimes \mathcal{B}, \mathfrak{h}, D)$ for the algebra $\mathcal{A} \otimes \mathcal{B}$.

On the level of homology, we want a map from the homology of the complex $\Omega_D^*(\mathcal{A})$ to the periodic cyclic cohomology of $\mathcal{B}$. It turns out that such a map can defined provided that one has certain relations, one of which is

(7) $$[[D, a], b] = 0 \quad \text{for all } a \in \mathcal{A} \text{ and } b \in \mathcal{B}.$$

We note that this relation is actually symmetric in a and b, as they commute.

The point of this general discussion is that it is natural to look at a K cycle for the tensor product $\mathcal{A} \otimes \mathcal{B}$ of two algebras, which satisfies (7). We can apply the constructions of section 1 to such a K cycle, and in particular the notions of a vector potential V and its Yang Mills action $YM(V)$. The gauge group of such a gauge theory would be the group $\mathcal{U}(\mathcal{A} \otimes \mathcal{B})$ of unitaries of $\mathcal{A} \otimes \mathcal{B}$. However, we can use the fact that we have two algebras to single out a class of vector potentials which is invariant under the action of the subroup $\mathcal{U}_{\mathcal{A}} \times \mathcal{U}_{\mathcal{B}}$.

Proposition 17

Let $\mathcal{V} = \mathcal{V}_{\mathcal{A}} + \mathcal{V}_{\mathcal{B}}$ be the subspace of the vector potentials $\mathcal{V}_{\mathcal{A}\otimes\mathcal{B}}$ which consists of sums of vector potentials relative to $\mathcal{A}$ and $\mathcal{B}$. Then $\mathcal{V}$ is invariant under the action of $\mathcal{U}_{\mathcal{A}} \times \mathcal{U}_{\mathcal{B}}$, and for every $V \in \mathcal{V}$, the operator $D + V$ still satifies equation (7).

To see this, recall that the action of the unitary group of $\mathcal{A} \otimes \mathcal{B}$ on vector potentials is determined by

$$g(D + V)\, g^* = D + \gamma_g(V).$$

Let us specialize this equation to elements $g = u\,v \in \mathcal{U}_{\mathcal{A}} \times \mathcal{U}_{\mathcal{B}}$.

Take $V = V^a + V^b \in \mathcal{V}_{\mathcal{A}} + \mathcal{V}_{\mathcal{B}}$. Then

$$uv(D + V^a + V^b)\, v^* u^* = u\, v\, D\, v^* u^* + u\, V^a u^* + v\, V^b v^*,$$

since by (7), every element V^a of $\mathcal{V}_{\mathcal{A}}$ (resp. V^b of $\mathcal{V}_{\mathcal{B}}$) commutes with $\mathcal{B}$ (resp. $\mathcal{A}$). Next,

$$u\, v\, D\, v^* u^* = u(D + v\,[D, v^*])\, u^* = D + u\,[D, u^*] + v[D, v^*],$$

again using (7). So we get :

$$\gamma_{uv}(V^a + V^b) = \gamma_u(V^a) + \gamma_v(V^b),$$

which shows that the space $\mathcal{V}$ is invariant under the action of $\mathcal{U}_{\mathcal{A}} \times \mathcal{U}_{\mathcal{B}}$. Finally, we compute:

$$[[D + V^a + V^b, a], b] = [[V^a, a], b] = [[V^a, b], a] = 0.$$

4. The standard U(1) × SU(2) × SU(3) model

In this section we shall build on the computation of the action functional in Example 5 i.e. in the case of the product of a continuum by a discrete 2-point space. We saw that almost by accident, we recovered the GWS model for leptons from a simple modification of the 4-D continuum. The question which we address in this section is: can one by a similar procedure incorporate the quarks as well as strong interactions?

Let us make some preliminary remarks. First, there is presently (1991) no doubt that the standard model of electroweak and strong interactions gives a remarkably accurate description of the known elementary particles. We refer to other works, such as [El], for a survey of the standard model, and will only give a skeleton description in order to fix notation.

The goal is to find a modification of the continuum spacetime geometry such that the bosonic part of the standard model becomes a pure gauge theory on this modified spacetime. (The fermionic part will be straightforward.) That is, we wish to find a new geometry such that the gauge fields and the Higgs fields of the continuum geometry become unified into a gauge field on the new geometry. In itself this is not a new idea, and most previous attempts to do such a unification used a new geometry consisting of $\mathbb{R}^4 \times F$, where F is a compact homogeneous space [CJ]. However, none of these attempts were able to succesfully reproduce realistic particle models, partly because of problems in producing chiral fermions on $\mathbb{R}^4$ from fermions on $\mathbb{R}^4 \times F$ [Wi]. What is new in our approach is to take F to be a finite set, albeit with a noncommutative geometry. Then the problems with producing chiral fermions immediately go away.

We wish, then, to find a finite space F such that when one computes the analog of the classical Lagrangian of electrodynamics, but instead on $\mathbb{R}^4 \times F$, one finds the classical Lagrangian of the standard model. Once the structure of this finite space F is given, we just apply our general method of computing the Yang-Mills action to $\mathbb{R}^4 \times F$, and will find the bosonic terms of the standard model action. The Higgs boson will be part of a gauge field, but coming from a finite difference, rather than a differential. The fermionic action will be straightforward to derive, and will give the fermionic terms of the standard model action.

a) *The standard model*

The Lagrangian density of the standard model contains five different terms:

$$\mathcal{L} = \mathcal{L}_G + \mathcal{L}_\Psi + \mathcal{L}_\phi + \mathcal{L}_Y + \mathcal{L}_V$$

which we now describe in a Euclidean version of the model.

1) *The pure gauge boson part* $\mathcal{L}_G$

$$\mathcal{L}_G = \frac{1}{4} g^{-2} \left(F_{\mu\nu} F^{\mu\nu}\right) + \frac{1}{4} g'^{-2} \left(G_{\mu\nu a} G_a^{\mu\nu}\right) + \frac{1}{4} g''^{-2} \left(H_{\mu\nu b} H_b^{\mu\nu}\right)$$

where $F_{\mu\nu}$ is the field strength tensor of a U(1)-gauge field A_μ, $G_{\mu\nu}$ is the field strength tensor of an SU(2)-gauge field W_μ, and $H_{\mu\nu}$ is the field strength tensor of an SU(3)-gauge field V_μ, the gluon field.

2) *The Fermion kinetic term* $\mathcal{L}_\Psi$

$$\mathcal{L}_\Psi = \overline{\Psi} (\gamma^\mu D_\mu) \Psi,$$

where Ψ is a spinorial field consisting of N copies, or generations, of a certain $\mathbb{C}^{15}$ representation of U(1) x SU(2) x SU(3). Here D_μ is the covariant derivative of the spinor field:

$$D_\mu \Psi = [\partial_\mu + \pi(A_\mu) + \pi'(W_\mu) + \pi''(V_\mu)] \Psi,$$

and π, π' and π'' are the respective representations of the Lie algebras of U(1), SU(2) and SU(3) on Ψ. The decomposition of the $\mathbb{C}^{15}$-representation into its irreducible components, listed by the particles of the first generation, is as follows:

Particle	$\pi \otimes \pi' \otimes \pi''$	Y
e_R	$\mathbb{C} \otimes \mathbb{C} \otimes \mathbb{C}$	-2
(e_L, ν_L)	$\mathbb{C} \otimes \mathbb{C}^2 \otimes \mathbb{C}$	-1
d_R	$\mathbb{C} \otimes \mathbb{C} \otimes \mathbb{C}^3$	-2/3
u_R	$\mathbb{C} \otimes \mathbb{C} \otimes \mathbb{C}^3$	4/3
(d_L, u_L)	$\mathbb{C} \otimes \mathbb{C}^2 \otimes \mathbb{C}^3$	1/3

The hypercharge Y, when multiplied by 3, labels the U(1) representation π. Hereafter we will write the fermion fields as N-vectors, labelled by their first-generation particle. For example, with the three known generations, using the standard particle names we have

$$\vec{e} = (e, \mu, \tau), \quad \vec{\nu} = (\nu_e, \nu_\mu, \nu_\tau), \quad \vec{u} = (u, c, t), \quad \vec{d} = (d, s, b).$$

3) *The kinetic terms for the Higgs fields*

$$\mathcal{L}_\phi = (D_\mu \phi)^* \; D^\mu \phi ,$$

where $\phi = \begin{pmatrix} \phi_1 \\ \phi_2 \end{pmatrix}$ is an SU(2) doublet of complex scalar fields with hypercharge Y = - 1.

4) *The Yukawa coupling of Higgs fields with Fermions*

$$\mathcal{L}_Y = \overline{e_R}\, M_e \phi_1 e_L + \overline{e_R}\, M_e \phi_2 \nu_L + \overline{d_R}\, M_d \phi_1 d_L + \overline{d_R}\, M_d \phi_2 u_L$$

$$+ \overline{u_R}\, M_u (-\overline{\phi}_2)\, d_L + \overline{u_R}\, M_u \overline{\phi}_1 u_L + \text{complex conjugate.}$$

Here M_e, M_d and M_u are N x N matrices whose singular values are, up to a constant, the masses of the fermions. Let us note that while $\phi = \begin{pmatrix} \phi_1 \\ \phi_2 \end{pmatrix}$ is involved in the Yukawa couplings to the electron e_R and down quark d_R, its conjugate

(8) $$\tilde{\phi} = \begin{pmatrix} -\overline{\phi}_2 \\ \overline{\phi}_1 \end{pmatrix}$$

is in the Yukawa coupling to the up quark u_R.

5) *The Higgs self interaction*

$$\mathcal{L}_V = -\mu^2 \phi^* \phi + \frac{1}{2}\lambda (\phi^* \phi)^2$$

has exactly the same form as in the GWS model.

Thus we see that there are essentially three new features of the complete standard model as compared to the GWS model:

A. The new SU(3) gauge symmetry, whose gauge fields are responsible for the strong interaction.

B. The new fermions, the quarks, with their new hypercharges.

C. The new Yukawa coupling terms involving the quarks.

We shall now briefly explain how these new features motivate a modification of Example 5, which lead us above to the GWS model for leptons. First, our model will still be a *product* of an ordinary Euclidean continuum by a finite space.

In example 4 β, our algebra $\mathcal{A}$ of functions on the finite space,was $\mathbb{C} \oplus \mathbb{C}$. But since we then considered a bundle on {a,b} with fiber $\mathbb{C}$ on a and $\mathbb{C}^2$ on b, we could have equally well used $\mathcal{A} = \mathbb{C} \oplus M_2(\mathbb{C})$, and taken the module $\mathcal{E}$ to be the same as $\mathcal{A}$. Let us see how point C. leads us to replace $\mathbb{C} \oplus M_2(\mathbb{C})$ by $\mathcal{A} = \mathbb{C} \oplus \mathbb{H}$ where $\mathbb{H}$ is the Hamilton algebra of quaternions. The point is simply that the equation (8) which relates ϕ and $\tilde{\phi}$ is the same as the unitary equivalence $2 \sim \bar{2}$ between the fundamental representation 2 of SU(2) and its complex conjugate or contragredient representation, i.e. one has :

$$g \in U(2)\ ,\ J\, g\, J^{-1} = \bar{g} \Leftrightarrow g \in SU(2),$$

where $J = \begin{pmatrix} 0 & -1 \\ 1 & 0 \end{pmatrix}$.

We remark that :

$$\{\, x \in M_2(\mathbb{C})\ ,\ J\, x\, J^{-1} = \bar{x} \,\}$$

defines an algebra, the quaternion algebra $\mathbb{H}$.

Next let us see how point A. leads us to the formalism of bimodules of Section 3. Indeed, look at any isodoublet of the form $\begin{pmatrix} u_L \\ d_L \end{pmatrix}$ of left handed quarks. It appears in 3 colors:

$$\begin{pmatrix} u_L^r & u_L^y & u_L^b \\ d_L^r & d_L^y & d_L^b \end{pmatrix}$$

which makes it clear that the corresponding representation of $SU(2) \times SU(3)$ is the tensor product $\mathbb{C}^2 \otimes \mathbb{C}^3$ of their fundamental representations. It is easy to convince oneself that even if one neglects the difference between U(n) and SU(n), there is no way to obtain such groups and representations from a single algebra and its unitary group. The solution that we found is to take $(\mathfrak{A}, \mathfrak{B})$ bimodules, with $\mathfrak{A} = \mathbb{C} \oplus \mathbb{H}$ and $\mathfrak{B} = \mathbb{C} \oplus M_3(\mathbb{C})$.

We are now ready to describe the geometric structure of a finite space F which, when crossed by $\mathbb{R}^4$, gives the standard model.

b) *Geometric structure of the finite space* F

The structure is given by an $\mathfrak{A}$ - $\mathfrak{B}$ bimodule $(\mathfrak{h}, D, \Gamma)$ where $\mathfrak{A}$ is the involutive algebra $\mathbb{C} \oplus \mathbb{H}$ while $\mathfrak{B}$ is the involutive algebra $\mathbb{C} \oplus M_3(\mathbb{C})$. Unlike $\mathfrak{B}$, the algebra $\mathfrak{A}$ is only an algebra over $\mathbb{R}$. The involutive representations π of $\mathfrak{A}$ in a finite dimensional Hilbert space are characterized (up to unitary equivalence) by three multiplicities: n_+, n_-, m, where

$$\mathfrak{h}_\pi = \mathbb{C}^{n_+} \oplus \mathbb{C}^{n_-} \oplus \mathbb{C}^{2m},$$

and if $a = (\lambda, q) \in \mathbb{C} \oplus \mathbb{H}$, $\pi(a)$ is the block diagonal matrix:

$$\pi(a) = \left(\lambda \otimes I_{n_+}\right) \oplus \left(\bar{\lambda} \otimes I_{n_-}\right) \oplus \left(\begin{bmatrix} \alpha & \beta \\ -\bar{\beta} & \bar{\alpha} \end{bmatrix} \otimes I_m\right).$$

Here we are writing the quaternion q as $q = \alpha + \beta j$, with $\alpha, \beta \in \mathbb{C} \subset \mathbb{H}$.

The representation of the complex involutive algebra $\mathcal{B}$ in $\mathfrak{h}$ gives a decomposition:

$$\mathfrak{h} = \mathfrak{h}_0 \oplus (\mathfrak{h}_1 \otimes \mathbb{C}^3)$$

in which $b = (b_0, b_1) \in \mathbb{C} \oplus M_3(\mathbb{C})$ acts by $\pi(b) = b_0 \oplus (1 \otimes b_1)$. It follows that the representation of $\mathcal{A}$ (which commutes with the representation of $\mathcal{B}$) is given by *a pair* π_0, π_1 of representations of $\mathcal{A}$ in Hilbert spaces $\mathfrak{h}_0$ and $\mathfrak{h}_1$. The $\mathcal{A}$ - $\mathcal{B}$ bimodule $\mathfrak{h}$ is thus completely described by the six multiplicities, namely (n_+^0, n_-^0, m^0) for π_0 and (n_+^1, n_-^1, m^1) for π_1. We shall take them to be of the form:

$$(n_0^+, n_0^-, m_0) = N(1,0,1)$$

$$(n_1^+, n_1^-, m_1) = N(1,1,1),$$

where N will eventually be the number of generators. That is,

$$\mathfrak{h} = [(\mathbb{C} \oplus \mathbb{H}) \oplus ((\mathbb{C} \oplus \mathbb{C} \oplus \mathbb{H}) \otimes \mathbb{C}^3)] \otimes \mathbb{R}^N.$$

We shall take the $\mathbb{Z}/2$ grading Γ in $\mathfrak{h}$ to be given by the element $\Gamma = (1, -1)$ of the center of $\mathcal{A}$. Finally, for D we shall take the most general selfadjoint operator in $\mathfrak{h}$ which anticommutes with Γ and commutes with $\mathbb{C} \oplus \mathcal{B}$, where $\mathbb{C} \subset \mathcal{A}$ is the diagonal subalgebra: $\{(\lambda, \lambda), \lambda \in \mathbb{C}\}$. (As we shall see, D encodes both the fermion masses and the Kobayaski-Maskawa mixing parameters.) It follows that the action of $\mathcal{A}$ and the operator D in $\mathfrak{h}_0$ (resp. $\mathfrak{h}_1$) have the following general form: (with $q = \alpha + \beta j \in \mathbb{H}$)

$$\pi_0(f,q) = \begin{bmatrix} f & 0 & 0 \\ 0 & \alpha & \beta \\ 0 & -\bar{\beta} & \bar{\alpha} \end{bmatrix} \qquad D_0 = \begin{bmatrix} 0 & M_e & 0 \\ M_e^* & 0 & 0 \\ 0 & 0 & 0 \end{bmatrix}$$

$$\pi_1(f,q) = \begin{bmatrix} f & 0 & 0 & 0 \\ 0 & \bar{f} & 0 & 0 \\ 0 & 0 & \alpha & \beta \\ 0 & 0 & -\bar{\beta} & \bar{\alpha} \end{bmatrix} \qquad D_1 = \begin{bmatrix} 0 & 0 & M_d & 0 \\ 0 & 0 & 0 & M_u \\ M_d^* & 0 & 0 & 0 \\ 0 & M_u^* & 0 & 0 \end{bmatrix}$$

where M_e, M_u, M_d are arbitrary complex $N \times N$ matrices.

c) *Gauge Theory on the finite space* F

We shall take the modules for $\mathfrak{A}$ and $\mathfrak{B}$ to be $\mathcal{E} = \mathfrak{A}$ and $\mathfrak{F} = \mathfrak{B}$ respectively. Then the Hilbert space $\mathcal{E} \otimes_{\mathfrak{A}} \mathfrak{h} \otimes_{\mathfrak{B}} \mathfrak{F}$ of gauged fermions is the same as $\mathfrak{h}$. The unitary gauge groups are $\mathcal{U}_{\mathfrak{A}} = U(1) \times SU(2)$ and $\mathcal{U}_{\mathfrak{B}} = U(1) \times U(3)$. The gauge fields are simply given by vector potentials, i.e. self-adjoint elements of $\Omega^1_D(\mathfrak{A})$ and $\Omega^1_D(\mathfrak{B})$. As $\mathfrak{B}$ commutes with D, $\Omega^1_D(\mathfrak{B})$ vanishes, and so $\mathfrak{B}$ will play no role in the finite geometry.

Let us look at $\Omega^1_D(\mathfrak{A})$. Write an element ρ of $\Omega^1(\mathfrak{A})$ as $\rho = \Sigma\, a_j\, da'_j$, with $a_j, a'_j \in \mathfrak{A}$; $a_j = (\lambda_j, q_j)$, $a'_j = (\lambda'_j, q'_j)$; $q_j = \alpha_j + \beta_j\, j$, $q'_j = \alpha'_j + \beta'_j\, j$. (One can simplify the calculations by noting that (π_0, D_0) is essentially the degenerate case $M_u = 0$ of (π_1, D_1)).

One finds

$$\pi_0(\rho) = \begin{pmatrix} 0 & \phi_1 M_e & \phi_2 M_e \\ \phi'_1 M_e^* & 0 & 0 \\ -\overline{\phi'_2} M_e^* & 0 & 0 \end{pmatrix},$$

$$\pi_1(\rho) = \begin{pmatrix} 0 & 0 & \phi_1 M_d & \phi_2 M_d \\ 0 & 0 & -\overline{\phi_2} M_u & \overline{\phi_1} M_u \\ \phi'_1 M_d^* & \phi'_2 M_u^* & 0 & 0 \\ -\overline{\phi'_2} M_d^* & \overline{\phi'_1} M_u^* & 0 & 0 \end{pmatrix}$$

where

$$\phi_1 = \Sigma\, \lambda_j\left(\alpha'_j - \lambda'_j\right) \qquad \phi_2 = \Sigma\, \lambda_j\, \beta'_j$$

$$\phi'_1 = \Sigma\, \alpha_j\left(\lambda'_j - \alpha'_j\right) + \beta_j \overline{\beta'_j} \qquad \phi'_2 = \Sigma\, \beta_j\left(\overline{\lambda'_j} - \overline{\alpha'_j}\right) - \alpha_j \beta'_j$$

Thus $\Omega^1_D(\mathfrak{A}) = \{(\phi_1, \phi_2, \phi_1', \phi_2') \in \mathbb{C}^4\}$, the differential $d\colon \mathfrak{A} \to \Omega^1_D(\mathfrak{A})$ being given by

$$d(\lambda, \alpha, \beta) = (\alpha - \lambda, \beta, \lambda - \alpha, -\beta).$$

If ρ is a vector potential then $\phi'_1 = \overline{\phi_1}$ and $\phi'_2 = -\phi_2$. Similarly, one computes that

$$\pi_0\left(d\rho\right) = \begin{pmatrix} \left(\phi_1 + \phi'_1\right)\ M_e M_e^* & 0 \\ 0 & P \end{pmatrix}$$

with $$P = 1/2 \begin{pmatrix} \phi_1+\phi'_1+Y & \phi_2+\phi'_2+Z \\ -\overline{\phi}_2-\overline{\phi}'_2+\overline{Z} & \overline{\phi}_1+\overline{\phi}'_1-\overline{Y} \end{pmatrix} \otimes M_e{}^* M_e,$$

and $$\pi_1(d\rho) = \begin{pmatrix} Q & 0 \\ 0 & R \end{pmatrix}$$

with $$Q = \begin{pmatrix} (\phi_1+\phi'_1) M_d M_d^* & (\phi_2+\phi'_2)\, M_d M_u^* \\ (-\overline{\phi}_2-\overline{\phi}'_2) M_u M_d^* & (\overline{\phi}_1+\overline{\phi}'_1)\, M_u M_u^* \end{pmatrix}$$

and $$R = 1/2 \begin{pmatrix} \phi_1+\phi'_1 & \phi_2+\phi'_2 \\ -\overline{\phi}_2-\overline{\phi}'_2 & \overline{\phi}_1+\overline{\phi}'_1 \end{pmatrix} \otimes (M_d{}^* M_d + M_u{}^* M_u)$$

$$+ 1/2 \begin{pmatrix} Y & Z \\ \overline{Z} & -\overline{Y} \end{pmatrix} \otimes (M_d{}^* M_d - M_u{}^* M_u).$$

Here Y and Z are new fields given by

$$Y = \Sigma(\alpha_j - \lambda_j)(\lambda'_j - \alpha'_j) - \beta_j \overline{\beta}'_j \quad \text{and} \quad Z = \Sigma - (\alpha_j - \lambda_j)\beta'_j + \beta_j(\overline{\alpha}'_j - \overline{\lambda}'_j).$$

We see that $\{\pi(d\rho): \rho \in J_0 \cap \Omega^1_D(\mathcal{A})\} = \{\pi(d\rho): (\phi_1, \phi_2, \phi_1', \phi_2') = 0\}$ consists of the Y and Z fields. Then the quotienting used to define $\Omega^2_D(\mathcal{A})$ amounts to quotienting out the Y and Z fields. Considering the products $\pi(\rho_1)\,\pi(\rho_2)$, we see that $\Omega^2_D(\mathcal{A}) \cong \mathbb{C}^4$, with the product $\Omega^1_D(\mathcal{A}) \times \Omega^1_D(\mathcal{A}) \to \Omega^2_D(\mathcal{A})$ given by

$$(\phi_1, \phi_2, \phi'_1, \phi'_2) \times (\eta_1, \eta_2, \eta'_1, \eta'_2) =$$

$$(\phi_1\eta_1' - \phi_2\overline{\eta}_2', \phi_1\eta_2' + \phi_2\overline{\eta}_1', \phi'_1\eta_1 - \phi'_2\overline{\eta}_2, \phi'_1\eta_2 + \phi'_2\overline{\eta}_1)$$

and the differential d: $\Omega^1_D(\mathcal{A}) \to \Omega^2_D(\mathcal{A})$ given by

$$d(\phi_1, \phi_2, \phi_1', \phi_2') = (\phi_1 + \phi_1', \phi_2 + \phi_2', \phi_1 + \phi_1', \phi_2 + \phi_2').$$

It follows that the curvature $\theta = dV + V^2$ of $V = (\phi_1, \phi_2, \overline{\phi}_1, -\phi_2)$ has image in $\Omega^2_D(\mathcal{A})$ given by

$$\pi_D(\theta) = \left(|1+\phi_1|^2 + |\phi_2|^2 - 1 \right) \ (1, 0, 1, 0) \in \Omega^2_D(\mathcal{A}).$$

Then the Yang-Mills action is

$$\mathrm{YM} = \langle\theta, \theta\rangle = \text{const.}\left(\left|1+\phi_1\right|^2 + \left|\phi_2\right|^2 - 1\right)^2.$$

Up to a shift of the ϕ_1 variable, this is the symmetry-breaking potential for the Higgs field, with $(\phi_1, \phi_2) = 0$ being a minimum.

Writing a fermion field $\Psi \in \mathfrak{h}$ as

$$\Psi = (e_R, e_L, \nu_L) \oplus (d_R, u_R, d_L, u_L),$$

the fermionic action

$$\mathfrak{L}_\Psi = \overline{\Psi}\,(D + V)\,\Psi$$

gives exactly the Yukawa couplings of the standard model, after a shift of the ϕ_1 field. When (ϕ_1, ϕ_2) is frozen at its minimum $(0, 0)$, $\mathfrak{L}_\Psi$ becomes

$$\overline{e_R}\, M_e\, e_L + \overline{d_R}\, M_d\, d_L + \overline{u_R}\, M_u\, u_L + \text{complex conjugate}$$

The "normal modes" of Ψ are given by the eigenstates of the matrices

$$(M_eM_e^*, M_e^*M_e, 0),\ (M_dM_d^*, M_uM_u^*, M_d^*M_d, M_u^*M_u),$$

and the corresponding fermion masses are the square roots of the eigenvalues. In this finite geometry, the fermion masses are the only physical information in the matrices M_e, M_d and M_u, but in the full standard model, to be described next, there is also a physically relevant N x N unitary matrix, the mixing matrix U. This matrix comes from the discrepancy between the eigenstates of the mass matrices and the weak curent interaction [El]. Explicitly, suppose that $M_u^*M_u$ and $M_d^*M_d$ are diagonalized by unitary matrices V_u and V_d:

$$M_u^*M_u = V_u\,\mathrm{Diag}_u\,V_u^{-1} \text{ and } M_d^*M_d = V_d\,\mathrm{Diag}_d\,V_d^{-1}.$$

Then $U = V_u^{-1}V_d$. We can easily describe U in terms of our finite-geometry. There are orthonormal bases for the vector spaces of d_L's and u_L's given by the eigenstates of the matrices $M_d^*M_d$ and $M_u^*M_u$. As $d_L + u_L j$ lies in $\mathbb{H}$, multiplication by the unit quaternion j maps the vector space of d_L's to the vector space of u_L's. Then U is simply the writing of this multiplication operator in terms of the preferred bases. As the eigenstates are only defined up to a phase, U is only defined up to right and left multiplication by $U(1)^N$.

Before leaving the finite-point geometry, we remark that there is a compact way to write its differential algebra. First, $\Omega_D^0(\mathfrak{A}) = \mathfrak{A} = \mathbb{C} \oplus \mathbb{H} \subset \mathbb{H} \oplus \mathbb{H}$. Next,

$$\Omega_D^1(\mathfrak{A}) = \{(\phi_1, \phi_2), (\phi_1', \phi_2') \in \mathbb{C}^2 \oplus \mathbb{C}^2\} \cong \{(q_1, q_2) \in \mathbb{H} \oplus \mathbb{H}\}.$$

With the identification $q_1 = \phi_1 + \phi_2\, j$ and $q_2 = \phi'_1 + \phi'_2\, j$, the bimodule structure on $\Omega^1_D(\mathcal{A})$ is given by

$$(\lambda, q)\,(q_1, q_2) = (\lambda q_1, q q_2) \qquad \forall q_1, q_2 \in \mathbb{H}$$

$$(q_1, q_2)\,(\lambda, q) = (q_1\, q, q_2\, \lambda) \qquad \lambda \in \mathbb{C}, q \in \mathbb{H}$$

and the differential d being again the *finite difference:*

$$d(\lambda, q) = (q - \lambda, \lambda - q) \in \mathbb{H} \oplus \mathbb{H}.$$

The involution on $\Omega^1_D(\mathcal{A})$ is given by:

$$(q_1, q_2)^* = (q_2^*, q_1^*) \quad \forall q_1, q_2 \in \mathbb{H}.$$

The space $\mathcal{V}$ of vector potentials is thus naturally isomorphic to $\mathbb{H}$.

Finally, $\Omega^2_D(\mathcal{A}) \equiv \mathbb{H} \oplus \mathbb{H}$ with an $\mathcal{A}$-bimodule structure given by:

$$(\lambda, q)\,(q_1, q_2)\,(\lambda', q') = (\lambda q_1\, \lambda', q q_2\, q') \quad \forall \lambda, \lambda' \in \mathbb{C}\ , q\text{'s} \in \mathbb{H}.$$

The product: $\Omega^1_D \times \Omega^1_D \to \Omega^2_D$ is given by:

$$(q_1, q_2)\ \ (q'_1, q'_2) = (q_1\, q'_2\, , q_2\, q'_1)$$

and the differential $d : \Omega^1_D \to \Omega^2_D$ is given by

$$d(q_1, q_2) = (q_1 + q_2, q_1 + q_2)\ .$$

The curvature θ of a vector potential $V = (\, q, q^*\,)$ is then

$$\theta = dV + V^2 = (q + q^* + qq^*\, , q + q^* + q^*q) = (|1 + q|^2 - 1)\ \ (1,1).$$

d) *Geometric structure of the standard model*

For the full standard model, we take the product geometry of the finite space (b) and the Riemannian geometry of a spin 4-manifold X, where the product is in the sense of Section 2. So we have an $\mathcal{A}$ - $\mathcal{B}$ bimodule $(\mathfrak{h}, D, \Gamma)$ with

$$\mathcal{A} = C^\infty(M, \mathbb{R}) \otimes (\mathbb{C} \oplus \mathbb{H}),\ \mathcal{B} = C^\infty(M, \mathbb{R}) \otimes (\mathbb{C} \oplus M_3(\mathbb{C})).$$

The corresponding unitary gauge groups are

$$\mathcal{U}_{\mathcal{A}} = \mathrm{Map}(X, U(1) \times SU(2)),\ \ \mathcal{U}_{\mathcal{B}} = \mathrm{Map}(X, U(1) \times U(3))$$

The Hilbert space is $\mathfrak{h} = \mathfrak{h}_0 \oplus (\mathfrak{h}_1 \otimes \mathbb{C}^3)$, with

$$\mathfrak{h}_0 = L^2(X,S) \otimes (\mathbb{C} \oplus \mathbb{H}) \otimes \mathbb{R}^N,\ \mathfrak{h}_1 = L^2(X,S) \otimes (\mathbb{C} \oplus \mathbb{C} \oplus \mathbb{H}) \otimes \mathbb{R}^N.$$

The representations π_0 and π_1 of $\mathcal{A}$ are the same as in (b). Letting z denote the element (1, -1) of the center of $\mathcal{A}$, the grading operator on $\mathfrak{h}$ is $\Gamma = \gamma_5 \otimes \pi(z)$. The self-adjoint operator $D = D_0 \oplus D_1$ is given by

$$D_0 = \begin{bmatrix} \partial_X \otimes I_N & \gamma_5 \otimes M_e & 0 \\ \gamma_5 \otimes M_e^* & \partial_X \otimes I_N & 0 \\ 0 & 0 & \partial_X \otimes I_N \end{bmatrix}$$

$$D_1 = \begin{bmatrix} \partial_X \otimes I_N & 0 & \gamma_5 \otimes M_d & 0 \\ 0 & \partial_X \otimes I_N & 0 & \gamma_5 \otimes M_u \\ \gamma_5 \otimes M_d^* & 0 & \partial_X \otimes I_N & 0 \\ 0 & \gamma_5 \otimes M_u^* & 0 & \partial_X \otimes I_N \end{bmatrix}$$

Here M_e, M_d and M_u are complex N x N matrices.

The computation of the gauge theory on this space is similar to that done in Example 5, so we shall only state the results. First, consider the $\mathcal{A}$ algebra. One finds that a universal 1-form ρ is represented by

$$\pi_0(\rho) = \begin{pmatrix} \gamma(A) \otimes I_N & \phi_1 \gamma_5 \otimes M_e & \phi_2 \gamma_5 \otimes M_e \\ \phi'_1 \gamma_5 \otimes M_e^* & \gamma(W_1) \otimes I_N & \gamma(W_2) \otimes I_N \\ -\overline{\phi'_2} \gamma_5 \otimes M_e^* & -\gamma(\overline{W}_2) \otimes I_N & \gamma(\overline{W}_1) \otimes I_N \end{pmatrix},$$

$$\pi_1(\rho) = \begin{pmatrix} \gamma(A) \otimes I_N & 0 & \phi_1 \gamma_5 \otimes M_d & \phi_2 \gamma_5 \otimes M_d \\ 0 & \gamma(\overline{A}) \otimes I_N & -\overline{\phi_2} \gamma_5 \otimes M_u & \overline{\phi_1} \gamma_5 \otimes M_u \\ \phi'_1 \gamma_5 \otimes M_d^* & \phi'_2 \gamma_5 \otimes M_u^* & \gamma(W_1) \otimes I_N & \gamma(W_2) \otimes I_N \\ -\overline{\phi'_2} \gamma_5 \otimes M_d^* & \overline{\phi'_1} \gamma_5 \otimes M_u^* & \gamma(-\overline{W}_2) \otimes I_N & \gamma(\overline{W}_1) \otimes I_N \end{pmatrix}$$

Here (A, W_1, W_2) are complex-valued 1-forms on X, and $(\phi_1, \phi_2, \phi'_1, \phi'_2)$ are complex-valued functions on X. So $\Omega^1_D(\mathcal{A}) \cong (\Lambda^1(X, \mathbb{C}))^3 \oplus (\Lambda^0(X, \mathbb{C}))^4$. The differential map $d: \mathcal{A} \rightarrow \Omega^1_D(\mathcal{A})$ is given by

$$d(\lambda, \alpha + \beta j) = (d\lambda, d\alpha, d\beta) \oplus (\alpha - \lambda, \beta, \lambda - \alpha, -\beta) \in \Omega^1_D(\mathcal{A}).$$

If ρ is a vector potential then A is u(1)-valued, W is su(2)-valued and $\phi'_1 = \bar{\phi}_1$, $\phi'_2 = -\phi_2$. Thus a vector potential consists of a u(1) gauge field A, an su(2) gauge field W and a Higgs doublet ϕ.

In order to compute $\Omega^2_D(\mathcal{A})$, it is enough to just consider $\pi_1(d\rho)$, as $\pi_0(d\rho)$ is then obtained by taking $M_u = 0$. Separating the various terms with respect to their differential-form grading on M, one finds

$$\pi_1(d\rho) = \begin{pmatrix} \gamma(dA)\otimes I_N & 0 & 0 & 0 \\ 0 & \gamma(d\bar{A})\otimes I_N & 0 & 0 \\ 0 & 0 & \gamma(dW_1)\otimes I_N & \gamma(dW_2)\otimes I_N \\ 0 & 0 & \gamma(-d\overline{W}_2)\otimes I_N & \gamma(d\overline{W}_1)\otimes I_N \end{pmatrix}$$

$$+ \begin{pmatrix} 0 & 0 & D\phi_1\gamma_5\otimes M_d & D\phi_2\gamma_5\otimes M_d \\ 0 & 0 & -\overline{D\phi_2}\gamma_5\otimes M_u & \overline{D\phi_1}\gamma_5\otimes M_u \\ D\phi'_1\gamma_5\otimes M_d^* & D\phi'_2\gamma_5\otimes M_u^* & 0 & 0 \\ -\overline{D\phi'_2}\gamma_5\otimes M_d^* & \overline{D\phi'_1}\gamma_5\otimes M_u^* & 0 & 0 \end{pmatrix}$$

$$+ \begin{pmatrix} \tilde{A}\otimes I_N & 0 & 0 & 0 \\ 0 & \bar{\tilde{A}}\otimes I_N & 0 & 0 \\ 0 & 0 & \tilde{W}_1\otimes I_N & \tilde{W}_2\otimes I_N \\ 0 & 0 & -\bar{\tilde{W}}_2\otimes I_N & \bar{\tilde{W}}_1\otimes I_N \end{pmatrix}$$

$$+ \begin{pmatrix} Q & 0 \\ 0 & R \end{pmatrix}$$

with
$$Q = \begin{pmatrix} (\phi_1 + \phi'_1) M_d M_d^* & (\phi_2 + \phi'_2) M_d M_u^* \\ (-\bar{\phi}_2 - \bar{\phi}'_2) M_u M_d^* & (\bar{\phi}_1 + \bar{\phi}'_1) M_u M_u^* \end{pmatrix}$$

and
$$R = 1/2 \begin{pmatrix} \phi_1 + \phi'_1 & \phi_2 + \phi'_2 \\ -\bar{\phi}_2 - \bar{\phi}'_2 & \bar{\phi}_1 + \bar{\phi}'_1 \end{pmatrix} \otimes (M_d^* M_d + M_u^* M_u)$$

$$+ 1/2 \begin{pmatrix} Y & Z \\ \bar{Z} & -\bar{Y} \end{pmatrix} \otimes (M_d^* M_d - M_u^* M_u).$$

Here $D(\phi_1, \phi_2) = d(\phi_1, \phi_2) + A(\phi_1, \phi_2) - (\phi_1, \phi_2)\begin{pmatrix} W_1 & W_2 \\ -\overline{W_2} & \overline{W_1} \end{pmatrix}$, and $\tilde{A}$, $\widetilde{W}$, Y and Z are new scalar fields. We see that

$$\{\pi(d\rho): \rho \in J_0 \cap \Omega^1(\mathfrak{A})\} = \{\pi(d\rho): A = W = (\phi_1, \phi_2, \phi_1', \phi_2') = 0\}$$

consists of the $\tilde{A}$, $\widetilde{W}$, Y and Z fields. Then the quotienting used to define $\Omega^2_D(\mathfrak{A})$ amounts to quotienting out the new scalar fields. Let us assume, for example, that the matrix $M_d^* M_d + M_u^* M_u$ is not a multiple of the N x N identity matrix; then we find that $\Omega^2_D(\mathfrak{A})$ comes from the tensor product of the differential algebra of the finite-point space by the exterior algebra of X. Namely, an element of $\Omega^2_D(\mathfrak{A})$ consists of

a. A $\mathbb{C}$-valued 2-form on X and an $\mathbb{H}$-valued 2-form on X.

b. Two $\mathbb{H}$-valued 1-forms on X.

c. Two $\mathbb{H}$-valued 0-forms on X.

If $V_{\mathfrak{A}}$ is a vector potential then its curvature $\theta_{\mathfrak{A}} = dV_{\mathfrak{A}} + V_{\mathfrak{A}}{}^2 \in \Omega^2_D(\mathfrak{A})$, a self-adjoint element, consists of the following components:

a. The curvature F_A of the u(1)-gauge field A and the curvature F_W of the su(2) gauge field W.

b. The covariant derivative $D\phi$ of the Higgs field ϕ, and its conjugate.

c. The function $\left(|1 + \phi_1|^2 + |\phi_2|^2 - 1 \right)$ times (1,1).

The story with the $\mathfrak{B}$ algebra is much simpler. As $\mathfrak{B}$ commutes with the off-diagonal terms of D, it is easy to see that $\Omega^*_D(\mathfrak{B})$ is just $\mathfrak{B} \otimes \Lambda^*(X, \mathbb{C})$, with the obvious multiplication and differentiation. Then a vector potential $V_{\mathfrak{B}}$ is the sum of a u(1) gauge field K and a u(3) gauge field V, and its curvature $\theta_{\mathfrak{B}} \in \Omega^2_D(\mathfrak{B})$ is same as the usual field strength $(dK, dV + V^2) \in (u(1) \oplus u(3)) \otimes \Lambda^2(X)$.

The gauge group of our theory is $\mathrm{Map}(X, U(1)_{\mathfrak{A}} \times SU(2)_{\mathfrak{A}} \times U(1)_{\mathfrak{B}} \times U(3)_{\mathfrak{B}})$. In order to correctly reduce the gauge fields to take values in $u(1) \oplus su(2) \oplus su(3)$, we must impose following condition on the gauge fields :

(9) $\quad A = K = - \mathrm{Tr}\, V.$

Then the contributions to the net hypercharges of the fermions are as shown:

	A	K	V	Y
e_R	-1	-1	0	-2

(e_L, ν_L)	0	-1	0	-1
d_R	-1	0	1/3	-2/3
u_R	1	0	1/3	4/3
(d_L, u_L)	0	0	1/3	1/3.

We show in the Appendix that (9) has a natural interpretation as a ***unimodularity*** condition on the gauge fields, of the same general type as the reduction from a U(N) gauge theory to an SU(N) gauge theory. In particular, equation (9) is an infinitesimal version of equation (12) of the Appendix.

We shall now compute the action. Let us start with the fermionic part. If we write a fermion field $\Psi \in \mathfrak{h}$ as

$$\Psi = (e_R, e_L, \nu_L) \oplus (d_R, u_R, d_L, u_L),$$

then $\overline{\Psi}\,(D + V_{\mathfrak{A}} + V_{\mathfrak{B}})\,\Psi$ becomes the terms $\mathcal{L}_\Psi$ and $\mathcal{L}_Y$ of the standard model, after a shift of the ϕ_1 field.

We now must compute the Yang-Mills action. If we were to follow the previous discussion, we would simply take

$$\text{(10)} \qquad YM = \langle\theta_{\mathfrak{A}}, \theta_{\mathfrak{A}}\rangle + \langle\theta_{\mathfrak{B}}, \theta_{\mathfrak{B}}\rangle.$$

However, this would be physically wrong, as our Hilbert spaces of fermions are not irreducible under the action of the gauge group. Consequently, using (10) would have the effect of artificially imposing relations among coupling constants. A more general gauge invariant bosonic action is given by

$$\text{(11)} \qquad \mathcal{L} = \mathrm{Tr}_\omega\,(z_1\, \theta_{\mathfrak{A}}{}^2\, D_\nabla{}^{-4}) + \mathrm{Tr}_\omega\,(z_2\, \theta_{\mathfrak{B}}{}^2\, D_\nabla{}^{-4}),$$

where z_1 and z_2 are arbitrary positive operators on $\mathfrak{h}$ which commute with the actions of $\mathfrak{A}$ and $\mathfrak{B}$, and with the operator D. With this freedom, the Lagrangian (11) reproduces the terms $\mathcal{L}_G + \mathcal{L}_\phi + \mathcal{L}_V$ of the standard model, with arbitrary constants in $\mathcal{L}_G$ and $\mathcal{L}_V$ (after a rescaling of the Higgs field). Thus we recover the standard model on the nose, with the same number of arbitrary coupling constants. On the other hand, one could require in addition that the operators z_1 and z_2 lie in the center of $\mathfrak{A} \otimes \mathfrak{B}$. In this case we find one relationship among the coupling constants of the standard model. We will not write out this relationship here, but will simply note that it gives the Higgs mass in terms of the W mass and the fermion masses. In particular, if the top quark mass is of the same order of magnitude as the W mass, then the relationship implies that the Higgs mass would be, also. However, this relationship is not preserved by the usual renormalization flow, and we do not know if it is physically meaningful.

Let us summarize some of the improvements of the present paper over our previous paper. In [CL] we had the following :

1. The complex conjugate of the up quark in the Hilbert space, and a charge conjugation in the operator D.

2. An $(\mathfrak{A}, \mathfrak{B})$ bimodule structure.

3. Equation (9) relating the U(1) factors.

In the present paper, we simplified the first point by changing the action of the $\mathfrak{A}$ algebra. (This simplification was noticed independently by D. Kastler.) We again have the bimodule structure, but Section 3 of the present paper puts this into a more general context. And equation (9) is now interpreted in the Appendix as a special case of a unimodularity condition which makes sense in noncommutative geometry.

5. Appendix

We discuss a notion of unimodularity which makes sense in a general algebraic setting. First, suppose that one has a C^*-algebra C and a self-adjoint trace τ on C. That is, $\tau(x^*) = \overline{\tau(x)}$ for all $x \in C$. Then one can define the phase of a unitary element of C by

$$\mathrm{Phase}_\tau(u) = \frac{1}{2\pi i}\int_0^1 \tau(u'(t)\, u(t)^{-1})\, dt,$$

where u(t) is a smooth path of unitaries joining 1 to u. So this phase is only defined in the connected component of the identity in the group $\mathcal{U}(C)$ of unitaries, and is ambiguous up to a countable subgroup of $\mathbb{R}$, namely the image $\langle\tau, K_0(C)\rangle$ of $K_0(C)$ by the trace [CK].

The condition $\mathrm{Phase}_\tau(u) = 0$ defines a normal subgroup of the connected component of the identity, which we will denote by $S_\tau(C)$.

Now let $\mathfrak{A}$ and $\mathfrak{B}$ be involutive algebras, and $(\mathfrak{h}, D)$ a (d, ∞)-summable bimodule over $\mathfrak{A}$ and $\mathfrak{B}$. We shall apply the above considerations to the C^*-algebra C generated by $\mathfrak{A}$ and $\mathfrak{B}$ in $\mathfrak{h}$, with a family of traces τ_ρ on C constructed from self-adjoint elements $\rho = \rho^*$ of the center of $\mathfrak{A}$:

$$\tau_\rho(x) = \mathrm{Tr}_\omega(\rho\, x\, |D|^{-d}) \qquad \text{for all } x \in C.$$

We thus get a normal subgroup $S_{\mathfrak{A}}(C)$ of the unitary group of C by intersecting all of the $S_{\tau_\rho}(C)$'s. Since $\mathcal{U}(\mathfrak{A}) \times \mathcal{U}(\mathfrak{B})$ is a subgroup of $\mathcal{U}(C)$, its intersection with $S_{\mathfrak{A}}(C)$ gives a normal subgroup $S(\mathfrak{A}, \mathfrak{B})$ of $\mathcal{U}(\mathfrak{A}) \times \mathcal{U}(\mathfrak{B})$.

Example 6: Let X be a Riemannian spin manifold. Take $\mathfrak{A} = M_N(C^\infty(X))$, $\mathfrak{B} = \mathbb{C}$, $\mathfrak{h} = (L^2(X,S))^N$ and D to be the Dirac operator. Then the space of self-adjoint elements of the center of $\mathfrak{A}$ is $\{f\, I_N : f \in C^\infty(X) \text{ real}\}$, and one finds $S(\mathfrak{A}, \mathfrak{B}) = \mathrm{Map}(X, SU(N))$. This is why in general, one can think of $S(\mathfrak{A}, \mathfrak{B})$ as a sort of unimodular unitary group.

Example 7: Take $\mathfrak{A}$, $\mathfrak{B}$, $\mathfrak{h}$ and D as in Section 4b above. A self-adjoint element of the center of $\mathfrak{A}$ can be written as $\lambda_1 e + \lambda_2(1 - e)$ for some real numbers λ_1 and λ_2, with $e = (1, 0) \in \mathbb{C} \oplus \mathbb{H}$ and $1 - e = (0, 1) \in \mathbb{C} \oplus \mathbb{H}$. It follows that

$$S(\mathfrak{A}, \mathfrak{B}) = (\mathcal{U}(\mathfrak{A}) \times \mathcal{U}(\mathfrak{B})) \cap (SU(e\mathfrak{h}) \times SU((1 - e)\mathfrak{h})).$$

Let then U be an element of $\mathcal{U}(\mathfrak{A}) \times \mathcal{U}(\mathfrak{B})$. It is given by a quadruple:

$$U = (\lambda, q), (u, v)); \qquad \lambda \in U(1),\ q \in SU(2),\ u \in U(1),\ v \in U(3).$$

We have $\mathfrak{h} = \mathfrak{h}_0 \oplus (\mathfrak{h}_1 \otimes \mathbb{C}^3)$, with the action of U given by

$$(\pi_0(\lambda, q) \otimes u) \oplus (\pi_1(\lambda, q) \otimes v).$$

This operator restricts to both $e\mathfrak{h}$ and $(1 - e)\mathfrak{h}$, and we must compute the determinants of these restrictions. We get

$$\det(U_e) = (u^2 (\det(v))^2)^N, \qquad \det(U_{1-e}) = (\lambda\, u\, (\det(v))^2)^N.$$

So the unimodularity condition is

(12) $\qquad \lambda = u = (\det(v))^{-1},$

and $S(\mathfrak{A}, \mathfrak{B}) = U(1) \times SU(2) \times SU(3)$.

Example 8: Take $\mathfrak{A}$, $\mathfrak{B}$, $\mathfrak{h}$ and D as in Section 4d above. Then it is easy to see that $S(\mathfrak{A}, \mathfrak{B})$ consists of maps from X to the unimodular unitary group of Example 7, that is Map(X, U(1) x SU(2) x SU(3)).

References

[CES] R. Coquereaux, G. Esposito-Farese and F. Scheck, "The Theory of Electroweak Interactions Described by SU(2|1) Algebraic Superconnetions", IHES preprint, to appear (1991)

[CJ] R. Coquereaux and A. Jadczyk, Riemannian Geometry, Fiber Bundles, Kaluza-Klein Theories and All That, World Scientific Press, Singapore (1988)

[CK] A. Connes and M. Karoubi, "Caractere Multiplicatif d'un Module de Fredholm", K-Theory 2, p. 431 (1988)

[CL] A. Connes and J. Lott, "Particle Models and Noncommutative Geometry", Nuc. Phys. B, Proc. Suppl. 18B, p. 29, North-Holland, Amsterdam (1990)

[Co1] A. Connes, "Noncommutative Differential Geometry", Publ. Math. IHES 62, p. 44 (1983)

[Co2] A. Connes, Noncommutative Differential Geometry, to appear

[CS] A. Connes and G. Skandalis, "The Longitudinal Index Theorem for Foliations", Publ. RIMS, Kyoto 20, p. 1139-1183 (1984)

[Di] J. Dixmier, "Existence de Traces Non-normales", C. R. Acad. Sci. Paris 262, p. 1107-1108 (1966)

[El] J. Ellis, "Phenomenology of Unified Gauge Theories", in Gauge Theories in High-Energy Physics, Part I, 1981 Les Houches Lectures, North-Holland, Amsterdam, p. 161 (1983)

[GWS] S. Glashow, Nucl. Phys. B22, p. 579 (1961)

S. Weinberg, Phys. Rev. Lett. 19, p. 1264 (1967)

A. Salam, Proc. 8th Nobel Symposium, ed. N. Svartholm, Almqvist and Wiksells, Stockholm, p. 367 (1968)

[Ka] G. Kasparov, "Equivariant KK-Theory and the Novikov Conjecture", Inv. Math. 91, p. 147 (1988)

[LM] H. Lawson and M.L Michelsohn, Spin Geometry, Princeton University Press, Princeton (1989)

[Wi] E. Witten, "Fermion Quantum Numbers in Kaluza-Klein Theory", in Proceedings of the 1983 Shelter Island Conference on Quantum Field Theory and the Foundations of Physics, eds. N. Khuri et al., MIT Press, p. 227 (1985)

INTERSECTION THEORY, INTEGRABLE HIERARCHIES AND TOPOLOGICAL FIELD THEORY

Robbert Dijkgraaf

School of Natural Sciences
Institute for Advanced Study
Princeton, NJ 08540

1. INTRODUCTION

The last two years have seen the emergence of a beautiful new subject in mathematical physics. It manages to combine a most exotic range of disciplines: two-dimensional quantum field theory, intersection theory on the moduli space of Riemann surfaces, integrable hierarchies, matrix integrals, random surfaces, and many more. The common denominator of all these fields is two-dimensional quantum gravity or, more general, low-dimensional string theory. Here the application of large-N techniques in matrix models, that are used to simulate fluctuating triangulated surfaces [1]-[3], has led to complete solvability [4]-[8]. (See *e.g.* the review papers [9], and also the lectures of S. Shenker in this volume.) Shortly after the onset of the remarkable developments in matrix models, Edward Witten presented compelling evidence for a relationship between random surfaces and the algebraic topology of moduli space [10, 11]. This proposal involved a particular quantum field theory, known as topological gravity [12], whose properties were further established in [13, 14] and generalized to the so-called multi-matrix models in [15]-[19]. This subject can, among many other things, be considered as a fruitful application of quantum field theory techniques to a particular problem in pure mathematics, and as such is a prime example of a much bigger program, also largely due to Witten, that has been taking shape in recent years. It is impossible to do fully justice to this subject within the confines of these lecture notes. I will however make an effort to indicate some of the more startling interconnections.

I think it is fair to say that one of the most exciting recent developments in this field has been the work of the Russian mathematician Maxim Kontsevich [20]-[22]. He did not only prove rigorously the first of a set of conjectures of Witten relating the calculus of intersection numbers on the moduli space of curves to integrable hierarchies of KdV type. In the process of that, he also derived a concrete 'matrix integral representation' of the string partition function. This is quite a curious result, since this matrix integral is in no obvious way related to the matrix models that initiated all these developments. In

New Symmetry Principles in Quantum Field Theory, Edited by
J. Fröhlich et al., Plenum Press, New York, 1992

fact, one could say that Kontsevich's model is used to triangulate moduli space, whereas the original models triangulated Riemann surfaces.

Let me first state the main results that will be the revolving point of these notes. Our starting point will be the particular two-dimensional quantum field theory know as topological gravity. Let $\mathcal{O}_n$ denote the observables in this theory and t_n the coupling constants to these operators. In the field theory we can consider correlation functions of the type

$$\left\langle \mathcal{O}_{n_1} \cdots \mathcal{O}_{n_s} \right\rangle_g, \tag{1.1}$$

where $\langle \cdots \rangle_g$ denotes the expectation value on a (connected) surface with g handles. These correlation functions represent, more or less by definition, characteristic numbers of the moduli space of Riemann surfaces. The string partition function $\tau(t)$ is defined as the generating functional of all possible correlation functions on all possible Riemann surfaces, not necessarily connected. That is, $\tau(t)$ has an asymptotic expansion of the form

$$\tau(t) = \exp \sum_{g=0}^{\infty} \left\langle \exp \sum_n t_n \mathcal{O}_n \right\rangle_g \tag{1.2}$$

Witten's conjecture states that, as anticipated in our notation,

$$\tau(t) \textit{ is a tau-function of the KdV hierarchy.} \tag{1.3}$$

We will explain the concept of a τ-function in great detail in section 3. It is a very useful and common notion in the study of integrable hierarchies. Essentially it implies that from $\tau(t)$ we can construct solutions to the famous Korteweg-de Vries equation

$$\frac{\partial u}{\partial t} = uu' + \frac{1}{6}u''' \tag{1.4}$$

and all its generalizations. The relevant point is that $\tau(t)$ is completely calculable. This integrable structure was first found in the matrix model description of random surfaces or Euclidean quantum gravity, and so Witten's conjecture can be restated as the equivalence of quantum and topological gravity. The intuition behind this remarkable equivalence is some kind of poorly understood universality in theories of two-dimensional gravity.*

The integral representation of $\tau(t)$ that emerges naturally in Kontsevich's solution involves the integral over a $N \times N$ Hermitian matrix Y of the form

$$\tau(Z) = \rho(Z)^{-1} \int dY \cdot \exp \, \mathrm{Tr} \left[-\frac{1}{2} Z Y^2 + \frac{i}{6} Y^3 \right] \tag{1.5}$$

Here Z is a second $N \times N$ Hermitian matrix, and $\rho(Z)$ is the one-loop integral

$$\rho(Z) = \int dY \cdot \exp -\frac{1}{2} \mathrm{Tr}\, Z Y^2 \tag{1.6}$$

*In the case of two-dimensional *gauge* theories the equivalence of the topological and the non-topological model (also known as Yang-Mills theory) is much better understood due to recent work of Witten [23].

It is clear that $\tau(Z)$ is conjugation invariant, and so only depends on the eigenvalues $z_1, \ldots, z_N$ of Z. The relation between (1.2) and (1.5) requires a map from the matrix Z to the coupling coefficients t_n. According to Kontsevich this map is given by

$$t_n = -\frac{1}{n} \operatorname{Tr} Z^{-n}. \tag{1.7}$$

As we will see, the matrix integral (1.5) has a natural expansion in Feynman diagrams and each diagram represents the integral over a particular cell in the moduli space of Riemann surfaces. It can therefore be considered as a finite-dimensional reduction of the path-integral of topological gravity.

Finally, there is a third and equivalent way to characterize $\tau(t)$, that naturally arises in topological field theory and in considerations of so-called loop operators. In this formulation a series of differential operators L_n — quadratic in the parameters t_k and their derivatives $\partial/\partial t_k$ — annihilate the string partition function [24,25]

$$L_n \cdot \tau = 0, \qquad n \geq -1. \tag{1.8}$$

These operators, that will be consider in greater detail in sections 4 and 7, form part of a Virasoro algebra—they satisfy the commutation relations

$$[L_n, L_m] = (n-m) L_{n+m}. \tag{1.9}$$

The constraints (1.8) can be used to derive recursion relations for the correlation functions and thus provide an alternative way to arrive at the asymptotic expansion for $\tau(t)$. From a computational perspective, (1.8) is perhaps the most practical characterization of $\tau(t)$.

There exist a whole hierarchy of generalizations of all these structures, where the results become more and more incomplete and conjectural as we move away from the original setting. The most simple generalization involves a choice of integer $p \geq 2$, with the case $p = 2$ corresponding to intersection theory on moduli space. In the case there are well-defined candidates for the corresponding problem in algebraic geometry, the integrable hierarchy, and the matrix integral. As we will see this can naturally be considered the A_{p-1} case in a series of models labeled by a simply-laced Lie group G, *i.e.*, G is of type A_n, D_n, E_6, E_7, or E_8. The most general case is based on an arbitrary two-dimensional topological field theory, and includes, among others, sigma-models with an almost complex target space. This field is known as *topological string theory* and its status as an integrable system is very unclear, to say the least.

Let me close this introduction with a short exposition of how this all is related to $c < 1$ string theory, where one studies minimal conformal field theories on fluctuating surfaces. These minimal models are labeled by two (relative prime) integers p and q—the labeling (p, q) is symmetric in p and q. The models can be considered as isolated points in an ill-defined space of all two-dimensional quantum field theories coupled to gravity, as is symbolically illustrated in *fig.* 1. A neighbourhood of a minimal model consists of massive QFT's obtained by perturbation. The counting of physical observables is rather complicated [26], but there is a simple subsector—which does not involve the reparametrization ghosts—and in this subsector we find

$$\#\ \text{states} = \tfrac{1}{2}(p-1)(q-1). \tag{1.10}$$

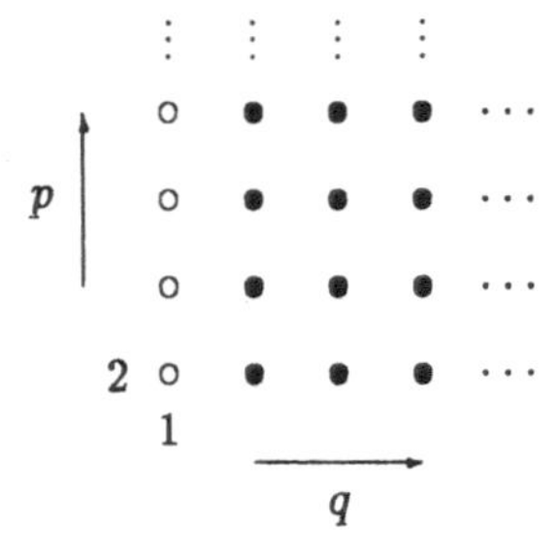

Fig. 1. *The space of $c < 1$ field theories coupled to quantum gravity contains as special point the (p, q) minimal models. The models of type $(p, 1)$, here indicated by open dots, should be considered as topological field theories, and are the main subject of these notes.*

This finite number of states is the defining property of minimal models. Through the work of Kazakov [3], one can use matrix models to represent both the random surfaces and the conformal field theory by choosing the right critical behavior of the matrix potential. This leads to a uniform description of all models with identical p, where a chain of $(p-1)$ matrices suffices. The second number q now labels the order of criticality. All models with fixed p are related through the pth KdV hierarchy [7], which moves horizontally in *fig.* 1. That is, their partition functions are given by the *same* τ-function, evaluated at different values of the coupling coefficients t_n.

Models of type $(p, 1)$ are strictly speaking not well-defined conformal field theories—according to the above formula they have zero physical fields. However, one can make sense out of these models as topological field theories, the so-called *topological minimal models*, and they will form the main topic of these lecture notes. Since the orbit of a topological model under the KdV flows will cover all other (p, q) models with fixed p, this is in fact a very elegant and powerful way to study the general case. We will see that the behavior of many quantities simplifies dramatically at the topological point $q = 1$, in particular the so-called Baker-Akhiezer function is exactly calculable, and this will lead to an explicit solution. As mentioned before this leads us again, through the work of Kontsevich, to matrix models, although of a very different kind. In fact, Kontsevich's generalized matrix model interpolates in a natural way between the models with different p—its allows 'vertical' deformations in *fig.* 1—and can thus be seen as complementary to the original double-scaled matrix models.

In the following sections we will try to explain the various aspects of these beautiful but at first sight highly mysterious results and their many generalizations. We will focus in particular on the question why matrix integrals arise naturally in integrable hierarchies, a point that is also addressed in a number of recent publications [22,27,28]. These notes are organized as follows: We start our discussion with (generalized) intersection theory on moduli space in §2. The integrable hierarchies of KdV type that emerge in that context are described in §3, which follows very closely the exposition of Segal and Wilson [29]. In §4 we show how the apparatus of Grassmannians can be applied to our specific problem and how all this is related to matrix integrals. The quantum field theories that have led us to these results and conjectures first enter in §5. Here we only consider the general properties of topological field theories that obey factorization conditions. This is made much more explicit in §6, where various two-dimensional models are constructed, with particular emphasis on the so-called Landau-Ginzburg models. In §7 we finally discuss topological gravity, and more generally topological string theory. Since we covered much of this material on a previous occasion [30], we will be here far less complete. In particular, the Virasoro constraints and their relation with loop equations will only be touched upon. Finally a point of caution: we have not always been extremely careful in our calculations with all 'irrelevant' numerical constants. We refer the reader to the literature for the factors of i and 2π.

2. INTERSECTION THEORY

We start from the quantum field theory end, or more precisely from the finite dimensional cohomology problem to which the quantum field theory, by construction, can be reduced to. We will return to the description of the full quantum field theories in the sections 5–7.

2.1. The moduli space of curves

Let us briefly recall some well-known facts about the moduli space of smooth complex curves or Riemann surfaces. A given topological surface Σ with g handles and s marked points $x_1, \ldots, x_s$ can be made into a complex manifold by endowing it with a complex structure J, that is, a local linear map on the tangent bundle that satisfies $J^2 = -1$ and the integrability condition $DJ = 0$. Two complex structures are considered equivalent if they are related by a diffeomorphism. This leaves us actually with a finite-dimensional space of inequivalent complex structures on the surface, known as the moduli space $\mathcal{M}_{g,s}$. By Riemann-Roch this is a space of complex dimension

$$\dim \mathcal{M}_{g,s} = 3g - 3 + s. \tag{2.1}$$

(The moduli spaces $\mathcal{M}_{0,s}$ are only well-defined for $s \geq 3$, the exceptional case $\mathcal{M}_{1,0}$ equals $\mathcal{M}_{1,1}$ and is thus one-dimensional.) Moduli spaces are not everywhere smooth, since surfaces with accidental higher symmetry give rise to so-called orbifold points.

One way to think about the complex structure on Σ is as the conformal class of a metric $h_{\mu\nu}$. Indeed, a metric defines a complex structure through

$$J_\mu{}^\nu = \sqrt{h}\, \epsilon_{\mu\lambda} h^{\lambda\nu}, \tag{2.2}$$

with $\epsilon_{\mu\nu}$ the Levi-Civita symbol, and it is a useful fact that all J's can be obtained in this way. Since the definition of J is independent of local rescalings of $h_{\mu\nu}$, we can represent $\mathcal{M}_{g,s}$ alternatively as the space of metrics modulo local rescalings and diffeomorphisms.

The moduli space $\mathcal{M}_{g,s}$ has a boundary and there exists a direct, intuitive interpretation of the points at the boundary — they represent degenerate surfaces. There are basically two ways in which a surface can degenerate. If we think in terms of a conformal class of metrics, the surface can either form a node—or, equivalently, a long neck—or two marked points can collide. The boundary of $\mathcal{M}_{g,s}$ can be thought to lie at infinity. One would like to compactify the moduli space by adding points at infinity, not unlike how one compactifies the plane $\mathbf{R}^n$ to the n-dimensional sphere. In this case the points at infinity represent particular Riemann surfaces. The Deligne-Mumford or 'stable' compactification [31] adds three types of these points, as is illustrated in *fig.* 2.

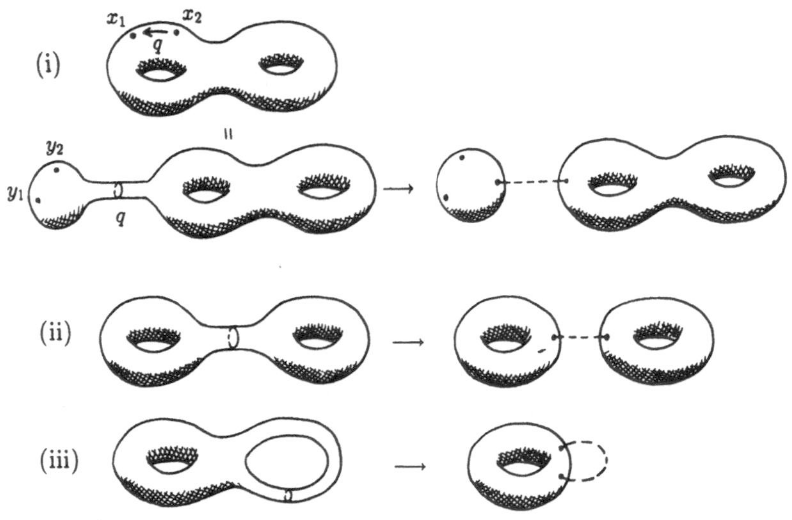

Fig. 2. *The three possible degenerations of a Riemann surface. A surface degenerates if (i) two marked points come close together, which is conformally equivalent to the formation of a long thin neck, (ii) a dividing cycle pinches, or (iii) a non-trivial homology cycle pinches. We have also indicated the surfaces that are added in the stable compactification of* $\mathcal{M}_{g,s}$.

(i) The process in which two points x_1 and x_2 'collide' if $q = x_1 - x_2$ tends to zero can (after a coordinate transformation $x \to x/q$) alternatively be described as the process in which a sphere, that contains x_1 and x_2 at fixed distance, pinches off the surface by forming a neck of length $\log q$. These two descriptions are fully equivalent, but the latter is actually more in the spirit of conformal field theory, since we see in an obvious way the operator product expansion emerge. It is also suggestive of another end point configuration. Here the natural final configuration is not simply the surface with $x_1 = x_2$, but it consists of a separate sphere containing the points x_1, x_2 and a third point where the infinite long tube was attached, together with the original surface with one marked point less. In the stable compactification we add this configuration as limit point, which symbolically can be written as the process

$$(g, s) \to (g, s-1) + (0, 3). \tag{2.3}$$

Recall that the thrice-punctured sphere has a unique complex structure, so there are no moduli associated to it and this boundary component has codimension one. The crucial property of this compactification is that the points x_i never come together. The two other degeneration modes are more straightforward.

(ii) If a cycle of non-trivial homology pinches, we replace the surface by a surface with one handle less and two extra marked points—the attachment points of the infinitely thin handle,

$$(g, s) \to (g-1, s+2). \tag{2.4}$$

(iii) In a similar spirit: in case a dividing cycle pinches, the resulting surface consists

of two disconnected surfaces of genus g' and $g - g'$, each having one extra puncture

$$(g, s) \to (g, s' + 1) + (g - g', s - s' + 1). \tag{2.5}$$

It can be shown that this prescription makes $\mathcal{M}_{g,s}$ into a compact smooth (orbifold) space $\overline{\mathcal{M}}_{g,s}$. We now want to consider its cohomology ring $H^*(\overline{\mathcal{M}}_{g,s})$. A particular set of elements has been considered by Mumford, Morita, and Miller [32]-[34]. These classes are constructed as follows [10]. There exist s natural line bundles $L_1, \ldots, L_s$ on the moduli space. The fiber of the bundle L_i at a point $\Sigma \in \mathcal{M}_{g,s}$ is the cotangent space to the point x_i on the surface Σ. These line bundles have first Chern classes that in de Rham cohomology are represented by the curvature F_i of an arbitrary $U(1)$ connection on L_i. This defines for us the $2n$-dimensional classes

$$\sigma_n(i) = c_1(L_i)^n \in H^{2n}(\mathcal{M}_{g,s}), \qquad \sigma_n(i) = \underbrace{F_i \wedge \ldots \wedge F_i}_{n}. \tag{2.6}$$

It is a non-trivial property that these are stable classes, *i.e.*, that their definition actually extends to the *compactified* moduli space $\overline{\mathcal{M}}_{g,s}$. The reason is basically that the line bundles L_i remain well-defined in the stable compactification, since the points x_i will never collide. At infinity they remain at finite distance while being separated from the bulk of the surface on a 'frozen' 3-punctured sphere, which carries no moduli.

We can now define 'correlation functions' by pairing the wedge product of the σ_n's with the fundamental class of the moduli space

$$\left\langle \sigma_{n_1} \cdots \sigma_{n_s} \right\rangle_g \equiv \langle \sigma_{n_1}(1) \cdots \sigma_{n_s}(s), \overline{\mathcal{M}}_{g,s} \rangle \tag{2.7}$$

Of course, these intersection numbers are only non-zero if the degree of the total class equals the dimension of moduli space, which gives the 'charge conservation' condition

$$\sum_{i=1}^{s} (n_i - 1) = 3g - 3. \tag{2.8}$$

Although the Chern classes $c_1(L_i)$ are in principle defined in integer cohomology, their intersection numbers will be in general rational numbers, due to the orbifold nature of the moduli space.

The amazing result, first conjectured by Witten [10,11] on the basis of the matrix model results and later proved rigorously by Kontsevich [20], states that all these intersection numbers are explicitly computable because their generating function obeys the equations of the KdV hierarchy. To state this result more precisely, it is convenient to rescale and relabel the classes σ_n as

$$\mathcal{O}_{2n+1} = (2n+1)!! \cdot \sigma_n, \qquad n \geq 0, \tag{2.9}$$

After this substitution the theorem states that the string partition function

$$\tau(t) = \exp \sum_{g=0}^{\infty} \left\langle \exp \sum_{k \text{ odd}} t_k \mathcal{O}_k \right\rangle_g \tag{2.10}$$

is a τ-function of the KdV hierarchy. In this sense intersection theory on the (universal) moduli space is completely integrable. We note that due to the charge conservation condition (2.8), it is very easy to introduce a string coupling constant λ that keeps track of the genus g of the different surfaces that appear in the partition function $\tau(t)$. Indeed, if we rescale the coupling constants by

$$t_k \rightarrow \lambda^{\frac{k}{3}-1} t_k, \tag{2.11}$$

we find

$$\tau(t) \rightarrow \exp \sum_{g=0}^{\infty} \lambda^{2g-2} \Big\langle \exp \sum_{k \text{ odd}} t_k \mathcal{O}_k \Big\rangle_g \tag{2.12}$$

Before we discuss the proof of the integrability of $\tau(t)$, let us first discuss some weaker results related to the 'trivial' nature of the classes σ_0 and σ_1.

2.2. The puncture and dilaton equations

The only consequence of the insertion of the 'puncture operator' σ_0 in a correlation function is that the same cohomology class is now integrated over $\overline{\mathcal{M}}_{g,s+1}$ instead of $\overline{\mathcal{M}}_{g,s}$. Since there is a natural projection map $\pi : \overline{\mathcal{M}}_{g,s+1} \rightarrow \overline{\mathcal{M}}_{g,s}$, where one simply forgets the position of the extra point, one might tend to conclude that the classes can be simply pull-backed and that thus the correlation function including σ_0's vanish by dimensional reasons. This is actually not true, due to subtleties at the divisor at infinity, where the forgotten point comes close to the other positions. This leads to extra corrections that can be expressed through the so-called puncture equation, due to Deligne [35], which is explained in more detail in [15],

$$\big\langle \sigma_0 \cdot \sigma_{n_1} \cdots \sigma_{n_s} \big\rangle_g = \sum_{i=1,\, n_i \neq 0}^{s} \big\langle \sigma_{n_1} \cdots \sigma_{n_i - 1} \cdots \sigma_{n_s} \big\rangle_g \tag{2.13}$$

This relation can be used to eliminate all σ_0's from a particular correlation function. It is of course only valid if both sides of the equation are well-defined, which is not the case in genus zero with less than three insertions. In fact, there we have the additional relation

$$\big\langle \sigma_0 \sigma_0 \sigma_0 \big\rangle_0 = 1, \tag{2.14}$$

a simple consequence of $\mathcal{M}_{0,3} =$ point.

A similar result can be obtained for the insertions of the next operator in line—the 'dilaton operator' σ_1. Since σ_1 is the first Chern class $c_1(L)$ it can be shown to calculate the degree of the canonical line bundle of a genus g surface with s punctures [11]. The dilaton equation exactly expresses this result

$$\big\langle \sigma_1 \cdot \sigma_{n_1} \cdots \sigma_{n_s} \big\rangle_g = (2g - 2 + s) \big\langle \sigma_{n_1} \cdots \sigma_{n_s} \big\rangle_g \tag{2.15}$$

Here again we must be careful if the right-hand side is ill-defined. This happens in genus one with no punctures. Here one finds

$$\big\langle \sigma_1 \big\rangle_1 = \frac{1}{24}. \tag{2.16}$$

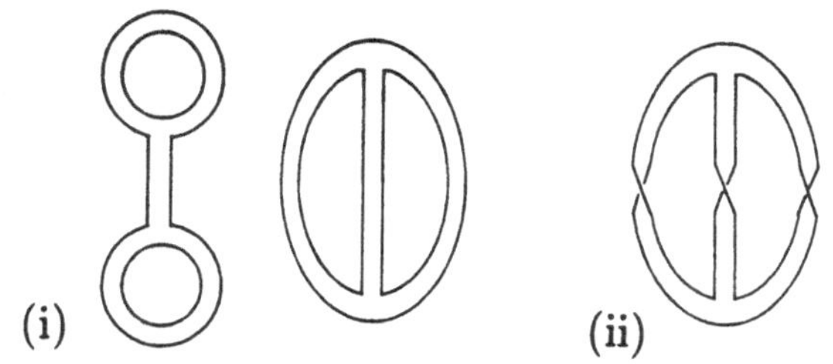

Fig. 3. *The realization of a three-punctured sphere (i) and a once-punctured torus (ii) as ribbon graphs.*

The above equations allow one to eliminate all operators σ_0 and σ_1. Actually this, together with the selection rule (2.8), suffices to reduce any intersection number in genus zero or one to (2.14) and (2.16) respectively. On the sphere the final result reads

$$\langle\sigma_{n_1}\cdots\sigma_{n_s}\rangle_0 = \frac{(n_1+\ldots+n_s)!}{n_1!\ldots n_s!} \tag{2.17}$$

On the torus it is more difficult to give a closed form expression, but one way to state the result is by means of generating functions

$$\left\langle \exp\sum_n t_n\sigma_n\right\rangle_1 = \frac{1}{24}\log\left\langle\sigma_0^3\cdot\exp\sum_n t_n\sigma_n\right\rangle_0 \tag{2.18}$$

In this fashion one can verify the KdV relations by hand at low genus. Similar verification have been done by Witten [11] in genus two and Horne [36] in genus three, using results of Mumford [32] and Faber [37] respectively.

2.3. Kontsevich's calculation

There exists a very explicit realization of the moduli spaces $\mathcal{M}_{g,s}$ for $s>0$ due to the work of Strebel [38], Penner [39], and others, that naturally arises in open string field theory. Recall that we can think of complex structures in terms of conformal classes of metrics. In open string field theory Riemann surfaces are built out of flat strips (propagators) with fixed width and flexible lengths ℓ_a by glueing them together with a three-point vertex, in such a way that all the curvature is localized in the mid-point of the vertex. With $3k$ propagators and $2k$ vertices one produce a surface of Euler number $\chi = 2-2g-s=-k$. The corresponding Feynman diagram is called a ribbon or fat graph. Some examples are sketched in *fig.* 3. These surface have by construction at least one boundary. We can associate a closed surface with punctures to a ribbon graph Γ by glueing onto the boundary components $C_1,\ldots,C_s$ infinitely long cylinders—which are conformally equivalent to punctured disks. We have $6g-6+3s$ real variables ℓ_a parametrizing the surface, s of which correspond to the total lengths of the boundaries while the other variables parametrize a top-dimensional simplex in the moduli space $\mathcal{M}_{g,s}$. It is a powerful result that we can actually obtain any point of the moduli space once and only once by considering all connected graphs Γ of the appropriate Euler number.

We can define a 'path-integral' whose expansion in Feynman diagrams naturally produces the graphs Γ. The fundamental idea, basically due to 't Hooft [40], is to use an

Hermitian $N \times N$ matrix Y to represent the double line propagator, appropriate for a ribbon graph. Indeed, the matrix integral

$$Z = \int dY \cdot \exp \frac{1}{\lambda} \mathrm{Tr} \, [-\frac{1}{2} Y^2 + \frac{1}{3} Y^3] \tag{2.19}$$

has an expansion

$$\log Z = \sum_{\Gamma} \frac{1}{n(\Gamma)} \lambda^{2g-2+s} N^s \tag{2.20}$$

where we sum over all connected graphs Γ; $n(\Gamma)$ is the order of the automorphism group of the graph Γ, λ is the string coupling constant. Z simply counts all top-dimensional cells exactly once, and thus represents an 'integral' over moduli space.

Kontsevich has used this 'open string field theory' description of moduli space to calculate the intersection numbers of the classes σ_n. The first step is to write the Chern classes $c_1(L_i)$ in terms of the lengths ℓ_a's. Let us concentrate on one boundary component C_i of the surface. It has total length

$$p_i = \sum_{a \in I_i} \ell_a, \tag{2.21}$$

where I_i labels all propagators that contribute to the boundary C_i. According to Kontsevich the subdivision of p_i in the ℓ_a can be used to define a natural connection on the circle bundle, or actually the polygon bundle, L_i associate to the boundary. The first Chern class of this line bundle L_i is the given by the curvature of this connection, and takes the form

$$c_1(L_i) = \sum_{a,b \in I_i} d\theta_a \wedge d\theta_b, \qquad \theta_a = \ell_a / p_i. \tag{2.22}$$

We now want to perform the integral of these Chern classes over moduli space. We do not have the space here to reproduce the complete calculation, which has many deep subtleties. Let us briefly indicate the crucial steps. First, one considers not the classes σ_n but the 'loop operators'

$$w(z) = \int dp \; e^{-zp} \exp \tfrac{1}{2} p^2 c_1(L) = \sum_{k \text{ odd}} \frac{1}{k} z^{-k} \mathcal{O}_k. \tag{2.23}$$

One then evaluates the correlation function of these operators $w(z)$ for fixed genus g. This is an integral over both $\mathcal{M}_{g,s}$ and the boundary lengths $p_1, \ldots, p_s$. According to Kontsevich this integral becomes elementary when expressed in terms of the coordinates ℓ_a. With the two-form Ω defined by

$$\Omega = \sum_{i=1}^{s} \tfrac{1}{2} p_i^2 c_1(L_i), \tag{2.24}$$

we see that the integral contains indeed the simple measure factor

$$e^{\Omega} \cdot dp_1 \wedge \ldots \wedge dp_s = c \cdot \prod_a d\ell_a. \tag{2.25}$$

Surprisingly the constant c does not depend on the graph Γ. One simply finds $c = 2^k$ for a surface with Euler characteristic $\chi = -k$ (see Appendix C of [22]), and the calculation reduces to

$$\langle w(z_1)\cdots w(z_s)\rangle_g = \sum_\Gamma \frac{2^k}{n(\Gamma)} \int [d\ell] \prod_{i,a\in I_i} e^{-z_i \ell_a} \tag{2.26}$$

which can be directly evaluated to give

$$\langle w(z_1)\cdots w(z_s)\rangle_g = \sum_\Gamma \frac{2^{-2k}}{n(\Gamma)} \prod_{prop} \frac{2}{z_i + z_j}. \tag{2.27}$$

Here the summation is over all diagrams Γ that contribute to $\mathcal{M}_{g,s}$ and the product is over all propagators. A factor $2/(z_i + z_j)$ is included when the two sides of the propagator form part of the loops C_i and C_j. This is but a small modification of the Feynman rules of the integral (2.19). We only have to change the quadratic part. The factor 2^{-2k} can be absorbed in the weight of the cubic vertex, or in the string coupling constant, if one wishes. In fact, we arrive in this way directly at a representation of the generating functional of the loop operator correlation functions

$$\tau(z_1,\ldots,z_N) = \exp \sum_{g=0}^{\infty} \langle \exp \sum_{i=1}^{N} w(z_i)\rangle_g \tag{2.28}$$

by means of the matrix integral

$$\tau(Z) = \rho(Z)^{-1} \int dY \cdot \exp\, Tr[-\frac{1}{2} Z Y^2 + \frac{1}{6} Y^3]. \tag{2.29}$$

Here $z_1, \ldots, z_N$ are the eigenvalues of the $N \times N$ Hermitian matrix Z. We have divided by the one-loop integral

$$\rho(Z) = \int dY \cdot \exp -\frac{1}{2} \mathrm{Tr}\, Z Y^2 = \prod_{i,j} (z_i + z_j)^{-\frac{1}{2}} \tag{2.30}$$

in order to obtain the correct asymptotic expansion. We see from (2.23) that (2.29) corresponds to the following parametrization of the coupling constants t_k

$$t_k = \sum_{i=1}^{N} \frac{1}{k} z_i^{-k} = \frac{1}{k} \mathrm{Tr}\, Z^{-k}. \tag{2.31}$$

We will learn in §3 that this is a very natural parametrization from the point of view of the KdV hierarchy. The matrix integral can be made absolutely convergent by putting a factor of i in front of the cubic term in the action. One easily sees that, by charge conservation, this is equivalent to a redefinition $t_k \to -t_k$. In this way we recover Kontsevich's result [20] that we mentioned in the Introduction.

2.4. Witten's generalizations

Witten has formulated in [18,19] a generalization of the above intersection theory, which involves the choice of an integer $p \geq 2$, and extra quantum numbers $k_i = 0, \ldots, p-2$ for each of the points x_i. Our observables are now of the form $\sigma_{n,k}$, and their correlation functions are defined as

$$\left\langle \sigma_{n_1,k_1} \cdots \sigma_{n_s,k_s} \right\rangle_g \equiv p^{-g} \left\langle \sigma_{n_1}(1) \cdots \sigma_{n_s}(s) \cdot c_D(V), \overline{\mathcal{M}}'_{g,s} \right\rangle \tag{2.32}$$

Let us explain the different objects that enter in this definition. We have already met the Mumford classes σ_n. The cohomology class $c_D(V)$ is the top dimensional Chern class of a complex D-dimensional vector bundle V over moduli space. The definition of V depends on the numbers p and k_i as follows [18,19]. Let S be the line bundle $K^{p-1} \otimes_i \mathcal{O}(z_i)^{k_i}$. That is, its section are $p-1$ forms that can have poles of order k_i at the point x_i. The degree of S is given by

$$\deg(S) = (p-1)(2g-2) + \sum_{i=1}^{s} k_i. \tag{2.33}$$

If this degree is a multiple of p we can define another line bundle T as the pth root of S

$$T^{\otimes p} = S, \quad \text{if} \quad \deg(S) \equiv 0 \pmod{p}. \tag{2.34}$$

Actually there are p^{2g} choices, and this defines a branched covering of $\mathcal{M}_{g,s}$ that is denoted as $\mathcal{M}'_{g,s}$. The fiber V_Σ of the vector bundle V at a point $\Sigma \in \mathcal{M}_{g,s}$ is defined as the space of holomorphic section of T

$$V_\Sigma = H^0(\Sigma, T), \quad \text{if} \quad H^1(\Sigma, T) = 0. \tag{2.35}$$

This is true if the degree of T is large enough. (See [19] for a more precise definition that takes into account that $H^1(\Sigma, T)$ may not always vanish.) In that case Riemann-Roch tells us that the fiber of V has complex dimension

$$D = d(g-1) + \sum_{i=1}^{s} q_i, \tag{2.36}$$

where we introduced the new notations

$$d = \frac{p-2}{p}, \qquad q_i = \frac{k_i}{p}. \tag{2.37}$$

By construction D is always a non-negative integer. Combining all these ingredients we see that the correlation functions are only non-vanishing if the following charge conservation holds

$$\sum_{i=1}^{s} (n_i + q_i - 1) = (3-d)(g-1). \tag{2.38}$$

It is sometimes convenient to denote the operators $\sigma_{n,k}$ slightly different. From the definition it is clear that $\sigma_{n,k}$ is more or less the tensor product of a Chern class and 'fractional class' related to the bundle T. That is, we would like to write symbolically

$$\sigma_{n,k} \sim \sigma_n \otimes \phi_k \tag{2.39}$$

with ϕ_k an operator of fractional charge $q = k/p$. We will make more sense out of this notation later on, but at this moment we only want to emphasize the special nature of the fields $\sigma_{0,k}$ that we also write as ϕ_k. These operators do not depend on the Mumford classes, only on the top Chern class $c_D(V)$, and we will refer to these operators as 'primary.' The primary correlation functions

$$\left\langle \phi_{k_1} \cdots \phi_{k_s} \right\rangle_g \equiv p^{-g} \langle c_D(V), \overline{\mathcal{M}}'_{g,s} \rangle \tag{2.40}$$

can be studied in their one right, and this is actually what we will undertake in §6. The primary fields ϕ_k can be considered as the observables in a separate topological field theory, and the combinations with the Mumford classes can be described as the coupling to topological gravity. This line of thought suggests many other generalizations, as we will see in the second half of these notes.

To state the generalized conjectures about the general correlation functions (2.32) we have again to renormalize our operators, this time as

$$\mathcal{O}_{np+k+1} = (np+k+1)((n-1)p+k+1)\cdots(k+1)\cdot\sigma_{n,k}. \tag{2.41}$$

So, the operators $\mathcal{O}_n$ in this case are labeled by a positive integer n that satisfies $n \neq 0 \pmod p$. The claim is that $\tau(t)$ as defined by

$$\tau(t) = \exp \sum_{g=0}^{\infty} \left\langle \exp \sum_n t_n \mathcal{O}_n \right\rangle_g \tag{2.42}$$

is now a τ-function of the pth generalized KdV hierarchy. Actually there is also a simple candidate matrix integral representation of $\tau(t)$ that we will meet in the next section—we basically replace the cubic interaction by a vertex of order $p+1$. In §§6 and 7 we will consider the quantum field theory that lies behind this conjecture.

It is not difficult to find the generalizations of the puncture and dilaton equations in this setup. These are again related to the special role of the operators $\sigma_{0,0}$ and $\sigma_{1,0}$. For instance, one easily obtains the generalized puncture equation

$$\left\langle \sigma_{0,0} \cdot \sigma_{n_1,k_1} \cdots \sigma_{n_s,k_s} \right\rangle_g = \sum_{i=1,\, n_i \neq 0}^{s} \left\langle \sigma_{n_1,k_1} \cdots \sigma_{n_i-1,k_i} \cdots \sigma_{n_s,k_s} \right\rangle_g \tag{2.43}$$

The genus zero correction reads in this case

$$\left\langle \sigma_{0,0}\sigma_{0,i}\sigma_{0,p-i-2} \right\rangle_0 = 1 \tag{2.44}$$

or, in the notation (2.41),

$$\langle \mathcal{O}_1 \mathcal{O}_i \mathcal{O}_j \rangle_0 = ij \cdot \delta_{i+j,p} \tag{2.45}$$

A dilaton equation, completely analogous to (2.15), can also be derived, with as only new ingredient the expectation value [19]

$$\langle \mathcal{O}_{p+1} \rangle_1 = (p+1)\langle \sigma_{1,0} \rangle_1 = \frac{p^2-1}{24}. \tag{2.46}$$

As a little warm-up for our discussion in §4.4, we mention here that the puncture equation (2.43) can be written as a linear partial differential equation for the string partition function $\tau(t)$

$$L_{-1} \cdot \tau = 0, \tag{2.47}$$

where L_{-1} is the differential operator

$$L_{-1} = -\frac{\partial}{\partial t_1} + \sum_{k=p+1}^{\infty} k\, t_k \frac{\partial}{\partial t_{k-p}} + \tfrac{1}{2} \sum_{i+j=p} ij\, t_i t_j. \tag{2.48}$$

This identity follows immediately, once it is realized that in a generating functional a derivative $\partial\tau/\partial t_k$ represent an insertion of the operator $\mathcal{O}_n$ in each correlation function, whereas a multiplication $t_n \cdot \tau$ eliminates the same operator. The last term on the RHS represents the correction for the exceptional case (2.45). A little further experimentation, which we leave as an exercise to the reader, shows that a particular linear combination of the dilaton equation and the charge conservation condition (2.38) leads to a similar differential equation

$$L_0 \cdot \tau = 0, \tag{2.49}$$

with L_0 given by

$$L_0 = -\frac{\partial}{\partial t_{p+1}} + \sum_{k=1}^{\infty} k\, t_k \frac{\partial}{\partial t_k} + \frac{p^2-1}{24}. \tag{2.50}$$

These two equations are but the beginning of a full tower of equations, as we alluded to in the Introduction, and we will discus this at greater lengths in §4.4.

Let us end this brief description with some specializations. The case $p = 2$ should reduce to 'standard' intersection theory on moduli space, which is indeed the case, since the bundle V is now zero-dimensional by (2.36). Another interesting case, although somewhat outside our range of definition, is the case $p = -1$ [18], where (with all $m_i = 0$) V can be identified with the bundle of quadratic differentials, *i.e.*, the cotangent bundle $T^*\mathcal{M}_{g,0}$ to moduli space. In this case the top Chern class $c_D(V)$ actually represents, up to a sign, the Euler class of moduli space and we can calculate

$$\langle 1 \rangle_g = \langle c_D(V), \overline{\mathcal{M}}_{g,0} \rangle = (-1)^{3g-3} \chi_{g,0}, \tag{2.51}$$

where we used the notation $\chi_{g,s}$ for the (virtual) Euler characteristic of the moduli space $\mathcal{M}_{g,s}$. Actually due to the work of Harer and Zagier [41] and, in particular, Penner [39] we know a concrete realization of the generating functional of these Euler numbers.

If Y is again an $N \times N$ matrix then the following matrix integral generates all Euler characteristics

$$\log \int dY \cdot \exp \frac{1}{\lambda} \mathrm{Tr}\left[\log(1-Y) - Y\right] = \sum_{g,s} N^s \lambda^{2g-2+s} \chi_{g,s}. \tag{2.52}$$

The proof of this result is based on the observation that this matrix integral simply counts all k-simplices in the cell decomposition of $\mathcal{M}_{g,s}$ weighted with a factor $(-1)^k$.

3. INTEGRABLE HIERARCHIES

In this section we will change our point of view and investigate in greater detail the integrable hierarchies that have emerged in solvable string theories and intersection theory. The following is basically a self-contained elementary exposition of a number of familiar techniques in the theory of integrable systems, and some readers may wish to skip this part. A good survey of these matters is given in [29], which we will follow closely. We gladly acknowledge that a much higher level of sophistication can be found in the extensive literature on integrable systems. All these techniques will find their due applications in §4.

3.1. The generalized KdV and KP hierarchies

We will first describe the so-called generalized KdV hierarchy in its scalar Lax formulation. Let x be a real variable and let D denote the operator

$$D = \frac{\partial}{\partial x}. \tag{3.1}$$

For a given integer $p \geq 2$ we will consider the general differential operator of order p

$$L = D^p + \sum_{i=0}^{p-2} u_i(x) D^i. \tag{3.2}$$

Here the coefficients $u_i(x)$ are *a priori* arbitrary functions in the variable x. Note that by conjugation the operator L can always be put in the above form with vanishing term of order $p-1$. We now consider the flow of L in an infinite set of 'times' $t_1, t_2, \ldots$ as generated by Hamiltonians $H_1, H_2, \ldots$

$$\frac{\partial L}{\partial t_n} = [H_n, L]. \tag{3.3}$$

In order to be able to consider the operator $L(t)$ as a simultaneous functions of all parameters t_n the Hamiltonians—which in general are time-dependent themselves—are required to generate commuting flows, and therefore satisfy the zero-curvature relation, or Zakharov-Shabat equations,

$$\frac{\partial H_m}{\partial t_n} - \frac{\partial H_n}{\partial t_m} + [H_m, H_n] = 0. \tag{3.4}$$

There is a unique basis for such commuting Hamiltonians; they are given by the non-negative part of the fractional powers of L

$$H_n = (L^{n/p})_+. \tag{3.5}$$

Let us explain this notation. We can define fractional powers of L as Laurent series in the differential D. So in general these are so-called *pseudo*-differential operators, *i.e.*, operators of the form

$$A = \sum_{i=-\infty}^{n} a_i(x) D^i. \tag{3.6}$$

The restriction of this sum to only the non-negative powers of D is denoted by A_+

$$A_+ = \sum_{i=0}^{n} a_i(x) D^i. \tag{3.7}$$

One similarly defines $A_- = A - A_+$. We take this occasion to introduce another useful concept: the ***residue*** of a pseudo-differential operator. This is defined as the coefficient of D^{-1} in the Laurent expansion (3.6) of A

$$\operatorname{res} A = a_{-1}(x). \tag{3.8}$$

We will apply this notion in a moment.

It is not difficult to explicitly verify that the flows (3.5) indeed commute. These hierarchies, describing the evolution of an pth order differential operator, are known as the generalized KdV or Gelfand-Dikii [42] hierarchies. Since the operator L satisfies (3.3), its coefficients can be considered to be functions of both the coordinate x and the times t_n. In fact, this is a somewhat redundant parametrization, because x can be identified with t_1, since we have

$$H_1 = (L^{1/p})_+ = D. \tag{3.9}$$

Furthermore, if n is a multiple of p, the flows t_n are trivial

$$\frac{\partial L}{\partial t_n} = 0, \qquad \text{if } n = q \cdot p, \tag{3.10}$$

since in that case $H_n = L^q$ which clearly commutes with L.

As an example we can consider the simplest case $p = 2$, where we have

$$L = D^2 + u(x), \tag{3.11}$$

the Hamiltonian of a particle in a potential $u(x)$. One arrives at the original Korteweg-de Vries equation by considering the first non-trivial flow of this operator, as generated by $H_3 = D^3 + \frac{3}{2}uD + \frac{3}{4}u'$,

$$\frac{\partial L}{\partial t_3} = \frac{\partial u}{\partial t_3} = [H_3, L] = u\frac{\partial u}{\partial x} + \frac{1}{6}\frac{\partial^3 u}{\partial x^3}. \tag{3.12}$$

All the KdV flows can be considered special cases of the more general KP hierarchy which is defined as follows. Let Q denote the unique p^{th} root of the differential operator L

$$Q = L^{1/p}, \tag{3.13}$$

that has an expansion of the form

$$Q = D + \sum_{i=1}^{\infty} q_i D^{-i}. \tag{3.14}$$

Then Q satisfies a similar set of evolution equations as L

$$\frac{\partial Q}{\partial t_n} = [H_n, Q], \qquad H_n = Q_+^n. \tag{3.15}$$

These can be considered as the defining equations of the KP hierarchy, if Q is now taken to be an ***arbitrary*** pseudo-differential operator with an expansion (3.14). One specializes to the pth KdV case by imposing that $Q^p = L$ is a pure differential operator, that is, by requiring

$$Q^p_- = 0. \tag{3.16}$$

The KP hierarchy can be slightly rewritten by introducing yet another pseudo-differential operator K through

$$Q = KDK^{-1}, \tag{3.17}$$

where K has an expansion of the form

$$K = 1 + \sum_{i=1}^{\infty} a_i D^{-i}. \tag{3.18}$$

It satisfies the equation

$$\frac{\partial K}{\partial t_n} = -Q_-^n \cdot K. \tag{3.19}$$

We end this long list of definitions with a side remark. It is rather straightforward to take the 'classical' limit, also known as the *dispersionless* KdV hierarchy [43,44] (see also the lectures of I. Krichever in these proceedings.) We simply put $\hbar$ in all the obvious places and take the $\hbar \to 0$ limit. In this fashion the differential D becomes the classical momentum y conjugate to x, and the differential operator L becomes a function, polynomial in y, on the phase space (y, x) with symplectic form $dy \wedge dx$,

$$L(y,x) = y^p + u_{p-2}(x) y^{p-2} + \ldots + u_0(x). \tag{3.20}$$

All commutators reduce to Poisson brackets, in particular we have

$$\frac{\partial L}{\partial t_n} = \{H_n, L\} = \frac{\partial H_n}{\partial y}\frac{\partial L}{\partial x} - \frac{\partial H_n}{\partial x}\frac{\partial L}{\partial y}. \tag{3.21}$$

We will see that, in terms of the string partition function, $\hbar$ corresponds to the string coupling constant λ, and that the classical limit implies the restriction to genus zero. We will return to these matters in §6.4.

3.2. The τ-function and the Baker-Akhiezer function

To a solution $L(t)$ of the KdV equations (3.3) one can associate a so-called *tau-function* $\tau(t)$. It is in several ways related to the differential operator L. A particularly concrete connection, though perhaps not the most insightful, is given in terms of the operator K defined in (3.17)

$$\text{res}\, K = -\frac{\partial}{\partial x} \log \tau, \tag{3.22}$$

or, equivalently, in terms of the Lax operator L

$$\text{res}\, L^{n/p} = \frac{\partial^2}{\partial x \partial t_n} \log \tau. \tag{3.23}$$

Note that if $i = 1, \ldots, p-1$,

$$\text{res}\, L^{i/p} = \frac{i}{p} \cdot u_{p-i-1} + \cdots \tag{3.24}$$

where the ellipses indicate polynomials in the coefficients u_j and there derivatives $u'_j, u''_j, \ldots$ with respect to $x = t_1$, all satisfying $j > p - i - 1$. Thus the transformations from the coefficients u_{i-1} to the residues $\text{res}\, L^{i/p}$ is seen to be upper-triangular and thus invertible. In this sense the operator L can be reconstructed from the τ-function.

The opposite problem, how to construct τ out of L, will be considered in greater detail in §3.4. A crucial ingredient in this construction is played by the eigenfunctions ψ of L. Since the operator L depends itself on the times t_n, these eigenfunctions depend on the coordinate $x = t_1$, the other variables t_n and the eigenvalue that we choose to parametrize as z^p. With the understanding that x can be identified with t_1 we will write the eigenfunction as $\psi(t, z)$. Its defining equation is

$$L\,\psi(t, z) = z^p\, \psi(t, z). \tag{3.25}$$

The function $\psi(t, z)$ is called the *Baker-Akhiezer function*. It satisfies the Schrödinger equations

$$\frac{\partial \psi}{\partial t_n} = H_n \psi. \tag{3.26}$$

We will assume it is normalized such that

$$\psi(t, z) = g(t, z) \cdot \eta(t, z), \qquad g(t, z) = \exp \sum_n t_n z^n, \tag{3.27}$$

where $\eta(t, z)$ has an asymptotic expansion in z^{-1} of the form

$$\eta(t, z) = 1 + \sum_{i=1}^{\infty} a_i(t) z^{-i}. \tag{3.28}$$

Note that $\eta = 1$ corresponds to the trivial case $L = D^p$. In fact, since in the general case $L = K D^p K^{-1}$, one can write ψ as

$$\psi(t, z) = K \cdot g(t, z) \tag{3.29}$$

which shows that the coefficients a_i in (3.17) and (3.28) are indeed identical.

There is a beautiful way to generate solutions to the KP hierarchy using Grassmannians or equivalently two-dimensional free (chiral) fermions. This method, originally due to Sato [45,46], is further explored and exposed in the beautiful paper by Segal and Wilson [29].

The basic idea is the following. For every operator $L(t)$ we can consider its Baker-Akhiezer function $\psi(t,z)$ as defined in (3.25). For fixed value of the t_n's, and after an appropriate analytic continuation in z, the function $\psi(z)$ can be seen as a wave function in the Hilbert space $H = L^2(S^1)$ with z the coordinate on the unit circle. When the function $\psi(t,z)$ evolves in time it will sweep out a linear subspace W of H

$$\psi(t,z) \in W \subset H. \tag{3.30}$$

The method of Grassmannians constructs the solution $\psi(t,z)$ out of a given linear space $W \subset H$.

The wave functions in H can be given the interpretation of one particle fermion states, with Hamiltonian $z\partial_z$. We have a basis of eigenstates $\{z^n\}$ in H with energy $n \in \mathbf{Z}$, and a natural polarization

$$H = H_+ \oplus H_- \tag{3.31}$$

in terms of positive and negative energy states. That is, H_+ has a basis $z^0, z^1, \ldots,$ whereas H_- is spanned by $z^{-1}, z^{-2}, \ldots$. The Grassmannian $Gr(H)$ is defined as the space of subsets W that (in some precise sense) are comparable to H_+. The Grassmannian can be seen as the second quantized fermion Fock space. Indeed, to the subspace H_+ we can associate the semi-infinite wedge product

$$|0\rangle = z^0 \wedge z^1 \wedge z^2 \wedge \ldots \tag{3.32}$$

which is just the Dirac vacuum, where—with an unfortunate choice of conventions—we filled all 'positive energy states.' If $w_0, w_1, \ldots$ is a basis for the subspace W, we can similarly construct the state

$$|W\rangle = w_0 \wedge w_1 \wedge w_2 \wedge \ldots \tag{3.33}$$

Recall that in this notation the fermion fields have an expansion

$$\psi(z) = \sum_n \psi_n z^n, \qquad \psi^*(z) = \sum_n \psi_n^* z^n, \tag{3.34}$$

and act on the states as

$$\psi_n = \frac{\partial}{\partial z^n}, \qquad \psi_n^* = z^{-n} \wedge . \tag{3.35}$$

These operators obviously satisfy the canonical anti-commutation relations

$$\{\psi_n, \psi_m\} = \{\psi_n^*, \psi_m^*\} = 0, \qquad \{\psi_n, \psi_m^*\} = \delta_{n+m,0}. \tag{3.36}$$

We want the Grassmannian $Gr(H)$ to consist of subspaces W such that the inner product of the states corresponding to H_+ and W is well-defined. Recall that

$$\langle 0|W\rangle = \det\langle w_i, z^j\rangle \tag{3.37}$$

in terms of the one-particle inner product $\langle\cdot,\cdot\rangle$, so W and H_+ must be comparable in size. This will certainly be the case if W is *transverse* to H_-. By this we mean the following. Let w be the map $H_+ \to W$ that sends the basis elements z^i of H_+ to the basis elements w_i of W

$$z^i \to w_i = \sum_j (w_+)_{ij} z^j + \sum_k (w_-)_{ik} z^{-k}. \tag{3.38}$$

This defines for us the projections w_+ and w_-. We now have

$$\langle 0|W\rangle = \det w_+ \tag{3.39}$$

We will require that w_+ is invertible. In this case W can be seen as the graph of the map $A = w_+^{-1} w_-$, and we can choose a basis

$$w_i = z^i + \sum_{j=1}^{\infty} A_{ij} z^{-j}. \tag{3.40}$$

Clearly the case $A = 0$ corresponds to $W = H_+$. In terms of matrix A_{ij} the state $|W\rangle$ can be written as the Bogoliubov transform

$$|W\rangle = h|0\rangle, \qquad h = \exp \sum_{i,j} A_{ij} \psi_i \psi_j^*, \tag{3.41}$$

with h an element of a subgroup of $GL(\infty, \mathbf{C})$ that has a natural action on the Grassmannian $Gr(H)$.

There is a smaller subgroup $\Gamma_+ \subset GL(\infty, \mathbf{C})$ defined by multiplication with a function $g(z)$ that is holomorphic on the unit disk D_0. A map $g(z) \in \Gamma_+$ simply multiplies every element $w \in W$. With respect to the polarization $H_+ \oplus H_-$ of H it has a decomposition of the form

$$g = \begin{pmatrix} a & b \\ 0 & c \end{pmatrix} \tag{3.42}$$

We can now consider the orbit $\Gamma_+ \cdot W$ of a subspace $W \in Gr(H)$ under the action of Γ_+. If we parametrize $g(z)$ as

$$g(t, z) = \exp \sum_{n=1}^{\infty} t_n z^n \tag{3.43}$$

the evolution operator $U(t)$ that acts in the fermion Fock space is defined by

$$U(t)|W\rangle = |gW\rangle. \tag{3.44}$$

Multiplication by z^n simply shifts $z^k \to z^{k+n}$, so $U(t)$ is realized in terms of the fermion fields as

$$U(t) = \exp \sum_{n=1}^{\infty} t_n J_n. \tag{3.45}$$

Here J_n are the modes of the fermion current $J(z) = \psi\psi^*(z)$, with the fundamental property $[J_n, \psi_k] = \psi_{k+n}$. We can write $U(t)$ as

$$U(t) = \exp \oint \frac{dz}{2\pi i} \log g(z) \cdot J(z). \tag{3.46}$$

We are now finally in a position to explain how all this is related to the Lax pair formulation of the integrable hierarchies that we considered in the previous subsections. The subspace W has, by definition, a unique element $\eta = w_0$ satisfying

$$\eta(z) = 1 + \sum_{i=1}^{\infty} a_i z^{-i}. \tag{3.47}$$

It can be described formally as the intersection

$$\eta(z) = W \cap (1 + H_-). \tag{3.48}$$

Similarly the transform $g^{-1}W$ contains an element with that property

$$\eta(t,z) = g^{-1}W \cap (1 + H_-). \tag{3.49}$$

Recall here that g is parametrized as in (3.43). But this last equation implies directly that

$$\psi(t,z) = g(t,z) \cdot \eta(t,z) \in W. \tag{3.50}$$

Since this is true for *all* times t_n and W is a linear space, we have the property that all (multiple) derivatives of ψ still lie in W. In particular we have

$$g^{-1}\frac{\partial \psi}{\partial t_n} = z^n + \ldots \in g^{-1}W, \tag{3.51}$$

and

$$g^{-1}\frac{\partial^k \psi}{\partial x^k} = z^k + \ldots \in g^{-1}W. \tag{3.52}$$

Since $g^{-1}W$ has a unique basis of the form (3.40), this implies that ψ satisfies an equation

$$\frac{\partial \psi}{\partial t_n} = H_n \psi. \tag{3.53}$$

where H_n is an nth order differential operator in x. By construction the H_n commute, and thus we have found a solution to the KP hierarchy. The reduction to KdV comes about when W satisfies the extra constraint

$$z^p \cdot W \subset W. \tag{3.54}$$

In that case one finds the additional relation

$$L\psi = z^p \psi, \tag{3.55}$$

which defines the differential operator $L = H_p$.

3.4. τ-functions and determinants

The tau-function $\tau(t)$ associated to a point W in the Grassmannian is now defined as

$$\tau(t) = \frac{\det(g^{-1}w)_+}{g^{-1}\det w_+} \tag{3.56}$$

with g parametrized as in (3.43), or in our notation (3.40)-(3.42)

$$\tau(t) = \det(1 + a^{-1}bA). \tag{3.57}$$

In terms of the fermion Fock space we have the representation

$$\tau(t) = \frac{\langle 0|U(t)|W\rangle}{g^{-1}\langle 0|W\rangle} \tag{3.58}$$

Since the state

$$\langle t| = \langle 0|U(t) \tag{3.59}$$

can be considered as a coherent state in the Hilbert space, the τ-function can alternatively be regarded as the wave-function of the state $|W\rangle$.

The bosonization formulas

$$J = \partial\phi, \quad \psi = e^{\phi}, \quad \psi^* = e^{-\phi}, \tag{3.60}$$

lead us to consider the operators

$$\gamma_+(z) = \exp\sum_{n=1}^{\infty}\frac{1}{n}z^{-n}\frac{\partial}{\partial t_n}. \tag{3.61}$$

Insertion of the fermi field $\psi(z)$ corresponds to action on $\tau(t)$ with the differential operator

$$\gamma(z) \equiv \gamma_+(z)\cdot g(z), \tag{3.62}$$

together with an increase in fermi number of one unit. If we denote the N-particle ground state as

$$|N\rangle = z^N \wedge z^{N+1} \wedge z^{N+2} \wedge \ldots \tag{3.63}$$

then we evidently have

$$\gamma(z_1)\cdots\gamma(z_N)\cdot\tau(0) = \langle N|\psi(z_1)\cdots\psi(z_N)|W\rangle \tag{3.64}$$

where the notation on the LHS indicates that we first take derivatives and then put the arguments $t_n = 0$ in the τ-function. This implies in particular that

$$\gamma_+(z_1)\cdots\gamma_+(z_N)\cdot\tau(0) = \Delta(z)^{-1}\langle N|\psi(z_1)\cdots\psi(z_N)|W\rangle \tag{3.65}$$

with

$$\Delta(z) = \langle N|\psi(z_1)\cdots\psi(z_N)|0\rangle = \prod_{i>j}(z_i - z_j) = \det z_i^{j-1}, \tag{3.66}$$

the Vandermonde determinant. In fact, (3.65) can be easily evaluated, since only the first N one-particle states $w_1, \ldots, w_N$ contribute to give

$$\langle N|\psi(z_1)\cdots\psi(z_N)|W\rangle = \det(w_{j-1}(z_i)) \equiv \Delta(w; z). \tag{3.67}$$

Here we introduced the 'generalized Vandermonde' determinant $\Delta(w; z)$ for an arbitrary basis $\{w_k(z)\}$. Note that in this language the Baker-Akhiezer function is recovered as the fermion one-point function

$$\psi(t, z) = \gamma(z) \log \tau(t) \tag{3.68}$$

By expanding ψ in powers of z^{-1} we find as a corollary that our definition of τ coincides with the one given in §3.2.

These fermionic correlation functions occur naturally if we consider special elements $g(z) \in \Gamma_+$ such that g^{-1} is a Nth order polynomial with zeroes z_i outside D_0

$$g(z)^{-1} = \prod_{i=1}^{N}(1 - z/z_i). \tag{3.69}$$

We can regard g^{-1} to be the characteristic polynomial

$$g(z)^{-1} = \det(1 - z \cdot Z^{-1}), \tag{3.70}$$

where Z is an Hermitian matrix with eigenvalues $z_1, \ldots, z_N$. This corresponds to the choice of parameters t_n as

$$t_n = \frac{1}{n}\sum_{i=1}^{N} z_i^{-n} = \frac{1}{n}\,\mathrm{Tr}\, Z^{-n}, \tag{3.71}$$

also known as 'Miwa's coordinates' [47]. We will write the tau-function in this parametrization as

$$\tau(t) = \tau(Z). \tag{3.72}$$

For this particular choice of g we find immediately with (3.46) that

$$\tau(Z) = \gamma_+(z_1)\cdots\gamma_+(z_N)\cdot\tau(0), \tag{3.73}$$

which we already evaluated to be ratio

$$\tau(Z) = \Delta(w; z)/\Delta(z). \tag{3.74}$$

This is our main results of this section. It gives a closed expression for the τ-function in the specific parametrization (3.71) once we know an explicit basis $w_k(z)$ for the point $W \in Gr(H)$.

4. MATRIX INTEGRALS

We will now turn to the application of the results obtained in the previous section to the specific solutions of the KdV hierarchies that appear in theories of quantum gravity. In this section we will perform the 'Wick rotation' $x \to ix$, $t_n \to it_n$, in order to achieve better convergence properties of many of our quantities. This change is of course completely irrelevant on the level of asymptotic expansions. Note that the derivative $D = -i\partial/\partial x$ is now an Hermitian operator. We will assume that the Lax operator L has (possibly complex) coefficients such that it is Hermitian too. It consequently has real eigenvalues z^p.

4.1. The string equation and Airy functions

We can think of the many different solutions to the KdV hierarchies, such as the famous soliton solutions, as being related to different initial conditions $L(0)$ for the Lax operator $L(t)$. For the solutions that appeared in §2 these initial conditions can be found as follows. Recall that in these cases the τ-function has a very concrete interpretation: it is given by the string partition function of the $(p,1)$ minimal topological model. Stated otherwise, the logarithm of τ is a generating functional for all correlation functions on connected surfaces of arbitrary genus g, and consequently has an expansion

$$\log \tau = \sum_{g=0}^{\infty} \left\langle \exp \sum_n t_n \mathcal{O}_n \right\rangle_g \tag{4.1}$$

If we think of the variables t_n as describing a background, then $\log \tau$ becomes indeed the 'string free energy'

$$\log \tau = F = \langle 1 \rangle_t \tag{4.2}$$

where the subscript $\langle \cdot \rangle_t$ simply indicates the insertion of the '$\exp t_n \mathcal{O}_n$' term and a summation over all genera is understood. In this notation equations (3.22)–(3.23) relate the residues with special two-point functions

$$\operatorname{res} K = -\langle \mathcal{O}_1 \rangle_t \qquad \operatorname{res} L^{n/p} = \langle \mathcal{O}_1 \mathcal{O}_n \rangle_t \tag{4.3}$$

We now want to determine the initial condition $L(0)$, where we have put all times t_n to zero for $n > 1$. Putting the coupling constant $t_1 = x$ to a non-zero value and expanding $\log \tau$ up to first order in the other coupling constants, we find for $i = 1, \ldots, p-1$ with (2.44)

$$\operatorname{res} L^{i/p} = p \cdot \delta_{i,p-1}. \tag{4.4}$$

(The factor p that appears here can be absorbed in the string coupling constant, and we will drop it in the consequent.) This translates in the following initial value for the differential operator L

$$L(0) = D^p + x. \tag{4.5}$$

That is, only a linear potential $u_0(x) = x$ and no higher derivative terms appear. The equation

$$u_i(x) = x \cdot \delta_{i,0}, \tag{4.6}$$

or, equivalently,

$$[D, L] = 1, \tag{4.7}$$

is known as the *string equation* for the $(p, 1)$ model, in the notation of the introduction.* It is this extremely trivial form of the string equation that leads to the complete solvability. Indeed, we will now be able to explicitly solve for the τ-function, because in this case the Baker-Akhiezer function

$$L\,\psi(x,z) = z^p\,\psi(x,z) \tag{4.8}$$

is simply given by the generalized Airy function

$$\psi(x,z) = c(z)\cdot\int dy\,\exp i[(x - z^p)y + \frac{y^{p+1}}{p+1}]. \tag{4.9}$$

This should be contrasted with the much more complicates solutions for an arbitrary (p, q) model [48]. The coefficient $c(z)$ is fixed by requiring the asymptotic expansion (3.28)

$$\psi(x,z) = e^{ixz}\left(1 + \sum_{i=1}^{\infty} a_i(x) z^{-i}\right). \tag{4.10}$$

The normalization can be chosen as

$$c(z) = e^{\frac{ip}{p+1}z^{p+1}}\sqrt{z^{p-1}}. \tag{4.11}$$

The asymptotic expansion is now verified by shifting $y \to y + z$ in the integral (4.9), so that one obtains the representation

$$\psi(x,z) = e^{ixz}\sqrt{z^{p-1}}\int dy\cdot e^{ixy + iS(y,z)} \tag{4.12}$$

with

$$S(y,z) = \frac{1}{p+1}\left[(y+z)^{p+1}\right]_{\geq 2}. \tag{4.13}$$

Here the subscript '≥ 2' implies that we only keep terms in y of second order or higher. By using the stationary phase approximation to such integrals one then easily deduces the expansion (4.10).

We now wish to determine the function $\tau(Z)$ for this special initial value condition

$$L = D^p + x. \tag{4.14}$$

According to the general discussion of section 3, the Baker-Akhiezer function $\psi(t, z)$ is an element of W for all values of t. This is in particular true if we restrict to the parameter $t_1 = x$. So, all the functions $v_k(z)$ that appear in the Taylor expansion

$$\psi(x,z) = \sum_{k=0}^{\infty} v_k(z)\frac{x^k}{k!}. \tag{4.15}$$

*For general (p, q) it reads $[H_q, L] = 1$ [7], which in the case $(p, q) = (2, 3)$ gives the celebrated Painlevé equation $u^2 + \frac{1}{6}u'' = x$ of [4].

are elements of W:

$$v_k(z) \in W. \tag{4.16}$$

Furthermore, the functions $v_k(z)$ form a basis, since they are given by

$$v_k(z) = \sqrt{z^{p-1}} \int dy \cdot (y+z)^k \cdot e^{iS(y,z)}, \tag{4.17}$$

and thus have to property

$$v_k(z) = z^k(1 + O(z^{-1})). \tag{4.18}$$

They therefore define an element of the Fock space by

$$|W\rangle = v_0 \wedge v_1 \wedge v_2 \wedge \ldots \tag{4.19}$$

and the τ-function is, according to the main result of §3, simply obtained as the following ratio of determinants

$$\tau(Z) = \Delta(v; z)/\Delta(z). \tag{4.20}$$

4.2. More general models: the small phase space

One can slightly generalize the previous case by considering an operator of the form

$$L = W(D) + x, \tag{4.21}$$

with W an arbitrary polynomial of order p with *constant* coefficients

$$W(y) = y^p + g_{p-2}y^{p-2} + \ldots + g_0. \tag{4.22}$$

If we introduce a potential V by the relation

$$W(y) = V'(y), \tag{4.23}$$

and a function $\zeta(z)$ through

$$V'(\zeta) = z^p, \quad \zeta = z + O(z^{-1}), \tag{4.24}$$

the Baker-Akhiezer function reads in this case

$$\psi(x,z) = e^{ix\zeta}\sqrt{V''(\zeta)} \int dy \cdot e^{ixy + iS(y,\zeta)} \tag{4.25}$$

with S defined as

$$S(y,\zeta) = [V(y+\zeta)]_{\geq 2} = V(y+\zeta) - V(\zeta) - yV'(\zeta). \tag{4.26}$$

In this case the moments $v_k(z)$ are given by

$$v_k(z) = \sqrt{V''(\zeta)} \int dy \cdot (y+\zeta)^k \cdot e^{iS(y,\zeta)} \tag{4.27}$$

This case is actually of great interest, since an operator of the form (4.21) is naturally obtained if we let the operator $L = D^p + x$ evolve in the 'primary' times $t_1, \ldots, t_{p-1}$. The subspace parametrized by these t_i's is known as the 'small phase space' [15]. In intersection theory the restriction to the small phase space implies that we only consider the top Chern class of the vector bundle V, not the Mumford-Morita-Miller classes σ_n. In the corresponding quantum field theory the primary operators $\mathcal{O}_1, \ldots, \mathcal{O}_{p-1}$ are the physical fields of a $(p,1)$ topological minimal model. The small phase space describes the moduli space of topological field theories that can be reached by perturbation of the minimal models. When we consider $\tau(t)$ as a function on the small phase space, it receives furthermore, by charge conservation, only contributions at genus zero.

In the context of integrable hierarchies of KdV type, these special properties of the first $p-1$ flows are reflected in the fact that the corresponding Hamiltonians $H_i = L_+^{1/p}$ do not depend on x. Therefore the dispersionless, or spherical, approximation is exact, and we have the classical equations

$$\frac{\partial L}{\partial t_i} = \{H_i, L\} = \{L_+^{i/p}, x\} = \frac{\partial L_+^{i/p}}{\partial y} \tag{4.28}$$

with y the classical momentum corresponding to D. These equations fully determine the t-dependence of L. The initial value

$$W(y) = y^p \tag{4.29}$$

evolves to a general potential of type (4.22), where the coefficients $g_i(t)$ are given implicitly by the relation

$$i(p-i)t_{p-i} = \operatorname{res} L^{i/p} = \oint \frac{dy}{2\pi i} \, L(y)^{i/p}, \qquad (i = 1, \ldots, p-1). \tag{4.30}$$

This equation follows immediately from the results (4.3) and (2.44).

4.3. Matrix integrals

We will now show following Kontsevich [22] how the previous results are related to matrix integrals. First recall the wonderful result of Harish-Chandra [50], Mehta [51], Itzykson and Zuber [52], for the following integral over the unitary group $U(N)$

$$\int dU \, \exp i \operatorname{Tr}[UXU^\dagger Y] = c \cdot \frac{\det e^{ix_i y_j}}{\Delta(x)\Delta(y)}, \tag{4.31}$$

with dU the Haar measure and x_i, y_i the eigenvalues of the Hermitian matrices X and Y. This result can be obtained in many different ways, perhaps most elegantly as an application of the Duistermaat-Heckman localization formula [53].

As a special application of the above equation consider the matrix Fourier transform, *i.e.*, the integral over a hermitian matrix Y in a external field X, both $N \times N$ matrices, of the form

$$\tau(X) = \int dY \, \exp i \operatorname{Tr}[XY + V(Y)]. \tag{4.32}$$

By conjugation invariance, this is only a function of the eigenvalues $x_1, \ldots, x_N$. We can use (4.31) to integrate out the angular variables U in the decomposition

$$Y = U \cdot \operatorname{diag}(y_1, \ldots, y_N) \cdot U^\dagger, \tag{4.33}$$

which also introduces a Jacobian

$$dY = \Delta(y)^2 \cdot dU \cdot [dy]. \tag{4.34}$$

This leaves us with an integral over the eigenvalues y_i of the form

$$\tau(X) = \int [dy] \, \Delta(y) \Delta(x)^{-1} \exp \sum_j i[x_j y_j + V(y_j)] \tag{4.35}$$

Now the Vandermonde determinant $\Delta(y)$ is a sum of terms of the form

$$\pm \, y_1^{i_1} \cdots y_N^{i_N}, \tag{4.36}$$

and for each of these terms the integral $\tau(X)$ factorizes in separate integrals over the individual eigenvalues y_i. If we introduce the function

$$w(x) = \int dy \cdot e^{ixy + iV(y)} \tag{4.37}$$

and its derivatives

$$w_k(x) = \int dy \cdot y^k \cdot e^{ixy+iV(y)}, \tag{4.38}$$

the contribution of (4.36) is simply

$$\pm \, w_{i_1}(x_1) \cdots w_{i_N}(x_N). \tag{4.39}$$

So we can evaluate $\tau(X)$ straightforwardly as

$$\tau(X) = \det(w_{j-1}(x_i))/\Delta(x) = \Delta(w; x)/\Delta(x) \tag{4.40}$$

in the notation of the previous section. This very much suggests that $\tau(X)$ is a τ-function for the KP hierarchy with one-particle wave-functions $w_0, w_1, \ldots$. This will be the case if these functions have appropriate asymptotic expansions, and so depends on the choice of potential V and the normalization of the integral.

We first wish to apply this result using the functions $v_k(z)$ that appeared in §4.1. Recall that these where the derivatives (at $x = 0$) of the Baker-Akhiezer function of the operator $L = D^p + x$. In this case one simply puts

$$V(Y) = \frac{1}{p+1} Y^{p+1}, \qquad X = -Z^p, \tag{4.41}$$

and

$$\tau(Z) = c(Z) \cdot \int dY \, \exp i \, \mathrm{Tr} \, [-Z^p \cdot Y + \frac{Y^{p+1}}{p+1}] \tag{4.42}$$

with a proper choice of normalization $c(Z)$. This constant is basically the product of the individual constants $c(z_i)$ of (4.11), together with a correction replacing the Vandermonde determinant $\Delta(z^p)$ by $\Delta(z)$

$$c(Z) = \det Z^{\frac{p-1}{2}} \cdot e^{\frac{ip}{p+1} \mathrm{Tr} \, Z^{p+1}} \frac{\Delta(z^p)}{\Delta(z)}. \tag{4.43}$$

With all these ingredients the integral (4.42) reproduces indeed our τ-function

$$\tau(Z) = \Delta(v; z)/\Delta(z) \tag{4.44}$$

The integral can be written in yet another fashion. Let $S(Y, Z)$ be given as in (4.13)

$$S(Y, Z) = \frac{1}{p+1} \mathrm{Tr} \, [(Y + Z)^{p+1}]_{\geq 2}, \tag{4.45}$$

and let $S_2(Y, Z)$ denote the part of the function $S(Y, Z)$ of second order in Y

$$S_2(Y, Z) = \sum_{k=0}^{p-1} \mathrm{Tr} \, [Y Z^k Y Z^{p-1-k}], \tag{4.46}$$

One easily evaluates the Gaussian integral

$$\rho(Z) \equiv \int dY \, e^{iS_2(Y,Z)} = \prod_{i,j} \sqrt{\frac{z_i - z_j}{z_i^p - z_j^p}} = \det Z^{\frac{1-p}{2}} \frac{\Delta(z)}{\Delta(z^p)}. \tag{4.47}$$

This can be used to rewrite $\tau(Z)$ as

$$\tau(Z) = \rho(Z)^{-1} \int dY \, e^{iS(Y,Z)}. \tag{4.48}$$

This matrix integral has been suggested as the relevant one for the $(p, 1)$ model by several authors, see *e.g.* [22,49,27]. Here we have shown that it follows naturally from the Lax pair formulation of the string equation, by which the general (p, q) model is characterized.

The integral (4.48) has an obvious asymptotic expansion in matrix Feynman diagrams. In this way we recover the cell decomposition used by Kontsevich to derive his results for $p = 2$. This elegant geometrical interpretation is less clear for higher values of p. First of all one must include higher order vertices (up to order $p + 1$), which implies the consideration of cells in $\mathcal{M}_{g,s}$ of non-zero codimension. Secondly, the Feynman rules associated to these vertices become Z-dependent, and generally quite complicated. It would be very interesting if this prescription could nevertheless be directly related to the intersection formulas of §2.4.

As is also observed in [22,27], the formula (4.48) makes sense for a much bigger class of potentials. We can consider the general case

$$\tau(Z) = c(Z) \cdot \int dY \, \exp i \operatorname{Tr}[-W(Z) \cdot Y + V(Y)] \tag{4.49}$$

If $V(Y)$ has a well-defined Taylor expansion around $Y = Z$, we can put

$$W(Y) = V'(Y). \tag{4.50}$$

After a shift in the integration variable Y, the matrix integral $\tau(Z)$ can be written in the form (4.48), if we define the action to be

$$S(Y, Z) = \operatorname{Tr}[V(Y + Z) - V(Z) - Y \cdot V'(Z)], \tag{4.51}$$

and $S_2(Y, Z)$ similarly as in (4.46) as the quadratic part of S. One easily calculates

$$\rho(Z) = (\det V''(Z))^{-1/2} \frac{\Delta(z)}{\Delta(V'(z))} \tag{4.52}$$

and the matrix integral (4.48) becomes again a ration of determinants

$$\tau(Z) = \Delta(v; z)/\Delta(z). \tag{4.53}$$

In this more general case the functions $v_k(z)$ are given by

$$v_k(z) = \sqrt{V''(z)} \int dy \cdot (y + z)^k \cdot e^{iS(y,z)}, \tag{4.54}$$

and, again under some mild restriction on the potential V, this produces a solution of the KP hierarchy.

One might wonder if all these solutions are completely independent. This brings us to the following point. The most general matrix integral that naturally leads to a τ-function of the KP hierarchy depends both on the choice of potential V and the external field Z. This might be a redundant parametrization, in the sense that a particular variation δV corresponds to a flow in the times t_n encoded by Z. We meet a simple example of this phenomenon if $V(Y)$ is a polynomial of order $p + 1$. In this case the functions v_k that we constructed above are identical to the ones obtained in §4.2, after we perform a reparametrization

$$z \to \zeta(z), \qquad \zeta^p = W(z). \tag{4.55}$$

Recall that we considered at that point the τ-function associated to the Lax operator $L = W(D) + x$. A Lax operator of that form could be obtained, if we start with $W(Y) = Y^p$ and flow in the primary times $t_1, \ldots, t_{p-1}$. Indeed, as we showed, in that case one ends up with a general polynomial $W(Y)$ of order p whose coefficients are determined through the relation

$$t_{p-i} = \frac{1}{i(p-i)} \operatorname{res} W^{i/p}, \qquad i = 1, \ldots, p-1. \tag{4.56}$$

We now see that the correct identification of the KP times t_n is given by

$$t_n = \frac{1}{n}\mathrm{Tr}\, W(Z)^{-n/p}, \tag{4.57}$$

which just expresses the reparametrization (4.55) of the coordinate z. Summarizing, the change in potential can be absorbed in the redefinition in the coupling coefficients (4.57), together with the extra term (4.56) for the primary couplings. Of course, it is possible to consider more general flows, where the order of the potential changes. This cannot be absorbed in a shift of the KdV times, and in this way one can interpolate between different values of p.

As a final application, we can consider the limit $p \to -1$. In that case we find

$$W(Y) = Y^{-1}, \qquad V(Y) = \log Y. \tag{4.58}$$

One easily verifies that the Kontsevich matrix model now becomes identical to Penner's model that computes the Euler characteristic of moduli space, in accordance with our considerations of §2.4

4.4. The Virasoro constraints

We will make here a few brief comments on the occurrence of Virasoro algebras (and more general W-algebras) in the context of these integrable hierarchies. We already mentioned this characterization of the tau-function in the introduction and we will see in §7 that it occurs naturally in the context of topological string theory. There exists by now a quite extensive literature on this subject, see *e.g.* our selection [54], and we will be very brief at this point.

Let us start with a few definitions. In the free fermion theory there is a holomorphic stress tensor

$$T(z) = \tfrac{1}{2}(\partial_z\psi^*\psi - \psi^*\partial_z\psi) = \tfrac{1}{2}J^2, \tag{4.59}$$

which describes the behaviour of the quantum theory under reparametrizations $z \to w(z)$ of the unit circle. Its modes L_n, defined by

$$T(z) = \sum_n L_n z^{-n-2}, \tag{4.60}$$

satisfy the $c = 1$ Virasoro algebra

$$[L_n, L_m] = (n-m)L_{n+m} + \frac{1}{12}n(n^2-1)\delta_{n+m,0}, \tag{4.61}$$

and are the infinitesimal generators of Diff(S^1). They can be represented as differential operators acting on the τ-function through the relation

$$L_{n_1}\cdots L_{n_k}\cdot\tau(t) = \langle t|L_{n_1}\cdots L_{n_k}|W\rangle. \tag{4.62}$$

This gives, with $n > 0$, the explicit realizations

$$\begin{aligned} L_{-n} &= \sum_{k=n+1}^{\infty} k\, t_k \frac{\partial}{\partial t_{k-n}} + \tfrac{1}{2} \sum_{i+j=n} ij\, t_i t_j, \\ L_0 &= \sum_{k=1}^{\infty} k\, t_k \frac{\partial}{\partial t_k}, \\ L_n &= \sum_{k=1}^{\infty} k\, t_k \frac{\partial}{\partial t_{k+n}} + \tfrac{1}{2} \sum_{i+j=n} \frac{\partial^2}{\partial t_i \partial t_j}. \end{aligned} \tag{4.63}$$

In a τ-function of the pth KdV hierarchy all t_k's with $k \equiv 0 \pmod p$ can consistently be put to zero, and only the modes of the form $L_{q\cdot p}$ act non-trivially.

In the previous sections we have constructed a special solution of the pth KdV hierarchy, that turns out to have remarkably simple properties with respect to the Virasoro generators. The results can be formulated as follows. Let us introduce the redefined operators L'_n by

$$p \cdot L'_n = L_{n\cdot p} - \frac{\partial}{\partial t_{1+(n+1)p}} + \delta_{n,0} \frac{p^2 - 1}{24}, \qquad n \geq -1. \tag{4.64}$$

One easily verifies that they satisfy the algebra

$$[L'_n, L'_m] = (n - m) L'_{n+m}. \tag{4.65}$$

The τ-function of the $(p, 1)$ model can now be shown to obey the equations

$$L'_n \cdot \tau = 0, \qquad n \geq -1. \tag{4.66}$$

A derivation of this result is given in [24]. These equations have a straightforward interpretation as recursion relations for correlation functions, as we will see explicitly for the case $p = 2$ in §7.

There exists actually a much bigger algebra of currents bilinear in the fermions—the so-called $W_{1+\infty}$ algebra [56], generated by currents of the form

$$W(z) = \partial_z^i \psi^* \partial_z^j \psi. \tag{4.67}$$

The modes of this fields can also be realized as differential operators, and can be considered as the infinitesimal generators of the group $GL(\infty, \mathbf{C})$ that acts naturally on Sato's Grassmannian. The τ-function of the $(p, 1)$ model can be shown to satisfy the more general constraints

$$W_n^{(s)} \cdot \tau = 0, \qquad s = 2, \ldots, p, \ n \geq 1 - s, \tag{4.68}$$

where $W_n^{(s)}$ are the modes of a spin s generator $W^{(s)}(z)$. See *e.g.* [55,56] for more details. The fields $W^{(s)}$ generate a so-called W-algebra associated to the Lie group $SL(p)$.

These constraints can actually be found directly in the matrix model representation of the τ-function [57]. A very elegant way [21,58,59] proceeds as follows. Recall that for the $(p,1)$ models the τ-function was, up to normalization, given by the generalized matrix Airy function

$$A(X) = \int dY \cdot \exp i \operatorname{Tr}[XY + \frac{Y^{p+1}}{p+1}]. \tag{4.69}$$

This function satisfies the 'matrix Airy equation'

$$\left(\frac{\partial^p}{\partial X^p} + X\right) A = 0, \tag{4.70}$$

a reflection in matrix terms of the 'string equation' $L = D^p + x$. When this equation is written out in terms of the coordinates t_n, its gives rise to a set of equations of the form

$$Q \cdot \tau = 0. \tag{4.71}$$

where Q is a differential operators of order p or less. The claim is, that these equations reproduces exactly the Virasoro and W-constraints.

5. TOPOLOGICAL FIELD THEORY

We now shift our perspective completely and start our discussion of topological field theories [60,61]. The models we consider in this and the next section can be considered as 'matter theories' and only describe the primary fields $\mathcal{O}_1, \ldots, \mathcal{O}_{p-1}$ that featured in our previous models. Consequently, we will only recover the KdV equations on the small phase space. In order to extend our discussion to the full integrable hierarchies these models have to be coupled to (topological) gravity—a subject that we will postpone until §7, and then only discuss marginally.

5.1. The stress-energy tensor

Let us start with some remarks of a rather general nature. For other introductions see *e.g.* [62]-[64]. Given a compact manifold M of dimension D and a quantum field theory with some set of fundamental fields $\phi(x)$, one can consider the vacuum amplitude or partition function

$$Z(M) = \int [d\phi]\, e^{-S[\phi]} \tag{5.1}$$

Here we formally defined $Z(M)$ by the path-integral over all field configurations on M weighted with some action S. In general the partition function $Z(M)$ will depend on many geometrical data. For instance, in almost all quantum field theories we need a Riemannian structure on M, *i.e.*, a metric $g_{\mu\nu}(x)$. This metric enters both in the definition of the action S and in the definition of the measure $[d\phi]$. Other possible ingredients might be an orientation and, if fermions are involved, a choice of spin structure, or in the case of gauge fields a choice of fiber bundle.

Although through its definition the quantum field theory can seem to depend on all information encoded in the metric $g_{\mu\nu}$, it might be the case that the actual amplitudes have some invariance under particular changes $\delta g_{\mu\nu}$ in the metric. An example is the invariance under reparametrizations

$$\delta g_{\mu\nu} = D_\mu \epsilon_\nu + D_\nu \epsilon_\mu. \tag{5.2}$$

As is well-know, this leads to the conservation law

$$D^\mu T_{\mu\nu} = 0 \tag{5.3}$$

for the expectation value of the stress-energy tensor

$$T_{\mu\nu} = \frac{1}{\sqrt{g}} \frac{\delta S}{\delta g^{\mu\nu}}, \tag{5.4}$$

that encodes the reaction of the quantum field theory to metric deformations. Another example of a symmetry is conformal invariance: the invariance under local rescalings of the metric

$$\delta g_{\mu\nu} = \epsilon \cdot g_{\mu\nu}, \tag{5.5}$$

with $\epsilon(x)$ an arbitrary function. This implies that the trace of $T_{\mu\nu}$ vanishes

$$T^\mu{}_\mu = 0. \tag{5.6}$$

Quantum field theories with this property are known as conformal field theories.

A quantum field theory will be called *topological*, when it is invariant under arbitrary smooth deformation of the metric:

$$\delta g_{\mu\nu} = \epsilon_{\mu\nu}. \tag{5.7}$$

In this fashion only the topology of the manifold M will matter, and the partition function $Z(M)$ will be a topological invariant. In this case the stress tensor will have to vanish completely

$$T_{\mu\nu} = 0. \tag{5.8}$$

This discussion can be generalized from vacuum amplitudes to physical correlation functions. In a generic quantum field theory, the correlation functions depend on the position of the operators. But in a topological field theory the physical correlation functions of local operators are just numbers,

$$\langle \phi_{i_1}(x_1) \cdots \phi_{i_s}(x_s) \rangle_M \equiv \int [d\phi]\, \phi_{i_1}(x_1) \cdots \phi_{i_s}(x_s) \cdot e^{-S} = constant. \tag{5.9}$$

This is again a manifestation of the metric-independence of physical observables, and a consequence of the vanishing of the stress-energy tensor, now within correlation functions

$$\langle T_{\mu\nu}(x) \phi_{i_1}(x_1) \cdots \phi_{i_s}(x_s) \rangle_M = 0 \tag{5.10}$$

In §5.4 we will see how these conditions can be realized in practice.

We will at this point not be interested in actually evaluating the path-integral for a given quantum field theory. We will rather, following Atiyah [65], deduce from the above definition certain properties of partition functions, and more generally correlation functions, that can be lifted to the status of 'axioms.' The most important property is the concept of factorization.

Let us in our imagination cut the manifold M along a codimension one subspace B, so that M splits into two parts M_1 and M_2. In this case we can do the path-integral in two steps. Let us first fix the values of the fields $\phi(x)$ to some given configuration ϕ' on B, and do the separate integrals over M_1 and M_2 with these boundary conditions. If we write

$$\Psi_{M_i}(\phi') = \int_{\phi|_B=\phi'} [d\phi]\, e^{-S[\phi]} \tag{5.11}$$

for the path-integral over the component M_i with fixed values ϕ' at B, it is clear that we will have

$$Z(M) = \int [d\phi']\, \Psi_{M_1}(\phi') \cdot \Psi_{M_2}(\phi') \tag{5.12}$$

To make this more precise we have to consider general space-times M with boundaries. To each connected component B of the boundary ∂M of M a quantum field theory will associate a Hilbert space of states $\mathcal{H}_B$

$$B \to \mathcal{H}_B. \tag{5.13}$$

In explicit examples this Hilbert space will be constructed by considering canonical quantization on the cylinder $B \times \mathbf{R}$, where B is considered space-like and $\mathbf{R}$ the time direction. States $|\Psi\rangle \in \mathcal{H}_B$ can be represented as wave functions $\Psi(\phi) = \langle\phi|\Psi\rangle$ of the fundamental field variables $\phi(x)$ restricted to B. In our examples the vector spaces $\mathcal{H}_B$ will actually be finite-dimensional.

The assignment of vector spaces (5.13) should satisfy some intuitively clear axioms. If the boundary ∂M has several disconnected components $B_1, \ldots, B_s$ we define

$$\mathcal{H}_{\partial M} = \mathcal{H}_{B_1} \otimes \cdots \otimes \mathcal{H}_{B_s}. \tag{5.14}$$

Furthermore, if the boundary is empty, $\mathcal{H}$ is one-dimensional

$$\mathcal{H}_\emptyset = \mathbf{C}. \tag{5.15}$$

The definition of $\mathcal{H}_B$ will in general depend on the orientation of B. If $-B$ denotes the manifold with reversed orientation, we have

$$\mathcal{H}_{-B} = \mathcal{H}_B^*, \tag{5.16}$$

since the path-integral on the cylinder $B \times \mathbf{R}$ gives a canonical isomorphism

$$\mathcal{H}_B \otimes \mathcal{H}_{-B} \to \mathbf{C}. \tag{5.17}$$

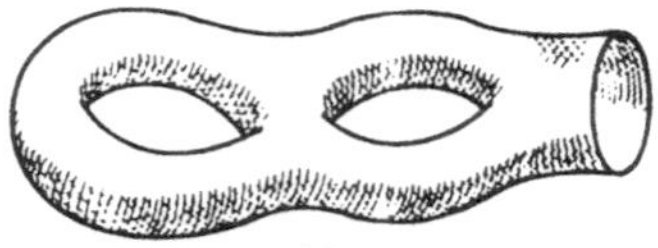

Fig. 4. *If a space-time M has a single boundary B, a quantum field theory will associate to M a state $|M\rangle$ in the Hilbert space $\mathcal{H}_B$.*

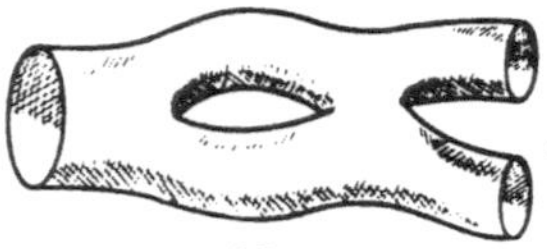

Fig. 5. *The path-integral on a space-time M with boundaries B and B' will define a transition amplitude $\Phi_M : \mathcal{H}_B \to \mathcal{H}_{B'}$.*

The space-time M will defines a certain state in the Hilbert space associated to its own boundary $\partial M = B$

$$|M\rangle \in \mathcal{H}_B, \tag{5.18}$$

where the orientation of M and B agree. Intuitively, the wave function $\Psi_M(\phi')$ representation of this state is defined as the path-integral over all field configurations ϕ on M that restrict to ϕ' at the boundary as in (5.11), see *fig.* 4.

In general we can consider the case that we give part of ∂M (say B') an orientation that agrees with that of M, and the other part (B) the reverse orientation, as in *fig.* 5. In that case the path-integral on M will define an element in $\mathcal{H}_B^* \otimes \mathcal{H}_{B'}$, or equivalently a transition amplitude

$$\Phi_M : \mathcal{H}_B \to \mathcal{H}_{B'}. \tag{5.19}$$

Factorization is now the property that if we cut M in two parts M_1 and M_2 along an intermediate slice B'', the transition amplitudes satisfy

$$\Phi_M = \Phi_{M_2} \circ \Phi_{M_1}. \tag{5.20}$$

This property is illustrated in *fig.* 6. Since we compose linear operators, the process includes a sum over states in the intermediate Hilbert space $\mathcal{H}_{B''}$.

We should stress that this type of factorization is characteristic of 'matter theories.' Theories of gravity, that somehow include an integral over all metrics on M, behave characteristically differently. This is due to the fact that in quantum gravity the path-integral only becomes well-defined if one integrates over all metrics modulo diffeomorphisms. It is clear that not all diffeomorphisms of a manifold M respect the decomposition $M = M_1 \cup M_2$, so factorization of the path-integral is not obvious. Another

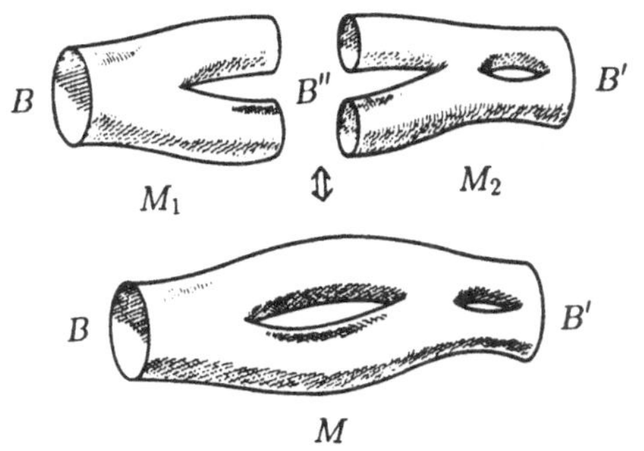

Fig. 6. *The factorization axiom presumes that if a space-time M is sliced into two parts M_1 and M_2 the corresponding transition amplitudes can be composed $\Phi_M = \Phi_{M_2} \circ \Phi_{M_1}$.*

important modification in theories of quantum gravity is the natural and unavoidable occurrence of singular configurations, where the topology of the manifold actually changes. In Polyakov's word, the space of metrics has a boundary of degenerate manifolds, and the path-integral naturally receives contributions of this boundary. We met this phenomenon already when we discussed the stable compactification of the moduli space of Riemann surfaces in §2.1.

5.3. Topological field theory in two dimensions

In this subsection we will restrict our investigations to two dimensions ($D = 2$) where our manifolds M are surfaces of genus g. Since the only connected compact one-dimensional manifold is the circle S^1, we have only one vector space to consider

$$\mathcal{H} \equiv \mathcal{H}_{S^1}, \tag{5.21}$$

and for convenience we will assume it to be finite dimensional

$$\dim \mathcal{H} = N < \infty. \tag{5.22}$$

The data of a two-dimensional topological field theory are now obtained by considering respectively the sphere with one, two, and three holes [66]. Let us briefly run through the argument.

(i) The disk gives rise to a particular state

$$\mathbf{1} \in \mathcal{H}, \tag{5.23}$$

that we will denote as the identity, for reasons that become obvious in a moment.

(ii) The cylinder gives a bilinear map

$$\eta : \mathcal{H} \otimes \mathcal{H} \to \mathbf{C}, \tag{5.24}$$

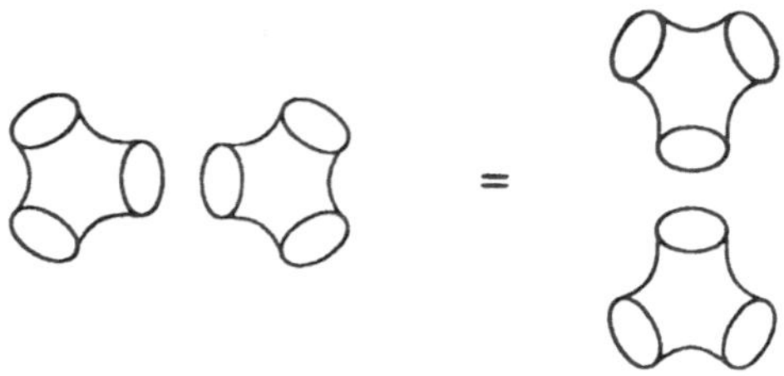

Fig. 7. *Duality of the four-point function on the sphere leads to associativity of the operator product algebra.*

which we notate as

$$\eta(a, b) = \langle a, b \rangle. \tag{5.25}$$

By factorization, this inner product η will be non-degenerate, but not necessarily positive. It allows us to identify the 'incoming states' in $\mathcal{H}$ with the 'outgoing states' in the dual space $\mathcal{H}^*$. Note that here both boundary components have an orientation that is opposite to that of the cylinder, and the map η should be sharply distinguished from the identity map (5.17).

(iii) Finally the pair of pants,—or the sphere with three holes—corresponds, again with the appropriate choice of orientations, to a bilinear map

$$c : \mathcal{H} \otimes \mathcal{H} \to \mathcal{H}, \tag{5.26}$$

If we introduce the notation

$$c(a, b) = a \times b, \tag{5.27}$$

this makes $\mathcal{H}$ into an algebra, the operator product algebra of the topological field theory. Indeed, this allow us to identify states with operators in a very simple way. It is convenient to choose an explicit basis $\{\phi_0, \ldots, \phi_{N-1}\}$ for $\mathcal{H}$, with $\phi_0 = \mathbf{1}$, in terms of which we have in component notation

$$\eta_{ij} = \langle \phi_i, \phi_j \rangle, \qquad \phi_i \times \phi_j = \sum_k c_{ij}{}^k \phi_k. \tag{5.28}$$

All these data suffice to calculate any partition or correlation function, since every surface can by factorization be reduced to a collection of three-holed spheres. Of course, there are many inequivalent ways to factorize a surface. The final answer should however not depend on the particular choice of factorization. This concept, known as duality, gives further constraint on the data N, η and c. For instance, a simple consequence of the symmetry of the 3-punctured sphere is the compatibility of the metric η with the algebra $\mathcal{H}$

$$\langle a \times b, c \rangle = \langle a, b \times c \rangle. \tag{5.29}$$

If we consider the sphere with four holes, there are two inequivalent ways of factorization as illustrated in *fig.* 7. This leads to so-called s-t duality, which translates in associativity of the algebra $\mathcal{H}$

$$(a \times b) \times c = a \times (b \times c). \tag{5.30}$$

It can be easily checked that no further conditions will be found when we consider more complicated surfaces.

Fig. 8. *The operator $H = c_i{}^{ij}\phi_j$ creates an handle on the surface Σ.*

With the concept of factorization, it is extremely easy to calculate higher genus partition and correlation functions. In fact, we can introduce an operator H that creates an handle [10]. It is defined as the state associated to the torus with one puncture (see *fig.* 8), and has the representation

$$H = \sum_{a \in \text{basis}} (\operatorname{Tr} a^*) \cdot a = \sum_{i,j} c_i{}^{ij} \phi_j. \tag{5.31}$$

In this fashion a genus g partition function Z_g can be written as a genus zero correlation function

$$Z_g = \big\langle \underbrace{H \cdots H}_{g} \big\rangle_0 \tag{5.32}$$

From this expression it is immediately clear that for topological field theories, in which the factorization axiom holds, partition functions grow polynomial for large genus

$$Z_g \sim c^g, \tag{5.33}$$

with some constant c. This behaviour contrasts starkly with results in two-dimensional quantum gravity and string theory, where the size of the partition function is related to the volume of moduli space and has the famous factorial growth [67]

$$Z_g \sim (2g)! \tag{5.34}$$

This already indicates to us, that the simple kind of general covariance that we are studying in this and the following section, is not the one relevant for gravity theories. It has to be appropriately modified—we will have to give up factorization—and we will return to these modifications in section 7. However, before we discuss all this, we will first have to explain how topological invariance can be realized in practice. Hereto we will consider again the general D-dimensional case.

5.4. *The Q-cohomology*

It might not be obvious how all of the above structures can be realized in a quantum field theory. Indeed, one does not expect in general such a simple structure; finite dimensional Hilbert spaces are already exceptional in quantum mechanics, let alone quantum field theory. The crucial ingredient is to introduce a cohomology principle into the game.

The metric independence independence of a quantum field theory was expressed by the vanishing of the energy-momentum tensor

$$T_{\mu\nu} = 0. \tag{5.35}$$

In theories of quantum gravity this general covariance is achieved by making the metric $g_{\mu\nu}$ a dynamic variable and integrating over all metrics in the path-integral. The stress tensor of the matter is now balanced by the contribution of the gravity sector

$$T_{\mu\nu} = T_{\mu\nu}^{(matter)} + T_{\mu\nu}^{(gravity)} = 0. \tag{5.36}$$

However, the introduction of a fluctuating metric can be avoided if the quantum field theory contains a fermionic charge Q that is nilpotent

$$Q^2 = 0. \tag{5.37}$$

One can think about Q as a generating a BRST-like symmetry, like in gauge theories, and define a physical Hilbert space $\mathcal{H}$ as the cohomology ring of Q in the full Hilbert space $\mathcal{H}'$ of the original field theory

$$\mathcal{H} = H_Q^*(\mathcal{H}') = \ker Q/\operatorname{im} Q. \tag{5.38}$$

That is, physical states are annihilated by Q and identified modulo Q-exact states. It does not matter which particular operator we choose to represent a class, since a well-known argument shows that spurious states (states of the form $\{Q,\lambda\}$) decouple inside physical correlation functions

$$\langle\{Q,\lambda\}\mathcal{O}_1\cdots\mathcal{O}_s\rangle = 0, \qquad \text{if } \{Q,\mathcal{O}_i\} = 0. \tag{5.39}$$

The bracket $\{\cdot,\cdot\}$ represents here either a commutator or an anti-commutator, whichever is appropriate. The existence of the operator Q by itself is of course not enough to warrant general covariance. To that end we need another essential ingredient, namely that the energy-momentum tensor $T_{\mu\nu}$ is spurious itself

$$T_{\mu\nu} = \{Q, G_{\mu\nu}\}, \tag{5.40}$$

and thus cohomologically trivial. The operator $G_{\mu\nu}$ must be a fermionic rank two tensor. The conserved charges related to $T_{\mu\nu}$ and $G_{\mu\nu}$ are

$$P_\mu = \int T_{\mu 0}, \qquad G_\mu = \int G_{\mu 0}, \tag{5.41}$$

where the integrals are taken over a $D-1$ dimensional spacelike surface. These charges form an anti-commuting extension of the translation group

$$P_\mu = \{Q, G_\mu\}, \qquad \{G_\mu, G_\nu\} = 0, \tag{5.42}$$

that can be viewed as a supersymmetry algebra with charges of spin zero and one, instead of the usual spin one-half.

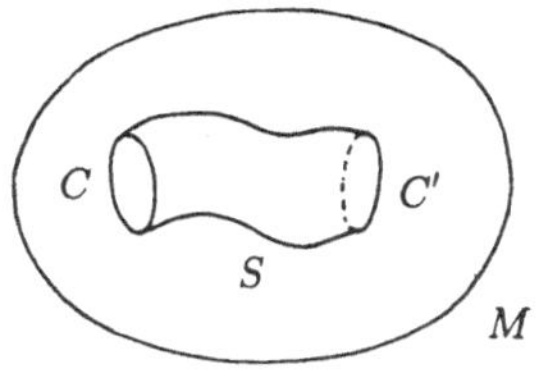

Fig. 9. *If two k-dimensional submanifolds C and C' bound an $k+1$ dimensional manifold S in the space-time M, the non-local operators $\phi(C)$ and $\phi(C')$ are equivalent as physical observables.*

5.5. The descent equation

Regarding the Q symmetry as a spin zero supersymmetry is a very fruitful analogy. In fact, it is convenient to go to a 'superspace' formulation of the theory, since this leads naturally to the non-local observables in the model. In addition to the previous space-time coordinates x^μ, we will now have D Grassmannian coordinates θ^μ. Starting from a physical field $\phi(x)$, that we will sometimes denote as $\phi^{(0)}(x)$ to indicate that it is a local operator, the superfield $\Phi(x,\theta)$ is given by

$$\begin{aligned}\Phi(x,\theta) &= \exp \theta^\mu G_\mu \cdot \phi(x) \\ &= \phi^{(0)}(x) + \phi^{(1)}_\mu(x)\theta^\mu + \ldots + \phi^{(D)}_{\mu_1\ldots\mu_D}(x)\theta^{\mu_1}\cdots\theta^{\mu_D}.\end{aligned} \tag{5.43}$$

Here the fields $\phi^{(k)}$ are generated from $\phi^{(0)}$ by repeated application of G_μ

$$\phi^{(k)}_{\mu_1\ldots\mu_k}(x) = \{G_{\mu_1}, \{G_{\mu_2}, \ldots, \{G_{\mu_k}, \phi^{(0)}(x)\}\ldots\}\}. \tag{5.44}$$

Since $\phi^{(k)}_{\mu_1\ldots\mu_k}$ is antisymmetric in all its indices, it represents a k-form, and we can write (5.44) symbolically as

$$\{G, \phi^{(k)}\} = \phi^{(k+1)}. \tag{5.45}$$

Since by (5.42) we have $\{Q, G\} = d$, these differential forms satisfy the important *descent equation*

$$d\phi^{(k)} = \{Q, \phi^{(k+1)}\}. \tag{5.46}$$

We can draw two conclusions from this equation. First, it suggests a new class of *non-local* physical observables. If C is a k-dimensional closed submanifold, the descent equation shows that

$$\phi(C) \equiv \int_C \phi^{(k)}(x) \tag{5.47}$$

is a physical observable, since

$$\{Q, \phi(C)\} = \int_C d\phi^{(k-1)} = \int_{\partial C} \phi^{(k-1)} = 0. \tag{5.48}$$

Secondly, if C and C' represent the same class in $H_k(M)$, we have $C - C' = \partial S$ and

$$\phi(C) - \phi(C') = \int_S d\phi^{(k)} = \{Q, \int_S \phi^{(k+1)}\}. \tag{5.49}$$

So the physical observable depends only on the homology class of C, as is illustrated in *fig.* 9. That is, for each class in $H_k(M)$ and each element in $\mathcal{H}$ we can construct a non-local operator. In the case of the zero form fields, we have

$$\phi^{(0)}(x) - \phi^{(0)}(y) = \{Q, \int_x^y dx^\mu \, \phi_\mu^{(1)}\}, \tag{5.50}$$

which shows explicitly our claim that the correlators of the local physical operators do not depend on the positions.

5.6. Perturbations of topological field theories

One very important non-local operator, that always exists, is the top-form

$$\phi(M) = \int_M \phi^{(D)}(x). \tag{5.51}$$

The importance of this particular operators lies in the following fact. Since $\phi^{(D)}$ is a volume form, it can be integrated over the space-time itself. In this way we can make a modification of the action of the topological quantum field theory by adding (5.51) with some coupling coefficient t to the action S

$$S \to S - t \cdot \int_M \phi^{(D)}, \tag{5.52}$$

while preserving the general covariance. So, the top-dimensional partner of any local observable defines a perturbation of the topological field theory. We can in this way define a multi-parameter family of topological field theories whose partition functions are given by

$$Z(M,t) = \langle \exp \sum_k t_k \cdot \int_M \phi_k^{(D)} \rangle. \tag{5.53}$$

The coupling coefficients t_k can be seen as a moduli in the space of topological field theories.

If we specialize to the case of two dimensions where M is a surface Σ, we see that apart from our local observables

$$\phi(P) \in \mathcal{H}, \tag{5.54}$$

associated to a point $P \in \Sigma$, we also can consider the operators

$$\phi(S^1) = \oint \phi^{(1)}, \qquad \phi(\Sigma) = \int_\Sigma \phi^{(2)}, \tag{5.55}$$

that correspond to a (non-trivial) cycle and the surface Σ itself. Temporarily discarding the one-forms, the most general correlation function in the model is of the form

$$\left\langle \phi_{i_1} \cdots \phi_{i_s} \int \phi_{j_1}^{(2)} \cdots \int \phi_{j_r}^{(2)} \right\rangle_g \tag{5.56}$$

We have already given away how these correlators can in principle be calculated. One simply considers the more general object

$$\left\langle \phi_{i_1} \cdots \phi_{i_s} \exp \sum_k t_k \int \phi_k^{(2)} \right\rangle_g \tag{5.57}$$

The expansion of this object in the coupling constants t_k clearly generates all possible inclusions of two-form operators. However, according to the above reasoning it can alternatively be seen as a correlation function of only the local operators ϕ_i in a new, perturbed topological field theory, described by the coupling constants t_k. Since this new theory satisfies all the axioms of §5.3, all quantities can again be calculated by factorization. That is, all essential information is encoded in the perturbed three-point functions

$$c_{ijk}(t) = \left\langle \phi_i \phi_j \phi_k \exp \sum_k t_k \int \phi_k^{(2)} \right\rangle_0 \tag{5.58}$$

which define a family of multiplications on the vector space $\mathcal{H}$.

So we finally arrive at the following beautiful picture of (a family of) two-dimensional topological field theories. It is given by a bundle of commutative, associative algebras $\mathcal{H}$ fibered over a moduli space X. The data N, η, and c vary continuously and satisfy the relations of §5.3. Moreover, there is a well-defined map from a state $\phi_k \in \mathcal{H}$ to a vector field $\partial/\partial t_k$ on X. Furthermore, these vector fields commute, and can be integrated to give local coordinates t_k. In the following section we will give some very explicit examples of all this.

6. TOPOLOGICAL CONFORMAL FIELD THEORY

We have seen that a general two dimensional topological field theory is solved in terms of the perturbed three-point functions $c_{ijk}(t)$. Our aim will now be to calculate these quantities in some examples. To this end we have to introduce some auxiliary structure: we will impose the condition that the topological field theory be conformally invariant. This might seem a bit superfluous. All *physical* quantities in a topological field theory are automatically invariant under conformal transformations, since they are independent of the metric. However, we demand that even the non-physical sector—the 'big' Hilbert space $\mathcal{H}'$—be conformally invariant. We will call these models *topological conformal field theories* (TCFT), see also [17,30].

6.1. Twisting $N = 2$ superconformal models

Conformal invariance requires that the trace of the stress-energy tensor vanishes

$$T_\mu{}^\mu = 0, \tag{6.1}$$

not just within physical correlation functions, but identically as an operator. The stress tensor still has a Q partner $G_{\mu\nu}$

$$T_{\mu\nu} = \{Q, G_{\mu\nu}\}, \tag{6.2}$$

which is also assumed to be traceless

$$G_\mu{}^\mu = 0. \tag{6.3}$$

In complex coordinates $(z, \overline{z})$ the two remaining components $T = T_{zz}$ and $\overline{T} = T_{\bar{z}\bar{z}}$ are respectively holomorphic and anti-holomorphic fields. We will for the moment restrict our attention only to the left-moving fields, *i.e.*, fields holomorphic in z. Let us make a list of the different fields that appear in this sector.

First of all we have the stress tensor $T(z)\,dz^2$ and its Q-partner $G(z)\,dz^2$, both quadratic differentials. These have the standard decompositions into modes

$$T(z) = \sum L_n z^{-n-2}, \qquad G(z) = \sum G_n z^{-n-2}. \tag{6.4}$$

We will make the further important assumption, that the Hilbert space $\mathcal{H}'$ will decompose in highest weight modules of the algebra generated by the L_n and G_n, which is an extension of the $c = 0$ Virasoro algebra

$$[L_m, L_n] = (m - n)L_{m+n}, \tag{6.5}$$

by a vector supersymmetry:

$$[L_m, G_n] = (m - n)G_{m+n}, \qquad \{G_m, G_n\} = 0. \tag{6.6}$$

The conformal central charge c, that in principle could appear in the Virasoro algebra, must necessarily vanish, since a conformal anomaly is not allowed in a general covariant theory, *cf* critical string theory. We have primary fields ϕ_i satisfying

$$L_n|\phi_i\rangle = 0, \qquad G_n|\phi_i\rangle = 0, \quad (n > 0). \tag{6.7}$$

Notice the second relation, which is not a consequence of the first. Actually in this respect conformal invariance is somewhat of a misnomer, since we really require invariance under the larger algebra generated by both the stress tensor T and its anti-commuting partner G.

There is a second multiplet of fields related to the (anti-commuting) BRST-current $Q(z)\,dz$, whose conserved charge is the left-moving part of the scalar Q-operator

$$Q = Q_0 + \overline{Q}_0, \tag{6.8}$$

with

$$Q_0 = \oint Q(z)\,dz, \qquad Q(z) = \sum Q_n z^{-n-1}. \tag{6.9}$$

We will use the notations Q and Q_0 interchangeably if no confusion arises. Since the BRST-current is a spin one field, it is—by an argument that we will sketch in a moment—necessarily exact:

$$Q(z) = \{Q, J(z)\}, \qquad J(z) = \sum J_n z^{-n-1}, \tag{6.10}$$

Table 1. *The conserved currents of a topological conformal field theory with their conformal dimensions, charges, and statistics.*

current	spin h	charge q	$h+\frac{1}{2}q$	statistics
$T(z)$	2	0	2	boson
$G(z)$	2	-1	$\frac{3}{2}$	fermion
$Q(z)$	1	1	$\frac{3}{2}$	fermion
$J(z)$	1	0	1	boson

where $J(z)\,dz$ is a bosonic field of spin one, that is, a honest abelian current. So we arrive at the multiplet of holomorphic fields listed in table 1.

All fields $\phi_i(z,\bar{z})$ in the theory can be chosen to have some definite conformal dimensions h_i and charges q_i

$$L_0|\phi_i\rangle = h_i|\phi_i\rangle, \qquad J_0|\phi_i\rangle = q_i|\phi_i\rangle. \tag{6.11}$$

Since we have

$$L_0 = \{Q, G_0\}, \qquad [Q, L_0] = 0, \tag{6.12}$$

we see immediately that all physical states, satisfying $Q|\phi\rangle = 0$, are ground states, with $h=0$. Indeed, suppose $L_0|\phi\rangle = h|\phi\rangle$ with $h \neq 0$, then by a familiar argument this state is exact

$$|\phi\rangle = \frac{1}{h}L_0|\phi\rangle = Q|\lambda\rangle, \qquad |\lambda\rangle = \frac{1}{h}G_0|\phi\rangle. \tag{6.13}$$

The multiplet $\{T,G,Q,J\}$ is very reminiscent of the generators of the $N=2$ superconformal algebra, where we would have a stress tensor T, two spin $3/2$ supercurrents Q^+, Q^- and a $U(1)$ current J. The two sets of currents are related by the 'twist' [60,61,68]

$$T \to T + \tfrac{1}{2}\partial J, \tag{6.14}$$

which gives in particular $L_0 \to L_0 + \frac{1}{2}J_0$ and accordingly adds the charge to the conformal dimensions (see table 1)

$$h \to h + \tfrac{1}{2}q. \tag{6.15}$$

We can now consider the algebra generated by the modes L_n, G_n, Q_n, J_n of the four currents. If they form a *closed* algebra—which need not be the case, of course—their commutators are completely fixed, due to the uniqueness of the $N=2$ superconformal algebra, that we would obtain after twisting. In this way we arrive at the following algebra

$$\begin{aligned}
[L_m, L_n] &= (m-n)L_{m+n}, & [J_m, J_n] &= d\cdot m\,\delta_{n+m,0},\\
[L_m, G_n] &= (m-n)G_{m+n}, & [J_m, G_n] &= -G_{m+n},\\
[L_m, Q_n] &= -n\,Q_{m+n}, & [J_m, Q_n] &= Q_{m+n},
\end{aligned}$$

$$
\begin{aligned}
\{G_m, Q_n\} &= L_{m+n} + nJ_{m+n} + \tfrac{1}{2}d \cdot m(m+1)\delta_{n+m,0}, \\
[L_m, J_n] &= -nJ_{m+n} - \tfrac{1}{2}d \cdot m(m+1)\delta_{m+n,0}.
\end{aligned}
\tag{6.16}
$$

Most of these relations are self-evident; they simply express the conformal dimensions and charges of the currents. We note the appearance of a central charge d for the $U(1)$ symmetry. In the case of a topological sigma-model d will correspond to the complex dimension of the target space, which explains the choice of symbol. The same constant features as an anomaly in the commutator of the stress tensor with the $U(1)$ current; we will return to this point momentarily. The central charge for the corresponding $N = 2$ supersymmetric theory is related to the anomaly in the topological model by $c = 3d$.

The anomaly in the current $J(z)$ leads to selection rules for the correlation functions. $J(z)$ does not transform as a proper current under coordinate transformations, as we read off from the commutation relations of L_n with J_m in (6.16). If $z \to w(z)$ denotes a holomorphic coordinate transformation, then

$$
J_0 \to J_0 + d \cdot \oint \partial_z(\log \partial_z w), \tag{6.17}
$$

where the additional term measures the winding number of the map $\partial_z w$. So, under the inversion $w = 1/z$ the zero mode J_0 goes to $J_0 - d$. This results in a background charge of $-d$ on the sphere. The background charge for arbitrary topology can be found from the covariant expression of the $U(1)$ anomaly

$$
\nabla_\mu J^\mu = -\frac{d}{8\pi}\sqrt{h}\, R, \tag{6.18}
$$

where $h_{\mu\nu}$ is the world-sheet metric and R is the associated scalar curvature. Consequently, since the curvature density integrates to the Euler character $\chi = 2 - 2g$ on a genus g surface, there is a background charge of $d \cdot (g - 1)$ present that has to be compensated. Correlation functions

$$
\langle \phi_{i_1} \cdots \phi_{i_s} \rangle_g \tag{6.19}
$$

therefore obey the selection rule

$$
\sum_{i=1}^{s} q_i = d(1 - g). \tag{6.20}
$$

In relation to this we note that the two-form operators

$$
\phi_i^{(2)} = G_{-1}\overline{G}_{-1}\phi_i^{(0)} \tag{6.21}
$$

have (*left, right*) charges $(q_i - 1, \overline{q}_i - 1)$.

By the above twisting *any* $N = 2$ superconformal model produces a topological CFT. This not only provides a wealth of examples, but also allows us to use some very interesting results derived in the context of $N = 2$ supersymmetry. For instance, the unitary

$N = 2$ models come with a natural positive inner product, that we can carry over to the topological field theory. Because of the twisting, this inner product is not Lorentz-invariant. In fact, with this inner product we have the hermiticity property $Q^\dagger = G_0$, which relates a scalar and a vector quantity. By considering the relation

$$0 = \langle\phi|L_0|\phi\rangle = \|Q|\phi\rangle\|^2 + \|G_0|\phi\rangle\|^2, \tag{6.22}$$

one shows that in this case the ground states satisfy both

$$Q|\phi\rangle = 0, \qquad G_0|\phi\rangle = 0. \tag{6.23}$$

The latter condition picks out a unique representative for each Q-cohomology class. It should be compared to harmonic forms in ordinary de Rham cohomology theory. In the $N = 2$ context these fields are known as chiral primary fields [69].

6.2. Generalized $SL(2,C)$ invariance and its consequences

We now want to derive an important symmetry property of topological conformal field theories. To this end we recall that the non-local operators can be combined into one superfield (5.43)

$$\Phi(\theta,\overline{\theta}) = \phi^{(0)} + \phi^{(1,0)}\theta + \phi^{(0,1)}\overline{\theta} + \phi^{(2)}\theta\,\overline{\theta}. \tag{6.24}$$

Here we suppressed the $(z,\overline{z})$ dependence. Let us now consider a correlation function on the sphere of the form

$$\Big\langle\prod_{i=1}^{s}\int d^2z\, d^2\theta\; \Phi_i(z,\overline{z},\theta,\overline{\theta})\Big\rangle_0. \tag{6.25}$$

This expression is not well-defined, since it is invariant under a fermionic extension of the well-known $SL(2,C)$ symmetry, generated by operators L_0, L_1, L_{-1} and G_0, G_1, G_{-1}. We have to factor out the infinite volume of this group in order to obtain a finite answer. These symmetries correspond to the super-Möbius transformations

$$z \longrightarrow \frac{az+b}{cz+d}, \qquad \theta \longrightarrow \frac{\theta + \alpha z^2 + \beta z + \gamma}{(cz+d)^2}. \tag{6.26}$$

This extended $SL(2,C)$ symmetry can be used to fix three of the z-coordinates, say z_1, z_2, z_3, at 0, 1, and ∞, and put three of the θ-coordinates to zero. We choose these anti-commuting coordinates to be $\theta_1, \theta_2, \theta_3$. If we recall that

$$\int d^2\theta\cdot\Phi = \phi^{(2)}, \qquad \Phi\big|_{\theta=0} = \phi^{(0)}, \tag{6.27}$$

then we see that after gauge fixing we are left with a correlation function of the form

$$\Big\langle\phi^{(0)}_{i_1}\phi^{(0)}_{i_2}\phi^{(0)}_{i_3}\int\phi^{(2)}_{i_4}\cdots\int\phi^{(2)}_{i_s}\Big\rangle_0. \tag{6.28}$$

Since we started from an expression that was explicitly symmetric in *all* indices $i_1, \ldots, i_s$, this correlator also has this permutational symmetry. That is, it does not matter which three operators we represent as zero-forms—the generalized $SL(2,C)$ invariance tells us that we can interchange a zero and a two-form. An alternative derivation of this result uses the Ward identities associated to this global invariance [17].

In terms of the coefficients $c_{ijk}(t)$ the permutation symmetry of (6.28) gives the important integrability condition

$$\frac{\partial c_{ijk}}{\partial t_l} = \frac{\partial c_{ijl}}{\partial t_k}. \tag{6.29}$$

We stress that this is an additional condition imposed on the family of algebras $c(t)$, and a consequence of the topological invariance. We can integrate this relation three times—at least locally—to find that the three-point functions are actually the third derivatives of a function $F(t)$, the so-called free energy

$$c_{ijk}(t) = \frac{\partial^3 F(t)}{\partial t_i \partial t_j \partial t_k}. \tag{6.30}$$

Symbolically, $F(t)$ is defined as

$$F(t) = \left\langle \exp \sum_i t_i \int \Phi_i \right\rangle, \tag{6.31}$$

that is, $F = \log \tau$, with $\tau(t)$ the string partition function.

As a corollary to the result (6.29), consider the special identity operator $\phi_0 = \mathbf{1}$. There is no corresponding two-form, since G commutes with the identity. So the coupling coefficient t_0 does not exist—there is no modulus associated to $\mathbf{1} \in \mathcal{H}$. This fact, combined with integrability relation (6.29), shows that

$$0 = \frac{\partial c_{ijk}}{\partial t_0} = \frac{\partial c_{ij0}}{\partial t_k} = \frac{\partial \eta_{ij}}{\partial t_k}. \tag{6.32}$$

So we have shown that the metric or two-point function η_{ij} is independent of the deformation parameters t_k.

Summarizing, we have derived in this section three additional important results, that followed from the conformal invariance. First, all of the coefficients $c_{ijk}(t)$ are derived from a single function, the free energy $F(t)$. This function is well-defined on local patches of the moduli space of TCFT's. Secondly, the metric $\eta_{ij}(t)$ on the Hilbert space of states is in fact independent of the couplings t_k. This is equivalent to the statement that $\partial F/\partial t_0$ is a quadratic function. Finally, there exists an anomalous $U(1)$ symmetry, which assigns charge q_i to the local physical fields ϕ_i, with a background charge d, and which corresponds to a scaling relation for the free energy of the form

$$\sum_j (q_j - 1) t_j \frac{\partial}{\partial t_j} F(t) = (d-3) F(t). \tag{6.33}$$

Remarkably enough, these three ingredients will be sufficient to solve the theory in many special cases, in particular for all models with $d < 1$ [17].

At this point it might be helpful to consider some examples of TCFT's. The simplest example is a free field theory of d complex bosons and fermions. It can obtained by twisting d copies of the $c = 3$ free field realization of the $N = 2$ superconformal algebra. The bosons will be written as $x^i, x^{\bar{\imath}}$, with $i = 1, \ldots, d$. In order to describe the fermions we have to make a choice. This choice is related to the following ambiguity. Up to now we basically only discussed the left-moving sector. Here the topological and the superconformal model were related by the twist

$$T \to T \pm \tfrac{1}{2}\partial J. \tag{6.34}$$

Here the $\pm$ sign is completely irrelevant, since we have no absolute way to fix the overall sign of the $U(1)$ charge. However, if we combine left and right-movers we can consider either the $(+,+)$ or the $(+,-)$ version. Although the resulting TCFT's are still equivalent, since *quantum-mechanically* we can also not distinguish the signs of the separate left and right-moving currents, the interpretation in terms of the *classical* geometry of the original superconformal model might be vastly different. In all its generality this phenomenon is known as 'mirror symmetry' [70]. Even in case of our simple free field theory we can consider two different actions and symmetry realizations that lead to equivalent quantum field theories. The problem manifests itself here only in the spin assignments of the fermions. In the $N = 2$ model we have four different kinds of fermionic operators that we can denote as

$$\psi^i_L(z),\ \psi^i_R(\bar{z}),\ \psi^{\bar{\imath}}_L(z),\ \psi^{\bar{\imath}}_R(\bar{z}). \tag{6.35}$$

These fields, all of spin one-half, obtain the unusual spins of zero and one after twisting. The two possible twists identify as the spin zero fields either the pair $\psi^i_L, \psi^{\bar{\imath}}_R$ or the pair ψ^i_L, ψ^i_R. We will continue to work with the latter choice, and will combine the two fields into one field

$$\psi^i(z,\bar{z}) = \psi^i_L(z) + \psi^i_R(\bar{z}). \tag{6.36}$$

We denote the spin one fields as $\chi^{\bar{\imath}}_z, \chi^{\bar{\imath}}_{\bar{z}}$. They are the two components of a one-form $\chi^{\bar{\imath}}_\mu$. The action of these fields reads, with a summation over $i = \bar{\imath}$ understood,

$$S_0 = \int d^2z \left(\partial_\mu x^{\bar{\imath}} \partial^\mu x^i + \chi^{\bar{\imath}}_\mu \partial^\mu \psi^i \right). \tag{6.37}$$

The next step is to identify the nilpotent charge Q. With $\delta = \{Q, \cdot\}$ the non-vanishing transformations are simply

$$\delta x^i = \psi^i, \qquad \delta \chi^{\bar{\imath}}_\mu = -\partial_\mu x^{\bar{\imath}}. \tag{6.38}$$

With these field transformations we easily verify that the action is invariant. The stress-energy tensor is indeed Q-exact

$$T(z) = \partial_z x^{\bar{\imath}} \partial_z x^i + \chi^{\bar{\imath}}_z \partial_z \psi^i_L = \{Q, G\}, \qquad G(z) = -\chi^{\bar{\imath}}_z \partial_z x^i. \tag{6.39}$$

The set of currents is complemented in this example by

$$Q(z) = \partial_z x^{\bar{\imath}} \psi^i_L, \qquad J(z) = -\chi^{\bar{\imath}}_z \psi^i. \tag{6.40}$$

This trivial free field theory can be generalized in two interesting directions. First, we can consider non-linear sigma models, where the flat space-time of one-complex dimension is replaced by an arbitrary d-dimensional Ricci-flat Kähler manifold M. In superfield notation, with $X^i = x^i + \theta\psi^i + \ldots$, the action reads

$$S = \int d^2z\, d^4\theta \cdot K(X^i, X^{\bar{\imath}}). \tag{6.41}$$

where K is the Kähler potential, and $x^i, x^{\bar{\imath}}$ are the local coordinates on M. These models are well-known to possess $N = 2$ superconformal invariance, which translates into topological invariance after twisting. In the above model we simply have chosen the flat target space metric

$$K(X^i, X^{\bar{\imath}}) = X^i X^{\bar{\imath}}. \tag{6.42}$$

A second possible generalization is the inclusion of a superpotential $W(X)$. In this case we have a so-called topological Landau-Ginzburg model [71], see also [72]. The action reads

$$S = S_0 + \int d^2z\, d^2\theta \cdot W(X^i) + \text{c.c.} \tag{6.43}$$

or, after the integration over $(\theta, \bar{\theta})$ and eliminating the auxiliary fields,

$$\begin{aligned} S = \int d^2z \Big(& \partial_z x^{\bar{\imath}} \partial_{\bar{z}} x^i + \chi_z^{\bar{\imath}} \partial_{\bar{z}} \psi_L^i + \chi_{\bar{z}}^{\bar{\imath}} \partial_z \psi_R^i \\ & + \frac{\partial W}{\partial x^{\bar{\imath}}} \frac{\partial \overline{W}}{\partial x^i} + \psi_L^i \psi_R^j \frac{\partial^2 W}{\partial x^i \partial x^j} + \chi_z^{\bar{\imath}} \chi_{\bar{z}}^{\bar{\jmath}} \frac{\partial^2 \overline{W}}{\partial x^{\bar{\imath}} \partial x^{\bar{\jmath}}} \Big). \end{aligned} \tag{6.44}$$

In this case we have to modify the transformation rules by [71]

$$\delta\psi_L^i = \frac{\partial W}{\partial x^{\bar{\imath}}}, \qquad \delta\psi_R^i = -\frac{\partial W}{\partial x^{\bar{\imath}}}. \tag{6.45}$$

Landau-Ginzburg models become only conformal invariant at the renormalization group fixed point, if the superpotential is quasi-homogeneous [73,74]. That is, if there exists a scaling law, with charges q_i, such that (from now we will use lower index notation $x_i \equiv x^{\bar{\imath}}$ to avoid confusion with powers of the coordinates)

$$W(\lambda^{q_i} x_i) = \lambda W(x_i). \tag{6.46}$$

Landau-Ginzburg models have the wonderful property that the correlation functions of the chiral primary fields—or, after twisting, the physical fields— are completely determined by the superpotential W. In fact, for a quasi-homogeneous potential $W(x_i)$ the local physical fields are simply functions $f(x_i)$ of the invariant field $x_i \equiv x^{\bar{\imath}}$. Since we have

$$\frac{\partial W}{\partial x_i} = \{Q, \psi_L^i\}, \tag{6.47}$$

every function of the form $f = g^i \partial_i W$ is Q-exact, and the physical state space is accordingly given by

$$\mathcal{H} = \frac{\mathbf{C}[x]}{\partial W(x)}. \tag{6.48}$$

Table 2. *The simply-laced Lie groups G, their Dynkin diagrams Γ, the corresponding superpotentials W, and the central charge d of the topological minimal model.*

G	Γ	W	d
A_n	$1\ \bullet - \bullet - \cdots - \bullet - \bullet\, n$	x^{n+1}	$\frac{n-1}{n+1}$
D_n	$\begin{array}{c}\bullet \\ \mid \\ 1\ \bullet - \bullet\ - \bullet - \cdots - \bullet - \bullet\, n-1\end{array}$	$x^{n-1} + xy^2$	$\frac{n-2}{n-1}$
E_6	$\begin{array}{c}\bullet \\ \mid \\ \bullet - \bullet - \bullet - \bullet - \bullet\end{array}$	$x^3 + y^4$	$\frac{4}{5}$
E_7	$\begin{array}{c}\bullet \\ \mid \\ \bullet - \bullet - \bullet - \bullet - \bullet - \bullet\end{array}$	$x^3 + xy^3$	$\frac{8}{9}$
E_8	$\begin{array}{c}\bullet \\ \mid \\ \bullet - \bullet - \bullet - \bullet - \bullet - \bullet - \bullet\end{array}$	$x^3 + y^5$	$\frac{14}{15}$

Since there are no singularities in the operator product of two function $f(x_i)$, the operator algebra does not receive corrections, and is also given by (6.48) now considered as a polynomial ring, where $\mathbf{C}[x]$ is the set of polynomials in the x_i, and we have factored out the ideal generated by

$$\frac{\partial W}{\partial x_i} = 0. \tag{6.49}$$

The charge of an element $\phi(x)$ is determined by assigning charge q_i to the fundamental field x_i.

In general we can consider deformations of W by an element in $\mathcal{H}$ that preserve the quasi-homogeneity, *i.e.*, an operator of charge $q = 1$, a modulus or marginal operator. A simple example is the potential

$$W = x^3 + y^3 + z^3 + a \cdot xyz. \tag{6.50}$$

This clearly introduces a modulus—for any value of a the potential is homogeneous, whereas the different rings $\mathcal{H}$ are non-isomorphic. There is a special set of potentials that have no moduli. The corresponding field theories are known as topological minimal models—the twisted versions of the minimal $N = 2$ superconformal models, and as such were first considered by Eguchi and Yang [68] and, in relation with solvable string theories, by Li [16],. In this way the classification of rational singularities enters the subject, since these potentials have an ADE classification, as given in table 2. This is of course the same ADE labeling that features in the classification of modular invariant partition functions in minimal conformal field theories [75].

It is not difficult to give the explicit forms of the rings $\mathcal{H}$. For instance, the A_{p-1} ring with

$$W(x) = x^p \tag{6.51}$$

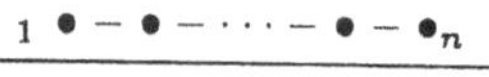

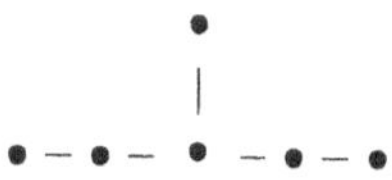

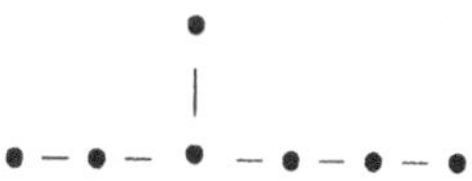

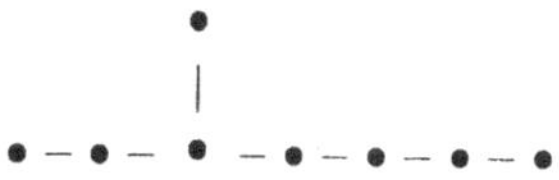

is simply given by

$$x^{p-1} = 0. \tag{6.52}$$

Thus, a basis is furnished by the elements

$$\left\{1, x, x^2, \ldots, x^{p-2}\right\}. \tag{6.53}$$

Quite generally the fields $\phi_i \in \mathcal{H}$ are in one-to-one correspondence with the exponents of the simply-laced Lie group G. The field ϕ_i has charge $q_i = i/p$ with $i+1$ an exponent, and $p = h_G$, the dual Coxeter number of G. The 'central charge' d of the minimal model is given by

$$d = \frac{p-2}{p}, \tag{6.54}$$

see also table 2. The metric η is determined through charge conservation, and is of the skew-diagonal form

$$\eta_{ij} = \delta_{i+j,p-2}. \tag{6.55}$$

6.4. The perturbed operator algebra

It is clear that a choice of superpotential $W(x)$ determines the operator algebra (6.48) and therefore all correlation functions of the local operators $\phi^{(0)}$. As we have explained, by factorization, it suffice to calculate the perturbed three-point function

$$c_{ijk}(t) = \left\langle \phi_i \phi_j \phi_k \exp \sum_n t_n \int \phi_n^{(2)} \right\rangle_0 \tag{6.56}$$

in order to include also the non-local two form $\phi^{(2)}$. We will be able to calculate this algebra as a result of the existence of a flat metric on the moduli space of TCFT's. Much of the structure we will meet, in particular the existence of 'flat coordinates' associated to a rational singularity, has already appeared in the mathematical literature through the work of K. Saito [76]. Recently, these techniques have been applied in the context of topological field theory [77,78]. We will sketch here the less sophisticated argument of [17].

For definiteness we will consider only the models in the A-series, so that $W(x) = x^p$, and the basis elements of $\mathcal{H}$ are given by

$$\phi_i = x^i, \qquad i = 0, 1, \ldots, p-2. \tag{6.57}$$

The crucial idea is that the perturbed algebra is still of the form (6.48), but with a more general superpotential

$$W = x^p + \sum_{i=0}^{p-2} g_i(t) x^i. \tag{6.58}$$

Intuitively, a result of this form is to be expected, since we add to the action the operators

$$\phi_i^{(2)} = \int d^2\theta \cdot X^i. \tag{6.59}$$

However, the extra terms on the RHS of (6.58) are not of the simple form $t_i x^i$. This is the case because the operators are only given by $\phi_i = x^i$ at the conformal ($t = 0$) point, and become more complicated t-dependent polynomials after a general perturbation. We therefore should allow for more general functions $g_i(t)$ that behave as $g_i = t_i + \mathcal{O}(t^2)$. For general values of the coupling constants the fields ϕ_i are defined by

$$\phi_i(x,t) = \frac{\partial W}{\partial t_i} = \sum_{j=0}^{p-2} \frac{\partial g_j}{\partial t_i} x^j. \tag{6.60}$$

We want to calculate the functions $g_i(t)$, or equivalently the fields $\phi_i(x,t)$.

Of course, one can also consider the g_i's as coordinates on the space of topological field theories, and this is the perfect legitimate point of view taken in [71]. In that case one should worry about contact terms that are quite generally associated to reparametrizations of moduli [80]. The preferred coordinates t_k, that naturally follow from the conformal invariance underlying the model, are referred to as 'flat coordinates.' The reason for this—not quite appropriate—nomenclature is the following. In §6.2 we proved that in a general topological conformal field theory the metric

$$\eta_{ij} = \langle \phi_i \phi_j \rangle_0 \tag{6.61}$$

is constant, when consider a function of the moduli t_i. That is, we always have a flat metric on the space of perturbed TCFT's and we can choose coordinates t_i such that this metric is constant. We will use this crucial ingredient to determine the correlation functions. Notice that for a *given* superpotential W the operator products are completely determined—they are given by the polynomial ring

$$\phi_i(x)\,\phi_j(x) = \sum_k c_{ij}{}^k \phi_k(x) \qquad (\mathrm{mod}\ W'(x)). \tag{6.62}$$

The inner product should be compatible with this multiplication, and is thereby uniquely determined to be

$$\langle \phi_i \phi_j \rangle = \oint \frac{dx}{2\pi i} \frac{\phi_i(x)\,\phi_j(x)}{W'(x)}. \tag{6.63}$$

(Here we used in an essential way the fact that W allows no moduli. Otherwise this metric would only be determined up to a conformal factor. See for instance [79] for an example where this phenomenon occurs.) The above metric clearly respects the operator algebra (6.62), since terms proportional to $W'(x)$ do not contribute. It can also be directly derived from the path-integral representation [71]. We know that the fields ϕ_i should be orthogonal with respect to this inner product. These orthogonal polynomials can be defined in the following elegant way as derivatives of fractional powers of the superpotential

$$\phi_i(x) = \frac{1}{i+1} dW_+^{(i+1)/p}. \tag{6.64}$$

Here $d = \partial/\partial x$, and the $+$ indicates again a truncation to positive powers of x, similar as in our discussion of the KdV hierarchy in §3.1. One easily verifies, with $z = W^{1/p}$ that indeed

$$\begin{aligned} \langle \phi_i \phi_j \rangle &= \frac{1}{2\pi i} \oint \frac{(z^i dz)_+ (z^j dz)_+}{z^{p-1} dz} \\ &= \frac{1}{2\pi i} \oint z^{i+j-p+1} dz = \delta_{i+j,p-2}. \end{aligned} \tag{6.65}$$

The reader is encouraged to verify that the +'s have legitimately been dropped in this derivation. The solution of the model results if we combine the equations (6.60) and (6.64), and obtain the important result

$$(i+1)\frac{\partial W}{\partial t_i} = \frac{\partial}{\partial x} W_+^{(i+1)/p}. \tag{6.66}$$

After a relabelling $t_i \to (i+1)t_{i+1}$ this is exactly the equation (4.28) that described the first few 'primary' flows in the pth KdV hierarchy. We can identify the fundamental Landau-Ginzburg field x with the classical momentum $y = D$, and the superpotential $W(x)$ with the classical version of the Lax operator $L(y)$. In this way the operators are related by

$$\mathcal{O}_k = k \cdot \phi_{k-1} = dW_+^{k/p}. \tag{6.67}$$

Some more work is needed to show that also the string partition function and the τ-function coincide. In this way we have recovered a crucial part of the quantum field theory underlying the structures of §§2–4. The twisted $N = 2$ minimal models where first proposed as the relevant quantum field theories for the $(p, 1)$ string theory by Keke Li [16]. As explained in [18,19] these minimal models naturally give rise to the top Chern classes that appeared in the intersection numbers defined in §2.4.

In case the reader is wondering: equation (6.66) is sufficient to explicitly solve the operators $\phi_i(x)$ and the superpotential $W(x)$. The result reads

$$\phi_i(x) = (-1)^i \det \begin{pmatrix} -x & 1 & 0 & \cdots & 0 \\ t_{p-2} & -x & 1 & \ddots & \vdots \\ t_{p-3} & t_{p-2} & \ddots & & 0 \\ \vdots & & \ddots & & 1 \\ t_{p-i} & \cdots & t_{p-3} & t_{p-2} & -x \end{pmatrix} \tag{6.68}$$

The superpotential is obtained by putting $\partial W/\partial x = \phi_{p-1}$ in the above equation.

7. TOPOLOGICAL STRING THEORY

In this final section we will start our discussion of topological gravity and topological string theory. Through these vastly more complicated quantum field theories we will be able to make contact with the intersection theory and integrable hierarchies that featured in the first few sections.

There exists a general recipe to associate to a given moduli space a topological quantum field theory whose correlation functions reduce to intersection numbers of cohomology classes of the moduli space [62]. In the case of the moduli space of Riemann surfaces this field theory is known as topological gravity [12]. The fundamental multiplet consists of the spin connection ω_μ, its anti-commuting partner ψ_μ, and a bosonic scalar field ϕ. The BRST transformations read

$$\delta\omega_\mu = \psi_\mu, \qquad \delta\psi_\mu = \partial_\mu\phi, \qquad \delta\phi = 0. \tag{7.1}$$

The cohomology classes σ_n correspond to the physical observables

$$\sigma_n = \phi^n. \tag{7.2}$$

The relation with intersection theory is in this formulation conceptually most clear. However, it does not lead easily to explicit calculations. Therefore we choose here a slightly more complicated, but equivalent, representation due to Erik and Herman Verlinde [14,30]. We will refer to this theory in all its generality as *topological string theory*.

As in conventional string theory, topological strings will consist of a matter theory coupled to Liouville and ghost fields. The matter sector will be a topological CFT as was studied in the previous section. We have seen that the existence of the nilpotent Q-charge simplified things enormously. Our aim will be to introduce a dynamical metric, or, after gauge fixing, a Liouville-like field and ghosts, while preserving the Q-symmetry at all stages. The fundamental symmetries, generated by the stress-tensor $T_{\mu\nu}$ and its partner $G_{\mu\nu}$, that we encountered in the previous section, will now be lifted to the status of gauge symmetries. In our presentation we will follow very closely [14,30]. We will explain the construction starting with our most simple example.

7.1. *Action and symmetries*

Let us reconsider the $d = 1$ topological sigma model of §6.3 whose action we now write for a general world-sheet metric $g_{\mu\nu}$ as

$$S = \int d^2z\, \sqrt{g}g^{\mu\nu}(\partial_\mu x \partial_\nu \overline{x} + \chi_\mu \partial_\nu \psi). \tag{7.3}$$

We want to couple this model to gravity, so that the metric $g_{\mu\nu}$ becomes dynamical, without ruining the topological invariance guaranteed by the Q-symmetry. Therefore, each field must have a fermionic partner. So, we introduce the metric's fermionic partner $\psi_{\mu\nu}$ together with the transformation rules

$$\delta g_{\mu\nu} = \psi_{\mu\nu}, \qquad \delta\psi_{\mu\nu} = 0. \tag{7.4}$$

Just as variations of the action with respect to the metric give the stress-energy tensor, variations with respect to $\psi_{\mu\nu}$ give its partner $G_{\mu\nu}$

$$T^{\mu\nu} = \frac{1}{\sqrt{g}}\frac{\delta S}{\delta g_{\mu\nu}}, \qquad G^{\mu\nu} = \frac{1}{\sqrt{g}}\frac{\delta S}{\delta \psi_{\mu\nu}}. \tag{7.5}$$

We already found that $G_{\mu\nu}$ was given by $\chi_\mu \partial_\nu \psi$. These functional derivatives can be integrated to get the full action describing the coupling of topological matter to topological gravity

$$S_m = \int d^2z\, \sqrt{g}g^{\mu\nu}(\partial_\mu x \partial_\nu \overline{x} + \chi_\mu \partial_\nu \psi + \psi_{\mu\rho}\chi^\rho \partial_\nu x). \tag{7.6}$$

We will work in the conformal gauge

$$g_{\mu\nu} = e^\phi \delta_{\mu\nu}, \qquad \psi_{\mu\nu} = \rho\, e^\phi \delta_{\mu\nu}, \tag{7.7}$$

which requires us to introduce the bosonic (commuting) ghosts β and γ in addition to the fermionic (anticommuting) ghosts b and c. The ghosts b and β have spin two, whereas c and γ have spin -1. Of course, the central charge of this ghost multiplet is again $c = 0$, by our familiar theme of 'bosons cancel fermions.' It is an easy exercise to write down the generators T, G, Q, and J for the ghost system, and verify that they obey the relations we have come to expect.

In terms of the conformal factors ϕ and ρ the Q-transformation of the metric reads

$$\delta\phi = \rho, \qquad \delta\rho = 0. \tag{7.8}$$

Since the total central charge vanishes, we can fix the Weyl transformations that shift these modes by the conditions $\partial\overline{\partial}\phi = \widehat{R}$ and $\partial\overline{\partial}\rho = 0$, where $\widehat{R}$ is the world-sheet curvature of a fiducial background metric $\hat{g}_{\mu\nu}$. These constraints will be implemented with the aid of Lagrange multipliers π and λ. Taking all together, the total action for the theory is

$$\begin{aligned} S &= S_m + S_{gh} + S_L, \\ S_{gh} &= \int d^2z\,(b\overline{\partial}c + \beta\overline{\partial}\gamma + \text{c.c.}), \\ S_L &= \int d^2z\,(\pi(\partial\overline{\partial}\phi - \widehat{R}) - \lambda\partial\overline{\partial}\rho). \end{aligned} \tag{7.9}$$

In order to find the physical spectrum we have to know the correct BRST charge. In fact, here we have two candidates. First, we have a topological field theory that comes naturally with a nilpotent Q charge—we have been very careful to preserve this feature. Secondly, we are dealing with a gauge theory complete with ghosts and its own BRST charge, since we gauged the symmetries generated by the stress tensor T and its fermionic partner G. This BRST charge is of the form

$$Q_{BRST} = \oint \frac{dz}{2\pi i} :c\left(T_m + T_L + \tfrac{1}{2}T_{gh}\right) + \gamma\left(G_m + G_L + \tfrac{1}{2}G_{gh}\right): \tag{7.10}$$

and is a simple generalization of the well-known Q_{BRST} of bosonic string theory. A precise analysis shows that in this case the correct BRST charge, whose cohomology determines the spectrum of the theory, is the sum of these two charges

$$Q_{total} = Q + Q_{BRST} \tag{7.11}$$

An analysis, that we will not repeat here, shows that the physical operators with respect to Q_{total} are of the form

$$\mathcal{O} = \phi_i \cdot \tilde{\sigma}_n, \tag{7.12}$$

where ϕ_i is a matter field with charge q_i and $\tilde{\sigma}_n$ is a combination of ghost and Liouville fields with charge n, a non-negative integer. These gravitational fields $\tilde{\sigma}_n$ are a new feature of topological string theory—they will of course correspond to the Mumford-Morita-Miller classes of §2.

We now must prescribe how to calculate correlation functions of these operators. As in critical string theory, we define the higher genus correlation functions so that they become volume forms on the moduli space of Riemann surfaces. These forms are then integrated to get the final correlation functions, using the Beltrami differentials $\mu_1, \ldots, \mu_{3g-3}$. The precise definition is

$$\langle \mathcal{O}_1 \ldots \mathcal{O}_n \rangle_g \equiv \int_{\mathcal{M}_g} \langle \int \mathcal{O}_1^{(2)} \ldots \int \mathcal{O}_n^{(2)} \prod_{a=1}^{3g-3} \int \mu_a G \int \overline{\mu}_a \overline{G} \rangle_g . \tag{7.13}$$

The operators are of the form $\mathcal{O}_k = \phi_{i_k} \cdot \tilde{\sigma}_{n_k}$, and satisfy the selection rule

$$\sum_k (n_k + q_{i_k} - 1) = (d-3)(1-g). \tag{7.14}$$

The total background charge on the RHS is composed of the matter contribution $d(1-g)$ and a term needed to cancel the charge of the $3g-3$ G-insertions. The correlation function (7.13) is a generalization of the one for critical string theory, with the anti-ghosts $b(z)$ are replaced by the supercurrents $G(z)$, see [30] for further comments on this analogy.

7.2. Pure Topological Gravity

We will proceed to solve topological string theory in the case of pure topological gravity. That is, the case that no matter sector is present ($d = 0$). Since we work with a cohomological theory, there are many possible representations of the operators. We will follow here [14] and choose a representation in which the correlation functions receive no contribution from the interior of moduli space.

Let us start with the definition of the operators $\tilde{\sigma}_n$

$$\tilde{\sigma}_n = 3^n e^{\frac{2}{3}(n-1)\pi} (\tilde{\gamma})^n, \qquad \text{with} \quad \tilde{\gamma} = \overline{\partial} \left(\tfrac{1}{2} \partial c + c \partial \phi - \text{c.c.} \right). \tag{7.15}$$

In order to understand this definition a bit more intuitively, introduce $Q = Q_0 + \overline{Q}_0$ and $Q_- = Q_0 - \overline{Q}_0$. Since these operators have an interpretation as Dolbeault operators ∂, $\overline{\partial}$ on the world-sheet, one can verify that

$$\tilde{\gamma} = \{Q, \{Q_-, \phi\}\} \sim \partial \overline{\partial} \phi = \widehat{R} \tag{7.16}$$

So $\tilde{\gamma}$ is roughly the curvature, and $\tilde{\sigma}_n$ corresponds to the nth power of $\widehat{R}$. A much precise analysis [14,30] shows that, in the notation of §2, the operator $\tilde{\sigma}_n$ is related to the cohomology class $\sigma_n \in H^*(\mathcal{M}_{g,s})$ by

$$\tilde{\sigma}_n = (2n+1)!! \cdot \sigma_n = \mathcal{O}_{2n+1}. \tag{7.17}$$

We note that $\tilde{\gamma}$ naively appears to decouple because it is BRST exact. However, the model is not well-defined as it stands. To define the correlation functions unambiguously, the fields must be restricted to a subspace $\mathcal{H}_0$ of the 'big' Hilbert space $\mathcal{H}'$

$$\mathcal{H}_0 \equiv \left\{ |\phi\rangle \in \mathcal{H}' \; ; \; (b_0 - \overline{b}_0)|\phi\rangle = 0 \right\}; \tag{7.18}$$

otherwise, there is an obstruction to globally define correlation function over $\mathcal{M}_{g,s}$ [81]. So the physical Hilbert space $\mathcal{H} = H_Q^*(\mathcal{H}_0)$ is composed of Q cohomology classes equivariant with respect to $b_0 - \overline{b}_0$. This insures that local coordinates can be consistently defined at the punctures. It can easily be seen that, even though $\tilde{\gamma}$ is BRST exact in $\mathcal{H}'$, it is not in $\mathcal{H}_0$ and therefore does not decouple.

As was shown in [14] the correlation functions for pure topological gravity are completely determined from the boundary of moduli space. In fact, contact terms, which occur whenever two operators collide and are quite generally of the form

$$\mathcal{O}_i(z) \cdot \mathcal{O}_j(0) \sim \mathcal{O}_k(0)\delta^2(z), \tag{7.19}$$

suffice to solve the model. To see this we must work out the contact term algebra explicitly. We will sketch the argument of [14].

Define the states $|\tilde{\sigma}_m\rangle = \tilde{\sigma}_m(0)|0\rangle$, and consider the contribution of the operator $\tilde{\sigma}_n$ when it is close to the operator $\tilde{\sigma}_m$, located at $z = 0$ in some coordinates

$$\int d^2z \cdot \tilde{\sigma}_n(z,\bar{z}) \cdot \tilde{\sigma}_m(0)|0\rangle = \int_{|q|<\epsilon} \frac{d^2q}{|q|^2} G_0 \overline{G}_0 \, q^{L_0}\bar{q}^{\overline{L}_0} \tilde{\sigma}_n(1)|\tilde{\sigma}_m\rangle. \tag{7.20}$$

Now recall that $\tilde{\sigma}_n \sim \tilde{\gamma}^n$ with $\tilde{\gamma} = \{Q, \{Q_-, \phi\}\}$. Since L_0 is Q exact, we find a contribution due to

$$\int d^2q \; \partial_q \partial_{\bar{q}} \, q^{L_0}\bar{q}^{\overline{L}_0} \phi(0)|\tilde{\sigma}_{n+m-1}\rangle, \tag{7.21}$$

where $q^{L_0}\bar{q}^{\overline{L}_0}\phi(0) \sim \log|q|^2 +$ regular, because of the anomalous covariance of the field ϕ. This clearly gives a δ-function contribution, and with the inclusion of all factors one finds in the end

$$\int d^2z \, \tilde{\sigma}_n(z,\bar{z})\tilde{\sigma}_m(0) = (2m+1)\tilde{\sigma}_{n+m-1}(0). \tag{7.22}$$

Note that this expression is not symmetric in n and m,

$$\int [\tilde{\sigma}_n, \tilde{\sigma}_m] = 2(m-n)\tilde{\sigma}_{m+n-1}, \tag{7.23}$$

where we see the appearance of a Virasoro algebra.

Let us now consider a general correlator of the form

$$\langle \tilde{\sigma}_n \tilde{\sigma}_{k_1} \ldots \tilde{\sigma}_{k_s} \rangle_g. \tag{7.24}$$

We want to study the contributions to this amplitude that we obtain if we integrate the two-form $\tilde{\sigma}_n$ over the Riemann surface. In general there are three different contributions: the contact terms that we examined above, factorization terms, and, finally, bulk terms. The factorization terms arise when the surface itself degenerates—the cases *(ii)* and *(iii)* of *fig.* 2. They may be expressed as sums over factorized correlation functions with $\tilde{\sigma}_i$ insertions, and are, by charge conservation, of the general form

$$\sum_{i+j=n-2} A_{ij}^{(n)} \langle \tilde{\sigma}_i \tilde{\sigma}_{k_1} \ldots \tilde{\sigma}_{k_r} \rangle_{g'} \langle \tilde{\sigma}_j \tilde{\sigma}_{k_{r+1}} \ldots \tilde{\sigma}_{k_s} \rangle_{g-g'} \tag{7.25}$$

or

$$\sum_{i+j=n-2} B_{ij}^{(n)} \langle \tilde{\sigma}_i \tilde{\sigma}_j \tilde{\sigma}_{k_1} \ldots \tilde{\sigma}_{k_s} \rangle_{g-1} \tag{7.26}$$

with some coefficients $A_{ij}^{(n)}, B_{ij}^{(n)}$. The bulk terms describe the contributions from the interior of moduli space. Quite remarkably, one can show that these contributions are absent due to the asymmetry in the contact terms. Also the factorization terms are completely determined by consistency: $A_{ij}^{(n)} = B_{ij}^{(n)} = \frac{1}{2}$. So, in the end we are left with the following recursion relations for the correlation functions of topological gravity.

$$\begin{aligned}\Big\langle \tilde{\sigma}_n \prod_{k\in S} \tilde{\sigma}_k \Big\rangle_g &= \sum_{k\in S} (2k+1) \Big\langle \tilde{\sigma}_{n+k-1} \prod_{k'\neq k} \tilde{\sigma}_{k'} \Big\rangle_g + \sum_{i+j=n-2} \tfrac{1}{2} \Big\langle \tilde{\sigma}_i \tilde{\sigma}_j \prod \tilde{\sigma}_k \Big\rangle_{g-1} \\ &+ \sum_{\substack{g';\ S=X\cup Y \\ i+j=n-2}} \tfrac{1}{2} \Big\langle \tilde{\sigma}_i \prod_{k\in X} \tilde{\sigma}_k \Big\rangle_{g'} \Big\langle \tilde{\sigma}_j \prod_{k\in Y} \tilde{\sigma}_k \Big\rangle_{g-g'}. \end{aligned} \tag{7.27}$$

Here $S = \{k_1, \ldots, k_s\}$, and the last summation is over all partitions of S into two subsets X, Y. The three terms on the RHS are associated to the three inequivalent degenerations of *fig.* 2: contact terms and factorization terms of a homology cycle or dividing cycle. The contributions always come from the boundary of moduli space, so the recursion relation lowers the genus g and/or the number of operators s, and ultimately gives factors of the three-point function on the sphere,

$$\langle \tilde{\sigma}_0^3 \rangle_0 = \langle \sigma_0^3 \rangle_0 = 1, \tag{7.28}$$

or the one-point function on the torus

$$\langle \tilde{\sigma}_1 \rangle_1 = 3 \langle \sigma_1 \rangle_1 = \frac{1}{8}. \tag{7.29}$$

Therefore the recursion relations suffice to calculate any intersection number. In the case $n = 0$ the relation (7.27) is nothing but the puncture equation (2.13); the case $n = 1$ is a linear combination of the dilaton equation (2.15) and the ghost charge conservation (2.8). Note that in these two cases the factorization terms are absent.

Although these relations are manifestly non-linear in terms of the free energy F, they turn out to be linear when written in terms of the partition function $\tau = e^F$, due to the wonderful relation

$$\tau^{-1} \partial^2 \tau = \partial^2 F + (\partial F)^2. \tag{7.30}$$

Indeed, consider the partition function

$$\tau(\tilde{t}) = \exp \sum_{g=0}^{\infty} \Big\langle \exp \sum_n \tilde{t}_n \tilde{\sigma}_n \Big\rangle_g \tag{7.31}$$

where the coupling coefficients $\tilde{t}_n$ are related to the KdV times by $\tilde{t}_n = t_{2n+1}$. Now, if we recall that a derivative $\partial\tau/\partial\tilde{t}_n$ corresponds to an insertion of $\tilde{\sigma}_n$, whereas a multiplication $\tilde{t}_n \cdot \tau$ removes the same operator from the generating function, the recursion relations (7.27) can be expressed as linear, homogeneous differential equations for the τ-function

$$L_n \cdot \tau = 0, \qquad (n \geq -1). \tag{7.32}$$

where L_n denote the differential operators

$$
\begin{aligned}
L_{-1} &= -\tfrac{1}{2}\frac{\partial}{\partial \tilde{t}_0} + \sum_{k=1}^{\infty}(k+\tfrac{1}{2})\tilde{t}_k\frac{\partial}{\partial \tilde{t}_{k-1}} + \tfrac{1}{4}\tilde{t}_0^2 \\
L_0 &= -\tfrac{1}{2}\frac{\partial}{\partial \tilde{t}_1} + \sum_{k=0}^{\infty}(k+\tfrac{1}{2})\tilde{t}_k\frac{\partial}{\partial \tilde{t}_k} + \tfrac{1}{16} \quad , \\
L_n &= -\tfrac{1}{2}\frac{\partial}{\partial \tilde{t}_{n-1}} + \sum_{k=0}^{\infty}(k+\tfrac{1}{2})\tilde{t}_k\frac{\partial}{\partial \tilde{t}_{k+n}} + \tfrac{1}{4}\sum_{i=1}^{n}\frac{\partial^2}{\partial \tilde{t}_{i-1}\partial \tilde{t}_{n-i}}.
\end{aligned}
\tag{7.33}
$$

These are exactly the equations that according to §4.4 characterize the τ-function of the KdV hierarchy that we studied in such great detail before. So, the above manipulations amount to a sketch of a quantum field theory proof of Witten's original conjecture.

In the more general case of a topological minimal model coupled to gravity, these Virasoro constraints are replaced by the W-constraints

$$W_n^{(s)} \cdot \tau = 0, \qquad s = 2, \ldots, p, \ n \geq 1 - s, \tag{7.34}$$

The differential operators $W_n^{(s)}$ form a W-algebra associated to the Lie group G that labeled the minimal model. They are spin s currents that generalize the Virasoro algebra. We have one W-generator for each Casimir in the group G. The constraints now obtain an interpretation as multiple contact and factorization terms [24,16]. A full discussion of these relations is however completely outside the scope of these lectures.

Acknowledgements

I wish to thank the organizers of this wonderful school for inviting me to present these lectures and for creating such a stimulating and enjoyable atmosphere. Needless to say I benefitted greatly from countless insightful discussion with my collaborators Erik Verlinde, Herman Verlinde, and Edward Witten. I further thank Rob Rudd for his assistance in preparing a preliminary version of these notes. This research was supported by the W.M. Keck Foundation.

References

[1] V. Kazakov, Phys. Lett. **159B** (1985) 303; F. David, Nucl. Phys. **B257** (1985) 45; V. Kazakov, I. Kostov, and A. Migdal, Phys. Lett. **157B** (1985) 295; J. Fröhlich, *The statistical mechanics of surfaces,* in *Applications of Field Theory to Statistical Mechanics*, L. Garrido ed. (Springer, 1985).

[2] F. David, Phys. Lett. **159B** (1985) 303; D. Boulatov, V. Kazakov, and A. Migdal, Nucl. Phys. **B275 [FS117]** (1986) 543; A. Billoire and F. David, Phys. Lett. **186B** (1986) 279; J. Jurkievic, A. Krzywicki, and B. Peterson, Phys. Lett. **186B** (1986) 273; J. Ambjorn, B. Durhuus, J. Fröhlich, and P. Orland, Nucl. Phys. **B270 [FS16]** (1986) 457.

[3] V. Kazakov, Mod. Phys. Lett. **A4** (1989) 2125.

[4] E. Brézin and V. Kazakov, *Exactly solvable field theories of closed strings,* Phys. Lett. **B236** (1990) 144.
M. Douglas and S. Shenker, *Strings in less than one dimension,* Nucl. Phys. **B335** (1990) 635.
D.J. Gross and A. Migdal, *Nonperturbative two dimensional quantum gravity,* Phys. Rev. Lett. **64** (1990) 127.

[5] D.J. Gross and A. Migdal, *A nonperturbative treatment of two dimensional quantum gravity,* Nucl. Phys. **B340** (1990) 333.
T. Banks, M. Douglas, N. Seiberg, and S. Shenker, *Microscopic and macroscopic loops in nonperturbative two dimensional gravity,* Phys. Lett. **238B** (1990) 279.

[6] E. Brézin, M. Douglas, V. Kazakov, and S. Shenker, *The Ising model coupled to 2d gravity: a nonperturbative analysis,* Phys. Lett. **237B** (1990) 43
D.J. Gross and A. Migdal, *Nonperturbative solution of the Ising model on a random surface,* Phys. Rev. Lett. **64** (1990) 717.
C. Crnković, P. Ginsparg, and G. Moore, *The Ising model, the Yang-Lee edge singularity, and 2d quantum gravity,* Phys. Lett. **237B** (1990) 196.

[7] M. Douglas, *Strings in less than one dimension and the generalized KdV hierarchies,* Phys. Lett. **238B** (1990) 176.

[8] E. Brézin, V. Kazakov, and Al.B. Zamolodchikov, *Scaling violation in a field theory of closed strings in one physical dimension,* Nucl. Phys. **B338** (1990).
D.J. Gross and N. Miljković, *A nonperturbative solution of $D = 1$ string theory,* Phys. Lett. **238B** (1990) 217.
P. Ginsparg and J. Zinn-Justin, *2-d Gravity and 1-d matter,* Phys. Lett. **240B** (1990) 333.
D.J. Gross and I.R. Klebanov, *One-dimensional string theory on a circle,* Nucl. Phys. **B344** (1990) 475.

[9] *Random Surfaces, Quantum Gravity and Strings,* proceedings of the 1990 Cargèse workshop, O. Alvarez, E. Marinari, and P. Windey eds. (plenum Press, 1991).
D.J. Gross, *Nonperturbative string theory,* in the Proceedings of the *International Colloquium on Modern Quantum Field Theory,* Bombay, 1990.
I.R. Klebanov, *String theory in two dimensions,* lectures at 1991 Trieste Spring School on *String Theory and Quantum Gravity,* Princeton preprint PUPT-1271 (July, 1991).
P. Ginsparg, *Matrix models of 2d gravity,* lectures at 1991 Trieste Summer School, Los Alamos preprint (December, 1991).

[10] E. Witten, *On the topological phase of two dimensional gravity,* Nucl. Phys. **B340** (1990) 281.

[11] E. Witten, *Two dimensional gravity and intersection theory on moduli space,* Surveys In Diff. Geom. **1** (1991) 243.

[12] J. Labastida, M. Pernici, and E. Witten, *Topological gravity in two dimensions,* Nucl. Phys. **B310** (1988) 611.
D. Montano and J. Sonnenschein, *The topology of moduli space and quantum field theory,* Nucl. Phys. **B313** (1989) 258; Nucl. Phys. **324** (1990) 348.
R. Myers and V. Periwal, *Topological gravity and moduli space,* Nucl. Phys. **333** (1990) 536.

[13] J. Distler, *2d quantum gravity, topological field theory and the multicritical matrix models,* Nucl. Phys. **B342** (1990) 523.

[14] E. Verlinde and H. Verlinde, *A solution of two-dimensional topological quantum gravity,* Nucl. Phys. **B348** (1991) 457.

[15] R. Dijkgraaf and E. Witten, *Mean field theory, topological field theory, and multi-matrix models,* Nucl. Phys. **B342** (1990) 486.

[16] K. Li, *Topological gravity with minimal matter; Recursion relations in topological gravity with minimal matter,* Nucl. Phys. **354** (1991) 711, 725.

[17] R. Dijkgraaf, E. Verlinde, and H. Verlinde, *Topological strings in $d < 1$,* Nucl. Phys. **B352** (1991) 59.

[18] E. Witten, *The N matrix model and gauged WZW models,* IAS preprint IASSNS-HEP-91/26 (June, 1991).

[19] E. Witten, *Algebraic geometry associated with matrix models of two dimensional gravity,* IAS preprint IASSNS-HEP-91/74 (October, 1991).

[20] M. Kontsevich, *Intersection theory on the moduli space of curves,* Funk. Anal. i Pril. **25** (1991) 50 (in Russian).

[21] M. Kontsevich, *Intersection theory on the moduli space of curves and the matrix Airy function,* 30 Arbeitstagung Bonn, Max-Planck-Institut preprint MPI/91-47.

[22] M. Kontsevich, *Intersection theory on the moduli space of curves and the matrix Airy function,* Max-Planck-Institut preprint MPI/91-77.

[23] E. Witten, Lectures at IAS, Fall 1991.

[24] R. Dijkgraaf, E. Verlinde, and H. Verlinde, *Loop equations and Virasoro constraints in non-perturbative 2d quantum gravity,* Nucl. Phys. **B348** (1991) 435.

[25] M. Fukuma, H. Kawai, and R. Nakayama, *Continuum Schwinger-Dyson equations and universal structures in two-dimensional quantum gravity,* Int. J. Mod. Phys. **A6** (1991) 1385.

[26] B.H. Lian and G.J. Zuckerman, *New selection rules and physical states in 2d gravity: conformal gauge,* Phys. Lett. **254** (1991) 417.

[27] S. Kharchev, A. Marschakov, A. Mironov, A. Morozov, A. Zabrodin, *Unification of all string models with* $c<1$, Lebedev Institute preprint FIAN/TD-9/91.

[28] C. Itzykson and J.-B. Zuber, *Combinatorics of the Modular Group II: The Kontsevich Integrals,* Saclay preprint SPhT/92-001, submitted to Int. J. Mod. Phys.

[29] G. Segal and G. Wilson, *Loop groups and equations of KdV type,* Publ. Math. I.H.E.S. **61** (1985) 1.

[30] R. Dijkgraaf, E. Verlinde, and H. Verlinde, *Notes on topological string theory and 2d quantum gravity,* in *String Theory and Quantum Gravity*, Proceedings of the Trieste Spring School 1990, M. Green *et al.* eds. (World-Scientific, 1991).

[31] P. Deligne and M. Mumford, *The irreducibility of the space of curves of given genus,* Publ. I.H.E.S. **45** (1969) 75.

[32] D. Mumford, *Towards an enumerative geometry of the moduli space of curves,* in *Arithmetic and Geometry,* M. Artin and J. Tate eds. (Birkhäuser, Basel, 1983).

[33] S. Morita, *Characteristic classes of surface bundles,* Invent. Math. **90** (1987) 551.

[34] E. Miller, *The homology of the mapping class group,* J. Diff. Geom. **24** (1986) 1.

[35] P. Deligne, Letter to E. Witten (October, 1989).

[36] J. Horne, *Intersection theory and two dimensional gravity at genus 3 and 4,* Mod. Phys. Lett. **A5** (1990) 2127.

[37] C. Faber, *Chow rings of moduli spaces of curves: I and II,* Ann. Math. **132** (1990) 331; 421.

[38] K. Strebel, *Quadratic Differentials* (Springer-Verlag, 1984).

[39] R.C. Penner, *The decorated Teichmüller space of punctured surfaces,* Commun. Math. Phys. **113** (1987) 299; *Perturbative series and the moduli space of Riemann surfaces,* J. Diff. Geom. **27** (1988) 35.

[40] G. 't Hooft, Nucl. Phys. **B72** (1974) 461.

[41] J. Harer and D. Zagier, *The Euler characteristic of the moduli space of curves,* Invent. Math. **185** (1986) 457.

[42] I.M. Gelfand and L.A. Dikii, *The resolvent and Hamiltonian systems,* Funct. Anal. Appl. **11:2** (1977) 93.

[43] V.E. Zakharov, Funk. Anal. i Pril. **14** (1980) 89.

[44] I. Krichever, *Topological minimal models and dispersionless Lax equation,* Turin preprint, Commun. Math. Phys. to appear (1991); *Topological minimal models and soliton equations,* Landau Institute preprint (1991).

[45] M. Sato, *Soliton equations as dynamical systems on infinite dimensional Grassmann manifolds,* RIMS Kokyuroku **439** (1981) 30.
M. Sato and Y. Sato, *Soliton equations as dynamical systems in an infinite dimensional Grassmannian,* in *Nonlinear Partial Differential Equations in Applied Sciences,* P.Lax, H. Fujita, and G. Strang eds. (North-Holland, Amsterdam, 1982).

[46] E. Date, M. Jimbo, M. Kashiwara, T. Miwa, *Transformation groups for soliton equations,* RIMS Symp. *Nonlinear Integrable Systems—Classical Theory and Quantum Theory* (World Scientific, Singapore, 1983).

[47] T. Miwa, Proc. Jap. Acd. **58A** (1982) 9.

[48] G. Moore, *Geometry of the string equations,* Commun. Math. Phys. 133 (1990) 261; *Matrix models of 2d gravity and isomonodromic deformation,* in *Random Surfaces, Quantum Gravity and Strings,* Proceedings of the 1990 Cargèse workshop, O. Alvarez, E. Marinari, and P. Windey eds. (Plenum Press, 1991).

[49] M. Adler and P. van Moerbeke, *The W_p-gravity version of the Witten-Kontsevich model,* Brandeis preprint (September, 1991).

[50] Harish-Chandra, *Differential operators on a semisimple Lie algebra,* Am. J. Math. **79** (1987) 87.

[51] M.L. Mehta, *A method of integration over matrix variables,* Commun. Math. Phys. **79** (1981) 327.

[52] C. Itzykson and J.-B. Zuber, *The planar approximation II,* J. Math. Phys. **21** (1980) 411.

[53] R.F. Picken, *The Duistermaat-Heckman integration formula on flag manifolds,* J. Math. Phys. **B31** (1990) 616.

[54] A. Miranov, A. Morozov, Phys. Lett. **B252** (1990) 47. Yu. Makeenko, A. Marshakov, A. Mironov, A. Morozov, Nucl. Phys. **356** (1991) 574. H.S. La, Commun. Math. Phys. **140** (1991) 569. A.M. Semikhatov, Lebedev Institute preprints.

[55] J. Goeree, *W constraints in 2-d quantum gravity,* Nucl. Phys. **B358** (1991) 73.

[56] M. Fukuma, H. Kawai, and R. Nakayama, *Infinite dimensional Grassmannian structure of two-dimensional quantum gravity,* University of Tokyo preprint UT-572 (November, 1990).

[57] E. Witten, *On the Kontsevich model and other models of two dimensional gravity,* IAS preprint IASSNS-HEP-91/24 (June, 1991).

[58] D.J. Gross and M. Newman, *Unitary and Hermitian matrices in an external field II: the Kontsevich model and continuum Virasoro constraints,* Princeton preprint (December, 1991).

[59] A. Marschakov, A. Mironov, and A. Morozov, *On equivalence of topological and quantum 2d gravity,* Lebedev Institute preprint FIAN/TD/04-91.

[60] E. Witten, *Topological quantum field theory,* Commun. Math. Phys. **117** (1988) 353.

[61] E. Witten, *Topological sigma models,* Commun. Math. Phys. **118** (1988) 411.

[62] E. Witten, *Introduction to cohomological field theories,* Int. J. Mod. Phys. **A6** (1991) 2775.

[63] P. van Baal, *An introduction to topological Yang-Mills theory,* Acta Phys. Polon. **B21** (1990) 73.

[64] D. Birmingham, M. Blau, M. Rakowski, and G. Thompson, *Topological field theory,* preprint CERN-TH 6045/91, to be published in Physics Reports.

[65] M.F. Atiyah, *Topological quantum field theories,* Publ. Math. I.H.E.S. **68** (1988) 175.

[66] R. Dijkgraaf, *A geometrical approach to two-dimensional conformal field theory*, Ph.D. Thesis (Utrecht, 1989).

[67] S. Shenker, *The strength of nonperturbative effects in string theory*, in Random Surfaces, Quantum Gravity and Strings, Proceedings of the 1990 Cargèse workshop, O. Alvarez, E. Marinari, and P. Windey eds. (Plenum Press, 1991).

[68] T. Eguchi and S.-K. Yang, *$N = 2$ superconformal models as topological field theories*, Mod. Phys. Lett **A5** (1990) 1693.

[69] W. Lerche, C. Vafa, and N.P. Warner, *Chiral rings in $N = 2$ superconformal theories*, Nucl. Phys **B324** (1989) 427.

[70] B. Greene and R. Plesser, *Duality in Calabi-Yau moduli space*, Nucl. Phys. **B338** (1990) 15.

[71] C. Vafa, *Topological Landau-Ginzburg models*, Mod. Phys. Lett. **A6** (1990) 337.

[72] K. Ito, *Topological phase of $N = 2$ superconformal field theory and topological Landau-Ginzburg field theories*, Phys. Lett. **250** (1990) 91.

[73] E. Martinec, *Algebraic geometry and effective lagrangians*, Phys. Lett. **217B** (1989) 431; *Criticality, catastrophe and compactifications*, in V.G. Knizhnik memorial volume, 1989.

[74] C. Vafa and N. Warner, *Catastrophes and the classification of conformal field theories*, Phys. Lett. **218B** (1989) 51.

[75] A. Cappelli, C. Itzykson, and J.-B. Zuber, *The ADE classification of minimal and $A_1^{(1)}$ conformal theories*, Commun. Math. Phys. **113** (1987) 1.

[76] K. Saito, Publ. RIMS **19** (1983) 1231.
M. Saito, Ann. Inst. Fourier **39:1** (1989) 27.

[77] B. Blok and A. Varchenko, *Topological conformal field theories and the flat coordinates*, IAS preprint IASSNS-HEP-91/5 (January, 1991).

[78] W. Lerche, D.J. Smit, and N.P. Warner, *Differential equations for periods and flat coordinates in two-dimensional matter theories*, Preprint UCB-PTH-91-39 (July, 1991).

[79] E. Verlinde and N.P. Warner, *Topological Landau-Ginzburg matter at $c = 3$*, IAS preprint IASSNS-HEP-91-16 (March, 1991).

[80] D. Kutasov, *Geometry on the space of conformal field theories and contact terms*, Phys. Lett **B220** (1989) 153.

[81] J. Distler and P. Nelson, *Topological couplings and contact terms in 2d field theory*, Commun. Math. Phys. **138** (1991) 273.

QUANTUM SYMMETRY IN CONFORMAL FIELD THEORY BY HAMILTONIAN METHODS

L.D. Faddeev

Steklov Mathematical Insitute, St. Petersburg and
Research Institute in Theoretical Physics
Helsinki University, Helsinki

INTRODUCTION

I shall discuss the generalization of the group symmetry which was found recently in two-dimensional field theory, in particular in the conformal field theory. My approach will be dual to that of Professor Mack. I shall consider a particular example and gradually develop general insight by studying it. This example will be the $WZNW$ model for the group $SU(2)$, so that the pair – a compact group and an integer mentioned by Professor Bott will define the model.

I shall use the conservative language of Hamiltonian field theory. So I shall not refer to the basic notions and tools of the bootstrap approach to CFT pioneered by Belavin, Polyakov and Zamolodchikov [1], such as primary fields, vertex operators, operator expansions, etc. Also I shall not mix operators and states.

Instead, I shall consider only the multiples of fields, directly entering into the lagrangian, and investigate their operator content by means of quantization, based on the experience I got together with my collaborators in St. Petersburg working on the integrable magnetic chains, (for review see [2]).

1. CLASSICAL FORMULATION

First I describe the formulation of the WZNW model as a classical field theory. The space time M is a cylinder $S^1 \times R^1$ with coordinates x, t, $\quad 0 \leq x \leq 2\pi$, $\quad -\infty < t < \infty$ with Minkowskian metric. The field $g(x,t)$ is a unitary unimodular 2×2 matrix

$$g : M \longrightarrow su(2)\,.$$

The action functional looks as follows [3]

$$A = -\frac{1}{8\gamma}\int \mathrm{tr}(\partial g g^{-1})^2 dx dt + \frac{1}{12\gamma}\int (d^{-1}\mathrm{tr}(dgg^{-1})^3)_* \tag{1}$$

where the pullback of the integral of the closed but not exact 3-form $\mathrm{tr}(dg\, g^{-1})^3$ is used to define the WZ term [4] in action. This nonunivalent action (in the terminology of

New Symmetry Principles in Quantum Field Theory, Edited by
J. Fröhlich et al., Plenum Press, New York, 1992

Novikov [5]) can be used in quantum theory only if the coupling constant γ is quantized, namely

$$l = \frac{\pi}{\gamma}$$

must be an integer.

The phase space of the model is spaned by the Cauchy data $g(x), J_0(x)$ for fixed time, where we use the notation J_μ for the right invariant currents

$$J_\mu = \partial_\mu g g^{-1} .$$

To specify their Poisson brackets it is convenient to use the first order formalism when g and J_0 are considered as independent variables. The action, equivalent to (1) can be taken in the form

$$A = -\frac{1}{4\gamma} \int \mathrm{tr} \left[\partial_0 g g^{-1} J_0 - \frac{1}{2} \left(J_0^2 + J_1^2 \right)_{-1} \right] dx dt + \mathrm{WZ}$$

where WZ is a WZ-term, i.e. the second term in the RHS of (1). It is considered as a functional of g alone as it contains $\partial_0 g$ only linearly. Furthermore,

$$J_1 = \partial_x g g^{-1} .$$

The variation of the canonical 1-forms in A leads to the sympletic form

$$\Omega = \frac{1}{4\gamma} \int \mathrm{tr} \left(dgg^{-1} \wedge dJ_0 + J_1 dgg^{-1} \wedge dgg^{-1} \right) dx dt .$$

In derivation one is to use the property of WZ

$$\delta \mathrm{WZ} = \frac{1}{4\gamma} \int \mathrm{tr}(J_1 \delta g g^{-1} \partial_0 g g^{-1}) dx dt$$

which follows from

$$\delta \mathrm{tr}(dgg^{-1})^3 = 3 \ d \ \mathrm{tr} \left(\delta g g^{-1} (dgg^{-1})^2 \right) .$$

Now we can find the Poisson brackets by inverting Ω. We find

$$\begin{aligned}
&\{g^1(x), g^2(y)\} = 0 \\
&\{J_0^1(x), g^2(y)\} = -2\gamma C g^2(y) \delta(x-y) \\
&\{J_0^1(x) J_0^2(y)\} = -\gamma \left[J_0^1(x) - J_1^1(x) - J_0^2(x) + J_1^2(x), C \right] \ \delta(x-y) .
\end{aligned}$$

Here $\{A^1(x), B^2(y)\}$ stands for the 4×4 matrix of all Poisson brackets of 2×2 matrices A and B, arranged in the same fashion, as in the product of matrices

$$A^1 = A \otimes I$$

and

$$B^2 = I \otimes B ;$$

furthermore C is a constant 4×4 matrix given by

$$C = \sum_a \sigma^a \otimes \sigma^a ,$$

where σ^a , $a = 1,2,3$ are Pauli matrices. These notations and their quantum counterpart will be used throughout all my lectures.

Now it is easy to see that the combination

$$L = \frac{1}{2}(J_0 + J_1)$$

and

$$R = \frac{1}{2}g^{-1}(J_0 - J_1)g$$

have Poisson brackets

$$\begin{aligned} \{L^1(x), L^2(y)\} &= \gamma/2[C, L^1(x) - L^2(y)]\delta(x-y) + \gamma C\delta'(x-y) ; \\ \{R^1(x), R^2(y)\} &= \gamma/2[C, R^1(x) - R^2(y)]\delta(x-y) - \gamma C\delta'(x-y) \end{aligned}$$

and

$$\{L^1(x), R^2(y)\} = 0 .$$

Thus we see that the phase space of the WZNW model contains two exemplars of the Kac-Moody algebra; the coupling constant γ defines the corresponding central charge.

The dynamics is trivial in terms of L and R: the equations of motion are linear

$$\partial_t L - \partial_x l = 0 ; \quad \partial_t R + \partial_x R = 0 ;$$

so only the reformulation of dynamics in terms of the original field $g(x,t)$ is needed. However the most interesting things happen in this passage. To get g we are to solve the differential equation

$$\partial_x g = Lg - gR .$$

The general solution is given by

$$g(x) = u_0(x) K v_0(x) ,$$

where

$$\begin{aligned} u_0(x) &= \overleftarrow{\exp}\left\{\int_0^x L(y)dy\right\} ; \\ v_0(x) &= \overrightarrow{\exp}\left\{-\int_0^x R(y)dy\right\} \end{aligned}$$

are solutions of equations

$$\partial_x u_0 = Lu_0 ; \quad \partial_x v_0 = -v_0 R$$

and

$$K = g(0)$$

plays the role of the integration constant. However it cannot be arbitrary. Indeed, introducing the monodromies M_L and M_R for u_0 and v_0

$$\begin{aligned} u_0(x + 2\pi) &= u_0(x) M_L ; \\ v_0(x + 2\pi) &= M_R v_0(x) , \end{aligned}$$

or

$$M_L = u_0(2\pi) ; \quad M_R = v_0(2\pi)$$

we have from the periodicity of $g(x)$

$$K = M_L K M_R$$

or

$$M_R^{-1} = K^{-1} M_L K \ .$$

Thus M_L and M_R could be diagonalized simultaneously

$$\begin{aligned} M_L &= Z_L D Z_L^{-1} \ , \\ M_R &= Z_R^{-1} D^{-1} Z_R \ , \end{aligned}$$

where

$$D = \begin{pmatrix} e^{ip} & 0 \\ 0 & e^{-ip} \end{pmatrix}$$

and p must be confined to the half-period $0 \le p \le \pi$ to kill the Weyl invariance. Now we have the relation on K

$$[Z_L^{-1} K Z_R^{-1}, D] = 0 \ ,$$

so that

$$K = Z_L Q Z_R$$

and Q is diagonal

$$Q = \begin{pmatrix} e^{iq} & 0 \\ 0 & e^{-iq} \end{pmatrix} \ , \quad 0 \le q \le 2\pi \ .$$

The coordinate q is the only one which enters $g(x)$, but is absent in currents L and R. It is a cyclic coordinate. Its role is not evident until both chiralities L and R, or truly canonical fields g and J_0 are considered.

This concludes the discussion of the maps

$$(g(x), J_0(x)) \leftrightarrow (L(x), R(x), q) \ .$$

Alternatively we can use new chiral fields

$$\begin{aligned} u_F(x) &= u_0(x) Z_L Q \ , \\ u_F(x) &= Q Z_R v_0(x) \ . \end{aligned}$$

They have diagonal monodromy

$$\begin{aligned} u_F(x + 2\pi) &= u_F(x) D \ , \\ u_F(x + 2\pi) &= D^{-1} v_F(x) \end{aligned}$$

and the local field $g(x)$ is expressed through them as follows:

$$g(x) = u_F(x) Q^{-1} v_F(x) \ .$$

We have the algebra of Poisson brackets for $L(x)$ and $R(x)$. To understand the underlying structure better one has to get the corresponding brackets for $u_F(x)$ and $v_F(x)$. To get an idea how they could look, it is instructive to consider first a more simple model.

2. CLASSICAL AND DEFORMED TOP

We shall forget about the field-theoretic WZNW-model for the time being and consider a mechanical model, which plays the role of that for the zero models of WZNW, namely the classical isotropic top.

The phase space of the top is T^*G. The natural coordinates ω and a_L (or $a_R = \omega^{-1}a_L\omega$) have values in the group and Lie algebra, correspondingly. The lagrangian

$$\begin{aligned} l &= \mathrm{tr}(a_L\dot{\omega}\omega^{-1} - \frac{1}{2}\omega^2) = \\ &= \mathrm{tr}(a_R\dot{\omega}^{-1}\dot{\omega} - \frac{1}{2}\omega^2) \end{aligned}$$

leads to the equations of motion

$$\dot{a}_L = 0 \; ; \quad \dot{\omega} = a_L\omega$$

and Poisson brackets

$$\{\omega^1, \omega^2\} = 0 \; ; \tag{2.a}$$

$$\{a_L^1, \omega^2\} = -C\omega^2 \; ; \tag{2.b}$$

$$\{a_L^1, a_L^2\} = -\frac{1}{2}[C, a_L^1 - a_L^2] \, , \tag{2.c}$$

or

$$\begin{aligned} \{a_R^1, \omega^2\} &= -\omega^2 C \; ; \\ \{a_R^1, a_R^2\} &= -\frac{1}{2}[C, a_R^1 - a_R^2] \, , \end{aligned}$$

and

$$\{a_L^1, a_R^2\} = 0 \, .$$

Introducing

$$\begin{aligned} a_L &= u_0 \begin{pmatrix} p & 0 \\ 0 & -p \end{pmatrix} u_0^{-1} \; ; \\ a_R &= v_0^{-1} \begin{pmatrix} p & 0 \\ 0 & -p \end{pmatrix} v_0 \, , \end{aligned}$$

where $u_0, v_0 \in G/H$, H being the Cartan subgroup of diagonal matrices, we have

$$\omega = u_0 Q v_0 \, ,$$

where Q is diagonal

$$Q = \begin{pmatrix} e^{iq} & 0 \\ 0 & e^{-iq} \end{pmatrix} .$$

The analogy with the formulas above is striking. We are now looking for the Poission brackets for u_0 and v_0. Let us parametrize them by means of Euler angles

$$\begin{aligned} u_0 &= e^{i\alpha\sigma_3} e^{i\beta\sigma_2} \; ; \\ v_0 &= e^{i\gamma\sigma_2} e^{i\delta\sigma_3} \, . \end{aligned}$$

The absence of a diagonal part on the right of u_0 and left of v_0 corresponds to the choice of the representatives in G/H. The canonical 1-form looks as follows

$$\mathrm{tr}(a_L d\omega\omega^{-1}) = p dq + p\cos 2\beta d\alpha + p\cos 2\gamma d\delta \, ,$$

so that we can calculate all the brackets we want. In particular, as was shown in [6], for the matrices

$$u = u_0 Q \,, \quad v = Q v_0$$

we get the quadratic algebra

$$\begin{aligned}
\{u^1, u^2\} &= u^1 u^2 r_0(p) \,; && (3.a)\\
\{v^1, v^2\} &= -r_0(p) v^1 v^2 \,; && (3.b)\\
\{u_0^1, v_0^2\} &= 0 \,; && (3.c)\\
\{u^1, D^2\} &= u^1 D^2 s \,; && (3.d)\\
\{v^1, D^2\} &= s D^2 v^1 \,; && (3.e)
\end{aligned}$$

where

$$r_0(p) = \frac{i}{p} \begin{pmatrix} 0 & 0 & 0 & 0 \\ 0 & 0 & 1 & 0 \\ 0 & -1 & 0 & 0 \\ 0 & 0 & 0 & 0 \end{pmatrix}$$

and

$$s = \mathrm{diag}\frac{1}{2}(1, -1, -1, 1) \,.$$

Now we turn to the quantum version. Moreover we shall consider the t-deformation of the classical top in the general spirit of quantum groups. We shall need such a deformation for the interpretation of some formulas for the WZNW model. At the moment we shall introduce it without discussion.

The existing experience of work with the quadratic Poisson algebras gives a natural proposal for t-deformation of the formulas (3). We are compelled to use the letter t instead of the usual q which is already used for the coordinate conjugate to p. It is enough to write down consistent quadratic relations and check that they turn into (3) in the limit $t \to 1$. We shall follow this tactic and propose formulas in already-quantized form. Consider the quadratic relations

$$\begin{aligned}
R_t u^1 u^2 &= u^2 u^1 R_t(p) \\
R_t(p) v^1 v^2 &= v^2 v^1 R_t
\end{aligned} \qquad (4)$$

for the operator-valued matrices u and v. Here R_t is the usual $sl(2)$ R-matrix

$$R_t = \begin{pmatrix} t^{1/2} & 0 & 0 & 0 \\ 0 & t^{-1/2} & 0 & 0 \\ 0 & t^{1/2} - t^{-3/2} & t^{-1/2} & 0 \\ 0 & 0 & 0 & t^{1/2} \end{pmatrix}$$

and $R_t(p)$ the corresponding $6j$ symbol

$$R_t(p) = \begin{pmatrix} t^{1/2} & 0 & 0 & 0 \\ 0 & t^{-1/2}\sqrt{1 - \left(\frac{t - t^{-1}}{e^{ip} - e^{-ip}}\right)^2} & \frac{t^{1/2} - t^{-3/2}}{1 - e^{2ip}} & 0 \\ 0 & \frac{t^{1/2} - t^{-3/2}}{1 - e^{-2ip}} & t^{-1/2}\sqrt{1 - \left(\frac{t - t^{-1}}{e^{ip} - e^{-ip}}\right)^2} & 0 \\ 0 & 0 & 0 & t^{1/2} \end{pmatrix}$$

(see i.e. Ref. [7]).

The self-consistency of these relations follows from the Yang-Baxter relation for R_t

$$R_t^{12}R_t^{13}R_t^{23} = R_t^{23}R_t^{13}R_t^{12}$$

and its generalization for $R_t(p)$

$$(Q^1)^{-1}R_t^{23}(p)Q^1R^{13}(p)(Q^3)^{-1}R_t^{12}(p)Q^3 = R_t^{12}(p)(Q^2)^{-1}R_t^{13}(p)Q^2R_t^{23}(p)\,.$$

Here the matrix Q

$$Q = \begin{pmatrix} e^{iq} & 0 \\ 0 & e^{-iq} \end{pmatrix}$$

consists of the shift operators for variable p, i.e.

$$e^{-iq}f(p)e^{iq} = f\left(p + \frac{1}{i}\ln t\right)\,.$$

The relation for $R_t(p)$ was first introduced in Ref. [8]. The classical relations (3) follow from (4) in the contraction limit. If we write

$$t = e^{i\gamma\hbar}$$

where $\hbar$ is the ordinary Planck constant and γ measures the t-deformation and take into account, that

$$\{\cdot\,,\,\cdot\} = \lim_{\hbar\to 0} i\frac{[\cdot\,,\,\cdot]}{\hbar}\,,$$

we shall get (3.a), (3.b) in the limit $\gamma = 0$ and $\hbar = 0$, if we renormalize p as

$$p \to \gamma p\,.$$

Other relations (3) are either unchanged, i.e.

$$u_0^1v_0^2 = v_0^2u_0^1$$

or trivially modified, i.e.

$$D^2u^1 = u^1D^2\sigma \tag{5}$$

where

$$\sigma = t^{\sigma_3\otimes\sigma_3} = \begin{pmatrix} t & & & \\ & t^{-1} & & \\ & & t^{-1} & \\ & & & t \end{pmatrix}\,.$$

Now we can introduce the analogues of the original variable ω and a. As always after deformation, the Lie-algebraic variable a_L is to be substituted by the Lie-group like one. So we introduce

$$\omega = uQ^{-1}v$$

and

$$A = uDu^{-1}\,.$$

The classical variable a appears in the limit

$$A = 1 - 2\gamma a + \cdots\,.$$

Let us find the relations for ω and A.

Beginning with the relations for ω we have a chain of commutations:

$$\begin{aligned}\omega^1\omega^2 &= u^1(Q^1)^{-1}v^1u^2(Q^2)^{-1}v^2\\ &= u^1v_0^1u_0^2v^2 = u^1u_0^2v_0^1v^2\\ &= u^1u^2(Q^1Q^2)^{-1}v^1v^2\\ &= R_t^{-1}u^2u^1R_t(p)(Q^1Q^2)^{-1}R_t(p)^{-1}v^2v^1R_t\,.\end{aligned}$$

Now from the explicit form of Q^1Q^2

$$Q^2Q^2 = \begin{pmatrix} e^{2iq} & & & \\ & 1 & & \\ & & 1 & \\ & & & e^{-2iq}\end{pmatrix} = Q^2Q^1$$

it follows that $R_t(p)$ commutes with Q^1Q^2, so that $R_t(p)$ cancels in the last line. We continue

$$\begin{aligned}R_t\omega^1\omega^2 &= u^2u^1(Q^1)^{-1}(Q^2)^{-1}v^2v^1R_t\\ &= u^2u_0^1v_0^2v^1R_t = u^2v_0^2u_0^1v^1R_t\\ &= \omega^2\omega^1R_t\,.\end{aligned}$$

Thus we get the defining relation of the quantum group in the sense of [9]

$$R_t\omega^1\omega^2 = \omega^2\omega^1R_t\,. \tag{6}$$

Let us come now to the relation between ω and A. We have

$$\begin{aligned}\omega^1A^2 &= u^1v_0^1u^2D^2(u^2)^{-1} = u^1u^2D^2(u^2)^{-1}v_0^1\\ &= R_t^{-1}u^2u^1R_t(p)D^2(u^2)^{-1}v_0^1\,.\end{aligned}$$

Now introduce $1 = \mathcal{D}^2(\mathcal{D}^2)^{-1}$ between u^2 and u^1, use (5) in the form

$$(\mathcal{D}^2)^{-1}u^1 = u^1\sigma^{-1}(\mathcal{D}^2)^{-1}$$

and the fact that the matrix

$$\tilde{R}_t(p) = (\mathcal{D}^2)^{-1}\sigma^{-1}R_t(p)\mathcal{D}^2$$

has essentially the same properties as $R_t(p)$. More explicitly

$$\tilde{R}_t(p) = \Pi R_t(p)^{-1}\Pi$$

when Π – permutation matrix, which can be checked directly. The relation (4) can be rewritten as

$$\tilde{R}_tu^1u^2 = u^2u^1\tilde{R}_t(p)$$

or

$$(u^2)^{-1}\tilde{R}_tu^1 = u^1\tilde{R}_t(p)(u^2)^{-1}$$

where

$$\tilde{R}_t = \Pi R_t^{-1}\Pi$$

so that we can continue

$$\begin{aligned} R_t\omega^1 A^2 &= u^2 D^2 u^1 \tilde{R}_t(p)(u^2)^1 v_0^{-1} \\ &= u^2 D^2 (u^2)^{-1} \tilde{R}_t u^1 v_0^1 \\ &= A^2 \tilde{R}_t \omega^1 . \end{aligned}$$

Once more the $6j$ symbols disappear in the final formula. With more symmetric notations

$$R_t^+ = \tilde{R}_t \ ; \quad R_t^- = R_t$$

we have the relation

$$R_t^- \omega^1 A^2 = A^2 R_t^+ \omega^1 . \tag{7}$$

The relation between A can be calculated in a similar way and we present the final result

$$A^1 (R_t^-)^{-1} A^2 R_t^- = (R_t^+)^{-1} A^2 R_t^+ A^1 . \tag{8}$$

Thus we constructed all commutation relations defining $(T^*A)_t$. The variables of the base ω constitute quantum group (relations (6)). The variables of the fiber A have commutation relations (8) of the corresponding quantum Lie algebra in the form of Reshetikhin and Semenov-Tjan-Shansky [10]. The cross-relations (7) between ω and A define the deformation of the sympletic structure. It is instructive to check how the classical relations (2) follow in the contraction limit. To this end one is to observe that the Casimir C can be written

$$C = r^- - r^+$$

where

$$R_t^\pm = 1 + i\hbar\gamma r^\pm + \cdots . \tag{9}$$

With this comment we finish the considering of the top and return to our main goal.

3. EXCHANGE ALGEBRA

Let us return to the problem of the Poisson brackets (or commutation relations) for the chiral components $u_0(x), v_0(x)$ or $u_F(x), v_F(x)$ of the local field $g(x)$ in the WZNW model. The example of section 2 shows that one could expect the quadratic Poisson algebra for $u_F(x)$ and $v_F(x)$ with a suitable generalization of matrix $r_0(p)$. It is indeed the case and a corresponding algebra was proposed in 1989 [11], [7], [12]. It was an exchange algebra with the classical R-matrix from the theory of quantum groups. The interrelation of CFT and quantum groups which was already seen previously in connection with the Knizhnik-Zamolodchikov equation [13], [14], became more straightforward. However the derivations or reasonings in [7] were not completely satisfactory and looked rather mysterious. Because of that Volkov and I, in the beginning of 1990, made an effort to derive the exchange algebra directly from the lagrangian formulation. Evidently the same spirit prompted the recent publications [15], [16].

The important new idea appeared in the end of 1990, in the conversations between Alekseev, Volkov, Semenov-Tjan-Shansky and me. It led to the connection of the exchange algebra and R-matrix so straightforwardly that we called our publication [17] "The unraveling of the quantum group structure in WZNW model". Here follows the idea of our proposal.

If the coefficient $L(x)$ in the equation

$$u' = Lu$$

is gauge transformed

$$L(x) = T(x)s(x)T^{-1}(x) + \partial T(x)T^{-1}(x)$$

then the solution can be written as

$$u(x) = T(x)\overleftarrow{\exp}\left\{\int_0^x s(y)dy\right\}.$$

The idea is to find the entries $T(x), s(x)$ in such a way, that they have only ultralocal Poisson brackets (i.e. without $\delta'(x-y)$ terms) leading to the usual Kac-Moody brackets for $L(x)$. As soon as $s(x)$ is ultralocal, the Poisson brackets of the transport matrix $\exp\{\int_0^x s(y)dy\}$ are easily calculable, as is well known in the inverse scattering method [18].

The realization of this idea is quite simple and beautiful. Consider $T(x)$ – the upper triangular matrix and $s(x)$ – the lower triangular matrix with the Poisson brackets

$$\begin{aligned}
\{T^1(x), T^2(y)\} &= 0\,;\\
\{T^1(x), s^2(y)\} &= \gamma T^1(x) r \delta(x-y)\,;\\
\{s^1(x), s^2(y)\} &= \gamma[s^1(x) + s^2(y), r]\delta(x-y)\,.
\end{aligned}$$

Here r is a classical r-matrix ($r = -r^+$ from (§2)

$$r = \frac{1}{2}\sigma^3 \otimes \sigma^3 + 2\sigma^+ \otimes \sigma^-\,,$$

or

$$r = \begin{pmatrix} 1/2 & 0 & 0 & 0\\ 0 & -1/2 & 2 & 0\\ 0 & 0 & -1/2 & 0\\ 0 & 0 & 0 & 1/2 \end{pmatrix}.$$

Then

$$L(x) = T(x)s(x)T^{-1}(x) + \partial T(x)T^{-1}(x)$$

has the Poisson brackets of the left current of the WZNW model.

The relations for $T(x)$ and $s(x)$ have a nice geometrical meaning. As they are ultralocal we can drop the variable x, returning from loops to the finite dimensional objects. Then the relations are nothing but those on T^*B, where B is a Borel subgroup of G, namely the subgroup of upper triangular matrices. Thus $T(x)$ and $s(x)$ constitute the loops of T^*B.

Of course, T and s contain 4 variables where $L(x)$ has only 3. The resolution consists in the observation, that one can construct from the diagonal elements of T and s a current $l(x)$, such that

$$\begin{aligned}
\{l(x), l(y)\} &= \gamma\delta'(x-y)\,;\\
\{L(x), l(y)\} &= 0\,.
\end{aligned}$$

In other words, we can write symbolically

$$KM = LT^*B/LH\,,$$

where LH is a loop of the Cartan subgroup, generated by $l(x)$.

Note also the appearance of the r-matrix in an entirely classical setting – it is nothing but a structure constants matrix for the Borel subalgebra.

The ultralocal expression for $L(x)$ allows us to calculate the Poisson brackets for $u(x)$. We shall present the corresponding formulas in an already quantized form.

Quantization can be performed in the spirit of the Quantum Inverse Scattering Method [2]. The circle S^1, over which x runs is to be substituted by a finite chain with running a discrete variable $n, n = 1, \ldots, N$. The continuous limit $N \longrightarrow \infty$, or $\Delta \longrightarrow 0$, where

$$\Delta = 2\pi / N$$

is obtained in the usual way

$$x = n\Delta \, .$$

The current $L(x)$ is substituted now by a set of matrices L_n; in continuous limit we have

$$L_n = I + \Delta L(x) + \cdots \, .$$

The differential equation

$$u' = Lu$$

turns into

$$u_{n+1} = L_n u_n$$

with the solution

$$u_n = \overleftarrow{\prod_{k \leq n}} L_k \, .$$

The gauge transformation now looks as

$$L_n = T_{n+1} S_n T_n^{-1} \tag{10}$$

where in continuous limit

$$T_n \longrightarrow T(x) \, ; \quad S_n \longrightarrow I + \Delta s(x) + \cdots \, .$$

We introduce now the main commutation relations

$$\begin{aligned} R^+ S_n^1 S_n^2 &= S_n^2 S_n^1 R^+ \\ R^+ T_n^1 T_n^2 &= T_n^2 T_1^n R^+ \\ T_n^1 S_n^2 &= S_n^2 T_n^1 R^+ \, . \end{aligned} \tag{11}$$

Here R^+ is a quantum R-matrix, introduced in section 2. In the first two formulas one can use also R^- instead of R^+. The last one can also be written as

$$S_n^1 T_n^2 = T_n^2 S_n^1 R^- \, .$$

These quantum relations turn into the classical continuous ones in the limit $\Delta \longrightarrow 0$, $\hbar \longrightarrow 0$. Note that the group variables T_n are not commuting now – this is a first manifestation of the noncommutative geometry, more of which will be seen in what follows.

The commutation relations for T_n, S_n for a fixed n define the regular representation of "quantized" or rather deformed group B. In other words, it is quantization of the deformed phase space $(T^* B)_t$.

The combination (10), defining the Kac-Moody generators on the chain, enters solely into the corresponding commutation relations. It can be checked by means of (11) that the following relations have place

$$\begin{aligned} R^+ L_n^1 L_n^2 R^- &= L_n^2 L_n^1 \\ L_n^1 L_{n+1}^2 &= L_{n+1}^2 R^+ L_n^1 \, . \end{aligned} \tag{12}$$

In the first formula one can substitute $R^{\pm} \longrightarrow R^{\mp}$.

Formulas (12) can be rewritten into one

$$L_m^1 (R_{m-n-1}^-)^{-1} L_n^2 R_{m-n}^- = (R_{m-n}^+)^{-1} L_n^2 R_{m-n+1}^+ L_m^1$$

using the notation

$$R_n = \begin{cases} R & n = 0 \\ I & n \neq 0 \end{cases} .$$

It was proposed by M. Sememonv-Tjan-Shansky and published in [19]. It looks like a natural generalization of the commutation relation in the quantum Lie algebra of [10] but accounts for the central extension.

Now the problem of the calculation of the commutation relations for the chiral components u_n is reduced to a simple algebra. It follows from (12) that L_n commutes with the product of $L_{n+1} L_n L_{n-1}$

$$L_n^1 L_{n+1}^2 L_n^2 L_{n-1}^2 = L_{n+1}^2 L_n^2 L_{n-1}^2 L_n^1$$

and, of course,

$$L_n^1 L_m^2 = L_m^2 L_n^1 , \quad |n - m| \geq 2 .$$

So commuting u_m and u_n one must take into account only the very first $n (n = 1, 2)$ or the last ones if $n = m$. We get for $m, n \leq N - 1$

$$\begin{aligned} u_m^1 u_n^2 R^{\pm} &= u_n^2 u_m^1 \quad m \gtrless n \\ R^+ u_n^1 u_n^2 R^- &= u_n^2 u_n^1 . \end{aligned}$$

Note that the fields u_n do not commute when their arguments coincide.

For monodromy ($n = N$), we are to take into account that L_N and L_1 are not commuting. We get

$$M_L^1 (R^-)^{-1} M_L^2 R^- = (R^+)^{-1} M_L^2 R^+ M_L^1 . \tag{13}$$

We already had such a relation in Section 2. Thus we see the proper place of the quantum group (rather quantum Lie algebra) in CFT – it is a monodromy of a chiral field with a trivial normalization at some point ($u_0(0) = I$). However it is not the whole story of quantum groups.

Indeed, let us turn to the local field

$$g_n = u_n K v_n ,$$

where K plays the role of the initial condition

$$K = g_0 = g_N .$$

In section 1 we constructed K classically through the matrices Z_L, Z_R, diagonalizing M_L and M_R and an additional variable q. This process could be repeated here. However we shall propose instead some characterization of K in terms of simple commutation relations and check their consistency.

We require that the following relations take place

$$\begin{aligned} K^1 L_1^2 &= L_1^2 R^+ K^1 \\ R^- K^1 L_N^2 &= L_N^2 K^1 \end{aligned}$$

and K commutes with L_n, $n = 2, \ldots, N-1$. This is the quantum counterpart of the commutation relation between $J_0(x)$ and $g(y)$ (at $y = 0$). Moreover, K itself constitutes a quantum group

$$RK^1K^2 = K^2K^1R\,, \tag{14}$$

where R stands for R^+ or R^-. It follows that K has similar relations with u_n

$$K^1u_n^2 = u_n^2R^+K^1\,, \quad n = 1, \ldots, N-1$$

and monodromy

$$R^-K^1M_L^2 = M_L^2R^+K^2\,. \tag{15}$$

Corresponding relations with v_n (which we do not discuss explicitly) look as follows:

$$K^1R^+v_n^2 = v_n^2K^1\,.$$

(One reads the relations for u from right to left.) Now by simple algebra one can check that the local field g_n satisfies the relations

$$\begin{aligned} g_m^1g_n^2 &= g_n^2g_m^1\,; \quad n \neq m \\ Rg_n^1g_n^2 &= g_n^2g_n^1R \end{aligned}$$

among itself and

$$\begin{aligned} g_n^1L_{n+1}^2 &= L_{n+1}^2R^+g_n^1\,, \\ R^-g_n^1L_n^2 &= L_n^2g_n^1\,, \\ g_n^1L_m^2 &= L_m^2g_n^1\,, \quad m \neq n, n+1\,, \end{aligned}$$

between g_n and L_m.

These relations turn into those of Section 1 in the classical continuous limit.

The consistency with the periodicity condition is now checked in the following manner - M_LKM_R has exactly the same relations with K, L_n and M_L and itself (and their right counterparts) as K. It follows that

$$K = \alpha M_LKM_R$$

where α commutes with all dynamical variables, so it is a constant and must be equal to 1 by comparison of determinents.

The relations (13), (14), (15) exactly coincide with the relations (6),(7),(8) of Section 2 if we substitute

$$K \longrightarrow \omega\,; \quad M_L \longrightarrow A\,.$$

Thus the local field at a fixed point and the monodromy of a chiral component around the circle from this point give us a representation of $(T^*A)_t$. These data comprise zero modes of the WZNW model and essentially define its full structure. This establishes the intimate connection of Quantum Groups and Kac-Moody algebras.

We can now turn to the quantum analogues of the quasiperiodic chiral components $u_F(x)$ and $v_F(x)$. They were defined classically in Section 1 by

$$u_F(x) = u(x)N_L\,; \quad v_F(x) = N_Rv(x)\,,$$

where

$$N_L = Z_LQ\,; \quad N_R = QZ_R$$

and

$$K = g(0) = Z_L Q Z_R = N_L Z_R = Z_L N_R \, .$$

We shall use the same definitions in the quantum case, i.e.

$$u_F(n) = u_n N_L \, .$$

As Z_R commutes with all left variables, N_L has the same commutation relations with L_n as K, namely

$$\begin{aligned} N^1 L_1^1 &= L_1^2 R^+ N^1 \, , \\ N^1 L_n^2 &= L_n^2 N^1 \, , \quad n = 2, \ldots, N-1 \, . \end{aligned}$$

For N_L itself we are to take the relations (4)

$$R N_L^1 N_L^2 = N_L^2 N_L^1 R(p) \, ,$$

where p is defined by the diagonalization of M_L. From this and commutation relations for u_n we get

$$\begin{aligned} u_F^1(m) u_F^2(n) &= u_F^2(m) u_F^1(n) R^{\pm}(p) \, ; m \lessgtr n \\ R u_F^1(n) u_F^2(n) &= u_F^2(n) u_F^1(n) R(p) \, . \end{aligned}$$

The continuous analogue of these relations was proposed in [7]. They are consistent with the quasiperiodicity of $u_F(n)$

$$u_F(N) = u_F(0) \mathcal{D}$$

due to the property of $R(p)$

$$R^+(p) = (\mathcal{D}^2)^{-1} \sigma^{-1} R^-(p) \mathcal{D}^2$$

mentioned in Section 2 and the commutation relation

$$N^1 \mathcal{D}^2 = \mathcal{D}^2 N^1 \sigma^{-1} \, .$$

With this we end the discussion of the exchange algebra.

4. DISCUSSION AND CONCLUSION

The connection of the WZNW model with the deformed top ($(T^*G)_t$) found above allows us to say a lot about the former. In fact one can translate everything known for the top into a statement on the WZNW model. The crucial observation, made in [19], is that the centers of the algebra $\mathfrak{A}$, generated by L_n and that generated by A, i.e. $\mathfrak{B} = U_t(\mathfrak{A})$ coincide. This allows us to say that the irreducible representations of $\mathfrak{A}$ are labeled by those of $\mathfrak{B}$.

More exactly, it is known that the compact quantum Lie algebra $SU(2)_t$ for $t = e^{i\pi/l}$ has $l-1$ representations $V_j, j = 0, \frac{1}{2}, \ldots, \frac{l}{2} - 1$ of dimension $2j+1$ with highest weight and infinite series of cyclic representations of dimension l. One can forget about the latter in the quantization of a top, if one restrict values of p and q to the reals. Thus the Hilbert space of quantized (and deformed) top is

$$\mathcal{H}_{\text{top}} = \sum_j V_j \otimes V_j \, .$$

Now let us return to WZNW. The quantized $L(x)$ gives a representation of the Kac-Moody algebra of the level $k = l - 2$ (the appearance of the famous shift is explained in terms nearest to this text in [19]). There are $k + 1 = l - 1$ such representations $\mathcal{H}_j$ and due to the observation above

$$\mathcal{H}_j = \mathcal{H}_0 \otimes V_j \, .$$

In particular, the fusion rules for $\mathcal{H}_j$ are generated by the tensor algebra of V_j. The Hilbert space of the WZNW model has the form

$$\mathcal{H}_{WZNW} = \sum \mathcal{H}_j \otimes \mathcal{H}_j \, .$$

The local field at a fixed point can be used as an "intertwinner" between different $\mathcal{H}_j$ in the sense of the generalized approach to superselection sectors (see e.g. [20] and lectures of Prof. Mack and Friedenhagen).

The representation of the local field through the chiral components

$$g(x) = u_0(x) K v_0(x)$$

could prove to be very useful for the calculations of the correlation functions. Indeed, the reconstruction of the conformal block

$$\left\langle u_0^1(x_1 + t_1) \cdots u_0^n(x_n + t_n) \right\rangle$$

can be reduced to the Riemann-Hilbert problem similar (and simpler) to that considered by Smirnov [21], see also [22]. However this program was not realized yet.

I turn now to the general conclusions.

1. Change of the nature of symmetry. The relations (12) for L_n are invariant to the external gauge transformation

$$L_n \longrightarrow h_{n+1} L_n h_n^{-1}$$

only if h_n constitute a deformed loop group

$$\begin{aligned} h_n^1 h_m^2 &= h_m^2 h_n^1 \, , \quad n \neq m \\ R h_n^1 h_n^2 &= h_n^2 h_n^1 R \, , \end{aligned}$$

commuting with L_n

$$h_n^1 L_m^2 = L_m^2 h_n^1 \, .$$

So the symmetry itself becomes deformed – it is a symmetry with respect to an action of the quantum loop group.

2. Noncommutative geometry. I have already made several comments to the effect that the ideology of noncommutative geometry naturally enters my exposition. Let us reiterate that the natural object for the values of the local field is a deformed loop group in the sense just explained. Having in mind the connection of the σ-model and string it is natural to speculate that the target space for the string could be noncommutative.

3. Dimensional reduction. We have seen that due to the monodromy map the full WZNW (1+1 dim) model was reduced to the finite-dimensional (1 dim) dynamical system on $(T^*G)_\gamma, \gamma = \frac{\pi}{k+2}$. In its turn WZNW is a reduction of a Chern-Simons model (1+2 dim) by transgression. Thus our example prompts the investigation of a more general situation, where such reductions are possible.

4. Equivalent deformations. The same thing can be put in a different fashion. Beginning with a finite dimensional Lie algebra G we can construct an affine algebra $(\hat{G}_k)$ or quantized algebra G_t. The first is still Lie algebra, but infinite dimensional, the second is finite dimensional, but has only deformed Lie algebra structure. Each modification is parametrized by one parameter: k and t, correspondingly. We have seen that $\hat{G}_k$ and G_t are essentially equivalent if $t = e^{i\pi/k+2}$ (for $SU(2)$).

One can draw a diagram

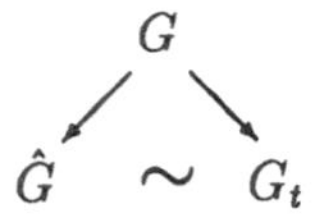

and ask a natural question if this diagram has a continuation. (It was mentioned to me that I. Frenkel raised the same question.) The continuation must be a two-parameter family, which may be a local Lie-algebra (in two dimensions), or a deformed Lie algebra (in one dimension), or a finite dimensional algebra with more complicated relations. There are candidates for all three – extended the current algebra on a disk, deformed $K-M$ algebra and Sklyanin algebra. To establish the relation between their representations is an interesting unsolved problem.

References

1. A. Belavin, A. Polyakov, A. Zamolodchikov: Nucl. Phys. **B241** 333 (1984).
2. L. Faddeev: Les Hauches lecture in "Recent Advances in Field Theory", R. Stora, J.-B. Zuber (eds) Amsterdam, North- Holland (1984).
3. E. Witten: Commun. Math. Phys. **92** 451 (1984).
4. J. Wess, B. Zumino: Phys. Rev. **B37** 95 (1971).
5. S. Novikov: Uspehi Math. Sci. **37** 3 (1982).
6. A. Alekseev, L. Faddeev: MIT preprint CTP N1957 (1991) (to be published in Commun. Math. Phys.).
7. L. Faddeev: Commun. Math. Phys. **132** (1990).
8. J.-L. Gervais, A. Neveu: Nucl. Phys. **B238** 125 (1984).
9. N. Reshetikhin, L. Takhtajan, L. Faddeev: Algebra and Analysis, **1** 147 (1989) (in Russian).
10. N. Reshetikhin, M. Semenov-Tjan-Shansky: Lett. Math. Phys. **19** 133 (1990).
11. B. Blok: Phys. Lett. **233B** 359 (1989).
12. A. Alekseev, S. Statashvili: Commun. Math. Phys. **133** 353 (1990).
13. A. Tsuchiya, Y. Kanie: Advance Stud. Pure Math. **16** 297 (1988).
14. T. Kohno: Ann. Inst. Fourier (Grenoble) **37** 4 (1987).
15. K. Gawedsky: preprint IHES (1990).
16. M. Chu, P. Goddared, L. Halliday, D. Olive, A. Schwimmer: Phys. Lett. **266B** 71 (1991).

17. A. Alekseev, L. Faddeev, M. Semenov-Tjan-Shansky, A. Volkov: preprint CERN-TH-5981/91 (1991).

18. L. Faddeev, L. Takhtajan: "Hamiltonian Methods in the Theory of Solitons" Springer (1987).

19. A. Alekseev, L. Faddeev, M. Semenov-Tjan-Shansky: LOMI preprint E-5-91 (1991).

20. G. Mack, V. Schomerus: preprint DESY 91-037 (1991).

21. F. Smirnov: "Formfactors in Completely Integrable Models of Quantum Field Theory", World Scientific (to be published).

22. F. Smirnov: preprint RIMS-773 (1991).

OBSERVABLES, SUPERSELECTION SECTORS AND GAUGE GROUPS

Claus Fredenhagen

II. Institut für Theoretische Physik
Universität Hamburg

1. Introduction

Symmetries play an important role in quantum physics from its very beginning. There are two types of symmetry which must be distinguished, namely internal and external symmetries. External symmetries have an active interpretation; they transform states and observables such that certain structures are respected, e.g. transition probabilities, the Hamiltonian, the S-matrix etc.. Typical examples are the space time symmetries in a situation where space time is a priori given in the sense of classical physics. Internal symmetries, on the contrary, have only a passive interpretation; they do not transform the states of the system but merely change the description of the system. Examples are gauge transformations in gauge theories and diffeomorphisms in general relativity. Hence internal symmetries occur only in cases where the formulation of the theory contains some redundancy. It is tempting to remove this redundancy and to base the theoretical description exclusively on observables. Then the question arises whether the internal symmetry has an intrinsic meaning and may be recovered from the observables. Conversely, one may interpret the observed structure of the state space in terms of an internal symmetry and develop a redundant description where this symmetry acts explicitely. Such a redundant description might be useful for the investigation of perturbed theories where the symmetry is slightly broken, either spontaneously or dynamically.

The case of spontaneous breakdown of symmetry is especially interesting since it shows how internal symmetries might turn into external ones. E.g. in the BCS-model, the normally unobservable phase of the Fermi field becomes observable (modulo π) in the supraconducting phase and shows up for instance in the Josephson effect. A similar transmutation of internal symmetries into external ones occurs in elementary particle physics, and in a possible scenario for quantum gravity a spontaneous breakdown of diffeomorphism symmetry might lead to a classical notion of space time.

In my lecture I want to describe the present status of an analysis of possible symmetries starting from observables only. Its point of departure is the algebra of observables in quantum field theory with its local structure satisfying spacelike

New Symmetry Principles in Quantum Field Theory, Edited by
J. Fröhlich et al., Plenum Press, New York, 1992

commutativity. One then selects the set of states one is interested in. There are two selection criteria which were analyzed. One is the Doplicher-Haag-Roberts (DHR) criterion which singles out those states which have a trivial structure at spacelike infinity [1]; states which can be obtained by applying local fields to the vacuum belong to this set. The second criterion, due to Borchers [2], admitts only states of positive energy. In the absence of massless particles such states have a trivial structure on topologically trivial (i.e. contractible) parts of spacelike infinity [3]. Examples are obtained by applying Mandelstam's path dependent field operators to the vacuum (for a rigorous construction in the framework of lattice gauge theories see [4], cf. also [5,6]).

The set of pure states satisfying either one of these criteria decomposes into superselection sectors. There is a (commutative, associative) product on the set of finite direct sums of sectors which can be lifted to a tensor-like product structure on the corresponding Hilbert spaces. This lift induces some kind of curvature which can physically be interpreted as particle statistics. It is characterized by a unitary representation of the permutation group in $D \geq 3$ (criterion I)[1] respectively $D \geq 4$ space time dimensions [3] and a unitary representation of the braid group in $D = 2$ (criterion I) resp. $D = 3$ (criterion II) space time dimensions [7,8,9]. In $D = 2$ space time dimensions one gets in the case of criterion II a more involved structure, due to the fact that spacelike infinity is not connected which describes the way soliton sectors may be composed to multisoliton sectors (see [10] for an early approach abstracted from models and [11] for a recent proposal which works in the general case).

In the permutation group case, Doplicher and Roberts [12,13] recently succeeded in constructing a unique compact group G whose dual is the set of sectors and a larger algebra $\mathcal{F}$ (called the field algebra) on which G acts covariantly such that the invariant elements of $\mathcal{F}$ are the observables. $\mathcal{F}$ again has a local structure, but in contrast to the observable algebra it contains also fermionic elements which anticommute at spacelike distances.

Whether a similar construction is possible in the braid group case is unknown up to now. In models of conformal field theory quantum groups at roots of unity play an important role. In the lectures of Professor Mack we heard how the structure of quantum groups must be modified in order to fit to the simplest models.

2. The Algebraic Approach to Quantum Field Theory

The structure of quantum field theory may be summarized in the following three principles:

(i) one is dealing with a quantum theory, i.e. the observables are selfadjoint elements of an operator algebra $\mathcal{A}$ which is isomorphic to a selfadjoint algebra of Hilbert space operators.

(ii) locality: observables are measurable in certain subregions of space time, and this association of regions of measurement is compatible with the algebraic structure. So for each space time region one considers the local algebras of observables

$$\mathcal{A}(\mathcal{O}) = \{A \in \mathcal{A} \mid A \text{ is measurable within } \mathcal{O}\}. \tag{2.1}$$

Einstein causality is then incorporated into the scheme by the requirement that observables which can be measured in causally disjoint regions are represented by commuting operators.

(iii) stability: the precize formulation is still under discussion. It is usually incorporated by the selection of a preferred set of states, e.g. the states inducing positive energy representations.

For the following we need some facts from the mathematics of operator algebras. Field theory is normally formulated in terms of pointlike localized fields $\varphi(x)$ which may be smeared with test functions f in order to give bona fide operators

$$\varphi(f) = \int dx \varphi(x) f(x) \quad . \tag{2.2}$$

These operators, however, are unbounded in general, and are thus defined only on a dense subspace of Hilbert space. Much more convenient for a structural analysis are bounded operators which can be defined on the whole Hilbert space. One therefore studies those bounded operators which can be constructed from smeared fields, e.g. $\exp(i\varphi(f))$ for $\varphi(f)$ selfadjoint, and defines the local algebras of observables as being generated by bounded functions of smeared fields with test functions having support in the region considered. It is, however, a major problem whether commutation properties of fields induce the corresponding properties for the bounded operators. An analogous problem in mathematics is the integrability of Lie algebra representations to representations of the associated Lie groups. (For recent progress in the field theoretical problem see [14] and references therein.)

A bounded operator A on a Hilbert space $\mathcal{H}$ is a linear mapping $A : \mathcal{H} \to \mathcal{H}$ such that

$$||A|| := \sup_{\Phi \in \mathcal{H}, ||\Phi||=1} ||A\Phi|| < \infty \quad . \tag{2.3}$$

$||A||$ is called the norm of A; it is a norm on the linear space $\mathcal{B}(\mathcal{H})$ of bounded operators on $\mathcal{H}$. Moreover, since $||AB|| \leq ||A|| \, ||B||$, $\mathcal{B}(\mathcal{H})$ is a normed algebra. There is an important additional structure on $\mathcal{B}(\mathcal{H})$, namely the involution $A \mapsto A^*$ where A^* is the adjoint operator. It satisfies

$$||A^*|| = ||A|| \tag{2.4}$$

and

$$||A^*A|| = ||A||^2. \tag{2.5}$$

One can now abstract these properties from $\mathcal{B}(\mathcal{H})$ and define a C*-algebra to be a complete normed *-algebra with a norm satisfying (2.5). Actually, every C*-algebra is isomorphic to a norm closed algebra of Hilbert space operators. There is, however, no distinguished realization.

States on a C*-algebra are defined as expectation functionals, i.e. linear functionals

$$\omega : \mathcal{A} \to \mathbb{C} \tag{2.6}$$

such that

$$\omega(A^*A) \geq 0 \quad \text{(positivity)} \tag{2.7}$$

and

$$||\omega|| := \sup_{A \in \mathcal{A}, ||A||=1} \omega(A) \quad \text{(normalization)}. \tag{2.8}$$

Provided $\mathcal{A}$ has a unit $\mathbb{1}$ (which we will assume in the following) one has

$$||\omega|| = \omega(\mathbb{1}) \quad . \tag{2.9}$$

For a selfadjoint bounded operator A the probability distribution of measured values in the state ω is given in terms of a probability measure $\mu_{\omega,A}$ which is determined by

$$\int d\mu_{\omega,A}(a)f(a) = \omega(f(A)) \tag{2.10}$$

for all continuous functions f. (Note that a C*-algebra contains with a selfadjoint operator A all polynomials of A and because of completeness all continuous functions of A.)

Examples for states are vector or density matrix states in a representation of $\mathcal{A}$. A (unital) representation of a C*-algebra $\mathcal{A}$ in a Hilbert space $\mathcal{H}$ is a *-homomorphism π of $\mathcal{A}$ into $\mathcal{B}(\mathcal{H})$ with $\pi(1) = 1$. Each vector $\Phi \in \mathcal{H}$ with norm 1 induces a state (a "vector state") on $\mathcal{A}$ by

$$\omega_\Phi(A) = (\Phi, \pi(A)\Phi) \quad . \tag{2.11}$$

Similarly, a positive trace class operator ρ on $\mathcal{H}$ with $tr\rho = 1$ induces the state (a "density matrix state")

$$\omega_\rho(A) = tr\rho\pi(A) \quad . \tag{2.12}$$

Actually, by the famous GNS (Gelfand-Naimark-Segal)-construction every state is a vector state in a suitable representation, namely associated to each state ω there is a representation π_ω on a Hilbert space $\mathcal{H}_\omega$ and a unit vector $\Phi_\omega \in \mathcal{H}_\omega$ such that

(i) $\omega(A) = (\Phi_\omega, \pi_\omega(A)\Phi_\omega) \quad , \quad A \in \mathcal{A}$
(ii) Φ_ω is cyclic for $\pi_\omega(\mathcal{A})$ (i.e. $\pi_\omega(\mathcal{A})\Phi_\omega$ is dense in $\mathcal{H}_\omega$),

and the GNS-triple $(\pi_\omega, \mathcal{H}_\omega, \Phi_\omega)$ is unique up to unitary equivalence.

As an illustration we consider a density matrix state of the form (2.12), where

$$\rho = \sum_{i \in \mathbb{N}} \lambda_i |\Phi_i\rangle\langle\Phi_i| \tag{2.13}$$

with $\lambda_i \geq 0, \sum \lambda_i = 1$ and $||\Phi_i|| = 1$. Then we define a representation π_ρ of $\mathcal{A}$ on the Hilbert space of square summable sequences $\Psi = (\Psi_n)_{n \in \mathbb{N}}$ with values in $\mathcal{H}$ by

$$(\pi_\rho(A)\Psi)_n = \pi(A)\Psi_n \tag{2.14}$$

The vector $\Omega = (\Omega_n)_{n \in \mathbb{N}}$, $\Omega_n = \lambda^{\frac{1}{2}}\Phi_n$ then induces the state ω_ρ, $\mathcal{H}_{\omega_\rho}$ is the closure of $\pi_\rho(\mathcal{A})\Omega$, and $\pi_{\omega_\rho}(A)$ is the restriction of $\pi_\rho(A)$ to $\mathcal{H}_{\omega_\rho}$.

There is a special class of C*-algebras, the so called von Neumann algebras (or W*-algebras). They are *-isomorphic to weakly closed selfadjoint subalgebras of $\mathcal{B}(\mathcal{H})$. Alternatively, they can be characterized by von Neumann's bicommutant theorem: the commutant $\mathcal{N}'$ of a set $\mathcal{N} \subset \mathcal{B}(\mathcal{H})$ is the set

$$\mathcal{N}' = \{A \in \mathcal{B}(\mathcal{H}) | AB = BA \quad \forall A \in \mathcal{N}\}. \tag{2.15}$$

Then for $\mathcal{N}$ selfadjoint (i.e. $\mathcal{N}$ contains the adjoints of all its elements) the bicommutant $\mathcal{N}''$ (the commutant of the commutant) is the smallest von Neumann algebra containing $\mathcal{N}$. A third characterization uses the fact that the density matrix states ω_ρ, ρ density matrix in $\mathcal{H}$, on a C*-algebra $\mathcal{N} \subset \mathcal{B}(\mathcal{H})$ generate by finite linear combinations a norm closed subspace of the dual $\mathcal{N}^*$ of $\mathcal{N}$. $\mathcal{N}$ is a von Neumann algebra if and only if it coincides with the dual of this subspace which is then called the predual $\mathcal{N}_*$ of $\mathcal{N}$. States in $\mathcal{N}_*$ are called normal. The σ-topology on a von

Neumann algebra $\mathcal{N}$ is the coarsest topology for which all elements of the predual are continuous.

We close this mathematical disgression by some remarks on equivalence of representations. Two representations π_1, π_2 of a C*-algebra $\mathcal{A}$ are unitarily equivalent if there is a unitary operator $U : \mathcal{H}_{\pi_1} \to \mathcal{H}_{\pi_2}$ such that

$$U\pi_1(A) = \pi_2(A)U \quad , A \in \mathcal{A} \quad . \tag{2.16}$$

They are quasiequivalent when they are unitarily equivalent up to multiplicity, or, equivalently, when their state spaces coincide. Here the state space of a representation is the set of states induced by density matrices in the representation space. Since von Neumann algebras have a distinguished set of states, the set of normal states, they have up to equivalence only one normal (i.e. σ-continuous) faithful representation. Non normal representations of von Neumann algebras are pathological and are usually disregarded. On the contrary, generic C*-algebras have many inequivalent faithful representations which are to be treated on an equal footing.

We return now to the description of the algebraic approach to quantum field theory. Let $\mathcal{K}$ denote the set of open double cones $\mathcal{O}$ ("diamonds") in Minkowski space M,

$$\mathcal{O} = V_+ + x \cap V_- + y \tag{2.17}$$

where $V_\pm$ denote the interior of the forward, resp. backward light cone and x, y are points in Minkowski space with $x - y \in V_+$. The theory is now formulated in terms of a family of von Neumann algebras acting irreducibly in some Hilbert space $\mathcal{H}$,

$$\mathcal{A} = (\mathcal{A}(\mathcal{O}))_{\mathcal{O}\in\mathcal{K}} \quad , \quad \bigcap_{\mathcal{O}\in\mathcal{K}} \mathcal{A}(\mathcal{O})' = \mathbb{C}\mathbb{1} \tag{2.18}$$

such that $\mathcal{A}(\mathcal{O}_1) \subset \mathcal{A}(\mathcal{O}_2)$ for $\mathcal{O}_1 \subset \mathcal{O}_2$ and

$$\mathcal{A}(\mathcal{O}_1) \subset \mathcal{A}(\mathcal{O}_2)' \tag{2.19}$$

if $\mathcal{O}_1$ is contained in the spacelike complement of $\mathcal{O}_2$ (locality). $\mathcal{A}$ is called the local net or the Haag-Kastler net [15]. Since $\mathcal{K}$ is a directed set one can define the algebra

$$\mathcal{A}_0 = \bigcup_{\mathcal{O}\in\mathcal{K}} \mathcal{A}(\mathcal{O}) \quad . \tag{2.20}$$

$\mathcal{A}_0$ is independent of the way the local net is represented in $\mathcal{H}$. It has a unique norm inherited from the local algebras and therefore a unique completion to a C*-algebra which also will be denoted by $\mathcal{A}$.

As a reminiscence on the construction of $\mathcal{A}(\mathcal{O})$ from pointlike localized fields we require

$$\mathcal{A}(\mathcal{O}) \subset \bigvee_{i\in I} \mathcal{A}(\mathcal{O}_i) \tag{2.21}$$

whenever $\mathcal{O} \subset \bigcup_{i\in I} \mathcal{O}_i$ (weak additivity). Here the symbol $\bigvee$ means the generated von Neumann algebra.

3. The State Space of Positive Energy Representations

The C*-algebra $\mathcal{A}$ of local observables constructed in the preceding section has, in general, a tremendous number of states from which one has to select the interesting ones. The leading idea here is the principle of stability. We may realize it as the concept of positive energy representations, a point of view first emphasized by Borchers [16]. We assume that the translations x of Minkowski space act on $\mathcal{A}$ by automorphisms such that the local structure is respected,

$$\alpha_x(\mathcal{A}(\mathcal{O})) = \mathcal{A}(\mathcal{O}+x) \quad . \tag{3.1}$$

A positive energy representation is then a representation π of $\mathcal{A}$ in some Hilbert space $\mathcal{H}$ together with a strongly continuous representation $x \mapsto U(x) = e^{ixP}$ of the translation group implementing α_x,

$$\pi \circ \alpha_x = Ad\, U(x) \circ \pi \tag{3.2}$$

such that $sp\, P \subset \overline{V_+}$. Positive energy representations have remarkable properties:

(i) U can be chosen such that the operators $U(x)$ are in the von Neumann algebra $\pi(\mathcal{A})$" [17]. This may be considered as an abstract Sugawara construction and justifies a posteriori the spectrum condition as a condition on observables. Moreover, the choice can be made such that $sp\, P$ is Lorentz invariant [18,19].

(ii) Let $x \in V_+$ and $\Phi \in e^{-xP}\mathcal{H}$. Then one finds the Reeh-Schlieder property (provided the net satisfies weak additivity):
Φ is cyclic and separating for $\pi(\mathcal{A}(\mathcal{O}))$ for all $\mathcal{O} \in \mathcal{K}$.
Here separating means that $A \in \pi(\mathcal{A}(\mathcal{O}))$ is uniquely determined by the vector $A\Phi$.

(iii) Assume again that weak additivity is satisfied in the representation π. Let $E \in \pi(\mathcal{A}(\mathcal{O}))$" be a nonzero projection, and let $\mathcal{O}_1 \in \mathcal{K}$ such that $\mathcal{O}+x \subset \mathcal{O}_1$ for all x in some neighbourhood of the origin in the translation group. Then there is some isometry $W \in \pi(\mathcal{A}(\mathcal{O}_1))$" such that $WW^* = E$ (Borchers Property [20]).

(iv) Provided the defining representation of $\mathcal{A}$ is a positive energy represensation, any positive energy representation π has normal restrictions to the local algebras $\mathcal{A}(\mathcal{O})$, $\mathcal{O} \in \mathcal{K}$ (local normality [21]). In particular, weak additivity holds in every positive energy representation.

In applications of the last property one has to be cautious since in practice one usually knows the representation only on a σ-dense subalgebra. Property (iv) conforms with the statement that non normal representation of von Neumann algebras can be disregarded.

In order to abstract the relevant structure we invoke Haag's principle of local definiteness [22]:
For all $\mathcal{O} \in \mathcal{K}$, $\mathcal{A}(\mathcal{O})$ is a factor (i.e. has trivial center), and every admissible state of $\mathcal{A}$ has a normal restriction to $\mathcal{A}(\mathcal{O})$.

In examples, and under certain reasonably looking conditions also in the general case one can determine $\mathcal{A}(\mathcal{O})$ up to isomorphy [23]. It turns out to be isomorphic to the unique hyperfinite type III_1 factor $\mathcal{R}$ in the classification of Connes. $\mathcal{R}$ occurs everywhere in quantum physics of infinite systems. In field theory the interesting structure is the relative position of all the factors $\mathcal{A}(\mathcal{O}) \simeq \mathcal{R}$. For the inclusion of two factors there are now interesting results available, due to the work of Jones, Ocneanu, Popa, Wenzl and others [24-27].

A classification of nets seems to be far beyond the reach in the moment. However, a partical classification can be done in terms of properties of the state space. The space of all admissable states

$$\mathcal{S}_{\text{ad}} = \{\omega \in \mathcal{S} | \, \omega|_{\mathcal{A}(\mathcal{O})} \text{ is normal} \, \forall \mathcal{O} \in \mathcal{K}\} \tag{3.3}$$

($\mathcal{S}$ denotes the set of all states of $\mathcal{A}$) is much too large for a reasonable classification due to the noncompactness of spatial sections in Minkowski space. Therefore one may compactify the spatial sections. The simplest possibility is the one point compactification by adding a point ∞ at spacelike infinity. One then enlarges also the set of double cones by adding "double cones at infinity"

$$\mathcal{O}' = \bigcup_{\substack{\mathcal{O}_1 \in \mathcal{K} \\ \mathcal{O}_1 \text{ spacelike to } \mathcal{O}}} \mathcal{O}_1 \quad . \tag{3.4}$$

Let $\mathcal{K}_\infty$ be the set of all these double cones. One now may define for $\mathcal{O} \in \mathcal{K}$ the algebra $\mathcal{A}(\mathcal{O}')$ to be the maximal algebra compatible with locality

$$\mathcal{A}(\mathcal{O}') = \mathcal{A}(\mathcal{O})' \quad . \tag{3.5}$$

It is, however, not clear whether this extension of the system of local algebras still satisfies weak additivity. The minimal algebra which one may associate to $\mathcal{O}'$ is

$$\mathcal{A}_{\min}(\mathcal{O}') = \bigvee_{\substack{\mathcal{O}_1 \in \mathcal{K} \\ \mathcal{O}_1 \text{ spacelike to } \mathcal{O}}} \mathcal{A}(\mathcal{O}_1) \quad . \tag{3.6}$$

One says that the local net fulfils Haag dulity if $\mathcal{A}_{\min}(\mathcal{O}') = \mathcal{A}(\mathcal{O}')$. Weak additivity is then automatically satisfied. In cases where Haag duality fails one may introduce the dual net

$$\mathcal{A}^d(\mathcal{O}) = \mathcal{A}_{\min}(\mathcal{O}')' \quad . \tag{3.7}$$

Under quite general conditions the dual net is again local [28]. Then $\mathcal{A}^d$ automatically satisfies Haag duality. Another possibility is to require weak additivity directly: Let

$$\bigcup \mathcal{O}_i \cup \bigcup \hat{\mathcal{O}}_j' \supset \mathcal{O}' \quad , \quad \mathcal{O}_i, \hat{\mathcal{O}}_j, \mathcal{O} \in \mathcal{K} \quad . \tag{3.8}$$

Then we require

$$\mathcal{A}(\mathcal{O})' \subset \bigvee_i \mathcal{A}(\mathcal{O}_i) \vee \bigvee_j \mathcal{A}(\hat{\mathcal{O}}_j)' \quad . \tag{3.9}$$

The nice feature of this requirement is that it is actually independent of the original representation; namely by taking the commutants we get the equivalent condition

$$\mathcal{A}(\mathcal{O}) \supset \bigcap_i \mathcal{A}(\mathcal{O}_i)' \cap \bigcap_j \mathcal{A}(\hat{\mathcal{O}}_j) \quad . \tag{3.10}$$

Let us denote the enlarged family of local algebras by $\mathcal{A}_\infty = (\mathcal{A}(\mathcal{O}))_{\mathcal{O} \in \mathcal{K}_\infty}$. Since $\mathcal{K}_\infty$ is not directed the union of the local algebras is not an algebra. A global algebra $\mathcal{A}_\infty$ may implicitly be defined as follows:

Let $(\pi_\mathcal{O})_{\mathcal{O} \in \mathcal{K}}$ be any family of representation $\pi_\mathcal{O} : \mathcal{A}(\mathcal{O}) \to \mathcal{B}(\mathcal{H})$ such that $\pi_\mathcal{O}|_{\mathcal{A}(\mathcal{O}_1)} = \pi_{\mathcal{O}_1}$ for $\mathcal{O}_1 \subset \mathcal{O}$. Then there is a unique representation π of $\mathcal{A}_\infty$ such that $\pi|_{\mathcal{A}(\mathcal{O})} = \pi_\mathcal{O}$.

The compatibility conditions are simpler for representations than for states. Hence we look at all representations π of $\mathcal{A}_\infty$ on separable Hilbert spaces such that $\pi|_{\mathcal{A}(\mathcal{O})}$ is normal for all $\mathcal{O} \in \mathcal{K}_\infty$. But for a type III factor π is normal if and only if it is untarily equivalent to the identical representation. Hence we find the DHR selection criterion: $\pi|_{\mathcal{A}(\mathcal{O})}$ is unitarily equivalent to the identity of $\mathcal{A}(\mathcal{O})$ for all $\mathcal{O} \in \mathcal{K}_\infty$.

A similar criterion had previously been considered by Borchers. It is clear that the DHR criterion does not cover all representations of interest; for instance, in QED electrically charged states have a rather complicated structure at spacelike infinity. But even in completely massive theories the DHR criterion is too restrictive.

Let us consider an irreducible positive energy representation π of $\mathcal{A}$ where the energy momentum spectrum contains an isolated mass shell. Such a representation describes states of a massive particle in a theory without massless particles. One finds the following result [3]:

Theorem *There is a unique vacuum representation π_0 of $\mathcal{A}$ such that*

$$\pi|_{\mathcal{A}(\mathcal{C}')} \simeq \pi_0|_{\mathcal{A}(\mathcal{C}')}$$

where $\mathcal{C}$ denotes a spacelike cone, i.e. a set of the form $\mathcal{C} = a + \bigcup_{\lambda>0} \lambda\mathcal{O}$ for some double cone of spacelike directions ($r_+^2 = r_-^2 = -1$, $r_+ - r_- \in V_+$)

$$\mathcal{O} = \mathcal{O}_{r_+,r_-} = \{r \in M, r^2 = -1, r \in r_+ + V_- \cap r_- + V_+\}$$

and $\mathcal{A}(\mathcal{C}')$ is the C^*-algebra generated by all algebras $\mathcal{A}(\mathcal{O})$ with $\mathcal{O}$ spacelike to $\mathcal{C}$.

In the following we will restrict ourselves to the conceptually simplest situation where $\mathcal{A}$ satisfies Haag duality and where only representations fulfilling the DHR criterion are considered.

4. The Trace of an Internal Symmetry

Can you hear the shape of a drum? Kac' famous question of whether knowledge of the eigenvalues of a Laplacian contains enough information about the geometry of a compact Riemanncian space has an analogue in field theory: can the internal symmetries be detected from the observable data?

In order to investigate this question we look at a system with a compact internal symmetry group G. The system is described by a net of von Neumann algebras $\mathcal{F} = (\mathcal{F}(\mathcal{O}))_{\mathcal{O}\in\mathcal{K}}$ together with a faithful action $g \mapsto \alpha_g$ of G by automorphismus of $\mathcal{F}$ with

$$\alpha_g(\mathcal{F}(\mathcal{O})) = \mathcal{F}(\mathcal{O}) \quad ,\mathcal{O} \in \mathcal{K}, g \in G \tag{4.1}$$

and such that $g \mapsto \alpha_g(F)$ is σ-continnous for every $F \in \mathcal{F}(\mathcal{O})$ and every $\mathcal{O} \in \mathcal{K}$. $\mathcal{F}$ is assumed to satisfy graded spacelike commutativity, i.e. there is a $k \in G$ with $k^2 = e$ inducing a grading of $\mathcal{F}$ such that the graded commutator between spacelike separated elements of $\mathcal{F}$ vanishes.

The observable algebra $\mathcal{A}(\mathcal{O})$ is defined as the set of fixed points of $\mathcal{F}(\mathcal{O})$ under G,

$$\mathcal{A}(\mathcal{O}) = \{A \in \mathcal{F}(\mathcal{O}), \alpha_g(A) = A\,\forall g \in G\} \quad . \tag{4.2}$$

We consider an irreducible faithful representation of $\mathcal{F}$ on a separable Hilbert space $\mathcal{H}$ together with a strongly continnous representation U of G which implements α. Let $\mathcal{H}_0$ be the subspace of G invariant vectors and π_0 the corresponding subrepresentation of $\mathcal{A}$ on $\mathcal{H}_0$ ("vacuum representation"). We assume

(i) $\mathcal{F}(\mathcal{O})\mathcal{H}_0$ is dense in $\mathcal{H}$ $\forall \mathcal{O} \in \mathcal{K}$
(ii) $\pi_0(\mathcal{A}(\mathcal{O}'))' = \pi_0(\mathcal{A}(\mathcal{O}))$ (**Haag duality**)
(iii) $E \in \mathcal{A}(\mathcal{O})$ a projection, $\overline{\mathcal{O}} \subset \mathcal{O}_1$, $\mathcal{O}, \mathcal{O}_1 \in \mathcal{K} \Rightarrow \exists W \leftarrow \mathcal{A}(\mathcal{O}_1)$ such that $W^*W = 1$, $WW^* = E$ (**Borchers property**).

The representation of $\mathcal{A}$ on $\mathcal{H}$ decomposes according to the irreducible subrepresentation of U, and since $\mathcal{F}$ is a *-algebra with a faithful action of G, every irreducible representation of G occurs; moreover, there are nonzero irreducible tensors in $\mathcal{F}(\mathcal{O})$ transforming according to these representations.

Now let $F_1, ..., F_n$ be a tensor transforming according to a unitary representation τ of G which is equivalent to some subrepresentation of U. We consider the mapping $\chi : \mathcal{A} \to \mathcal{B}(\mathcal{H})$,

$$\chi(A) = \sum_{i=1}^{n} F_i^* A F_i \quad . \tag{4.3}$$

For $A \in \mathcal{A}(\mathcal{O}_1)$ with $\mathcal{O}_1 \supset \mathcal{O}$ $\chi(A)$ is a G-invariant element of $\mathcal{F}(\mathcal{O}_1)$, hence $\chi(A) \in \mathcal{A}(\mathcal{O}_1)$ and χ is a positive mapping from $\mathcal{A}$ into $\mathcal{A}$. Actually, χ is completely positive, i.e. it is of the form

$$\chi(A) = F^* \pi(A) F \tag{4.4}$$

with $\pi(A) = A \otimes 1_n \in M_n(\mathcal{B}(\mathcal{H}))$ and $F^* = (F_1^*, ..., F_n^*)$.

One can now exploit the fact that in quantum field theory all multiplicities are equal (Borchers property). As shown by Doplicher and Roberts [29] there is a τ-tensor $(\psi_1^*, ..., \psi_n^*)$ in $\mathcal{F}(\mathcal{O})$ such that

$$\psi_i^* \psi_j = \delta_{ij} \quad , \quad \sum \psi_i \psi_i^* = 1 \quad , \tag{4.5}$$

and all τ-tensors $(F_1, ..., F_n)$ in $\mathcal{F}(\mathcal{O})$ are of the form

$$F_i = \psi_i^* B \tag{4.6}$$

for some $B \in \mathcal{A}(\mathcal{O})$. Unfortunately, the proof of this remarkable fact is rather indirect. I will not try to reproduce it here and refer to [29,13]. I will instead discuss the rich and surprising consequences of this fact.

First one may define as in (4.3) the associated positive mapping

$$\rho(A) = \sum \psi_i A \psi_i^* \quad . \tag{4.7}$$

But due to (4.5), ρ is not merely a positive mapping but an endomorphism of $\mathcal{A}$ (i.e. it respects also the multiplicative structure of $\mathcal{A}$). For $n > 1$ ρ is not surjective; but there exists a positive mapping ϕ from $\mathcal{A}$ into $\mathcal{A}$ which is a left inverse of ρ,

$$\phi(A) = \frac{1}{n} \sum_{i=1}^{n} \psi_i^* A \psi_i \quad . \tag{4.8}$$

The multiplet $(\psi_1, ..., \psi_n)$ is of course not unique. Let $U \in \mathcal{A}(\mathcal{O})$ be unitary and $\psi_i' = U\psi_i$. Then $(\psi_1', ..., \psi_n')$ is another multiplet with the same algebraic properties, and

$$U = \sum \psi_i' \psi_i^* \quad . \tag{4.9}$$

Conversely, any multiplet $(\psi_1', ..., \psi_n')$ is of the form $U\psi_i$, $i = 1, ..., n$ for some unitary $U \in \mathcal{A}(\mathcal{O})$.

The operators $\psi_1, ..., \psi_n$ generate a unique *-algebra, the so called Cuntz algebra $\mathcal{O}_n^0$ [30]. Any G-invariant element of $\mathcal{O}_n^0$ belongs to $\mathcal{A}(\mathcal{O})$, hence we may try to detect the invariant subalgebra of $\mathcal{O}_n^0$,

$$\mathcal{O}_G^0 = \{A \in \mathcal{O}_n^0 \quad , \alpha_g(A) = A \ \forall g \in G\} \tag{4.10}$$

within $\mathcal{A}(\mathcal{O})$.

Now $\mathcal{O}_G^0$ inherits a $\mathbb{Z}$-grading from $\mathcal{O}_n^0$ (number of ψ's-number of ψ^*'s), and every homogenous $T \in \mathcal{O}_G^0$ with degree $k \in \mathbb{Z}$ may be written in the form

$$T = \sum_{i,j} t_{ij} \psi_{i_1} ... \psi_{i_{n+k}} \psi_{j_1}^* ... \psi_{j_n}^* \tag{4.11}$$

with $t_{ij} \in \mathbb{C}$ and $n \in \mathbb{N}$ sufficiently large. The G-invariance of T means that (t_{ij}) is an intertwiner between the n-th and the $(n+k)$-th tensor power of the representation τ of G. Inside of $\mathcal{A}(\mathcal{O})$ where G-invariance is automatically satisfied intertwiners T of the form (4.11) can be characterized by the intertwining property $(m = n + k)$

$$T\rho^n(A) = \rho^m(A)T \quad . \tag{4.12}$$

Hence we find find the equality

$$\mathcal{O}_G^0 = \bigcup_{n,m \geq 0} (\rho^n, \rho^m) =: \mathcal{O}_\rho^0 \tag{4.13}$$

where (ρ^n, ρ^m) denotes the set of all operators $T \in \mathcal{A}$ satisfying (4.12).

$\mathcal{O}_\rho^0$ is defined exclusively in terms of observables. Moreover, for certain intertwiners between tensor powers of τ the corresponding elements of $\mathcal{O}_\rho^0$ can be directly specified. Consider e.g. the intertwiners arising from permutations of factors in tensor products. They correspond to the elements

$$\hat{\varepsilon}(p) = \sum_i \psi_{p(i_1)} ... \psi_{p(i_n)} \psi_{i_n}^* ... \psi_{i_1}^* \quad , p \in S_n \tag{4.14}$$

of $\mathcal{O}_{U(n)}^0 \subset \mathcal{O}_G^0 \subset \mathcal{O}_n^0$. Now let $\mathcal{O}_1 \subset \mathcal{O}'$, $\mathcal{O}_1 \in \mathcal{K}$ and $\psi'^* = (\psi_1'^*, ..., \psi_n'^*)$ be a τ-tensor in $\mathcal{F}(\mathcal{O}_1)$ satisfying (4.5). Then from (4.9)

$$\psi_i' = U\psi_i \,\text{with}\, U = \sum \psi_j' \psi_j^* \in \mathcal{A} \tag{4.15}$$

and

$$\rho'(A) := \sum \psi_i' A \psi_i'^* = U\rho(A)U^{-1} \tag{4.16}$$

i.e. $U \in (\rho, \rho')$. Hence for the transposition $p = (12)$ we get

$$\begin{aligned} \hat{\varepsilon}(12) &= \sum_{i,j} \psi_i \psi_j \psi_i^* \psi_j^* \\ &= \sum_{i,j} \psi_i U^{-1} \psi_j' \psi_i^* \psi_j^* \\ &= \pm \sum_{i,j} \psi_i U^{-1} \psi_i^* \psi_j' \psi_j^* \\ &= \pm \rho(U)^{-1} U =: \pm \varepsilon_\rho \end{aligned} \tag{4.17}$$

where the third equality sign comes from graded locality, and the intrinsically defined operator ε_ρ is the so called statistics operator (see below).

In a similar way in the case $\tau(G) \subset SU(n)$ the operator corresponding to the determinant

$$R = \frac{1}{\sqrt{d!}} \sum_{p \in S_n} \mathrm{sign}(p) \psi_{p(1)} \ldots \psi_{p(n)} \tag{4.18}$$

can be intrinsically characterized as an isometric intertwiner in (id, ρ^n). In sections 6 and 7 we will see that these data can be extracted from a generic quantum field theory in $D \geq 3$ space time dimensions and that they suffice to recover the group G and the field algebra $\mathcal{F}$.

5. Composition of Sectors

We now return to the general situation where only the observable algebra is given, and the internal symmetries are to be detected.

Let $\mathcal{A} \subset \mathcal{B}(\mathcal{H})$ and $\Omega \in \mathcal{H}$ with $\omega_0(\cdot) = (\Omega, \cdot\Omega)$ denoting the vacuum state, and consider the set of positive linear functionals

$$\mathcal{S} = \{\omega \in \mathcal{A}_+^* \mid \pi_\omega \, \text{satisfies the DHR criterion}\} \quad . \tag{5.1}$$

A norm dense subset $\mathcal{S}_0$ of $\mathcal{S}$ is formed by those functionals which are dominated at spacelike infinity by $\omega_0, \mathcal{S}_0 = \bigcup_{\mathcal{O} \in \mathcal{K}} \mathcal{S}(\mathcal{O})$, where

$$\begin{aligned} \mathcal{S}(\mathcal{O}) = \{\omega \in \mathcal{S} \mid \exists T \in \mathcal{A}(\mathcal{O}), T \geq 0 \, \text{such that} \\ \omega(A') = \omega_0(TA') \forall A' \in \mathcal{A}(\mathcal{O}')\} \quad . \end{aligned} \tag{5.2}$$

Let $\omega \in \mathcal{S}(\mathcal{O})$ and let $(\pi_\omega, \mathcal{H}_\omega, \Phi_\omega)$ denote the GNS triple associated to ω. By

$$A'\Omega \mapsto \pi_\omega(A')\Phi_\omega \quad , A' \in \mathcal{A}(\mathcal{O}') \tag{5.3}$$

one defines a mapping S_ω from $\mathcal{H}_0$ into $\mathcal{H}_\omega$ which intertwines the restrictions of $id \equiv \pi_0$ and π_ω to $\mathcal{A}(\mathcal{O}')$. We then introduce the completely positive mapping on $\mathcal{A}$,

$$\chi_\omega(A) = S_\omega^* \pi_\omega(A) S_\omega \quad , A \in \mathcal{A} \quad . \tag{5.4}$$

χ_ω satisfies $\omega_0 \circ \chi_\omega = \omega$ and

$$\chi_\omega(A'BC') = A'\chi_\omega(B)C' \quad , A', C' \in \mathcal{A}(\mathcal{O}'), B \in \mathcal{A} \quad , \tag{5.5}$$

and is uniquely determined by these properties. It follows that the association $\omega \mapsto \chi_\omega$ respects positive linear combinations. By Haag duality, χ_ω maps $\mathcal{A}(\mathcal{O}_1)$ into $\mathcal{A}(\mathcal{O}_1)$ for $\mathcal{O}_1 \supset \mathcal{O}$, hence $\mathcal{A}$ into $\mathcal{A}$. Therefore we can introduce a product of states by

$$\begin{cases} \mathcal{S}_0 \times \mathcal{S}_0 \to \mathcal{S}_0 \\ \omega_1, \omega_2 \mapsto \omega_1 \times \omega_2 := \omega_0 \circ \chi_{\omega_1} \chi_{\omega_2} \quad . \end{cases} \tag{5.6}$$

In general, the product of pure states is not pure. This phenomenon is connected with the failure of Haag duality in the induced representation. Let

$$d(\omega) = \mathrm{Ind}[\pi_\omega(\mathcal{A}(\mathcal{O}'))' : \pi_\omega(\mathcal{A}(\mathcal{O}))]^{\frac{1}{2}} \tag{5.7}$$

where Ind$[M:N]$ means the Jones index [24] of the inclusion of von Neumann algebras $N \subset M$. The index can be characterized in terms of a conditional expectation $\mathcal{E}: M \to N$. It is the best possible constant in the Pimsner-Popa-inequality [26]

$$\mathcal{E}(A) \geq \mathrm{Ind}[M:N]^{-1}A \quad , A \geq 0 \quad . \tag{5.8}$$

$d(\omega)$ is called the statistical dimension of π_ω in the DHR theory.

Let $\mathcal{S}_{\mathrm{fin}}(\mathcal{O}) = \{\omega \in \mathcal{S}(\mathcal{O}) | \, d(\omega) < \infty\}$. States in $\mathcal{S}_{\mathrm{fin}}(\mathcal{O})$ are finite convex combinations of pure states, and products of states in $\mathcal{S}_{\mathrm{fin}}(\mathcal{O})$ are again in $\mathcal{S}_{\mathrm{fin}}(\mathcal{O})$. For the statistical dimensions one finds the relations

$$\begin{aligned} d(\sum \omega_i) &= \sum d(\omega_i) \\ d(\omega_1 \times \omega_2) &\leq d(\omega_1) d(\omega_2) \quad . \end{aligned} \tag{5.10}$$

There is also a conjugation $\omega \mapsto \overline{\omega}$ in $\mathcal{S}_{\mathrm{fin}}(\mathcal{O})$ with $d(\omega) = d(\overline{\omega})$ which satisfies the inequality

$$\begin{aligned} \omega \times \overline{\omega} &\geq d(\omega)^{-1} \, \chi_\omega(1) \omega_0 \chi_\omega(1) \\ \overline{\omega} \times \omega &\geq d(\omega)^{-1} \, \chi_{\bar{\omega}}(1) \omega_0 \chi_{\overline{\omega}}(1) \end{aligned} \tag{5.11}$$

where we used the convention

$$(A\omega B)(C) = \omega(ACB) \tag{5.12}$$

for left and right multiplication of linear functionals on an algebra by elements of the algebra.

6. Statistics

States which are localized spacelike to each other commute, i.e. if $\omega_i \in \mathcal{S}(\mathcal{O}_i), i = 1, 2$ and $\mathcal{O}_1 \subset \mathcal{O}_2'$ then

$$\omega_1 \times \omega_2 = \omega_2 \times \omega_1 \quad . \tag{6.1}$$

This follows from the fact that restricted to any subalgebra $\mathcal{A}(\mathcal{O})$ there are states $\hat{\omega}_i$ in the vacuum sector coinciding with ω_i on $\mathcal{A}(\mathcal{O})$,

$$\hat{\omega}_i = S_i^* \omega_0 S_i \, , \, S_i \in \mathcal{A}(\hat{\mathcal{O}}_i) \tag{6.2}$$

such that $\hat{\mathcal{O}}_1 \subset \hat{\mathcal{O}}_2'$, $\mathcal{O}_i \subset \hat{\mathcal{O}}_i, i = 1, 2$. Then for $A \in \mathcal{A}(\mathcal{O})$ we find

$$\begin{aligned} \omega_1 \times \omega_2(A) &= \omega_0(S_1^* S_2^* A S_2 S_1) \\ &= \omega_0(S_2^* S_1^* A S_1 S_2) \\ &= \omega_2 \times \omega_1(A) \quad . \end{aligned} \tag{6.3}$$

Now we may lift the product structure from the states to the vectors which induce the states. Here we have to fix conventions by choosing suitable representations. It is convenient to use as representations endomorphismus ρ of $\mathcal{A}$ which are special cases of the positive mappings χ_ω. We find

$$\chi_\omega(A) = S^* \rho(A) S \, , \, A \in \mathcal{A} \tag{6.4}$$

where $\rho \in \mathrm{End}(\mathcal{A})$ and $S \in \mathcal{A}$ are not uniquely determined by ω. For ρ irreducible the freedom consists in replacing S by US and ρ by $AdU \circ \rho$ for unitaries $U \in \mathcal{A}$. The state ω is induced by the vector $S\Omega$ in the representation ρ.

Now let $\chi_{\omega_i}(\cdot) = S_i^* \rho_i(\cdot) S_i$, $i = 1, 2$, then $\omega_1 \times \omega_2$ is induced by the vector $\rho_1(S_2)S_1\Omega$ in the representation $\rho_1\rho_2$. Similarly, $\omega_2 \times \omega_1$ is induced by the vector $\rho_2(S_1)S_2\Omega$ in the representation $\rho_2\rho_1$. For mutually spacelike localized states one finds

$$\rho_2(S_1)S_2\Omega = \varepsilon(\rho_1, \rho_2)\rho_1(S_2)S_1\Omega \tag{6.5}$$

where $\varepsilon(\rho_1, \rho_2) \in (\rho_1\rho_2, \rho_2\rho_1)$ is unitary.

In $D \geq 3$ dimensions $\varepsilon(\rho_1, \rho_2)$ turns out to be uniquely determined by ρ_1 and ρ_2 where as in $D = 2$ dimensions one obtains two possibly different operators $\varepsilon(\rho_1, \rho_2)_<$, $\varepsilon(\rho_1, \rho_2)_>$, dependent on whether $\mathcal{O}_1$ is to the left or to the right of $\mathcal{O}_2$. In the special case $\rho_1 = \rho_2 = \rho$ the operator $\varepsilon_\rho = \varepsilon(\rho, \rho)$ $(D \geq 3)$ resp. $\varepsilon_\rho = \varepsilon(\rho, \rho)_<$ $(D = 2)$ is called the statistics operator. It satisfies the relation

$$\varepsilon_\rho \rho(\varepsilon_\rho)\varepsilon_\rho = \rho(\varepsilon_\rho)\varepsilon_\rho\rho(\varepsilon_\rho) \tag{6.6}$$

and in $D \geq 3$ dimensions

$$\varepsilon_\rho^2 = 1 \tag{6.7}$$

and this induces by $\sigma_n \mapsto \varepsilon^{(\rho)}(\sigma_n) := \rho^{n-1}(\varepsilon_\rho)$ a unitary representation $\varepsilon^{(\rho)}$ of the braid group B_∞ which degenerates in $D \geq 3$ dimensions to a representation of the permutation group S_∞.

For an analysis of this representation one uses the left inverse ϕ of ρ. It is the positive mapping determined by the conjugate of the state $\omega_0 \circ \rho$,

$$\phi = d(\omega)\chi_{\overline{\omega_0 \circ \rho}} \quad . \tag{6.8}$$

By iterating ϕ we get a function φ of positive type on B_∞,

$$\lim_{n \to \infty} \phi^n \circ \varepsilon^{(\rho)}(b) = \varphi(b)1 \tag{6.9}$$

which is a Markov trace in the sense of Jones [24], i.e. it is a trace and satisfies

$$\varphi(b_n\sigma_n^{\pm 1}) = \varphi(b_n)\varphi(\sigma_n^{\pm 1}) \quad , b_n \in B_n \quad . \tag{6.10}$$

$\lambda_\rho = \varphi(\sigma_n) = \kappa(\rho)d(\rho)$ is called the statistics parameter in the DHR theory and the Markov parameter in the Jones theory, $d(\rho)$ is the statistical dimension and $\kappa(\rho)$ the statistical phase.

The Markov trace φ induces a ribbon invariant and by renormalization a link invariant. If for instance ρ^2 is irreducible $\varepsilon(\rho)$ is an abelian representation. If ρ^2 is the direct sum of two irreducible representations one obtains the representation of the braid group discovered by Jones, Ocneanu and Wenzl and the corresponding link invariants (Jones and HOMFLY invariant). If ρ^2 is the direct sum of three inequivalent representations one of which has statistical dimension 1 one gets the Murakami-Wenzl representations and the Kauffman invariant. In any case, the statistical dimensions must be the square roots of possible Jones indices and are therefore restricted to the set $\{2\cos\frac{\pi}{m}, m \geq 3\} \cup [2, \infty)$. Further restrictions have been found by Longo [31] and Rehren [32].

7. Reconstruction of the Symmetry Group

In the case where $\varepsilon^{(\rho)}$ is a representation of the symmetric group the statistical dimension d must be a natural number and the statistical phase κ can only assume the values ± 1. One then can define a bosonized representation by

$$\hat{\varepsilon}(p) = \begin{cases} \varepsilon^{(\rho)}(p) & , \quad \kappa = 1 \\ \text{sign}(p)\varepsilon^{(\rho)}(p) & , \quad \kappa = -1 \quad . \end{cases} \tag{7.1}$$

The representation ρ^d contains a subrepresentation with statistical dimension 1 which corresponds to the projection

$$E_d = \frac{1}{d!} \sum_{p \in S_d} \text{sign}(p)\hat{\varepsilon}(p) \quad . \tag{7.2}$$

We consider the case where this representation is equivalent to the vacuum representation. Since ρ is not assumed to be irreducible one may satisfy this property by replacing ρ by $\rho \oplus \overline{\rho}$ if necessary ($\bar{\rho}$ is an endomorphism conjugate to ρ, i.e. it is unitarily equivalent to $\pi_{\overline{\omega_0 \circ \rho}}$). By Borchers property there is an isometry $R \in \mathcal{A}$ with $RR^* = E_d$.

The reconstruction of the symmetry group due to Doplicher and Roberts now proceeds in the following steps. First one shows that the operators $\hat{\varepsilon}(p)$ and R generate a unique *-algebra which is isomorphic to $\mathcal{O}^0_{SU(d)}$. One then uses the fact that $\mathcal{A}$ as well as $\mathcal{O}^0_d$ are bimodules over $\mathcal{O}^0_{SU(d)}$, and one can perform the bimodule tensor product

$$\mathcal{F}^0 = \mathcal{A} \otimes_{\mathcal{O}^0_{SU(d)}} \mathcal{O}^0_d \tag{7.3}$$

with natural embeddings

$$\mathcal{O}^0_d \to \mathcal{F}^0 \quad , \quad \psi_i \mapsto 1 \otimes \psi_i \tag{7.4}$$

and

$$\mathcal{A} \to \mathcal{F}^0 \quad , \quad A \mapsto A \otimes 1 \quad . \tag{7.5}$$

$\mathcal{F}^0$ gets the structure of an associative *-algebra by the definitions

$$\begin{aligned} (A \otimes 1)(1 \otimes \psi) &= A \otimes \psi \\ (1 \otimes \psi_i)(A \otimes 1) &= \rho(A) \otimes \psi_i \\ (1 \otimes \psi_i^*)(A \otimes 1) &= R^* \rho^{n-1}(A) \otimes \hat{\psi}_i \end{aligned} \tag{7.6}$$

for $A \in \mathcal{A}$, $\psi \in \mathcal{O}^0_d$, $i = 1, ..., n$ and with

$$\hat{\psi}_i = \frac{1}{\sqrt{(d-1)!}} \sum_{p \in S_n, p(1)=i} \text{sign}(p)\psi_{p(2)} \cdots \psi_{p(d)} \quad . \tag{7.7}$$

$\mathcal{F}^0$ is not yet the field algebra, and $SU(d)$ is not yet the gange group, in general, as we used only special intertwiners in $\mathcal{O}^0_\rho$. This becomes visible in the nontriviality of the relative commutant of $\mathcal{A}$ in $\mathcal{F}^0$. It is gratisfying that the relative commutant coincides with the center of $\mathcal{F}^0$ which is an abelian *-algebra. $SU(d)$ acts ergodically on the center and hence transitive on its spectrum.

The final step in the DR reconstruction is now done by choosing a point z in the spectrum. The stabilizer G_z of z is a subgroup of $SU(d)$ which can be identified (up to conjugation within $SU(d)$) with the gauge group, and the field algebra is

$$\mathcal{F} = \mathcal{F}^0 / \mathcal{I}_z \tag{7.8}$$

where $\mathcal{I}_z$ is the ideal generated by the kernel of z.

8. "Quantum Symmetry"

At present it is not known what the appropriate generalization of the concept of an internal symmetry group is which is applicable in theories with braid group statistics. I want to discuss several attempts.

The most direct one from the point of view of the theory of superselection sectors is an investigation of the intertwiner algebra $\mathcal{O}^0_\rho$. Following Ocneanu [25] one can describe it as a "path algebra". One introduces a graph whose vertices are the irreducible representations occuring as subrepresentations of powers of ρ, and where $N_{\alpha\beta}$ directed edges point from ρ_α to ρ_β if $\rho\rho_\alpha$ contains ρ_β $N_{\alpha\beta}$ times. The algebra is now generated by formal linear combinations (ξ, η) where ξ and η are paths on the graph which start at the identity and have a common endpoint with the identification

$$(\xi, \eta) = \sum_e (\xi \circ e, \eta \circ e) \tag{8.1}$$

where the sum goes over all edges e emanating from the joint end point of ξ and η. The product is Witten's string product

$$(\xi, \eta)(\xi', \eta') = \delta_{\eta\xi'}(\xi, \eta') \tag{8.2}$$

where (8.1) has been used to make η and ξ' equally long. The intertwiners $\varepsilon^{(\rho)}(\rho_n) = \rho^{n-1}(\varepsilon_\rho)$ have the form

$$\rho^{n-1}(\varepsilon_\rho) = \sum R^{(n,n+1)}_{\xi\eta}(\xi, \eta) \tag{8.3}$$

where the coefficients $R^{(n,n+1)}_{\xi\eta}$ vanish unless the paths ξ and η have length $n+1$ and coincide up to length $n-1$. The matrices $R^{(n,n+1)}$ satisfy the constant Yang Baxter equations,

$$R^{(n,n+1)} R^{(n-1,n)} R^{(n,n+1)} = R^{(n-1,n)} R^{(n,n+1)} R^{(n-1,n)} \quad . \tag{8.4}$$

A complete list of conditions which $\mathcal{O}^0_\rho$ has to satisfy in order to occur in the theory of superselection sectors is not known at the moment.

Another approach which is near to the formulation of the DR reconstruction uses terms of the theory of categories. The objets are the endomorphisms of the DHR theory and the arrows are the intertwiners between endomorphisms. Fröhlich and Kerler showed how the category of indecomposable representations of a quantum group at a root of unity can be used to define a new category by factoring over the degenerate representations such that the new category is equivalent to the category of positive energy representations of certain conformal field theories. For more details I refer to the contribution of Dr. Kerler to these proceedings.

In the Fröhlich-Kerler approach one does not try to construct covariant fields which exhibit the quantum symmetry directly. A construction of covariant fields for special models has been performed by Todorov et al. [33] in an indefinite metric framework. Mack and Schomerus [34] recently succeeded in such a construction in a Hilbert space (i.e. with positive definite metric). Their symmetry is no longer a quantum group as in [33] but a quasi quantum group in the sense of Drinfeld (see the lectures of Prof.Mack).

There is also a proposal for the general case with a finite number of sectors ("rational theories"). Rehren [35] constructs a field algebra which is similar to the field algebra of Doplicher and Roberts. He starts from an endomorphism ρ which is equivalent to the direct sum of all sectors. Then he introduces the algebra

$$M = \bigcup_{n \geq 0} (\rho^n, \rho^n) \tag{8.5}$$

and the vector space

$$K = \bigcup_{n \geq 0} (\rho^{n+1}, \rho^n) \tag{8.6}$$

which has a natural structure as an M-bimodule

$$m \cdot k = \rho(m)k \quad , \quad k \cdot m = km \quad . \tag{8.7}$$

Now K may be decomposed into a direct sum of irreducible M-bimodules

$$K = \bigoplus_{\alpha} K_\alpha \tag{8.8}$$

where $K_\alpha = E_\alpha K$, $E_\alpha \in \rho(\mathcal{A})'$, $E_\alpha \rho(\cdot) \simeq \rho_\alpha(\cdot)$ and $\dim_M K_\alpha = d(\rho_\alpha)$. The tensor product of these bimodules can be decomposed into irreducible ones

$$K_\alpha \otimes_M K_\beta = \bigoplus N^\gamma_{\alpha\beta} K_\gamma \tag{8.9}$$

where $N^\gamma_{\alpha\beta}$ is the multiplicity of ρ_γ in $\rho_\alpha \rho_\beta$. He then constructs a field algebra $\mathcal{F}$ which is generated by $\mathcal{A}$ and ψ_k, $k \in K$ together with an embedding $\tau : M \to \mathcal{F} \cap \mathcal{A}'$ such that

(i) $k \to \psi_k$ is a bimodule homomorphism
(ii) $\psi_k A = \rho(A)\psi_k$, $A \in \mathcal{A}$
(iii) $\psi_k^* = R^* \psi_{\overline{k}}$

where R is an isometry in (id,ρ^2) with $R^*\rho(R) = d(\rho)^{-1}U$, U unitary, and $\overline{k} = d(\rho)^{\frac{1}{2}} \rho(k^*)R$.

The analogy of Rehren's approach to more conventional formulations of symmetry can be seen by choosing a basis of K as a right module over M. Let $b_i \in K$, $i = 1, ..., n$ with $\sum b_i b_i^* = 1$ and $b_i^* b_j = \delta_{ij} p_i$ where p_i are projections in M with $\sum \varphi(p_i) = d(\rho)$, φ being the Markov trace associated to ρ. Then the left action of M on K can be described by a homomorphism $\hat{\rho} : M \to M_n(M)$,

$$\rho(m) b_i = \sum b_j \hat{\rho}_{ji}(m) \quad , \quad \hat{\rho}_{ji}(m) \in M \tag{8.10}$$

and the covariance of the fields ψ_i under the symmetry assumes the form

$$\tau(m)\psi_i = \sum \psi_j \tau \circ \hat{\rho}_{ji}(m) \quad . \tag{8.11}$$

References

1. S. Doplicher, R. Haag, J.E. Roberts: Local Observables and Particle Statistics. Commun. Math. Phys.**23**,199(1971) and **35**,49(1974)
2. H.-J. Borchers: Local Rings and the Connection of Spin with Statistics. Commun. Math. Phys. **1**,281(1965)
3. D. Buchholz, K. Fredenhagen: Locality and the Structure of Particle States. Commun. Math. Phys.**84**,1(1982)
4. K. Fredenhagen, M. Marcu: Charged States in $\mathbb{Z}_2$ Gauge Theories. Commun. Math. Phys. **92**, 81 (1983)
5. J. Fröhlich, P. A. Marchetti: Soliton Quantization in Lattice Field Theories. Commun. Math. Phys. **112**, 343 (1987)
6. K. Szlachanyi: Non-Local Fields in the Z(2) Higgs Model: The Global Gauge Symmetry Breaking and the Confinement Problem. Commun. Math. Phys. **108**, 319 (1987)
7. K. Fredenhagen, K.H. Rehren, B. Schroer: Superselection Sectors with Braid Group Statistics and Exchange Algebras. Commun. Math. Phys. **125**,201(1989)
8. R. Longo: Index for Subfactors and Statistics of Quantum Fields. Commun. Math. Phys.**126**,217(1989) and **130**,285(1990)
9. J. Fröhlich, Gabbiani: Braid Statistics in Local Quantum Field Theory. Rev. Math. Phys. **2**,251(1990)
10. J. Fröhlich: New Superselection Sectors ("Soliton States") in Two Dimensional Bose Quantum Field Theory Models. Commun. Math. Phys. **47**,269(1976)
11. K. Fredenhagen: Generalizations of the Theory of Superselection Sectors. In "The Algebraic Theory of Superselection Sectors. Introduction and Recent Results", D. Kastler (ed.), World Scientific 1990
12. S. Doplicher, J.E. Roberts: Endomorphisms of C^*-algebras, Cross Products and Duality for Compact Groups. Ann. Math. **130**, 75 (1989)
13. S. Doplicher, J.E. Roberts: Why there is a Field Algebra with a Compact Gauge Group Describing the Superselection Structure in Particle Physics. Commun. Math. Phys.**131**,51(1990)
14. H.J. Borchers, J. Yngvason: Positivity of Wightman functions and the Existence of Local Nets. Commun. Math. Phys. **127**, 607 (1990)
15. R. Haag, D. Kastler: An Algebraic Approach to Quantum Field Theory. J. Math. Phys.**5**,848(1964)
16. H.J. Borchers: On the Vacuum State in Quantum Field Theory. Commun. Math. Phys. **1**, 57 (1965)
17. H.J. Borchers: Energy and Momentum as Observables in Quantum Field Theory. Commun. Math. Phys. **2**, 49 (1966)
18. H.-J. Borchers D. Buchholz, : The Energy Momentum Spectrum in Local Field Theories with Broken Lorentz Symmetry. Commun. Math. Phys. **97**, 169 (1985)
19. H.-J. Borchers: Locality and Covariance of the Spectrum. Bielefeld 1984
20. H.-J. Borchers: A Remark on a Theorem of B. Misra. Commun. Math. Phys. **4**, 315 (1967)
21. M. Takesaki, W. Winnink: Local Normality in Quantum Statistical Mechanics. Commun. Math. Phys. **30**, 129 (1973)
22. R. Haag, H. Narnhofer, U. Stein: On Quantum Field Theory in Gravitational Background. Commun. Math. Phys.**94**, 219 (1984)
23. D. Buchholz, C. D'Antoni, K. Fredenhagen:The Universal Structure of Local Algebras. Commun. Math. Phys. **111**, 123 (1987)
24. V.F. Jones: Index of Subfactors. Invent.Math.**72**,1 (1983)

25. A. Ocneanu: Quantized Groups, String Algebras and Galois Theory for Algebras. Lond.Math.Soc., Lect. Notes Vol. **136**, Evans, Takesaki (eds.),119-172(1989)
26. M. Pimsner, S. Popa: Entropy and Index for Subfactors. Ann. Sci. Ec. Norm. Sup. **19**, 57 (1986)
27. H. Wenzl: Hecke Algebras of Type A_n and Subfactors. Invent. Math. **92**, 349 (1988)
28. J.J. Bisognano, E.H. Wichmann: On the Duality Condition for a Hermitean Scalar Field. J. Math. Phys **17**, 303 (1975)
29. S. Doplicher, J.E. Roberts: Fields, Statistics and Non Abelian Gauge Groups. Commun. Math. Phys. **28**, 331 (1972)
30. J. Cuntz: Simple C^*-Algebras Generated by Isometries. Commun. Math. Phys. **57**, 173 (1977)
31. R. Longo: Minimal Index and Braided Subfactors (preprint)
32. K.H. Rehren: Braid Group Statistics and their Superselection Rules. In "The Algebraic Theory of Superselection Sectors. Introduction and Recent Results", D. Kastler (ed.), World Scientific 1990
33. L.K. Hadjiivanov, R.R. Paunov, I.T. Todorov:Quantum Group Extended Chiral p-Models. INRNE-TH-90-7 (preprint)
34. G. Mack, V. Schomerus: Quasi Quantum Group Symmetry and Local Braid Relations in the Conformal Ising Model. DESY 91-060 (preprint)
35. K.H. Rehren: Field Operators for Anyons and Plektons. DESY 91-043 (preprint)

INCOMPRESSIBLE QUANTUM FLUIDS, GAUGE-INVARIANCE, AND CURRENT ALGEBRA

Jürg Fröhlich and Urban M. Studer

Theoretische Physik, ETH – Hönggerberg
CH-8093 Zürich

1. INTRODUCTION

This rather long paper is a tale of non-relativistic quantum theory summarizing research that has been conducted during the last one and a half years, and the main results of which have been sketched in two lectures presented at the Cargèse summer school of 1991, as well as in lectures at several other institutions. Coworkers in our endeavor have been, or are, Thomas Kerler, Pieralberto Marchetti and Tony Zee. Basic help and guidance were generously provided by Rudolf Morf. We are deeply grateful to these colleagues without whom our enterprise would have suffered premature shipwreck. We also thank J. Avron and G. Felder for very helpful discussions.

After some basic ideas underlying our approach had been developed during a students seminar on the quantum Hall effect at ETH organized by Rudolf Morf and J.F., we became aware of independent, but slightly prior work of X.G. Wen [1,2] that bears much resemblance with ours [3,4,5]. A 1982 paper of B.I. Halperin [6], supplemented by more recent results on current algebra [7,8,9] and on Chern-Simons gauge theory [10,11,12], has been instrumental in triggering the work in [2,3]. J.F. should also like to acknowledge some very stimulating discussions with Paul Wiegmann, in spring of 1989, whose remarks turned out to be much to the point.

Work vaguely or closely related to Wen's and ours has been carried out by several people and can be found in [13], and refs. indicated therein.

The task assigned to J.F. at the Cargèse school was to lecture on low-dimensional quantum theory with braid statistics and quantum symmetries. This task could have been fulfilled by lecturing on the beautiful *mathematics* of braid statistics and quantum symmetries that involves operator algebra theory, quantum groups and their subtle representation theory, holomorphic vector bundles over Riemann surfaces, and, perhaps most importantly, the theory of tensor categories. However, as physicists,

New Symmetry Principles in Quantum Field Theory, Edited by
J. Fröhlich et al., Plenum Press, New York, 1992

we may have a feeling of loosing ground in this world of mathematics. In any event, other people essentially took over that task, and it appeared desirable to lecture about *physical systems* with braid statistics and quantum symmetires. Fortunately, such systems exist in nature! A two-dimensional electron gas in a strong transverse magnetic field can exhibit quasi-particle excitations of *fractional charge* and *fractional (abelian braid) statistics*, the famous Laughlin vortices. One can imagine two-dimensional systems of condensed matter which will actually exhibit quasi-particle excitations with *non-abelian* braid statistics and quantum symmetries; see e.g. [14]. But it is likely that such systems have not been realized in the laboratory, yet. [Candidate systems are 2D systems with broken reflection – and time reversal invariance made of particles of spin ≥ 1.]

The phenomena of braid statistics and quantum symmetries in a two-dimensional quantum system appear to be intimately related to the property of local gauge invariance of the system. One of the key ideas underlying the work described in this paper is that one can acquire a surprisingly rich amount of information on a system of non-relativistic matter by studying how it reacts when coupled to external gauge fields. In Sect. 2, we therefore study how systems of non-relativistic quantum mechanical particles with spin interact with external electromagnetic fields, with "tidal gauge fields" providing a quantum-mechanical description of Coriolis forces and spin precession in moving coordinates, and to a variable metric on space. Our formalism can be applied to systems in one, two, and three space dimensions. It reveals a basic $U(1)_{\text{em}} \times SU(2)_{\text{spin}}$-gauge-invariance of non-relativistic quantum theory which gives rise to powerful *Ward identities.*

In Sect. 3, we review and "explain" a number of classic effects in non-relativistic quantum theory from the point of view of its $U(1)_{\text{em}} \times SU(2)_{\text{spin}}$ gauge invariance, (supplemented by certain assumptions concerning the structure of states that minimize the energy – , or free energy density). Included are the Aharonov-Bohm effect and its $SU(2)_{\text{spin}}$-variant, the Aharonov-Casher effect, flux quantization in superconductors and vorticity quantization in superfluids, the London equation for the supercurrent density in a superconductor and the related Anderson-Higgs mechanism, and different variants of the Einstein-de Haas (-Barnett) effect.

It turns out that the celebrated quantum Hall effect (and the related quantum Hall effect for spin currents [5]) encountered in two-dimensional electron gases (realized, for example, in heterojunctures) subject to a strong, transverse, external magnetic field is yet another phenomenon reflecting the $U(1) \times SU(2)$-gauge-invariance of non-relativistic quantum theory. In Sect. 4, we therefore study two-dimensional, *incompressible* electron fluids in external electromagnetic fields. The notion of incompressibility that we are using is the following: A system at zero temperature (but positive density) is incompressible if the energy of all physical states describing extended (as opposed to localized) excitations of the groundstate is *strictly* above the ground state energy. Incompressible systems are free of dissipation, and therefore the longitudinal resistance vanishes. Experimentally, this is found to be the case when the Hall conductivity is on a plateau [15].

By using $U(1)\times SU(2)$-Ward identities we show that two-dimensional, incompressible quantum fluids have *universal* properties. For example, their effective action as a functional of small perturbations in the external electromagnetic field has a universal form which we determine explicitly. The notion of universality that emerges here is very much the same as the one encountered in the theory of critical phenomena associated with continuous phase transitions.

Our results on the effective action, summarized in Sect. 4, imply the general equations describing the Hall effects for the electric charge – and current density and for the spin – and spin-current density in systems with vanishing longitudinal resistances. Moreover, they yield a proof of the Goldstone theorem for non-abelian symmetries.

In Sect. 4, we also use our expression for the effective action to find the spectrum of charge-, flux- and spin-carrying excitations of an incompressible quantum fluid, and we discuss the possible values of their electric charge and spin, and their statistics. Our analysis provides first insights into why the Hall conductivity and various other quantities characterizing the system, e.g., its magnetic susceptibility, are quantized. But our reasoning is somewhat heuristic, mathematically.

In order to bring more rigour into that analysis, we derive and discuss, in Sect. 5, algebras of chiral currents circulating in an incompressible quantum fluid along domain boundaries across which the value of the Hall conductivity jumps, in particular along its edges. The electric edge currents form chiral $U(1)$-current algebras, the edge spin-currents form $SU(2)$-Kac-Moody algebras. These results can be derived from $U(1)\times SU(2)$-gauge-invariance by using well known results on the $(1+1)$-dimensional chiral gauge anomalies and their relation to $(2+1)$-dimensional Chern-Simons theory [16]. [An alternative derivation of the existence of algebras of chiral edge currents in incompressible Hall fluids from quantized Chern-Simons theory, based on results in [10,11,12], is given in [3].]

The well known representation theory of chiral current algebras, combined with some physically natural requirements, then leads us to find *discrete* sets of possible values of the Hall conductivity, (certain rational multiples of $\frac{e^2}{h}$), of the fractional charges of excitations, and of other interesting quantities, which are compatible with the incompressibility of the Hall fluid. Our results can be viewed as "*gap-labelling theorems*": The energy spectrum of a two-dimensional electron fluid in an external magnetic field can have a *positive gap* above the groundstate energy (reflecting its *incompressibility*) only if its Hall conductivity belongs to a certain discrete set.

We also find the statistics of fractionally charged excitations (Laughlin vortices) from the representation theory of the algebras of chiral edge currents.

A complete discussion of edge spin-currents and currents associated with internal symmetries of the system would take too much space and is therefore deferred to another paper [17]. However, a few basic ideas are provided in Sect. 5.

Most of this paper was written during a two-weeks' stay of J.F. at I.H.É.S, Bures-sur-Yvette. J.F. thanks the director of I.H.É.S, M. Berger, his colleages and the staff at the Institut for their very friendly hospitality during a period that was quite hectic for him.

2. NON-RELATIVISTIC QUANTUM MECHANICS OF SPINNING PARTICLES COUPLED TO EXTERNAL METRICS AND ELECTROMAGNETIC FIELDS

In this section we recall the formulation of non-relativistic quantum mechanics in general, including moving, coordinates on a Riemannian space. We consider systems of spinning particles coupled to the space metric and to external electromagnetic fields. [For mathematical background see, e.g. [18].] Since we are interested in time-dependent many-particle systems, it will be convenient to use a second quantized Lagrangian formalism [19].

Physical space is a two-, or three-dimensional manifold, M, with possibly time-dependent metric, space-time is given by $N := \mathbf{R} \times M$. The system is confined to the interior of a space-time cyclinder $\Lambda \subset N$. The intersection of Λ with a fixed-time slice is denoted by Ω_t where t is time. In local coordinates, points in M are denoted by $\mathbf{x}, \mathbf{y}, \cdots$, points in N by $x = (t, \mathbf{x}), y = (t, \mathbf{y}), \cdots$. The Riemannian metric on M is denoted by $g_{ij}(t, \mathbf{x})$; space-time N carries the metric $\eta_{\mu\nu}(x)$, where $\eta_{00}(x) = 1$, $\eta_{0i}(x) = \eta_{i0}(x) = 0, \eta_{ij}(x) = -g_{ij}(t, \mathbf{x})$. In the tangent space at a point $\mathbf{x} \in M$ we also have the flat, Cartesian metric, δ_{AB}. [Similarly, in the tangent space at a space-time point $x \in N$ we have the usual Lorentz metric $\eta^0_{\alpha\beta}$.]

If the dimension of M is two we imagine that M is a surface in a three-dimensional Riemannian manifold L with metric also denoted by g_{ij}, and the metric on M is the induced metric. In physical applications L will usually be three-dimensional Euclidean space $\mathbf{E}^3$, and M will be some surface in $\mathbf{E}^3$.

So far, time is merely a parameter, and we temporarily omit it from our notations. In the cotangent bundle to L we choose local sections of orthonormal frames $(e^A(\mathbf{x}))^3_{A=1}$. The components of $e^A(\mathbf{x})$ in the basis $(dx^j)^3_{j=1}$ of $T^*_{\mathbf{x}}(L)$ are denoted by ${e^A}_i(\mathbf{x})$ and are called *"dreibein* (fields)". If $\dim M = 2$ we choose $(e^A(\mathbf{x}))^3_{A=1}$ such that, for $\mathbf{x} \in M \subset L$, $e^3(\mathbf{x})$ is orthogonal to $T^*_{\mathbf{x}}(M)$ in the metric of $T^*_{\mathbf{x}}(L)$. The metric on L can be expressed in terms of the dreibein as follows:

$$g_{ij}(\mathbf{x}) = \delta_{AB} {e^A}_i(\mathbf{x}) {e^B}_j(\mathbf{x}) \, . \tag{2.1}$$

If $\dim M = 2$ we choose local coordinates on L in a neighborhood of M such that the metric on M at a point $\mathbf{x}$ is given by

$$g_{ij}(\mathbf{x}) = \sum_{A,B=1}^{2} \delta_{AB} e^{A}{}_{i}(\mathbf{x}) e^{B}{}_{j}(\mathbf{x}) \ , \quad i,j = 1,2, \tag{2.2}$$

i.e., the coordinate x^3 is transversal to M.

The inverse of the dreibein $e^A{}_i$ is given by

$$\mathcal{E}_A{}^i(\mathbf{x}) = \delta_{AB} g^{ij}(\mathbf{x}) e^B{}_j(\mathbf{x}) \ , \tag{2.3}$$

where (g^{ij}) is the inverse of (g_{ij}). Clearly

$$\mathcal{E}_A{}^i e^B{}_i = \delta^B_A \ , \quad \delta^{AB} \mathcal{E}_A{}^i \mathcal{E}_B{}^j = g^{ij} \ . \tag{2.4}$$

The dreibein $e^A{}_i$ is the matrix which transforms the coordinate basis (dx^i) of $T^*_{\mathbf{x}}(L)$ to an orthonormal basis, $(e^A(\mathbf{x}))$, of $T^*_{\mathbf{x}}(L)$,

$$e^A(\mathbf{x}) = e^A{}_i(\mathbf{x}) dx^i \ . \tag{2.5}$$

Similarly, $\mathcal{E}_A{}^i$ transforms the basis $\left(\frac{\partial}{\partial x^i}\right)$ of $T_{\mathbf{x}}(L)$ to an orthonormal basis, $(\mathcal{E}_A(\mathbf{x}))$, of $T_{\mathbf{x}}(L)$,

$$\mathcal{E}_A(\mathbf{x}) = \mathcal{E}_A{}^i(\mathbf{x}) \frac{\partial}{\partial x^i} \ . \tag{2.6}$$

On every cotangent space $T^*_{\mathbf{x}}(L)$, $\mathbf{x} \in L$, we have a three-dimensional (spin-1) representation, $(R(\mathbf{x}) \in SO(3))$ of the rotation group, acting on the dreibein $e^A{}_i$ as follows

$$^{R}e^A{}_i(\mathbf{x}) = R(\mathbf{x})^A{}_B e^B{}_i(\mathbf{x}) \ . \tag{2.7}$$

We require that parallel transport on L be given by the Levi-Cività connection $\Gamma^i{}_{jl}$, so that the torsion, T, vanishes. Then we may define Cartan's *spin connection* $\lambda^A{}_B$ through Cartan's first structure equation

$$de^A + \lambda^A{}_B \wedge e^B \equiv T^A = 0 \ . \tag{2.8}$$

These equations enable us to express $\lambda_i{}^A{}_B$ in terms of the dreibeins $e^A{}_i$, their derivatives, and their inverses $\mathcal{E}_A{}^i$; (see [18]).

The curvature 2-form $\mathcal{R}^A{}_B$ of L is defined by Cartan's second structure equation

$$\mathcal{R}^A{}_B = d\lambda^A{}_B + \lambda^A{}_C \wedge \lambda^C{}_B \ . \tag{2.9}$$

It is easy to deduce from (2.8) and (2.9) how λ and R transform under the "gauge-transformations" (2.7) of the dreibein:

$$\begin{aligned} {}^{R}\lambda(\mathbf{x}) &= R(\mathbf{x})\lambda(\mathbf{x})R^T(\mathbf{x}) + R(\mathbf{x})dR^T(\mathbf{x}) \ , \\ {}^{R}\mathcal{R}(\mathbf{x}) &= R(\mathbf{x})\mathcal{R}(\mathbf{x})R^T(\mathbf{x}) \ . \end{aligned} \tag{2.10}$$

We now assume that the manifold L admit a spin structure. Then we may introduce *spinor bundles* over L. Let $s = 0, 1/2, 1, \cdots$ denote the spin, i.e., $2s+1$ is the dimension of an irreducible representation of $SU(2) = \widetilde{SO(3)}$ with spin s. The fibre of the spin-s spinor bundle, $E^{(s)}$, over L is isomorphic to the $(2s+1)$-dimensional Hilbert space, $\mathcal{D}^{(s)}$, carrying the spin-s representation of $SU(2)$. Sections of the spin-s spinor bundle are denoted by $\psi^{(s)}(\mathbf{x})$. From now on, we choose the gauge transformations $(R(\mathbf{x}))$ to be $SU(2)$-valued. The action of these gauge transformations on the cotangent bundle is given by their adjoint (spin-1) representation, usually also denoted by $R(\mathbf{x})$[1]. Under a gauge transformation $(R(\mathbf{x}))$, a section $\psi^{(s)}$ of $E^{(s)}$ transforms as follows:

$$\psi^{(s)}(\mathbf{x}) \mapsto {}^{R}\psi^{(s)}(\mathbf{x}) := U^{(s)}(R(\mathbf{x}))\psi^{(s)}(\mathbf{x}) , \tag{2.11}$$

where $U^{(s)}$ is the spin-s representation of $SU(2)$. The transition functions of the spin-s spinor bundle are inherited from the transition functions of the cotangent bundle, $T^*(L)$, by lifting them to the spin-s representation of $SU(2)$. [Since we have assumed that L have a spin structure this is possible even if s is half-integer.]

Physically, what is meant by "spin up" or "spin down" is now a *local notion*, depending on the point $\mathbf{x} \in L$ at which the spin is located and determined by the frame $(e^A(\mathbf{x}))_{A=1}^3$.

We intend to develop non-relativistic quantum mechanics on Hilbert spaces of sections of these spinor bundles. In non-relativistic quantum mechanics, wave functions are complex-valued. We therefore tensor the fibre space $\mathcal{D}^{(s)}$-real when s is integer – by $\mathbf{C}$. The structure group of the resulting bundle, still denoted by $E^{(s)}$, is then $U(1) \times SU(2)$. The factor $U(1)$ (phase transformations of $\psi^{(s)}$) is connected to electromagnetism, as recognized by Weyl more than sixty years ago.

In order to keep our notations simple, it is advantageous to formulate quantum mechanics by using the language of second quantization. The sections $\psi^{(s)}(\mathbf{x})$ of $E^{(s)}$ are then interpreted as operator-valued distributions acting on Fock space and subject to *equal-time canonical (anti-) commutation relations*

$$\left[\psi_\alpha^{(s)\#}(\mathbf{x}), \psi_\beta^{(s)\#}(\mathbf{y})\right]_{\pm} = 0 , \tag{2.12}$$

$$\left[\psi_\alpha^{(s)}(\mathbf{x}), \psi_\beta^{(s)*}(\mathbf{y})\right]_{\pm} = \frac{1}{\sqrt{g(\mathbf{x})}}\delta_{\alpha\beta}\delta(\mathbf{x}-\mathbf{y}) , \quad \alpha, \beta = 1, \cdots, 2s+1 ,$$

where $[\ ,\ ]_+$ denotes the anti-commutator and $[\ ,\ ]_-$ the usual commutator, $\psi^{(s)\#} = \psi^{(s)}$ or $\psi^{(s)*}$; $\psi^{(s)*}$, the *creation operator*, is the adjoint (on Fock space) of $\psi^{(s)}$, the *annihilation operator*; $g(\mathbf{x})$ denotes the determinant of $(g_{ij}(\mathbf{x}))$. The usual connection between spin and statistics is to choose anti-commutators in (2.12), corresponding to Fermi statistics, when s is half-integer, and commutators, corresponding to Bose statistics, when s is an integer.

Our purpose is now to specify some nonrelativistic dynamical laws for the operators $\psi^{(s)\#}$ in the Heisenberg picture. Let $\psi^{(s)\#}(x) = \psi^{(s)\#}(t, \mathbf{x})$ denote the Heisenberg

[1] There is little danger of confusion.

- picture creation - and annihilation operators with initial conditions $\psi^{(s)\#}(0,\mathbf{x}) = \psi^{(s)\#}(\mathbf{x})$. In order to formulate local dynamical laws for $\psi^{(s)\#}(x)$, we need to be able to differentiate these fields in t and $\mathbf{x}$. This necessitates introducing a notion of parallel displacement in $E^{(s)}$. Parallel displacement in $E^{(s)}$ is defined with the help of a $U(1) \times SU(2)$-connection, (a vector potential with values in $\mathbf{R} \oplus su(2)$, where $su(2)$ is the Lie algebra of $SU(2)$). Once such a connection is fixed, derivatives of sections $\psi^{(s)\#}$ are defined as *covariant derivatives.* Setting $x^0 := ct, (x^\mu) := (x^0, \mathbf{x})$, the covariant derivative in the μ-direction is given by

$$D_\mu = \frac{\partial}{\partial x^\mu} + ia_\mu(x) + w_\mu^{(s)}(x) \, , \tag{2.13}$$

where $a(x) := a_j(x)dx^j$ is the $U(1)$-connection (i.e., $a_j(x)$ is the jth component of a real-valued vector potential), and $a_0(x)$ is the scalar potential, $w^{(s)}(x) := w_j^{(s)}(x)dx^j$ is the $SU(2)$-connection, and $w_0^{(s)}(x)$ is the "Zeeman potential" in the spin-s representation of $su(2)$, i.e.,

$$w_\mu^{(s)}(x) = i \sum_{A=1}^{3} w_{\mu A}(x) L_A^{(s)} \, , \tag{2.14}$$

where $(L_A^{(s)})_{A=1}^3$ are Hermitian - we are physicists - generators of $su(2)$ in the spin-s representation, normalized such that $L_A^{(1/2)} = \sigma_A$, where σ_1, σ_2 and σ_3 are the usual Pauli matrices. We shall see that we should identify a with the electromagnetic vector potential, up to multiplication by a constant of nature. What about $w^{(s)}$? Clearly the spin connection $\lambda^A{}_B$, introduced in (2.8), must enter the definition of $w^{(s)}$. But we can add to λ a one-form, ρ, transforming under the adjoint representation of the $SU(2)$-gauge group. The sum is then still an $SU(2)$-connection. Hence

$$w_\mu^{(s)}(x) = \lambda_\mu^{(s)}(x) + \rho_\mu^{(s)}(x) \, , \tag{2.15}$$

where

$$\lambda_\mu^{(s)}(x) = \frac{i}{2} \sum_{A,B,C=1}^{3} \varepsilon_A{}^{BC} \lambda_\mu{}^A{}_B(x) L_C^{(s)} \, , \tag{2.16}$$

$\varepsilon_A{}^{BC} = \varepsilon^{ABC}$ is the sign of the permutation (ABC) of $(1\ 2\ 3)$, and where

$$\rho_\mu{}^{(s)}(x) = i \sum_{A=1}^{3} \rho_{\mu A}(x) L_A^{(s)} \, . \tag{2.16'}$$

Under an $SU(2)$-gauge-transformation of the cotangent bundle, $\rho_\mu^{(s)}$ transforms as follows:

$$\rho_\mu{}^{(s)}(x) \mapsto {}^R\rho_\mu{}^{(s)}(x) = U^{(s)}(R(x))\rho_\mu{}^{(s)}(x)U^{(s)}(R(x))^* \, . \tag{2.17}$$

The transformation law of $\lambda_\mu{}^{(s)}$ can be inferred from (2.10).

If the dreibein $(e^A{}_i)$ is time-independent λ_0 vanishes, but after a time-dependent $SU(2)$-gauge - transformation λ_0 may be different form zero. In general, ρ_0 will be different from zero.

We shall see that, physically, (ρ_0, ρ) describes *Zeeman* - and *spin-orbit couplings* of the magnetic moments carried by the particles to the electromagnetic field. Geometrically, the part (ρ_0, ρ) of the $SU(2)$-connection w yields non-trivial *torsion.*

Having introduced a $U(1)\times SU(2)$-connection and defined covariant differentiation of $\psi^{(s)\#}$, we are now in a position to formulate local dynamical laws. It is convenient to use the *Lagrangian formalism*, but we could also work in the Hamiltonian formalism; see [3]. Let us consider a system of non-relativistic particles of fixed spin s and, to simplify our notations, drop the superscript $^{(s)}$. Our ansatz for the action of the system is ($dx = dt\, d\mathbf{x}$)

$$S_\Lambda(\psi^*, \psi; a, w, g) := \int_\Lambda \sqrt{g(t,\mathbf{x})}dx \Big[i\hbar c(\psi^* D_0 \psi)(x) \\ - \frac{g^{kl}(t,\mathbf{x})}{2m}(-i\hbar D_k\psi)^*(x)(-i\hbar D_l\psi)(x) - U(\psi^*,\psi)(x)\Big] , \qquad (2.18)$$

where the covariant derivatives are given in (2.13), m is the effective mass of the particles, and $U(\psi^*, \psi)$ is a $U(1) \times SU(2)$-invariant functional of ψ^* and ψ, e.g.,

$$\begin{aligned} U(\psi^*,\psi)(x) &= \int \frac{1}{2}\sqrt{g(t,\mathbf{y})}d\mathbf{y} : (\psi^*(t,\mathbf{x})\psi(t,\mathbf{x}) - n)V(\mathbf{x},\mathbf{y}) \\ &\times \ (\psi^*(t,\mathbf{y})\psi(t,\mathbf{y}) - n) : + v(t,\mathbf{x})\psi^*(t,\mathbf{x})\psi(t,\mathbf{x}) \, . \end{aligned} \qquad (2.19)$$

The double colons indicate Wick ordering, V is some repulsive pair potential, n is the background density of the system, and $v(t, \mathbf{x})$ is a possibly time-dependent one-body (background) potential.

We recall that $\Lambda \subset \mathbf{R} \times M$ is a cylindrical region to which the system is confined. At fixed time t we impose Dirichlet boundary conditions at the boundary, $\partial\Omega_t$, of the region Ω_t to which the system is confined.

The field equations (or Euler-Lagrange equations) for $\psi(x)$ and $\psi^*(x)$ follow by setting the variation of S_Λ with respect to $\psi^*(x)$ and $\psi(x)$, respectively, to zero. The resulting equations are reminiscent of the *Pauli equations* for ψ and ψ^*.

In order to interpret these equations physically we start with a simple situation: We choose space M to be given by $\mathbf{E}^2$ (the $x-y$ plane in $L = \mathbf{E}^3$) or by $\mathbf{E}^3; g_{ij}(t, \mathbf{x}) = \delta_{ij}$, for all times t and all $\mathbf{x} \in M$, $\Lambda = \mathbf{R} \times \Omega$, where Ω is some time-independent open set in M. The field equation for $\psi(x)$ obtained by varying the action S_Λ defined in (2.18) with respect to $\psi^*(x)$ then essentially reduces to the Pauli equation found in standard text books of quantum mechanics [20], with a minor modification of order $1/m(m_0c)^2$ discussed in [5], *provided* we identify the $U(1)$-connection a with the electromagnetic vector potential

$$a_j(x) = \frac{e}{\hbar c}A_j(x) \, , \quad \text{and} \quad a_o(x) = -\frac{e}{\hbar c}\phi(x) \, , \qquad (2.20)$$

where $-e$ is the charge of the particle, ϕ is the electrostatic potential, and the coefficients of the $su(2)$-valued components ρ_μ are expressed in terms of the electromagnetic field $(\vec{E}, \vec{B})$ as follows:

$$\rho_{oA}(x) = -\frac{\mu}{2c} B_A(x) \,, \tag{2.21}$$

where $B_A(x)$ is the A-component of the magnetic field $\vec{B}(x)$ in the basis $(e^1(x), e^2(x), e^3(x))$, and

$$\rho_{kA}(x) = -\frac{\mu}{4c} \sum_{C=1}^{3} \varepsilon_{kAC} E_C(x) \,, \tag{2.22}$$

with E_C the C-component of the electric field $\vec{E}$. In these equations μ is the magnetic moment of the particles, (up to a factor $\frac{\hbar}{2}$). For electrons, $\mu \approx -\frac{e}{m_o c}$, where m_o is the electron mass in empty space. [In standard situations of solid state physics, the effective mass m can be considerably smaller than m_0.] The symbol ε_{kAC} is defined by

$$\varepsilon_{kAC} = \varepsilon_{kAC}(x) := e^D{}_k(x) \varepsilon_{DAC} \,, \tag{2.23}$$

where ε_{DAC} is the sign of the permutation (DAC) of (1 2 3). Of course, in the present case $e^D{}_k(x) = \delta_k^D$, but formula (2.23) is valid in general. Formulas (2.20)-(2.22) have been derived in [5] by comparing the Euler-Lagrange equations corresponding to the action S_Λ with the usual Pauli equation, including the Zeeman term and spin-orbit couplings.

It is now straightforward to find the correct physical interpretations of the connections a and w for spaces M which are arbitrary Riemannian spin manifolds. The $U(1)$-connection a is still expressed in terms of the electromagnetic vector potential $A = (-\phi, \vec{A})$ by formula (2.20). The $SU(2)$-connection w is given by

$$w_\mu = \lambda_\mu + \rho_\mu \,, \tag{2.24}$$

where λ_μ is the affine spin connection corresponding to the dreibein $e^A{}_i(x)$, see (2.8), and the coefficients of ρ_μ are given by

$$\rho_{oA}(x) = -\frac{\mu}{2c} \mathcal{E}_A{}^l(x) B_l(x) \equiv -\frac{\mu}{2c} B_A(x) \tag{2.25}$$

where $(\mathcal{E}_A{}^l(x))$ is the inverse of the dreibein $(e^A{}_i(x))$ and B_l is the l-component of $\vec{B}$ in the basis (dx^1, dx^2, dx^3); moreover

$$\begin{aligned} \rho_{jA}(x) &= -\frac{\mu}{4c} e^D{}_j(x) \varepsilon_{DAC} \mathcal{E}_C{}^l(x) E_l(x) \\ &\equiv -\frac{\mu}{4c} \varepsilon_{jA}{}^l(x) E_l(x) \,, \end{aligned} \tag{2.26}$$

where E_l is the l-component of $\vec{E}$ in local coordinates. Note that (2.25) and (2.26) are consistent with the transformation law (2.17) of ρ_μ under $SU(2)$-gauge-transformations.

We recall that the potential V in (2.19) is a pair potential (e.g. Coulomb, for charged particles, or van der Waals, for neutral atoms or molecules), and v is a potential created by the background in which the particles are moving; (v might depend on the scalar curvature of M).

We now suppose that the background of the system is moving according to some *classical flow* $\phi(t,\cdot)$. Here $\phi(t,\mathbf{y})$ is the position in M of a point particle at time t starting at position $\mathbf{y}$ at time 0. Then, in the x-coordinates, the one-body potential $v(x)$ and the magnetic and electric fields $\vec{B}_c(x)$ and $\vec{E}_c(x)$, created by the background are *time-dependent*. This implies that, in the time-independent $\mathbf{x}$-coordinates on M, the Hamiltonian of the system is time-dependent which complicates the mathematical analysis of the system, in particular the analysis of its *thermal equilibrium properties*. It is quite clear, physically, that thermal equilibrium in such a system will be reached locally in regions moving with the background, (according to the flow $\phi(t,\cdot)$). Thus, we ought to formulate quantum mechanics in "*moving coordinates*", (y^1, y^2, y^3), where

$$\mathbf{x} = \phi(t,\mathbf{y}) \,, \quad \text{i.e.,} \quad \mathbf{y} = \phi^{-1}(t,\mathbf{x}) \,. \tag{2.27}$$

Time will not be transformed. In the new coordinates (y^1, y^2, y^3), the one-body potential $v(t,\mathbf{y})$ and the background fields $\vec{B}_c(t,\mathbf{y})$ and $\vec{E}_c(t,\mathbf{y})$ might now be *time-independent*. In this case, the Hamiltonian for spinless particles ($s = 0$) will be time-independent, and we can apply the rules of Gibbsian statistical mechanics to study thermal equilibrium.

Unfortunately, for spinning particles ($s = 1/2, 1, \cdots$), the situation is not quite as neat, because, in the y-coordinates, the dreibeins $\tilde{e}^A{}_i(y)$ are now *time-dependent*:

$$\tilde{e}^A{}_i(y) = e^A{}_j(\phi(t,\mathbf{y}))\frac{\partial \phi^j(t,\mathbf{y})}{\partial y^i} \,. \tag{2.28}$$

In order to eliminate as much of this undesirable time-dependence as possible, we may try to perform a suitable $SU(2)$-gauge transformation on the new dreibeins $\tilde{e}^A{}_i(y)$. What is the optimal choice? The answer is, perhaps, somewhat ambiguous, in general. But the following choice tends to be quite optimal: Let $(f^j(t,\mathbf{x}))$ be the vector (velocity) field generating the flow $\phi(t,\cdot)$, i.e.,

$$\frac{\partial}{\partial t}\phi(t,\mathbf{y}) = f(t, \phi(t,\mathbf{y})) \,. \tag{2.29}$$

Let

$$f^A(t,\mathbf{x}) := e^A{}_j(\mathbf{x}) f^j(t,\mathbf{x}) \,.$$

Then the infinitesimal rotation of an orthonormal frame carried along by the flow ϕ, at the point $\mathbf{x}$ and at time t, is given by

$$\Omega^A{}_B(t,\mathbf{x}) = \frac{1}{2}\{(\partial_B f^A)(t,\mathbf{x}) - (\partial_A f^B)(t,\mathbf{x})\} \,, \tag{2.30}$$

where $\partial_A = \mathcal{E}^i_A(\mathbf{x})\frac{\partial}{\partial x^i}$; see (2.6). The vector $\vec{\Omega}(t,\mathbf{x})$ dual to the antisymmetric matrix $(\Omega^A{}_B(t,\mathbf{x}))$ is called the *vorticity* of the vector field f and is the local angular velocity of the rotation induced by ϕ of a frame at the point $\mathbf{x}$, at time t.

We define a rotation matrix $R(t,\mathbf{x})^A{}_B$ by setting

$$R(t,\mathbf{x})^A{}_B := T\left[\exp\int_0^t dt'\Omega(t',\mathbf{x})\right]^A{}_B , \tag{2.31}$$

where T denotes time ordering. [The r.h.s. of (2.31) can be defined, for example, by a convergent Dyson series if $\vec{\Omega}(t,\mathbf{x})$ is uniformly bounded in t.] We now define

$$\hat{e}^A{}_i(y) := {}^R\tilde{e}^A{}_i(y) = R(t,\phi(t,\mathbf{y}))^A{}_B\tilde{e}^B{}_i(y) , \tag{2.32}$$

where $\tilde{e}^B{}_i(y)$ is given by (2.28). We also define the following transformed quantities:

$$\begin{aligned}
\hat{g}^{kl}(t,\mathbf{y}) &:= \frac{\partial y^k}{\partial x^m}\frac{\partial y^l}{\partial x^n}g^{mn}(t,\phi(t,\mathbf{y})) ,\\
\hat{\psi}(t,\mathbf{y}) &:= U^{(s)}(t,\mathbf{y})\psi(t,\phi(t,\mathbf{y})) ,\\
\hat{w}_k^{(s)}(t,\mathbf{y}) &:= \frac{\partial x^l}{\partial y^k}\Big\{U^{(s)}(t,\mathbf{y})w_l^{(s)}(t,\phi(t,\mathbf{y}))U^{(s)}(t,\mathbf{y})^*\\
&\qquad +U^{(s)}(t,\mathbf{y})\left(\frac{\partial}{\partial x^l}U^{(s)}\right)^*(t,\mathbf{y})\Big\} ,\\
\hat{w}_0^{(s)}(t,\mathbf{y}) &:= U^{(s)}(t,\mathbf{y})\left[w_0^{(s)}(t,\phi(t,\mathbf{y}))+\frac{\partial x^l}{\partial y^0}w_l^{(s)}(t,\phi(t,\mathbf{y}))\right]U^{(s)}(t,\mathbf{y})\\
&+ U^{(s)}(t,\mathbf{y})\frac{1}{c}\frac{\partial}{\partial t}U^{(s)}(t,\mathbf{y})^* ,
\end{aligned} \tag{2.33}$$

where

$$U^{(s)}(t,\mathbf{y}) := U^{(s)}(R(t,\phi(t,\mathbf{y}))) ,$$

and, for $l=1,2,3$,

$$\left(\frac{\partial}{\partial x^l}U^{(s)}\right)(t,\mathbf{y}) := U^{(s)}\left(\frac{\partial}{\partial x^l}R(t,\mathbf{x})\right)\Big|_{\mathbf{x}=\phi(t,\mathbf{y})} . \tag{2.34}$$

Finally, we have

$$\hat{a}_k(t,\mathbf{y}) := \frac{\partial x^l}{\partial y^k}a_l(t,\phi(t,\mathbf{y})) ,$$

and

$$\hat{a}_0(t,\mathbf{y}) := a_0(t,\phi(t,\mathbf{y})) + \frac{\partial x^l}{\partial y^0}a_l(t,\phi(t,y)) . \tag{2.35}$$

Our aim is now to rewrite the action S_Λ introduced in (2.18), (2.19) in the moving y-coordinates, using the transformations (2.32)–(2.35). By (2.33), (2.34),

$$\psi(t,\mathbf{x}) = U^{(s)}(R(t,\mathbf{x}))^*\hat{\psi}(t,\phi^{-1}(t,\mathbf{x})) . \tag{2.36}$$

Hence

$$\begin{aligned}
U^{(s)}(R(t,\mathbf{x}))\frac{\partial}{\partial t}\psi(t,\mathbf{x}) &= \frac{\partial}{\partial t}\hat{\psi}(t,\mathbf{y}) - \hat{f}^j(t,\mathbf{y})\frac{\partial}{\partial y^j}\hat{\psi}(t,\mathbf{y})\\
&-\frac{i}{4}\sum_{A,B,C}\varepsilon_A{}^{BC}\hat{\Omega}^A{}_B(t,\mathbf{y})L_C^{(s)}\hat{\psi}(t,\mathbf{y}) ,
\end{aligned} \tag{2.37}$$

where $-\hat{f}^j(t,\mathbf{y})$ is the jth component of the vector field generating $\phi^{-1}(t,\cdot)$ in $\mathbf{y}$-coordinates, and $\hat{\Omega}^A{}_B(t,\mathbf{y})$ is the vorticity of f in $\mathbf{y}$-coordinates with respect to the dreibein $(\hat{e}^A(t,\mathbf{y}))$. By comparing (2.37) with the last equation in (2.33) and with (2.34) and (2.35) we see that

$$U^{(s)}(R(t,\mathbf{x}))\left(\frac{1}{c}\frac{\partial}{\partial t}+ia_0(x)+w_0^{(s)}(x)\right)\psi(x) \tag{2.38}$$
$$=\left(\frac{1}{c}\frac{\partial}{\partial t}+i\hat{a}_0(y)+\hat{w}_0^{(s)}(y)\right)\hat{\psi}(y)-\frac{1}{c}\hat{f}^j(y)\left(\frac{\partial}{\partial y^j}+i\hat{a}_j(y)+\hat{w}_j^{(s)}(y)\right)\hat{\psi}(y)\,.$$

We now define the new covariant derivatives

$$\begin{aligned}\hat{D}_0 &:= \frac{1}{c}\frac{\partial}{\partial t}+i\hat{a}_0(y)+\hat{w}_0{}^{(s)}(y)\\ \hat{D}_j &:= \frac{\partial}{\partial y^j}+i\hat{a}_j(y)-i\frac{m}{\hbar}\hat{f}_j(y)+\hat{w}_j^{(s)}(y)\end{aligned} \tag{2.39}$$

where $\hat{f}_j=\hat{g}_{jl}\hat{f}^l$, and the new one-body potential

$$\begin{aligned}\hat{v}(t,\mathbf{y}) &:= v(t,\phi(t,\mathbf{y}))-\frac{m}{2}\hat{f}^j(y)\hat{f}_j(y)\\ &\quad -i\frac{\hbar}{2}\frac{1}{\sqrt{\hat{g}}}\frac{\partial}{\partial y^j}\left(\sqrt{\hat{g}}\hat{f}^j\right)(y)\,,\end{aligned} \tag{2.40}$$

as well as the two-body potential

$$\hat{V}(t,\mathbf{y},\mathbf{y}'):=V(\phi(t,\mathbf{y}),\phi(t,\mathbf{y}'))\,. \tag{2.41}$$

After these preparations, one verifies easily that

$$\begin{aligned}&S_\Lambda(\psi^*,\psi;a,w,g)\\ &=\hat{S}_{\hat{\Lambda}}(\hat{\psi}^*,\hat{\psi};\hat{a},\hat{w},\hat{f},\hat{g})\\ &=\int_{\hat{\Lambda}}\sqrt{\hat{g}(t,\mathbf{y})}dy\,\Big[i\hbar c(\hat{\psi}^*\hat{D}_0\hat{\psi})(y)\\ &\quad-\frac{\hat{g}^{kl}(t,\mathbf{y})}{2m}(-i\hbar\hat{D}_k\hat{\psi})^*(y)(-i\hbar\hat{D}_l\hat{\psi})(y)\\ &\quad-\hat{U}(\hat{\psi}^*,\hat{\psi})(y)\Big]\,,\end{aligned} \tag{2.42}$$

where in the definition of $\hat{U}$ the potentials $\hat{v}$ and $\hat{V}$ of (2.40) and (2.41) are used, and $\hat{\Lambda}:=\{(t,\mathbf{y}):(t,\mathbf{x}=\phi(t,\mathbf{y}))\in\Lambda\}$. To prove (2.42), one expands the r.h.s. of (2.42) in powers of $\hat{f}$, integrates by part, and compares the resulting expression to (2.38), using (2.40), (2.39) and the fact that $(U^{(s)}\psi)^*(U^{(s)}\psi)=\psi^*\psi$.

Let us pause to interpret the result (2.42). By (2.39), $-\frac{m}{\hbar}\hat{f}_j$ enters the action $\hat{S}$ as a contribution to the $U(1)$-connection. By (2.20), $-m\hat{f}_j$ and $\frac{e}{c}\hat{A}_j$ play analogous roles, i.e.,

$$-m\vec{f}\leftrightarrow\frac{e}{c}\vec{A}\,. \tag{2.43}$$

The vector potential $\vec{A}$ gives rise to the *Lorentz force* in the classical limit. The Lorentz force has the same form as the *Coriolis force* if one replaces $\frac{e}{c}\vec{B}$ by $-2m\vec{\Omega}$, where $\vec{\Omega}$ is the local angular velocity which is precisely half the curl of the vector field $\vec{f}$. Thus, $\vec{f}$ is the vector potential that gives rise to the Coriolis force in the classical limit.

By (2.37) and (2.38), the new action $\hat{S}$ contains a term

$$\hat{\psi}^* \left(\hat{\vec{\Omega}} \cdot \frac{\hbar}{2} \vec{L}^{(s)} \right) \hat{\psi} , \tag{2.44}$$

where $\hat{\vec{\Omega}}$ is the curl of $\hat{\vec{f}}$. This has the form of the *Zeeman term*

$$- \mu \hat{\psi}^* \left(\hat{\vec{B}} \cdot \frac{\hbar}{2} \vec{L}^{(s)} \right) \hat{\psi} \tag{2.45}$$

which, by (2.25), (2.24) and (2.39), also appears in $\hat{S}$. Of course, $\mu \hat{\vec{B}}$ is precisely the angular velocity of spin precession in a magnetic field.

Next, we must analyze the one-body potential $\hat{v}$ in moving coordinates. By (2.40), $\hat{v}$ is *complex-valued, unless*

$$\frac{1}{\sqrt{\hat{g}}} \frac{\partial}{\partial y^j} \left(\sqrt{\hat{g}} \hat{f}^j \right) = 0 , \tag{2.46}$$

i.e., *unless the vector field $\hat{\vec{f}}$ is divergence-free.* A divergence-free vector field generates a *volume-preserving* flow ϕ, hence

$$\hat{g}(t, \mathbf{y}) \equiv \det(\hat{g}_{kl}(t, \mathbf{y})) = g(t, \phi(t, \mathbf{y})) . \tag{2.47}$$

Thus, for volume-preserving (i.e., *incompressible*) flows, and *only* for such flows, $\hat{v}$ is again real-valued. [This is, because if volume is preserved by ϕ then, by (2.47), the quantum mechanical time-evolution in the moving coordinate system preserves probabilities with respect to the volume element $\sqrt{g(t, \phi(t, \mathbf{y}))} d\mathbf{y}$, and hence is generated by a *Hermitian* (selfadjoint) Hamiltonian!] But $\hat{v}$ contains an additional term, $-\frac{m}{2} \hat{f}^j(y) \hat{f}_j(y)$, that was not present in the original one-body potential. What does it correspond to physically? It is the potential of the *centrifugal force*, (because $\frac{m}{2} \frac{\partial}{\partial y^i} (\hat{f}^j(t, \mathbf{y}) \hat{f}_j(t, \mathbf{y}))$ is precisely the i-component of the centrifugal force at the point $\mathbf{y}$, at time t; note, incidentally, that $\frac{m}{2} \hat{\vec{f}} \cdot \hat{\vec{f}}$ is the classical kinetic energy of the particle in the rest frame which must be subtracted in the $\mathbf{y}$-frame).

In conclusion, we find that quantum mechanics in moving coordinates is Hamiltonian, with a *Hermitian* (but possibly still time-dependent) Hamiltonian operator, iff the flow ϕ defining the moving coordinate system is *volume-preserving*, or *incompressible.* Henceforth this property is usually required. It is worthwhile recalling that in *two* space dimensions, incompressible flows are automatically symplectic (Hamiltonian) flows, because the vector fields generating them are divergence-free and hence are dual to the gradient of some (scalar) Hamiltonian function.

Let us consider, as an example, a system of particles of charge $-e$ and magnetic moment $\mu = -\frac{e}{m_0 c}$, with $m_0 = m$, (e.g., electrons, neglecting their anomalous magnetic moment). For such a system, we can eliminate, to order $\max(\vec{B}^2, |\partial_j \vec{B}|)$ the effect of an external magnetic field $\vec{B}$ by choosing moving coordinates with vorticity field $2\vec{\Omega} = -\mu\vec{B}$ and velocity field $\vec{f} = -\mu\vec{A} = \frac{e\vec{A}}{mc}$, where the electromagnetic vector potential $\vec{A}$ is chosen in the Coulomb gauge, div $\vec{A} = 0$, in order for $\vec{f}$ to be divergence-free, up to a modification of the one-body potential v by the potential $-\frac{m}{2}\vec{f}\cdot\vec{f}$ of the centrifugal force and additional spin-orbit couplings, proportional to derivatives of $\vec{B}$ (if $\vec{B}$ is not homogeneous). This theorem follows directly from (2.43) - (2.45). It is the quantum-mechanical version of *Larmor's theorem.* (This theorem can be generalized, in order to take anomalous magnetic moments into account, by suitably changing the definition of $R(t, \mathbf{x})$ in eq. (2.31).)

Before we turn to some applications of the formalism presented in this section, we wish to emphasize once more that it applies equally well to (one-), two- and three-dimensional systems. It often happens in solid state physics, e.g. in two-dimensional heterojunctures used in measurements of the quantized Hall effect, that the system exhibits an approximate internal symmetry described by some compact group G. The spinors $\psi^{(s)}$ then transform according to some non-trivial representation, π, of G. A breaking of G might be described as the effect of coupling $\psi^{(s)\#}$ to an external gauge field with values in the representation $d\pi$ of the Lie algebra of G. Let us denote this gauge field by Z. By modifying the covariant derivatives,

$$D_\mu \mapsto D'_\mu := D_\mu + d\pi(Z_\mu) , \tag{2.48}$$

we may easily extend the entire formalism developed in this section to systems with gauged internal symmetries. This is important in applications, (e.g., to the quantum Hall effect).

Note that the action S_Λ introduced in eq. (2.18) is $U(1)_{\rm em} \times SU(2)_{\rm spin} \times G_{\rm internal}$ gauge-invariant: It does not change if, for an arbitrary real-valued function χ, an $SU(2)$-valued function R and a G-valued function g, the following substitutions are made:

$$\begin{aligned} \psi^{(s)} &\mapsto e^{i\chi}U^{(s)}(R) \otimes \pi(g)\psi^{(s)} , \\ a_\mu &\mapsto a_\mu - \partial_\mu \chi , \\ w_\mu &\mapsto R w_\mu R^* + R\partial_\mu R^* , \end{aligned} \tag{2.49}$$

and

$$Z_\mu \mapsto g Z_\mu g^{-1} + g\partial_\mu g^{-1} .$$

Thus, barring gauge anomalies, (which actually cannot appear in systems of finitely many non-relativistic particles), the non-relativistic quantum mechanics of such systems is $U(1)_{\rm em} \times SU(2)_{\rm spin} \times G_{\rm internal}$ gauge-invariant. Ward identities expressing this gauge-invariance turn out to play an important role in establishing certain universal properties of such systems; see [5] and Sect. 4.

3. SOME KEY EFFECTS RELATED TO THE $U(1)_{em} \times SU(2)_{spin}$-GAUGE-INVARIANCE OF NON- RELATIVISTIC QUANTUM MECHANICS

Before we turn to our main topic, the analysis of two-dimensional, incompressible quantum fluids and their relation to one-dimensional chiral current algebras, we wish, in this section, to sketch some effects in quantum mechanics related to its $U(1)_{em} \times SU(2)_{spin}(\times G_{internal})$-gauge invariance. Most of the material reviewed here is well known, but our perspective, emphasizing gauge-invariance, may be somewhat novel in a few instances.

(1) The Aharonov-Bohm effect [21]

A key effect reflecting Weyl's $U(1)_{em}$-gauge principle realized in quantum theory is the *Aharonov-Bohm effect*: Consider the scattering of quantum mechanical particles at a magnetic solenoid; (the wave functions of the particles are required to vanish inside the solenoid). Then the diffraction pattern seen on a screen depends non-trivially on the magnetic flux, Φ, through the solenoid in a periodic fashion, with period $\frac{h}{q}$ (or $\frac{hc}{q}$, in the units used in Sect. 2), where q is the charge of the particles. This is a consequence of the fact that the vector potential $\vec{A}$ outside the solenoid cannot be gauged away, globally, in spite of the fact that there is no electromagnetic field, thus leading to non-integrable $U(1)$-phases of quantum-mechanical wave functions which change the diffraction pattern.

The Aharonov-Bohm effect explains the possibility of fractional (or θ–, or abelian braid–) statistics of anyons [22] in two-dimensional systems: Anyons are particles carrying electric charge q and magnetic flux Φ ($= \sigma_H^{-1} q$, where σ_H is a "Hall conductivity") and hence give rise to Aharonov-Bohm phases which one can interpret as statistical phases.

After what we have learned in Sect. 2 on the $U(1)$-vector potential of Coriolis forces, it is clear that there should also exist a *"tidal" Aharonov-Bohm effect:* Consider a mass-current conducting superfluid in a large container penerated by some straight cylindrical tube that excludes the quantum fluid. Now set the fluid in circular motion around the axis of the tube with velocity field $\vec{f}$, where $|\vec{f}(r)| = \frac{V}{2\pi r}$ at a distance r from the axis of the tube, and V is a quantity of dimension cm^2/sec, the total vorticity. [We note that $V = \frac{2\pi \bar{L}_z}{NM}$, where M is the mass of the particles constituting the quantum fluid, $\bar{L}_z$ is the expectation value of the component of the total angular momentum operator parallel to the tube in the state of the system, and N is the particle number.] Small mass-currents excited in this system, scattered at the tube, will exhibit an Aharonov-Bohm effect depending periodically on V, with period $\frac{h}{m}$, where m is the mass of the particles constituting the current; see (2.43).

While this effect may be somewhat difficult to test experimentally, it is important theoretically: Consider a superfluid film with manifestly (e.g., by rotating it) or spontaneously broken time reversal and reflection-in-lines invariance. Such a two-dimensional superfluid will, in general, exhibit *vortex excitations* of vorticity $V = n\frac{h}{M}$, $n = 0, \pm1, \pm2, \cdots$, where M is the mass of the constituent particles in the superfluid, and *fractional mass* (rather than fractional charge) $\sigma_H^{-1}V$, where $\sigma_H = \frac{M^2}{h}\sigma$ is the "tidal" Hall conductivity. Such excitations give rise to Aharonov-Bohm phases and hence are anyons if σ is not an integer, i.e., if the superfluid shows a fractional "tidal" Hall effect. The presence of such excitations may be tested experimentally by measuring fluctuations in the longitudinal resistance of superfluid current conduction; (see [23] for an analogous experiment).

In superfluids of particles with magnetic moments there are mixed "tidal" and electromagnetic effects (e.g., binding electric charge or magnetization to vorticity). See also [24] for a discussion of various effects encountered in superfluids.

(2) Flux quantization [25]

A superconductor exhibits the *Meissner effect*: A magnetic field cannot penetrate into the bulk of a superconducting material. However, in a type II superconductor, thin magnetic field tubes can thread through the bulk. They have the property that they carry a magnetic flux Φ which is an integer multiple of $\frac{hc}{q}$, where q is the charge of the particles in the condensate, (e.g., $q = -2e$, for BCS pairs of electrons). These tubes are called *Abrikosov vortices.* The quantization of Φ is explained by requiring that outside an Abrikosov vortex the quantum mechanical properties of the system, in particular its superconducting nature, remain unchanged. From what we have said about the Aharonov-Bohm effect it follws that this requirement is fulfilled precisely if Φ is an integer multiple of $\frac{hc}{q}$.

The formalism developed in Sect. 2 makes it clear that the Meissner effect and flux quantization for Abrikosov vortices have their partners in the theory of superfluidity: Consider a superfluid in some container. Now set the container in uniform rotation. The superfluid inside the container abhors angular velocity which would destroy the superfluidity and does, therefore, *not* follow the rotation of the container's walls. However, just like there can be Abrikosov vortices in a type II superconductor, the superfluid can eventually be set in motion, and the motion is generated by a velocity field $\vec{f}$, whose curl, $2\vec{\Omega}$, is localized along thin tubes. The tidal variant of the Aharonov-Bohm effect then predicts that the total vorticity in such a tube is quantized to be an integer multiple of $\frac{h}{M}$, where M is the mass of the particles (e.g. ^3He-pairs) constituting the superfluid. [This can also be understood by appealing to the quantization of orbital angular momentum.] If, in such a superfluid, one can excite mass-currents of quantum mechanical particles of mass $m < M$ one may be able to test the tidal Aharonov-Bohm effect.

Our conclusions survive a more detailed theoretical anaysis (see e.g. [26]) and are apparently tested experimentally. The phenomena described here may also be relevant in the astrophysics of neutron stars which are apparently superfluid.

(3) The Aharonov-Casher effect [27]

Consider a system of quantum mechanical particles with spin s, electric charge 0, but with a magnetic moment $\mu \neq 0$, in a plane or in three-dimensional space. [The paritcles could be neutrons, or neutral atoms,....] Following Aharonov and Casher, we would like to study the influence of an external electric field on the dynamics of such particles. As a consequence of *relativistic effects* rapidly moving particles will, in their rest frame, feel a magnetic field that interacts with their magnetic moment.

In the formalism of Sect. 2, this effect should be described as follows: We choose the dreibein $(e^A(\mathbf{x}))_{A=1}^3$ to be the obvious one, namely $e^A{}_i(\mathbf{x}) = \delta_i^A$, for all $\mathbf{x}$, (with e^3 perpendicular to the plane of the system, in the case of a two-dimensional system). By equations (2.15), (2.21) and (2.22), the $SU(2)$-connection w on the spin-s spinor bundle $E^{(s)}$ is given by

$$\begin{aligned} w_\mu^{(s)}(\mathbf{x}) &= i\sum_{A=1}^{3} w_{\mu A}(\mathbf{x}) L_A^{(s)}\,, \quad \text{with} \\ w_{oA}(\mathbf{x}) &= -\frac{\mu}{2c} B_A(\mathbf{x}) = 0\,, \quad \text{and} \\ w_{iA}(\mathbf{x}) &= -\frac{\mu}{4c}\varepsilon_{iAD} E_D(\mathbf{x})\,. \end{aligned}$$

For general electic fields, the curvature, $dw(\mathbf{x}) + (w\wedge w)(\mathbf{x})$, of the $SU(2)$-connection w will not vanish on full-measure sets of space, and so we are not surprised to find that the electric field $\vec{E}(\mathbf{x})$ gives rise to non-trivial spin-orbit interactions. However, if we consider a system of particles confined to the $x-y$ plane in E^3 moving in the electric field of a charged wire placed along the z-axis with constant charge Q per unit of length we encounter an $SU(2)$-version of the Aharonov-Bohm effect: The electric field $\vec{E}(\mathbf{x})$ is then given by $\vec{E}(\mathbf{x}) = \frac{Q}{2\pi r^2}(x,y,0)$, where $r=\sqrt{x^2+y^2}$. The coefficients of the $SU(2)$-connection w are given by

$$w_{13}(\mathbf{x}) = \frac{\mu}{4c} E_2(\mathbf{x}) = \frac{\mu Q}{8\pi c r^2} y\,, \tag{3.1}$$

$$w_{23}(\mathbf{x}) = -\frac{\mu}{4c} E_1(\mathbf{x}) = -\frac{\mu Q}{8\pi c r^2} x\,, \tag{3.2}$$

$w_{i1} = w_{i2} \equiv 0$, for $i = 1,2$, and $w_{3i}(\mathbf{x})$ – which does *not* vanish – does not enter the dynamics of a system confined to the $x-y$ plane. One then checks easily that, for the two-dimensional system in the $x-y$ plane,

$$dw(\mathbf{x}) + (w\wedge w)(\mathbf{x}) = -\frac{\mu Q}{4c}\delta(\mathbf{x})\sigma_3 dx^1\wedge dx^2\,, \tag{3.3}$$

i.e., w is *flat* outside the wire.

The quantum mechanics of this system is described by the action S_Λ introduced in (2.18), with $\Lambda = \mathsf{R}\times(\mathsf{E}^2\backslash\{0\})$, $a_\mu = 0$, and $w_\mu^{(s)} = i w_{\mu 3} L_3^{(s)}$, with $w_{o3} = 0$ and

w_{i3} as given in (3.1) and (3.2). The point is that the scattering of the particles at the charged wire depends on its charge per unit of length, Q, because, although w is flat except at the origin, it cannot be gauged away globally! Therefore, w gives rise to "non-integrable $SU(2)$-phase factors" in the wave functions of the particles which affect their interference patterns. These patterns are periodic in Q with a period given by $\frac{4hc}{\mu}$, as follows easily from (3.1), (3.2) and (3.3). The effect described here was first described by Aharonov and Casher [27] in a somewhat more classical language.

Next, let us consider a *two-dimensional* system on a cone with tip at $\mathbf{x} = 0$. The system consists of particles with non-zero spin. Then the spin connection $\lambda_k{}^A{}_B$, although flat for $\mathbf{x} \neq 0$, cannot be gauged away globally, although $\rho_\mu = 0$ if there are no electromagnetic fields. The $SU(2)$-connection w has the same form as in the previous example, but Q is now given by the defect angle. Scattering of particles at the tip of the cone now yields interference patterns depending on the defect angle Q. This is the "geometrical version" of the Aharonov-Casher effect which is presumably better known than its electromagnetic cousin, see e.g. [28]. What might be more surprising is that we could consider spinning particles on a two-dimensional crystal lattice with disclinations. The scattering at a disclination should also display a "geometrical Aharonov-Casher effect".

Do spinless particles "see" the tip of the cone, or is spin important? The answer depends on our choice of a quantum-mechanical state space: We must impose some "boundary conditions" on the wave functions: $\psi(r, \varphi + 2\pi - Q) = e^{i\theta}\psi(r, \varphi)$, where φ is the polar angle, and θ is some phase to be specified; besides some boundary condition at $r = 0$. But no matter how we choose θ, we can make the tip of the cone *"invisible"* to spinless particles by threading a magnetic flux through $\mathbf{x} = 0$. If the particles have spin *and* a non-zero magnetic moment then, in addition, we would have to put a charge at $\mathbf{x} = 0$, in order to make the tip invisible.

Recall that the Aharonov-Bohm effect explains why two-dimensional quantum theory can describe anyons with fractional statistics, namely particles carrying charge and flux (or mass and vorticity, ...). It is natural to ask whether the Aharonov-Casher effect also has something to do with exotic statistics in two-dimensional quantum theory. The answer is yes! The Aharonov-Casher effect is closely related to the existence of particles in two-dimensional quantum theory with *non-abelian braid statistics* [29]. Such particles can have topological interactions that can be described by some $SU(2)$-Knizhnik-Zamolodchikov connection [30]. Consider, for example, a two-dimensional chiral spin liquid made of particles with spin $s_0 \geq 1$ – if such systems exist. An incompressible chiral spin liquid of this type will most likely exhibit excitations of arbitrary spin $s = 1/2, \cdots, s_0$. The claim is that an excitation of non-zero spin $s < s_0$ will exhibit non-abelian braid statistics, as pointed out in [14]. This will be discussed again in the following section.

We would like to finally remark that there is also an analogue of the Aharonov-

Casher effect where $SU(2)_{\text{spin}}$ is replaced by a gauged internal symmetry group G. This effect can, perhaps, be tested in inhomogeneous heterojunctures. It is related, physically and mathematically, to the existence of particles in two-dimensional quantum theory with topological pair interactions described by a G-Knizhnik-Zamolodchikov connection that, just as in the case of $SU(2)_{\text{spin}}$, may give rise to non-abelian braid statistics.

(4) Einstein-de Haas (-Barnett) effect [31]

Consider a cylinder of iron or some other ferromagnetic material suspended at a wire in such a way that it can freely rotate around its axis. Let us suppose that, initially, it is demagnetized and at rest. Now, imagine that the cylinder is set into rapid rotation around its axis. As explained in Sect. 2, the quantum mechanics of the electrons in this material should now be described in a uniformly rotating coordinate system fixed to the background. In this coordinate system, the electronic Hamiltonian will be time-independent, but it now contains a Zeeman term

$$\vec{\Omega} \cdot \frac{\hbar}{2}\vec{\sigma}\,, \quad (\vec{\Omega} = \text{ angular velocity})\,, \tag{3.4}$$

a tidal vector potential $\vec{f} = \vec{\Omega} \wedge \vec{x}$, and a potential $-\frac{m}{2}|\vec{\Omega} \wedge \vec{x}|^2$ of centrifugal forces; see (2.44), (2.39) and (2.30), and (2.40), respectively. These terms can be combined into $\vec{\Omega} \cdot \vec{J}$, where $\vec{J}$ is the total angular momentum operator [32]. The centrifugal forces will be balanced by the chemical potential of the background. Thus the Hamiltonian is essentially equivalent to the one for the cylinder at rest in a magnetic field $\vec{B} = -\mu^{-1}\vec{\Omega}$. The result is, in both cases, that the cylinder is *magnetized*, because the spins will be aligned with $-\vec{\Omega}, \pm\vec{B}$, respectively. Conversely, if one turns on a magnetic field, $\vec{B}$, antiparallel to the spontaneous magnetization of a magnetized piece of iron, thereby *increasing* the *free energy* of the system, the system reacts by starting to rotate around the axis of the external magnetic field so as to offset the effect of $\vec{B}$ on the electrons by rotation. It thereby returns to a state corresponding to a local minimum of the free energy. A similar effect is observed when one tries to magnetize a paramagnet.

It would appear interesting to test a *local version* of this effect in a "ferro-fluid". If the magnetic field acting on a highly mobile ferro-fluid, locally in thermal equilibrium, is modified locally the fluid reacts by starting to flow with a velocity field that optimally offsets the change in the magnetic field so as to restore local equilibrium. The particle – and magnetic current densities induced are given by $n\vec{f}$ and $\vec{M} \otimes \vec{f}$, respectively, where $\vec{f}$ is the velocity field, n the particle density and $\vec{M}$ the magnetization density. A somewhat analogous effect for quantum Hall fluids will be discussed in the next section.

There is another variant [32] of the Einstein-de Haas effect: consider a beam of non-relativistic particles, e.g. heavy ions, with spin, rotating in a storage ring with some mean angular velocity $\vec{\Omega}$. Then they experience a tidal Zeeman energy, given in (2.44), in addition to the usual magnetic Zeeman energy (2.45). After relaxation to a steady state, the tidal Zeeman energy obviously affects the ratio of "spin-up" to "spin-down" ions in the beam!

Similar considerations are important e.g. in the study of electronic spectra of rotating molecules in the Born-Oppenheimer approximation; see [32].

(5) Supercurrents [25]

Consider a superconducting condensate of charged bosons, e.g. electron pairs, of charge q and mass M, in equilibrium. Imagine that a magnetic field, $\vec{B}$, is turned on inside the bulk of this system. Since the superconducting state minimizes the free energy of the system, the condensate reacts to turning on $\vec{B}$ by developing a *flow* with velocity field $\vec{f}$ in such a way as to offset the effect of $\vec{B}$. Neglecting the centrifugal potential, $-\frac{m}{2}\vec{f}\cdot\vec{f}$, and the magnetic field created by the resulting current, it follows from eqs. (2.42), (2.43) and (2.46) that the optimal velocity field $\vec{f}$ is given by

$$\vec{f} = \frac{q}{Mc}\vec{A}^{\,T} ,$$

where $\vec{A}^{\,T}$ is the vector potential of $\vec{B}$ in the Coulomb gauge (i.e., div $\vec{A}^{\,T} = 0$). Thus the system exhibits a supercurrent density, $\vec{j}_s$, given in our approximation by

$$\vec{j}_s = qn\vec{f} = \frac{q^2 n}{Mc}\vec{A}^{\,T} , \tag{3.5}$$

where n is the density of the condensate. This is the London equation for type II superconductors. Recalling that $\vec{j}(x) = \delta S_{\text{eff.}}(\vec{A})/\delta\vec{A}(x)$ – see also Sect. 4 – one may proceed from eq. (3.5), fairly easily, to the Anderson-Higgs mechanism. Note that, by eq. (3.5), a supercurrent $\vec{j}_s$ is really a sign for the presence of a *vector potential*, $\vec{A}^T$, and thus can be used for experimental tests of the Aharonov-Bohm effect![2]

There is an $SU(2)_{\text{spin}}$-analogue of these effects in condensates of *neutral* bosons with *magnetic moments*. For example, in principle, one encounters an "$SU(2)$-Anderson-Higgs mechanism" and, for spin-polarized condensates, spin supercurrents induced by electric fields.

We have already alluded to the *Hall effect* earlier in this section. Just as the Aharonov-Bohm effect reflects the $U(1)_{\text{em}}$-gauge-invariance of quantum theory, so does the Hall effect for the electric current, as emphasized by Laughlin [33]. Actually one might view the Hall effect as a time-dependent version of the Aharonov-Bohm effect [21]. In the same vein, both the Aharonov-Casher effect and the Hall effect for the *spin current* reflect the $SU(2)_{\text{spin}}$-gauge-invariance of non-relativistic quantum theory, as emphasized in [5]. In the next section, we attempt to unravel the universal aspects of the quantum Hall effect in two-dimensional, incompressible electron fluids with broken parity and time reversal invariance.

---

[2] N. Byers and C.N. Yang, Phys. Rev. Lett. **7**, 46 (1961).

4. "SCALING LIMIT" OF THE EFFECTIVE ACTION OF A TWO-DIMENSIONAL, INCOMPRESSIBLE QUANTUM FLUID

In this section we study the generating ("partition") function of two-dimensional non-relativistic quantum systems coupled to electromagnetic fields:

$$Z_\Lambda(a,w) := \int \mathcal{D}\psi^* \mathcal{D}\psi e^{iS_\Lambda(\psi^*,\psi;a,w)/\hbar} , \tag{4.1}$$

where the gauge potentials a and w have been introduced in (2.13)–(2.16), and S_Λ is the action of the system given in (2.18); see also (2.39), (2.40) and (2.42). The integration variables ψ^* and ψ are Grassmann variables (anti-commuting c-numbers) for Fermi statistics, and complex c-number fields for Bose statistics.

We have not displayed the metric, g_{ij}, of space explicitly, since it will be kept fixed, and usually $M = \mathsf{E}^2$ with $g_{ij} = \delta_{ij}$, for simplicity. We realize that, for the study of the *stress tensor*, pressure – and density fluctuations and curvature effects, we would have to choose a variable external metric (or, at least, a variable conformal factor in g_{ij}). This would be important for an understanding of density waves, in particular surface density waves (which are interesting in two-dimensional quantum fluids), and of critical phenomena. But, unfortunately, we cannot cover everything that is interesting; confer e.g. to [17]. We note, however, that curvature effects can be studied by analyzing the dependence of $Z_\Lambda(a,w)$ on w which contains the spin connection, λ; see (2.15).

We define the electric charge – and current densities, j^0 and $\vec{j}$, by

$$\begin{aligned} j^0(x) &= \psi^*(x)\psi(x) , \\ j^k(x) &= -\frac{i\hbar}{2mc} g^{kl}(x) \left[(D_l\psi)^*(x)\psi(x) - \psi^*(x)(D_l\psi)(x)\right] , \end{aligned} \tag{4.2}$$

and the spin – and spin current densities, $\vec{s}^\mu(x)$, by

$$\begin{aligned} \vec{s}^{\,0}(x) &= \psi^*(x)\vec{L}^{(s)}\psi(x) , \\ \vec{s}^{\,k}(x) &= -\frac{i\hbar}{2mc} g^{kl}(x) \left[(D_l\psi)^*(x)\vec{L}^{(s)}\psi(x) - \psi^*(x)\vec{L}^{(s)}(D_l\psi)(x)\right] , \end{aligned} \tag{4.3}$$

where $(L_1^{(s)}, L_2^{(s)}, L_3^{(s)})$ are the generators of the spin-s representation of $su(2)$. Similarly, we can define the currents associated with internal symmetries; but, for simplicity, we shall not consider them here. The electric current is conserved (continuity equation holds), but the spin current is, in general, *not* conserved, because it couples to a non-abelian vector potential. It is, however, *covariantly conserved*; see (4.11).

It is straightforward to infer from (4.1), (2.18), (4.2) and (4.3) that the *time-ordered current Green functions* of the system are given, *at non-coinciding arguments*, by

$$\begin{aligned} &\left\langle T\left[\prod_{i=1}^{n} j^{\mu_i}(x_i) \prod_{l=1}^{m} s^{\nu_l}_{A_l}(y_l)\right]\right\rangle^c_{a,w} \\ &= i^{n+m} \prod_{i=1}^{n} \frac{\delta}{\delta a_{\mu_i}(x_i)} \prod_{l=1}^{m} \frac{\delta}{\delta w_{\nu_l A_l}(y_l)} \ln Z_\Lambda(a,w) , \end{aligned} \tag{4.4}$$

where $\langle(\cdot)\rangle^c_{a,w}$ denotes the connected expectation functional of the system in an external gauge field configuration, (a, w), (with "ground state asymptotic conditions", as $t \to \pm\infty$, to be specific), and T indicates time-ordering. At *coinciding arguments*, eq. (4.4) is modified by *Schwinger terms*, (but this will not be very important).

We define the effective gauge field action by

$$S_\Lambda^{\text{eff}}(a, w) := \frac{\hbar}{i} \ln Z_\Lambda(a, w) . \tag{4.5}$$

The idea is to try to calculate the "leading terms" in $S_\Lambda^{\text{eff}}(a, w)$ which, via (4.4), will provide us with information on the current Green functions. By "leading terms" we mean those terms which dominate at large distance scales and low frequencies. The calculation of the leading terms in S_Λ^{eff} may look like a fairly vast problem. Actually, making a single assumption on the excitation spectrum of the system, "*incompressibility*", and using the $U(1)_{\text{em}} \times SU(2)_{\text{spin}}$-*gauge-invariance* of the system, that calculation can be carried out, [5].

Let χ be a real-valued function and R an $SU(2)$-valued function on space-time $N = \mathbf{R} \times M$. Consider the gauge transformations in eq. (2.49), i.e.,

$$a \mapsto {}^\chi a , \quad {}^\chi a_\mu := a_\mu - \partial_\mu \chi , \tag{4.6}$$

and

$$w \mapsto {}^R w , \quad {}^R w_\mu := R w_\mu R^* + R \partial_\mu R^* . \tag{4.7}$$

Changing integration variables,

$$\psi \mapsto {}^{\chi, R}\psi := e^{i\chi} U^{(s)}(R) \psi \tag{4.8}$$

in the functional integral (4.1), and using the gauge-invariance of S_Λ under the transformations (4.6) – (4.8) and the fact that the Jacobian of (4.8) is unity, we find the *Ward identity*

$$S_\Lambda^{\text{eff}}(a, w) = S_\Lambda^{\text{eff}}({}^\chi a, {}^R w) , \tag{4.9}$$

for all χ and R. [For a system of finitely many particles in a bounded region of space, (4.9) can be proven rigorously. This identity is stable under passing to limits, for χ's and dR's of compact support.]

By differentiating (4.9) in χ or R and setting $\chi = 0, R = 1$, we find, using (4.4) (for $n + m = 1$), that

$$\frac{1}{\sqrt{g}} \partial_\mu \left(\sqrt{g} \langle j^\mu \rangle_{a,w}\right) = 0 , \tag{4.10}$$

and

$$\frac{1}{\sqrt{g}} D_\mu \left(\sqrt{g} \langle s^\mu \rangle_{a,w}\right)_A = 0 , \quad A = 1, 2, 3 ,$$

or

$$\frac{1}{\sqrt{g(x)}} \partial_\mu \left(\sqrt{g(x)} \langle \vec{s}^\mu(x) \rangle_{a,w}\right) = 2\vec{w}_\mu(x) \wedge \langle \vec{s}^\mu(x) \rangle_{a,w} , \tag{4.11}$$

for *arbitrary* a and w. These "infinitesimal" Ward identities play an important role in determining the general form of S_Λ^{eff}. They can be generalized, in an obvious way, to systems with internal symmetries.

We now proceed to determine the form of S_Λ^{eff} in the scaling limit. We need to consider ever larger systems and ever slower variations in time. Let $1 \leq \theta < \infty$ be a scale parameter. We set

$$\begin{aligned} g_{ij}(x) &\equiv g_{ij}^{(\theta)}(x) := \gamma_{ij}\left(\frac{x}{\theta}\right), \quad \text{and} \\ \Lambda &\equiv \Lambda^{(\theta)} := \theta\Lambda_0 , \end{aligned} \tag{4.12}$$

where γ_{ij} is a fixed metric on M (e.g. $\gamma_{ij} = \delta_{ij}$), and Λ_0 is a fixed space-time cylinder;

$$x = (x^0, \mathbf{x}) = \theta(\xi^0, \boldsymbol{\xi}) = \theta\xi , \quad \xi \in \Lambda_0 . \tag{4.13}$$

Then

$$\frac{\partial}{\partial x^\mu} = \theta^{-1}\frac{\partial}{\partial \xi^\mu} . \tag{4.14}$$

We propose to study the reaction of the system to a small change in the external gauge potentials a and w. We choose fixed background potentials, $a_c(x)$ and $w_c(x)$, defined on all of space-time, and set

$$a_\mu{}^{(\theta)}(x) := a_{c,\mu}(x) + \theta^{-1}\tilde{a}_\mu\left(\frac{x}{\theta}\right) , \tag{4.15}$$

and

$$w_\mu{}^{(\theta)}(x) := w_{c,\mu}(x) + \theta^{-1}\tilde{w}_\mu\left(\frac{x}{\theta}\right) , \tag{4.16}$$

where $\tilde{a}_\mu(\xi)$ and $\tilde{w}_\mu(\xi)$ are *fixed* functions defined on Λ_0. If m is the effective mass of the particles and μ their magnetic moment in physical $(t, \mathbf{x})$-coordinates then the mass $m^{(\theta)}$ and magnetic moment $\mu^{(\theta)}$ in *rescaled coordinates*, $\tau \equiv \frac{\xi^0}{c}, \boldsymbol{\xi}$, are given by

$$m^{(\theta)} = m \cdot \theta , \quad \text{and } \mu^{(\theta)} = \mu\theta^{-1} , \tag{4.17}$$

as follows from eqs. (2.18), (2.25) and (2.26), (i.e., in the *rescaled* system the particles are heavy and have small magnetic moments. Moreover, the range of the two-body potential, in the *rescaled* system, becomes shorter and shorter, as θ becomes large).

One *basic assumption* underlying our analysis is that $S_{\theta\Lambda_0}^{\text{eff}}(a^{(\theta)}, w^{(\theta)})$ is *four times continuously differentiable* in $\tilde{a}_\mu^{(\theta)}(x) \equiv \theta^{-1}\tilde{a}_\mu\left(\frac{x}{\theta}\right)$ and $\tilde{w}_\mu^{(\theta)}(x) \equiv \theta^{-1}\tilde{w}_\mu\left(\frac{x}{\theta}\right)$ at $\tilde{a}_\mu^{(\theta)} = \tilde{w}_\mu^{(\theta)} = 0$, for a *suitable choice of background potentials*, a_c and w_c, and for $\tilde{a}_\mu$ and $\tilde{w}_\mu$ constrained to belong to *suitable spaces*, $\mathcal{A}$ and $\mathcal{W}$, *of fluctuation potentials*, to be specified later. We may then expand $S_{\theta\Lambda_0}^{\text{eff}}$ to third order in $\tilde{a}^{(\theta)}$ and $\tilde{w}^{(\theta)}$, with a fourth order remainder term. Among the terms thus generated we shall only retain the leading terms in θ, namely those scaling with a *non-negative power* of θ which are commonly called *relevant* and *marginal* terms. The sum of these terms will be denoted by $S_{\Lambda_0}^*(\tilde{a}, \tilde{w})$, a functional that we call the *scaling limit of the effective action.*

Using identity (4.4) to find the Taylor coefficients of $S^{\text{eff}}_{\theta\Lambda_0}(a^{(\theta)}, w^{(\theta)})$, plugging (4.15) and (4.16) into the resulting expressions, and finally passing to $(\xi^0, \boldsymbol{\xi})$-coordinates, we find that the coefficient of the term of nth order in $\tilde{a}$ and of mth order in $\tilde{w}$ in $S^{\text{eff}}_{\theta\Lambda_0}$ is given by a distribution

$$\varphi_\theta^{\mu_1\cdots\nu_1\cdots}{}_{A_1\cdots A_m}(\xi_1, \cdots, \xi_n, \eta_1, \cdots, \eta_m)$$

which, at non-coinciding arguments, is given by

$$(-i)^{n+m}\left\langle T\left[\prod_{i=1}^{n}(\theta^2 j^{\mu_i}(\theta\xi_i))\prod_{l=1}^{m}(\theta^2 s^{\nu_l}_{A_l}(\theta\eta_l))\right]\right\rangle^c_{a_c, w_c}, \tag{4.18}$$

in accordance with the circumstance that, in *three* space-time dimensions, the scaling dimension of currents is 2!

We may now formulate our *basic assumption of incompressibility*: We imagine that, for *certain choices of the background potentials a_c and w_c*, the excitation spectrum of the system above its groundstate (energy) is such that connected Green functions of its *currents* have *"good" cluster properties* (better than in a system with Goldstone bosons), in such a way that the limits of the distributions $\varphi_\theta^{\mu_1\cdots\nu_1\cdots}{}_{A_1\cdots A_m}$, as $\theta \to \infty$, are *local distributions*, i.e., sums of products of derivatives of δ-functions.

This incompressiblity assumption is by no means a mild or minor assumption. It tends to be a really hard analytical problem of many-body theory to show that, for a concrete system, it is satisfied. [For some recent ideas about how to establish it for quantum Hall fluids at certain filling factors see [34,35,14].] What we propose to do here is to use it to calculate the general form of the action $S^*_{\Lambda_0}$ in the scaling limit. We only sketch some ideas; for the details see [5].

Our calculation is based on the following four principles:

(A) *Incompressibility*: $\varphi_\theta^{\mu_1\cdots\mu_n\nu_1\cdots\nu_m}{}_{A_1\cdots A_m}$ converge to local distributions, as $\theta \to \infty$, for all n and m.

(B) $U(1)_{\text{em}} \times SU(2)_{\text{spin}}$-*gauge-invariance*: Ward identities (4.9) – (4.11).

(C) *Only relevant and marginal terms are kept in $S^*_{\Lambda_0}$*.

(D) *Extra symmetries of the system*, e.g., for $a_{c0} = 0$, $w_{c\mu A} = \delta_{A3} w_{c\mu 3}$, global rotations around the 3-axis in spin space are a continuous, global symmetry of the system with an associated *conserved* Noether current $s_{\mu 3}(x)$; or translation invariance in the scaling limit $(\theta \to \infty)\cdots$, *are exploited to reduce the number of terms.*

From (A) and eqs. (4.15) and (4.16) it immediately follows that all terms contributing to $S^{\text{eff}}_{\theta\Lambda_0}$ of order 4 or higher in $\tilde{a}$ and $\tilde{w}$ are *irrelevant*, (scaling like $\theta^{-D}, D >$

0). In particular, a fourth-order remainder term does *not* contribute to $S^*_{\Lambda_0}$, (principle (C)). We now present the final result, in the special case of systems which are incompressible for a choice of $w_{c\mu}$ satisfying

$$w_{c\mu A}(x) = \delta_{A3} w_{c\mu 3}(x) \, , \tag{4.19}$$

or, in view of eqs. (2.25) and (2.26), for a background electromagnetic field $(\vec{E}_c, \vec{B}_c)$ with

$$\vec{B}_c(x) = (0,0,B_c(x)) \, , \; \vec{E}_c(x) = (E_1(x), E_2(x), 0) \, , \tag{4.20}$$

and a spin connection

$$\left(\lambda_\mu{}^A{}_B\right) = \begin{pmatrix} 0 & \lambda_\mu & 0 \\ -\lambda_\mu & 0 & 0 \\ 0 & 0 & 1 \end{pmatrix} \, , \tag{4.21}$$

in the coordinate system $(e^1(x), e^2(x), e^3(x))$. In this situation, the scaling limit of the effective action is given by

$$\begin{aligned} -\frac{1}{\hbar} S^*_{\Lambda_0}(\tilde{a}, \tilde{w}) &= \int_{\Lambda_0} j^\mu_c \tilde{a}_\mu dv + \int_{\Lambda_0} m^\mu_3 \tilde{w}_{\mu 3} dv \\ &+ \sum_{A=1}^{2} \int_{\Lambda_0} \tau_1^{\mu\nu} \tilde{w}_{\mu A} \tilde{w}_{\nu A} dv + \sum_{A,B=1}^{2} \int_{\Lambda_0} \tau_2^{\mu\nu} \varepsilon_{AB} \tilde{w}_{\mu A} \tilde{w}_{\nu B} dv \\ &+ \sum_{A,B,C=1}^{3} \int_{\Lambda_0} \eta^{\mu\nu\rho}_{ABC} \tilde{w}_{\mu A} \tilde{w}_{\nu B} \tilde{w}_{\rho C} dv \\ &+ \frac{\sigma}{4\pi} \int_{\Lambda_0} \tilde{a} \wedge d\tilde{a} + \frac{\chi}{2\pi} \int_{\Lambda_0} \tilde{a} \wedge d\tilde{w}_3 + \frac{\sigma_s}{4\pi} \int_{\Lambda_0} \tilde{w}_3 \wedge d\tilde{w}_3 \\ &+ \frac{k}{4\pi} \int_{\Lambda_0} \mathrm{tr} \left(w \wedge dw + \frac{2}{3} w \wedge w \wedge w \right) + B.T. \end{aligned} \tag{4.22}$$

where j^μ_c is an electric – and m^μ_3 a magnetic supercurrent circulating in the system when $a = a_c$, $w = w_c$; $\tau_1^{\mu\nu}$ is a function symmetric in μ and ν, while $\tau_2^{\mu\nu}$ is antisymmetric in μ and ν; the function $\eta^{\mu\nu\rho}_{ABC}$ is symmetric under interchanges of $(\mu A), (\nu B)$ and (ρC) and vanishes if two or more of the indices A, B, C are equal to 3; $dv = \sqrt{\gamma(\xi)} d\xi$ is the volume element on space-time; σ, χ, σ_s and k are real constants, whose possible values will be studied in Sect. 5; $w(\xi) = w_c^{(\theta)}(\xi) + \tilde{w}(\xi)$ is the total $SU(2)$ connection, with $w_c^{(\theta)}(\xi) = \theta w_c(\theta\xi)$, by (4.16); and "B.T." are boundary terms only depending on $a|_{\partial\Lambda_0}$, $w|_{\partial\Lambda_0}$ which will be studied in Sect. 5. In the last four terms on the r.h.s. of (4.22) we are using a *new notation:*

$$\begin{aligned} \tilde{a} &= \sum_{\mu=0}^{2} \tilde{a}_\mu d\xi^\mu \, , \quad d\tilde{a} = \sum_{\mu,\nu=0}^{2} \partial_\mu \tilde{a}_\nu \, d\xi^\mu \wedge d\xi^\nu \\ \tilde{w}_3 &= \sum_{\mu=0}^{2} \tilde{w}_{\mu 3} d\xi^\mu \, , \quad w = \sum_{\mu=0}^{2} \sum_{A=1}^{3} w_{\mu A} \sigma_A \, d\xi^\mu \, . \end{aligned} \tag{4.23}$$

See [5] for more details.

In Sect. 5, we shall use results on $U(1)$ – and $SU(2)$ chiral current algebra to determine the possible values of σ, χ, σ_s and k and find some relations between them.

Here we wish to point out that the functions $j_c^\mu, m_3^\mu, \tau_1^{\mu\nu}, \tau_2^{\mu\nu}$ and $\eta_{ABC}^{\mu\nu\rho}$ are not all independent, but are constrained by the infinitesimal Ward identities (4.10) and (4.11): By (4.4)

$$\langle j^\mu(\xi)\rangle_{a^{(\theta)},w^{(\theta)}} = \frac{\delta S^*_{\Lambda_0}(\tilde{a},\tilde{w})}{\delta \tilde{a}_\mu(\xi)} + \cdots , \tag{4.24}$$

$$\langle s_A^\mu(\xi)\rangle_{a^{(\theta)},w^{(\theta)}} = \frac{\delta S^*_{\Lambda_0}(\tilde{a},\tilde{w})}{\delta \tilde{w}_{\mu A}(\xi)} + \cdots . \tag{4.25}$$

The dots stand for contributions from irrelevant terms in the effective action. We calculate the r.h.s. of these equations by using (4.22) and plug the result into eqs. (4.10) and (4.11). As a result we obtain the following constraints (see [5]).

$$(a) \qquad \frac{1}{\sqrt{g}}\partial_\mu(\sqrt{g} j_c^\mu) = 0 .$$

$$(b) \qquad \frac{1}{\sqrt{g}}\partial_\mu(\sqrt{g} m_3^\mu) = 0 .$$

$$(c) \qquad \sum_{B=1}^{2} \varepsilon_{AB}\left\{m_3^\mu - 2\tau_1^{0\mu} w_{c03}^{(\theta)}\right\}\tilde{w}_{\mu B} + 2 w_{c03}^{(\theta)} \sum_{j=1}^{2} \tau_2^{0j}\tilde{w}_{jA} = 0 , \quad A = 1,2 .$$

$$(d) \qquad \frac{1}{\sqrt{g}}\partial_\mu\left\{\sqrt{g}\tau_1^{\mu\nu}\tilde{w}_{\nu A} + \sqrt{g}\sum_{B=1}^{2}\varepsilon_{AB}\tau_2^{\mu\nu}\tilde{w}_{\nu B}\right\} = -2\left\{\sum_{B=1}^{2}\varepsilon_{AB}\tau_1^{\mu\nu}\tilde{w}_{\nu B} - \tau_2^{\mu\nu}\tilde{w}_{\nu A}\right\}\tilde{w}_{\mu 3} - 3\sum_{B=1}^{2}\varepsilon_{AB}\sum_{C,D=1}^{3}\eta_{BCD}^{0\nu\rho} w_{c03}^{(\theta)}\tilde{w}_{\nu C}\tilde{w}_{\rho D} , \quad A = 1,2 . \tag{4.26}$$

Constraints (a) and (b) just express the conservation of the supercurrents j_c^μ and m_3^μ when $\tilde{a} = \tilde{w} = 0$.

If we impose (4.26)-(c) and (d), for *arbitrary* smooth fluctuation potentials $\tilde{w}$, then it follows that

$$m_3^\mu = \tau_1^{\mu\nu} = \tau_2^{\mu\nu} = 0 , \quad \text{for all } \mu \text{ and } \nu , \tag{4.27}$$

in particular, the system *cannot* be *magnetized* ($m_3^0 = 0$) and cannot support persistent spin currents. This may seem rather strange, because we would expect that if $w_{c03} = -\frac{\mu}{2c}B_c$, for some large magnetic field $\vec{B}_c = (0,0,B_c)$, then the system would be magnetized in the 3-direction. What has gone wrong? The point is that the assumed properties that S_Λ^{eff} is four times continuously differentiable in $\tilde{a}$ and $\tilde{w}$ and that the system remains incompressible in an *arbitrary* function-space neighborhood of (a_c, w_c) of sufficiently small diameter must fail for magnetized systems! The reason is that an *arbitrarily small* fluctuation field $\tilde{w}$ which oscillates rapidly in time can destroy the incompressibility of the system, and hence our estimate on the fourth order remainder in the Taylor expansion of S_Λ^{eff} breaks down.

We thus assume, for example, that, for a *time-independent* background field w_c, the system remains incompressible and S_Λ^{eff} is four times continuously differentiable in $(\tilde{a}, \tilde{w})$ on the function-space sets

$$\begin{aligned} \mathcal{A} &= \{\tilde{a}_\mu \in \mathcal{S}\} , \\ \mathcal{W} &= \{\tilde{w}_{\mu A} \in \mathcal{S} : \tilde{w}_{\mu A} \text{ is time-independent}\} , \end{aligned} \tag{4.28}$$

where $\mathcal{S}$ is some Schwartz space neighbourhood of 0. Then constraints (c) and (d) of (4.26) imply that

$$\tau_1^{00}(\xi) = \frac{m_3^0(\xi)}{2w_{c03}(\xi)} , \quad \tau_1^{0i} = \tau_1^{ij} = 0 , \tag{4.29}$$

$$\begin{aligned} \tau_2^{\mu\nu} &= 0 , \quad \eta_{AA3}^{000}(\xi) = -\frac{\tau_1^{00}(\xi)}{3w_{c03}(\xi)} , \quad \text{for } A = 1,2; \\ &\text{all other } \eta_{ABC}^{\mu\nu\rho} \text{ vanish.} \end{aligned} \tag{4.30}$$

Hence $(m_3^\mu) = (m_3^0, \mathbf{0})$. Under somewhat more restrictive assumptions on $\mathcal{W}$, imposing e.g. relations (2.25) and (2.26) on $\tilde{w}$ which couple $\tilde{w}$ to $\tilde{a}$, a non-zero spin current $\mathbf{m}_3 = (m_3^1, m_3^2)$ is possible, too. For a more detailed discussion see [5].

A corollary of our derivation of $S_{\Lambda_0}^*$, in particular of (4.29), using gauge invariance and incompressibility, is the *Goldstone theorem*, [36]: If the magnetization, $M = \mu\frac{\hbar}{2}m_3^0$, does *not* tend to 0, as $\vec{B}_c = (0,0,B_c)$ tends to 0 (with $w_{c03} = -\frac{\mu}{2c}B_c$) then the system *cannot* be incompressible at $\vec{B}_c = 0$, i.e., there are gapless extended modes, the *Goldstone bosons*, coupled to the groundstate by the spin current; see [5]. Our proof also works for systems with continuous non-abelian internal symmetries.

Next, let us briefly discuss the *linear response equations* (4.24) and (4.25) that follow from our expression (4.22) for the effective action $S_{\Lambda_0}^*$ in the scaling limit, for systems characterized by conditions (4.28) – (4.30). It is a simple exercise to verify that

$$\begin{aligned} \sqrt{g(\xi)}\,\langle j^\mu(\xi)\rangle_{a,w} &= \sqrt{g(\xi)}j_c^\mu(\xi) + \frac{\sigma}{2\pi}\varepsilon^{\mu\nu\rho}(\partial_\nu\tilde{a}_\rho)(\xi) \\ &\quad + \frac{\chi}{2\pi}\varepsilon^{\mu\nu\rho}(\partial_\nu\tilde{w}_{\rho 3})(\xi) + \cdots \end{aligned} \tag{4.31}$$

and

$$\begin{aligned} \sqrt{g(\xi)}\,\langle s_A^\mu(\xi)\rangle_{a,w} &= \sqrt{g(\xi)}\delta_{A3}\delta_0^\mu m_3^0(\xi) + \delta_{A3}\frac{\chi}{2\pi}\varepsilon^{\mu\nu\rho}(\partial_\nu\tilde{a}_\rho)(\xi) \\ &+ \delta_{A3}\frac{\sigma_s}{2\pi}\varepsilon^{\mu\nu\rho}(\partial_\nu\tilde{w}_{\rho 3})(\xi) \\ &- \frac{k}{\pi}\varepsilon^{\mu\nu\rho}\{(\partial_\nu w_{\rho A})(\xi) - \varepsilon_{ABC}w_{\nu B}(\xi)w_{\rho C}(\xi)\} \\ &+ \sqrt{g(\xi)}2(1-\delta_{A3})\delta_0^\mu\tau_1^{00}(\xi)\tilde{w}_{0A}(\xi) + \cdots , \end{aligned} \tag{4.32}$$

where the dots stand for terms coming from irrelevant terms in the effective action, or from terms of order two in $\tilde{w}$ (e.g. a term proportional to η_{ABC}^{000}) which are of little interest in linear response theory. Furthermore, $w_{\mu A} = w_{c\mu A}^{(\theta)} + \tilde{w}_{\mu A}$.

In order to understand the physical contents of these equations, we must remind ourselves of the physical meaning of the connections a and w elucidated in Sect. 2: From eqs. (2.20), (2.39) and (2.43) we know that

$$a_j(x) = \frac{e}{\hbar c} A_j(x) - \frac{m}{\hbar} f_j(x) , \tag{4.33}$$

where $\vec{A}$ is the electromagnetic vector potential, $-e$ is the charge and m the effective mass of the particles in the quantum fluid, and $\vec{f}$ is a divergence-free velocity field generating some incompressible superfluid flow. Furthermore, by (2.20),

$$a_0(x) = -\frac{e}{\hbar c} \phi(x) , \tag{4.34}$$

where ϕ is the electrostatic potential.

Since we are studying two-dimensional incompressible quantum fluids on a surface M imbedded in E^3, it is natural to choose an $SU(2)_{\text{spin}}$-gauge with the property that $e^3(t, \mathbf{x})$ is orthogonal to the tangent space of M at $\mathbf{x}$, for all times t, as discussed at the beginning of Sect. 2. Then the $SU(2)$-spin connection $\lambda^{(1/2)}$ has the form

$$\lambda_j^{(1/2)} = i\lambda_j \sigma_3 \, , \; j = 1,2; \quad \text{and } \lambda_{0B}^{\;A} = \frac{1}{2} \left[\delta^{AC} \delta_{BD} \mathcal{E}_C^l \partial_0 e_l^D - \mathcal{E}_B^l \partial_0 e_l^A \right] . \tag{4.35}$$

It then follows from (2.24), (2.25) and (2.39), (2.44) that

$$w_{0A}(x) = -\frac{\mu}{2c} B_A(x) + \delta_{A3} \Omega(x) + \lambda_{0A} \tag{4.36}$$

where $\vec{\Omega}(x) = (0, 0, \Omega(x))$ is the curl of $\vec{f}$ in the dreibein basis $(e^A(x))_{A=1}^3$, and μ is the magnetic moment of the particles. Finally, by (2.15), (4.35), (2.26) and (2.33),

$$w_{jA}(x) = \delta_{A3}(\lambda_j(x) + \cdots) - \frac{\mu}{4c} \sum_{C=1}^{3} \varepsilon_{jAC}(x) E_C(x) , \tag{4.37}$$

where the dots correspond to terms proportional to derivatives of $\Omega(t', \mathbf{x}), t' \leq t$, (and are generated by the $SU(2)_{\text{spin}}$-gauge-transformation defined in (2.31)).

Finally, we define the charge density operator, in physical units

$$\rho(\xi) := e\sqrt{g(\xi)} j^0(\xi) , \tag{4.38}$$

the electric current density by

$$\mathbf{J}(\xi) = ec\sqrt{g(\xi)} \mathbf{j}(\xi) , \tag{4.39}$$

the spin density by

$$\vec{S}^0(\xi) = \frac{\hbar}{2} \sqrt{g(\xi)} \vec{s}^{\,0}(\xi) , \tag{4.40}$$

and the spin current density by

$$\vec{S}^j(\xi) = \frac{\hbar c}{2} \sqrt{g(\xi)} \vec{s}^{\,j}(\xi) . \tag{4.41}$$

Then equation (4.31) for the 0-component reads

$$\begin{aligned}\langle\rho(\xi)\rangle_{a,w} &= \rho_c(\xi)+\frac{\sigma_H}{c}\tilde{B}_3(\xi)-2\sigma\frac{em}{h}\tilde{\Omega}(\xi)\\ &- \frac{\chi}{2\pi}\left(\frac{e\mu}{4c}\boldsymbol{\nabla}\cdot\tilde{\boldsymbol{E}}(\xi)-e\mathcal{R}(\xi)\right)+\cdots,\end{aligned} \tag{4.42}$$

where $\sigma_H=\frac{e^2}{h}\sigma$ is the Hall conductivity, $\boldsymbol{\nabla}\cdot(\quad)$ denotes the divergence , $\mathcal{R}(\xi)=$ curl $\boldsymbol{\lambda}(\xi)$ is the scalar curvature of M at ξ, and the dots stand for contributions from irrelevant terms. It will turn out that

$$\chi_\perp:=\frac{e\mu}{4\pi c}\chi \tag{4.43}$$

is the magnetic susceptibility of the system in the 3-direction normal to the surface. In (4.42) and the following formulas the tildes $\tilde{}$ indicate contributions from $\tilde{a}$ and $\tilde{w}$; (we have absorbed the spin connection λ into $\tilde{w}$, but without decorating it with a $\tilde{}$). Next, one verifies that

$$\begin{aligned}\left\langle J^i(\xi)\right\rangle_{a,w} &= J^i_c(\xi)+\sigma_H\varepsilon^{ij}\tilde{E}_j(\xi)\\ &+ \frac{\sigma em}{h}\varepsilon^{ij}\frac{\partial}{\partial\tau}\tilde{f}_j(\xi)\\ &- \frac{\chi}{2\pi}\left(\frac{e\mu}{2}\varepsilon^{ij}\partial_j\tilde{B}_3(\xi)-ec\varepsilon^{ij}\partial_j\tilde{\Omega}(\xi)\right)\\ &+ \frac{\chi}{2\pi}\left(\frac{e\mu}{4c}\frac{\partial}{\partial\tau}\tilde{E}^i(\xi)-e\varepsilon^{ij}\frac{\partial}{\partial\tau}\lambda_j(\xi)\right)\\ &+ \cdots,\end{aligned} \tag{4.44}$$

where $\tau=\xi^0/c$ is the rescaled time variable.

From (4.32) we find, for example, that

$$\begin{aligned}\mu\langle S^0_3(\xi)\rangle_{a,w} &= M(\xi)+\sigma_H^{\text{spin}}\left(\frac{\mu}{2c}\boldsymbol{\nabla}\cdot\tilde{\boldsymbol{E}}(\xi)-2\mathcal{R}(\xi)\right)+k\frac{\mu^2\hbar}{8\pi c}\boldsymbol{\nabla}\cdot\mathbf{E}_c(\xi)\\ &+ \chi_\perp\left(\tilde{B}_3(\xi)-\frac{2c}{\mu}\tilde{\Omega}(\xi)\right)+\cdots,\end{aligned} \tag{4.45}$$

where M is the magnetization at (a_c,w_c), $\chi_\perp$ is the magnetic susceptibility at (a_c,w_c) given in (4.43), and

$$\sigma_H^{\text{spin}}=\mu\hbar\frac{k}{4\pi}-\mu\hbar\frac{\sigma_s}{8\pi} \tag{4.46}$$

is the *Hall conductivity for the spin current.* As eq. (4.45) shows, σ_H^{spin} is a *pseudoscalar.* Next

$$\begin{aligned}\langle S^i_3(\xi)\rangle_{a,w} &= \sigma_H^{\text{spin}}\Bigg(\varepsilon^{ij}\partial_j\tilde{B}_3(\xi)-\frac{2c}{\mu}\varepsilon^{ij}\partial_j\tilde{\Omega}(\xi)-\frac{1}{2c}\frac{\partial}{\partial\tau}\tilde{E}^i(\xi)\\ &+ \frac{2}{\mu}\frac{\partial}{\partial\tau}\lambda^i(\xi)\Bigg)+k\frac{\mu\hbar}{4\pi}\varepsilon^{ij}\partial_jB_c(\xi)\\ &+ \chi_\perp\left(c\mu^{-1}\varepsilon^{ij}\tilde{E}_j(\xi)+\frac{mc}{\mu e}\frac{\partial}{\partial\tau}\tilde{f}^i(\xi)\right)+\cdots\end{aligned} \tag{4.47}$$

where the dots stand for terms proportional to λ_0 and further irrelevant and higher-order terms. A similar story could be told for $\langle S^{\mu}_{A}(\xi)\rangle_{a,w}$, but we refrain from telling it and refer the reader to his drawing board, or to [5]. [We do not guarantee all signs and factors of 2π in our formulas!]

We encourage the reader to notice how neatly our formulas summarize the laws of the Hall effect, including effects due to *tidal forces* coming from *superfluid flow* and due to the *curvature* of the sample. [We believe that the tidal terms might be relevant in the study of the transition from one plateau of σ_H to the next one in very pure samples.]

Our next topic concerns the analysis of some quasi-particle excitations above the groundstate in a two-dimensional, incompressible quantum fluid, whose effective action in the scaling limit is given by the action $S^*_{\Lambda_0}$ computed above; see (4.22). For simplicity, we start by considering a flat, two-dimensional system of charged fermions with vanishing magnetic moment, so that the $SU(2)_{\text{spin}}$-connection w vanishes identically in an appropriate $SU(2)$-gauge, $(e^1(x), e^2(x), e^3(x))$ are chosen to be time-independent, so that there is no tidal Zeeman term; see Sect. 2). We suppose that, in a small neighborhood of a suitably chosen background potential a_c – typically $a_{c0} = 0, b_c = da_c$ constant and of suitable magnitude – the system is incompressible. Then the action in the scaling limit is given by

$$-\frac{1}{\hbar}S^*_{\Lambda_0}(\tilde{a}) = \int_{\Lambda_0} j^{\mu}_{c}\tilde{a}_{\mu}dv + \frac{\sigma}{4\pi}\int_{\Lambda_0} \tilde{a} \wedge d\tilde{a} \ , \tag{4.48}$$

up to boundary terms. The first term on the r.h.s. is unimportant in the following discussion, and we set $j^{\mu}_{c} = 0$.

Let us produce a *"Laughlin vortex"* [37] in this system by turning on a magnetic field $\tilde{b}(\xi) = \partial_1\tilde{a}_2(\xi) - \partial_2\tilde{a}_1(\xi)$ in a small disc. [Actually, $\tilde{b}(\xi)$ could be a vorticity field of a superfluid flow if, instead of a quantum Hall fluid, we consider a superfluid film. We shall nevertheless use "magnetic language" in the following discussion.] From our discussion of the Aharonov-Bohm effect in Sect. 3–(1) we know that this excitation only disturbs the system locally, and thus may have a finite energy difference to the groundstate energy, if

$$\frac{1}{2\pi}\int \tilde{b}(t,\boldsymbol{\xi})d^2\boldsymbol{\xi} = n \ , \quad n \in \mathbf{Z} \ . \tag{4.49}$$

By eq. (4.31) for $\mu = 0$, we have that

$$\langle j^0(\xi)\rangle_{a,w} = \frac{\sigma}{2\pi}\tilde{b}(\xi) \ , \tag{4.50}$$

and hence the charge of the excitation (background charge normalized to 0) is given by

$$q = \int \langle j^0(t,\boldsymbol{\xi})\rangle_{a,w} d^2\boldsymbol{\xi} = \sigma n \ .$$

If σ is not an integer then q will be fractional, in general. Now, consider two such excitations localized in two disjoint small disks and interchange them along some paths oriented anti-clock-wise. According to Sect. 3–(1), the Aharonov-Bohm phase picked up in this process is given by

$$e^{2\pi i\theta} := e^{i\pi qn} = e^{i\pi\sigma n^2} , \tag{4.51}$$

where we have normalized the statistical phase θ such that $\theta = 1/2$ corresponds to *Fermi statistics*; $\theta = 0$ corresponds to *bosons*, and $\theta \neq 0, 1/2$ (mod 1) to *anyons* [22]. Thus, Laughlin vortices are anyons, unless σn^2 is an integer.

Among the excitations that one can produce in this fashion there should be the particles constituting the system. Let us suppose that the state of the system is fully spin-polarized, (as is the case for filling factors $\nu \overset{\text{e.g.}}{=} \frac{1}{3}, \frac{1}{5}$ in quantum Hall fluids). Suppose a magnetic flux of n_0 produces a state of N electrons. From (4.50) we then infer that

$$\sigma = \frac{N}{n_0} . \tag{4.52}$$

If N is *odd* this state is composed of N fermions and hence describes a fermion, so that, by (4.51),

$$e^{i\pi N n_0} = -1 . \tag{4.53}$$

Thus n_0 must be *odd*, too. In fact, one may show that if N and n_0 have no common divisor then n_0 is odd. In particular, for $N = 1$ we conclude that

$$\sigma = 1/n_0 \text{ , with } n_0 \text{ odd} . \tag{4.54}$$

This is the famous odd-denominator rule; see e.g. [38]. An excitation with vorticity 1 then has fractional charge $q = 1/n_0$ and is an anyon, for $n_0 > 1$.

Note that the vector potential, $\tilde{a}$, created by a pointlike excitation of charge q located at $\xi = 0$ is given by

$$\tilde{a}_i(\xi) = -\frac{q}{\sigma}\varepsilon_{ij}\frac{\xi^j}{|\boldsymbol{\xi}|^2} , \tag{4.55}$$

as follows from (4.49) and (4.50) for $\langle j^0(\xi)\rangle_{a,w} = q\delta_0(\xi)$. This is the "$U(1)$-Knizhnik-Zamolodchikov connection".

Next, we consider another "in vitro" system, namely a "chiral spin liquid". [It is not entirely clear that such systems exist in nature.] A chiral spin liquid is a system of neutral particles of spin $s > 0$ and with non-zero magnetic moment ($\mu = 1$, in present units) having a spin-singlet groundstate for some non-zero, constant magnetic field, $\vec{B}_c$. It is assumed, here, to be incompressible and to exhibit breaking of parity and time reversal, but no spontaneous magnetization. In our formalism, the effective action of such a system in the scaling limit is given by

$$-\frac{1}{\hbar}S^*_{\Lambda_0}(w) = \frac{k}{4\pi}\int_{\Lambda_0} \mathrm{tr}(w \wedge dw + \frac{2}{3}w \wedge w \wedge w) , \tag{4.56}$$

up to boundary terms. Under reflections in lines, w_i transforms as a vector, w_0 as a pseudoscalar and k as a pseudoscalar. Let us consider an excitation created by turning on an $SU(2)$-gauge field w with field strength, g, given by

$$g(\xi) = dw(\xi) + w(\xi) \wedge w(\xi) \ .$$

For example, we may choose g to be given by

$$\vec{g}_{12}(\xi) = -\vec{e} g_0(\boldsymbol{\xi}) \ ,$$

where $\vec{e}$ is some unit vector in $\mathbf{R}^3$ and g_0 is time-independent, $\vec{g}_{0i}(\xi) = 0$. By eq. (4.32), the spin density of this excitation is given by

$$\langle \vec{s}^{\,0}(\xi) \rangle_w = \vec{e} \frac{k}{\pi} g_0(\boldsymbol{\xi}) \ , \tag{4.57}$$

so that the expectation value of its total spin operator, $\vec{S}$, is given by

$$\langle \vec{S} \rangle_w = \vec{e} \frac{k}{\pi} \int g_0(\boldsymbol{\xi}) d^2 \boldsymbol{\xi} \ .$$

Such an excitation is commonly called a "*spinon*". Quantum mechanically, spin is quantized: $\vec{S} \cdot \vec{S} = 4l(l+1)$, $l \in \frac{1}{2}\mathbf{Z}$. Consider a spinon of spin l located at the point $\boldsymbol{\xi} = \boldsymbol{\xi}_1$. Then eq. (4.57) says that $\vec{g}_{12}(\boldsymbol{\xi})$ is the solution of the equation

$$\langle \vec{L}^{(l)} \rangle_w \delta(\boldsymbol{\xi} - \boldsymbol{\xi}_1) = -\frac{k}{\pi} \vec{g}_{12}(\xi) \ , \tag{4.58}$$

where $\vec{L}^{(l)}$ is the spin operator, $\vec{S}$, in the spin-l representation; (see Sect. 2). A connection $\vec{w}$ for the field strength $\vec{g}$ satisfying (4.58), with $\vec{g}_{0i}(\xi) = 0$, is given by

$$\vec{w}_0(\xi) = 0 \ , \quad \vec{w}_i(\boldsymbol{\xi}) = \frac{2}{k} \langle \vec{L}^{(l)} \rangle_w \varepsilon_{ij} \frac{\xi^j - \xi_1^j}{|\boldsymbol{\xi} - \boldsymbol{\xi}_1|^2} \ . \tag{4.59}$$

Suppose, we now create a second spinon of spin l' moving in the background gauge field $\vec{w}$ excited by the first spinon. Its dynamics is coupled to $\vec{w}$ through the covariant derivatives (see Sect. 2, eqs. (2.13), (2.14)):

$$D_\mu = \partial_\mu + i \vec{w}_\mu \cdot \vec{L}^{(l')} \ , \tag{4.60}$$

with $\vec{w}_\mu$ as in (4.59). Let us imagine that it makes sense to do "two-spinon quantum mechanics" on a Hilbert space $\mathcal{H}^{(l)} \otimes \mathcal{H}^{(l')}$, with

$$\mathcal{H}^{(l)} = \mathcal{D}^{(l)} \otimes L^2(M, dv) \ ,$$

where $\mathcal{D}^{(l)}$ carries the spin-l representation of $SU(2)$. By (4.59) and (4.60), the covariant derivatives on $\mathcal{H}^{(l)} \otimes \mathcal{H}^{(l')}$ are then given by

$$D_0^1 = \frac{\partial}{\partial \xi_1^0} \ , \quad D_j^1 = \frac{\partial}{\partial \xi_1^j} + \frac{2i}{k} \varepsilon_{jn} \frac{\xi_1^n - \xi_2^n}{|\boldsymbol{\xi}_1 - \boldsymbol{\xi}_2|^2} \sum_{A=1}^{3} L_A^{(l)} \otimes L_A^{(l')} \ ,$$

and

$$D_0^2 = \frac{\partial}{\partial \xi_2^0} \ , \quad D_j^2 = \frac{\partial}{\partial \xi_2^j} + \frac{2i}{k} \varepsilon_{jn} \frac{\xi_2^n - \xi_1^n}{|\boldsymbol{\xi}_1 - \boldsymbol{\xi}_2|^2} \sum_{A=1}^{3} L_A^{(l)} \otimes L_A^{(l')} \ , \tag{4.61}$$

These are the covariant derivatives associated with the celebrated *Knizhnik-Zamolodchikov connection*, [30]. For the "two-spinon quantum mechanics" with parallel transport given by (4.61) to be consistent with unitarity, it is necessary that

$$k = \pm(\kappa + 2) , \quad \kappa = 1, 2, \cdots . \tag{4.62}$$

This follows from results in [30,39]. Recalling what we have said in Sect. 3–(3) about the Aharonov-Casher effect, we observe that the "phase factor" arising in the parallel transport of a quantum mechanical spinon in the field excited by a classical spinon with spin orthogonal to the plane of the system is an Aharonov-Casher phase factor.

Let us now exchange the positions of two quantum mechanical, pointlike spinons along anti-clockwise oriented paths. Then the "Aharonov-Casher phase factor" multiplying the wave function is given by a matrix

$$R_{ll'}^{(\kappa)} : \mathcal{D}^{(l)} \otimes \mathcal{D}^{(l')} \to \mathcal{D}^{(l')} \otimes \mathcal{D}^{(l)}$$

which is the *braid matrix* for exchanging a chiral vertex of spin l with a chiral vertex of spin l' in the chiral *Wess-Zumino-Novikov-Witten model* [30] at level κ. It is given by

$$R_{ll'}^{(\kappa)} = T\pi_l \otimes \pi_{l'}(\mathcal{R}^{(\kappa)}) , \tag{4.63}$$

where $\mathcal{R}^{(\kappa)}$ is the universal R-matrix of the quantum group $U_q(sl_2)$, with $q = \exp i\pi/(\kappa+2)$, and T is the flip (transposition of factors). All this can be extended to "n-spinon quantum mechanics". The matrices $R_{ll'}^{(\kappa)}$ determine an exotic quantum statistics related to non-abelian (for $\kappa > 1$, $l, l' < \frac{\kappa}{2}$) representations of the braid groups (more precisely, the groupoids of coloured braids) which is commonly called *non-abelian braid statistics* [40,29]. We wish to note that l and l' are forced to be $\leq \frac{\kappa}{2}$, i.e., there are no spinons of spin $> \frac{\kappa}{2}$. One might call this phenomenon *"spin screening"*. If the particles of spin s constituting the chiral spin liquid appear as spinon excitations above the groundstate then

$$\kappa \geq 2s , \tag{4.64}$$

since these particles carry spin s. One can argue that the statistics of these particles must be *abelian* braid statistics, i.e., they are anyons. In fact, it then follows that they are semions ($\theta = 1/4$). Now, for a given level κ, the matrices $R_{ll}^{(\kappa)}$ define an abelian representation of the braid groups if and only if $2l = \kappa$. It follows that, for a chiral spin liquid made of particles of spin s

$$\kappa = 2s . \tag{4.65}$$

Any spinon-excitation of spin $l < s$ then has *non-abelian* braid statistics!

The reader may feel that our "derivation" of "spinon quantum mechanics" from the effective action $S^*_{\Lambda_0}(w)$ given in (4.56) is based on idealizations – see (4.58) – and jumps in the logics – reasoning between (4.60) and (4.61) – that might make it appear to be quite problematic. Actually, it turns out that our conclusions concerning spinon statistics, in particular eqs. (4.63) and (4.65), are perfectly correct. This follows from an analysis of the mysterious boundary terms, "B.T.", in the effective action; see [17], and Sect. 5 for the example of anyons.

In order to understand spin-singlet quantum Hall fluids, one must glue the Laughlin vortices described in (4.49) – (4.53) to the spinons discussed above. One checks that for $\sigma = 2/n_0$, n_0 odd, and $\kappa = 2s = 1$, a Laughlin vortex of vorticity $n = -\frac{n_0}{2}$ (!) glued to a spinon of spin $s = 1/2$ is an excitation of charge $q = -1$, spin $1/2$ and *Fermi statistics*, [3,17]. These are the properties of an electron. In an electronic quantum Hall fluid (*without* any *very exotic* internal symmetries) one does not find any excitations with non-abelian braid statistics. However, if one could manufacture a quantum Hall fluid made of charge carriers of spin $s = \frac{3}{2}, \frac{5}{2}, \cdots$, with a spin-singlet ground state it would display excitations with non-abelian braid statistics [14]. It may appear difficult to build such a system, in practice. But, perhaps, one can think of incompressible superfluid films of particles of higher spin, with broken parity and time reversal, which would also exhibit excitations with non-abelian braid statistics.

The analysis sketched above extends, in a straightforward way, to systems with continuous internal symmetries and corresponding gauge fields; see [17].

It may be worthwhile emphasizing that in quantum Hall fluids with non-vanishing magnetic susceptibility (spin-polarized Hall fluids) the fractional statistics of Laughlin vortices always appears as a consequence of a *combination* of the Aharonov-Bohm – *and* the Aharonov-Casher effect; (but notice that, for spin-polarized quantum Hall fluids, the Aharonov-Casher phase factors are automatically abelian). This is a consequence of the fact that electrons have a non-vanishing magnetic moment and follows from eq. (4.42).

Finally, we come to a brief comment concerning the relation of our definition of the Hall conductivity $\sigma_H = \frac{e^2}{h}\sigma$ as the coefficient of a Chern-Simons term, $\frac{\sigma}{4\pi}\int \tilde{a} \wedge d\tilde{a}$, in the effective gauge field action $S^*_{A_0}$, see (4.22), of an incompressible quantum Hall fluid to the more conventional definition via the *Kubo formula* [41]. It follows easily from eqs. (4.4), (4.5) and (4.22) that σ appears in the following *current sum rules:* For every choice of a permutation $(\mu\nu\rho)$ of (012),

$$i\frac{\sigma}{\pi} = \text{sign}\,(\mu\nu\rho) \int (x-y)^\mu \,\langle T[j^\nu(x) j^\rho(y)]\rangle^c_{a_c,w_c}\, d^3y\,. \tag{4.66}$$

These are *three* equations for one and the same quantity σ. The equation for $(\mu\nu\rho) =$ (012) is

$$i\frac{\sigma}{\pi} = \int (t-s)\left\langle T[j^1(t,\mathbf{x}) j^2(s,\mathbf{y})]\right\rangle^c_{a_c,w_c} ds\; d^2\mathbf{y} \tag{4.67}$$

which is just the Kubo formula (in "mathematical units", with no guarantee for signs and factors of π); compare e.g. to [41]. The other two equations are an automatic consequence of $U(1)_{\text{em}}$-gauge-invariance. See [5] for a more systematic study of current sum rules and "proofs".

Thouless and coworkers [42], and followers [43], have derived from the Kubo formula that

$$\sigma = \frac{1}{n_0} c_1 , \tag{4.68}$$

where n_0 is the groundstate degeneracy and c_1 is the *first Chern number* of a vector bundle over a two-dimensional torus of magnetic fluxes (ϕ_1, ϕ_2). So, c_1 is an integer which, in formula (4.52), was called $N = \#$ of electrons created when one turns on a local magnetic field of total flux n_0. Does our formulation "know" that n_0 is the degeneracy of the groundstate? Yes, it does! This follows e.g. from the material in Sect. 5 and has been noted in [1]; (see also [17] for a more precise derivation).

Bellissard [44] and Avron, Seiler and Simon [45] have also given a definition of σ as an *index*. Their definition is equivalent to ours, too, and the proof follows from the material in Sect. 5; see Sect. 6 of ref. [3].

We finally note that σ_H^{spin} (for $k = 0$, i.e., spin-polarized quantum Hall fluids) can be shown to be given by a Kubo formula involving spin currents and can then be shown to be proportional to a first Chern number of a vector bundle over a two-dimensional torus of electric charges per unit length (Q_1, Q_2).

In a fairly precise sense one finds that the Hall effect for the electric current is a time-dependent form of the *Aharanov-Bohm effect*, while the Hall effect for the spin current corresponds to the time-dependent *Aharonov-Casher effect.*

5. ANOMALY CANCELLATION AND ALGEBRAS OF CHIRAL EDGE CURRENTS IN TWO-DIMENSIONAL, INCOMPRESSIBLE QUANTUM FLUIDS

In this last section we outline some ideas on the origin of the quantization of the values of the constants σ, χ, σ_s and k which appear as the coefficients of the Chern-Simons terms in the effective action $S^*_{\Lambda_0}$ of incompressible quantum fluids in the scaling limit; see (4.22). This topic is intimately connected with the so far mysterious boundary terms, "B.T", on the r.h.s of eq. (4.22). Since this is a somewhat technical topic, we have to limit our review to a few basic aspects and refer the reader to [4,17] for more details.

Briefly, our analysis of the boundary terms in $S^*_{\Lambda_0}$ and of the quantization of the coefficients σ, χ, σ_s and k relies upon the following two *key ideas*:

(i) The "important" - more precisely the *anomalous* - part of the boundary terms in the action $S^*_{\Lambda_0}$ is completely determined by the Chern-Simons terms in $S^*_{\Lambda_0}$ by invoking $U(1)_{\text{em}} \times SU(2)_{\text{spin}}(\times G_{\text{internal}})$-gauge-invariance of the *total* effective action of non-relativistic quantum theory.

(ii) This anomalous part of the boundary terms of $S^*_{\Lambda_0}$ turns out to be the generating functional of the connected Green functions of chiral current operators which generate $U(1)$-, $SU(2)$- (and G-) current (Kac-Moody) algebras [9]. Some physical and mathematical principles concerning the representation theory of these current algebras then constrain the values of the coefficients σ, χ, σ_s and k to belong to certain discrete sets.

Remark on (ii). We already have found constraints on the values of $\sigma, (\chi, \sigma_s)$ and k in Sect. 4 by analyzing the statistics of Laughlin vortices and "spinons" and imposing the constraint that, among excitations composed of Laughlin vortices glued to spinons, one should find excited states of the particles constituting the incompressible quantum fluid – in the case of a quantum Hall fluid, the *electrons* or *holes.* In Sect. 4, it turned out that if one imposes the principle of unitarity on the quantum mechanics of spinons then k must be an integer. Our analysis of Laughlin vortices predicted σ to be a rational number (with an *odd* denominator for quantum Hall fluids composed of spinless, charged fermions).

Let us start our analysis by recalling a well known lemma that shows how $SU(2)$-gauge-invariance forces k to be an integer: Let g be an $SU(2)$-gauge-transformation with the property that

$$g(\tau, \boldsymbol{\xi}) \to \mathbf{1}, \quad \text{continuously as } (\tau, \boldsymbol{\xi}) \to \partial\Lambda_0 \ , \tag{5.1}$$

or $\tau \to \pm\infty$. For Λ_0 a cylinder, the family of all such $SU(2)$-gauge transformations splits into disjoint homotopy classes labelled by an integer winding number, $n(g)$; (recall that $\pi_3(SU(2)) = \mathbb{Z}!$). Let g be a gauge transformation with winding number $n(g) \neq 0$. The gauge-transformed $SU(2)$-connection, ${}^g w$, is given by

$$w \mapsto {}^g w = g w g^{-1} + g d g^{-1} \ . \tag{5.2}$$

Let us study how the $SU(2)$-Chern-Simons term

$$S_{CS}(w) := \frac{k}{4\pi} \int_{\Lambda_0} tr(w \wedge dw + \frac{2}{3} w \wedge w \wedge w) \tag{5.3}$$

in $S^*_{\Lambda_0}$ transforms under the transformation (5.2). The well known answer is that

$$S_{CS}({}^g w) = S_{CS}(w) + 2\pi k n(g) \ . \tag{5.4}$$

Now, non-relativistic quantum theory is fully gauge-invariant under local $SU(2)_{\text{spin}}$-gauge-transformations, including time-dependant ones. Therefore, the generating (partition) function

$$Z_\Lambda(a^{(\theta)}, w^{(\theta)}) \equiv \exp \frac{i}{\hbar} S^{\text{eff}}_\Lambda(a^{(\theta)}, w^{(\theta)}) \sim \exp \frac{i}{\hbar} S^*_{(\Lambda/\theta)}(\tilde{a}, \tilde{w}) \tag{5.5}$$

must be invariant under the transformation (5.2). Asymptotically, as $\theta \to \infty$, the *only* gauge-variance of $Z_{\theta\Lambda_0}(a^{(\theta)}, w^{(\theta)})$ comes from the Chern-Simons term (5.3) in

$S^*_{\Lambda_0}(\tilde{a},\tilde{w})$. **Hence** we must **require** that

$$\begin{aligned} \exp\frac{i}{\hbar}S^*_{\Lambda_0}({}^g w) &= \exp\frac{i}{\hbar}\left[S^*_{\Lambda_0}(w) - 2\pi k\hbar n(g)\right] \\ &\stackrel{!}{=} \exp\frac{i}{\hbar}S^*_{\Lambda_0}(w)\,, \end{aligned} \tag{5.6}$$

for arbitrary integers $n(g)$. Thus

$$k \in \mathbf{Z}\,. \tag{5.7}$$

The same result could have been deduced by considering the transformation properties of the Chern-Simons term $S_{CS}(w)$ under gauge transformations, g, not vanishing at the boundary $\partial\Lambda_0$. The non-invariance of $S_{CS}(w)$ under such gauge transformations actually determines one of the *boundary terms* in $S^*_{\Lambda_0}$, [2,17]. What about the values of σ, σ_s and χ? Consider, for example, the abelian Chern-Simons term

$$S'_{CS}(\tilde{a}) := \frac{\sigma}{4\pi}\int_{\Lambda_0} \tilde{a}\wedge d\tilde{a} \tag{5.8}$$

in the effective action $S^*_{\Lambda_0}$. As explained in [5], $S^*_{\Lambda_0}(\tilde{a},\tilde{w})$ must be invariant under $U(1)_{\text{em}}$-gauge-transformations

$$\tilde{a} \mapsto {}^{\chi}\tilde{a} = \tilde{a} + d\chi\,, \tag{5.9}$$

(in spite of the fact that $\tilde{a}$ is only a fluctuation potential, i.e., $\tilde{a} = a - a_c$!) Since $\pi_3(U(1)) = 0$, the local $U(1)$-gauge-transformations on Λ_0 do not split into different homotopy classes, and hence there is no a-priori quantization of σ. However, by considering the transformation properties of $S'_{CS}(\tilde{a})$ under gauge transformations, χ, which do *not* vanish at the boundary $\partial\Lambda_0$, we shall be able to infer some constraints on the possible values of σ.

In order not to get lost in many technicalities, we refrain from studying a general quantum Hall fluid here; but see [3,17]. Rather, we shall confine our analysis of boundary terms and edge currents to idealized quantum Hall fluids of *spinless fermions*, so that $w = 0$, from now on. This is an important special case for coming to grips with the general case (which also involves $SU(2)_{\text{spin}}$ and, possibly, G_{internal}). But the general case would lead us into a little orgy of "*branching rules*" for representations of subalgebras of Kac-Moody algebras which is deferred to another paper, although, physically, the general case is important for understanding quantum Hall fluids with spin-singlet groundstates (e.g., for a filling factor $\nu = \frac{8}{5}$ [46]), or with internal symmetries, (e.g. certain hierarchy states of the electron fluid, or, perhaps, the fluid corresponding to $\nu = \frac{5}{2}$; [3,47]).

If the particles in a two-dimensional, incompressible quantum fluid are spinless fermions then $w = 0$, and its effective action in the scaling limit is given by

$$-S^*_{\Lambda_0}(\tilde{a}) = \int_{\Lambda_0} j^\mu_c(\xi)\tilde{a}_\mu(\xi)dv + S'_{CS}(\tilde{a}) + \text{B.T.}(\tilde{a})\,, \tag{5.10}$$

where $S'_{CS}(\tilde{a})$ is given in (5.8), and B.T. stands for the celebrated boundary terms, and, for the rest of this section, $\hbar = 1$. Let us now perform a gauge transformation (5.9) on $\tilde{a}$, with χ not vanishing at $\partial\Lambda_0$. Then

$$\begin{aligned} S^*_{\Lambda_0}(\tilde{a} + d\chi) &= S^*_{\Lambda_0}(\tilde{a}) - \int_{\partial\Lambda_0} j_{c,n}\chi do \\ &+ \frac{\sigma}{4\pi}\int_{\partial\Lambda_0} d\chi \wedge \tilde{a} - \text{B.T.}(\tilde{a} + d\chi) + \text{B.T.}(\tilde{a}) , \end{aligned} \tag{5.11}$$

where $j_{c,n}$ is the component of $j_c{}^{\mu}$ normal to the boundary $\partial\Lambda_0$ of Λ_0, and do is the surface element. Note that $S^*_{\Lambda_0}(\tilde{a} + d\chi)$ would be equal to $S^*_{\Lambda_0}(\tilde{a})$, and we could set B.T.$= 0$, if

$$j_{c,n} \propto \text{dual of } d\tilde{a}\,|_{\partial\Lambda_0} \; .$$

However, j_c is the current supported by the quantum fluid when $a = a_c, \tilde{a} = 0$, and $\tilde{a}$ is an arbitrary fluctuation potential. Therefore such a relation between j_c and $d\tilde{a}$ does not make sense for arbitrary $\tilde{a}$. Experimentally, for the electron fluid in a heterojuncture, for example, $\tilde{a}$ can be tuned in a fairly arbitrary way, and the boundary $\partial\Lambda_0$ is such that there is no leakage of electric charge through $\partial\Lambda_0$, i.e.,

$$j_{c,n} = 0 \; . \tag{5.12}$$

In this case, the second term on the r.h.s. of (5.11) vanishes, but the third term is different from 0, for suitable choices of χ and $\tilde{a}$. Imposing gauge invariance of the effective action $S^*_{\Lambda_0}$ thus yields the following equation for the boundary terms:

$$\text{B.T.}(\tilde{a} + d\chi) - \text{B.T.}(\tilde{a}) = \frac{\sigma}{4\pi}\int_{\partial\Lambda_0} d\chi \wedge \tilde{a} \; , \tag{5.13}$$

for arbitrary $\tilde{a}$ and χ. This equation is well known from the study of the (1+1)-*dimensional chiral anomaly* [16]. To solve it, it is convenient to use light-cone coordinates on $\partial\Lambda_0$. We set

$$u_\pm = \frac{1}{\sqrt{2}}(v\tau \pm \theta) \tag{5.14}$$

where τ is a time-like and θ a space-like coordinate on $\partial\Lambda_0$, and v is some velocity. Since the term $\int_{\partial\Lambda_0} \tilde{a} \wedge d\chi$ is topological, eq. (5.13) does not impose any specific choice of τ, θ and v. Mathematically, it is convenient to set $v = 1$ and choose θ to be an angle ranging over the interval $[0, 2\pi]$. However, if τ and θ are measured in physical units then v would be the *propagation speed of surface charge density waves.* The value of this physically interesting quantity will not be determined by $S^*_{\Lambda_0}$. [It would only be computable from a more microscopic analysis of the system.] We now set

$$\tilde{a}\,|_{\partial\Lambda_0} = A_+ du_+ + A_- du_- \; , \tag{5.15}$$

where

$$A_\pm := \frac{1}{\sqrt{2}}\left(\tilde{a}_\tau\,|_{\partial\Lambda_0} \pm \tilde{a}_\theta|_{\partial\Lambda_0}\right) \; . \tag{5.16}$$

In light cone coordinates, the r.h.s. of eq. (5.13) is given by

$$\frac{\sigma}{4\pi}\int_{\partial\Lambda_0} d\chi \wedge \tilde{a} = \frac{\sigma}{4\pi}\int_{\partial\Lambda_0}(A_+\partial_-\chi - A_-\partial_+\chi)d^2u \;, \tag{5.17}$$

where

$$\partial_\pm\chi \equiv \frac{\partial}{\partial u_\pm}\chi \;.$$

We note that, in light-cone coordinates, the d'Alembertian, $\square = \frac{\partial^2}{\partial\tau^2} - \frac{\partial^2}{\partial\theta^2}$, is given by

$$\square = 2\partial_+\partial_- \;. \tag{5.18}$$

After these preparations, it is a simple exercise to verify that the solution of the functional equation (5.13) is given by

$$\begin{aligned} \text{B.T.}(\tilde{a}) &= -\frac{\sigma}{4\pi}\Delta_R(A) + W(A) \\ &= \frac{\sigma}{4\pi}\Delta_L(A) + W(A) \;, \end{aligned} \tag{5.19}$$

where

$$\Delta_{L/R}(A) = \int_{\partial\Lambda_0}\left\{A_\mp A_\pm - 2A_\mp\frac{\partial^2_\pm}{\square}A_\mp\right\}d^2u \;, \tag{5.20}$$

and $W(A)$ is an arbitrary *gauge-invariant* functional of the boundary vector potential given by A_+ and A_-. Note that replacing σ by $-\sigma$ corresponds to replacing L (left) by R (right)! Readers, who still remember the basic formulas arising in the study of the $(1+1)$-dimensional, chiral $U(1)$-anomaly will recognize $\Delta_L(A)$ as the effective gauge field action of a chiral (left-moving) relativistic fermion minimally coupled to a $U(1)$-gauge field A, in two space-time dimensions. One checks that

$$\begin{aligned} \frac{i}{4\pi}\Delta_L(A) &= \ln\det\left[\partial\!\!\!/ + i\,A\!\!\!/\left(\frac{1-\gamma_5}{2}\right)\right] \\ &= \ln\det(\partial\!\!\!/ + i\,A\!\!\!/)\Big|_{A_+=0} + \frac{i}{4\pi}\int A_+A_-d^2u \;. \end{aligned} \tag{5.21}$$

Let ψ be a $(1+1)$-dimensional two-component Dirac spinor, and $\bar{\psi} = \psi^*\gamma_0$ its conjugate. The expression for the left-moving current, j^μ_L, is given by

$$j^\mu_L = N\left(\bar{\psi}\gamma^\mu\left(\frac{1-\gamma_5}{2}\right)\psi\right) \;, \tag{5.22}$$

where N indicates normal ordering. This current generates a chiral $U(1)$-current algebra. Comparing (5.22) to (5.21) and recalling the basics of Berezin integration, we observe that $\frac{i}{4\pi}\Delta_L(A)\Big|_{A_+=0}$ is the generating functional for the connected Green functions of the left moving current j^μ_L.

We conclude that *if σ were an integer* we could cancel the anomaly, $\frac{\sigma}{4\pi}\int_{\partial\Lambda_0} d\chi\wedge\tilde{a}$, of the Chern-Simons term, $\frac{\sigma}{4\pi}\int_{\Lambda_0}\tilde{a}\wedge d\tilde{a}$, in the effective action of the quantum fluid under a gauge transformation, $\tilde{a} \mapsto \tilde{a} + d\chi$, *by $|\sigma|$ bands of left- or right-moving* (depending on the sign of σ) *free, relativistic complex fermions* minimally coupled to the boundary vector potential. An analysis due to Halperin [6] and elaborated upon in [3] shows that this describes precisely the physics of boundary degrees of freedom of an integer (non-interacting) quantum Hall fluid *with* $|\sigma|$ *filled Landau levels.* Actually, the logics can be turned around: If we consider a non-interacting quantum Hall fluid with N filled Landau bands coupled to a small fluctuation vector potential $\tilde{a}$ then those quantum mechanical degrees of freedom which are localized near the boundary of the system produce a $U(1)$-gauge-anomaly corresponding to the action $\frac{iN}{4\pi}\Delta_{L/R}(A)$, (where the choice of L or R depends on the sign of the external magnetic field). For this anomaly to be canceled – as required by the $U(1)$-gauge invariance of non-relativistic quantum theory – it is necessary that the effective gauge field action of the *bulk degrees of freedom* contain a Chern-Simons term $\pm\frac{iN}{4\pi}\int_{\Lambda_0}\tilde{a}\wedge d\tilde{a}$. As shown in eqs. (4.42) and (4.44), this term reproduces the basic equations of the quantum Hall effect, with a quantized Hall conductivity $\sigma_H = \frac{e^2}{h}N$, $N = 0, 1, 2\cdots$.

So we understand the integral quantum Hall effect for non-interacting electrons pretty well – although there are actually still plenty of interesting analytical (*spectral*) problems for systems with a large amount of disorder and for systems of spinning electrons with spin-orbit interactions which should be studied more carefully!

But what if σ is *not* an integer? Then the $U(1)$-anomaly of the Chern-Simons term in the effective action is cancelled by the term $\pm\frac{i\sigma}{4\pi}\Delta_{L/R}(A)$, as shown above. Of course $\frac{i}{4\pi}\Delta_{L/R}(A)$ remains the generating functional of a chiral $U(1)$-current algebra of left- or right-moving currents. What kind of a system does the corresponding chiral $U(1)$- current, $J^{\pm} = j^{\mu=0}_{L/R}$, describe physically? Of course, it still describes chiral electric charge density waves circulating around the boundary edges of the system. But what are the basic charge carriers like? Here a little general culture on current algebra (see e.g. [9]) helps: Let us start by considering free, massless Dirac fermions in $1+1$ dimensions. By (5.22)

$$j^{\mu}_{L/R} = \frac{1}{2}(j^{\mu} \mp j^{\mu}_5) \, . \tag{5.23}$$

Let us first suppose that the external gauge field A is set to 0. Then j^{μ} and j^{μ}_5 are conserved currents, i.e.,

$$\partial_{\mu}j^{\mu} = \partial_{\mu}j^{\mu}_5 = 0 \, . \tag{5.24}$$

The general solution of eqs. (5.24) is

$$j^{\mu} = \varepsilon^{\mu\nu}\partial_{\nu}\varphi \, , \quad j^{\mu}_5 = \varepsilon^{\mu\nu}\partial_{\nu}\varphi_5 \, , \tag{5.25}$$

where φ and φ_5 are scalar fields of scaling dimension 0. However, in two space-time dimensions, $j^{\mu}_5 = -\varepsilon^{\mu\nu}j_{\nu}$, with $\varepsilon^{01} = -\varepsilon^{10} = 1$. Therefore $j^{\mu}_5 = \partial^{\mu}\varphi$, and (5.24) implies that

$$\partial_{\mu}\partial^{\mu}\varphi \equiv \Box\varphi = 0 \, , \tag{5.26}$$

i.e., φ is a free, massless scalar field. Any solution of (5.26) has the form

$$\varphi = \sqrt{2}(\varphi_L(u_+) + \varphi_R(u_-)) \, . \tag{5.27}$$

By (5.23),

$$J^+ = -\partial_- \varphi_R \, , \quad J^- = \partial_+ \varphi_L \, , \tag{5.28}$$

with $\partial_\pm J^\pm = 0$. These formulas hold at the level of quantized fields and are at the origin of abelian bosonization in two space-time dimensions. Now, any sum of free fields is again a free field. Thus, let us write, for fun,

$$\varphi = \frac{1}{2\pi}(\phi_1 + \cdots + \phi_N) \, , \tag{5.29}$$

where $\phi_1, \cdots, \phi_N$ are distinct, free, massless scalar fields. We set

$$\hat{\phi} = \begin{pmatrix} \phi_1 \\ \vdots \\ \phi_N \end{pmatrix} . \tag{5.30}$$

For $A = 0$, the action of $\hat{\phi}$ is given by

$$S_{WZW}(\hat{\phi}) = \frac{1}{4\pi} \int_{\partial \Lambda_0} \partial_+ \hat{\phi} \cdot K \partial_- \hat{\phi} d^2 u \, , \tag{5.31}$$

where K is a positive $N \times N$ matrix, and $\hat{a} \cdot \hat{b} := \sum_{i=1}^{N} a_i b_i$. If one wishes to describe chiral left- (or right-) moving free fields one supplements the action (5.31) by the constraints

$$\partial_- \hat{\phi} = 0 \quad (\partial_+ \hat{\phi} = 0, \text{ resp.}) \, . \tag{5.32}$$

The matrix K describes linear couplings between the fields $\phi_1, \cdots, \phi_N$ and fixes their normalization when one uses a standard path-integral quantization.

Let us now study what happens when one attempts to couple the fields $\phi_1, \cdots \phi_N$ to the vector potential A. The first problem one encounters is that expressions like $\partial_\pm \hat{\phi}$, for example the chiral constraint $\partial_- \hat{\phi} = 0$, are *not* invariant under $U(1)$-gauge-transformations. We must find out how $\hat{\phi}$ transforms under gauge transformations. For $N = 1$ and $K = 1$, it is well known and easy to check that the fermion operators ψ_L and ψ_R are given by vertex operators, $\psi_{L/R} =: e^{i\varphi_{L/R}} :$, where the double colons indicate Wick ordering. Hence φ and $\phi_1, \cdots, \phi_N$ transform like angular variables under $U(1)$-gauge transformations. An adequate ansatz is

$$\phi_j \mapsto {}^{\chi}\phi_j = \phi_j + \sum_{i=1}^{N} (K^{-1})_{ji} \chi \, . \tag{5.33}$$

A gauge-invariant form of the chiral constraint is then given by

$$\partial_- \phi_j - \sum_{i=1}^{N} (K^{-1})_{ji} A_- = 0 \, . \tag{5.34}$$

We set

$$\hat{\chi} := \begin{pmatrix} \chi \\ \vdots \\ \chi \end{pmatrix} , \quad \hat{A} = \begin{pmatrix} A \\ \vdots \\ A \end{pmatrix} , \tag{5.35}$$

with N components each. An action reducing to (5.31) for $A = 0$ is given by

$$\begin{aligned} S_{WZW}(\hat{\phi}, \hat{A}) &= \frac{1}{4\pi} \int_{\partial\Lambda_0} \partial_+\hat{\phi} \cdot K \partial_- \hat{\phi} d^2u \\ &- \frac{1}{2\pi} \int_{\partial\Lambda_0} \left\{ \hat{A}_- \cdot \partial_+ \hat{\phi} - (\partial_- \hat{\phi} - K^{-1} \hat{A}_-) \cdot \hat{A}_+ \right\} d^2u \\ &+ \frac{k}{4\pi} \int_{\partial\Lambda_0} A_- A_+ d^2u , \end{aligned} \tag{5.36}$$

where

$$k := \sum_{i,j} (K^{-1})_{ij} . \tag{5.37}$$

Note that expression (5.36) is symmetric in "+" and "−". If we want to describe chiral fields we supplement the dynamics determined by the action (5.36) by the *gauge-invariant* chiral constraint (5.34). Let us now check how $S_{WZW}(\hat{\phi}, \hat{A})$ transforms under the $U(1)$-gauge-transformations (5.33) and $A_\pm \mapsto {}^\chi A_\pm = A_\pm + \partial_\pm \chi$. After a fairly brief calculation we find that

$$\begin{aligned} S_{WZW}({}^\chi\hat{\phi}, {}^\chi\hat{A}) &= S_{WZW}(\hat{\phi}, \hat{A}) \\ &+ \frac{k}{4\pi} \int_{\partial\Lambda_0} (A_+ \partial_- \chi - A_- \partial_+ \chi) d^2u \\ &+ \frac{1}{2\pi} \int_{\partial\Lambda_0} (\partial_- \hat{\phi} - K^{-1} \hat{A}_-) \cdot \partial_+ \hat{\chi} d^2u . \end{aligned} \tag{5.38}$$

The last term on the r.h.s. of (5.38) vanishes when the chiral constraint (5.34) is imposed. We observe that the second term on the r.h.s. of (5.38) is precisely the anomaly (5.17) of the Chern-Simons action if $k = \sigma$.

Let us also note that

$$\zeta_L(A) := \int \mathcal{D}\hat{\phi} e^{-iS_{WZW}(\hat{\phi}, \hat{A})} \delta(\partial_- \hat{\phi} - K^{-1} \hat{A}_-) \tag{5.39}$$

is, for $A_+ = 0$, the generating function for the current $\frac{1}{2\pi} \sum_{i=1}^N \partial_+ \phi_i = \partial_+ \varphi$ which, by eq. (5.28), is precisely the left-handed current J^-. Since the integration measure is gauge-invariant, i.e., $\mathcal{D}\hat{\phi} = \mathcal{D}^\chi \hat{\phi}$, it follows from (5.38) and (5.17), (5.20) that

$$\zeta_L(A) = \exp\left(-\frac{ik}{4\pi} \Delta_L(A) \right) . \tag{5.40}$$

Thus, we conclude that if the coefficient σ of the Chern-Simons term, $\frac{\sigma}{4\pi} \int_{\Lambda_0} \tilde{a} \wedge d\tilde{a}$, in the effective action $S^*_{\Lambda_0}$ of an incompressible quantum fluid satisfies

$$\sigma = k \equiv \sum_{ij=1}^N (K^{-1})_{ij} \tag{5.41}$$

then

$$\exp(-iS'_{CS}(\tilde{a})) \zeta_L(A = \tilde{a}|_{\partial\Lambda_0}) \tag{5.42}$$

is $U(1)$-gauge-invariant, i.e., *anomaly-free*. For $\sigma = -k$, the same holds if "+" and "−" and "left (L)" and "right (R)" are interchanged.

Next, we must investigate the physics of the $\hat{\phi}$-system on the boundary of an incompressible quantum fluid. In particular, we must find physical constraints on the matrix K. When the gauge field A is zero, the electric charge operator Q is given by

$$Q = \oint j^0 d\theta = \oint (\partial_\theta \varphi) d\theta \, , \tag{5.43}$$

by (5.25). By (5.14), (5.29) and (5.34), this yields

$$\begin{aligned} Q &= \sum_{j=1}^{N} Q_j \, , \quad \text{with} \\ Q_j &= \frac{1}{2\pi} \oint (\partial_+ \phi_j) d\theta \, . \end{aligned} \tag{5.44}$$

Unfortunately, these expressions are *not* $U(1)$-gauge-invariant. But it is clear how to render them gauge-invariant. The correct definition of the charge operator associated with ϕ_j is

$$Q_j = \frac{1}{2\pi} \oint (\partial_+ \phi_j - \sum_i (K^{-1})_{ji} A_+) d\theta \tag{5.45}$$

which is manifestly gauge invariant. Let us replace

$$\hat{A} = \begin{pmatrix} A \\ \vdots \\ A \end{pmatrix} \mapsto \hat{\alpha} = \begin{pmatrix} \alpha_1 \\ \vdots \\ \alpha_N \end{pmatrix} , \tag{5.46}$$

where under a $U(1)$-gauge transformation χ

$$a_j \mapsto {}^{\chi}\alpha_j = \alpha_j + d\chi \, , \tag{5.47}$$

for $j = 1, \cdots, N$. We call α_j the "vector potential of the jth band", in accordance with the structure of the couplings $\mp\frac{1}{2\pi}\int_{\partial\Lambda_0} \hat{\alpha}_\mp \cdot \partial_\pm \hat{\phi}$ in the action $S_{WZW}(\hat{\phi}, \hat{\alpha})$ given in (5.36). Imagine that we now increase the magnetic flux inside the system by n_i units in the ith band, e.g. by creating n_i Laughlin vortices in the ith band, $i = 1, \cdots, N$. Then

$$\frac{1}{2\pi} \oint \alpha_{i\theta} d\theta = n_i \, , \tag{5.48}$$

and the charge in the jth band, Q_j, changes by an amount

$$\Delta Q_j = \sum_i (K^{-1})_{ji} n_i \, , \tag{5.49}$$

as follows from (5.45), with $\hat{A}$ replaced by $\hat{\alpha}$. In vector notation,

$$\Delta \hat{Q} = K^{-1} \hat{n} \, , \quad \text{or} \quad \hat{n} = K \Delta \hat{Q} \, . \tag{5.50}$$

We now imagine that every band admits excitations with the quantum numbers of an electron or hole, i.e., for every $j = 1, \cdots, N$, there are excitations changing the total charges by $\Delta Q_i^{(j)} = \pm\delta_{ij}$ and having Fermi statistics. By formula (5.36) for the action S_{WZW}, such excitations are created by the vertex operators

$$\exp \frac{i}{2\pi} \int \hat{\alpha}_- \cdot \partial_+ \hat{\phi} d\theta \; , \tag{5.51}$$

with

$$-\frac{\sqrt{2}}{2\pi} \oint \hat{\alpha}_- d\theta = \frac{1}{2\pi} \oint \hat{\alpha}_\theta d\theta = \hat{n} \; , \tag{5.52}$$

and $\hat{n}$ is given by

$$n_i = \sum_{l=1}^{N} K_{il} \Delta Q_l^{(j)} = K_{ij} \; , \quad i = 1, \cdots, N \; ; \tag{5.53}$$

see (5.50). The statistics of the vertex operator (5.51) is described by the phase

$$\exp 2\pi i h(\hat{\alpha}) \; , \tag{5.54}$$

where $h(\hat{\alpha})$ is its conformal dimension. By (5.36), $h(\hat{\alpha})$ turns out to be given by

$$\begin{aligned} h(\hat{\alpha}) &= \frac{1}{2} \hat{n} \cdot K^{-1} \hat{n} \\ &= \frac{1}{2} \Delta \hat{Q} \cdot K \Delta \hat{Q} \; , \end{aligned} \tag{5.55}$$

and the second equation follows from (5.50). Thus, for an electron or hole in the jth band,

$$h(\hat{\alpha}) = \frac{1}{2} K_{jj} \; .$$

By (5.54), this excitation has Fermi statistics iff

$$K_{jj} = 2l^{(j)} + 1 \; , \quad l^{(j)} = 0, 1, 2, \cdots \; , \tag{5.56}$$

for $j = 1, \cdots, N$. Clearly, electrons and holes are excitations which are *relatively local* to each other, (meaning that microscopic electronic wave functions are *single-valued*). Hence a vertex operator creating an electron or hole in the ith band must commute or anti-commute with a vertex operator creating an electron or hole in the jth band, for all i and j. One readily checks that this will be the case iff

$$\exp 2\pi i \Delta \hat{Q}^{(i)} \cdot K \Delta \hat{Q}^{(j)} = \exp 2\pi i K_{ij} = 1 \; ,$$

hence

$$K_{ij} \in \mathbf{Z} \; , \quad \text{for all } i \text{ and } j \; . \tag{5.57}$$

Actually, if one assumes that two vertex operators creating electrons in *different* bands must commute – as one normally would – then it follows that

$$K_{ij} \in 2\mathbf{Z} \; , \quad \text{for} \quad i \neq j \; . \tag{5.57'}$$

Plugging results (5.56) and (5.57) into formula (5.41) one observes that

$$\sigma = \pm \sum_{i,j=1}^{N} (K^{-1})_{ij} \tag{5.58}$$

is a *rational* number, and hence the Hall conductivity $\sigma_H = \frac{e^2}{h}\sigma$ is a *rational* multiple of $\frac{e^2}{h}$, for every *incompressible* quantum fluid of scalar (spin-polarized) electrons! A similar conclusion holds if spin and internal symmetries are included, but one obtains different sets of rational numbers as the possible values of σ compatible with incompressibility; see [17].

We note that, for $\hat{\alpha} = \hat{A}$ (see (5.46)) and $\Delta Q = \sum_{j=1}^{N} \Delta Q_j$, eq. (5.49) implies that

$$\Delta Q = \frac{1}{2\pi} \sum_{j,i=1}^{N} (K^{-1})_{ji} \oint A_\theta d\theta = \frac{\sigma}{2\pi} \int \tilde{b}(t, \boldsymbol{\xi}) d^2\xi \ , \tag{5.59}$$

by Stokes' theorem. This is an integrated form of eq. (4.31), see also (4.42) and (4.50), for a quantum fluid with vanishing magnetic susceptibility, as is the case for spinless electrons.

Clearly, for a given rational value of σ, formula (5.58), along with the constraints (5.56) and(5.57), does not determine the "band coupling matrix" K uniquely. This is an intrinsic weakness of our very general approach. A given rational value of σ corresponding to a plateau of the Hall conductivity can, in general, be reproduced by many different systems of chiral boundary currents corresponding to distinct K-matrices. In order to find out which K-matrix is the most likely candidate corresponding to a given plateau of σ, one must invoke additional information on the quantum Hall fluid, in particular stability properties against small perturbations, whose elucidation requires analytical or numerical work, or *symmetries.* As a first step towards reducing the plethora of possible K-matrices we propose to study what kind of *invariant* information is coded into a matrix K. For this purpose one should try to find the full spectrum of charged excitations of a system corresponding to a given matrix K satisfying (5.56) and (5.57). For an excitation of a quantum Hall fluid to have a finite energy difference to the groundstate energy (called a finite-energy excitation), it should perturb the groundstate only *locally.* As a corollary of our discussion of the Aharonov-Bohm effect it follows that the magnetic flux $\hat{n}$ of a finite-energy excitation must be quantized, i.e.,

$$n_j \in \mathbb{Z} \ , \quad \text{for} \quad j = 1, \cdots, N \ .$$

[If an electron in the jth band is transported around such an excitation it picks up a statistical phase $\exp 2\pi i n_j$ which is unity if $n_j \in \mathbb{Z}$.] We conclude that finite-energy excitations of an incompressible quantum Hall fluid can be labelled, in part, by their magnetic flux quantum numbers $\hat{n}$ which are the sites of the lattice $\Phi := \mathbb{Z}^n$. Eq. (5.50) then says that the electric charges corresponding to an excitation with magnetic flux $\hat{n}$ are given by $\Delta\hat{Q} = K^{-1}\hat{n}$ and form the sites of a lattice $\Gamma := K^{-1}\Phi$. The lattice Γ contains the sublattice $\mathbf{Z}^n$ of excitations with integer charge, i.e., of multi-electron, multi-hole excitations. The quotient space, $\Gamma/\mathbf{Z}^n$, is an abelian group with n generators. It tells us everything about the possible *fractional* charges of finite-energy excitations.

We now observe that what we are calling the "j^{th} band", $j = 1, \cdots N$, is based on a somewhat arbitrary convention of how the fields ϕ_j are coupled to the external electromagnetic vector potential A, (i.e., on the electric charges assigned to these fields). If S is some integral $N \times N$ matrix of determinant 1, i.e., $S \in SL(N, \mathbf{Z})$, then S leaves Φ invariant. Two systems corresponding to matrices K and K', with

$$K' = S^T K S \tag{5.60}$$

describe the same lattices (Φ and Γ) of excitations and correspond to *equivalent* quantum Hall fluids which only differ in the assignment of electric charges to the fields $\phi_1, \cdots, \phi_N$. This observation poses the problem of defining and then finding *normal forms* for the integral, positive quadratic forms K on the lattice Φ, (with respect to conjugation by $SL(N, \mathbf{Z})$); see [4]. This is known to be a subtle mathematical problem which is not solved in general; (see [48]).

But let us return to the problem of symmetries of quantum Hall fluids. A natural symmetry of such a fluid at small values of the filling factor ν is likely to be invariance under arbitrary permutations of the bands. This symmetry would imply that

$$K_{ij} = K_{\pi(i)\pi(j)}\ , \quad i, j = 1, \cdots N\ , \tag{5.61}$$

for arbitrary permutations, π, of $\{1, \cdots, N\}$. Together with conditions (5.56) and (5.57) eq. (5.61) implies that

$$K_{ii} = 2l + 1\ , \quad i = 1, \cdots, N\ , \tag{5.62}$$

for some $l = 0, 1, 2, \cdots$ *independent* of i, and

$$K_{ij} = n \in \mathbf{Z}\ , \quad \text{for } i \neq j\ . \tag{5.63}$$

Thus

$$K = (2l + 1 - n)1_N + nNP_N\ , \tag{5.64}$$

where P_N is the orthogonal projection on the unit vector in $\mathbf{R}^N$ all of whose components are given by $1/\sqrt{N}$. Hence

$$K^{-1} = (2l + 1 - n)^{-1}\left(1_N - \frac{nN}{2l + 1 + n(N-1)}P_N\right)\ ,$$

and this equation and (5.58) yield

$$\sigma \equiv \sigma_K = \pm\frac{N}{2l + 1 + n(N-1)}\ . \tag{5.65}$$

Imposing constraint (5.57') we must assume that n is an *even* integer. This reproduces the *odd-denominator rule*. [In general, the odd-denominator rule only holds for an odd number of bands! See also [4].]

A "second generation hierarchy state" of a quantum Hall fluid might be defined as a system with a coupling matrix K_h given by

$$K_h = \begin{pmatrix} K & & 0 \\ & \ddots & \\ 0 & & K \end{pmatrix} + m(pN)P_{pN} \,, \tag{5.66}$$

where K is an $N \times N$ matrix of the form (5.64), the first matrix on the r.h.s. of (5.66) is a $(pN) \times (pN)$ matrix build from p matrices K, and m is an (even) integer. For the Hall conductivity of this system one finds [4]

$$\sigma_{K_h} = \pm \frac{1}{m + (1/p|\sigma_k|)} = \pm \frac{p|\sigma_k|}{mp|\sigma_K| + 1} \,. \tag{5.67}$$

One can now go on and define "third generation hierarchy states" , etc.

Next, one might ask what form the matrix K must have if the system exhibits a full unitary group, $U(N)$, of symmetries permuting its N bands of edge current; (an obvious example of such a system is an integer quantum Hall fluid of non-interacting electrons with $\sigma = \pm N$). The algebra of edge currents must then contain a Kac-Moody subalgebra $\widehat{su}(N)$ (at level 1). This is a much larger symmetry than the permutation symmetry discussed above. Correspondingly, the K-matrices compatible with this larger symmetry are more constrained: They have the form

$$K_{ii} = 2l + 1 \,,\; i = 1, \cdots, N, \quad K_{ij} = 2l \,,\; \text{for } i \neq j \,, \tag{5.68}$$

for some $l = 0, 1, 2, \cdots$. The corresponding Hall conductivity is found to be

$$\sigma_K = \pm \frac{N}{2lN + 1} \,. \tag{5.69}$$

The proof of (5.68) (see [4]) involves showing that there is a matrix $S \in SL(N, \mathbf{Z})$ such that

$$\begin{aligned} S^T K S &=: R \,, \quad \text{with} \\ R_{NN} &= 2l + 1 \,,\; R_{NN-1} = R_{N-1N} = -1 \,, \end{aligned} \tag{5.70}$$

and $(R_{ij})_{i,j=1}^{N-1}$ is the Cartan matrix of $su(N)$. In connection with quantum Hall fluids the matrix R first appeared in [49].

We note that quantum Hall fluids with K-matrices as in (5.68) correspond to Jain's states [50].

It may be good to consider the simplest example of a fractional quantum Hall fluid covered by our theory: We set $N = 1$ and $K = 2l + 1$. For $l = 0$, this is an integer quantum Hall fluid with $\sigma = \pm 1$. For $l = 1$, corresponding to $\sigma = \pm 1/3$, we find Laughlin's fluid [34]. There are also quantum Hall fluids corresponding to $l = 2$ and 3, ($\sigma = \pm 1/5, \pm 1/7$, respectively). There are no known quantum Hall fluids corresponding to $l = 4, 5, \cdots$, since they would correspond to electron gases of so low a density that they form a Wigner crystal and thereby loose their incompressibility.

The charged excitations in a fluid with $K = 2l + 1$ have vorticity $n = 1, \cdots, 2l + 1$ and charge $n/(2l + 1)$. For $n < 2l + 1$, their charge is thus *fractional*, and by (5.54), (5.55), they are *anyons*.

At the end of Sect. 4 (see (4.68)) we mentioned that, in the conventional approach to the quantum Hall effect, the denominator $n_0 = 2l + 1$ of the Hall conductivity σ is interpreted as the degeneracy of the ground state of the quantum Hall fluid. In our approach this has a straightforward explanation: The algebra of a chiral edge current of a quantum Hall fluid with $\sigma = \frac{1}{2l+1}$ ($K = 2l + 1, N = 1$) has $2l + 1$ inequivalent representations labelled by fluxes $1, 2, \cdots, 2l + 1$ which correspond to charges $\frac{1}{2l+1}, \frac{2}{2l+1}, \cdots, 1$. Every one of these representations corresponds to a groundstate of the quantum Hall fluid with a one-component boundary, in the thermodynamic limit which is approached when the scale parameter θ tends to ∞. In this limit, the $2l + 1$ distinct groundstates have the *same* energy per electron.

It is shown in [1,3,4] that, in the scaling limit, the groundstates of such a quantum Hall fluid are described by the conformal blocks of the free, massless field at level $2l+1$. On a Riemann surface of genus g with n punctures there are thus $(2l+1)^{g+n}$ degenerate groundstates. These results are best understood by studying the topological Chern-Simons gauge theory associated to the chiral edge currents [10,12]. The quantized gauge potential of this theory turns out to be the *vector potential of the conserved electric current density* $(j^\mu)^2_{\mu=0}$; see [3].

It is worthwhile to observe that the conformal blocks of the massless free field at level $2l + 1$ on the plane with n punctures are the *Laughlin wave functions* for n quasi-particles (characterized by their magnetic flux) of a quantum Hall fluid with $\sigma = \pm\frac{1}{2l+1}$. [This "coincidence" may partially justify some ansätze for hierarchy constructions based on Laughlin-type wave functions for quasi-particles. By and large, it may however have played rather a misleading role.]

We wish to note, furthermore, that if a vortex of strength $n = 2l + 1$ is created in the bulk of a quantum Hall fluid with $\sigma = \pm\frac{1}{2l+1}$ and a one-component boundary then, in the thermodynamic limit ($\theta \to \infty$), the total charge of the fluid changes by $\pm K^{-1}(2l + 1) = \sigma(2l + 1) = \pm 1$, as shown in eq. (5.59). More precisely, a charge of ± 1 is transferred from the place where the vortex is created to the boundary of the system; see also Sect. 6 of [3] for more details. This result relates our definition of the Hall conductivity σ to one where σ is defined as an index, [44,45].

The results reviewed here for the simple example of a quantum Hall fluid with $N = 1$ and $K = 2l + 1$ have straightforward extensions to fluids corresponding to arbitrary N and general K-matrices as discussed above.

Finally, the material in this section can be generalized to incompressible quantum fluids of particles with spin and internal symmetries. These generalizations are important in understanding quantum Hall fluids with $\sigma = \frac{8}{5}$ (spin singlet state) or $\sigma = \frac{5}{2}$, for example. But this is another story.

Now that we have reached the end of this paper life would just start to become interesting. We could now continue our tale by studying the *domain structure* of incompressible quantum fluids, in particular of quantum Hall fluids, some aspects of the *transition* of a quantum Hall fluid from *one plateau* of σ to a *neighbouring plateau*, presumably closely related to the problem of domain structure and *domain wandering*, the *stability* of plateaux and the *role of disorder* in the stability problem, $\cdots\infty$. But, most importantly, we should now finally address the analytical problem of *proving* that certain quantum fluids are indeed *incompressible*.

But all this must await another occasion – quite apart from the fact that much further more analytical work is needed!

REFERENCES

1. X.G. Wen, Phys. Rev. B**40**, 7387 (1989): Int. J. Mod. Phys. B**2**, 239 (1990).
2. X.G. Wen, Phys. Rev. B**43**, 11025 (1991); B. Blok and X.G. Wen, Phys. Rev. B**42**, 8133, 8145 (1990).
3. J. Fröhlich and T. Kerler, Nucl. Phys. B**354**, 369 (1991).
4. J. Fröhlich and A. Zee, Nucl. Phys. B**364**, 517 (1991).
5. J. Fröhlich and U.M. Studer, "$U(1) \times SU(2)$-gauge invariance of non-relativistic quantum mechanics, and generalized Hall effects", preprint ETH-TH/91-13 (September 1991).
6. B.I. Halperin, Phys. Rev. B**25**, 2185 (1982).
7. A.A. Belavin, A.M. Polyakov, and A.B. Zamolodchikov, Nucl. Phys. B**241**, 333 (1984).
8. V.G. Knizhnik and A.B. Zamolodchikov, Nucl. Phys. B**247**, 83 (1984); R. Dijkgraaf, "A geometrical approach to two-dimensional conformal field theory", Ph.D. thesis, Utrecht, 1989.
9. P. Goddard and D. Olive, Int. J. Mod. Phys. **1**, 303 (1986).
10. E. Witten, Commun. Math. Phys. **121**, 351 (1989).
11. J. Fröhlich and P.-A. Marchetti, Lett. Math. Phys. **16**, 347 (1988); J. Fröhlich and C. King, Commun. Math. Phys. **126**, 167 (1989).

12. G. Moore and N. Seiberg, Phys. Lett. B**220**, 422 (1989); S. Elitzur, G. Moore, A. Schwimmer, and N. Seiberg, Nucl. Phys. B**326**, 108 (1989).

13. M. Büttiker, Phys. Rev. B**38**, 9375 (1988); A.H. MacDonald, Phys. Rev. Lett. **64**, 220 (1990); M. Stone, Int. J. Mod. Phys. B**5**, 509 (1991); A.V. Balatsky, Phys. Rev. B**43**, 1257 (1991).

14. S.C. Zhang, T.H. Hansson, and S. Kivelson, Phys. Rev. Lett. **62**, 82, 980 (E) (1989); J. Fröhlich, T. Kerler, and P.-A. Marchetti, "Non-abelian bosonization in two-dimensional condensed matter pysics", preprint DFPD91/TH/16 (July 1991).

15. R.E. Prange and S.M. Girvin (eds.), *The quantum Hall effect*, Graduate Texts in Contemporary Physics (Springer, New York, 1987).

16. S.B. Treiman, R. Jackiw, B. Zumino, and E. Witten, *Current algebra and anomalies* (World Scientific, Singapore, 1985).

17. A.H. Chamsedine, J. Fröhlich, and U.M. Studer, to appear.

18. T. Eguchi, P.B. Gilkey, and A.J. Hanson, Phys. Rep. **66**, 213 (1980).

19. See e.g.: J.W. Negele and H. Orland, *Quantum many-particle systems*, Frontiers in Physics, vol. 68 (Addison-Wesley), New York, 1987).

20. G. Baym, *Lectures on quantum mechanics* (Benjamin/Cummings Publishing Company, Reading, Massachusetts, 1969); A. Messiah, *Quantum mechanics* (North-Holland Publishing Company, Amsterdam, 1961); C. Piron, *Mécanique quantique (base et applications)*, (Lausanne: Presses Polytechniques et Universitaires Romandes, 1990).

21. Y. Aharonov and D. Bohm, Phys. Rev. **115**, 485 (1959).

22. J.M. Leinaas and J. Myrheim, Nuovo Cimento **37** B, 1 (1977); G.A. Goldin, R. Menikoff, and D.H. Sharp, J. Math. Phys. **21**, 650 (1980); **22**, 1664 (1981); Phys. Rev. Lett. **51**, 2246 (1983); F. Wilczek, Phys. Rev. Lett. **48**, 1144 (1982); **49**, 957 (1982).

23. R.G. Clark et al., Phys. Rev. Lett. **60**, 1747 (1988); J.A. Simmons et al., Phys. Rev. Lett. **63**, 1731 (1989).

24. G.E. Volovik and V.M. Yakovenko, J. Phys. : Condens. Matter **1**, 5263 (1989); G.E. Volovik, "Exotic properties of superfluid ^3He", preprint 1991.

25. P.G. de Gennes, *Superconductivity of metals and alloys* (Benjamin, New York, 1966).

26. D. Pines and P. Nozières, *The theory of quantum liquids* (Addison-Wesley, Redwood City CA, 1989).

27. Y. Aharonov and A. Casher, Phys. Rev. Lett. **53**, 319 (1984).

28. G. t'Hooft, Commun. Math. Phys. **117**, 685 (1988); S. Deser and R. Jackiw, Commun. Math. Phys. **118**, 495 (1988); B.S. Kay and U.M. Studer, Commun. Math. Phys. **139**, 103 (1991).

29. J. Fröhlich, F. Gabbiani, and P.-A Marchetti, in *The algebraic theory of superselection sectors: Introduction and recent results*, ed. D. Kastler (World Scientific, Singapore, 1990); J. Fröhlich and P.-A. Marchetti, Nucl. Phys. B **356**, 533 (1991).

30. V.G. Knizhnik and A.B. Zamolodchikov, in ref. [8]; A. Tsuchiya and Y. Kanie, Lett. Math. Phys. **13**, 303 (1987).

31. L.D. Landau and E.M. Lifshitz, *Electrodynamics of continuous media*, vol. 8 of Course of theoretical physics (Pergamon Press, Oxford, 1960).

32. A.K. Kerman and N. Onishi, Nucl. Phys. A**361**, 179 (1981); J.S. Bell and J.M. Leinaas, Nucl. Phys. B**212**, 131 (1983); B**284**, 488 (1987).

33. R.B. Laughlin, Phys. Rev. B**23**, 5632 (1981); see also ref. [6].

34. R.B. Laughlin, Phys. Rev. Lett. **50**, 1395 (1983); B.I. Halperin, Helv. Phys. Acta **56**, 75 (1983); F.D.M. Haldane, Phys. Rev. Lett. **51**, 605 (1983); F.D.M. Haldane and E.H. Rezayi, Phys. Rev. Lett. **54**, 237 (1985); R. Morf and B.I. Halperin, Phys. Rev. B**33**, 2221 (1986).

35. D.H. Lee and S.C. Zhang, Phys. Rev. Lett. **66**, 1220 (1991); S.C. Zhang, "The Chern-Simons-Landau-Ginzburg theory of the fractional quantum Hall effect", preprint RJ 8364 (76087) (September 1991); N. Read, Phys. Rev. Lett. **62**, 86 (1989).

36. J. Goldstone, Nuovo Cimento **19**, 154 (1961); J. Goldstone, A. Salam, and S. Weinberg, Phys. Rev. **127**, 965 (1962).

37. B.I. Halperin, Phys. Rev. Lett. **52**, 1583, 2390(E) (1984); D. Arovas, J.R. Schrieffer, and F. Wilczek, Phys. Rev. Lett. **53**, 722 (1984); S. Girvin and A.H. MacDonald, Phys. Rev. Lett. **58**, 1252 (1987); R.B. Laughlin, in *Two-dimensional strongly correlated electronic systems*, eds. Z.-Z. Gan and Z.-B. Su (Gordon and Breach, London, 1989).

38. R. Tao and Y.-S. Wu, Phys. Rev. B**31**, 6859 (1985).

39. T. Kohno, Ann. de l'Inst. Fourier de l'Univ. de Grenoble **37**, 139 (1987); in *Contemporary mathematics: Artin's braid group*, eds. J.S. Birman and A. Libgober (American Math. Soc. , Providence RI, 1988).

40. K. Fredenhagen, K.-H. Rehren, and B. Schroer, Commun. Math. Phys. **125**, 201 (1989).

41. E. Fradkin, *Field theories of condensed matter systems*, Frontiers in Physics, vol. 82 (Addison-Wesley, Redwood City, California, 1991).

42. D.J. Thouless, M. Kohmoto, M.P. Nightingale, and M. den Nijs, Phys. Rev. Lett. **49**, 405 (1982); M. Kohmoto, Ann. Phys. (N.Y.) **160**, 343 (1985); Q. Niu and D.J. Thouless, Phys. Rev. B**35**, 2188 (1987).

43. J.E. Avron, R. Seiler, and B. Simon, Phys. Rev. Lett. **51**, 51 (1983); I. Dana, J.E. Avron, and J. Zak, J. Phys. C**18**, L679 (1985); J.E. Avron, R. Seiler, and L. Yaffe, Commun. Math. Phys. **110**, 33 (1987); H. Kunz, Commun. Math. Phys. **112**, 121 (1987).

44. J. Bellissard, in *Localization in disordered systems*, eds. W. Weller and P. Ziesche, Texts in Physics (Teubner, Leipzig, 1988).

45. J.E. Avron, R. Seiler, and B. Simon, Phys. Rev. Lett. **65**, 2185 (1990).

46. R.G. Clark et al., Phys. Rev. Lett. **62**, 1536 (1989); J.P. Eisenstein et al. , Phys. Rev. Lett. **62**, 1540 (1989); B.I. Halperin, in ref. [34].

47. J.P. Eisenstein et al., Phys. Rev. Lett. **61**, 997 (1988); T. Chakraborty and P. Pietiläinen, *The fractional quantum Hall effect: Properties of an incompressible quantum fluid* (Springer, Berlin Heidelberg New York, 1988); and refs. therein; M. Greiter, X.G. Wen, and F. Wilczek, Phys. Rev. Lett. **66**, 3205 (1991).

48. J.-P. Serre, *A course in arithmetic*, Graduate Texts in Mathematics, vol. 7 (Springer, New York, 1973).

49. N. Read, Phys. Rev. Lett. **65**, 1502 (1990).

50. J.K. Jain, Phys. Rev. Lett. **63**, 199 (1989); Phys. Rev. B**40**, 8079 (1989).

NON-COMPACT WZW CONFORMAL FIELD THEORIES[1,2]

Krzysztof Gawędzki

C.N.R.S., I.H.E.S.
91440 Bures-sur-Yvette, France

ABSTRACT

We discuss non-compact WZW sigma models, especially the ones with symmetric space $H^{\mathbf{C}}/H$ as the target, for H a compact Lie group. They offer examples of non-rational conformal field theories. We remind their relation to the compact WZW models but stress their distinctive features like the continuous spectrum of conformal weights, diverging partition functions and the presence of two types of operators analogous to the local and non-local insertions recently discussed in the Liouville theory. Gauging non-compact abelian subgroups of $H^{\mathbf{C}}$ leads to non-rational coset theories. In particular, gauging one-parameter boosts in the $SL(2,\mathbf{C})/SU(2)$ model gives an alternative, explicitly stable construction of a conformal sigma model with the euclidean 2D black hole target. We compute the (regularized) toroidal partition function and discuss the spectrum of the theory. A comparison is made with more standard approach based on the $U(1)$ coset of the $SU(1,1)$ WZW theory where stability is not evident but where unitarity becomes more transparent.

1. INTRODUCTION

The four years which passed since the previous Cargèse Institute of the series have brought a marked progress in the understanding of rational Conformal Field Theories (CFT's), a class of 2D massless quantum field models, see e.g. [1]. The simplest of those theories is the free field with values in a circle of rational radius, more complicated examples are provided by the Wess-Zumino-Witten (WZW) sigma models with a general compact Lie group G as the target [2],[3][4] or by the coset theories obtained by gauging a subgroup of G in the WZW theory [5],[6]. The characteristic property of the rational CFT's is

1. decomposition of the euclidean Green functions into a finite sum of products of holomorphic and antiholomorphic "conformal blocks".

---

[1]extended version of lectures read at the Summer Institute "New Symmetry Principles in Quantum Field Theory", Cargèse, July 16-27, 1991

[2]based on joint work in progress with Antti Kupiainen; these notes cover also some preliminary material far from being completely understood, reflecting there the present author's point of view which has been changing with time and has not yet reached the final form

New Symmetry Principles in Quantum Field Theory, Edited by
J. Fröhlich et al., Plenum Press, New York, 1992

This is accompanied by other simplifying features, which are more or less special from the point of view of the general quantum field theory, like

2. discreteness of finite volume energy spectrum,
3. one-to-one correspondence between states and operators,
4. simple structure of the operator product expansions,
5. finiteness of the partition functions,
6. simple factorization properties.

Our knowledge about the rational WZW and coset theories seems rather satisfactory today (although one might argue that a subtle cleaning of some fine mathematical points remains to be done; also the basic problem of classification of the rational CFT's has not been solved). It contains the exact solution for the spectrum and for the low genus Green functions (see e.g. [7]). It seems then reasonable to go beyond the study of relatively simple conformal theories where properties 1-6 hold, especially since examples of conformal models without those properties appear rather naturally. The best known instance is the Liouville theory describing the conformal mode of 2D gravity [8]. It is in this model, essential for the treatment of non-critical string theory, where the new features related to the failure of 1-6 where first discussed, see inspiring lectures [9].

Mathematically, the passage from rational to irrational CFT's involves a shift from purely algebraic treatment to more analysis. The parallel might be the passage from representation theory of compact Lie groups to the non-compact case. Indeed, the canonical WZW example of rational CFT is related to the representation theory of loop groups of compact groups [10] and it is expected that (largely non-existent, see however [11],[12]) theory of representations of loop groups of non-compact type will underlie an interesting class of irrational CFT's. The first candidates which come to mind are the WZW theories with non-compact groups as targets. These however, if quantized as in the compact case, have unbounded below energy and, consequently, no stability and no euclidean picture. A possible solution is to pass to their coset models where in some cases one may expect to recover stability, see Sec. 5 below. In the present course, we shall start instead from a different series of non-rational CFT's which have bounded below energy and stable euclidean picture but are non-unitary. These are the WZW-type sigma models with non-compact target spaces $H^{\mathbf{C}}/H$ where H is a compact (simple, connected, simply connected) Lie group. We shall call them shortly $H^{\mathbf{C}}/H$ WZW models. It should be stressed that, contrary to what the name might suggest, this is a different class of models than the coset G/H theories in the CFT sense. The latter are obtained by gauging a subgroup $H \subset G$ (or more generally $H \subset G_{\text{left}} \times G_{\text{right}}$) in the group G WZW model and correspond rather to conformal sigma models with orbit space of the left-right action of H on G as the target. To avoid the terminological confusion, we shall label them as G mod H coset theories[3]. In fact, a coset theory G mod H factorizes into the group G WZW theory times the $H^{\mathbf{C}}/H$ one, decoupled in the planar topology, and, in general, coupled only via zero modes. This is how the general $H^{\mathbf{C}}/H$ WZW theories manifested themselves for the first time [13],[14]. The $SL(2,\mathbf{C})/SU(2)$ model has been discussed earlier in [15]. The Green functions of the $H^{\mathbf{C}}/H$ models which appear in this context have also a 3D interpretation: they compute the scalar product of Schrödinger picture states of the 3D Chern-Simons [16],[17],[7] field theory with gauge group H. In more geometric terms, they give the hermitian structure which pairs conformal blocks of the (rational) group H WZW model into its Green

[3]we are fully aware that this arrogant attempt to change accepted terminology is bound to be futile

functions. In this guise, the $H^{\mathbf{C}}/H$ theories may be thought of as models dual to the ones with the compact group H as the target. All this is briefly recalled in Sec. 2.

In Sec. 3, we discuss free field representations of the $H^{\mathbf{C}}/H$ WZW models on the simplest example with $H = SU(2)$. We compute explicitly (the finite part of) the partition function of the model and discuss its spectrum and the relation between the space of states and the operators of the theory.

The coset scenario for producing new CFT's may also work in the case of the $H^{\mathbf{C}}/H$ WZW models if one gauges out a non-compact abelian subgroup $N \in H^{\mathbf{C}}$ (the result will be called an $H^{\mathbf{C}}/H$ mod N theory). In Sec. 4, using free field representations, we show that the $SL(2,\mathbf{C})/SU(2)$ mod $\mathbf{R}$ model where $\mathbf{R}$ is embedded into $SL(2,\mathbf{C})$ by $t \longmapsto e^{t\sigma^3}$ gives a conformal sigma model with the recently found [18],[19] 2D euclidean black hole as the target. We discuss the partition functions and the spectrum of this model. A comparison is made between the $SL(2,\mathbf{C})/SU(2)$ mod $\mathbf{R}$ theory and the rational parafermionic $SU(2)$ mod $U(1)$ model.

Finally, in Sec. 5, we contrast our approach to the black hole conformal sigma model with Witten's original proposal [18] based on the $SU(1,1)$ mod $U(1)$ coset theory, see also [20]-[29]. The free field calculation of the partition functions of the black hole model may be also repeated within Witten's scenario giving the same result but it requires complex shifts and rotations of the fields in the functional integral. One may reasonably expect that both models coincide, the two approaches being complementary: the $SL(2,\mathbf{C})/SU(2)$ mod $\mathbf{R}$ picture provides an explicitly stable construction whereas the $SU(1,1)$ mod $U(1)$ approach should be more useful for demonstrating unitarity of the theory.

2. ORIGIN OF THE $H^{\mathbf{C}}/H$ WZW MODELS

2.1. From the coset G mod H theories

Let us start by recalling the formulation of a coset G mod H theory as a partially gauged, group G WZW model (with compact G). The basic fields on the closed Riemann surface Σ are $G^{\mathbf{C}}$-valued functions g and gauge fields $A = A_z dz + A_{\bar z} d\bar z$ with values in the complexified Lie algebra $\mathcal{H}^{\mathbf{C}}$ of a group $H \subset G$. H is supposed to be embedded into G in two possibly different ways: $\iota_{l,r} : H \hookrightarrow G$. We shall denote by "tr" the invariant form on the Lie algebra $\mathcal{G}$ ($\mathcal{H}$) of G (H) normalized to give 2 as the length squared of the longest roots. We assume that via embeddings $\iota_{l,r}$, tr on $\mathcal{G}$ induces a single invariant form on $\mathcal{H}$ equal η times tr on $\mathcal{H}$. The euclidean action of the coset model takes the form [14]

$$kS(g,A) \;=\; kS(g) + \tfrac{ik}{\pi} \int_\Sigma \mathrm{tr}\,[A^r_z(g^{-1}\partial_{\bar z} g) + (g\partial_z g^{-1})A^l_{\bar z} \\ + i\mathrm{Ad}_g(A^r_z)A^l_{\bar z} - i\eta A_z A_{\bar z}\, d^2 z \tag{1}$$

where the superscripts "l,r" refer to the embeddings of H into G. $kS(g)$ is the pure WZW action

$$S(g) \;=\; -\tfrac{1}{2\pi} \int_\Sigma \mathrm{tr}\,(g^{-1}\partial_z g)(g^{-1}\partial_{\bar z} g)\, d^2 z + \tfrac{1}{24\pi i} \int_\Sigma d^{-1}\mathrm{tr}\,(g^{-1}dg)^{\wedge 3} \tag{2}$$

where we have used a shorthand notation for the Wess-Zumino topological term [2]. Coupling constant k ("level") is a positive integer. Under the complex ($H^{\mathbf{C}}$-valued) chiral gauge transformations

$$\begin{aligned} g &\longmapsto h_1^l g h_2^{r\dagger} \\ A_{\bar z} &\longmapsto {}^{h_1}A_{\bar z} \equiv \mathrm{Ad}_{h_1}(A_{\bar z}) - i h_1 \partial_{\bar z} h_1^{-1} , \\ A_z &\longmapsto {}^{h_2}A_z \equiv \mathrm{Ad}_{h_2^{\dagger-1}}(A_z) - i h_2^{\dagger-1} \partial_z h_2^{\dagger} , \end{aligned}$$

action (1) transforms like follows:

$$S(h_1^l g h_2^{r\dagger}, {}^{h_2}A_z dz + {}^{h_1}A_{\bar z} d\bar z) \;=\; S(g, A) + \eta S(h_1 h_2^{\dagger}, {}^{h_2}A_z dz + {}^{h_1}A_{\bar z} d\bar z) \; . \tag{3}$$

In particular, it is invariant under the unitary gauge transformations with H-valued $h_1 = h_2 = h$.

The Green functions of the coset model are formally given by the functional integral

$$\int - \mathrm{e}^{-kS(g,A)} \, Dg \, DA \tag{4}$$

over G-valued fields g and real (i.e. $\mathcal{H}$-valued) gauge fields A. As the insertion, we should take an expression invariant under the unitary gauge transformations. An example is provided by

$$\prod_\alpha \mathrm{tr}_{R_\alpha} g(\xi_\alpha) n_\alpha \tag{5}$$

where "tr_R" stands for the trace in representation R of G (in vector space V_R) and $n_\alpha \in G$ satisfy

$$u^l n_\alpha u^{r\dagger} = n_\alpha \tag{6}$$

for $u \in H$. For example, if $u^l = u^r$, we may take $n_\alpha = 1$.

On the Riemann sphere, we may parametrize real gauge fields A by $H^{\mathbf{C}}$-valued gauge transformations by putting $A_{\bar z}(h) = h^{-1}\partial_{\bar z} h$. Action (1) becomes then

$$kS(g, A(h)) \;=\; kS(h^l g h^{r\dagger}) - \eta k S(h h^{\dagger}) \; . \tag{7}$$

The Jacobian of the change of variables is (we ignore the zero modes for the moment)

$$\frac{\partial(A(h))}{\partial(h)} \;=\; \det(\bar\partial_h^* \bar\partial_h) = \; \mathrm{e}^{2h^\vee S(hh^\dagger)} (\det(-\Delta))^{\dim H} \tag{8}$$

where $\bar\partial_h = d\bar z(\partial_{\bar z} + \mathrm{ad}_{A_{\bar z}(h)})$, $h^\vee$ is the dual Coxeter number of H and Δ is the scalar Laplacian. More exactly, the change of variables $A \mapsto h$ gives the following expression for the Green functions (4) with insertion (5):

$$C \int \Big(\prod_\alpha \mathrm{tr}_{R_\alpha} g(\xi_\alpha) \, (h^l n_\alpha h^{r\dagger})^{-1}(\xi_\alpha) \Big) \, \mathrm{e}^{-kS(g)} \, \mathrm{e}^{(\eta k + 2h^\vee) S(hh^\dagger)} \, Dg \, \delta(h(\xi_0)) \, Dh \tag{9}$$

where $C = (\det'(-\Delta)/\mathrm{area})^{\dim H}$ with the determinant without the zero mode contribution. Expression (9) combines the Green functions of the compact group G WZW model

$$\Gamma \;=\; \int (\bigotimes_\alpha g_{R_\alpha}(\xi_\alpha)) \, \mathrm{e}^{-kS(g)} \, Dg \;\in\; \bigotimes_\alpha \mathrm{End}\, V_{R_\alpha} \tag{10}$$

(where g_R denotes the representation R matrix of g) with those of a field theory with fields $hh^\dagger$

$$\int (< \Gamma, \otimes (h^l n_\alpha h^{r\dagger})^{-1}_{R_\alpha}(\xi_\alpha)) > \; e^{\kappa S(hh^\dagger)} \; \delta(hh^\dagger(\xi_0)) \; D(hh^\dagger) \; . \tag{11}$$

In the last expression Γ may be any tensor in $\otimes \mathrm{End} V_{R_\alpha}$ such that

$$(\otimes \gamma^l_{R_\alpha}) \Gamma (\otimes \gamma^{r\dagger}_{R_\alpha}) = \Gamma \tag{12}$$

for $\gamma \in H^{\mathbf{C}}$. This condition guarantees that the integral is independent of point ξ_0 in the δ-function in (11) fixing the global $H^{\mathbf{C}}$ invariance. Green functions (10) certainly satisfy condition (12). $< \cdot, \cdot >$ in (11) stands for the scalar product induced from that of spaces V_R. Fields $hh^\dagger$ may be viewed as taking values in the non-compact symmetric space $H^{\mathbf{C}}/H$ and functional integral (11) as defining the (euclidean) Green functions of the $H^{\mathbf{C}}/H$ WZW theory (also in a general world-sheet topology). The euclidean action $-\kappa S(hh^\dagger)$ of the model is unambiguously defined[4] and real, non-negative [14]. We shall see that it leads to functional integrals of type (11) which are stable for any real $\kappa > h^\vee$. On the other hand, the Minkowskian action is not real: the Wess-Zumino term is purely imaginary so that we should not expect the theory to be unitary. We shall return to these issues below.

On a higher genus Riemann surface a similar treatment of the coset theory Green functions produces again a combination of the G and $H^{\mathbf{C}}/H$ WZW Green functions but this time both twisted by coupling to an external flat gauge field A_{flat} and the result contains an integral over the moduli of A_{flat} [14], essentially coinciding with the moduli of complex $H^{\mathbf{C}}$-bundles.

2.2. From the scalar product of the Chern-Simons theory states

The Schrödinger picture states of the 3D Chern-Simons theory with gauge group H on manifold $\Sigma \times \mathbf{R}$ and in the presence of the Wilson lines $\{\xi_\alpha\} \times \mathbf{R}$ in representations R_α are functionals

$$\psi : \mathcal{A} \longrightarrow \bigotimes_\alpha V_{R_\alpha} \tag{13}$$

on space $\mathcal{A}$ of real gauge fields A [30]. $\mathcal{A}$ has a natural complex structure obtained by identifying it with the space of forms $A_{\bar z} d\bar z$. Functionals ψ are required to be holomorphic and to transform covariantly under the complex gauge transformations:

$$\psi({}^h A) \;=\; e^{kS(h^{-1}) + \pi^{-1} ik \int \mathrm{tr}(h^{-1}\partial_z h) A_{\bar z} \, d^2 z} \bigotimes_\alpha h_{R_\alpha}(\xi_\alpha) \; \psi(A) \; . \tag{14}$$

k this time denotes the coupling constant of the Chern-Simons theory. The space of states defined as above is finite-dimensional. The scalar product of the states is formally given by the functional integral

$$\|\psi\|^2 \;=\; \int < \psi(A), \psi(A) > \; e^{-\pi^{-1} k \int \mathrm{tr} A_z A_{\bar z} \, d^2 z} \; DA \; . \tag{15}$$

On $\Sigma = \mathbf{C}P^1$, upon the change of variables $A \mapsto h$, eq. (15) becomes

$$\|\psi\|^2 \;=\; (\det{}'(-\Delta)/\mathrm{area})^{\dim H}$$
$$\cdot \int < \psi(0) \otimes \overline{\psi(0)}, \otimes (hh^\dagger)^{-1}_{R_\alpha}(\xi_\alpha) > \; e^{(k+2h^\vee) S(hh^\dagger)} \; \delta(hh^\dagger(\xi_0)) \; D(hh^\dagger) \tag{16}$$

[4] $H^{\mathbf{C}}/H$ is topologically trivial

which is a Green function of type (11) (for $G = H$ and $n_\alpha \equiv 1$).

2.3. From the hermitian structure coupling conformal blocks of the group H WZW theory

Green functions of the group H WZW model in an external $\mathcal{H}$-valued field A

$$\int (\bigotimes_\alpha g(\xi_\alpha)_{R_\alpha})\ \mathrm{e}^{-kS(g,A)}\ Dg \tag{17}$$

can be expressed as

$$\sum_{a,b} \Omega^{ab}\ \psi_a(A) \otimes \overline{\psi_b(A)}\ \mathrm{e}^{-\pi^{-1}k \int \mathrm{tr}\, A_{\bar z} A_z d^2z} \tag{18}$$

where (ψ_a) is a basis of the Chern-Simons states considered above and the inverse matrix

$$(\Omega^{-1})_{ab} \;=\; (\psi_a, \psi_b) \tag{19}$$

in the scalar product of (15), see [7],[31]. In the planar or toroidal geometry, the dependence of the basis vectors ψ_a on the insertion points and the complex structure may be chosen analytic and such that the scalar products (ψ_a, ψ_b) remain constant. Expression (18) gives then the decomposition of the Green functions into sum of combinations of conformal blocks demonstrating the rational character of the WZW theories with compact targets. As we see, scalar product (15) given by the Green functions of the $H^{\mathbf{C}}/H$ theory determines the way the conformal blocks of group H WZW theory are put together to build the complete Green functions.

The WZW theories with targets H and $H^{\mathbf{C}}/H$ may be considered as dual to each other. An elegant way to express this duality is to consider the coset H mod H model. This is a topological theory in the sense of [32]: its Green functions

$$\int (\prod_\alpha \mathrm{tr}_{R_\alpha} g(\xi_\alpha))\ \mathrm{e}^{-kS(g,A)}\ Dg\, DA \tag{20}$$

are independent of the location of the insertions and of the complex structure of the surface [14]. Integrating representation (18) over gauge fields A, one infers [33] that they are in fact equal to the dimensions of the spaces of states ψ known explicitly due to [34]. On the other hand, the coset Green functions factorize, as we have seen, into a combination of products of those of the group H and of the symmetric space $H^{\mathbf{C}}/H$ WZW models. This is the precise expression of the duality between both theories.

In the planar case, the $H^{\mathbf{C}}/H$ theory with Green functions (11) may be also viewed as an analytic continuation of those of the H theory to negative levels. This relation becomes more complicated on higher genera as, for example, a look into the respective partition functions shows. It is not excluded, however, that both models describe different aspects of the same structure analytic in k.

3. FREE FIELD REPRESENTATION OF THE $H^{\mathbf{C}}/H$ WZW THEORY

Functional integral (2.11) defining the Green functions of the $H^{\mathbf{C}}/H$ WZW theory may be computed by iterative Gaussian integration. This was noticed in [15] for the $H = SU(2)$ case and was implemented in the present context and for general H in [13],[14] for the twisted toroidal partition function and in [7],[35] for the planar Green functions. One can also compute toroidal Green functions. Free field representation for the model on a surface of genus > 2 is still an open problem. Below, we shall stick to the $SU(2)$ case, for simplicity.

Symmetric space $SL(2,\mathbf{C})/SU(2)$ coincides with the upper sheet H_3^+ of 3D mass hyperboloid. Convenient global coordinate system on H_3^+ is provided by the parametrization

$$hh^\dagger = \begin{pmatrix} e^{\phi}(1+|v|^2)^{1/2} & v \\ \bar{v} & e^{-\phi}(1+|v|^2)^{1/2} \end{pmatrix} \tag{1}$$

with ϕ real and v complex. The $SL(2,\mathbf{C})$-invariant measure on H_3^+, $d(hh^\dagger) = d\phi\, d^2v$. In coordinates (1),

$$S(hh^\dagger) = -\tfrac{1}{\pi}\int [(\partial_z\tilde{\phi})(\partial_{\bar z}\tilde{\phi}) + (\partial_z + \partial_z\tilde{\phi})\bar{v}\,(\partial_{\bar z} + \partial_{\bar z}\tilde{\phi})v)]\, d^2z \tag{2}$$

where[5] $\tilde{\phi} \equiv \phi - \frac{1}{2}\log(1+|v|^2)$. We shall also need a gauged version of the action. If we gauge the $U(1)$ group embedded into $SU(2)$ asymmetrically by $\iota_l(e^{i\theta}) = e^{i\theta\sigma^3}$, $\iota_r(e^{i\theta}) = e^{-i\theta\sigma^3}$, then the transformation law (2.3) implies, for $h_1 = e^{\lambda\sigma^3}$ and $h_2 = e^{-\lambda\sigma^3}$, that

$$S(e^{\lambda\sigma^3} g e^{\lambda\sigma^3}, A + id\lambda) = S(g, A) \tag{3}$$

for any $SL(2,\mathbf{C})$-valued g (in particular for $g = hh^\dagger$) and for any complex 1-form A. Consequently, taking A purely imaginary may be interpreted as gauging of subgroup $\mathbf{R} \hookrightarrow \{e^{\lambda\sigma^3} \mid \lambda \text{ real}\}$ in $SL(2,\mathbf{C})$ (which is the global symmetry group of the H_3^+ WZW model). $\mathbf{R}$ corresponds to the boosts in the third direction under the standard relation between $SL(2,\mathbf{C})$ and the Lorentz group. A direct computation gives

$$\begin{aligned} S(hh^\dagger, \tfrac{1}{2i}A) &= -\tfrac{1}{\pi}\int [(\partial_z\tilde{\phi} + A_z)(\partial_{\bar z}\tilde{\phi} + A_{\bar z}) \\ &\quad + (\partial_z + \partial_z\tilde{\phi} + A_z)\bar{v}\,(\partial_{\bar z} + \partial_{\bar z}\tilde{\phi} + A_{\bar z})v]\, d^2z\,. \end{aligned} \tag{4}$$

Invariance (3) becomes obvious in (4) since transformation $hh^\dagger \longmapsto e^{\lambda\sigma^3}hh^\dagger e^{\lambda\sigma^3}$ translates in coordinates (1) into $(\phi, v) \longmapsto (\phi + 2\lambda, v)$.

3.1. Toroidal partition function

First, let us describe the calculation [14] of the twisted partition function $\mathcal{Z}^{H_3^+}(\tau, U)$ of the H_3^+ WZW theory on torus $T_\tau \equiv \mathbf{C}/(2\pi\mathbf{Z} + 2\pi\tau\mathbf{Z})$, $\tau = \tau_1 + i\tau_2$, $\tau_2 > 0$. It is given by the functional integral:

$$\mathcal{Z}^{H_3^+}(\tau, U) = \int e^{\kappa S(\gamma_U hh^\dagger \gamma_U^\dagger)}\, D(hh^\dagger) \tag{5}$$

where $\gamma_U = \exp[-\frac{1}{4\tau_2}U(z-\bar{z})\sigma^3]$, $U \equiv U_1 + iU_2$, satisfies

$$\gamma_U(z + 2\pi) = \gamma(z) \quad\text{and}\quad \gamma_U(z + 2\pi\tau) = e^{-\pi i U\sigma^3}\gamma_U(z)$$

and the action is extended to twisted field configurations [14] by putting

$$S(\gamma_U hh^\dagger \gamma_U^\dagger) = S(hh^\dagger, \tfrac{1}{4i}(\tau_2^{-1}\bar{U}dz + \tau_2^{-1}Ud\bar{z})) + \tfrac{\pi}{\tau_2}U_1^2\,. \tag{6}$$

[5] it will become clear below why we use ϕ and not $\tilde{\phi}$ in parametrizing H_3^+

Using the explicit form (4) of the action, we obtain

$$\begin{aligned} \mathcal{Z}^{H_3^+}(\tau,U) &= e^{\pi\kappa\tau_2^{-1}U_1^2}\int e^{-\pi^{-1}\kappa\int(\partial_z\phi+\tau_2^{-1}\bar{U}/2)(\partial_{\bar z}\phi+\tau_2^{-1}U/2)\,d^2z} \\ &\quad\cdot e^{-\pi^{-1}\kappa\int(\partial_z+\partial_z\phi+\tau_2^{-1}\bar{U}/2)\bar{v}\,(\partial_{\bar z}+\partial_{\bar z}\phi+\tau_2^{-1}U/2)v\,d^2z}\,D(hh^\dagger)\,. \end{aligned} \tag{7}$$

where we have shifted $\tilde{\phi}\mapsto\phi$. The v-integral is gaussian and produces

$$\begin{aligned} &\det\left((\bar\partial+\bar\partial\phi+\tfrac{1}{2}\tau_2^{-1}U\,d\bar z)^*\,(\bar\partial+\bar\partial\phi+\tfrac{1}{2}\tau_2^{-1}U\,d\bar z)\right)^{-1} \\ &= e^{2\pi^{-1}\int(\partial_z\phi)(\bar\partial_{\bar z}\phi)\,d^2z+(2\pi i)^{-1}\int\phi\mathcal{R}}\,\det\left((\bar\partial+\tfrac{1}{2}\tau_2^{-1}U\,d\bar z)^*\,(\bar\partial+\tfrac{1}{2}\tau_2^{-1}U\,d\bar z)\right)^{-1} \end{aligned} \tag{8}$$

where $\mathcal{R}$ denotes the metric curvature form. Rather surprisingly, the resulting effective ϕ theory is the free field with the background charge so that we obtain again a calculable functional integral. Eq. (8) follows by the standard chiral anomaly calculation and does not depend on the regularization scheme used to define the determinants, within a large class. In particular, the absence of the Liouville $\sim\int e^{\phi}$ term in the effective action, of the type appearing in the conformal anomaly calculation, is here not an artifact of the choice of the zeta function regularization.

The presence of the (generic) twist U breaks the global $SL(2,\mathbf{C})$ symmetry of the theory to the diagonal $U(1)^{\mathbf{C}}$. The remaining symmetry results, however, in the divergence of the ϕ-integral (and, consequently, of the partition function) due to the zero mode contribution. This divergence may be extracted in the usual way as the infinite volume of $U(1)^{\mathbf{C}}$ leading to the insertion of $\delta(\phi(0))$ fixing the ϕ zero mode under the integral. The total central charge of the theory is easily computable from the standard dependence of the resulting determinants on the conformal factor of the metric. It is equal $c_{-\kappa}\equiv 3\kappa/(\kappa-2)$ or $\kappa\dim H/(\kappa-h^\vee)$ for general H. The determinants are well known [36]. The final result is (in the flat metric; $q\equiv e^{2\pi i\tau}$):

$$\begin{aligned} \mathcal{Z}^{H_3^+}(\tau,U) &= C\tau_2^{-1/2}(q\bar q)^{-1/8}\exp\left[-\pi(\kappa-2)U_2^2/\tau_2\right]|\sin(\pi U)|^{-2} \\ &\quad\cdot\left|\prod_{n=1}^{\infty}(1-e^{2\pi iU}q^n)(1-q^n)(1-e^{-2\pi iU}q^n)\right|^{-2}. \end{aligned} \tag{9}$$

3.2. Quantum-mechanical model

It will be useful to interpret expression (9) in the hamiltonian language. Let us first do it in the simpler quantum-mechanical case obtained from field theory by taking field configurations independent of the space coordinate (this approximation, ignoring the contributions of stringy oscillations, has been widely used in 2D gravity where it goes under the catchy name of "mini superspace"). The quantum-mechanical system that we obtain here is the geodesic motion on H_3^+ with the euclidean action

$$-S_{\text{mini}}(hh^\dagger)=\tfrac{\kappa}{4}\int\operatorname{tr}\left((hh^\dagger)^{-1}\partial_t(hh^\dagger)\right)^2dt\,. \tag{10}$$

Unlike in the 2D theory, in the "mini" case also the real time action is real and unitarity is recovered. The space of states is $L^2(H_3^+,d(hh^\dagger))\cong L^2(\mathbf{R}\times\mathbf{C},d\phi\,d^2v)$ and it carries the unitary representation of $SL(2,\mathbf{C})$ defined by

$$(gf)(hh^\dagger)=f(g^{-1}hh^\dagger g^{\dagger-1})\,. \tag{11}$$

On the infinitesimal level, this action may be described by generators of $sl(2,\mathbf{C})\oplus sl(2,\mathbf{C}) \cong sl(2,\mathbf{C})^{\mathbf{C}}$:

$$\begin{aligned} J^1 &= \tfrac{i}{4}(1+|v|^2)^{-1/2}(v\mathrm{e}^{\phi}-\bar{v}\mathrm{e}^{-\phi})\partial_\phi - \tfrac{i}{2}(1+|v|^2)^{1/2}(\mathrm{e}^{\phi}\partial_{\bar{v}}+\mathrm{e}^{-\phi}\partial_v)\ , \\ J^2 &= \tfrac{i}{4}(1+|v|^2)^{-1/2}(v\mathrm{e}^{\phi}+\bar{v}\mathrm{e}^{-\phi})\partial_\phi - \tfrac{i}{2}(1+|v|^2)^{1/2}(\mathrm{e}^{\phi}\partial_{\bar{v}}-\mathrm{e}^{-\phi}\partial_v)\ , \\ J^3 &= -\tfrac{1}{2}\partial_\phi - \tfrac{1}{2}v\partial_v + \tfrac{1}{2}\bar{v}\partial_{\bar{v}}\ , \end{aligned}$$

satisfying $[J^a, J^b] = i\epsilon^{abc}J^c$ and by $\bar{J}^a$'s given by the complex-conjugate vector fields. $J^{a*} = -\bar{J}^a$ so that $J^a - \bar{J}^a$ and $i(J^a+\bar{J}^a)$ are the hermitian generators of $sl(2,\mathbf{C})$. The Hamiltonian may be taken as $-2\kappa^{-1}\Delta$ where Δ denotes the Laplace-Beltrami operator on H_3^+ with the $SL(2,\mathbf{C})$-invariant metric.,

$$\begin{aligned} \Delta = \vec{J}^{\,2} = \vec{\bar{J}}^{\,2} &= \tfrac{1}{4}\partial_\phi^2 - \tfrac{1}{4}|v|^2(1+|v|^2)^{-1}\partial_\phi^2 \\ &+ (1+|v|^2)\partial_v\partial_{\bar{v}} + \tfrac{1}{4}(v\partial_v - \bar{v}\partial_{\bar{v}})^2 + \tfrac{1}{2}(v\partial_v + \bar{v}\partial_{\bar{v}})\ . \end{aligned} \tag{12}$$

$-\Delta$ has continuous bounded below spectrum starting from $\frac{1}{4}$ and induces the decomposition

$$L^2(H_3^+) \cong \int_{\rho>0}^{\oplus} \mathcal{H}_\rho\, \rho^2 d\rho \tag{13}$$

into the direct integral of irreducible unitary representations of $SL(2,\mathbf{C})$ from the principal continuous series [37],[38] on which $-\Delta$ acts as multiplication by $(1+\rho^2)/4$. $\mathcal{H}_\rho$ may be realized as the space of homogeneous functions of degree $-1+i\rho$ on non-negative matrices $h'h'^\dagger$ with determinant zero, i.e. on the upper light cone V_3^+. The parametrization by (ϕ, v) together with all the formulae concerning the action of $SL(2,\mathbf{C})$ pass to the case of V_3^+ provided that we replace everywhere[6] factor $1+|v|^2$ by $|v|^2$. The scalar product in $\mathcal{H}_\rho$ is that of $L^2(\,\delta(2-\mathrm{tr}\,h'h'^\dagger)\,d(h'h'^\dagger)\,)$. Operators $J^3-\bar{J}^3 = -v\partial_v+\bar{v}\partial_{\bar{v}}$ and $i(J^3+\bar{J}^3) = -i\partial_\phi$ may be diagonalized at the same time as Δ and their joint spectrum is $\mathbf{Z}\times\mathbf{R}$ in each $\mathcal{H}_\rho$ which, consequently, is very different from the highest- or lowest-weight representation spaces of $sl(2,\mathbf{C})\oplus sl(2,\mathbf{C})$: both J^3 and $\bar{J}^3$ have continuous unbounded spectrum here!

The heat kernel on H_3^+ is known explicitly and it has a simple form:

$$\mathrm{e}^{t\Delta}(h_1h_1^\dagger, h_2h_2^\dagger) = (\pi t)^{-3/2}\tfrac{d}{\sinh d}\,\mathrm{e}^{-t/4-d^2/t} \tag{14}$$

where d is the hyperbolic distance between $h_1h_1^\dagger$ and $h_2h_2^\dagger$ or between $hh^\dagger$ and 1 where $h = h_2^{-1}h_1$. In the more standard parametrization of H_3^+

$$hh^\dagger = (1+\vec{x}^{\,2})^{1/2} + \vec{x}\cdot\vec{\sigma}\ , \tag{15}$$

$d = \sinh^{-1}(|\vec{x}|)$. Operator $\mathrm{e}^{t\Delta}$ is certainly not of trace class since $-\Delta$ has continuous spectrum and moreover of infinite multiplicity. In the formal expression

$$\int \mathrm{e}^{t\Delta}(\mathrm{e}^{-\pi i U\sigma^3}hh^\dagger\mathrm{e}^{\pi i\bar{U}\sigma^3},\, hh^\dagger)\, d(hh^\dagger) \tag{16}$$

for $\mathrm{tr}\,\mathrm{e}^{t\Delta}\,\mathrm{e}^{2\pi i(UJ^3-\bar{U}\bar{J}^3)}$, the integral diverges due to the $U(1)^{\mathbf{C}}$ symmetry of the integrated kernel. That is the familiar problem which we have encountered already in the two-dimensional theory. We solve it again by fixing the $U(1)^{\mathbf{C}}$ invariance in the standard

[6]this is why formula (12) was written in a clumsy way

fashion. This leads to the insertion of $\delta(\phi)$ under the integral of the right hand side of (16) which renders it finite (for $U \notin \mathbf{Z}$). The hyperbolic distance between $e^{-\pi i U\sigma^3} hh^\dagger e^{\pi i \bar{U}\sigma^3}$ and $hh^\dagger$

$$d = \cosh^{-1}\Big((1+|v|^2)\cosh(2\pi U_2) - |v|^2\cos(2\pi U_1)\Big) \tag{17}$$

for $hh^\dagger = \begin{pmatrix} (1+|v|^2)^{1/2} & v \\ \bar{v} & (1+|v|^2)^{1/2} \end{pmatrix}$ and an easy calculation gives

$$\begin{aligned} \mathrm{tr}_{\mathrm{ren}}\, e^{4\pi\tau_2\kappa^{-1}\Delta}\, e^{2\pi i(UJ^3 - \bar{U}\bar{J}^3)} &\equiv \int e^{4\pi\tau_2\kappa^{-1}\Delta}(e^{-\pi i U\sigma^3} hh^\dagger e^{\pi i\bar{U}\sigma^3}, hh^\dagger)\, \delta(\phi)\, d(hh^\dagger) \\ &= \frac{\kappa^{1/2}}{8\pi\tau_2^{1/2}} e^{-\pi\tau_2/\kappa - \pi\kappa U_2^2/\tau_2} |\sin(\pi U)|^{-2}\,. \end{aligned} \tag{18}$$

On the other hand, the quantum-mechanical partition function

$$\mathcal{Z}^{H_3^+}_{\mathrm{mini}}(\tau, U) = \int e^{\kappa S_{\mathrm{mini}}(hh^\dagger)}\, \delta(\phi(t_0))\, D(hh^\dagger)$$

over twisted paths on $[0, 2\pi\tau_2]$ satisfying $hh^\dagger(2\pi\tau_2) = e^{-\pi i U\sigma^3} hh^\dagger(0) e^{\pi i\bar{U}\sigma^3}$ may be again computed by iterative gaussian integration. Not too surprisingly, one finds

$$\mathcal{Z}^{H_3^+}_{\mathrm{mini}}(\tau, U) = C\tau_2^{-1/2} e^{-\pi\kappa U_2^2/\tau_2} |\sin(\pi U)|^{-2}\,. \tag{19}$$

Comparing eqs. (18) and (19), we find that

$$\mathcal{Z}^{H_3^+}_{\mathrm{mini}}(\tau, U) = C\,\mathrm{tr}_{\mathrm{ren}}\, e^{4\pi\tau_2\kappa^{-1}(\Delta + 1/4)}\, e^{2\pi i(UJ^3 - \bar{U}\bar{J}^3)} \tag{20}$$

which establishes a Feynman-Kac type formula for the hyperbolic space H_3^+. Similar formulae may be produced for other symmetric spaces $H^{\mathbf{C}}/H$.

3.3. Space of states

Let us return now to the interpretation of expression (9) for the 2D partition function which becomes now straightforward. Using eq. (19) and (20), we obtain

$$\begin{aligned} \mathcal{Z}^{H_3^+}(\tau, U) &= C\,(q\bar{q})^{-c-\kappa/24}\, \mathrm{tr}_{\mathrm{ren}}\, e^{4\pi\tau_2(\kappa-2)^{-1}\Delta}\, e^{2\pi i(UJ^3 - \bar{U}\bar{J}^3)} \\ &\quad \cdot \Big| \prod_{n=1}^{\infty} (1 - e^{2\pi i U} q^n)(1 - q^n)(1 - e^{-2\pi i U} q^n) \Big|^{-2}\,. \end{aligned} \tag{21}$$

The first term on the right is the familiar prefactor with the central charge. Next comes essentially the mini-space contribution with $\kappa \mapsto \kappa - 2$ and then, multiplicatively, the contribution of the oscillatory degrees of freedom. By studying the canonical quantization of the H_3^+ WZW theory, one may infer that its space of states should carry a representation of the affine algebra $\hat{sl}(2,\mathbf{C}) \oplus \hat{sl}(2,\mathbf{C})$ of level $-\kappa$, extending the mini-space representation of $sl(2,\mathbf{C}) \oplus sl(2,\mathbf{C})$. Let $\hat{b}_\pm$ ($\hat{n}_\pm$) denote the subalgebras of $\hat{sl}(2,\mathbf{C})$ generated by J^a_n with $\pm n \geq 0$ ($\pm n > 0$). The action of $sl(2,\mathbf{C}) \oplus sl(2,\mathbf{C})$ in $L^2(H_3^+)$ may be extended to a representation of $\hat{b}_+ \oplus \hat{b}_+$ by making J^a_n and $\bar{J}^a_n$ for $n > 0$ act trivially (the bar refers to the second copy). Let us choose a dense invariant subdomain in $L^2(H_3^+)$ like the space $\mathcal{S}(H_3^+)$ of fast decreasing functions (in $\vec{x}$ of (15)). $\hat{sl}(2,\mathbf{C}) \oplus \hat{sl}(2,\mathbf{C})$ acts then in the space

$$\hat{\mathcal{H}}^{H_3^+} = \Big(\mathcal{U}(\hat{sl}(2,\mathbf{C})) \otimes \mathcal{U}(\hat{sl}(2,\mathbf{C}))\Big) \otimes_{\mathcal{U}(\hat{b}_+)\otimes\mathcal{U}(\hat{b}_+)} \mathcal{S}(H_3^+) \tag{22}$$

where $\mathcal{U}$ denotes the enveloping algebra. This gives the representation of $\hat{sl}(2,\mathbf{C}) \oplus \hat{sl}(2,\mathbf{C})$ induced from the action of $sl(2,\mathbf{C}) \oplus sl(2,\mathbf{C})$ in $L^2(H_3^+)$. In plain English, space $\hat{\mathcal{H}}^{H_3^+}$ is spanned by $\mathcal{S}(H_3^+)$ and by the descendents obtained by repeated action of J_n^a and $\bar{J}_n^b$ with $n < 0$ on the states in $\mathcal{S}(H_3^+)$. As a vector space,

$$\hat{\mathcal{H}}^{H_3^+} \cong \mathrm{Sym}(\hat{n}_-) \otimes \mathrm{Sym}(\hat{n}_-) \otimes \mathcal{S}(H_3^+)$$

where Sym denotes the symmetric algebra. As usually, the Sugawara construction allows to define the action in $\hat{\mathcal{H}}^{H_3^+}$ of two commuting Virasoro algebras (of central charge $c_{-\kappa}$):

$$L_n = -\frac{1}{\kappa-2} \sum_{m,a} : J_m^a J_{n-m}^a : \tag{23}$$

and similarly for $\bar{L}_n$. It is then the standard result that the contribution of the descendent states to $\mathrm{tr}\, q^{L_0} \bar{q}^{\bar{L}_0} e^{4\pi i (U J_0^3 - \bar{U} \bar{J}_0^3)}$ is the infinite product factor in (21). Since

$$q^{L_0} \bar{q}^{\bar{L}_0} \Big|_{L^2(H_3^+)} = e^{4\pi \tau_2 (\kappa-2)^{-1} \Delta} \, , \tag{24}$$

also the (renormalized) zero-level states contribution is recovered in (21).

The hamiltonian interpretation of the field-theoretic partition function may be then summarized in the following (Feynman-Kac type) formula:

$$\mathcal{Z}^{H_3^+}(\tau, U) \;=\; (q\bar{q})^{-c_{-\kappa}/24}\, \mathrm{tr}_{\mathrm{ren}}\, q^{L_0} \bar{q}^{\bar{L}_0} e^{2\pi i (U J_0^3 - \bar{U} \bar{J}_0^3)} \tag{25}$$

where on the right hand side the (renormalized) trace is taken over the space $\hat{\mathcal{H}}^{H_3^+}$ carrying the representation of $\hat{sl}(2,\mathbf{C}) \oplus \hat{sl}(2,\mathbf{C})$ induced from $L^2(H_3^+)$. The structure of the partition function of (21) and of the space of states appears to be much simpler here than in the case of compact WZW models. The probable reason is that $\hat{\mathcal{H}}^{H_3^+}$ may be decomposed into a direct integral of representations induced from $\mathcal{H}_\rho$, which we expect to be irreducible, at least in a suitable sense and for almost all ρ. Similar decomposition in the compact case (into a finite direct sum) yields representations which should be further reduced. $\hat{\mathcal{H}}^{H_3^+}$ carries a natural hermitian form (,) extending the scalar product of $L^2(H_3^+)$. It may be characterized by the conjugacy relation $J_n^{a*} = -\bar{J}_{-n}^a$. It is certainly non-positive since for $\chi \in L^2(H_3^+)$

$$((J_{-1}^1 - \bar{J}_{-1}^1)\chi), (J_{-1}^1 - \bar{J}_{-1}^1)\chi) = -\tfrac{\kappa}{2}(\chi, \chi) \, . \tag{26}$$

We expect however that (,) is non-degenerate.

3.4. Green functions

In Sec. 2.1 and 2.2, we have seen that the matrix elements $hh^\dagger(\xi)_j$ of spin $j = 0, \frac{1}{2}, 1, \ldots$ representations appear as natural insertions in the $SL(2,\mathbf{C})/SU(2)$ WZW theory, provided that they are arranged into combinations invariant under the global $SL(2,\mathbf{C})$ symmetry $hh^\dagger \mapsto \gamma h h^\dagger \gamma^\dagger$ (this is like the neutrality condition in the 2D Coulomb gas correlations). The corresponding Green functions are calculable by the iterative gaussian integration in parametrization (1). Let us explain how this works on the simplest example of the planar spin $\frac{1}{2}$ two-point function [7]

$$\begin{aligned} &\int \Big(\mathrm{tr}_{1/2}\, hh^\dagger(\xi_1)\,(hh^\dagger)^{-1}(\xi_2)\Big)\; e^{\kappa \int S(hh^\dagger)}\, \delta(hh^\dagger(\xi_0))\, D(hh^\dagger) \\ &= \int \big(|(e^\phi v)(\xi_1) - (e^\phi v)(\xi_2)|^2 + e^{\phi(\xi_1)-\phi(\xi_2)} + e^{\phi(\xi_2)-\phi(\xi_1)}\big) \\ &\qquad \cdot e^{-\pi^{-1}\kappa \int [(\partial_z \phi)(\partial_{\bar z}\phi) + (\partial_{\bar z} + \partial_{\bar z}\phi)\bar{v}\,(\partial_z + \partial_z \phi) v)]\, d^2 z} \\ &\qquad \cdot \delta(\phi(\xi_0))\, \delta^2(v(\xi_0))\, D\phi\, Dv \end{aligned} \tag{27}$$

where we have already shifted $\tilde{\phi} \mapsto \phi$. The v-integral is gaussian. It produces the partition function

$$e^{2\pi^{-1}\int(\partial_z\phi)(\bar{\partial}_{\bar{z}}\phi)\,d^2z+(2\pi i)^{-1}\int\phi\mathcal{R}}\left(\det{}'(\bar{\partial}^*\bar{\partial})/\text{area}\right)^{-1} \tag{28}$$

(which changes the coupling constant of the effective ϕ-integral from κ to $\kappa-2$, compare eq. (8)) and the normalized expectation

$$\begin{aligned} &< |(e^{\phi}v)(\xi_1) - (e^{\phi}v)(\xi_2)|^2 > \\ &= (\pi\kappa)^{-1}|\xi_1-\xi_2|^2\int e^{2\phi(\zeta)}|\xi_1-\zeta|^{-2}|\xi_2-\zeta|^{-2}d^2\zeta\,. \end{aligned} \tag{29}$$

Notice the appearance of the linear term $\sim \int\phi\mathcal{R}$ in the effective ϕ-action and of the $e^{2\phi(\zeta)}$ insertion corresponding, respectively, to the background and screening charges in the Coulomb gas interpretation of the resulting ϕ-field theory. The integral over ϕ is again gaussian but requires a renormalization of the polynomial in $e^{\pm\phi(\xi_\alpha)}$ and $e^{2\phi(\zeta)}$ to render it finite. If we extract the most divergent factor multiplicatively, the terms with milder divergences will not survive the renormalization. In the case at hand, these are terms $e^{\phi(\xi_1)-\phi(\xi_2)}+e^{\phi(\xi_2)-\phi(\xi_1)}$ on the right hand side of (27). They drop out leaving us with the result

$$\begin{aligned} &\text{const.}\,|\xi_1-\xi_2|^{2-1/(\kappa-2)}\int|(\xi_1-\zeta)(\xi_2-\zeta)|^{-2+2/(\kappa-2)}d^2\zeta \\ &\qquad = \text{const.}\,|\xi_1-\xi_2|^{3/(\kappa-2)} \end{aligned} \tag{30}$$

(in the flat metric). Replacing $\text{tr}_{\frac{1}{2}}$ in (27) by tr_j for higher spins, we obtain a ϕ-integral with $2j$ screening charges and finally

$$\text{const.}\,|\xi_1-\xi_2|^{4j(j+1)/(\kappa-2)} \tag{31}$$

provided that $2j+1<\kappa-2$. Otherwise, the integral over the positions of the screening charges diverges[7]. Higher Green functions may be computed similarly [7],[39], also for the general $H^{\mathbf{C}}/H$ theories [35].

From the form of the general Green functions (also with the current and energy-momentum insertions) one infers that fields $hh^\dagger(\xi)$ are primary, both for the $\hat{sl}(2,\mathbf{C})\oplus\hat{sl}(2,\mathbf{C})$ and Vir$\oplus$Vir algebras. Their conformal weights $\Delta_j=\bar{\Delta}_j$ are, as read from eq. (31), $-\frac{j(j+1)}{\kappa-2}<0$. Occurrence of fields with negative dimensions, so with Green functions growing with the distance, might seem incompatible with the stability although not necessarily in a non-unitary theory as ours. The point, however, lies elsewhere. Such fields ($:e^{\alpha\phi}:$ for α real) are clearly present for the massless free (uncompactified) field ϕ which gives a stable unitary theory and are also expected in the Liouville theory [9], believed to be stable and unitary (there, they correspond to the local operators in terminology of [9]). These operators escape the standard relation between the spectrum of energy and of conformal weights since they correspond to eigenfunctions of the Hamiltonian outside the generalized eigenspaces. This may be seen already in the "mini-space" quantum-mechanical picture which is stable and unitary for the H_3^+ theory: although

$$-\tfrac{1}{\kappa-2}\Delta\, hh_j^\dagger = -\tfrac{j(j+1)}{\kappa-2}\,hh_j^\dagger\,,$$

[7]this is the dual manifestation of the restriction to spins $j\leq k/2$ in the $SU(2)$ WZW model or, more generally, of its fusion rules, see [7]

the matrix elements of $hh_j^\dagger$ are not the generalized eigenfunctions of $-\Delta$ due to their too rapid growth at infinity. Appearance of operators with negative conformal dimensions may be typical for irrational theories with continuous spectrum of $L_0, \bar{L}_0$. Notice nevertheless that in the $H^{\mathbf{C}}/H$ WZW model they come in a finite number whereas for the massless free field and for the Liouville theory, there is a continuous family of such fields.

Besides fields with negative dimensions which do not correspond neither to true nor to generalized states of the theory, it is natural to expect existence of fields with positive dimensions corresponding to the states in the spectrum of $L_0, \bar{L}_0$. The natural candidates for such fields are given by $f_{\rho,m_l,m_r}(hh^\dagger(\xi))$ where f_{ρ,m_l,m_r} is a joint generalized eigenfunction of $-\Delta, J^3, \bar{J}^3$ corresponding to eigenvalues $\frac{1}{4}(1+\rho^2), m_l = \frac{1}{2}(n+i\omega), m_r = \frac{1}{2}(-n+i\omega)$ where $\rho \geq 0$, $n \in \mathbf{Z}$ and $\omega \in \mathbf{R}$. In the space $\mathcal{H}_\rho$ (of homogenous functions on V_3^+), the corresponding eigenfunction is

$$e^{-i\omega\phi - in\arg(v)}\,|v|^{-1+i\rho}\ . \tag{32}$$

Eigenfunction f_{ρ,m_l,m_r} on H_3^+ is obtained by applying to (32) the Gelfand-Graev integral transformation [37] realizing the isomorphism (13):

$$\begin{aligned} f_{\rho,m_l,m_r}(\phi, v) \;&=\; e^{-i\omega\phi - in\arg(v)}\,(1+|v|^2)^{i\omega/2} \\ &\cdot \int_0^{2\pi} d\theta \int_0^{\infty} dr\; e^{in\theta}\, r^{i\rho+i\omega}\,[1+2|v|r\cos\theta + (1+|v|^2)r^2]^{-1-i\rho}\ . \end{aligned} \tag{33}$$

For example, for $\rho = m_l = m_r = 0$, we obtain the elliptic integral

$$f_{0,0,0}(v) \;=\; \pi \int_0^{2\pi} (1+|v|^2\sin^2\theta)^{-1/2}\, d\theta\ . \tag{34}$$

Unfortunately, we were not able to compute the Green functions of fields f_{ρ,m_l,m_r} exactly. It remains then to be seen if they indeed give rise, upon multiplicative renormalization, to primary fields with conformal weights $\Delta_\rho = \bar{\Delta}_\rho = \frac{1+\rho^2}{4(\kappa-2)}$.

4. $SL(2,\mathbf{C})/SU(2)$ mod $\mathbf{R}$ COSET THEORY

4.1 2D black hole sigma model

In Sec. 3, we have coupled the $SL(2,\mathbf{C})/SU(2) \equiv H_3^+$ WZW model to an abelian gauge field A in the way which rendered the action invariant under the non-compact gauge transformations:

$$S(e^{\lambda\sigma^3/2}hh^\dagger e^{\lambda\sigma^3/2}, \tfrac{1}{2i}(A-d\lambda)) \;=\; S(hh^\dagger, \tfrac{1}{2i}A)\ , \tag{1}$$

see (3.3). Following the scenario for producing coset theories from compact WZW models, let us consider the functional integral

$$\begin{aligned} &\int - \; e^{\kappa S(hh^\dagger, (2i)^{-1}A)}\, D(hh^\dagger)\, DA \\ = &\int - \; e^{-\pi^{-1}\kappa \int[(\partial_z\phi + A_z)(\partial_{\bar z}\phi + A_{\bar z}) + (\partial_z + \partial_z\phi + A_z)\bar{v}\,(\partial_{\bar z} + \partial_{\bar z}\phi + A_{\bar z})v]\,d^2z}\, D\phi\, Dv\, DA \end{aligned} \tag{2}$$

with gauge invariant insertions. First notice that, by the gauge invariance, the integral over ϕ factors as the (infinite) volume of the gauge group. Since A enters quadratically into the action, it may be integrated out (for appropriate, e.g. A-independent, insertions) giving

$$C \int - \, e^{-\pi^{-1}\kappa \int (1+|v|^2)^{-1}(\partial_z \bar{v})(\partial_{\bar{z}} v)\, d^2 z} \prod_{\xi} \frac{d^2 v(\xi)}{1+|v(\xi)|^2} \,. \tag{3}$$

The effective action for v:

$$\begin{aligned} S_{\text{eff}}(v) \;&\equiv\; \tfrac{\kappa}{\pi} \int (1 + |v|^2)^{-1} (\partial_z \bar{v})(\partial_{\bar{z}} v)\, d^2 z \\ &= \tfrac{\kappa}{\pi} \sum_{a=1,2} \int (1 + |v|^2)^{-1} (\partial_z v^a)(\partial_{\bar{z}} v^a)\, d^2 z \end{aligned}$$

if we integrate by parts. $v = v^1 + iv^2$. It is the action of a sigma model with the complex plane with metric

$$(1 + |v|^2)^{-1} (dz \otimes d\bar{z} + d\bar{z} \otimes dz) \tag{4}$$

as the target. It was noticed recently [18],[19] that this target metric (together with the dilaton field $\Phi = \log(1 + |v|^2))$ forms a euclidean black hole solution of equations of 2D gravity (with unit mass). It describes an infinite cigar becoming asymptotically a cylinder (the scalar curvature goes down as $|v|^{-2}$ as $v \to \infty$). The Minkowskian counterpart of this solution is the metric

$$(1 - v^+ v^-)^{-1} (dv^+ dv^- + dv^- dv^+) \tag{5}$$

with the asymptotically flat region $\pm v^{\pm} > 0$ with future horizon $v^- = 0,\ v^+ > 0$ and past horizon $v^+ = 0,\ v^- < 0$, another such region for $v^+ \leftrightarrow v^-$, and future and past singularities at $v^+ v^- = 1$.

4.2 Toroidal partition function

As it stands, functional integral (3) for the black hole target is difficult to compute directly. Instead, we may go back to expression (2) and integrate first over $hh^\dagger$ and then over A. Let us illustrate this on the example of the twisted toroidal partition function

$$\mathcal{Z}^{\text{bh}}(\tau, U) \;=\; \int e^{\kappa S(\gamma_U h h^\dagger \gamma_U^\dagger,\, (2i)^{-1} A)}\, D(hh^\dagger)\, DA \tag{6}$$

where the action for the twisted field configurations is coupled to the gauge field by putting

$$\begin{aligned} S(\gamma_U h h^\dagger \gamma_U^\dagger, \tfrac{1}{2i} A) \;&=\; S(hh^\dagger, \tfrac{1}{2i}(A + \tfrac{1}{2}\tau_2^{-1} \bar{U} dz + \tfrac{1}{2}\tau_2^{-1} U d\bar{z})) \\ &\quad + \tfrac{1}{2\pi\tau_2} U_1 \int (A_z + A_{\bar{z}}) d^2 z \;+\; \tfrac{\pi}{\tau_2} U_1^2 \,. \end{aligned} \tag{7}$$

The parametrization of A by the Hodge decomposition

$$A = d\mu + {*d\nu} + \tau_2^{-1} (\bar{u} dz + u d\bar{z})/2 \tag{8}$$

(μ, ν real functions, $u = u_1 + iu_2$) gives for the volumes

$$DA \;=\; C \tau_2^{-2} \det{}'(\bar{\partial}^* \bar{\partial})\ \delta(\mu(\xi_0))\ \delta(\nu(\xi_0))\ d^2 u\ D\mu\ D\nu \,.$$

Due to the gauge invariance of the action, the integral over μ factors out as the (infinite) volume of the gauge group. The ν-integral also factors out after unitary rotation $v \mapsto$

$e^{-i\nu}v$ so that the v- and ϕ-integrals produce the twisted partition function $\mathcal{Z}^{H_3^+}(\tau,u)$ of the H_3^+ WZW theory. As the result, we obtain

$$\mathcal{Z}^{\rm bh}(\tau,U) \;=\; C\tau_2^{-2}\det{}'(\bar\partial^*\bar\partial)\int \mathrm{e}^{-\pi^{-1}\kappa\int(\partial_z\nu)(\partial_{\bar z}\nu)\,d^2z-\pi\kappa\tau_2^{-1}(U_1-u_1)^2} \cdot \mathcal{Z}^{H_3^+}(\tau,u)\,d^2u\,\delta(\nu(\xi_0))\,D\nu\,. \tag{9}$$

The ν-integral is straightforward and for $\mathcal{Z}^{H_3^+}(\tau,u)$ we have expression (3.9). Hence

$$\begin{aligned}\mathcal{Z}^{\rm bh}(\tau,U) \;&=\; C\tau_2^{-1/2}\int \mathrm{e}^{-\pi\kappa\tau_2^{-1}(U_1-u_1)^2}\,\mathcal{Z}(\tau,u)\,|\eta(\tau)|^2\,d^2u\\ &=\; C\tau_2^{-1}(q\bar q)^{-1/12}\int \mathrm{e}^{-\pi\kappa\tau_2^{-1}(U_1-u_1)^2-\pi(\kappa-2)\tau_2^{-1}u_2^2}\,|\sin(\pi u)|^{-2}\\ &\qquad\cdot\left|\prod_{n=1}^{\infty}(1-\mathrm{e}^{2\pi iu}q^n)(1-\mathrm{e}^{-2\pi iu}q^n)\right|^{-2}d^2u\end{aligned} \tag{10}$$

where $\eta(\tau)\equiv q^{1/24}\prod\limits_{n\geq1}(1-q^n)$ is the Dedekind function. The u-integral diverges logarithmically due to the singularity $\sim|u|^{-2}$ at zero. This singularity is repeated on the lattice $\mathbf{Z}+\tau\mathbf{Z}$, as follows immediately from the bi-periodicity of expression $\mathrm{e}^{2\pi\tau_2^{-1}u_2^2}|\sin(\pi u)|^{-2}\left|\prod\limits_{n=1}^{\infty}(1-\mathrm{e}^{2\pi iu}q^n)(1-\mathrm{e}^{-2\pi iu}q^n)\right|^{-2}$. Let us explain this divergence of a relatively simple nature.

4.3 Mini-space partition function

It is instructive to start with the mini-space case (we remind that this means taking fields $hh^\dagger$ and A_z, $A_{\bar z}$ independent of the space variable). For

$$\mathcal{Z}^{\rm bh}_{\rm mini}(\tau,U) \;=\; \int \mathrm{e}^{\kappa S_{\rm mini}(\gamma_U hh^\dagger\gamma_U^\dagger,\,(2i)^{-1}A)}\,D(hh^\dagger)\,DA\,, \tag{11}$$

we may also proceed as before integrating first over A to get the twisted partition function for the quantum-mechanical particle moving on the euclidean black hole:

$$\mathcal{Z}^{\rm bh}_{\rm mini}(\tau,U) \;=\; C\int \mathrm{e}^{-(\kappa/2)\int_0^{2\pi\tau_2}(1+|v|^2)^{-1}|(\partial_t-i\tau_2^{-1}U_1)v|^2\,dt}\prod_{\xi}\frac{d^2v(\xi)}{1+|v(\xi)|^2}\,. \tag{12}$$

On the other hand, integrating first over $hh^\dagger$ and then over A, we obtain:

$$\mathcal{Z}^{\rm bh}_{\rm mini}(\tau,U) \;=\; C\tau_2^{-1}\int \mathrm{e}^{-\pi\kappa\tau_2^{-1}((U_1-u_1)^2+u_2^2)}\,|\sin(\pi u)|^{-2}\,d^2u\,. \tag{13}$$

The right hand side of eq. (13) may be rewritten, with the use of eqs. (3.18)-(3.20), as

$$\begin{aligned}&C\tau_2^{-1/2}\int {\rm tr}_{\rm ren}\left(\mathrm{e}^{4\pi\tau_2\kappa^{-1}(\Delta+1/4)}\,\mathrm{e}^{2\pi i(uJ^3-\bar u\bar J^3)}\right)\mathrm{e}^{-\pi\kappa\tau_2^{-1}(U_1-u_1)^2}\,d^2u\\ &=C\tau_2^{-1/2}\int \mathrm{e}^{4\pi\tau_2\kappa^{-1}(\Delta+1/4)}(2\pi u_2,\mathrm{e}^{-2\pi iu_1}v;0,v)\,\mathrm{e}^{-\pi\kappa\tau_2^{-1}(U_1-u_1)^2}\,d^2u\,d^2v\,.\end{aligned} \tag{14}$$

Notice that

$$\int \mathrm{e}^{t\Delta}(2\pi u_2,v;0,v')\,du_2 \;=\; \tfrac{1}{2\pi}\,\mathrm{e}^{t\Delta_{\omega=0}}(v;v') \tag{15}$$

where $\Delta_{\omega=0}$ is the restriction of Laplacian Δ to the generalized eigensubspace of operator $i(J^3+\bar J^3)=-i\partial_\phi$ corresponding to eigenvalue 0. From the expression (3.12) for Δ, we infer that

$$\Delta_{\omega=0} \;=\; (1+|v|^2)\partial_v\partial_{\bar v}+\tfrac14(v\partial_v-\bar v\partial_{\bar v})^2+\tfrac12(v\partial_v+\bar v\partial_{\bar v}) \tag{16}$$

and is a selfadjoint operator in $L^2(d^2v)$. Moreover,

$$\begin{aligned}
&(\kappa/\tau_2)^{1/2} \int e^{4\pi\tau_2\kappa^{-1}\Delta_{\omega=0}}(e^{-2\pi i u_1}v;\, v')\, e^{-\pi\kappa\tau_2^{-1}(U_1-u_1)^2}\, du_1 \\
&= (\kappa/\tau_2)^{1/2} \int e^{4\pi\tau_2\kappa^{-1}\Delta_{\omega=0} + 2\pi i u_1(J^3-\bar{J}^3)}(v;\, v')\, e^{-\pi\kappa\tau_2^{-1}(U_1-u_1)^2}\, du_1 \\
&= e^{4\pi\tau_2\kappa^{-1}\Delta_{\omega=0} - \pi\tau_2\kappa^{-1}(J^3-\bar{J}^3)^2 + 2\pi i U_1(J^3-\bar{J}^3)}(v;\, v') \;=\; e^{4\pi\tau_2\kappa^{-1}\Delta^{\mathrm{bh}}}(e^{-2\pi i U_1}v;\, v') \qquad (17)
\end{aligned}$$

where we have introduced

$$\begin{aligned}
-\Delta_{\omega=0} + (J^3)^2 &= -\Delta_{\omega=0} + (\bar{J}^3)^2 \\
= -\tfrac{1}{2}\partial_v(1+|v|^2)\partial_{\bar{v}} - \tfrac{1}{2}\partial_{\bar{v}}(1+|v|^2)\partial_v &\equiv -\Delta^{\mathrm{bh}}\,. \qquad (18)
\end{aligned}$$

It is a Laplacian quantizing the classical Hamiltonian $p_v p_{\bar{v}}(1+|v|^2)$ of the particle on the (euclidean) black hole, with a specific choice of ordering prescription (different from the Laplace-Beltrami operator which would correspond to $(1+|v|^2)^{1/2}\partial_v\partial_{\bar{v}}(1+|v|^2)^{1/2}$). We may finally rewrite the mini-space partition function as

$$\mathcal{Z}^{\mathrm{bh}}_{\mathrm{mini}}(\tau, U) \;=\; C \int e^{4\pi\tau_2\kappa^{-1}(\Delta^{\mathrm{bh}}+1/4)}(e^{-2\pi i U_1}v;\, v)\, d^2v\,. \qquad (19)$$

The integral is divergent but the nature of this divergence is quite simple. For $v \to \infty$, where the metric becomes cylindrical in variable $\log v$, $\exp[t\Delta^{\mathrm{bh}}](e^{-2\pi i U_1}v;\, v)|v|^2$ approaches a constant (equal to the free heat kernel between the points on the cylinder of constant difference). Hence the divergence due to the infinite volume of the black hole cigar. It may be easily regularized by cutting integral over v to $|v| \le R$. Going back to integral (3.8), it is easy to see that such cutoff results in the replacement

$$e^{-\pi\kappa\tau_2^{-1}u_2^2} \;\longmapsto\; e^{-\pi\kappa\tau_2^{-1}u_2^2} - e^{-(4\pi\tau_2)^{-1}\kappa d_R^2} \qquad (20)$$

in the integrand of (13). Here $d_R = \cosh^{-1}\left(\cosh(2\pi u_2) + 2R^2|\sin(\pi u)|^2\right)$ stands for the hyperbolic distance between $e^{-\pi i U\sigma^3}hh^\dagger e^{\pi i\bar{U}\sigma^3}$ and $hh^\dagger = \begin{pmatrix}(1+R^2)^{1/2} & R \\ R & (1+R^2)^{1/2}\end{pmatrix}$. Such a replacement makes the integral in (13) convergent but behaving as $\mathcal{O}(\log R)$ (or more generally as $\mathcal{O}(\log(M^{-1/2}R))$ where M is the black hole mass; we consider here only the case $M = 1$). We could define the finite part of $\mathcal{Z}^{\mathrm{bh}}_{\mathrm{mini}}$ by subtracting this logarithmic divergence, i.e. by comparing it to half the partition function of a particle on the cylinder.

Let us go back to the interpretation of the result (10). As compared to expression (13) for the mini-space case, the main differences in (10) are the partial shift $\kappa \mapsto \kappa - 2$ and the presence of the big product inherited from the oscillatory modes of the H_3^+ theory. The shift of κ is easy: if we drop the infinite product from the right hand side of (10) to get the level zero (i.e. zero mode) contribution, we obtain, proceeding as for the mini-space case,

$$\begin{aligned}
\mathcal{Z}^{\mathrm{bh}}_{\mathrm{level\,0}}(\tau, U) &= C(q\bar{q})^{-(c_{-\kappa}-1)/24}\, \mathrm{tr}|_{\omega=0}\, e^{4\pi\tau_2(\kappa-2)^{-1}\Delta - 2\pi\tau_2\kappa^{-1}((J^3)^2+(\bar{J}^3)^2) + 2\pi i U_1(J^3-\bar{J}^3)} \\
&= C\,(q\bar{q})^{-(c_{-\kappa}-1)/24}\, \mathrm{tr}|_{\substack{\mathrm{level}\,0 \\ m_l+m_r=0}}\, q^{L_0^{\mathrm{cs}}}\,\bar{q}^{\bar{L}_0^{\mathrm{cs}}}\, e^{2\pi i(UJ_0^3-\bar{U}\bar{J}_0^3)} \qquad (21)
\end{aligned}$$

with the coset Virasoro generators

$$L_0^{\mathrm{cs}} = L_0 + \tfrac{1}{\kappa}\sum_n :J_n^3 J_{-n}^3:\,, \quad \bar{L}_0^{\mathrm{cs}} = \bar{L}_0 + \tfrac{1}{\kappa}\sum_n :\bar{J}_n^3 \bar{J}_{-n}^3:\,. \qquad (22)$$

The contribution of the higher level oscillatory modes is, however, less transparent than one may naively think if we want to interpret it in terms of gauge invariant states.

4.4 Asymmetric parafermions

Let us compare the situation to a somewhat similar case of a variant of rational parafermionic theory which may be described as the $SU(2)$ WZW model with the axial gauging of the $U(1)$ subgroup, i.e. with the diagonal $U(1)$ gauged asymmetrically. The twisted toroidal partition function for such parafermions is [40],[14]

$$\mathcal{Z}^{\mathrm{pf}}(U,\tau) = \int e^{-kS(\gamma_U g \gamma_U^\dagger, A)}\, Dg\, DA\,. \tag{23}$$

The integration is now over real A. Parametrizing A as before by the Hodge decomposition, one arrives at the formula

$$\mathcal{Z}^{\mathrm{pf}}(U,\tau) = C\tau_2^{-1/2} \int\limits_{\mathbf{C}/(\mathbf{Z}+\tau\mathbf{Z})} e^{\pi k \tau_2^{-1}(U_1 - iu_2)^2}\, \mathcal{Z}^{SU(2)}(\tau,u)\, |\eta(\tau)|^2\, d^2u \tag{24}$$

where $\mathcal{Z}^{SU(2)}(\tau,u)$ is the asymmetrically twisted partition function of the rational $SU(2)$ WZW model:

$$\mathcal{Z}^{SU(2)}(\tau,u) = (q\bar{q})^{-c_k/24}\, \mathrm{tr}\; q^{L_0}\bar{q}^{\bar{L}_0}\, e^{2\pi i (uJ_0^3 + \bar{u}\bar{J}_0^3)}\,. \tag{25}$$

The trace is taken over the space of states

$$\hat{\mathcal{H}}^{SU(2)} = \bigoplus_{j\leq k/2} \hat{\mathcal{H}}_j \otimes \overline{\hat{\mathcal{H}}_j} \tag{26}$$

where $\hat{\mathcal{H}}_j$ carries the irreducible spin j level k representation of the Kac-Moody algebra $\hat{sl}(2,\mathbf{C})$. Notice the sign in front of $\bar{u}\bar{J}_0^3$ in (25). The integrand on the right hand side of eq. (24) is a function on $\mathbf{C}/(\mathbf{Z}+\tau\mathbf{Z})$ only if $U_1 \in k^{-1}\mathbf{Z}$ and only such twists should be allowed. For other twists there is a global gauge anomaly: the ungauged global $U(1)$ symmetry is broken in the parafermionic theory to $\mathbf{Z}_k$. The spaces $\hat{\mathcal{H}}_j$ may be decomposed into the weight spaces according to the integral or half-integral eigenvalue m of J_0^3 and at the same time with respect to the level k representations of the $\hat{U}(1)$ affine algebra (similarly for the complex conjugates):

$$\hat{\mathcal{H}}_j \cong \bigoplus_m \hat{\mathcal{H}}^{\mathrm{sing}}_{j,m} \otimes \hat{\mathcal{H}}'_m \tag{27}$$

where $\hat{\mathcal{H}}^{\mathrm{sing}}_{j,m}$ is the subspace of $\hat{\mathcal{H}}_j$ where $J_0^3 = m$ and $J_n^3 = 0$ for $n > 0$. $\hat{\mathcal{H}}'_m$ is the space of the level k $J_0^3 = m$ irreducible representation of the $\hat{U}(1)$ algebra. The Sugawara Virasoro generator L_0 decomposes into the sum of $L_0^{\mathrm{cs}} \equiv L_0 - \frac{1}{k}\sum_n{:}J_n^3 J_{-n}^3{:}$ acting on spaces $\hat{\mathcal{H}}^{\mathrm{sing}}_{j,m}$ and $L'_0 \equiv \frac{1}{k}\sum_n{:}J_n^3 J_{-n}^3{:}$ acting on $\hat{\mathcal{H}}'_m$ (in fact on $\hat{\mathcal{H}}^{\mathrm{sing}}_{j,m}$, $L_0^{\mathrm{cs}} = L_0 - \frac{1}{k}m^2$). Accordingly, we obtain for the partition function of the $SU(2)$ WZW theory:

$$\mathcal{Z}^{SU(2)}(\tau,u) = (q\bar{q})^{-(c_k-1)/24} \sum_{m_l,m_r} \mathcal{Z}^{\mathrm{sing}}_{m_l,m_r}\, q^{m_l^2/k}\bar{q}^{m_r^2/k}\, |\eta(\tau)|^{-2}\, e^{2\pi i(um_l + \bar{u}m_r)} \tag{28}$$

where

$$\mathcal{Z}^{\mathrm{sing}}_{m_l,m_r} = \mathrm{tr}|_{\hat{\mathcal{H}}^{\mathrm{sing}}_{m_l,m_r}}\, q^{L_0^{\mathrm{cs}}}\, \bar{q}^{\bar{L}_0^{\mathrm{cs}}} \tag{29}$$

with

$$\hat{\mathcal{H}}^{\text{sing}}_{m_l,m_r} \equiv \bigoplus_j \hat{\mathcal{H}}^{\text{sing}}_{j,m_l} \otimes \overline{\hat{\mathcal{H}}}^{\text{sing}}_{j,m_r} . \tag{30}$$

$\mathcal{Z}^{\text{sing}}_{m_l,m_r}$ depends only on m_l and m_r mod $k/2$ [40] (essentially due to the compact nature of the gauged symmetry). More exactly,

$$\text{tr}|_{\hat{\mathcal{H}}^{\text{sing}}_{j,m}}\, q^{L_0^{cs}} = \text{tr}|_{\hat{\mathcal{H}}^{\text{sing}}_{j,m+k}}\, q^{L_0^{cs}} = \text{tr}|_{\hat{\mathcal{H}}^{\text{sing}}_{k-j,-m}}\, q^{L_0^{cs}} ,$$

see [40]. Upon the insertion of (28) into the right hand side of (24), the u_1-integral will enforce equality $m_l = -m_r \equiv m$. The sum over m may be reduced mod $k/2$, with the sum over the integral part of $2m/k$ used to extend the integration over u_2 to a gaussian one over the entire real line. Finally we get

$$\mathcal{Z}^{\text{pf}}(\tau, U) = C(q\bar{q})^{-(c_k-1)/24} \sum_{m=0,\frac{1}{2},\ldots,\frac{k}{2}} \mathcal{Z}^{\text{sing}}_{m,-m}\, \text{e}^{-4\pi i m U_1} . \tag{31}$$

As we see, the parafermionic partition function is consistent (modulo multiplicity) with the space of states of the coset theory obtained by imposing the gauge conditions

$$J^3_0 + \bar{J}^3_0 = 0, \quad J^3_n = \bar{J}^3_n = 0 \ \text{ for } \ n > 0 \tag{32}$$

in the space of states of the ungauged WZW theory with the Virasoro algebra given by the coset construction. On the other hand, we could replace the first gauge condition by $J^3_0 + \bar{J}^3 = kn$ for $n \in \mathbf{Z}$ or by $J^3 = -\bar{J}^3$ and obtain equivalent theory. The latter means that the asymmetric parafermions are indistinguishable from the symmetric ones.

4.5 Space of states

The level zero contribution (21) to the black hole partition function is fully consistent with the gauge conditions (32) imposed on states of the H_3^+ WZW theory (for zero modes, only the first condition of (32) restricts the states). The problem appears on the excited levels of the space of states $\hat{\mathcal{H}}^{H_3^+}$ of the ungauged theory. Let us consider, as an example, the first excited level with states of the form

$$\sum_{a=\pm,3} (J^a_{-1}\psi_a + \bar{J}^a_{-1}\bar{\psi}_a) \tag{33}$$

where ψ_a, $\bar{\psi}_a$ are level zero states, i.e. functions on H_3^+. The $J^3_0 + \bar{J}^3_0 = 0$ condition translates into

$$(J^3_0 + \bar{J}^3_0 \pm 1)\psi_\pm = 0 , \quad (J^3_0 + \bar{J}^3_0 \pm 1)\bar{\psi}_\pm = 0 , \quad (J^3_0 + \bar{J}^3_0)\psi_3 = 0 . \tag{34}$$

The other conditions of (32) give

$$\psi_3 = \tfrac{2}{\kappa}(J^+_0\psi_+ - J^-_0\psi_-) , \quad \bar{\psi}_3 = \tfrac{2}{\kappa}(\bar{J}^+_0\bar{\psi}_+ - \bar{J}^-_0\bar{\psi}_-) . \tag{35}$$

Notice, however, that in $L^2(H_3^+)$, $J^3_0 + \bar{J}^3_0$ is antihermitian so it has imaginary spectrum. Thus non-trivial solutions of (34) and (35) are not only out of $L^2(H_3^+)$ but do not belong to the generalized eigenspaces of J^3_0, $\bar{J}^3_0$ (they have $\text{e}^{\pm\phi}$ dependence on ϕ). At best, we have to change the Hilbert space. Notice how the situation here differs from the case of parafermions where no such problems arise. We may understand the above difficulty also by looking at the level one contribution to the partion function (10) which involves integrals

$$\tau_2^{-1}\int e^{-\pi\kappa\tau_2^{-1}(U_1-u_1)^2-\pi(\kappa-2)\tau_2^{-1}u_2^2}\,|\sin(\pi u)|^{-2}\,e^{\pm 2\pi i u}\,d^2u$$
$$= C\tau_2^{-1/2}\int e^{4\pi\tau_2(\kappa-2)^{-1}(\Delta+1/4)}(2\pi u_2, e^{-2\pi i u_1}v;\,0,v)$$
$$\cdot\, e^{\pm 2\pi(iu_1-u_2)}\,e^{-\pi\kappa\tau_2^{-1}(U_1-u_1)^2}\,d^2u\,d^2v\,. \tag{36}$$

By spectral analysis, we may decompose operators $e^{t\Delta}$ into the heat kernels acting in the generalized eigenspaces of J_0^3, $\bar{J}_0^3$:

$$e^{t\,(\Delta+1/4)}(2\pi u_2, e^{-2\pi i u_1}v';\,0,v) = \sum_n\int \mathcal{K}_{n,\omega}(t;|v'|,|v|)\,e^{2\pi i n u_1-2\pi i\omega u_2}\,d\omega\,. \tag{37}$$

This allows to rewrite integrals (36) as

$$C\sum_n \mathcal{K}_{n,\mp i}(4\pi\tau_2\kappa^{-1};|v|,|v|)\,e^{-\pi\tau_2\kappa^{-1}(n\pm1)^2+2\pi i(n\pm1)U_1}\,d|v|^2 \tag{38}$$

involving the analytic continuation of heat kernels $\mathcal{K}_{n,\omega}$ to imaginary values of ω. The question is whether such an analytic continuation (which exists) corresponds to a heat kernel in a different Hilbert space.

Summarizing. the gauge conditions (32) do not determine unambiguously the space of states. We have to supplement them with regularity conditions specifying domains of the operators that they involve (the same applies to the BRST definition of gauge invariant states). Ultimately, we should be able to build a Hilbert space of states at each level and to compute the contribution to the partition function as a trace of a heat kernel in such a space. We shall discuss a candidate solution of this problem in Sec. 5.

On top of the above difficulties with the interpretation of the partition function (but not unrelated to them) comes the fact that, as it stands, the integral on the right hand side of eq. (10) diverges. The source of this divergence is, as in the mini-space approximation, the infinite volume of the target space. This may be regularized for example by defining

$$\tilde{\mathcal{Z}}^{\rm bh}_{\rm reg}(\tau,U;R) = C\tau_2^{-1}\int e^{-\pi\kappa\tau_2^{-1}|U-u|^2}\,\mathcal{S}(\tau,u)\left(1-e^{-R^2\mathcal{S}(\tau,u)^{-1}}\right)d^2u \tag{39}$$

where

$$\mathcal{S}(\tau,u) \equiv (q\bar{q})^{-1/12}\,e^{2\pi\tau_2^{-1}u_2^2}\,|\sin(\pi u)|^{-2}\left|\prod_{n=1}^{\infty}(1-e^{2\pi i u}q^n)(1-e^{-2\pi i u}q^n)\right|^{-2}. \tag{40}$$

The partition function $\tilde{\mathcal{Z}}^{\rm bh}_{\rm reg}(\tau,U)$ is finite and when $R\to\infty$ and for $U_2=0$, we recover the infinite integral (10) (we have put the twists along both homology cycles in $\tilde{\mathcal{Z}}^{\rm bh}_{\rm reg}(\tau,U)$ so that in the limit $R\to\infty$ it corresponds to the black hole functional integral with boundary conditions $v(z+2\pi)=e^{-2\pi i\Phi}v(z)$, $v(z+2\pi\tau)=e^{-2\pi i\Theta}v(z)$ where $U=\Theta-\tau\Phi$); for $\Phi=0$, we recover $Z^{\rm bh}(\tau,U)$ with twist only along one cycle). $\mathcal{S}(\tau,U)$ is invariant under translations $U\longmapsto U+n+\tau m$ for n,m integers and is modular invariant. As a result, under $SL(2,\mathbf{Z})$ transformations,

$$\tilde{\mathcal{Z}}^{\rm bh}_{\rm reg}\left(\tfrac{a\tau+b}{c\tau+d},\tfrac{U}{c\tau+d};R\right) = \tilde{\mathcal{Z}}^{\rm bh}_{\rm reg}(\tau,U;R)\,, \tag{41}$$

i.e. the regularized partition function is modular covariant. Again the divergence is logarithmic in R and we could subtract it to define the renormalized partition function measuring the difference between the theories with the black hole and (half-)cylinder targets.

4.6 Partition functions at higher genera

On a higher genus Riemann surface Σ with the homology basis (a_α, b_β), $\alpha,\beta = 1,\ldots,$genus, and with the basic holomorphic forms ω^α, $\int_{a_\alpha}\omega^\beta = \delta^{\alpha\beta}$, $\int_{b_\alpha}\omega^\beta = \tau^{\alpha\beta} \equiv \tau_1^{\alpha\beta} + i\tau_2^{\alpha\beta}$, let us define the multivalued field

$$\tilde{\gamma}_U(P) = \mathrm{e}^{\pi\sigma^3\int_{P_0}^{P}(U^t\tau_2^{-1}\bar{\omega} - \bar{U}^t\tau_2^{-1}\omega)/2} \tag{42}$$

with values in the Cartan subgroup of $SU(2)$. Along the basic cycles

$$\tilde{\gamma}_U(a_\alpha P) = \mathrm{e}^{-\pi i\Phi_\alpha\sigma^3}\,\tilde{\gamma}_U(P)\,,$$
$$\tilde{\gamma}_U(b_\alpha P) = \mathrm{e}^{-\pi i\Theta_\alpha\sigma^3}\,\tilde{\gamma}_U(P)$$

where $U = \Theta - \tau\Phi$. The twisted partition function on Σ is given by

$$\tilde{\mathcal{Z}}^{bh}(\tau,U) \;=\; \int \mathrm{e}^{\kappa S(\tilde{\gamma}_U\tilde{\gamma}_U^\dagger,(2i)^{-1}A)}\,D(hh^\dagger)\,DA \tag{43}$$

with

$$S(\tilde{\gamma}_U hh^\dagger\tilde{\gamma}_U^\dagger,(2i)^{-1}A) \;=\; S(hh^\dagger,\tfrac{1}{2i}(A + \pi\bar{U}^t\tau_2^{-1}\omega + \pi U^t\tau_2^{-1}\bar{\omega}))$$
$$+\tfrac{1}{2i}\int A\wedge(\bar{U}^t\tau_2^{-1}\omega - U^t\tau_2^{-1}\bar{\omega}) \;+\; \pi U^t\tau_2^{-1}U\,. \tag{44}$$

It defines the higher genus partition function for the black hole with twists of the v-field by $\mathrm{e}^{-2\pi i\Phi_\alpha}$ factors along the a_α cycles and by $\mathrm{e}^{-2\pi i\Theta_\beta}$ along the b_β ones. We decompose again the gauge field according to Hodge:

$$A = d\mu + *d\nu + \pi\bar{u}^t\tau_2^{-1}\omega + \pi u^t\tau_2^{-1}\bar{\omega} \tag{45}$$

and integrate over the v-field (of $hh^\dagger$), ν and μ (the latter integral gives the volume of the gauge group). What is left is the ϕ functional integral and the integral over twists u:

$$\tilde{\mathcal{Z}}^{\mathrm{bh}}(\tau,U) \;=\; C\left(\frac{\det'(-\bar{\partial}^*\bar{\partial})}{\mathrm{area}}\right)^{1/2}\det\tau_2^{-1}\int \mathrm{e}^{-\pi\kappa(\bar{U}-\bar{u})^t\tau_2^{-1}(U-u) + (2\pi i)^{-1}\kappa\int(\partial\phi)(\bar{\partial}\phi)}$$
$$\cdot\det\left(\bar{\partial} + \bar{\partial}\phi + \pi u^t\tau_2^{-1}\bar{\omega})^*(\bar{\partial} + \bar{\partial}\phi + \pi u^t\tau_2^{-1}\bar{\omega})\right)^{-1}\delta(\phi(\xi_0))\,D\phi\,d^{2\,\mathrm{genus}}u\,. \tag{46}$$

By the chiral anomaly (compare the genus one formula (3.8)),

$$\det\left(\bar{\partial} + \bar{\partial}\phi + \pi u^t\tau_2^{-1}\bar{\omega})^*(\bar{\partial} + \bar{\partial}\phi + \pi u^t\tau_2^{-1}\bar{\omega})\right)^{-1} \;=\; \mathrm{e}^{i\pi^{-1}\int(\partial\phi)(\bar{\partial}\phi) + (2\pi i)^{-1}\int\phi\mathcal{R}}$$
$$\cdot\;\left(\det_{\alpha,\beta}(\textstyle\int \mathrm{e}^{2\phi}\overline{\eta_{u\alpha}}\eta_{u\beta}) \,/\, \det_{\alpha,\beta}(\int\overline{\eta_{u\alpha}}\eta_{u\beta})\right)^{-1}\det\left(\bar{\partial}_u^*\bar{\partial}_u\right)^{-1} \tag{47}$$

where $\bar{\partial}_u \equiv \bar{\partial} + \pi u^t\tau_2^{-1}\bar{\omega}$ and $\eta_{u\alpha}$, $\alpha = 1,\ldots,\mathrm{genus}-1$, form a basis of the 01-forms in the kernel of $\bar{\partial}_u^*$. Using eq. (47), we may rewrite the partition function as

$$\tilde{\mathcal{Z}}^{\mathrm{bh}}(\tau,U) \;=\; C\left(\frac{\det'(-\bar{\partial}^*\bar{\partial})}{\mathrm{area}}\right)^{1/2}\det\tau_2^{-1}\int \mathrm{e}^{-\pi\kappa(\bar{U}-\bar{u})^t\tau_2^{-1}(U-u) + (2\pi i)^{-1}(\kappa-2)\int(\partial\phi)(\bar{\partial}\phi)}$$
$$\cdot\;\mathrm{e}^{(2\pi i)^{-1}\int\phi\mathcal{R} - \int\bar{\eta}_u\exp(2\phi)\eta_u}\det\left(\bar{\partial}_u^*\bar{\partial}_u\right)^{-1}\delta(\phi(\xi_0))\,D\phi\,d\eta_u\,du \tag{48}$$

where the gaussian integral over $\eta_u \in \ker \partial_u^*$ was used to express the $\eta_{u\alpha}$ determinants. The expression is obviously similar to the Liouville partition function although the real relation between two theories lies probably deeper. In any way, we expect the ϕ and η integrals to be finite and to lead to an expression regular in u except for the contribution of $\det\left(\bar{\partial}_u^*\bar{\partial}_u\right)^{-1}$ which around $u=0$ behaves as $|u|^{-2}$ which is integrable for genus >1 and diverges logarithmically for genus 1 ($\eta_{u\alpha}$ may be chosen regular in u around $u=0$). This singularity is repeated around other points of $\mathbf{Z}+\tau\mathbf{Z}$. Thus, similarly as for the Liouville theory coupled to free bosonic field, see [41],[42], we expect the partition functions at higher genera to be convergent reflecting the finite dimension of the region in the target space relevant for the stringy interaction.

4.7 Green functions

Since the coset theory is an instance of a gauge theory, its Green functions should be given by functional integral with gauge invariant insertions. Examples of gauge invariant fields are $f_{\rho,m_l,m_r}(v(\xi))$ of eq. (3.33) with $m_l=-m_r\equiv m$ whose conformal weights are

$$\Delta_{\rho,m}=\bar{\Delta}_{\rho,m}=\tfrac{1+\rho^2}{4(\kappa-2)}+\tfrac{m^2}{\kappa}\,. \tag{49}$$

If we instead used $f_{\rho,m_l,m_r}(\phi(\xi),v(\xi))$ with $m_l+m_r\neq 0$ as local fields, we could still maintain local gauge invariance by adding compensating currents, i.e. by considering insertions

$$I(hh^\dagger,\tfrac{1}{2i}A)\;=\;\prod_\alpha f_{\rho_\alpha,m_{l\alpha},m_{r\alpha}}(\phi(\xi_\alpha),v(\xi_\alpha))\;\mathrm{e}^{-\int\limits_c A} \tag{50}$$

where c is a chain such that $\delta c=\sum\limits_\alpha(m_{l\alpha}+m_{r\alpha})\xi_\alpha$. In the planar case, the functional integral over the gauge field may be easily done upon parametrization $A=d\mu+{*d\nu}$. The integral over μ drops out because of gauge invariance and the integral over ν gives expectation value of chiral vertex operators

$$\int \mathrm{e}^{\,i\int\limits_{c+c'}\partial\nu\,-\,i\int\limits_{c-c'}\bar{\partial}\nu\,-\,\pi^{-1}\kappa\int(\partial_z\nu)(\partial_{\bar{z}}\nu)\,d^2z}\;D\nu \tag{51}$$

where $\delta c'=\sum\limits_\alpha(m_{l\alpha}-m_{r\alpha})\xi_\alpha$ (compare [14] where similar calculation was done for the parafermions). Altogether, we obtain

$$\int I(hh^\dagger,\tfrac{1}{2i}A)\;\mathrm{e}^{\kappa S(hh^\dagger,\,(2i)^{-1}A)}\;D(hh^\dagger)\;DA$$

$$=\;\mathrm{const.}\;\prod_{\alpha\neq\alpha'}(\xi_\alpha-\xi_{\alpha'})^{m_{l\alpha}m_{l\alpha'}/\kappa}\,(\bar{\xi}_\alpha-\bar{\xi}_{\alpha'})^{m_{r\alpha}m_{r\alpha'}/\kappa}\;\int I(hh^\dagger,0)\;\mathrm{e}^{\kappa S(hh^\dagger)}\;D(hh^\dagger) \tag{52}$$

where the $\prod\limits_{\alpha\neq\alpha'}$ factors come from the (properly renormalized) free field integral (51). They modify the conformal dimensions of fields f_{ρ,m_l,m_r} of the H_3^+ WZW theory to

$$\Delta_{\rho,m_l}=\tfrac{1+\rho^2}{4(\kappa-2)}+\tfrac{m_l^2}{\kappa}\,,\qquad \bar{\Delta}_{\rho,m_r}=\tfrac{1+\rho^2}{4(\kappa-2)}+\tfrac{m_r^2}{\kappa}\,. \tag{53}$$

producing operators with imaginary spin and hence never local. It is possible, however, that correlations of fields coming from common eigenfunctions on H_3^+ of $\Delta, J^3, \bar{J}^3$ which do not correspond to the spectrum, for example for ω imaginary, may be given sense.

If in the left hand side of (52) we integrated out the A-field, we would obtain the black hole functional integral with insertions which for large values of $|v(\xi_\alpha)|$ take form

$$\prod_\alpha \left(|v(\xi_\alpha)|^{m_{l\alpha}+m_{r\alpha}} f_{p_\alpha,m_{l\alpha},m_{r\alpha}}(0,|v(\xi_\alpha|) \right) \mathrm{e}^{-i\int\limits_{c+c'} \partial \arg(v) + i \int\limits_{c-c'} \bar\partial \arg(v)} . \tag{54}$$

We recover then the chiral vertex operators of field $\arg(v)(\xi)$ which, for large $|v|$, becomes a compactified free field. If fields with real m_l+m_r existed, they would be mutually local for $m_l+m_r \in \kappa\mathbf{Z}$, as are their asymptotic versions. We shall return to the discussion of this possibility in the next section.

5. $SU(1,1)$ mod $U(1)$ COSET THEORY

5.1 Functional integral formulation

The original proposal [18] for the conformal sigma model with 2D black hole target was based on a coset construction starting with $SU(1,1) \cong SL(2,\mathbf{R})$ WZW model. The parametrization

$$g = \begin{pmatrix} \mathrm{e}^{i\psi}(1+|v|^2)^{1/2} & v \\ \bar v & \mathrm{e}^{-i\psi}(1+|v|^2)^{1/2} \end{pmatrix} \tag{1}$$

where ψ is in $\mathbf{R}/(2\pi\mathbf{Z})$ and v is complex gives global coordinates on $SU(1,1)$. Comparing to parametrization (3.1) of positive elements in $SL(2,\mathbf{C})$, we see that it passes to the present one by simple substitution $\phi \mapsto i\psi$. Consequently, for the WZW action with the $U(1) \subset SU(1,1)$ gauged asymmetrically (i.e. with the axial $U(1)$ gauge), we obtain from eq. (3.4)

$$\begin{aligned} S(g,\tfrac{1}{2i}A) &= -\tfrac{1}{\pi}\int [(i\partial_z\tilde\psi + A_z)(i\partial_{\bar z}\tilde\psi + A_{\bar z}) \\ &+ (\partial_z + i\partial_z\tilde\psi + A_z)\bar v\,(\partial_{\bar z} + i\partial_{\bar z}\tilde\psi + A_{\bar z})v]\, d^2z \,. \end{aligned} \tag{2}$$

where $\tilde\psi \equiv \psi + \frac{1}{2}i\log(1+|v|^2)$. The axial gauge invariance is

$$S(\mathrm{e}^{i\lambda\sigma^3} g\, \mathrm{e}^{i\lambda\sigma^3}, \tfrac{1}{2i}(A - 2id\lambda)) \,. \tag{3}$$

The euclidean action $\pm\kappa S(g)$ for the $SU(1,1)$ WZW theory is not bounded below. For the minus sign (and κ positive) this is due to the $\tilde\psi$-field contribution. As a result, the stable euclidean picture is missing for this theory. In the coset functional integral

$$\int - \mathrm{e}^{\kappa S(g,(2i)^{-1}A)}\, Dg\, DA \,,$$

however, the $\tilde\psi$-field may be gauged out and absorbed by a translation of A. If A is taken real then the A integral is stable and the translation of A is complex (the axial gauge invariance requires imaginary A). In this case, moreover, after the translation, we recover the same integral as before for the $SU(2,\mathbf{C})/SU(2)$ mod $\mathbf{R}$ coset theory. It seems that the two coset theories coincide[8]. On the quantum-mechanical level, the equivalence of both approaches may be seen clearly.

[8]this is the point on which the present author's opinion has wavered most and might continue to do so with the progress in the understanding of both theories

5.2 Particle on SU(1,1)

The classical mini-space system which corresponds to the 2D WZW theory with target $SU(1,1)$ is the geodesic motion in the invariant metric on $SU(1,1)$ of signature, say, $(-,+,+)$. We may quantize it taking $L^2(SU(1,1))$ with the Haar measure (equal $d\psi\, d^2v$ in parametrization (1)) as the space of states in which $SU(1,1)_{\text{left}} \times SU(1,1)_{\text{right}}$ acts unitarily. Infinitesimally, we get the action of $sl(2,\mathbf{C}) \oplus sl(2,\mathbf{C})$ generated by J^a's and $\bar{J}^a$'s given by the same formulae as in the case of $L^2(H_3^+)$ except for the substitution $\phi \mapsto i\psi$. The hermiticity relations change, however, and we obtain

$$\begin{aligned} J^{a*} &= -J^a\ , \quad \bar{J}^{a*} = -\bar{J}^a \quad \text{for } a = 1,2\ , \\ J^{3*} &= J^3\ , \quad \bar{J}^{3*} = \bar{J}^3 \quad . \end{aligned} \tag{4}$$

Also $-\Delta \equiv -\vec{J}^2 = -\vec{\bar{J}}^2$ is no more bounded below. It is again given explicitly by eq. (3.12) with ∂_ϕ^2 replaced by $-\partial_\psi^2$. In fact

$$L^2(SU(1,1)) \cong \int_{\substack{\rho>0\\ \epsilon=0,1/2}}^{\oplus} \mathcal{D}_{\rho,\epsilon} \otimes \bar{\mathcal{D}}_{\rho,\epsilon}\, d\nu(\epsilon,\sigma) \ \bigoplus \bigoplus_{\substack{j=-1,-3/2,\dots\\ \pm}} \mathcal{D}_j^{\pm} \otimes \bar{\mathcal{D}}_j^{\pm}\ . \tag{5}$$

$\mathcal{D}_{\sigma,\epsilon}$ carry unitary irreducible representations of $SU(1,1)$ of the principal continuous series which may be realized in the space of sections of a spin bundle on the circle ($SU(1,1)$ acts naturally on S^1, ϵ corresponds to two choices of the spin structure). The eigenvalue of $\vec{J}^2$ on $\mathcal{D}_{\rho,\epsilon}$ is $-\frac{1}{4}(1+\rho^2)$. Spaces $\mathcal{D}_{-j}^{\pm}$ carry the lowest- (highest)-weight representations of $sl(2,\mathbf{C})$ of spin j. They give the discrete series of unitary, irreducible representations of $SU(1,1)$ with eigenvalue of $\vec{J}^2$ equal to $j(j+1)$ which is ≥ 0. If, instead of $SU(1,1)$, we considered its simply-connected covering $\widetilde{SU(1,1)}$ (where ψ takes values in the non-compactified real line), the direct sums in decomposition (5) over ϵ and j would be replaced by direct integrals over $0 \leq \epsilon < 1$ and $j < -1/2$. Since $-\Delta$ plays the role of Hamiltonian, the energy is not bounded below (nor above). This problem with stability renders the above quantization physically not very satisfactory. Indeed, the way we proceeded here is not the one used for example to quantize a particle in Minkowski space where one recovers satisfactory solution of the stability problem passing to the second-quantized level. Finding a stable quantization of the particle on $SU(1,1)$ or, more importantly, of the $SU(1,1)$ WZW field theory remains an open and seemingly very interesting problem[9]. Here, however, we shall be interested only in the coset $SU(1,1)$ mod $U(1)$ theory where coupling to the gauge field removes the unstable $\tilde{\psi}$ field. On the quantum-mechanical level, the gauge condition $J^3 + \bar{J}^3 = i\partial_\psi = 0$, cuts out from $L^2(SU(1,1))$ the contribution of the discrete series (and more) making $-\Delta$ positive. Besides,

$$L^2(SU(1,1))|_{J^3+\bar{J}^3=0} \cong L^2(d^2v) \cong L^2(H_3^+)|_{J^3+\bar{J}^3=0}$$

in a natural way and this isomorphism preserves (restrictions of) Δ, J^3 and $\bar{J}^3$. This proves on the mini-space level the identity of the coset theories $SU(1,1)$ mod $U(1)$ and $SL(2,\mathbf{C})/SU(2)$ mod $\mathbf{R}$. The generalized eigenfunctions f_{ρ,m_l,m_r} of Δ, J^3, $\bar{J}^3$ on $SU(1,1)$, corresponding to eigenvalues $-\frac{1}{4}(1+\rho^2)$, m_l, m_r with $m_l \pm m_r \in \mathbf{Z}$, are given by Jacobi functions [38]. For $m_l + m_r = 0$ they are independent of ψ and, although given by different expressions, coincide with similar eigenfunctions on H_3^+. For example, from the harmonic analysis on $SU(1,1)$, we obtain

$$f_{0,0,0}(v) = \pi \int_0^{2\pi} (1 + 2|v|^2 + 2v(1+|v|^2)^{1/2}\cos\theta)^{-1/2}\, d\theta \tag{6}$$

[9] we thank G. Gibbons for attracting our attention to it

which should be compared with eq. (3.34). For both $m_l + m_r$ equal and different from zero, eigenfunctions f_{ρ,m_l,m_r} seem to generate primary fields of dimensions given by eq. (4.53) (for $m_l + m_r \neq 0$, they should be dressed with line integrals of the gauge field, like in (4.50)). If $m_l + m_r \in \kappa\mathbf{Z}$, the corresponding fields are mutually local.

5.3 Space of states, unitarity, duality, problems

On the level of 2D field theories, neither $SL(2,\mathbf{C})/SU(2)$ mod $\mathbf{R}$ nor $SU(1,1)$ mod $U(1)$ theory has been shown to exist, least solved completely, so comparison is more difficult. The computation of the partition function in the first case did not require complex rotations or shifts of the fields so it seems more trustable. Nevertheless, we have seen that the interpretation of the excited contributions to it required analytic continuation of the heat kernels on the eigensubspaces of J^3, $\bar{J}^3$ in $L^2(H_3^+)$ to imaginary eigenvalues ω of $\frac{1}{i}(J^3 + \bar{J}^3) = i\partial_\phi$. But this should be given by the heat kernels in the eigenspaces of J^3, $\bar{J}^3$ in $L^2(SU(1,1))$, or more generally in $L^2(\widetilde{SU(1,1)})$, obtained by the substitution $\phi \mapsto i\psi$. It is then possible that the partition function becomes a trace over the gauge-invariant states of the $SU(1,1)$ WZW theory. Superficially, the $U(1)$ coset of the latter has the same problem as the parafermionic model discussed in Sec. 4.4: the ungauged (vector) $U(1)$ symmetry has global anomaly which seems to break $U(1)$ to $\mathbf{Z}_k$. Here, this is a spurious problem, however: if we start from the $\widetilde{SU(1,1)}$ WZW theory rather than from the $SU(1,1)$ one, the coset theory is the same but the complete $U(1)$ symmetry is present. The space of states of the $\widetilde{SU(1,1)}$ WZW theory should be a subspace of

$$\int^{\oplus}_{\substack{\rho>0\\0\leq\epsilon<1}} \hat{\mathcal{D}}_{\rho,\epsilon} \otimes \hat{\bar{\mathcal{D}}}_{\rho,\epsilon}\, d\nu(\epsilon,\sigma) \;\bigoplus\; \int^{\oplus}_{\substack{j<-1/2\\\pm}} \hat{\mathcal{D}}_j^{\pm} \otimes \hat{\bar{\mathcal{D}}}_j^{\pm} \tag{7}$$

where "$\hat{}$" denotes the representation space of the Kac-Moody algebra $\hat{sl}(2,\mathbf{C})$ induced (in the sense of Sec. 3.3) from the representations of $\widetilde{SU(1,1)}$. What exactly should be the subspace taken does not seem to be clear yet. A possibility is the appearance of the fusion rule $-\frac{1}{2}(\kappa-1) < j$ in the discrete series, analogous to the rule $j \leq k/2$ of the $SU(2)$ WZW theory. Spaces $\hat{\mathcal{D}}$ may be provided with the hermitian form for which $J_n^{a*} = -J_{-n}^a$ for $a = 1,2$ and $J_n^{3*} = J_{-n}^3$ (this agrees at level zero with the scalar product induced from $L^2(SU(1,1))$). The encouraging sign is the important result of Dixon-Lykken-Peskin [12] (see also [43]) who proved that the gauge conditions $J_n^3 = 0$, $n > 0$, cut out, under the restriction $-\frac{1}{2}\kappa \leq j$ on the discrete series, the negative norm states from the induced representations. Notice, that the latter condition disposes of the representations with negative eigenvalues of L_0^{cs}. Their absence should then be assured by stability if the coincidence with the explicitly stable H_3^+ mod $\mathbf{R}$ model really takes place. In that case, the $SU(1,1)$ mod $U(1)$ approach should allow to show the unitarity of the euclidean black hole CFT. Moreover, we should be able to assemble the calculated partition functions from the characters of the induced representations $\hat{\mathcal{D}}$. This is not simple even on the quantum mechanical level where we know that it works.

The gauge condition $J_0^3 + \bar{J}_0^3 = 0$ leaves us with spin-less, $U(1)$-charge zero sector of the theory. The functional integral for the partition function, as in any gauge theory, should be given by the trace over this subspace of states, as is also clearly indicated by the U dependence of the result (4.10). The primary fields f_{ρ,m_l,m_r} with $m_l + m_r = 0$ correspond to vectors in this sector. On the other hand, gauge conditions $J_0^3 + \bar{J}_0^3 = l\kappa$ should give for $0 \neq l \in \mathbf{Z}$ sectors with the spin and the $U(1)$ charge different from zero.

Fields f_{ρ,m_l,m_r} with $m_l + m_r = l\kappa$ should correspond to states in these sectors. From the point of view of the asymptotic free field with the cylindrical part of the cigar as the target, these are the winding sectors, see formula (4.54)[10]. The partition functions corresponding to the winding sectors can be also computed, essentially by inserting a Polyakov line with charge $l\kappa$ into the functional integral. We plan to return to these issues elsewhere.

Another open problem in the black hole CFT is a relation between the coset models $SU(1,1)$ mod $U(1)$ obtained by gauging the axial and the vector $U(1)$ subgroup. The vector theory has a more serious stability problem than the axial one since the vector gauging does not seem to remove completely the unbounded below modes. On a rather formal level one can argue that both theories have the same spectrum of mutually local operators [20]-[22],[44]. It was expected that they give the same CFT. The vector coset results in a sigma model with singular metric on the target. In the asymptotic region, the target also looks like a half-cylinder and the identity of the models would become there that of free fields compactified on dual radia [45]. We have not been able, however, to stabilize the functional integral for the vector theory in a sensible way to show that it has the same partition function as the axial coset. The situation should be contrasted with the case of parafermions. There, as we have seen in Sec. 4.4, both gaugings give the same theory, in fact already on the mini-space level. In particular, both partition functions coincide. The duality between the two $U(1)$ cosets of the $SU(1,1)$ WZW theory requires, in our opinion, further study. It may be that the vector description may be maintained only in the asymptotically flat region. The issue is important for understanding whether the coupling to dynamical gravity washes out the singularity at $v^+v^- = 1$ of the classical Minkowskian metric (4.5) interchanged by the duality with the non-singular horizon $v^+v^- = 0$, see [20]. Even less clear is what sense we can make of the sigma model which Minkowskian 2D black hole as the target which formally comes from gauging non-compact subgroup in $SU(1,1)$ theory [18] and how all these theories fit together. We clearly touch here on the relation between the stability and unitarity of the CFT's and the signature of the effective targets. If a progress can be made in understanding such issues fundamental for quantum gravity, the effort invested in studying relatively simple non-rational theories may pay back.

References

[1] Moore, G., Seiberg, N.: *Classical and quantum conformal field theory.* Commun. Math. Phys. **123**, 177-254 (1989).

[2] Witten, E.: *Non-abelian bosonization in two dimensions.* Commun. Math. Phys. **92**, 455-472 (1984).

[3] Knizhnik, V., Zamolodchikov, A. B.: *Current algebra and Wess-Zumino model in two dimensions.* Nucl. Phys. **B 247**, 83-103 (1984).

[4] Gepner, D., Witten, E.: *String theory on group manifolds.* Nucl. Phys. **B 278**, 493-549 (1986).

[10]one should not confuse these sectors with the infinite volume superselection sectors obtained by sending some of the charges "behind the moon"; they are rather descendents of the charge sectors corresponding to infinitely heavy external charges - we thank E. Seiler for correcting some of the author's original misconceptions about this point

[5] Goddard, P., Kent, A., Olive, D.: *Unitary representations of the Virasoro and super-Virasoro algebras.* Commun. Math. Phys. **103**, 105-119 (1986).

[6] Balog, J., Fehér, L, Forgács, P., O'Raifeartaigh, L.: *Toda theory and W-algebra from a gauged WZNW point of view.* Ann. Phys. **203**, 76-136 (1990).

[7] Gawędzki, K.: *Quadrature of conformal field theories.* Nucl. Phys. **B 328**, 733-752 (1989).

[8] Polyakov, A. M.: *Quantum geometry of bosonic strings.* Phys. Lett. **103 B**, 207-210 (1981).

[9] Seiberg, N.: *Notes on quantum Liouville theory and quantum gravity,* RU-90-29 preprint.

[10] Pressley, A., Segal, G.: *Loop groups.* Oxford: Clarendon Press 1986.

[11] Jakobsen, H. P., Kac, V. G.: *A new class of unitarizable highest weight representations of infinite dimensional Lie algebras.* In: Non-linear equations in field theory, Lecture Notes in Physics 226, Berlin: Springer, 1985, pp. 1-20.

[12] Dixon, L. J., Lykken, J., Peskin, M. E.: *$N = 2$ superconformal symmetry and $SO(2,1)$ current algebra.* Nuclear Physics **B 325**, 329-355 (1989).

[13] Gawędzki, K., Kupiainen, A.: *G/H conformal field theory from gauged WZW model,* Phys. Lett. **B 215**, 119-123 (1988).

[14] Gawędzki, K., Kupiainen, A: *Coset construction from functional integral.* Nucl. Phys. **B 320**, 625-668 (1989).

[15] Haba, Z: *Correlation functions of sigma-fields with values in hyperbolic space.* Int. J. Mod. Phys. A4 267-286 (1989).

[16] Witten, E.: *Quantum field theory and Jones polynomial.* Commun. Math. Phys. **121**, 351-399 (1989).

[17] Elitzur, S., Moore, G., Schwimmer, A., Seiberg, N.: *Remarks on the canonical quantization of the Chern-Simons-Witten theory.* Nucl. Phys. **B 326**, 104-134 (1989).

[18] Witten, E.: *String theory and black holes.* Phys. Rev. **D 44**, 314-324 (1991).

[19] Mandal, G., Sengupta, A. M., Wadia, S. R.: *Classical solutions of 2-dimensional string theory.* Lett. Mod. Phys. **A 6**, 1685-1692 (1991).

[20] Dijkgraaf, R., Verlinde, E., Verlinde, H.: *String propagation in black hole geometry,* PUPT-1252 and IASSNS-HEP-91/22 preprint.

[21] Giveon, A.: *Target space duality and stringy black holes.* LBL-30671 preprint.

[22] Kiritsis, E. B.: *Duality in gauged WZW models.* LBL-30747 and UCB-PTH-91/21 preprint.

[23] Bars, I.: *String propagation on black holes.* USC-91/HEP-B3 preprint.

[24] Martinec, E. J., Shatashvili, S. L.: *Black hole physics and Liouville theory.* EFI-91-22 preprint.

[25] Spiegelglas, M.: *String winding in a black hole geometry.* Technion-PH-23-91 preprint.

[26] Bershadsky, M., Kutasov, D.: *Comment on gauged WZW theory.* Phys. Lett. **B 266**, 345-352 (1991).

[27] de Alwis, S. P., Lykken, J.: *2d gravity and the black hole solution in 2d critical string theory.* FERMILAB-PUB-91/198-T and COLO-HEP-258 preprint.

[28] Distler, J., Nelson, Ph.: *New discrete states of strings near a black hole*, UPR-0462T and PUPT-1262 preprint.

[29] Tseytlin, A. A.: *On the form of the "black hole" solution in D=2 string theory.* Phys. Lett. **B 268**, 175-178 (1991).

[30] Gawędzki, K.: *Wess-Zumino-Witten conformal field theory.* In: Constructive Quantum Field Theory II, eds. Wightman, A. S., Velo, G., New York: Plenum 1990, pp. 89-120.

[31] Gawędzki, K.: *Constructive conformal field theory.* In: Functional Integration, Geometry and Strings, ed. Haba, Z., Sobczyk, J., Basel: Birkhäuser 1989, pp. 277-302.

[32] Witten, E.: *Topological quantum field theory.* Commun. Math. Phys. **117**, 353-386 (1988).

[33] Verlinde, H.: *Conformal field theory, two-dimensional gravity and quantization of Teichmüller space.* Nucl. Phys. **B 337**, 652-680 (1990).

[34] Verlinde, E.: *Fusion rules and modular transformations in 2d conformal field theory.* Nucl. Phys. **B 300** [FS 22], 360-376 (1988).

[35] Falceto, F., Gawędzki, K., Kupiainen, A.: *Scalar product of current blocks in WZW theory.* Phys. Lett. **B 260**, 101-108 (1991).

[36] Ray, D. B., Singer, I. M.: *Analytic torsion for complex manifolds*, Ann. Math. **98**, 154-177 (1973).

[37] Gelfand, I. M., Graev, M. I., Vilenkin, N. Ya.: Generalized Functions, Vol. 5, New York: Academic Press, 1966.

[38] Vilenkin, N. Ya.: Fonctions Spéciales et Théorie de la Représentation des Groupes, Paris: Dunod, 1969.

[39] Gawędzki, K.: *Geometry of Wess-Zumino-Witten models of conformal field theory.* In: Recent Advances in Field Theory, eds. Binétruy, P., Girardi, G., Sorba, P., Nucl. Phys. **B** (Proc. Suppl.) **18 B**, 1990, pp. 78-91.

[40] Gepner, D., Qiu, Z.: *Modular invariant partition functions for parafermionic field theories*, Nucl. Phys. B 285 [FS 19], 423-453 (1987).

[41] David, F.: *Conformal field theories coupled to 2-D gravity in the conformal gauge.* Mod. Phys. Lett. **A 3**, 1651-1656 (1988).

[42] Distler, J., Kawai, H.: *Conformal field theory and 2-D quantum gravity or who's afraid of Joseph Liouville.* Nucl. Phys. **B 321**, 509-527 (1989).

[43] Lian, B. H., Zuckerman, G. J.: *BRST cohomology and highest weight vectors. I.* Commun. Math. Phys. **135**, 547-580 (1991).

[44] Roček, M., Verlinde, E.: *Duality, quotients and currents.* ITP-SB-91-53 and IASSNS-HEP-91/68 preprint.

[45] Kadanoff, L. P.: *Lattice Coulomb gas representations of two-dimensional problems.* J. Phys. **A 11**, 1399-1417 (1978).

S-MATRIX THEORY FOR BLACK HOLES

G. 't Hooft

Institute for Theoretical Physics
Princetonplein 5, P.O. Box 80.006
3508 TA Utrecht, The Netherlands

ABSTRACT

We explain the principles of the laws of physics that we believe to be applicable for the quantum theory of black holes. In particular, black hole formation and evolution should be described in terms of a scattering matrix. This way black holes at the Planck scale become indistinguishable from other particles. This *S*-matrix can be derived from known laws of physics. Arguments are put forward in favor of a discrete algebra generating the Hilbert space of a black hole with its surrounding space-time including surrounding particles.

1. INTRODUCTION

There is quite a bit of controversy (and confusion) regarding the nature of physical law governing a black hole. Some of the difficulties have their origin in the deceptively clean picture given by the "classical" (here this means "non-quantum mechanical") solutions of Einstein's equations of gravity in the case of gravitational collapse. The metric tensor describing the fabric of space-time appears to be smooth and well-behaved in the vicinity of a region we call the "horizon", a surface beyond which there are space-time points from which no information can reach the outside world. It seems that one should be able to apply standard techniques from particle theory here to derive what a distant observer can perceive and, naturally, this exercise has been done[1].

There is one innocent-looking assumption that most practitioners then make. One observes that clouds of particles may venture into the "forbidden region", from which they can no longer escape or even emit any signal towards the outside world, and so one *assumes* that the corresponding states in Hilbert space may be treated the way one always does in quantum mechanics: the unseen modes are averaged over. Operators describing observations in the outside world are assumed to be diagonal in the sector of Hilbert space that is not seen, and hence in all computations one is obliged to sum over all unseen modes.

The immediate consequence of this practice is that the outside world alone is not anymore described by a single wave function but by a density matrix[2]. Even if one starts with a "pure" wave function, sooner or later one finds the system to be in a mixed state. It is as if part

New Symmetry Principles in Quantum Field Theory, Edited by
J. Fröhlich et al., Plenum Press, New York, 1992

of the wave functions "disappeared into the wormhole"; information escaped, as if the system were linked to a heat bath.

With large black holes one can perform Gedanken experiments, and consider observers who move semiclassically in the neighborhood of a black hole horizon. If we attach some sense of "reality" to these observers the correctness of the above assumptions seems to be an inescapable conclusion.

But what if the hole is small, so that classical observers are too bulky to enter? Or let us ask a question that is probably equivalent to this: suppose one keeps track of *all* possible states a black hole can be in, is it then still impossible to describe the hole in terms of pure quantum states alone? Will the very tiny black holes evolve according to conventional evolution equations in quantum physics or is the loss of information a fundamental new feature, even for them?

There is a big problem with any theory in which the loss of "quantum information" is accepted as a fundamental item. This is the fact that all effective laws become fuzzy. It is not difficult to construct an example of a theory in which pure states evolve into mixed states. Consider a system with a Hamiltonian that depends on a free physical parameter α (for instance the fine structure constant). A state $|\psi\rangle_0$ at $t=0$ evolves into the state

$$|\psi\rangle_t = e^{-iH(\alpha)t}|\psi\rangle_0 \tag{1.1}$$

at time t . The expectation value for an operator $\mathcal{O}$ evolves into

$$\langle\mathcal{O}\rangle_{t,\alpha} = {}_0\langle\psi|e^{iH(\alpha)t}\mathcal{O}\, e^{-iH(\alpha)t}|\psi\rangle_0 \ , \tag{1.2}$$

and from the ψ dependence one can recover the information that the system remained in a pure quantum state. But now assume that there is an *uncertainty* in α . We only know the first few decimal places. There is a distribution of values for α , each with a probability $P(\alpha)$. Our theory now predicts for the "expectation value" of $\mathcal{O}$:

$$\langle\mathcal{O}\rangle_t = \int d\alpha\, \langle\mathcal{O}\rangle_{t,\alpha}\, P(\alpha) \equiv \mathrm{Tr}\, \rho(t)\mathcal{O} \ ; \tag{1.3}$$

$$\rho(t) = \int d\alpha\, P(\alpha)\, e^{-iH(\alpha)t}|\psi\rangle_0\, {}_0\langle\psi|e^{iH(\alpha)t} \ . \tag{1.4}$$

This is an impure density matrix, of the kind one obtains in doing calculations with black holes. The outcome of a by now standard calculation is a thermal distribution of outgoing particles. A thermal distribution is always a mixed state.

In our example we clearly see what the remedy is. the extra uncertainty had nothing to do with quantum mechanics; the Hamiltonian was not yet known because of our incomplete knowledge of the laws of physics (in this case the value of α). By doing extra experiments or by working harder on the theory we can establish a more precise value for α , and thus obtain a more precise prediction for $\langle\mathcal{O}\rangle_t$.

Returning with this wisdom to the black hole, what knowledge was incomplete? Here I think one has a situation that is common to all macroscopic systems: because of the large number of quantum mechanical states it was hopelessly difficult to follow the evolution of just one such state precisely. One was forced to apply thermodynamics. The outcome of our calculations with black holes got the form of thermodynamic expressions because of the impossibility, in practice, to follow in detail the evolution of any particular quantum state.

But this does mean that our basic understanding of black holes at present is incomplete. In a statistical system such as a vessel containing an ideal gas, we have *in principle* a quantum theory that is precise enough to study pure quantum states. In particular, if we dilute the gas so much that a single atom remains, the thermodynamic description will no longer be correct, and we must use the real quantum theory. Similarly, if we want to understand how a black hole behaves when it reaches the Planck mass, we expect the thermodynamic expressions to break down.

The importance of a good quantum mechanical description is that it would enable us to link black holes with ordinary particles. The Planck region may well be populated by a lot of different types of fundamental particles. their "high energy limit" will probably consist of particles small enough and heavy enough to possess a horizon and thus be indistinguishable from black holes. What we want is a consistent theory that covers all of this region. If we had a "conventional" Schrödinger equation in this region, it would be relatively straightforward (at least conceptually) to extrapolate to large distance scales using renormalization group techniques, and recover the "standard model" (or more!)

There have been many proposals concerning the nature of our physical world near the Planck length. we have seen "supergravity", "string theory", "heterotic strings", et cetera. My problem with these ideas is that they seem to be ad hoc. The models are "postulated" and then afterwards the authors try to argue why things have to be this way (basically the argument is that the new model is "more beautiful" than anything else known).

It would be a lot safer if we could *derive* the only possible correct setting of variables and forces, directly from the presently established laws of physics. In these lectures we will argue that it is possible to do this, or at least to make a good start, by doing Gedanken experiments with black holes. The reason why black holes should be used as a starting point in a theory of elementary particles is that *anything* that is tiny enough and heavy enough to be considered an entry in the spectrum of ultra heavy elementary particles (beyond the Planck mass), must be essentially a black hole.

Black holes are defined as solutions of the classical, i.e. unquantized, Einstein equations of General Relativity[3]. This implies that we only know how to describe them reliably when they are considerably bigger than the Planck length and heavier than the Planck mass. What was discovered by Hawking[1] in 1975 is that these objects radiate and therefore must decrease in size. It is obvious that they will sooner or later enter the domain that we presently do not understand.

Curiously, it is not easy to see why Hawking's derivation of the thermal black hole radiation would not be exactly correct. Even in a functional integral expression for this calculation one might still expect wormhole configurations through which quantum information leaks towards a mystical "other universe"[4]. We will now decide to be merciless: topologically non-trivial space-times are forbidden (until further notice) so that, at least at the microscopic level, pure quantum mechanics can be restored. More precisely, what we require is first of all some quantum mechanically pure evolution operator, and secondly that this operator be consistent with all we know of large scale physics, in particular general relativity.

At first sight these requirements are in conflict with each other. General relativity predicts unequivocally that gravitational collapse is possible, and this produces a horizon with all its difficulties. However, we claim that a pure quantum prediction that naturally blends into thermodynamic behavior in the large scale limit is not at all impossible, but it is true that the requirement for this to happen is extremely restrictive. Combining it with all we already know about large scale physics may well yield an unambiguous theory. Anyway, we know for sure that the amendments needed at the horizon all refer to Planck scale physics. As long as this physics is not completely understood it will also be impossible to refute our theories on the ground of inconsistencies with known physics.

So this is our program. We *assume* that, as for all spatially confined systems, there exists such a thing as a "scattering matrix". One then tries to reconcile this scattering matrix with the laws of physics already known. We will find that this scattering matrix, to some extent, can be derived. More precisely: *the exact quantum behavior at large distance scales* (the distance scales reached in present particle experiments) *can be derived uniquely*.

The problem is an apparent acausality. If we apply linearized quantum field theory in the black hole background it seems ununderstandable how information that is thrown into the black hole can reemerge as information in the outgoing states. This is because the outgoing radiation originates at $t = -\infty$ and the ingoing matter proceeds until $t = +\infty$, so the information had to go backwards in time. We simply claim that precisely for this reason linearized quantum field theory is inappropriate here. One *must* take gravitational (if not other) interactions between in- and outgoing matter into account. One way to interpret what happens then is to assume that there is a fundamental *symmetry principle*, because matter inside the horizon is unobservable. One can then perform a transformation that transforms away the singularity at $t{=}+\infty$, and produces one at $t{=}-\infty$.

It is of crucial importance to note that what we are deriving is not only the (quantum) behavior of the black hole itself. It is the entire system, black hole *plus* all surrounding particles, that we are talking about. Using our (assumed) knowledge of physics at large distance scales we derive the properties of the black hole *and all other forms of matter* at energies larger than the Planck energy.

In ordinary quantum field systems behavior at small distance, or equivalently, at high energies, determines the behavior at large distances and low energies. In the present case the interdependence goes both ways, or, in other words, the whole construction will be over-determined. We expect stringent constraints of consistency, which, one might hope, may lead to a single unique theory. The point is that the symmetry principle just mentioned affects matter in an essential way, and thus may perhaps continue to be of relevance at the low energy domain.

This is the motivation of this work. It may lead to "the unique theory". Even though our work is far from finished, we will be able to show that there will be a remarkable role for the old string theory[5]. The mathematical expressions we derive are so similar to those of string theory that perhaps some of its results will apply without any change. But both the physical interpretation and the derivations will be very different. As a consequence, the mathematics is not identical. One important difference is the string constant (determining the masses of the excitations), which in our case turns out to be imaginary[6].

In the usual string theory one uses the obvious requirements of unitarity and causality to derive that the string is governed by a local Lagrangean on the string world sheet. To derive similar requirements for the strings born from black holes is far from easy. This is presently what is holding us back from considerations such as tachyon elimination and anomaly cancellation that so successfully seem to have given us the superstring scenario. What we advertise is a careful though slow process establishing the correct demands for a full black hole/string theory. If successful, one will know exactly the rules of the game and the ways how to select good from false scenarios and models.

2. QUANTUM HAIR

Classical black holes are characterized by exactly three parameters[3]:

the *mass* M , the *angular momentum* L , and the *electric charge* Q . If magnetic monopoles exist in nature then there will be a fourth parameter, namely magnetic charge Q_m , and if besides electromagnetism there are other long range $U(1)$ gauge fields then also their charges correspond to parameters for the black hole.

However, the existence of long range $U(1)$ gauge fields other than electromagnetism seems to be rather unlikely. Then, since L , Q (and Q_m) are all quantized, the number of different values they can take is limited, and indeed one can argue convincingly (more about this in Ref[6]) that the black hole can be in much more different quantum states than the ones labeled by L and Q (and Q_m), or in other words, the mass M must be a function of much more variables than these quantum numbers alone.

An interesting attempt to formulate new quantum numbers for black holes was initiated by Preskill, Krauss, Wilczek and others[7]. They took as a model field theory a $U(1)$ gauge theory in which the local symmetry undergoes a Higgs mechanism *via* a Higgs field with charge Ne . In addition one postulates the presence of particles with charge e . In such a theory there exist vortices, much like the Abrikosov vortex in a super conductor. These vortices can be constructed as classical solutions with cylindrical symmetry, at which the Higgs field makes one full rotation if one follows it around the vortex.

The behavior near the vortex of particles whose charge is only e is more complicated. One finds that because of the magnetic flux in the Abrikosov vortex the fields of these particles undergo a phase rotation when they flow around the vortex, in such a way that an Aharonov-Bohm effect is seen. The Aharonov-Bohm phase is $2\pi/N$, or, if we take a particle with charge ne , this phase will be $2\pi n/N$.

The importance of this Aharonov-Bohm phase is that it will be detectable for any charged particle, at any distance from the vortex, in such a way that we will detect its charge *modulo* N . This is surprising because *there is no long range gauge field present*!

An observer who can only detect large scale phenomena may not be able to uncover the chemical composition of the particle, but he can determine its charge modulo N . All he needs is a vortex, which to him will look just like a Nambu-Goto string.

Even if a particle were absorbed by a black hole, its electric charge would still reveal itself. Thus, charge modulo N is a quantum number that will survive even for black holes. It must be a strictly conserved charge.

One can then formalize the argument using only strings and charges modulo N , without ever referring to the original gauge field. Then there may exist many kinds of strings/vortices, so that the black hole may have a rich spectrum of these pseudo-invisible but absolutely conserved charges.

Will this argument allow us to specify all quantum numbers for a black hole? There are several reasons to doubt this. One is that an extremely large number of different kinds of strings must be postulated, which seems to be a substantial departure from the Standard Model at large distance scales.

Secondly, it is not at all obvious that it will be possible to do Aharonov-Bohm experiments with black holes. One then has to assume *first* that black holes indeed occur in well-defined quantum states, just like atoms and molecules. So this argument that black holes have quantum hair is rather circular.

In my lectures there is no need for the mechanism advertised by Preskill et al. It is neither necessary nor likely that all quantum states can be distinguished by means of some conserved quantum number(s). In my other lectures I use just the assumption that quantum states exist, and nothing else. No large-scale strings are needed.

3. DECAY INTO SMALL BLACK HOLES

Due to Hawking radiation the black hole looses energy, hence also mass. The intensity of the radiation will be proportional to T^4 , where T is the temperature, and the total area of the horizon, which for the Schwarzschild black hole is $4\pi R^2$; $R = 2M$. Since one expects[#]

$$T = 1/8\pi M \quad , \tag{3.1}$$

the mass loss should obey

$$\frac{dM}{dt} = - C\, T^4 R^2 = - C'/ M^2 \quad . \tag{3.2}$$

The constants C , C' depend on the number of independent particle types at the corresponding mass scale, and this will vary slightly with temperature; the coefficients will however stay of order one (as long as M stays considerably larger than the Planck mass).

Ignoring this slight mass dependence of C' , one finds

$$M(t) = C'' (t_0-t)^{\frac{1}{3}} \quad , \tag{3.3}$$

where t_0 is a moment where the thing explodes violently. Conversely, the lifetime of any given Schwarzschild black hole with mass M can be estimated to be

$$t_1 = M^3/3C' \quad . \tag{3.4}$$

Now this is the time needed for the complete disappearance of the black hole. One may also ask for the average lifetime of a black hole in a given quantum mechanical state, i.e. the average time between two Hawking emissions.

A rough estimate reveals that the wavelength of the average Hawking particle is of the order of the black hole radius R , and that this is also the expected average spatial distance between two Hawking particles. Therefore the lifetime of a given quantum state is of order R , i.e. of order $1/M$ in Planck units.

In the language of particle physics this implies that the radiating black hole is a resonance state that in an S matrix would produce a pole at the complex energy value

$$E = M - C_3\, i/M \quad , \tag{3.4}$$

where C_3 is again a constant of order one. This corresponds to the value

$$E^2 = M^2 - 2C_3\, iM_{Pl}{}^2 \tag{3.5}$$

for the Mandelstam variable s . We see that all black hole poles are expected to be below the real axis of s at a universal average distance of order one in units of the Planck mass squared.

It is not altogether unreasonable to assume that a black hole is just a pole in the S matrix like any other tiny physical object.

[#]As was pointed out by this author[8], the derivation of this formula requires an assumption concerning the interpretation of quantum wave functions for particles disappearing into the black hole. Though plausible, one can imagine this assumption to be wrong, in which case the black hole temperature will be different from (3.1).

4. THE S-MATRIX ANSATZ AND THE SHIFTING HORIZON

The problem with linearised quantum field theory in the black hole background is that the ingoing particles then seem to be independent of the outgoing ones. Hilbert space is then a *product* space, $|\psi\rangle = |\psi\rangle_{in} \times |\psi\rangle_{out}$. If we were to describe a black hole that obeys an overall Schrödinger equation then these in- and out-spaces cannot be allowed to be independent of each other. In contrast, one would expect the existence of an S-matrix:

$$|\psi\rangle_{out} = S\,|\psi\rangle_{in} \,, \tag{4.1}$$

and with this mapping of in- to out-states the degrees of freedom pictured in Fig. 1*a* are replaced by the ones of Fig. 1*b* or Fig. 1*c*.

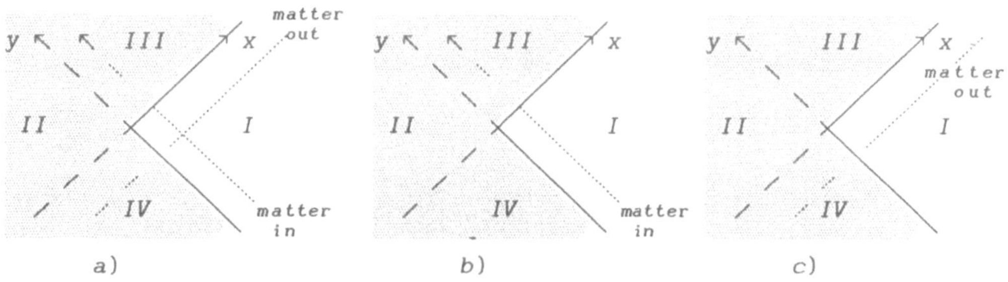

Fig. 1
a) Linearised quantum field theory produces a Hilbert space that is the product of two factors: $|\psi\rangle = |\psi\rangle_{in} \times |\psi\rangle_{out}$ Ingoing particles form the factor space $\{|\psi\rangle_{in}\}$ (*b*), and outgoing ones form $\{|\psi\rangle_{out}\}$ (*c*). x and y are the Kruskal coordinates for the Schwarzschild metric.

As stated earlier, the reason why superimposing in- and out-particles as in Fig. 1*a* is incorrect is the breakdown of linearised quantum field theory at distances closer than a Planck length from the horizon. Gravitational interactions there become super strong. We can obtain the black hole representations of Fig. 1*b* and Fig. 1*c* by adopting the following elementary procedure:

i) *Postulate* the existence of an S matrix, and

ii) take interactions between in- and out-states into account, in particular the gravitational ones.

We can look upon this procedure as a new and more precise formulation of the general coordinate transformation from Kruskal coordinates[3] to Schwarzschild coordinates, or from flat space-time to Rindler[9] space-time. There is no direct contradiction with anything we know about general relativity or quantum mechanics, but because of the crucial role attributed to the interactions the picture is only somewhat more complicated.

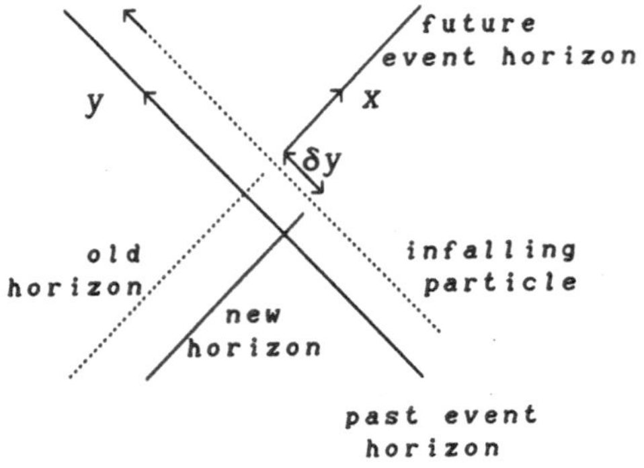

Fig. 2. The horizon displacement.

The most important ingredient of the gravitational interactions is the *horizon shift*[10]. Consider any particle falling into the black hole. Its gravitational field is assumed to be so weak that a linearised description of it, during the infall, is reasonable. The curvatures induced in the Kruskal frame are initially much *smaller* than the Planck length. Now perform a time translation (for the external observer). At the origin of Kruskal space this corresponds to a Lorentz transformation:

$$x \rightarrow \gamma^{-1} x \; ; \quad y \rightarrow \gamma\, y \; . \tag{4.2}$$

The γ factors here can grow very quickly, exponentially with external time. The very tiny initial curvatures soon become substantial, but are only seen as shifts in the y coordinate, because y has expanded such a lot.

The result is a representation of the space-time metric where the incoming particle enters along the y axis, with a velocity that has been boosted to become very close to the speed of light. Its energy in terms of the boosted coordinates has become so huge that the curvature became sizable. It is described completely by saying that two halves of the conventional Kruskal space are glued together along the y axis with a *shift* δy, depending explicitly on the angular coordinates ϑ and φ.

The calculation of the function $\delta y(\vartheta,\varphi)$ is elementary. In Rindler space, $\delta y(\tilde{x})$ is simply[11]

$$\delta y(\tilde{x}) \;=\; 4G_N p \, \log(\tilde{x}^2) + C \; , \tag{4.3}$$

where G_N is Newton's constant, p is the ingoing particle momentum, and C is an arbitrary constant.

In Kruskal space the angular dependence of δy is a bit more complicated than the $\tilde{x}$ dependence in Rindler space. It is found by inserting Einstein's equation, which is $R_{\mu\nu} = 0$, everywhere except where the particle comes in. Starting with an arbitrary δy as an Ansatz, one finds Einstein's equation to correspond to

$$(1 - \Delta_{\vartheta,\varphi})\, \delta y(\vartheta,\varphi) \;=\; 0 \; , \tag{4.4}$$

where $\Delta_{\vartheta,\varphi}$ is the angular Laplacian. At the angles ϑ_o,φ_o where the particle enters we simply compare with the Rindler result (4.3) to obtain

$$(1 - \Delta_{\vartheta,\varphi})\, \delta y(\vartheta,\varphi) \;=\; \kappa \, p_{in} \, \delta^2(\vartheta,\varphi;\vartheta_o,\varphi_o) \; , \tag{4.5}$$

where κ is a numerical constant related to Newton's constant.

This equation can be solved:

$$\delta y(\vartheta,\varphi) = f(\vartheta,\varphi;\vartheta_o,\varphi_o)\; p_{in}(\vartheta_o,\varphi_o) \quad ;$$

$$f = \kappa_1 \int_0^\infty \frac{\cos\left(\frac{\sqrt{3}}{2}\, s\right)\, ds}{\left(\cosh s - \cos\vartheta_1\right)^{\frac{1}{2}}} \quad , \tag{4.6}$$

where ϑ_1 is the angular separation between (ϑ,φ) and (ϑ_0,φ_o).

Other expressions for f are

$$f = \kappa_2 \int_{\vartheta_1}^{2\pi-\vartheta_1} dz\, \left(\cos\vartheta_1 - \cos z\right)^{-\frac{1}{2}} e^{-\frac{1}{2}\sqrt{3}\, z} \quad ; \tag{4.7}$$

and[12]

$$f = \frac{\pi\kappa_1}{\sqrt{2}} \frac{P_{-\frac{1}{2} + \frac{1}{2}i\sqrt{3}}\left(-\cos\vartheta_1\right)}{\cosh\left(\frac{1}{2}\pi\sqrt{3}\right)} \quad , \tag{4.8}$$

where P is a Legendre function with complex index (conical function). From (4.7) one sees directly that for all angles ϑ_1 f is positive.

5. SPACE-TIME SURROUNDING THE BLACK HOLE

The horizon shift discussed in the previous section is an essential ingredient in the S-matrix construction. Without it we would not be able to perform this task. Now we are. We will discuss this construction in the next chapter. First one has to understand what the relevant degrees of freedom are and where in space-time they live. Here, partly anticipating on our results, we observe that the outgoing configurations will depend on what goes in, and with a sensitivity that depends exponentially with $\delta t/4M$, where δt is the time interval as seen by the distant observer. However, the particles coming out later than a

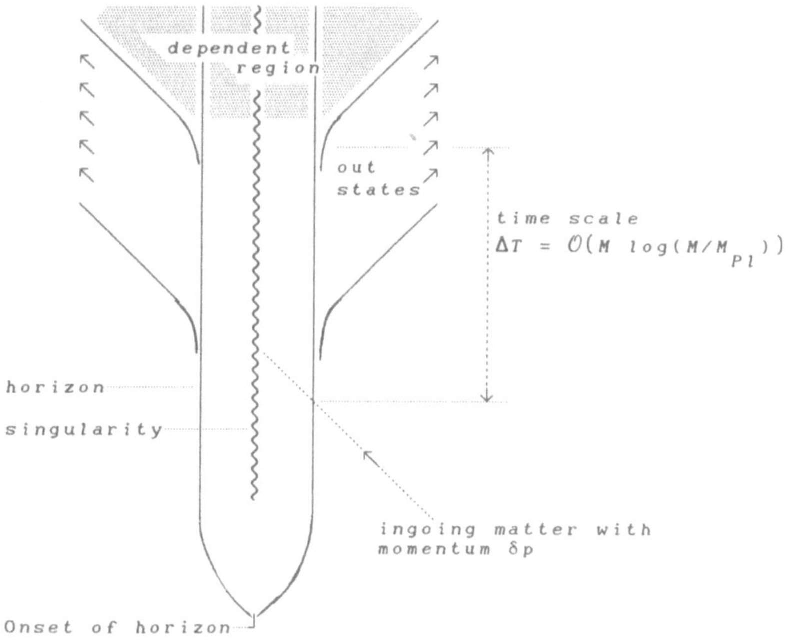

Fig. 3

lapse of time of the order of $\Delta t = 4M \log (M/M_{Pl})$, can no longer be described at all. We will assume, for simplicity, that these particles will be completely determined (in the quantum mechanical sense) by earlier events, or in other words, one is not allowed to choose these states any way one pleases. Simply counting states (as one can derive from the finite black hole entropy[14]) one notes that, given all incoming particles, the outgoing particles can be chosen freely only during an amount of time Δt , or *vise versa*.

Any incoming particle will not only affect the outgoing particles after a time lapse of order Δt , but also all that come out later than that. These limitations in choosing in- and outgoing states are just what they would be for any macroscopic system such as a finite size box containing a gas or liquid, connected to the outside world for instance via a tiny hole.

6. CONSTRUCTION OF THE S-MATRIX

The argument now goes as follows[6].

1. Consider one particular in-state and one particular out-state. Assume that someone gave us the amplitude defined by sandwiching the S-matrix between these two states:

$$\langle in | out \rangle \quad . \tag{6.1}$$

Both the in- and the out-state are described by giving all particles in some conveniently chosen wave packets. The ingoing wave packets look like

$$e^{-ip^i_{in}x} \; f^i_{in}(x,\vartheta,\varphi) \quad , \tag{6.2}$$

where i runs over all particles involved, and f^i_{in} are smooth functions. We assume them to be sharply peaked in the angular coordinates so that we know exactly where the particles enter into the horizon (so the *angular coordinates* and the *radial momenta* of all particles are sharply defined). Similarly the outgoing wave packets are

$$e^{-ip^i_{out}y} \; f^i_{out}(y,\vartheta,\varphi) \quad . \tag{6.3}$$

Now let us consider a small change in the ingoing state: $|in\rangle \rightarrow |in'\rangle$. This brings about a sharply defined small change in the distribution of the radial momenta $p_{in}(\vartheta,\varphi)$ on the horizon:

$$p_{in} \rightarrow p_{in} + \delta p_{in}(\vartheta,\varphi) \quad . \tag{6.4}$$

This δp_{in} now produces an (extra) horizon shift,

$$\delta y(\Omega) = \int f(\Omega-\Omega')\, \delta p_{in}(\Omega') \quad , \tag{6.5}$$

where f is the Green function computed in the previous section and Ω stands short for (ϑ,φ) ; $\Omega-\Omega'$ stands for the angle ϑ_1 between Ω and Ω' .

The horizon shift (6.5) does *not* affect the thermal nature of the Hawking radiation, but it *does* change the quantum states. All out-wave functions are shifted. (6.3) is replaced by

$$e^{-ip^i_{out}(y + \delta y(\Omega))} \; f^i_{out}(y,\Omega) \tag{6.6}$$

(the effect of the shift on f is of lesser importance). The shift δy

is assumed to be so small that the outgoing particle is not thrown over the horizon. This requires that we consider only those outgoing particles that are already sufficiently far separated from the horizon, or: they are not the ones that emerge later than the time interval Δt , as defined in the previous section, see Fig. 3. This is why the time interval Δt was necessary. Now such a restriction will also imply that we will have to reconsider the definition of inner products in our Hilbert space, and this will imply that the operator $\exp(-ip^i_{out}\delta y)$ might not be unitary. We will temporarily ignore this important observation.

We observe that the S-matrix element (6.1) is replaced:

$$\begin{aligned} \langle in' | out \rangle &= e^{-i\int p_{out}(\Omega)\ \delta y(\Omega) d^2\Omega} \langle in | out \rangle \\ &= e^{-i\int\int p_{out}(\Omega)\ f(\Omega-\Omega')\ \delta p_{in}(\Omega')\ d^2\Omega d^2\Omega'} \langle in | out \rangle \ . \end{aligned} \tag{6.7}$$

Here, $p_{out}(\Omega)$ is the total outgoing momentum at the angular coordinates Ω . What we have achieved is that we have been able to compute *another* matrix element of S . Now simply repeat this procedure many times. We then find *all* matrix elements of S to be equal to

$$\langle in | out \rangle = \mathcal{N}\, e^{-i\int\int p_{out}(\Omega)\ f(\Omega-\Omega')\ p_{in}(\Omega') d^2\Omega d^2\Omega'} \ , \tag{6.8}$$

where $\mathcal{N}$ is one common unknown factor. Apart from an overall phase, $\mathcal{N}$ should follow from unitarity.

The derivation of (6.8) ignores all interactions other than the gravitational ones. We will be able to do better than that, but let us first analyze this expression.

What is unconventional in the S-matrix (6.8) is the fact that the in- and out-states must have been characterized *exclusively* by specifying the *total* radial momentum distribution over the angular coordinates on the horizon. If there are more parameters necessary to characterize these states, these extra parameters will not figure in the S-matrix. But this would mean that two different states $|A\rangle$ and $|B\rangle$ could evolve into the same state $|\psi_{out}\rangle$, so these extra parameters will not be consistent with unitarity. *We cannot allow for other parameters than the total momentum distributions* (unless more kinds of interactions are taken into account).

Thus, if for the time being we only consider gravitational interactions, the in-states can be given as $|p_{in}(\Omega)\rangle$ and the out-states as $|p_{out}(\Omega)\rangle$. The operators $p_{in}(\vartheta,\varphi)$ commute for different values of ϑ and φ , and their representations span the entire Hilbert space; the same for $p_{out}(\vartheta,\varphi)$.

The canonically conjugated operators $u_{in}(\Omega)$, $u_{out}(\Omega)$ are defined by the commutation rules

$$[p_{in}(\Omega),u_{in}(\Omega')] = -i\delta^2(\Omega-\Omega') \tag{6.9}$$

(and similarly for the *out* operators), or,

$$\langle u_{in}(\Omega) | p_{in}(\Omega) \rangle = C \exp i\int d^2\Omega\ p_{in}(\Omega)u_{in}(\Omega) \ , \tag{6.10}$$

where C is a normalization constant.

In terms of the u operators the S-matrix is

$$\langle u_{out}(\Omega) | u_{in}(\Omega) \rangle = \int \mathcal{D}p_{out}\mathcal{D}p_{in} \exp\left(-ip_{in}u_{in} + ip_{out}u_{out} - ip_{out}fp_{in}\right) \ , \tag{6.11}$$

which is a Gaussian functional integral over the functions p_{out} and p_{in} . Since the inverse f^{-1} of f is $\kappa^{-1}(1 - \Delta_\Omega)$, the outcome of this functional integral is

$$\langle u_{out}(\Omega)|u_{in}(\Omega)\rangle = C\exp\left[-i\kappa^{-1}\int d^2\Omega\; u_{in}(\Omega)\,(1-\Delta_\Omega)\,u_{out}(\Omega)\right] =$$
$$C\exp\left[-i\kappa^{-1}\int d^2\Omega\;(\partial_\Omega u_{in}(\Omega)\,\partial_\Omega u_{out}(\Omega) + u_{in}(\Omega)\,u_{out}(\Omega))\right] \quad . \tag{6.12}$$

The last term in the brackets is something like a mass term and becomes subdominant if we concentrate on small subsections of the horizon. Therefore it will be ignored from now on. Eq. (6.12) seems to be more fundamental than (6.8) because it is local in Ω .

Fourier transforming back we get

$$\langle p_{out}(\Omega)|p_{in}(\Omega)\rangle =$$
$$\mathcal{N}\int \mathcal{D}u_{in}\mathcal{D}u_{out}\;\exp\int d^2\Omega\left[ip_{in}u_{in} - ip_{out}u_{out} - i\kappa^{-1}\partial_\Omega u_{in}(\Omega)\,\partial_\Omega u_{out}(\Omega)\right] \quad . \tag{6.13}$$

We reobtain (6.8) written as a functional integral.

It is illuminating to redefine

$$p_{out} = p_- \;\; ; \;\; p_{in} = p_+ \;\; ; \;\; u_{out} = x^- \;\; ; \;\; u_{in} = -x^+ \quad , \tag{6.14}$$

and to replace the angular coordinate Ω by a transverse coordinate $\tilde{x}$, so that if we define the transverse momentum components $\tilde{p} \cong 0$ one can write

$$\langle p_{out}(\tilde{x})|p_{in}(\tilde{x})\rangle = \mathcal{N}\int \mathcal{D}x^\mu(\tilde{x})\;\exp\int d^2\tilde{x}\left[ip_\mu(\tilde{x})x^\mu(\tilde{x}) - i\kappa^{-1}(\partial_\Omega x^\mu(\tilde{x}))^2\right] \quad . \tag{6.15}$$

Suppose now that both the in-state and the out-state are written as sets containing a finite number of particles, having not only fixed longitudinal momenta but also transverse momenta $\tilde{p}^i$. We then have to convolute the amplitude (6.15) with transverse wave functions $\exp(i\tilde{p}^i\tilde{x}^i)$ for each particle i . It becomes

$$\mathcal{N}\prod_i\int d\tilde{x}^i\int \mathcal{D}x^\mu(\tilde{x})\;\exp\left[i\sum_i p_\mu{}^i x^\mu(\tilde{x}^i) + \int d^2\tilde{x}\left[-i\kappa^{-1}(\partial_\Omega x^\mu(\tilde{x}))^2\right]\right] \quad . \tag{6.16}$$

This functional integral is very similar to the functional integral for a string amplitude, including the integration over Koba-Nielsen variables[13], except for the unusual imaginary value for the string constant:

$$T = 8\pi G_N i \quad . \tag{6.17}$$

7. ELECTROMAGNETISM

What happens if more interactions are included? The simplest to handle turn out to be the electromagnetic forces. Suppose that the particles that collapsed to form the black hole carried electric charges. The angular charge distribution was

$$\rho_{in}(\Omega) \quad . \tag{7.1}$$

As in the previous section, we consider a small change in this setting, so

$$\rho_{in}(\Omega) \rightarrow \rho_{in}(\Omega) + \delta\rho_{in}(\Omega) \quad . \tag{7.2}$$

The $\delta\rho_{in}(\Omega)$ produces an extra contribution to the vector potential at the horizon which is not difficult to compute[♩].

$$\delta A_\mu = \frac{1}{r_o^2}\,\delta_{\mu x}\,\delta(x)\,A(\Omega) \quad , \tag{7.3}$$

where r_o is the radius of the horizon, and $A(\Omega)$ must satisfy

$$\Delta_\Omega A(\Omega) = \delta\rho_{in}(\Omega) \quad . \tag{7.4}$$

The field (7.3) is only non-vanishing on the plane $x=0$, where it causes a sudden phase rotation for all wave packets that go through. An outgoing wave undergoes a phase rotation

$$e^{iQ\Lambda(\Omega)} \quad ; \quad \Lambda(\Omega) = r_o^{-1}\,A(\Omega) \ . \tag{7.5}$$

This rotation must be performed for all outgoing particles with charge Q . All together the outgoing wave is rotated as follows:

$$\begin{aligned} &|\,p_{out}(\Omega),\rho_{out}(\Omega)\rangle \ \rightarrow \\ &\exp -i\int d^2\Omega\int d^2\Omega'\ f_1(\Omega-\Omega')\ \rho_{out}(\Omega)\delta\rho_{in}(\Omega') \times |\,p_{out}(\Omega),\rho_{out}(\Omega)\rangle \quad , \end{aligned} \tag{7.6}$$

where $f_1(\Omega-\Omega')$ is a Green function that satisfies

$$\Delta_\Omega\, f_1(\Omega-\Omega') = -\kappa_e\delta^2(\Omega-\Omega') \quad . \tag{7.7}$$

κ_e is a numerical constant.

And using arguments identical to the ones of the previous section we repeat the infinitesimal changes to obtain the S-matrix dependence on $\rho_{in}(\Omega)$ and $\rho_{out}(\Omega)$:

$$\begin{aligned} &\langle p_{out}(\Omega),\rho_{out}(\Omega)\,|\,p_{in}(\Omega),\rho_{in}(\Omega)\rangle = \\ &\mathcal{N}\ e^{-i\int\int p_{out}(\Omega)\ f(\Omega-\Omega')\ p_{in}(\Omega')d^2\Omega d^2\Omega'} \times \\ &\qquad e^{-i\int\int \rho_{out}(\Omega)\ f_1(\Omega-\Omega')\ \rho_{in}(\Omega')d^2\Omega d^2\Omega'} \ . \end{aligned} \tag{7.8}$$

Now let us replace $\rho_{out}(\Omega)\rho_{in}(\Omega')$ by

$$-\tfrac{1}{2}\big(\rho_{out}(\Omega)-\rho_{in}(\Omega)\big)\big(\rho_{out}(\Omega')-\rho_{in}(\Omega')\big) \quad . \tag{7.9}$$

This differs from the previous expression by two extra terms in the exponent, one depending on $\rho_{out}(\Omega)$ only and the other depending on $\rho_{in}(\Omega)$ only. These would correspond to external "wave function renormalization factors" that do not describe interaction between the in- and the out-state. So we ignore them.

The electromagnetic contribution in (7.8) can then be written as a functional integral of the form

$$\int\mathcal{D}\Phi(\Omega)\ \exp\int d^2\Omega\Big(\frac{-i}{2\kappa_e}(\partial_\Omega\Phi)^2 + i\Phi(\rho_{out}-\rho_{in})\Big) \quad . \tag{7.10}$$

Now it may also be observed that the charge distribution ρ is actually a combination of Dirac delta distributions,

---

[♩]The unit e of electric charge of the ingoing particle is here included in ρ_{in} .

$$\rho(\Omega) = \sum Q_i \delta^2(\Omega-\Omega_i) \quad ; \quad Q_i = n_i e \quad . \tag{7.11}$$

Therefore, if we add an integer multiple of $2\pi/e$ to the field Φ the integrand does not change. In other words: Φ is a periodic variable.

Adding (7.10) to (6.15) we notice that the field Φ acts exactly as a fifth, periodic dimension. Hence, electromagnetism emerges naturally as a Kaluza-Klein theory.

8. HILBERT SPACE

In this section we briefly recapitulate the nature of the Hilbert space in which these S matrix elements are defined. As explained in Section 4, the states whose momentum and charge distribution over the horizon were given by $p(\Omega)$ and $\rho(\Omega)$ *include* all particles in the black hole's vicinity. But (if for simplicity we ignore electromagnetism) we can also form a complete basis in terms of states for which the canonical operators $x(\Omega)$ are given. These $x(\Omega)$ (eq. 6.12) may be interpreted as the coordinates of the horizon. Apparently, *the precise shape of the horizon determines the state of the surrounding particles!*

Furthermore, the *in*-horizon and the *out*-horizon do not commute. Therefore, the positions of the future event horizon and the past event horizon do not commute with each other. If we define a "black hole" as an object for which the location in space-time of the future event horizon is precisely determined, we can define a "white hole" as a state for which the past event horizon is precisely determined. *The white hole is a linear superposition of black holes* (and vice versa); operators for white holes do not commute with the ones for black holes. In our opinion this resolves the issue of white holes in general relativity.

Obviously, it is important that the horizon of the quantized black hole is not taken to be simply spherically symmetric. In a black hole with a history that is not spherically symmetric, the onset of the horizon, i.e. the point(s) in space-time (at the bottom of Fig. 3) where for the first time a region of space-time emerges from which no timelike geodesic can escape to $\mathcal{I}^+$, has a complicated geometrical structure. Its mathematical construction has the characteristics of a caustic. One might conjecture that the topological details of this caustic specify the quantum state a black hole may be in.

The fact that the geometry of the (future or past) horizon should determine the quantum state of the surrounding particles gives rise to interesting questions and problems. In ordinary quantum field theory the Hilbert space describing particles in a region of space-time is Fock space; an arbitrary, finite, number of particles with specified positions or momenta together define a state. But now, close to the horizon, a state must be defined by specifying the *total* momentum entering (or leaving) the horizon at a given solid angle Ω. Apparently we are not allowed to specify further how many particles there were, and what their other quantum numbers were. Together all these possibilities form just one state. So, our Hilbert space is set up differently from Fock space. The difference comes about of course because we have strong gravitational interactions that we are not allowed to ignore.

The best way to formulate the specifications of our basis elements here is to assume a lattice cut-off in the space of solid angles (one "lattice point" for each unit of horizon surface area somewhat bigger than $\delta\Sigma$ (the Planck distance squared), and then to specify that there should be *exactly one ingoing and one outgoing particle* at each $\delta\Sigma$. The momenta are given by the operators $p_{in}(\Omega)$ and $p_{out}(\Omega)$ (and the charges by $\rho_{in}(\Omega)$ and $\rho_{out}(\Omega)$). The in- and out-operators of course do not commute.

One may speculate that since $\delta\Sigma$ is extremely small, the totality of all these particles may be indistinguishable from an ordinary Dirac sea for the large-scale observers.

Also one may notice that the way conventional string theory deals with in- and outgoing particles is remarkably similar. Before integrating over the Koba-Nielsen variables the string amplitudes also depend exclusively on the distribution of total in- and outgoing momenta (see concluding remarks in Sect. 6).

If Hilbert space is constructed entirely from the operators $p(\tilde{x})$ and $x(\tilde{x})$ then these operators are hermitean by construction. But we have also seen that in terms of ordinary Fock space $p(\tilde{x})$, and hence also $x(\tilde{x})$ are probably not hermitean. There are different ways to approach this hermiticity problem, but we shall not elaborate here on this point.

9. RELATION BETWEEN TERMS IN THE HORIZON FUNCTIONAL INTEGRAL AND BASIC INTERACTIONS IN 4 DIMENSIONS

In principle one can pursue our doctrine to obtain more precise expressions for our black hole S matrix by including more and more interactions that we actually know to exist from ordinary particle theory. We should be certain to obtain a result that is accurate apart from a limitation in the angular resolution, because particle interactions are known only up to a certain energy. In this section we indicate some qualitative results.

The details of our "presently favored Standard Model" may well change in due time. We will denote anything used as an input regarding the fundamental interactions among in- and outgoing particles near the horizon, at whatever scale, by the words "standard model".

Suppose the standard model contains a scalar field. The effects of this field will be felt by slowly moving particles at some distance from the horizon. But at the horizon itself these effects are negligible. Consider namely a particle such as a nucleon, surrounded by a scalar field such as a pion field. Close to the horizon this particle will be Lorentz boosted to tremendous energies. The scalar field configuration will become more and more flattened. But unlike vector or tensor fields, its intensity will not be enhanced (it is Lorentz invariant). So the cumulated effect on particles traversing it will tend to zero.

However, one effect due to the scalar field will not go away. Suppose our standard model contains a Higgs field, rendering a $U(1)$ gauge boson massive. This means that the electromagnetic field surrounding a fast electrically charged particle will be of short range only. One can derive that the field equation (7.4) will change into

$$(\Delta_\Omega - M_A{}^2)A(\Omega) \quad = \quad \delta\rho_{in}(\Omega) \quad . \tag{9.1}$$

One may say that the incoming charge density $\rho_{in}(\Omega)$ is screened by charges coming from the Higgs particles.

This implies that the equations for the Φ field in Sect. 7 will obtain a mass term:

$$\int \mathcal{D}\Phi(\Omega)\ \exp \int d^2\Omega \Big(\frac{-i}{2\kappa_e}[(\partial_\Omega \Phi)^2 + M_A{}^2\Phi^2] + i\Phi\ \rho\Big) \quad . \tag{9.2}$$

Note that this mass term breaks explicitly the symmetry $\Phi \to \Phi + \Lambda$. This explicit symmetry breaking may be seen as a result of the finite and constant value of the Higgs field at the origin of Kruskal space-time.

Next, we may ask what happens if our standard model exhibits confinement. This means that at long distance scales no effect of the gauge field is seen and all allowed particles are neutral.

Confinement is usually considered to be the *dually opposite* of the Higgs mechanism: Bose condensation of magnetic monopoles. A magnetic monopole is an object to which the end point of a Dirac string is attached. A Dirac string is a singularity in a gauge transformation such that the gauge transformation makes one full rotation if we follow a loop around the string.

We must know how to describe the operator field of a monopole at the horizon. Suppose a monopole entered at the solid angle Ω_1 . This means that a Dirac string connects to the black hole at that point. The outgoing charged particles undergo a gauge rotation that rotates a full cycle if we follow a closed curve around Ω_1 (an anti-monopole may neutralize this elsewhere on the horizon).

The gauge jump for the vector potential field A can be identified with the periodic field Φ of Sect 7. So adding an entering monopole to the in-state implies that this field Φ is shifted by an amount $\Lambda(\Omega)$ where Λ makes a full cycle when followed over a loop around Ω_1. This is an operation that is called *disorder operator* in statistical physics and field theory. This operator, Φ_D , is dual to the original field Φ . We find that the dual transformation electricity $\leftrightarrow$ magnetism corresponds to the duality between Φ and Φ_D .

Thus, if we have confinement, a mass term will result in the equations for Φ_D . It explicitly breaks the symmetry $\Phi_D \rightarrow \Phi_D + C$. And this bars the transformation back to Φ . Therefore, *if confinement occurs, the field* Φ *is no longer well-defined, we have only* Φ_D . Its mass will be the glueball mass.

In Table 1 we list peculiarities of the mapping from 4 to 2 dimensions. The *generators* of local symmetry transformations in 4 dimensions correspond to the dynamic variables in 2 dimensions. Thus one expects that if the standard model includes a gravitino (requiring a supersymmetry generator of spin $\frac{1}{2}$) then a fermionic field variable will emerge in 2 dimensions.

But the above are merely qualitative features. They should be turned into precise quantitative rules and principles, for which further work is needed.

10. OPERATOR ALGEBRA ON THE HORIZON

A fundamental shortcoming of the procedure described above is that the dimensionality of Hilbert space is infinite from the start. The functions $p(\tilde{x})$ and $x(\tilde{x})$ generate an infinite set of basis elements. Yet the black hole entropy, as calculated from Hawking radiation, is finite. Indeed, we have not yet been able to reproduce Hawking radiation from our S-matrix. This is because we have ignored the *transverse* components of the gravitational shifts, and the string functionals we produced thus far only allow for infinitesimal string excitations. We shall now try to improve our description of the basis elements of Hilbert space. This we do by setting up an operator algebra. First we consider the algebra generated by the amplitudes we have.

Our starting point here is that states in Hilbert space are uniquely determined by specifying any one of the following four functions: the distribution of ingoing momenta $p_+(\tilde{x})$, the outgoing momenta $p_-(\tilde{x})$, the conjugated operators $x^+(\tilde{x})$, or $x^-(\tilde{x})$. They obey the algebra

$$[p_+(\tilde{x}),\ p_+(\tilde{x}')] = 0 \quad ; \quad [p_+(\tilde{x}),\ x^+(\tilde{x}')] = -i\delta^2(\tilde{x},\tilde{x}') \quad ; \tag{10.1}$$

$$[p_-(\tilde{x}),\ p_-(\tilde{x}')] = 0 \quad ; \quad [p_-(\tilde{x}),\ x^-(\tilde{x}')] = -i\delta^2(\tilde{x},\tilde{x}') \quad , \tag{10.2}$$

and we have the relation

$$x^-(\tilde{x}) = 4\pi G \int d^2\tilde{x}' \ f(\tilde{x},\tilde{x}') \ p_+(\tilde{x}') \quad . \tag{10.3}$$

This implies

$$[x^-(\tilde{x}),\ x^+(\tilde{x}')] = -4\pi i G \ f(\tilde{x},\tilde{x}') \quad , \tag{10.4}$$

so that we have also

$$x^+(\tilde{x}) = -4\pi G \int d^2\tilde{x}' \ f(\tilde{x},\tilde{x}') \ p_-(\tilde{x}') \quad . \tag{10.5}$$

Table 1

STANDARD MODEL IN 3+1 DIMENSIONS	INDUCED 2 DIMENSIONAL FIELD THEORY ON BLACK HOLE HORIZON
• *Spin 2:* $g_{\mu\nu}(\mathbf{x},t)$ local gauge generator: $u^\mu(\mathbf{x},t)$	String variables (*spin 1*): $x^\mu(\Omega)$
• *Spin 1:* $A_\mu(\mathbf{x},t)$ local gauge generator: $\Lambda(\mathbf{x},t)$ *mod* $2\pi/e$	Scalar variable (*spin 0*): $\Phi(\Omega)$ *mod* $2\pi/e$
• *Spin 0:* $\phi(\mathbf{x},t)$	No field at all
• Higgs mechanism: "spontaneous" mass M_A for vector field	explicit symmetry breaking; $\Phi(\Omega)$ gets same mass M_A .
• Confinement in vector field A_μ	Φ must be replaced by disorder op. Φ_D ; its symmetry broken.
• Non-Abelian gauge theory	only scalars Φ_i corresponding to Cartan subalgebra
• *Spin* $\frac{1}{2}$*:* fermions	no field at all
• *Spin* $\frac{3}{2}$*:* gravitino local gauge generator spin $\frac{1}{2}$	Spin $\frac{1}{2}$ fermion (?)

The algebraic relations among $\rho(\tilde{x})$ and $\phi(\tilde{x}')$ are slightly more subtle because of the quantization of electric charge and the ensuing periodic boundary conditions on ϕ . We will disregard these from here on.

The relations (10.1-5) are not infinitely accurate. This is because we neglected any gravitational curvature in the sideways directions. This is fine as long as transverse distance scales are kept considerably larger than the Planck scale. One may convince oneself that this implies neglecting higher orders in the derivatives $\partial x^\pm/\partial\tilde{x}$. Is there any way to obtain a more precise algebra? It is natural to search for an algebra that is invariant under Lorentz transformations. One might hope that such an algebra could generate the correct degrees of freedom at the Planck scale (in particular *quantized* degrees of freedom).

It was proposed in Ref[6] that Hilbert space on the horizon may be generated by the operator algebra of fundamental surface elements,

$$W^{\mu\nu}(\tilde{\sigma}) = \varepsilon^{ab}\frac{\partial x^\mu}{\partial \sigma^a}\frac{\partial x^\nu}{\partial \sigma^b} \quad . \tag{10.6}$$

where the transverse coordinates $\tilde{x}$ were replaced by more arbitrary surface coordinates σ^1, σ^2 . The relations (10.1-5) may be used in the case $\tilde{\sigma} = \tilde{x}$, when the derivatives are small. This means

$$W^{12} = 1 \quad ; \quad W^{1\mu} = \frac{\partial x^\mu}{\partial \sigma^2} \quad ; \quad W^{2\mu} = -\frac{\partial x^\mu}{\partial \sigma^1} \quad ; \quad W^{34} = \mathcal{O}(\partial x^\mu)^2 \ . \tag{10.7}$$

The commutation rules can then be rewritten in the form

$$\sum_\lambda [W^{\lambda\mu}(\tilde{\sigma}), W^{\lambda\nu}(\tilde{\sigma}')] = \tfrac{1}{2}T\ \varepsilon^{\mu\nu\kappa\lambda}W^{\kappa\lambda}(\tilde{\sigma})\delta^2(\tilde{\sigma}-\tilde{\sigma}') \quad , \tag{10.8}$$

which is written in such a way that it remains true in all coordinate frames. T is a constant ('string constant') equal to 8π in Planck units. In stead of (10.1-5) we can take this to be the equation that generalizes to arbitrary surfaces. It has the advantage of being linear in W .

Now (10.8) is not a closed algebra, because the left hand side still contains a summation. A complete algebra is obtained as follows.

Let K be i times the self dual part of W :

$$K^{\mu\nu} = i(W^{\mu\nu} + \tfrac{1}{2}\varepsilon^{\mu\nu\kappa\lambda}W^{\kappa\lambda}) \quad . \tag{10.9}$$

It has three independent components:

$$K_1 = i(W^{23} + W^{14}) \quad ; \quad K_2 = i(W^{31} + W^{24}) \quad ; \quad K_3 = i(W^{12} + W^{34}) \quad . \tag{10.10}$$

Now from (10.8) we derive that these obey a complete commutator algebra,

$$[K_a(\tilde{\sigma}), K_b(\tilde{\sigma}')] = iT\varepsilon_{abc}K_c(\tilde{\sigma})\delta^2(\tilde{\sigma}-\tilde{\sigma}') \quad . \tag{10.11}$$

Apart from a complication to be mentioned shortly, this is a local and complete algebra of the kind we were looking for. At first sight it seems to generate an infinite dimensional Hilbert space because the operators K , like the W , are distributions . But let us introduce test functions $f(\sigma)$, $g(\sigma)$ and define operators

$$L_a{}^{(f)} = T^{-1}\int K_a(\tilde{\sigma})f(\tilde{\sigma})\mathrm{d}^2\tilde{\sigma} \quad , \tag{10.12}$$

then these satisfy commutation rules:

$$[L_a{}^{(f)}, L_b{}^{(g)}] = i\varepsilon_{abc}L_c{}^{(fg)} \quad . \tag{10.13}$$

Let us now restrict to test functions $f(\tilde{\sigma})$ that can only take the values 0 or 1 . Then $L_a{}^{(f)}$ satisfy the commutation rules of ordinary angular momentum operators. Note that for such an f the integral (10.12) is nothing but a boundary integral:

$$L_1{}^{(f)} = iT^{-1}\oint_{\delta f}(x^2\mathrm{d}x^3 + x^1\mathrm{d}x^4) \quad , \quad \text{etc.,} \tag{10.14}$$

where δf stands for the boundary of the support of f . We conclude that for every closed curve δf on $\tilde{\sigma}$ space we have three 'angular momentum' operators $L_a{}^{(f)}$ that satisfy the usual commutation rules and addition rules for angular momenta. Given such a bunch of closed curves f_i we can characterize the contribution of that part of the horizon to Hilbert space by the usual quantum numbers l_i and m_i. These are discrete and so, in some sense, we seem to come close to our aim of realizing a discrete Hilbert space for black holes. We note an important resemblance with the loop variable approach to quantum gravity[15].

Unfortunately, there is a snag. The operators L_a are not hermitean. If we take x^i to be hermitean and x^4 anti-hermitean then in the definition (10.6) W^{ij} are hermitean and W^{i4} anti-hermitean. Therefore, $L_a^\dagger$ correspond to the *anti*-self dual parts of $W^{\mu\nu}$. The commutation rules between L_a and $L_a^\dagger$ are non-local (they follow from (10.1-5)). The operators L^2 are hermitean, but not necessarily positive (they are only nonnegative for *time-like* surface elements). If we may assume the smallest surface elements to be timelike we can still build our surface using quantum numbers l_i and m_i but the states we get are *not properly normalized* (it is for finding the norms of the states that we need hermitean conjugation). If

$$\psi\{l_i,m_i\}$$

are the basis elements constructed using the self dual operators L_i, and

$$\phi\{l_i,m_i\}$$

the basis elements generated by the anti-self dual $L_i^\dagger$, then we have

$$\langle\phi\{l_i',m_i'\}|\psi\{l_i,m_i\}\rangle = \prod_i \delta_{l_i,l_i'}\delta_{m_i,m_i'} , \tag{10.15}$$

but the ψ themselves, or the ϕ themselves, are not orthonormal.

Now remember our realization earlier that actually the operators $x^+(\tilde{x})$ and $x^-(\tilde{x})$ are *not* hermitean, when we pass from the "horizon Hilbert space" to ordinary Fock space, because the shift operators may move particles behind the horizon. It is conceivable that this will lead to hermiticity conditions altogether different from (10.15).

But it is far from clear whether or not we actually obtained a complete representation of our Hilbert space.

REFERENCES

1. S.W. Hawking, Commun. Math. Phys. **43** (1975) 199; J.B. Hartle and S.W. Hawking, Phys.Rev. **D13** (1976) 2188; W.G. Unruh, Phys. Rev. **D14** (1976) 870; R.M. Wald, Commun. Math. Phys. **45** (1975) 9
2. S.W. Hawking, Phys. Rev. **D14** (1976) 2460; Commun. Math. Phys. **87** (1982) 395; S.W. Hawking and R. Laflamme, Phys. Lett. **B209** (1988) 39; D.N. Page, Phys. Rev. Lett. **44** (1980) 301, Gen. Rel. Grav. **14** (1987) 299; D.J. Gross, Nucl. Phys. **B236** (1984) 349
3. C.W. Misner, K.S. Thorne and J.A. Wheeler, "Gravitation", Freeman, San Francisco, 1973; S.W. Hawking and G.F.R. Ellis, "The Large Scale Structure of Space-time", Cambridge: Cambridge Univ. Press, 1973; E.T. Newman et al, J. Math. Phys. **6** (1965) 918; B. Carter, Phys. Rev. **174** (1968) 1559; K.S. Thorne, "Black Holes: the Membrane Paradigm", Yale Univ. press, New Haven, 1986; S. Chandrasekhar, "The Mathematical Theory of Black Holes", Clarendon Press, Oxford University Press
4. S. Coleman, Nucl. Phys. **B310** (1988) 643; S.B. Giddings and A. Stroninger, Nucl. Phys. **B321** (1989) 481; *ibid.* **B306** (1988) 890
5. P. Goddard, J. Goldstone, C. Rebbi and C.B. Thorn, Nucl. Phys. **B56** (1973) 109; M.B. Green, J.H. Schwarz and E. Witten, "Superstring

Theory", Cambridge Univ. Press; D.J. Gross, et al, Nucl. Phys. **B 256** (1985) 253

6. G. 't Hooft, Phys. Scripta **T15** (1987) 143; *ibid.* **T36** (1991) 247; Nucl. Phys. **B335** (1990) 138; G. 't Hooft, "Black Hole Quantization and a Connection to String Theory" 1989 Lectures, Banff NATO ASI, Part 1, "Physics, Geometry and Topology, Series B: Physics Vol. 238. Ed. H.C. Lee, Plenum Press, New York (1990) 105-128; G. 't Hooft, "Quantum gravity and black holes", in: Proceedings of a NATO Advanced Study Institute on Nonperturbative Quantum Field Theory, Cargèse, July 1987, Eds. G. 't Hooft et al, Plenum Press, New York. 201-226
7. L. Kraus and F. Wilczek, Phys. Rev. Lett. **62** (1989) 1221; J. Preskill, L.M. Krauss, Nucl. Phys. **B341** (1990) 50; L.M. Krauss, Gen. Rel. Grav. **22** (1990); S. Coleman, J.Preskill and F. Wilczek, preprint IASSNS-91/17 CALT-68-1717/ HUTP-91-A016
8. G. 't Hooft, J. Geom. and Phys. **1** (1984) 45
9. W. Rindler, Am.J. Phys. **34** (1966) 1174
10. T. Dray and G. 't Hooft, Nucl Phys. **B253** (1985) 173
11. W.B. Bonner, Commun. Math. Phys. **13** (1969) 163; P.C. Aichelburg and R.U. Sexl, Gen.Rel. and Gravitation **2** (1971) 303
12. C. Lousto, private communication
13 Z. Koba and H.B. Nielsen, Nucl. Phys. **B10** (1969) 633 , *ibid.* **B12** (1969) 517; **B17** (1970) 206; Z. Phys. **229** (1969) 243
14. J.D. Bekenstein, Phys. Rev. **D7** (1973) 2333; R.M. Wald, Phys. Rev. **D20** (1979) 1271; G. 't Hooft, Nucl. Phys. **B256** (1985) 727; V.F. Mukhanov, "The Entropy of Black Holes", in "Complexity, Entropy and the Physics of Information, SFI Studies in the Sciences of Complexity, vol IX, Ed. W. Zurek, Addison-Wesley, 1990. *See also* M. Schiffer, "Black Hole Spectroscopy", São Paolo preprint IFT/P - 38/89 (1989)
15. A. Ashtekar, *Phys. Rev.* **D36** (1987) 1587; A. Ashtekar *et al*, *Class. Quantum Grav.* **6** (1989) **L**185; C. Rovelli, Class. Quant. Grav. **8** (1991) 297, *ibid.* **8** (1991) 317

NON-COMMUTATIVE GEOMETRY AND MATHEMATICAL PHYSICS

Arthur Jaffe

Harvard University
Cambridge, MA 02138, U.S.A.

1. MATHEMATICAL PHYSICS

What is the role of mathematics in physics?

We have come to accept, over the past few years, that physics can play a basic conceptual role in stimulating new points of view in mathematics. This can range, on the one hand, from giving technical insights into the proof of a theorem, to motivating, on the other hand, whole new subfields of investigation.

However, the flow of knowledge in the reverse direction, from mathematics to physics, has been more limited. There has been a prejudice among physicists that modern mathematics is unnecessary for the future evolution of physics. In the early 1960's, Res Jost introduced a series of lectures with the observation that for many physicists, an appropriate knowledge of mathematics comprised, "a familiarity with the Greek and Latin alphabets." This reference to the study of perturbation theory reflected the enormous distance between the mathematics and physics communities, just 30 years ago.

To put quantum field theory on a logical level, mathematics has been developed over these 30 years in the areas of partial differential equations, in probability theory, in analysis, and in geometry of infinite dimensional manifolds. In other words, one developed tools to study functions on the infinite dimensional configuration space or phase space of the quantum fields. This work was carried out by the "traditional" community of mathematical physicists, but it was more or less ignored by the majority of theoretical physicists.

Eventually, theoretical physicists themselves began to take an interest in abstract mathematics. Beginning with the mathematical study of gauge theories in the late 1970's, and carried on in a widespead way due to the popularity of string theory among theoretical physicists in the mid 1980's, the mathematical methods used by physicists became more sophisticated. Through string theory, links were made between problems worked on by physicists and frontier problems in algebraic geometry, number theory, and representation

New Symmetry Principles in Quantum Field Theory, Edited by
J. Fröhlich et al., Plenum Press, New York, 1992

theory. These subjects were related in many surprising mathematical forms to problems of interest to physicists.

However, string theory has not had the hoped-for link between speculation and the world of experiment. So some more phenomologically oriented physicists have begun to question whether string theory is physics at all.

In any case, many theoretical physicists have turned back toward the study of quantum fields – especially conformal field theories, or Chern-Simons theories. The former to have relevance in the theory of critical phenomena in physics, as well as to the new area in mathematics of representations of quantum groups. The latter have a link to electromagnetic phenomena in physics (see the lectures of Fröhlich), as well as with mathematical phenomena of knot theory and the Jones polynomial.

It is hard to predict the future, but one can hope that mathematics will provide the clue to bringing modern ideas about quantum physics in accord with experiment. Here one can only remark that the framework of non-commutative geometry seems so close to quantum physics, that I predict that it will play a fundamental role in the reformulation of physics for the future. Connes lectures here give us a set of insights at a classical level into the interplay between non-commutative space-time and the standard electro-weak theory. How such a theory can be quantized remains a very important question.

Which directions physics will turn remains a mystery to be revealed only by the future. Quantum fields on non-commutative spaces may emerge naturally as fundamental objects. It is clear that the old algebraic framework for quantum field theory is also close to non-commutative geometry, and so it may well provide new insights. The global theory of non-commutative geometry has been developed a great deal, but vast amount of structure in the local algebras of observables which has not yet been exploited. A mixture of ideas from algebraic field theory and ideas from non-commutative geometry provide a promising path for future development. I certainly hope that the whatever the future brings, the "new physics" will emerge on a time-scale sufficiently short that I will be able to share in its discussion.

These lectures are devoted to some work along the lines of non-commutative geometry. Already ideas from physics have influenced the mathematical developments of non-commutative geometry, including the discovery of the cocycle which I describe in these lectures, namely the JLO cocycle for entire cyclic cohomology. I also talk about the philosophy of quantization, and how quantum space (replacing ordinary commutative space) and quantization of fields have very much in common.

2. NON-COMMUTATIVE GEOMETRY

For a person in mathematical physics, notions of non-commutative geometry (NCG) seem very natural. Related ideas to those in NCG occur in quantum theory – especially supersymmetric quantum theory – and also in statistical mechanics. One can interpret NCG as a quantization of geometry, in the sense that quantum theory is a quantization of classical physics. Many basic notions of non-commutative geometry can be understood by thinking of NCG as a way to define and to integrate differentials $a_0 da_1 \cdots da_n$ in a framework more general than that of differential forms on manifolds.

The quantum functions $a_0, \ldots, a_n$ are operators; abstractly they may lie in a C^*-

algebra, or be realized as an algebra of operators acting on a Hilbert space. We assume here that the differential da is given by a (graded) commutator $da = [D, a]$. The integrals $\int a_0 da_1 \cdots da_n$ can be thought of as quantum mechanical expectation values $\langle a_0 da_1 \cdots da_n \rangle$. Connes has shown what results is a theory in which classical notions of geometry carry over. In particular, there is a natural interpretation of NCG in terms of a cohomology theory. This cohomology reduces to de Rham theory in the usual commutative case. In this sense NCG is a quantization of geometry. Quantum groups appear in NCG as examples of quantum spaces.

Non-commutative geometry has another aspect; it has a natural and straight-forward extension to an infinite dimensional setting. The concept of dimension of the underlying space of interest M arises indirectly, and can be studied in terms of the spectrum of an associated Laplace operator. In the infinite dimensional setting, the study of the cohomology problem has profited from recent ideas with their origins in the mathematical theory of quantum fields. Of course, this approach can also be used in the usual, finite dimensional theory. The infinite dimensional theory is called entire cyclic cohomology. It can provide the geometric framework for the physics of fields, or more generally for physics of an infinite number of degrees of freedom. However, the mathematical notions involved stand on their own, independent of the examples from physical theories.

In de Rham theory, one studies a manifold M by means of functions on M and the associated tangent, cotangent vector and spinor bundles. In NCG, the algebra $C_0^\infty(M)$ is replaced by an operator algebra $\mathcal{A}$ acting on a Hilbert space. Assume for the moment that $\mathcal{A}$ is the algebra of smooth functions on a compact manifold M, acting as an algebra of multiplication operators, i.e. $\mathcal{A} = C^\infty(M)$; then Connes has shown that NCG reduces to de Rham theory [C1]. Thus NCG is a bonafide generalization of ordinary geometry. It allows the study of more general spaces, and also it gives a more general setting for the study of ordinary spaces.

In de Rham theory, the space M can naturally be recovered from the algebra $\mathcal{A}$ of functions on M; it is given by the maximal ideals of $\mathcal{A}$. Likewise, in NCG the underlying space M is determined by $\mathcal{A}$. Granted this does not give a very intuitive notion of how M looks. However there are advantages to working in terms of $\mathcal{A}$, for a natural theory of invariants arises.

By invariants, we mean integrals of forms which are invariant under smooth changes of the differential d. In special cases, these invariants can be interpreted as the index (in the sense of Atiyah and Singer) of the differential d, and in this case these invariants of NCG reduce to standard integer invariants of manifolds which are unchanged by sufficiently smooth homotopies of M. Just as in the case of non-compact manifolds, or in the theory of von Neumann algebras, the index in NCG is not in general integer valued. But they arise as cohomology classes, as in the standard theory, and the invariants are stable under sufficiently smooth deformations of d.

The connection between NCG and ideas from quantum physics centers about the fact that the NCG differential da is given by a graded commutator. Here the algebra $\mathcal{A} = \mathcal{A}_+ \oplus \mathcal{A}_-$ can be decomposed into its even and odd parts under a $\mathbb{Z}_2$ grading, γ, where $\gamma(\mathcal{A}_\pm) = \pm\mathcal{A}_\pm$. Generally, the physics interpretation of the grading is that $\mathcal{A}_+$ and $\mathcal{A}_-$ denote bosonic and fermionic operators respectively. The graded commutator, given by an odd operator D, is

$$da = \begin{cases} Da - aD\,, & a \in \mathcal{A}_+ \\ Da + aD\,, & a \in \mathcal{A}_- \end{cases} .$$

We let $[D,a]$ denote the graded commutator with D. Clearly $d : \mathcal{A}_+ \to \mathcal{A}_-$ and $d : \mathcal{A}_- \to \mathcal{A}_+$. Thus for all a, sufficiently regular to define multiple commutators,

$$d^2 a = \big[D, [D,a]\big] = D^2 a - aD^2 \,.$$

We therefore infer that if $D^2 = \lambda I$ is a multiple of the identity, then

$$d^2 = 0\,.$$

In physics examples, D will often be linear in fermionic creation operators, so $D^2 = 0$. The operator d plays the role of the exterior differential.

A second natural differential is the adjoint d^* to the differential to d, defined by

$$d^* a = \begin{cases} D^* a - aD^* & a \in \mathcal{A}_+\,, \\ D^* a + aD^* & a \in \mathcal{A}_- \end{cases} .$$

So if $D^2 = \lambda I$ then also $D^{*2} = \bar{\lambda} I$ and therefore $d^{*2} = 0$. In physics examples, D^* often arises as a linear function of fermionic annihilation operators (namely the adjoint of D), so naturally $D^{*2} = 0$.

The signature-type operator

$$\delta = d + d^*$$

is the differential defined by the commutator with the self adjoint operator $Q = D + D^*$, namely

$$\delta a = [Q, a]\,,$$

where by $[Q,a]$ we denote the graded commutator. Then

$$\delta^2 a = Q^2 a - aQ^2\,.$$

It is natural to introduce a positive, self-adjoint Hamiltonian operator $H = Q^2$, for which

$$\delta^2 a = [H, a]\,.$$

If $D^2 = 0$, then the Hamiltonian H can be written in the form of a Laplacian,

$$H = DD^* + D^* D\,.$$

The differential δa can be thought of as an infinitesimal symmetry – namely super-symmetry. It is generated by Q, which in physics is called the supercharge operator. The energy operator H is the super-symmetric Hamiltonian, and it generates a one-parameter unitary group $\sigma_t(A) = e^{itH} A e^{-itH}$. Generally we wish $\mathcal{A}$ to be invariant under σ_t, and σ_t to act on $\mathcal{A}$ as a continuous, one-parameter group of $*$-automorphisms. The properties of $H = Q^2$ are reflected in three cases:

(i) *Compact, Finite Dimensional Space:* For p sufficiently large,

$$\mathrm{Tr}(\mathrm{H} + \mathrm{I})^{-\mathrm{p}/2} < \infty \, . \tag{1}$$

Then dim $= \inf p$ for which (1) is finite. This is the p-summable case of Connes [C1].

(ii) *Compact, Infinite Dimensional Space:* This is the Θ-summable case [C2, JLO1], for which

$$\mathrm{Tr}(\mathrm{e}^{-\mathrm{tH}}) < \infty \, , \quad \text{all } \mathrm{t} > 0 \, . \tag{2}$$

(iii) *Non-Compact Space, Finite or Infinite Dimensional:* In this case there is a functional ω on A which satisfies the super-KMS constraints [Ka, JLO2]:

$$\omega(\delta a) = 0 \qquad \text{infinitesimal invariance,}$$

and

$$\omega(ab) = \omega(\gamma(b)a(1)) \qquad \text{super-KMS condition.}$$

Here γ is the $\mathbb{Z}_2$ grading and $a(t) = e^{-tH} a e^{tH} = \sigma_{it}(a)$. Here σ_t is the group of $*$-automorphisms (isometries) above. Then σ_{it} is the analytic continuation of σ_t to imaginary time. In the C^* algebra setting, this continuation always exists on a dense subalgebra of $\mathcal{A}$.

One fundamental result is the existence of field theories corresponding to case (ii) and to case (iii):

Theorem 1. *Polynomial, $N = 2$ Supersymmetric quantum fields on a cylindrical space-time exist and are Θ-summable. (See [JL1].) The $\mathbb{R}^2$ limit of space-time yields a super-KMS functional [JWe].*

3. NCG AS COHOMOLOGY

The appropriate cohomology theory is cyclic cohomology. In the infinite-dimensional case, it is natural to work with a version called entire cyclic cohomology, which I describe here. It is striking how much the mathematical framework of entire cyclic cohomology looks like the mathematical framework of quantum field theory and statistical mechanics, when they are studied through expectation values. This has the consequence on the side of mathematics that representations and computational tools of mathematical physics, such as functional integrals, become natural tools in understanding cohomology. On the side of physics, this connection reveals a cohomological (geometric) content to quantum field theory. Finite volume theories fall in the Θ-summable class. Infinite volume ones are described by the super-KMS framework.

The actual similarity between NCG and physics can be illustrated through the study of cohomology through the study of dual space to $\mathcal{A}$, or rather to the dual space to the tensor algebra over $\mathcal{A}$,

$$\mathcal{A} = \mathbb{C} \oplus \mathcal{A} \oplus \mathcal{A}^2 \oplus \cdots ,$$

where $\mathcal{A}^n$ is the n-fold tensor product $\mathcal{A} \otimes \mathcal{A} \otimes \cdots \otimes \mathcal{A}$. The space $\mathcal{C}_n$ of $(n+1)$-multilinear functionals $f_n(a_0, a_1, \ldots, a_n)$ is the natural space of cochains of NCG. A cochain, evaluated on $a_0, da, \ldots, da_n$ can be interpreted as an integral of the differential $a_0 da_1 \cdots da_n$,

$$f(a_0, da, \ldots, da_n) = \int a_0 da_1 \cdots da_n \,. \tag{NCG}$$

We are familiar in the context of mathematical physics of studying the algebra $\mathcal{A}$ of obeservables (e.g. in a Haag-Kastler framework, or in the Wightman field framework) in terms of expectation values

$$f(a_0, da, \ldots, da_n) = \langle a_0 da_1 \cdots da_n \rangle \,. \tag{QFT}$$

From the expectation values in a cycle state $\langle \ \rangle$, we can reconstruct the algebra $\mathcal{A}$ (or the field operators) through a GNS construction.

In entire cyclic cohomology, one restricts the cochains (expectations) in the direct one $\mathcal{C} = \oplus_n \mathcal{C}_n$ to satisfy a growth condition in n. In fact, $\mathcal{C}_n$ has a natural norm, given by a norm $\|\cdot\|$ on $\mathcal{A}$, with

$$\|f_n\| = \sup_{\|a_j\|=1} |f_n(a_0, \ldots, a_n)| \,.$$

The entire growth condition of Connes is the assumption that every sequence $f = (f_0, f_1, f_2, \ldots) \in \mathcal{C}$ satisfies

$$n^{1/2} \|f_n\|^{1/n} \longrightarrow 0 \,. \tag{3}$$

Another way to describe a cochain f, in using language of quantum field theorists, is as an element of the dual to the algebra of fields. This approach was pursued in the 1960's by Borchers, and the algebra is often given his name. A cochain f is an element of the Borchers algebra restricted by a growth condition, namely: $\Sigma_n \|f_n\| |z|^n$ is bounded by a function of exponential order 2. Just as in the theory of analytic functions a growth condition ensures certain uniqueness of interpolations, this growth condition ensures a certain type of uniqueness of a normalized representation of cohomology classes described below.

On the space $\mathcal{C}$ there are two natural coboundary operators, the Hochschild "creation" operator

$$b : \mathcal{C} \longrightarrow \mathcal{C}\,, \quad b : \mathcal{C}_n \longrightarrow \mathcal{C}_{n+1}\,,$$

and Connes' "annihilation" operator

$$B : \mathcal{C} \longrightarrow \mathcal{C}\,, \quad B : \mathcal{C}_{n+1} \longrightarrow \mathcal{C}_n\,.$$

These operators satisfy

$$b^2 = B^2 = 0\,, \quad \text{as well as} \quad Bb + bB = 0\,.$$

These commutation conditions mean that the double complex described mean that the double complex described by b and B has a total cohomology, determined by the coboundary operator

$$\partial = b - B\,, \quad \text{which also satisfies} \quad \partial^2 = 0\,.$$

Explicitly

$$\begin{aligned}(bf_n)(a_0, a_1, \ldots, a_{n+1}) = \sum_{j=0}^{n} &(-1)^j f_n(a_0, \ldots, a_j a_{j+1}, \ldots, a_{n+1}) \\ &+ (-1)^{n+1} f_n(a_{n+1}a_0, a_1, \ldots, a_n)\end{aligned}\,,$$

and

$$\begin{aligned}(Bf_n)(a_0, \ldots, a_{n-1}) = f_n(I, a_0, \ldots, a_{n-1}) &- (-1)^n f_n(a_0, \ldots, a_{n-1}, I) \\ &+ \text{cyclic antisymmetrization}\end{aligned}\,.$$

4. THE COCYCLE EQUATION

The fundamental cohomology classes in entire cyclic cohomology, namely those with coboundary operator ∂, are given by solutions of the cocycle equation

$$\partial\tau = 0\,, \quad \tau \in \mathcal{C}\,. \tag{4}$$

This could also be written $b\tau = B\tau$. A solution is said to be trivial if τ is a coboundary, namely if $\tau = \partial G$ for some $G \in \mathcal{C}$. Thus the classes of cocycles τ are determined by the equivalence relation

$$\tau \cong \tau + \partial G\,.$$

The relation between solutions to the cocycle equation and geometric invariants are provided by an abstract result of Connes [C2] which pairs projections in $\mathcal{A}$ with cocycle classes τ:

Theorem 2. *Given a projection $e \in \mathcal{A}$ and a solution $\tau \in C$ to the cocycle equation (4), the pairing*

$$\langle \tau, e \rangle = \sum_{n=0}^{\infty} (-1)^n \frac{(2n)!}{n!} \tau_{2n}(e, e, \ldots, e) \tag{5}$$

depends only on the cohomology class of τ. In particular, $\langle \partial G, e \rangle = 0$.

The pairing (5) provides invariants $\langle \tau, e \rangle$ under a homotopy τ^λ of τ, for a family indexed by $\lambda \in [0,1]$. The problem of finding invariants of $\mathcal{A}$ (and hence of the underlying commutative or non-commutative space M) can be understood in terms of finding solutions to the cocycle equation. Thus what nontrivial τ's satisfy $\partial \tau = 0$?

At the present time, only one family of solutions for τ is known: this family is labelled by a differential δ or by its generator Q. Given a graded derivation $\delta a = [Q, a]$ on $\mathcal{A}$, there is a corresponding cohomology class τ^Q which is a solution to the cocycle equation. This class τ^Q is called the Chern character of Q.

Two different formulas for τ^Q have been proposed in the literature, namely τ_C [C2] and τ_{JLO} [JLO1]. Originally they appeared to define different solutions to the cocycle equation, but they have recently been shown to be cohomologous representations of the same solution [C3]. Each representation has its advantages; the JLO cocycle is given in a form which is amenable to computation. In particular, τ_{JLO} occurs as an expectation value of the type which often occurs in quantum mechanics. In this case, it can sometimes be written as a function space integral, yielding a beautiful integral representation of the cohomology class. The other representation τ_C has the advantage of being the unique cohomologous cocycle which is normalized in the sense of entire cyclic cohomology.

Here I describe the JLO cocycle. The formula for the nth component $\tau^Q_{JLO,n}$ of τ^Q_{JLO} is (for even n)

$$\begin{aligned}\tau^Q_{JLO,n}(a_0, a_1, \ldots, a_n) = \operatorname{Str} \int_{0 \le t_1 \le t_2 \le \cdots \le t_n \le 1} & \Big(a_0 e^{-t_1 Q^2} \delta a_1 e^{-(t_2 - t_1) Q^2} \\ & \cdots e^{-(t_n - t_{n-1}) Q^2} \delta a_n e^{-(1-t_n) Q^2} \Big) dt_1 \cdots dt_n \, . \end{aligned} \tag{JLO}$$

Here Str denotes the graded (super-) trace. The formula for odd n is similar. Note that the integration over the simplex $0 \le t_1 \le \cdots \le t_n \le 1$ has volume $1/n!$ and hence τ_n satisfies the bound $n \|\tau_n\|^{1/n} \le e$. Namely τ_n satisfies (3) and in fact has exponential growth of order 1.

There is an especially simple expression for $\tau^Q_{JLO,n}$ on the diagonal. Namely,

$$\tau^Q_{JLO,n}(a, a, \ldots, a) = \frac{d^n}{ds^n} \operatorname{Str}(a \exp(-Q^2 + s\delta a)) \Big|_{s=0} \, .$$

Theorem 3. *Let e^{-tQ^2} be trace class for $t > 0$. Then the sequence (JLO) is an element of C and it also is a solution to the cocycle equation (4).*

In summary, what we have shown, is that given a graded (super-) derivation Q, there exists a functional τ^{Q}_{JLO} on projections e which yields a cohomology invariant $\langle \tau^{Q}, e \rangle$, given by the pairing formula (5), namely

$$\langle \tau^{Q}_{JLO}, e \rangle = \sum_{n=0}^{\infty} (-1)^{n} \frac{(2n)!}{n!} \frac{d^{2n}}{ds^{2n}} \mathrm{Str}(e \exp(-Q^{2} + s\delta e)) \bigg|_{s=0} . \tag{6}$$

Since the pairing (6) is expressed as a graded trace of a heat kernel, it may have a functional integral representation. This can be extremely useful from a computational point of view. The examples of Theorem 1 do have the functional integral interpretation of $\langle \tau^{Q}_{JLO}, e \rangle$.

The pairing $\langle \tau^{Q}_{JLO}, e \rangle$ of (6) actually provides invariants which are stable under small deformations of Q. This is investigated in [EPJL] for perturbations $Q' = Q + q$ of Q, where a sufficient condition of q is given to yield stability of the invariants. If q is odd and bounded, this condition is satisfied. A sufficient condition is also given for a class of unbounded perturbations q, in terms of a relative bound between q and Q^{2}, which I do not state here.

Theorem 4. *Let q be odd and bounded, or more generally satisfy the condition described in [EPJL]. Then τ^{Q}_{JLO} and τ^{Q+q}_{JLO} are cohomologous, namely for some $G \in \mathcal{C}$,*

$$\tau^{Q}_{JLO} = \tau^{Q+q}_{JLO} + \partial G .$$

The cocycle τ^{Q}_{JLO} has a very simple geometric interpretation. If Q_{+} denotes the part of Q which maps $\mathcal{H}_{+}$ to $\mathcal{H}_{-}$ (even to odd subspaces of $\mathcal{H}$ with respect to the grading γ) then

$$\langle \tau^{Q}_{JLO}, e \rangle = \mathrm{Index}((eQe)_{+}) .$$

The formula (6) is a generalized McKean-Singer type of representation for the index. We pose the general question: Is there a second, noncohomologous solution to the cocycle equation? Such a solution might yield torsion or other invariants. Another cocycle might also arise in Chern-Simons type of situations.

5. QUANTUM SPACE

We have seen that the underlying geometric space M arising in a non-commutative geometry may not be a classical manifold. In fact, the notion of a quantum space M arises naturally in this context. We should then ask whether quantum spaces play (or will play) an important role in theoretical physics? If so, then in what way do they? It is best to approach these questions in terms of examples. Is quantum physics described by quantum field theory on a quantum space?

There are not so many concrete examples of quantum spaces. This is understandable, since the connection between the algebra $\mathcal{A}$ and the space M is not straightforward. Clearly it would be good to have intuitive pictures of these and of other examples. Let me mention several examples which do exist:

1) Connes' 2-sheeted modification of space-time which leads to a beautiful picture (at the classical level) of the standard electro-weak model [CL].

2) The quantum torus, studied by many authors. In this case the algebra $\mathcal{A}$ is generated by p, q where

$$pq = \gamma qp$$

and $\gamma \in \mathrm{C}\,, \quad |\gamma| = 1$.

3) The quantum unit disc, QD. This algebra $\mathcal{A}$ is generated by $z, \bar{z}$, and one can imagine functions given by power series

$$f(z, \bar{z}) = \sum a_{nm} z^n \bar{z}^m \ .$$

We consider now this third example. Classically z and $\bar{z}$ commute. In the quantum disc one has $\|z\| \leq 1$ and for the real parameter h, $0 < h < 1$. The quantum disc algebra (QD algebra) is generated by z, $\bar{z}$ such that

$$[z, \bar{z}] = h(1 - \bar{z}z)(1 - z\bar{z}) \ . \tag{7}$$

The QD was studied by Berezin [B] and by Klimek and Lesniewski [KL]. It follows that $z\bar{z}$ and $\bar{z}z$ commute, and that $[z, \bar{z}] \geq 0$. In fact, Klimek and Lesniewski show in [KL] that $SU(1,1)$ acts on this algebra $\mathcal{A}$, so that $z' = (az + b)(\bar{b}z + \bar{a})^{-1}$ and the corresponding $\bar{z}' = (\bar{a}\bar{z} + \bar{b})(b\bar{z} + a)^{-1}$, where $|a|^2 - |b|^2 = 1$, also satisfy (7).

Given the invariance of the QD algebra under $SU(1,1)$, one can use the QD algebra to define a quantum Riemann surface, QRS. In particular,

$$QRS = QD \text{ algebra/Discrete Subgroup of } SU(1,1) \ .$$

The QRS is specified by the deformation parameter h, and goes over continuously to the classical RS when $h \to 0$, as described in the following section.

Clearly the QRS as a family of deformations of a Poisson structure has many aspects in common with geometric quantization. But geometric quantization seems to yield the QRS only for a discrete subset of values of h. The relation between these methods of quantization is a problem to understand better in the future.

6. QUANTIZATION OF SPACE BY THE TOEPLITZ MAP

The QD algebra can also be represented by a quantization in terms of Toeplitz operators. Let $dw_h(\zeta)$ represent the measure on the classical unit disc, CD, given by

$$dw_h(\zeta) = \frac{i}{2}(1 - |\zeta|^2)^{-1+\frac{1}{h}} d\zeta d\bar{\zeta} \; .$$

Let us denote the L^2 Hilbert space

$$\mathcal{E} = L_2(CD, dw_h) \; ,$$

with the inner product

$$\langle f, g \rangle_h = \int_{CD} \bar{f} g dw_h \; .$$

Let $\mathcal{H} \subset \mathcal{E}$ denote the Hilbert subspace of functions holomorphic on CD, and let P be the orthogonal projection from $\mathcal{E}$ to $\mathcal{H}$. (The projection P is given by the Bergman kernel.) Then we can project the operators of multiplication by ζ on $\mathcal{E}$ and of multiplication by $\bar{\zeta}$ on $\mathcal{E}$, to operators z and $\bar{z}$ which act on $\mathcal{H}$. This projection is described by the commutative diagrams,

$$\begin{array}{ccc} \mathcal{E} & \xrightarrow{\zeta} & \mathcal{E} \\ P\downarrow & & \downarrow P \\ \mathcal{H} & \xrightarrow[z]{} & \mathcal{H} \end{array} \quad \text{and} \quad \begin{array}{ccc} \mathcal{E} & \xrightarrow{\bar{\zeta}} & \mathcal{E} \\ P\downarrow & & \downarrow P \\ \mathcal{H} & \xrightarrow[\bar{z}]{} & \mathcal{H} \end{array} \; . \qquad \text{(Toeplitz Quantization Map)}$$

Such operators given by multiplication followed by projection are called Toeplitz operators in the mathematics literature. The general Toeplitz operator Pf arises from multiplication by a bounded function $f(\zeta, \bar{\zeta})$ on $\mathcal{E}$ and from a projection P of $\mathcal{E}$ onto a subspace $\mathcal{H}$ of $\mathcal{E}$. We call the operator Pf acting on $\mathcal{H}$, the quantization of the bounded function $f(\zeta, \bar{\zeta})$ acting as a multiplication operator on L^2 of the classical unit disc.

Theorem 5. *[KL] For $0 < h < 1$, the Toeplitz operators $z, \bar{z}$ defined by the quantization map $z = P\zeta$ and $\bar{z} = P\bar{\zeta}$, satisfy $\bar{z} = z^*$ and provide a representation of the algebra (7).*

The quantization Pf has a classical limit given by the Poisson bracket $\{\; , \;\}$ on $\mathcal{E}$. In fact, to make this classical limit precise, remark that the parameter h of (7) enters both the quantization map $P = P_h$ and the definition of the space $\mathcal{H} = \mathcal{H}(h)$. Then in [KL] it is shown that the commutator, divided by h, converges to the Poisson bracket in the following sense:

Theorem 6. *Let f and g be bounded functions on the unit disk. As $h \to 0$, the quantization map P_h satisfies*

$$\|\frac{1}{h}[P_h f, P_h g] - P_h\{f, g\}\|_{\mathcal{H}(h)} \leq O(h^{1/2}) \; . \tag{8}$$

General properties of Toeplitz maps appear in the mathematics literature, see *e.g.* [DHK]. These similar structures in mathematics and in mathematical physics portend that future developments will profit from the convergence of these fields.

7. QUANTIZATION OF FIELDS BY THE TOEPLITZ MAP

In fact, the deformation quantization of space (a manifold) and the canonical quantization of a classical field are precisely the same in spirit. Both quantizations are Toeplitz operator quantizations. We obtained a quantum space from a classical space above (in the case of the quantum disc) by the Toeplitz quantization map. In quantum field theory, the usual procedure to pass from Euclidean (classical) fields to canonical (quantum) fields also uses a Toeplitz quantization map. In the latter case, there is also a twist in the inner product on $\mathcal{H}$ (yielding reflection positivity). The corresponding quantization by Toeplitz operators is known in mathematical physics as the Osterwalder-Schrader construction.

In the quantization of the disc in the previous section, the quantized coordinate function becomes the coordinate in quantum space. In field theory the classical fields are the coordinate functions on the "Euclidean" Hilbert space of classical field configurations. We obtain the quantum field operators on $\mathcal{H}$, as the Toeplitz projection of classical fields. In the case of Toeplitz quantization of the unit disc, the projection operator P is chosen to map from L^2-functions to holomorphic functions. In the case of field quantization we take the projection into the subspace of field configurations with a "reflection positive" inner product.

Consider $\mathcal{E} = L^2(\mathcal{S}', d\mu)$, where $d\mu$ is a measure on $\mathcal{S}'(\mathrm{R}^d)$ which satisfies reflection positively. Also let $\mathcal{H} \subset \mathcal{E}$ be the subspace of $\mathcal{E}$ obtained by first projecting onto functions of fields at positive times, and secondly projecting into the topology given by the reflection positive quadratic form. Since the reflection positive form is bounded by the inner product on $\mathcal{E}$, the operator P is a projection. The Toeplitz quantization of the coordinate function φ on $\mathcal{E}$ is the Osterwalder-Schrader quantization of a Euclidean quantum field, namely φ_{quantum} which acts on $\mathcal{H}$.

$$\begin{array}{ccc} \mathcal{E} & \xrightarrow{\varphi} & \mathcal{E} \\ P\big\downarrow & & \big\downarrow P \\ \mathcal{H} & \xrightarrow[\varphi_{\text{quantum}}]{} & \mathcal{H} \end{array}$$

Just as in the case of quantum space, the interacting quantized fields given in the constructive theory of fields [GJ2] have classical limits. In most of the constructive field theory literature, Planck's constant h is normalized to 1. However, in order to study the $h \to 0$ limit, it is necessary to study carefully the h-dependence of various quantities. The same methods which yield the construction for $h = 1$, also yield the construction for $0 < h < 1$. The h-dependence of the expectations of (unbounded) field operators has been studied, and one finds that the classical limit is well behaved. Unlike Theorem 6 which is stated for bounded functions, the classical limit of quantized fields has only been studied in the unbounded case.

The work by Eckmann [E] on the classical limit of quantum field theories concerns results about expectation values and scattering matrix elements. The theorems in this paper are actually translations of estimates in [GJS, GJ2] (where $h = 1$) in order to follow the dependence on h. I expect that the same cluster expansion methods could be developed, with the aid of techniques in [GJ1], to obtain estimates in an appropriate operator topology which would yield a quantum field theory version of Theorem 6. Meanwhile, we have the results of [E]:

Theorem 7. *In weakly coupled $\mathcal{P}(\phi)_2$ models the Euclidean expectations and their analytic continuations to scattering matrix elements are C^∞ functions of $h^{1/2}$ as $h \to 0+$.*

In quantum field theory, the measure $d\mu$ is invariant under a one-parameter group $T(t)$ which therefore acts as a unitary group of operators on $\mathcal{E}$. The physical interpretation of the variable t is the time parameter. The projection $PT(t)P$ of $T(t)$, for $t \geq 0$, onto $\mathcal{H}$, is the semigroup of e^{-tH} generated by the Hamiltonian H of quantum theory.

$$\begin{array}{ccc} \mathcal{E} & \xrightarrow{T(t)} & \mathcal{E} \\ {\scriptstyle P}\downarrow & & \downarrow{\scriptstyle P} \\ \mathcal{H} & \xrightarrow[e^{-tH}]{} & \mathcal{H} \end{array}$$

The self-adjointness of H follows from the fact that the inner product is $\mathcal{H}$ includes a time reflection. More generally, a group $\mathcal{G}$ may act invariantly on $d\mu$ and commute with the quantization map P, in which case the action of $\mathcal{G}$ on $\mathcal{H}$ is unitary.

The canonical quantization in field theory arises from the application of the quantization map P to ϕ and its time derivative. Let $q(f) = P\phi(f \otimes \delta)$ and $p(g) = P\phi(g \otimes \dot{\delta})$. Here $(f \otimes \delta)(x,t) = f(x)\delta(t)$ and $(g \otimes \dot{\delta})(x,t) = f(x)\frac{d}{dt}\delta(t)$. The resulting operators on $\mathcal{H}$ satisfy $[q(f), p(g)] = i\langle f, g\rangle_{L^2}$.

8. CONCLUSIONS

Ideas from physics have entered the development of non-commutative geometry. The JLO cocycle arises in this way. In fact, this cocycle even appears to have implications for mathematical problems in number theory [C4].

I am optimistic that ideas in the field of non-commutative geometry will also flow in the other direction, back into physics. Having remarked that similar quantization maps appear naturally both in the quantization of space and the quantization of fields, we can ask whether quantum fields on quantum space provide an appropriate physical theory? Here lie many open questions.

We can dream of the proper formulation for the next revolution in quantum physics. For 65 years quantum theory has remained central to our thinking, and there is no indication that this will change. It is tempting, however, when seeing the close connection between the ideas of quantum theory and those of non-commutative geometry, to speculate that non-commutative geometry will play a key role in the physics of the future.

REFERENCES

[B] F. Berezin, General concepts of quantization, *Commun. Math. Phys.*, **40** (1975) 153–174.

[C1] A. Connes, Non-commutative differential geometry, *Publ. Math. IHES,* **62** (1986), 257–360.

[C2] A. Connes, Entire cyclic cohomology of Banach algebras and characters of Θ-summable Fredholm modules, *K-Theory,* **1** (1988), 519–548.

[C3] A. Connes, On the Chern character of Θ-summable Fredholm modules, *Commun. Math. Phys.,* **139** (1991), 171–181.

[C4] A. Connes, Lectures at MIT, April 1991.

[CL] A. Connes and J. Lott, Non-commutative geometry and particle physics, Nuclear Physics B, 1990, and these proceedings.

[DHK] R. Douglas, S. Hurder, and J. Kaminker, Cyclic cocycles, renormalization and eta-invariants, *Invent. math.*, **103** (1991), 101–179.

[E] J.-P. Eckmann, Remarks on the classical limit of quantum field theories, *Lett. Math. Phys.,* **1** (1977), 387–394.

[EPJL] K. Ernst, P. Feng, A. Jaffe, and A. Lesniewski, Quantum K-Theory, II. Homotopy invariance of the Chern character, *Jour. Funct. Anal.*, **90** (1990), 355–368.

[GJ1] J. Glimm and A. Jaffe, The $\lambda(\phi^4)_2$ quantum field without cutoffs III: the physical vacuum, *Acta Math.,* **125** (1970), 203–267.

[GJ2] J. Glimm and A. Jaffe, "Quantum Physics," Springer Verlag 1987.

[GJS] J. Glimm, A. Jaffe, and T. Spencer, The Wightman axioms and particle structure in the $\mathcal{P}(\phi)_2$ quantum field model, *Ann. Math.,* **100** (1974), 585- 632.

[GS] E. Getzler and A. Szenes, On the Chern character of a theta-summable module, *J. Funct. Anal.,* **84** (1989), 343–357.

[JL1] A. Jaffe and A. Lesniewski, Supersymmetric field theory and infinite dimensional analysis, in "Nonperturbative Quantum Field Theory," Proceedings of the 1987 Cargèse Summer School, G. 't Hooft *et. al.*, Editors, Plenum Press 1988.

[JL2] A. Jaffe and A. Lesniewski, Geometry of supersymmetry, 1988 Erice Lectures, G. Velo and A. Wightman, Editors, by Plenum Press, New York, 1990.

[JL3] A. Jaffe and A. Lesniewski, 1990-1991 Harvard University Lectures on "Quantum fields, geometry, and analysis in infinite dimensions," to appear.

[JLO1] A. Jaffe, A. Lesniweski, and K. Osterwalder, Quantum K-Theory, I. The Chern character, *Commun. Math. Phys.,* **118** (1989), 1–14.

[JLO2] A. Jaffe, A. Lesniewski, and K. Osterwalder, On super-KMS functionals and entire cyclic cohomology, *K-Theory,* **2** (1989), 675–682.

[JLWi] A. Jaffe, A. Lesniewski and M. Wisniowski, Deformations of super-KMS functionals, *Commun. Math. Phys.,* **121** (1989), 527–540.

[JO] A. Jaffe and K. Osterwalder, Ward identities for non-commutative geometry, *Commun. Math. Phys.*, **132** (1990), 119–130.

[JWe] S. Janowsky and J. Weitsman, A vanishing theorem for supersymmetric quantum field theory and finite size effects in multi-phase cluster expansions, *Commun. Math. Phys.,* to appear.

[Ka] D. Kastler, Cyclic cocycles from graded KMS functionals, *Commun. Math. Phys.,* **121** (1989), 345–350.

[KL] S. Klimek and A. Lesniewski, Quantum Riemann surfaces, *Commun. Math. Phys.,* to appear.

WHITHAM THEORY FOR INTEGRABLE SYSTEMS AND TOPOLOGICAL QUANTUM FIELD THEORIES

I. Krichever

Landau Institute for Theoretical Physics
USSR Academy of Sciences

1. INTRODUCTION

During the last two years remarkable connections between the non-perturbative theory of two-dimensional gravity coupled with various matter fields, the theory of topological gravity coupled with topological matter fields, the theory of matrix models and, finally, the theory of integrable soliton equations with special Virasoro constraints have been found [1-11]. The main goal of these few lectures is to present the results of perturbation theory of algebraic-geometrical solutions of integrable equations which clarify some of this connections.

All the "integrable" partial differential equations, which are considered in the framework of the "soliton" theory, are equivalent to compatibility conditions of auxiliary linear problems. The general algebraic-geometrical construction of their exact periodic and quasi-periodic solutions was proposed in [12,13]. The cornerstone of this construction is a concept of the Baker-Akhiezer functions - the functions which are uniquely defined with the help of their analytical properties on auxiliary Riemann surfaces. The corresponding analytical properties are so specific that it follows from them and only from them that the Baker-Akhiezer functions are common eigenfunctions of an overdetermined system of linear equations. Consequently, coefficients of these equations are solutions of non-linear equations. The analytical properties of the Baker-Akhiezer functions are the generalization of properties of the Bloch solutions of the finite-gap Shturm-Liuville operators, which were found in a serious of papers by Novikov, Dubrovin, Matveev and Its (see the review of them in [14,15]; the reviews of further development of the finite-gap theory can be found in [16-20]).

Roughly speaking, the algebraic-geometrical construction gives a map from a set of algebraic-geometrical data to a space of solutions of integrable non-linear equations.

$$\{algebraic-geometrical\ data\} \longmapsto \{solutions\ of\ NLPDE\}$$

In a generic case the space of algebraic-geometrical data is a bundle over the moduli space $M_{g,N}$ of smooth algebraic curves Γ_g of genus g with N punctures $P_1, \ldots, P_N$. A bundle over $M_{g,N}$ corresponds to additional data which are: systems of local coordinates $k_\alpha^{-1}(Q)$, $k_\alpha^{-1}(P_\alpha) = 0$, in the neighborhoods of punctures and a set of g points $\gamma_1 \ldots, \gamma_g$ on Γ_g in the general position or equivalently a point on the Jacobian

New Symmetry Principles in Quantum Field Theory, Edited by
J. Fröhlich et al., Plenum Press, New York, 1992

$J(\Gamma)$ of the corresponding curve Γ. We shall denote this complete set of data by $\tilde{M}_{g,N}$ and intermediate bundle will be denoted by

$$\hat{M}_{g,N} = \{\Gamma_g, P_\alpha, k_\alpha^{-1}\}$$

It is to be mentioned that $\tilde{M}_{g,N}$ are "universal" data. For the given non-linear integrable equation the corresponding set of data have to be specified. For example, the solutions of the Kadomtsev-Petviashvili (KP) hierarchy can be obtained from the set of data $\tilde{M}_{g,N}$ for $N = 1$.

The algebraic-geometrical construction of [12,13] is by definition a sort of an "inverse transform" : from "spectral" (algebraic-geometrical) data to "solutions". A "direct" transform : from "solutions" to "spectral" data is an unavoidable step in a construction of the perturbation theory of integrable equations. Such theory has to contain answers to the following questions: "how many solutions can be obtained with the help of the algebraic-geometrical construction ?" and "what is a structure of the union of all algebraic-geometrical solutions corresponding to different genera ?". The last question is the question about the solutions of linearized equations on the background of given algebraic-geometrical solution of initial non-linear PDE because they form a tangent space to the manifold of all solutions.

In the next two chapters a brief review of the corresponding results using as an example the KP equation is given. In the fourth chapter the, so-called, Whitham equations are introduced. They are the equations on the moduli space $\hat{M}_{g,N}$ and are "averaging" integrable equations. It is remarkable that they are also "integrable". We shall begin the presentation of the construction of their solutions with the simplest case : the construction of the Whitham equations on $\hat{M}_{0,N}$ which are quasi-classical limit (dispersionless analogue) of soliton equations.

As it was found in [21], the particular solution of dispersionless Lax equations coincides with the perturbed superpotential of topological Landau-Ginzburg models [22]. It turns out, that dispersionless analogue of τ-function for the corresponding solution gives the solution of truncated Virasoro constraints. In the last chapter the relations between general Whitham equations and $1/N$-expansion for loop-equations for matrix models are discussed.

2. BAKER-AKHIEZER FUNCTIONS AND ALGEBRAIC-GEOMETRICAL SOLUTIONS OF THE KP EQUATION

Let Γ be a non-singular algebraic curve of genus g with N punctures P_α and fixed local parameters $k_\alpha^{-1}(Q)$ in the neighborhoods of punctures. For any set of the points $\gamma_1, \ldots, \gamma_g$ in the general position there exists a unique up to constant factor, $c(t_{\alpha,i})$, a function $\psi(t, Q)$, $t = (t_{\alpha,i})$,$\alpha = 1, \ldots, N$; $i = 1, \ldots,$ such that:

(i) the function ψ (as a function of the variable Q which is a point of Γ) is meromorphic everywhere except for the points P_α ant it has at most simple poles at the points $\gamma_1, \ldots, \gamma_g$ (if all of them are distinct).

(ii) at the neighborhood of the point P_α the function ψ has the form

$$\psi(t, Q) = \exp(\sum_{i=1}^{\infty} t_{\alpha,i} k_\alpha^i)(\sum_{s=0}^{\infty} \xi_{s,\alpha}(t) k_\alpha^{-s}), k_\alpha = k_\alpha(Q). \tag{2.1}$$

this is the most general definition of a scalar *multipoint* and *multivariable* Clebsh-Gordan-Baker-Akhiezer function (or simply the Baker-Akhiezer function). It depends on the variables $t = \{t_{1,i}, \ldots, t_{n,i}\}$ as on external parameters.

From the uniqueness of the Baker-Akhiezer function it follows that for each pair (α, n) there exists a unique operator $L_{\alpha,n}$ of the form

$$L_{\alpha,n} = \partial_{\alpha,1}^{n} + \sum_{j=1}^{n-1} u_j^{(\alpha,n)}(t)\partial_{\alpha,1}^{j} \tag{2.2}$$

(where $\partial_{\alpha,i} = \partial/\partial t_{\alpha,i}$) such that

$$(\partial_{\alpha,i} - L_{\alpha,n})\psi(t,Q) = 0. \tag{2.3}$$

The idea of the proof of theorems of this type which was proposed in [12,13] is universal.

For any formal serious of the form (2.1) their exists a unique operator $L_{\alpha,n}$ of the form (2.2) such that

$$(\partial_{\alpha,i} - L_{\alpha,n})\psi(t,Q) = O(k^{-1})\exp(\sum_{i=1}^{\infty} t_{\alpha,i}k_\alpha^i). \tag{2.4}$$

The coefficients of $L_{\alpha,n}$ are differential polynomials with respect to $\xi_{s,\alpha}$. They can be found after substitution of the serious (2.1) into (2.4).

It turns out that if the series (2.1) is not formal but is an expansion of the Baker-Akhiezer function in the neighborhood of P_α then the congruence (2.4) becomes an equality. Indeed, let us consider the function ψ_1

$$\psi_1 = (\partial_{\alpha,n} - L_{\alpha,n})\psi(t,Q). \tag{2.5}$$

It has the same analytical properties as ψ except only one. The expansion of this function in a neighborhood of P_α starts from $O(k^{-1})$. From uniqueness of the Baker-Akhiezer function it follows that $\psi_1 = 0$ and the equality (2.3) is proved.

Corollary. The operators $L_{\alpha,n}$ satisfy the compatibility conditions

$$[\partial_{\alpha,n} - L_{\alpha,n}, \partial_{\alpha,m} - L_{\alpha,m}] = 0. \tag{2.6}$$

Remark. The equations (2.6) are gauge invariant. For any function $g(t)$ operators

$$\tilde{L}_{\alpha,n} = gL_{\alpha,n}g^{-1} + (\partial_{\alpha,n})g^{(-1)} \tag{2.7}$$

have the same form (2.2) and satisfy the same operator equations (2.6). The gauge transformation (2.8) corresponds to the gauge transformation of the Baker-Akhiezer function

$$\psi_1(t,Q) = g(t)\psi(t,Q)$$

Example. One-point Baker-Akhiezer function.

In the one point case the Baker-Akhiezer function has an exponential singularity at a single point P_1 and depends on a single set of variables. Let us choose the normalization of the Baker-Akhiezer function with the help of the condition $\xi_{1,0} = 1$, i.e. an expansion of ψ in the neighborhood of P_1 equals

$$\psi(t_1, t_2, \ldots, Q) = \exp(\sum_{i=1}^{\infty} t_i k^i)(\sum_{s=0}^{\infty} \xi_s(t)k^{-s}). \tag{2.8}$$

In this case operator L_n has the form

$$L_n = \partial_1^n + \sum_{i=o}^{n-2} u_i^{(n)} \partial_1^i. \tag{2.9}$$

For example, for $n = 2, 3$ we have

$$L_2 = \partial_1^2 - u(t), L_3 = \partial_1^3 - \frac{3}{2} u \partial_1 + w.$$

Let us define the variables $x = t_1, y = \sigma^{-1} t_2, t = t_3$. Then from (2.6) it follows (for $n = 2, m = 3$) that $u(x, y, t, t_4, \ldots)$ satisfies the KP equation

$$\frac{3}{4} \sigma^2 u_{yy} + (u_t - \frac{3}{2} u u_x + \frac{1}{4} u_{xxx})_x = 0. \tag{2.10}$$

Remark. It should be emphasized that algebraic-geometrical construction is not a sort of abstract "existence" and "uniqueness" theorems. It provides the exact formulae for solutions in terms of the Riemann theta-functions. For example, the algebraic-geometrical solutions of the KP equation have the form

$$u(t_1, t_2, \ldots) = 2 \partial_1^2 \ln \theta(\sum_{i=1}^{\infty} U_i t_i + \Phi) + const \tag{2.11}$$

where $\theta(z_1, \ldots, z_g) = \theta(z_1, \ldots, z_g | B(\Gamma)$ is the Riemann theta-function which defined with the help of matrix $B(\Gamma)$ of b-periods of normalized holomorphic differentials on Γ. The vectors $J_i = \{U_{1,k}, \ldots, U_{g,i}\}$ are vectors of b-periods of normalized Abelian differentials with the only pole at P_1 (see details in [13]).

The equations (2.6) for $n = 2, m > 3$ describe evolutions of $u(x, y, t, t_4, \ldots)$ with respect to "higher times" or equivalently the whole KP hierarchy. Here it is necessary to make a few comments. In the original form the equations (2.6) are a set of non-linear equations on the coefficients $u_i^{(n)}$ and do not have the form of evolution equations. It can be shown (see [23]) that they are equivalent to evolution system in the form which was proposed by Kyoto group [24]. We shall show that for algebraic-geometrical solutions.

For any formal series of the form (2.8) there exists a unique pseudo-differential operator $\mathcal{L}$

$$\mathcal{L} = \partial_1 + \sum_{i=1}^{\infty} u_i(t_1, \ldots) \partial_1^{-i}. \tag{2.12}$$

such that

$$\mathcal{L} \psi(t, k) = k \psi(t, k). \tag{2.13}$$

Then the operators L_n which are uniquely defined from the congruence (2.4) are equal to

$$L_n = [\mathcal{L}^n]_+, \tag{2.14}$$

where $[\ldots]_+$ denotes a differential part of a pseudo-differential operator. From (2.3) it follows that if $\psi(t, k)$ is an expansion of the Baker-Akhiezer function then

$$(\partial_n - [\mathcal{L}^n]_+) \psi(t, Q) = 0. \tag{2.15}$$

The comparitability conditions of (2.13) and (2.15) imply the evolution equations

$$\partial_n \mathcal{L} = [[\mathcal{L}^n]_+, \mathcal{L}] \tag{2.16}$$

on the coefficients $u_i(t_1,\ldots)$ of $\mathcal{L}$. The equations (2.16) are the Sato form for the KP hierarchy.

At the end of this chapter we shall make a few comments about multi-point case. For each α the equation (2.6) up to gauge transformation are equivalent to the KP hierarchy corresponding to each set of variables $\{t_{\alpha,i}\}$. "Which is the interaction between two different "KP hierarchies"?

As it was found in [14,25] for the two-point case a full set of equations can be represented in the following form

$$[\partial_{\alpha,n} - L_{\alpha,n}, \partial_{\beta,n} - L_{\beta,n}] = D^{\alpha,\beta}_{N,m} H^{\alpha,\beta} , \tag{2.17}$$

where $H^{\alpha,\beta}$ is a two-dimensional Schrödinger operator in a magnetic field

$$H^{\alpha,\beta} = \frac{\partial^2}{\partial_{\alpha,1}\partial_{\beta,1}} + v_1^{\alpha,\beta}\partial_{\alpha,1} + v_2^{\alpha,\beta}\partial_{\alpha,2} + u^{\alpha,\beta} \tag{2.18}$$

and operators $D^{\alpha,\beta}_{N,m}$ are differential operators in variables $t_{\alpha,1}, t_{\beta,1}.\alpha, \beta$ are differential operators in variables $t_{\alpha,1}, t_{\beta,1}$.

The sense of (2.17) is as following. For the given operator $H^{\alpha,\beta}$ any differential operator D in the variables $t_{\alpha,1}, t_{\beta,1}$ why a number of can be uniquely represented in the form

$$D = D_1 H^{\alpha,\beta} + D_2 + D_3$$

where D_2 is a differential operator with respect to the variables $t_{\alpha,1}$ only and D_3 is a differential operator with respect to the variable $t_{\beta,1}$ only. The equation (2.17) implies that the second and the third term in the corresponding representation for the left hand side of (2.17) are equal to zero. This implies $n+m-1$ equations on $n+m$ unknown functions (the coefficients of operators $L_{\alpha,n}$ and $L_{\beta,m}$). The equations (2.17) are gauge invariant. That is why a number of equations equals a number of unknown functions.

3. PERIODIC PROBLEM FOR THE KP EQUATION

The KP hierarchy in the form (2.16) is a set of commuting evolution equations on a set of the coefficients $u_i(x), x = t_1$, of a pseudo-differential operator $\mathcal{L}$ (2.12). The original form of the KP equation (2.10) is a nonlinear equation on a single function $u(x,y,t)$ and, though it has not a purely evolution form, the Caushy problem can be formulated for it The Caushy data is a function of two variables $v(x,y) = u(x,y,t = 0)$. In the same sense all equations (2.6) for $N = 1, n = 2, m > 3$ can be formally considered as flows on a space of functions depending on the two variables x, y. These flows can be well-defined only on some subspaces of the space of functions on two variables. That's why the exact form of an equivalence between two forms of the KP hierarchy is a delicate subject. It is deeply connected with the problem of an invertibility of the algebraic-geometrical transform

$$\{\tilde{M}_{g,1}\} \longmapsto \{solutions\ of\ the\ KP\ hierarchy\} \tag{3.1}$$

The algebraic-geometrical solutions of the KP hierarchy are quasi-periodic functions of all variables, as it follows from the exact formula (2.11).

Let us consider periodic (in x and y) solutions. The corresponding subset of algebraic-geometrical data can be specified in the following way. For the given algebraic curve Γ with fixed point P_1 and local parameter in its neighborhood the differentials dp and dE can be defined as meromorphic differentials on Γ with the only poles at P_1 of the form

$$dp = idk(1 + O(k^{-2}), dE = i\sigma^{-1}dk^2(1 + O(k^{-3}) \tag{3.2}$$

and as differentials which have *real* periods along any cycle γ on Γ

$$\oint dp = U_\gamma, \mathrm{Im} U_\gamma = 0,$$

$$\oint dE = V_\gamma, \mathrm{Im} V_\gamma = 0. \tag{3.3}$$

If all the periods have the form

$$U_\gamma = \frac{2\pi n_\gamma}{a_1}, V_\gamma = \frac{2\pi m_\gamma}{a_2},$$

where n_γ and m_γ are integral numbers, then the corresponding solutions of the KP equation have periods a_1 and a_2 with respect to the variables x and y . The set of all "periodic" curves is dense in the whole moduli space for $a_1, a_2 \to \infty$.

Now we shall try to invert the map (3.1) in the case of periodic solutions. The solutions $\psi(x, y, w_1, w_2)$ of the non-stationary Shrodinger equation

$$(\sigma\partial_y - \partial_x^2 + u(x, y))\psi(x, y, w_1, w_2) = 0 \tag{3.4}$$

with a periodic potential $u(x, y)$ are called Bloch solutions, if they are eigenfunctions of the monodromy operators, i.e.

$$\psi(x + a_1, y, w_1, w_2) = w_1\psi(x, y, w_1, w_2)$$
$$\psi(x, y + a_2, w_1, w_2) = w_2\psi(x, y, w_1, w_2). \tag{3.5}$$

The set of pairs $Q = (w_1, w_2)$, for which their exists such a solution is called the Floque set and will be denoted by Γ. The multivalued functions $p(Q)$ and $E(Q)$ such that

$$w_1 = \exp(ipa_1), w_2 = \exp(iEa_2) \tag{3.6}$$

are called quasi-momentum and quasi-energy, respectively.

For the free operator with zero potential $u_0 = 0$, the Floque set is parametrized by the points of the complex k-plane

$$w_1^0 = \exp(ika_1), w_2^0 = \exp(-\sigma^{-1}k^2a_2) \tag{3.7}$$

and the Bloch solutions have the form

$$\psi_0(x, y, k) = \exp(ikx - \sigma^{-1}k^2y). \tag{3.8}$$

It turns out ([18]) that if *Re* $\sigma \neq 0$ then the Floque set of the operator (3.4) with the smooth potential $u(x, y)$ is isomorphic to the Riemann surface Γ (which has infinite genus in a generic case). The corresponding Riemann surfaces have such a specific structure that the theory of Abelian differentials, theta-functions and so on, can be constructed for them as well as for the finite genus case. (See detailed description of Γ in [18].) If Γ has a finite genus then u is an algebraic-geometrical potential. The last statement means that for periodic algebraic-geometrical solutions of the KP equation the corresponding curve in their algebraic-geometrical data coincides with the curve of the Bloch solutions. At the same time the Baker-Akhiezer function is the Bloch function. In [18] it was proved also that the algebraic geometrical solutions are dense in a space of all periodic (in x, y) solutions of the KP equation for $Re\ \sigma \neq 0$.

4. THE PERTURBATION THEORY OF THE ALGEBRAIC-GEOMETRICAL SOLUTIONS. WHITHAM EQUATIONS

The non-linear WKB (or Whitham) method can be applied to any non-linear equation which has a set of exact solutions of the form

$$u_0(x,y,t) = u_0(Ux + Vy + Wt + \Phi | I_1, \ldots, I_N), \tag{4.1}$$

where $u_0(z_1, \ldots | I_k)$ is a periodic function of the variables z_i depending on parameters (I_k). The vectors U, V, W are also functions of the same parameters : $U = U(I), V = V(I), W = W(I)$.

In the framework of the non-linear WBK-method (for (1 + 1)-systems see [26-28,31]) the asymptotic solutions of the form

$$u(x,y,t) = u_0(\varepsilon^{-1} S(X,Y,T) | I(X,Y,T)) + \varepsilon u_1 + \ldots \tag{4.2}$$

are constructed for the perturbed or non-perturbed initial equation. Here $X = \varepsilon x, Y = \varepsilon y, T = \varepsilon t$ are "slow" variables. If the vector $S(X,Y,T)$ is defined from the relations

$$\partial_X S = U(I(X,Y,T)) = U(X,Y,T),$$
$$\partial_Y S = V(X,Y,T), \quad \partial_T S = W(X,Y,T), \tag{4.3}$$

then the main term u_0 in the expansion (4.2) satisfies the initial equation up to the first order in ε. After that all the other terms of the series (4.2) are defined from the non-homogeneous linear equations. They can be easily solved if a full set of solutions for a homogeneous linear equation are known.

The KP equation linearized on the background u_0 has the form

$$\frac{3}{4}\sigma^2 v_{yy} + (v_t - \frac{3}{2}u_0 v_x - \frac{3}{2}u)_{0x} v + \frac{1}{4}v_{xxx})_x = 0. \tag{4.4}$$

The adjoined linear equation is

$$\frac{3}{4}\sigma^2 \Phi_{yy} - (\Phi_t + \frac{3}{2}u_0 \Phi_x + \frac{1}{4}\Phi_{xxx})_x = 0. \tag{4.5}$$

In the operator form the linearized KP hierarchy is linear equations on the pseudo-differential operator

$$\delta\mathcal{L} = \sum_{i=1}^{\infty} \delta u_i \partial_1^{-i}$$

and has the form

$$\partial_n \delta\mathcal{L} = [[\mathcal{L}_0^n]_+, \delta\mathcal{L}] + [\delta[\mathcal{L}^n]_+, \mathcal{L}_0].$$

Solutions of the linearized equations can be found from the following obvious observation. If a set of solutions of non-linear equation depending on some parameters is known, the derivatives (with respect to these parameters) are solutions of the linearized equation. In other words solutions of the linearized equation form a tangent space for a space of solutions of the non-linear equation.

In the case of algebraic-geometrical solutions there are deformations of algebraic-geometrical data which preserve genus of curves and there are deformations in "transversal" directions which increase genus of curves. The simplest type of the last deformations as follows. One has to take a pair of points Γ and "glue" between them a "thin" handle. For the periodic solutions u_0 of the KP equation (with $Re\ \sigma \neq 0$) it was shown [31] that such deformations together with deformations along $\tilde{M}_{g,1}$ give a full set of periodic solutions of the linearized equation (4.4) and the exact formulae were obtained for them in terms of the Riemann theta-functions . We shall not consider here a complete perturbation theory even for the KP equation. We restrict ourselves only to a part of this theory which contains a new features with respect to the usual perturbation theory.

The most essential part of it is the, so-called, Whitham equations, which define a "slow" dependence of parameters $I(X, Y, T)$ and, hence, define the main term of the series (4.2).

The asymptotic solutions of the form (4.2) can be constructed with an arbitrary dependence of the parameters I_k on slow variables. In this case the expansion (4.2) will be valid on a scale of order 1. The right hand side of the non-homogeneous linear equation for u_1 contains the first derivatives of the parameters I_k. Therefore, the choice of the dependence of I_k on slow variables can be used for the cancellation of the "secular" term in u_1. The necessary conditions of such cancellation are "ortogonality" conditions of the corresponding right hand side of the equation for u_1 to the cotangent bundle to $\hat{M}_{g,1}$. The cotangent bundle to a space of solutions of a non-linear equation is the space of solutions of an adjoint to the linearized equation. In [31] it was proved that the cotangent bundle for $\hat{M}_{g,1}$ corresponds to the product of the Baker-Akhiezer function and its "dual" $\psi^+(t, Q)$.

The dual Baker-Akhiezer function (in one point case) is a function which is meromorphic on Γ outside P_1 and has at most simple poles at the points $\gamma_1^+, \ldots, \gamma_g^+$. (If $\gamma_1, \ldots, \gamma_g$ are poles of the Baker-Akhiezer function then γ_s^+ are defined as a set of g points such that $\gamma_1, \ldots, \gamma_g$ and $\gamma_1^+, \ldots, \gamma_g^+$ are zeros of a differential $d\Omega$ on Γ with the only pole of order 2 at P_1.)

In the neighborhood of P_1 the function $\psi^+(t, Q)$ has the form

$$\psi^+(t_1, t_2, \ldots, Q) = \exp(\sum_{i=1}^{\infty} -t_i k^i)(\sum_{s=0}^{\infty} \xi_s^+(t) k^{-s}). \tag{4.6}$$

It turns out that if ψ satisfies the equations (2.13) and (2.15) then ψ^+ is a solution of the equations

$$\psi^+ \mathcal{L} = k\psi^+(t, k), \tag{4.7}$$

$$-\partial_n \psi^+ = \psi^+[\mathcal{L}^n]_+, \tag{4.8}$$

where the left action of differential operators is defined as usual

$$\psi^+(w\partial_1^i) = (-1)^i \partial_1^i (\psi^+ w). \tag{4.9}$$

From (2.13,2.15) and (4.7,4.8) it follows that if $\delta\mathcal{L}$ is a solution of the linearized equations then

$$\partial_n < \psi^+ \delta\mathcal{L}\psi > = < \psi^+(-[\mathcal{L}_0^n]_+ \delta\mathcal{L} + [\mathcal{L}_0^n]_+, \delta\mathcal{L}] + [\delta\mathcal{L}^n]_+, \mathcal{L}] + \delta\mathcal{L}[\mathcal{L}_0^n]_+)\psi > = 0. \tag{4.10}$$

(Here $< \ldots >$ denotes the mean value with respect to $x = t_1$). The equality (4.10) proves that $\psi^+(t, Q)\psi(t, Q)$ is a solution of adjoint equation.

From this statement it follows that if F is a right hand side of the equation for u_1 then the necessary conditions for uniform boundness of u_1 are the equalities

$$< \psi^+(t,Q)F(t)\psi(t,Q) > = 0, \tag{4.11}$$

which have to be fulfilled for all $Q \in \Gamma$.

In [31] it was shown that (4.11) can be represented in the following form. Let us consider the differentials dp, dE on Γ, which were defined by the conditions (3.2,3.3) and let us consider also a meromorphic differential $d\Omega$ with the only pole at P_1 of the form

$$d\Omega = idk^3(1 + O(k^{-4}), \tag{4.12}$$

which has *real* periods on Γ. The integrals $p(Q)$, $E(Q)$, $\Omega(Q)$ of these differentials are multivalued functions on the bundle M_g^* over $\hat{M}_{g,1}$

$$M_g^* = (\Gamma, P_1, k^{-1}, Q \in \Gamma).$$

If I_k are local coordinates on $\hat{M}_{g,1}$ and (λ, I_k) are local coordinates on M_g^*, then for any dependence of I_k on the variables X, Y, T the integrals p, E, Ω become functions of the variables X, Y, T: $p = p(\lambda, X, Y, T)$, $E = E(\lambda, X, Y, T)$, $\Omega = \Omega(\lambda, X, Y, T)$.

Theorem [31]. The necessary conditions for the existence of the asymptotic solutions of the equation

$$\frac{3}{4}\sigma^2 u_{yy} + (u_t - \frac{3}{2}uu_x + \frac{1}{4}u_{xxx})_x + \varepsilon K[u] = 0. \tag{4.13}$$

which has the form (4.2) with uniformly bounded first-order term are equivalent to the equation

$$\frac{\partial p}{\partial \lambda}(\frac{\partial E}{\partial T} - \frac{\partial \Omega}{\partial Y}) - \frac{\partial E}{\partial \lambda}(\frac{\partial p}{\partial T} - \frac{\partial \Omega}{\partial X}) + \frac{\partial \Omega}{\partial \lambda}(\frac{\partial p}{\partial Y} - \frac{\partial E}{\partial X}) = \frac{< \psi^+ K \psi >}{< \psi^+ \psi} \frac{\partial p}{\partial \lambda}. \tag{4.14}$$

Here $K[u]$ is an arbitrary differential polynomial.

The parameters of the algebraic-geometrical solutions of the whole KP hierarchy are points of the infinite dimensional manifold $\hat{M}_{g,1}$. If only a finite number of flows is considered then only a finite-dimensional part of $\hat{M}_{g,1}$ is essential.

Two local parameters k_1 and k would be called m-equivalent, if $k_1 = k + O(k^{-m})$. The class of m-equivalency of the local parameter is denoted by $[k^{-1}]_m$. From the definition of the Baker-Akhiezer function it follows that if it considered as a function of only m first "times" (i.e. $t_i = 0, i > m$) then ψ and the corresponding solutions of first equations (2.16) depend only on $[k^{-1}]_m$. The algebraic-geometrical solutions of the KP equation depend on $[k^{-1}]_3$. Hence, the number of there parameters equals $N = 3g + 2, I = (I_1, \ldots, I_N)$.

The equality (4.14) has to be fulfilled for any $Q \in \Gamma$, but they are not all independent. It turns out that they are equivalent to $3g + 2$ independent equations (i.e. the equations (4.14) are well-defined).

First of all, let us make an important remark. The equations (4.14) are invariant with respect to any change of local coordinate on Γ, $\lambda_1 = \lambda_1(\lambda, I)$. Hence, any particular choice of λ can be used. Let us choose a function $\lambda(Q)$ on Γ with the only pole of order $r = g + 3$ at P_1 such that

$$\lambda^{1/r}(Q) = k(Q) + O(k^{-3}(Q)) \tag{4.15}$$

(in a generic case such function exists and is unique). The function λ defines a local coordinate in the neighborhood of any point of Γ except points q_s where differential $d\lambda$ equals zero: $d\lambda(q_s) = 0$. The number of such points equals $N = 3g + 2$. The requirement that there are no poles in the left hand side of (4.14) implies (4.14).

Important remark. The Whitham equations are real equations because normalization conditions for differentials $dp, dE, d\Omega$ are real. Below we shall consider all times $\sqrt{-1}t_i$ as real variables (but the case where $a_i t_i$ are real for given complex numbers a_i can be considered in the same way).

Let us introduce meromorphic differentials $d\Omega_i$ with the only pole at P_1 of the form

$$d\Omega_i = d(k^i + O(k^{-1}))$$

and such that all their periods on Γ are *real*. (In our previous notations: $dp = d\Omega_1, dE = d\Omega_2, d\Omega = d\Omega_3$). Then the Whitham equations (in the case $K = 0$) can be written for any pair of times $i, j > 1$

$$\frac{\partial p}{\partial \lambda}(\partial_i \Omega_j - \partial_j \Omega_i) - \frac{\partial \Omega_j}{\partial \lambda}(\partial_i p - \partial_x \Omega_i) + \frac{\partial \Omega_i}{\partial \lambda}(\partial_j p - \partial_x \Omega_j) \; = \; 0. \qquad (4.16)$$

(we preserve here the same notation t_i for "slow" variables εt_i).

Let us consider now n-th reduction of the KP hierarchy which is the hierarchy of Lax equation. The Lax equations are equations on the coefficients of differential operator

$$L = \partial_1^n + u_{n-2}\partial_1^{n-2} + \ldots + u_0, \qquad (4.17)$$

which have the form

$$\partial_i L = [[L^{i/n}]_+, L], i = 1, 2, ..., \qquad (4.18)$$

They are the particular case of (2.16) corresponding to pseudo -differential operator $\mathcal{L}$ such that

$$L = \mathcal{L}^n. \qquad (4.19)$$

The subset $M_g(n)$ of algebraic geometrical data $\hat{M}_{g,1}$ which give the solutions of (4.18) is the following set of data: { curve Γ with puncture P_1 is such that there exists a function $E(Q)$ on Γ with the only pole at P_1 of order n; the local parameter in the neighborhood of P_1 is $E^{-1/n}$}. The corresponding solutions of (2.16) do not depend on $(t_n, \; t_{2n}, \; \ldots)$ and are solutions of (4.18).

The Whitham equations (4.16) for the choice $\lambda(Q) = E(Q)$ take the form

$$\partial_i p(t, E) \; = \; \partial_x \Omega_i(t, E). \qquad (4.20)$$

(for the KdV equation such form of the Whitham equations were obtained for the first time in [27]).

In [31] the construction of exact solutions for the Whitham equations was proposed. Only some particular cases of this construction would be considered below in detail. In the next chapter we shall start with simplest zero genus case.

5. DISPERSIONLESS LAX EQUATIONS

The algebraic-geometrical solutions of the KP hierarchy corresponding to zero genus case are simple constants. Nevertheless, the Whitham equations even in this case are not trivial.

The zero genus curve Γ_0 with puncture can be identified with usual complex p-plane where infinity corresponds to P_1. There are no moduli in this part of data, but there is still infinite number of parameters corresponding to a choice of local parameter K^{-1}

$$K(p) = p + \frac{u}{2}p^{-1} + \dots . \tag{5.1}$$

Hence, the Whitham equations in this case are the equations on the local parameter $K^{-1}(p)$. They have the form (4.16) where by definition

$$\Omega_i(p) = [K^i(p)]_+, \tag{5.2}$$

here $[...]_+$ denotes a non-negative part of Laurent series.

Example. From (5.2) we have that

$$\Omega_2 = k^2 + u, \quad \Omega_3 = k^3 + \frac{3}{2}uk + w$$

and equation (2.16) is equivalent to the dispersionless KP equation (which is also called the Khokhlov-Zabolotskaya equation).

$$\frac{3}{4}u_{yy} + (u_t - \frac{3}{2}uu_x) = 0. \tag{5.3}$$

The Whitham equations for genus zero case describe quasi-classical limit of the corresponding initial equations.(We would like to mention here two papers which were devoted to the dispersionless Lax and KP equations. In [29] the construction of integrals for dispersionless Lax equations was proposed. In [30] the construction of particular solution of (5.3) was presented. The dispersionless KdV equation is

$$u_t = \frac{3}{2}uu_x \tag{5.4}$$

and it is well known that all solutions of it can be obtained in the implicit form from the equation

$$x + tu = f(u), \tag{5.5}$$

where $f(u)$ is an arbitrary function. In that sense the dispersionless KdV equation is even "more integrable" than the KdV equation. The construction of the Whitham equations which was proposed in [31] is a deep generalization of (5.5).

Let us consider the dispersionless Lax equations. The corresponding hierarchy describes solutions of (4.18) which are slow functions of all the variables. They can be written as a system of evolution equation on the coefficients of a polynomial $E(p)$

$$E(p) = p^n + u_{n-2}p^{n-2} + \dots + u_0. \tag{5.6}$$

The analogues of (4.18) are the equations which are a particular case of the Whitham equations

$$\partial_i E = \frac{d\Omega_i}{dp}\partial_x E - \frac{dE}{dp}\partial_x \Omega_i, \tag{5.7}$$

here $\Omega_i(p)$ are the polynomials (5.2), where $K(p)$ is such that $K^n = E(p)$.

Let $p(E)$ be the inverse (multi-valued) function for (5.6)

$$p(E) = K + O(K^{-1}), K^n = E. \tag{5.8}$$

Then $\Omega_i(p) = \Omega_i(p(E)) = \Omega_i(E)$ can be also considered as multi-valued function of the variable E (or K)

$$\Omega_i(E) = K^i + \sum_{j=1}^{\infty} \chi_{i,j} K^{-j}. \tag{5.9}$$

The equations (5.7) are equivalent to the equations

$$\partial_i p(E) = \partial_x \Omega_i(E). \tag{5.10}$$

which is genus zero particular case of representation (4.10)

In this chapter we consider mainly only one special case of the construction [31], which provides solutions of the Whitham equations, corresponding to the loop-equations and topological minimal models. For the dispersionless Lax equations the corresponding construction looks as follows. Let us define the formal series

$$S_+(p) = \sum_{i=1}^{\infty} t_i \Omega_i(p) = \sum_{i=1}^{\infty} t_i K^i + O(K^{-1}) \tag{5.11}$$

(if only a finite number of t_i is not equal to zero, then $S_+(p)$ is a polynomial). The coefficients of S_+ are linear functions of t_i and polynomials on u_i. We introduce the dependence of u_j on the variables t_i with the help of the following relation: the ratio

$$B(p) = \frac{dS_+}{dE} = \sum_{i=0}^{\infty} b_i p^i \tag{5.12}$$

should contain only positive powers of the variable p. (If $t_i = 0$, for $i > N$, then it means that $B(p)$ *should* be a polynomial with respect to p.) The relation (5.12) defines $u_i(t_1, t_2, ...)$ as implicit function.

The defining relations (5.12) can be represented in another form. Let q_s be zeros of the polynomials

$$\frac{dE}{dp}(q_s) = 0. \tag{5.13}$$

then (5.12) are equivalent to the equalities

$$\frac{dS_+}{dp}(q_s) = 0. \tag{5.14}$$

Remark. From (5.14) it follows that u_i do not depend on the variables $t_n, t_{2n}, t_{3n}, \ldots$, because $\Omega_{pn} = E^p$, $p = 1, 2, 3, \ldots$.

Let us prove that if (5.12) or (5.14) is fulfilled, then

$$\partial_i S_+(E) = \Omega_i(E). \tag{5.15}$$

Consider the function $\partial_i S_+(E)$. From (5.11) it follows that

$$\partial_i S_+(E) = K^i + O(K^{-1}) = \Omega_i(E) + O(K^{-1}). \tag{5.16}$$

Hence, it is enough to prove that $\partial_i S_+(E)$ is a polynomial in p, because Ω_i is the only polynomial in p such that

$$\Omega_i(p) = K^i + O(K^{-1}). \tag{5.17}$$

The function $\partial_i S_+(E)$ is holomorphic everywhere except maybe at q_s. In the neighborhood of q_s the local coordinate is

$$(E - E_s)^{1/2} + \ldots \tag{5.18}$$

and the derivative $\partial_i S_+(E)$ might be singular at the points q_s. But the defining relations (5.14) imply that $\alpha_s = 0$. Therefore, $\partial_i S_+(E)$ is regular everywhere except at the infinity and is a polynomial. The equations (5.15) are proved.

The compatibility conditions of (5.15) imply (5.10) (because $p = \Omega_1, x = t_1$). Hence, (5.12) defines in the implicit form a solution of the dispersionless Lax equations.

Let us define a function

$$F(t_1, t_2, ...) = -\frac{1}{2} \operatorname{res}(S dS_+), \quad S = \sum_{i=1}^{\infty} t_i K^i. \tag{5.19}$$

From the following formulae it follows that the function

$$\tau_K(t_1, t_2, ...) = \exp F(t_1, t_2, ...) \tag{5.20}$$

is an analogue of the τ-function for usual Lax equations.

Remark. The functions F, τ do not depend on $t_n, t_{2n}, \ldots$.

In [21] it was proved, that

$$\partial_i F = -\operatorname{res}(K^i dS_+). \tag{5.21}$$

Hence, F is a homogeneous function of the variables t_i:

$$\sum_{i=1}^{\infty} t_i \partial_i F = 2F. \tag{5.22}$$

The function F contains all the information about Ω_i. For example, the coefficients $\chi_{i,j}$ of the expansions (5.9) equal:

$$\partial_i \partial_j F = -i\chi_{j,i} = -j\chi_{i,j}. \tag{5.23}$$

The function F satisfies the truncated version of the Virasoro constraints which were obtained in[10,11] for the partition function of two-dimensional gravity models. The dispersionless analogue of these constraints have the form ([21]):

$$\sum_{i=n+1}^{\infty} it_i \partial_{i-n} F + 12 \sum_j S d\delta_m S_+). \tag{5.29}$$

For $m = -n, 0, n, 2n, \ldots$ from the defining relations (5.14) it follows that the variation

$$\delta_m S_+ = n \frac{dS_+}{dE} E^{\frac{m+n}{n}} \tag{5.30}$$

is an entire function of the variable p. Therefore,

$$\delta_m S_+ = \sum_{i=1}^{\infty} t_i (i[K^{i+m}]_+ - \sum_{j=1}^{m} j\chi_{ij}[K^{m-j}]_+), \tag{5.31}$$

where χ_{ij} are the coefficients of the expansion (5.9).

The substitution of (5.31)into (5.29) gives for $m = 0, n, 2n, \ldots$

$$0 = \sum_{i=1}^{\infty} [it_i \partial_{i+m} F + \frac{1}{2} \sum_{j=1}^{m-1} (t_i \partial_i \partial_j F)(\partial_{m-j} F)]. \tag{5.32}$$

From (5.22) it follows that

$$\sum_{i=1}^{\infty}(t_i\partial_i\partial_j F) = \partial_j(\sum_{i=1}^{\infty} t_i\partial_i F) - \partial_j F = \partial_j F \tag{5.33}$$

and finally we obtain the equations (5.25, 5.26). For $m = -n$ we have

$$0 = -2 \sum_{i=-m+1}^{\infty} it_i\partial_{i+m}F + \mathrm{res}(\sum_{i=1}^{-m} it_i K^{i+m} dS_+). \tag{5.34}$$

The second term in (5.34) equals

$$\mathrm{res}(\sum_{i=1}^{-m} it_i K^{i+m} dS) = - \sum_{i+j=-m} ijt_it_j, \tag{5.35}$$

which proves (5.24).

Remark. A few comments on more general solutions of dispersionless Lax equations. The proof of the statement that the defining relations (5.14) implies that the corresponding polynomial $E(p)$ satisfies dispersionless Lax equations can be repeated without any changes if we introduce instead of S_+ a new function

$$\hat{S}_+ = S_+ + S_0,$$

where S_0 is piecewise holomorphic function with a constant jumps on some contours (see [31]).

6. THE TOPOLOGICAL MINIMAL MODELS

Topological minimal models were introduced in [32] and were considered in [33]. They are a twisted version of the discrete series of $N = 2$ superconformal Landau-Ginzburg (LG) models. A large class of the $N = 2$ superconformal LG models has been studied in [34-36]. It was shown, that a finite number of states are topological, which means that their operator products have no singularities. These states form a closed ring $\mathcal{R}$, which is called a primary chiral ring. It can be expressed in terms of the superpotential $W(p_i)$ of the corresponding model

$$\mathcal{R} = \frac{C[p_i]}{dW = 0}, \quad dW = \frac{\partial W}{\partial p_i} dp_i. \tag{6.1}$$

In topological models these primary states are the only local physical excitations.

In [22], it was shown that correlation functions of primary chiral fields and integrals of their second des can be expressed in terms of perturbed superpotentials $W(p_i, t_0, t_1, ...)$. For A_{n-1} model the unperturbed superpotential has the form:

$$W_0 = \frac{p^n}{n}. \tag{6.2}$$

(we use here the same normalization as in [22]). The equations which define its perturbation are as follows ([22]):

For any polynomial

$$W(p) = \frac{1}{n}(p^n + u_{n-2}p^{n-2} + ... + u_0) \tag{6.3}$$

polynomials Φ_i, $i = 0, 1, 2, ..., n-2$,

$$\Phi_i = p^i + O(p^{i-1}) \tag{6.4}$$

are defined with the help of the orthogonality conditions

$$\langle \Phi_i \Phi_j \rangle = \mathrm{res}(\frac{\Phi_i \Phi_j}{\partial_p W}) = \delta_{i,n-2-j}. \tag{6.5}$$

The equations which define the dependence of $W(p, t_0, t_1, ..., t_{n-2})$ on the "coupling constants" t_i have the form:

$$\partial_i W = \frac{\partial W}{\partial t_i} = -\Phi_i. \tag{6.6}$$

The solution of the equations (6.6) with the initial conditions

$$W(p, 0, 0, ...) = W_0(p) \tag{6.7}$$

were found in [22].

The relations of these results to the dispersionless Lax equations are given by the following theorem ([21]).

Theorem. *Let $E(p, t_1, t_2, ...t_{n-1}, t_{n+1}, ...)$ be the solution of the dispersionless Lax hierarchy which was constructed above. Then the superpotential of the perturbed A_{n-1} topological minimal model is equal to*

$$W(p, t_0, ...t_{n-2}) = \frac{1}{n} E(p, t_0, \frac{t_1}{2}, ..., \frac{t_{n-2}}{n-1}, \frac{1}{n+1}, 0, 0, ...). \tag{6.8}$$

The partition function of this model is equal to

$$F = F(t_0, \frac{t_1}{2}, ..., \frac{t_{n-2}}{n-1}, \frac{1}{n+1}, 0, 0, ...), \tag{6.9}$$

where $F(t_1, t_2, ...)$ is given by the formula (5.21).

By definition, the perturbed A_{n-1} model depends on $n-1$ variables. As it follows from the above-formulated theorem, after the natural extension as a function of an infinite number of "times" the partition function of this model satisfies the constraints (5.24-5.26). Now we are going to consider the relations of these constraints with the loop equations for matrix models.

7. LOOP EQUATIONS

The loop equations have been originally proposed for Yang-Mills theories. (The review of the latest works on the applications of the loop equations to matrix models and $2d$ quantum gravity can be found in [37].)

The hermitean matrix model is defined by the partition function

$$Z_N = \int DM \exp(-\mathrm{Tr}(V(M)), \tag{7.1}$$

where M is $N \times N$ hermitean matrix,

$$V(K) = \sum_{i=0}^{\infty} \tilde{t}_i K^i. \tag{7.2}$$

The Wilson loop corollator is by definition

$$\mathcal{W}(K) = \langle \mathrm{tr}\frac{1}{K-M}\rangle = -\sum_{i=0}^{\infty} K^{-i-1}\partial_i \log Z_N. \tag{7.3}$$

The loop equations are derived from the invariance of the integral (7.1) with respect to the infinitesimal shift of M and have the form

$$[\sum_{i=1}^{\infty} i\tilde{t}_i K^{i-1})\mathcal{W}(K)]_- = \mathcal{W}^2(K) + \frac{\delta}{\delta V}\mathcal{W}(K). \tag{7.4}$$

$[\ldots]_-$ denotes the negative part of Laurent series.The equation (7.4) has to be supplemented with the condition

$$\mathcal{W}(K) = \frac{N}{K} + O(K^{-2}). \tag{7.5}$$

The leading term of $1/N^{-2}$ expansion of a solution of (7.4) should be the solution of the truncated equation

$$(\sum_{i=1}^{\infty} i\tilde{t}_i K^{i-1})\mathcal{W}_0(K) = \mathcal{W}_0^2(K). \tag{7.6}$$

Below we consider only the "even" case $\tilde{t}_{2i+1} = 0$.

Let us consider the solution of the dispersionless KdV equation (the $n = 2$ case of the dispersionless Lax equations) which was constructed in the second section. If $F(t_1, t_3, ..., t_{2i+1}, ...)$ is defined by (5.19), the constraints (5.24-5.26) are equivalent to the equation

$$[(\sum_{i=1}^{\infty}(2i+1)t_{2i+1}K^{2i-1})(t_1K^{-1} + \sum_{j=1}^{\infty}(\partial_{2j-1}F)K^{-2j-1})]_- +$$

$$\frac{1}{2}(t_1K^{-1} + \sum_{j=1}^{\infty}\partial_{2j-1}FK^{-2j-1}) = 0. \tag{7.7}$$

From (5.21), it follows that

$$\mathcal{W}_0 = -\frac{d}{dE}(t_1K + S_-(K)), \tag{7.8}$$

where

$$S_- = S - S_+. \tag{7.9}$$

Therefore, if we define

$$\tilde{t}_{2i} = \frac{2i+1}{2i}t_{2i+1}, \quad N = -\frac{1}{2}t_1, \tag{7.10}$$

then the solution of the dispersionless KdV hierarchy gives the solution of (7.6) with the condition (7.5) with the help of the formula (7.8). This solution coincides with, so-called,"one-cut" solution [37]. At the end of this chapter we shall show that the Whitham equations (or more exactly, their special solutions) on the moduli space of hyperelliptic curves give "multi-cut" solutions of the equation (7.6).

Let us consider the Whitham equations for the genus g algebraic-geometrical solutions of the Lax equations (4.18). They are equations on the moduli space $M_g(n)$.

In full analogy with the genus zero case their particular solutions can be obtained with the help of the following defining relations

$$\frac{dS_+}{dE} \text{ is regular function on } \Gamma \text{ outside } P_1, \tag{7.11}$$

where dS_+ is defined with the help of formula (5.11) (the only difference with dispersionless case is the difference in the definition of the differentials $d\Omega_i$.

The relations (7.11) are equivalent to the equations

$$dS_+(q_s) = 0, \tag{7.12}$$

where q_s are zeros of the differential $dE(q_S) = 0$.

Example. $n = 2$ In this case the space $M_g(2)$ is the space of sets of distinct points $E_1, ..., E_{2g+1}$. The corresponding curves are hyperelliptic curves which are defined by the equation

$$y^2 = \prod_{i=1}^{2g+1} (E - E_i) = R(E). \tag{7.13}$$

The differentials $d\Omega_i$ has a'priori the form:

$$d\Omega_{2i+1} = \frac{Q_i(E)}{\sqrt{R(E)}} dE = \frac{2i+1}{2} \frac{E^{g+i} + ...}{\sqrt{R}} dE. \tag{7.14}$$

The coefficients of the polynomial $Q_i(E)$ are uniquely defined from the normalizing conditions for $d\Omega_i$, which are equivalent to a system of linear equations. They become the functions of E_i and can be expressed through complete hyperelliptic integrals. Therefore, the polynomial Q_i is also a function of the variables E_i, i.e.

$$Q_i(E) = Q_i(E|E_1, ...E_{2g+1}). \tag{7.15}$$

The defining relations (7.12) are equivalent to a set of non-linear transcendent (but not differential) equations

$$\sum_{i=1}^{\infty} t_i Q_i(E_m|E_1, E_2, ..., E_{2g+1}) = 0, m = 1, 2, ..., 2g+1. \tag{7.16}$$

They define E_m as functions of t_i, which are the solution of the Whitham equations on $M_g(n)$.

Let $F(t_1, ...)$ be the function given by the same formula (5.19) where S_+ corresponds to the solution of the Whitham equations on $M_g(n)$. Then all the relations (5.21-5.23) and the constraints (5.24-5.26) would be fulfilled as well. Hence, for $n = 2$, the formulae (7.8,7.9) after redefinition of "times" (7.10) provides "multi-cut solutions of the equation (7.6).

Important remark. These solutions are not analytical functions of t_i, because the normalizing conditions for Ω_i are real equations. Locally, analytical solutions of (7.6) can be obtained if we consider the Whitham-type equations on the Teichmüller space, which covers $M_g(n)$. Corresponding equations have the same form (4.20), but now normalizing conditions which define Ω_i, should be chosen in the form:

$$\oint_{a_i} d\Omega_i = 0, \tag{7.17}$$

where a_i, b_i is a canonical basis of cycles on Γ. After that all the previous statements will be fulfilled.

REFERENCES

1. E. Brézin, V. Kazakov, *Phys. Lett.* (1990),**B236**, 144.

2. M. Douglas, S. Shenker, *Nucl. Phys.* (1990) **B335**, 635.

3. D. J. Gross, A. Migdal, *Phys. Rev.* (1990) **64**, 127.

4. D. J. Gross, A Migdal, *Nucl. Phys.* (1990) **B340**, 333.

5. T. Banks, M. Douglas, N. Seiberg, S. Shenker, *Phys. Lett.* (1990) **238B**, 279.

6. M. Douglas , *Phys. Lett.* **238B** (1 B.Dubrovin, I. Krichever, S. Novikov, *Soviet Doklady* **229**, No 1 (1976), 15.

7. E. Witten, *Nucl. Phys.* **B340** (1990), 176.

8. R. Dijkgraaf, E. Witten, *Nucl. Phys.* **B342** (1990), 486.

9. J. Distler, *Nucl. Phys.* **B342** (1990), 523.

10. M. Fukuma, H. Kawai, Continuum Schwinger-Dyson equations and universal structures in two-dimensional quantum gravity, preprint Tokyo University UT-562, May 1990.

11. M. Fukuma, H. Kawai, Infinite dimensional Grassmanian structure of two-dimensional quantum gravity, preprint Tokyo University UT-572, November 1990.

12. I. Krichever, *Soviet Doklady* **227:2** (1976), 291.

13. I. Krichiver, *Funk. Anal. i Pril.* **11**, No 13 (1977), 15.

14. B. Dobrovin, V. Matveev, S. Novikov, *Uspekhi Mat. Nauk.]* **31**, No 1 (1976), 55–136.

15. V. Zakharov, S. Manakov, S. Novikov, L. Pitaevski, Soliton theory, Moscow, Nauka, 1980.

16. B. Dubrovin, *Uspekhi Mat. Nauk* **36**, No 2 (1981),11-80.

17. I.Krichever, S. Novikov, *Uspekhi Mat. Nauk,* **35**, No 6 (1980).

18. I. Krichever, *Uspekhi Mat. Nauk,* **44**, No 2 (1989), 121.

19. B. Dubrovin, I .Krichever, S. Novikov, Integrable systems, *VINITY AN USSR,* 1985.

20. I.Krichever, *Uspekhi Mat. Nauk* **32**, No 6 (1977), 180.

21. I. Krichever, Topological minimal models and dispersionless Lax equations, preprint ISI Turin (to appear in *Comm. Math. Phys.* (1991).

22. E. Verlinder, H. Verlinder, A solution of two-dimensional topological quantum gravity, preprint IASSNS-HEP-90/40, PUPT-1176 (1990).

23. M. Sato,Y. Sato Soliton equations as dynamical systems in an infinite dimensional Grassmann manifolds, in *Nonlinear Partial Differential equations in Applied Sciences* (North-Holland, Amsterdam, 1982).

24. E.Data, M. Kashivara, M. Jimbo, T. Miva Transformation groups for soliton equations, in *Nonlinear Integrable systems - Classical Theory and Quantum Theory* (World Scientific, Singapore, 1983).

25. B.Dubrovin, I. Krichever, *S. Novikov, Soviet Doklady* **229**, No 1 (1976), 15.

26. A. Gurevich, L. Pitaevskii, *JETP* **65**, No 3 (1973), 590.

27. H. Flashka, M. Forest, L.McLaughlin, *Comm. Pure and Appl. Math.* **33** (6).

28. S. Yu. Dobrokhotov, V. P. Maslov, *Soviet Scientific Reviews, Math. Phys. Rev.* OPA Amsterdam 3 (1982), 221-280.

29. V. E. Zakhavor, *Funk. Anal. i Pril.* **14** (1980), 89.

30. Y. Kodama, J. Gibbons, *Phys. Lett.* 135A (1989), 171.

31. I. Krichever, *Funk. Anal. i Pril.* **22(3)** (1988), 37–52.

32. T Eguchi, S.-K. Yang, $N = 2$ superconformal models as topological field theoreies, preprint of Tokyo University UT-564 (1990).

33. K.Li, Topological gravity with minimal matter , Caltech-preprint CALT-68-1662.

34. E. Martinec, *Phys. Lett.* **217B** (1989), 431.

35. C. Vafa, N. Warner, *Phys. Lett.* **218B** (1989), 51.

36. W. Lerche, C. Vafa, N.P. Warner, *Nucl. Phys.* **B324** (1989), 427.

37. Yu. Makeeno, Loop equations in matrix models and in 2D quantum gravity, Submitted to *Mod. Phys. Lett.* A.

QUANTUM SYMMETRY IN QUANTUM THEORY

Gerhard Mack and Volker Schomerus

II.Institut für Theoretische Physik
Universität Hamburg[1]

ABSTRACT

In quantum theory, internal symmetries more general than groups are possible. The generalization of the algebra $\mathcal{G}$ of functions on the group needs to be neither commutative nor associative, and it may involve truncation. With truncation, the dimensions of tensor products of representations of the dual algebra $\mathcal{G}^*$ may be smaller than the product of the dimensions of the original representations.

1. INTRODUCTION: SYMMETRY IN QUANTUM MECHANICS

Soon after the discovery of quantum mechanics, the exploitation of symmetry in atomic physics played an important role in working out the predictions of quantum mechanics. After that it was taken for granted that symmetries in quantum theory should be groups, as they are in classical physics. However, it need not be so. Supersymmetry was a first exception. Here we will discuss the most general algebraic structures which admit a physical interpretation as symmetries. We call them quasi quantum groups for short. There is some parallelism between the formulation of the theory and of its possible symmetries. In quantum mechanics, the points of phase space go away. But the "algebra of functions" on phase space remains, although it becomes noncommutative. Group elements may be regarded as points g in a space G. In classical mechanics, g can act as a map on points in phase space. When these points are no longer, the group elements are not needed either. It suffices to have an algebra $\mathcal{G}$ of "functions on the group G". This algebra may be noncommutative. It turns out that it need not be associative either. Quasiassociativity is enough.

To stay as close as possible to the traditional formulation of a symmetry we consider the dual algebra $\mathcal{G}^*$ of the algebra $\mathcal{G}$ of functions on a group. This is a generalization of the group algebra. We will assume that $\mathcal{G}^*$ has a unit element e which acts as the trivial transformation in the Hilbert space of physical states. $\mathcal{G}^*$ is called *cocommutative* if $\mathcal{G}$ is commutative, and *coassociative* when $\mathcal{G}$ is associative.

[1]E-mail: I02MAC@DHHDESY3.BITNET

New Symmetry Principles in Quantum Field Theory, Edited by
J. Fröhlich et al., Plenum Press, New York, 1992

The product in $\mathcal{G}$ yields a coproduct in $\mathcal{G}^*$ which is a homomorphism of algebras,

$$\Delta : \mathcal{G}^* \mapsto \mathcal{G}^* \otimes \mathcal{G}^* . \tag{1.1}$$

The various stages of generalization of the notion of symmetry can be characterized by the properties of the coproduct.

cocommutative	*coassociative*	$\Delta(e) = e \otimes e$	*name*
yes	*yes*	*yes*	*group algebra*
no	*yes*	*yes*	*Hopf algebra*
no	*no*	*yes*	*quasi Hopf algebra*
no	*no*	*no*	*weak quasi Hopf algebra*

There exists an extensive literature about quantum groups (i.e. quasitriangular Hopf algebras) [3]. Quasi-Hopf algebras were introduced by Drinfeld [4]. The present authors proposed to build in truncation, which leads to what we call weak quasi Hopf algebras [1].

The coproduct is used to define a tensor product $\tau \bigotimes \tau'$ of representations τ and τ' in vector spaces V and V' of $\mathcal{G}^*$.

$$(\tau \bigotimes \tau')(\xi) = (\tau \otimes \tau')(\Delta(\xi)) . \tag{1.2}$$

The right hand side involves the outer tensor product $\tau \otimes \tau'$ of representations, which is by definition a representation of $\mathcal{G}^* \otimes \mathcal{G}^*$. If $\Delta(e) \neq e \otimes e$ then there is *truncation*.This means that the "true" representation space of $\tau \bigotimes \tau'$ is in general a proper subspace of $V \otimes V'$. In this way the truncated tensor products of quantum group representations which appear in conformal field theory can be accomodated.

Among the representations of $\mathcal{G}^*$ there should occur a 1-dimensional representation ϵ. It furnishes a homomorphism of algebras

$$\epsilon : \mathcal{G}^* \mapsto \mathbf{C} . \tag{1.3}$$

ϵ defines the substitute for the trivial 1-dimensional representation of a group (and its group algebra).

We adopt the framework of second quantized quantum mechanics, so that there are field operators $\Psi(\mathbf{r}, t)$ which create particles or excitations. In order that $\mathcal{G}^*$ is a symmetry of a quantum mechanical system, its Hilbert space of physical states $\mathcal{H}$ should carry a unitary representation $\mathcal{U}$ of $\mathcal{G}^*$ with $\mathcal{U}(e) = 1$. Unitarity means that

$$\mathcal{U}(\xi^*) = \mathcal{U}(\xi)^* \quad . \tag{1.4}$$

The Hamiltonian H and the ground state $|0>$ should be invariant,

$$\mathcal{U}(\xi)|0>= |0> \epsilon(\xi) , \quad [\mathcal{U}(\xi), H] = 0 \text{ for all } \xi \in \mathcal{G}^* . \tag{1.5}$$

In addition, the tensor product is invoked to postulate a transformation law of field operators $\Psi^J_m(\mathbf{r}, t)$ [9]. We will use superscripts $I, J, K, ...$ to distinguish between field multiplets which transform according to some irreducible representations, and subscripts $i, j, k, ...$ to distinguish the members of a multiplet. Ψ^J is supposed to transform according to the finite dimensional representation τ^J of $\mathcal{G}^*$. In compact notation this says that

$$\mathcal{U}(\xi)\Psi^J_k(\mathbf{r}, t) = \Psi^J_l(\mathbf{r}, t)(\tau^J_{lk} \bigotimes \mathcal{U})(\xi) \quad . \tag{1.6}$$

More explicitly, if

$$\Delta(\xi) = \sum_{\sigma} \xi_{\sigma}^{1} \otimes \xi_{\sigma}^{2} , \tag{1.7}$$

then

$$\mathcal{U}(\xi)\Psi_{k}^{J}(\mathbf{r},t) = \sum_{\sigma} \Psi_{l}^{J}(\mathbf{r},t)\tau_{lk}^{J}(\xi_{\sigma}^{1})\mathcal{U}(\xi_{\sigma}^{2}) . \tag{1.8}$$

This can be used to move elements of $\mathcal{G}^*$ through field operators from left to right. In the special case where ξ is element of a symmetry group, $\Delta(\xi) = \xi \otimes \xi$, and the field transformation law reduces to the usual one.

From the transformation law of field operators one deduces that many particle states transform according to tensor products of representations.

Summing up, one needs a 1-dimensional representation ϵ in order to formulate the notion of invariance, a coproduct Δ in order to define a tensor product of representations and to formulate a transformation law of field operators, and a *-operation on $\mathcal{G}^*$ in order to state unitarity.

The desire to formulate $\mathcal{G}^*$-covariant local commutation relations for field operators will lead us to impose additional requirement. Neither commutativity nor accociativity of $\mathcal{G}$ (i.e. coassociativity and cocommutativity of $\mathcal{G}^*$) should be given up completely, but they can be weakened to quasi-associativity and braid-commutativity.

When coassociativity is weakened to quasi-coassociativity, the tensor product of representations is not associative, but there is still equivalence of representations

$$((\tau^{I} \bigotimes \tau^{J}) \bigotimes \tau^{K}) \cong (\tau^{I} \bigotimes (\tau^{J} \bigotimes \tau^{K})) . \tag{1.9}$$

Similarly, braid-commutativity will imply that

$$(\tau^{I} \bigotimes \tau^{J}) \cong (\tau^{J} \bigotimes \tau^{I})' . \tag{1.10}$$

The prime indicates interchange of indices ("the indices go with the matrices").

The formulation of the equivalence (1. 9) of triple tensor products of arbitrary representations involves a "reassociator" $\varphi \in \mathcal{G}^* \otimes \mathcal{G}^* \otimes \mathcal{G}^*$ with certain properties. Similarly, the formulation of braid-commutativity involves an element $R \in \mathcal{G}^* \otimes \mathcal{G}^*$. Relations involving ϵ, Δ and φ and R arise from the requirement that tensoring with the "trivial" 1-dimensional representation ϵ amounts to doing nothing. The existence of φ will enable us to define a $\mathcal{G}^*$-covariant product of field operators, so that one can write down $\mathcal{G}^*$-covariant operator product expansions etc.

In addition one requires the existence of an antipode $\mathcal{S}$. It is a (linear) antihomomorphism of algebras

$$\mathcal{S} : \mathcal{G}^* \mapsto \mathcal{G}^* . \tag{1.11}$$

The antihomomorphism property means that $\mathcal{S}(\xi\eta) = \mathcal{S}(\eta)\mathcal{S}(\xi)$ The inverse $\mathcal{S}^{-1}$ of the antipode is used to define the contragredient representation $\tilde{\tau}$ to a representation τ by

$$\tilde{\tau}(\xi) = {}^{t}\tau(\mathcal{S}^{-1}(\xi)) . \tag{1.12}$$

for $\xi \in \mathcal{G}^*$. ^{t} stands for the transpose of a matrix. When a suitable additional condition on $\mathcal{S}$ is imposed, the field transformation law (1. 8) can be inverted as follows

$$\Psi_{k}^{J}(\mathbf{r},t)\mathcal{U}(\xi) = \sum_{\sigma} \mathcal{U}(\xi_{\sigma}^{2})\tilde{\tau}_{kl}^{J}(\xi_{\sigma}^{1})\Psi_{l}^{J}(\mathbf{r},t) \equiv (\tilde{\tau}_{kl}^{J} \otimes \mathcal{U})(\Delta(\xi))\Psi_{l}^{J}(\mathbf{r},t) . \tag{1.13}$$

This tells us how to move elements of $\mathcal{G}^*$ through field operators from right to left. This relation serves as a starting point for the discussion of the transformation law of adjoints of field operators.

In quantum mechanics, it is very important that one can take the adjoint of operators. In other words, one has a *-operation. This is used to define the notion of a unitary representation through the requirement (1. 4). To give meaning to the left hand side of eq.(1. 4), a *-operation must be defined on $\mathcal{G}^*$. It should be an antilinear antihomomorphism of algebras $\mathcal{G}^* \mapsto \mathcal{G}^*$. In order that the *-operations preserve the algebraic structures, Δ and ϵ should be *-homomorphisms, i.e.

$$\Delta(\xi^*) = \Delta(\xi)^* , \quad (1.\ 14)$$
$$\epsilon(\xi^*) = \overline{\epsilon(\xi)} . \quad (1.\ 15)$$

But $\Delta(\xi) \in \mathcal{G}^* \otimes \mathcal{G}^*$. Therefore we must first say how the *-operation acts on $\mathcal{G}^* \otimes \mathcal{G}^*$. There are two possibilities

$$(\xi \otimes \eta)^* = \xi^* \otimes \eta^* , \quad \text{or} \quad (1.\ 16)$$
$$(\xi \otimes \eta)^* = \eta^* \otimes \xi^* . \quad (1.\ 17)$$

The best known examples of quantum group algebras $\mathcal{G}^*$ are the deformations $U_q(sl_2)$ of the universal envelopping algebra (or the group algebra) of $SU(2)$. They are coassociative but not cocommutative. They admit a *-operation, which is the deformation of the *-operation $\xi^* = \xi^{-1}$ in $SU(2)$, in two cases:

- q real, choice (1. 16) of *-operation,
- $|q| = 1$, choice (1. 17) of *-operation.

If q is a root of unity, the algebra $U_q(sl_2)$ is not semisimple. As a result, tensor products of irreducible representations are in general not fully reducible. But there is a weak quasi Hopf algebra which is obtained from $U_q(sl_2)$ by a process of truncation. It is semisimple (i.e. isomorphic to a sum of full matrix algebras). The truncation destroys not only $\Delta(e) = e \otimes e$, but coassociativity is weakened to quasi-coassociativity in addition.

Finally we come to a discussion of *statistics and locality.* In second quantized quantum theory, Bose and Fermi statistics are implemented through local commutation or anticommutation relations of field operators which create particles,

$$\Psi_i^I(\mathbf{r},t)\Psi_j^J(\mathbf{r}',t) = \pm\Psi_j^J(\mathbf{r}',t)\Psi_i^I(\mathbf{r},t) \quad \text{for } \mathbf{r}' \neq \mathbf{r} . \quad (1.18)$$

In a relativistic theory this must be true in every Lorentz frame. Consistency of a symmetry with Bose/Fermi statistics requires that this relation should be preserved by a symmetry transformation. This is indeed true for internal symmetry groups, for supersymmetry, and also for conformal transformations of space time [16].

In two and less space dimensions, Bose/Fermi statistics is not the most general possibility, but braid group statistics can also occur. This was explained in detail in the memorable lectures which Jürg Fröhlich presented at the last Cargèse school in 1987 [5]. He proposed that local commutation relations of fields should be be replaced by local braid relations as follows.

$$\Psi_i^I(\mathbf{r},t)\Psi_j^J(\mathbf{r}',t) = \omega\Psi_l^J(\mathbf{r}',t)\Psi_k^I(\mathbf{r},t)\hat{\mathcal{R}}_{kl,ij}^{IJ>} \quad (1.19)$$

if $\mathbf{r}' > \mathbf{r}$. We chose to pull out a phase factor ω which will not be determined by the symmetry. A similar relation is valid if $\mathbf{r}' < \mathbf{r}$. It involves the inverse of $\hat{\mathcal{R}}^{JI>}$. In one space dimension, $\mathbf{r}$ is a real number, and the condition $\mathbf{r}' > \mathbf{r}$ has the obvious meaning. In 2 space time dimensions, braid group statistics appears in situations where the wave function is multivalued, and the definition of > involves the positions of cuts. We will not go into an explanation of this here.

We take it as a defining feature of a quantum symmetry that it is consistent with local braid relations of this type. It will be seen, however, that braid matrices $\hat{\mathcal{R}}$ with entries in $\mathcal{G}^*$ should be admitted. (Originally it was proposed that they should be numbers.) We showed in [2] that the chiral Ising model - i.e. the conformal field theory with central charge $c = \frac{1}{2}$ - provides an example, with the truncated quantum group algebra $U_q(sl_2)$ with $q = \pm i$ as a symmetry. To the best of our knowledge, this was the first time that the consistency of nonabelian local braid relations (1. 19) has been demonstrated through the construction of a model. Originally it had been proposed that minimal conformal models have quasitriangular Hopf algebras ("quantum groups") as symmetries [7, 8], but this identification is not quite satisfactory, because the local braid relations, which should come with the symmetry, are not satisfied [10]. A hint where to go came from the work of Dijkgraaf, Pasquier and Roche [6].They considered correlation functions of an orbifold model with a quasi Hopf symmetry.

The $\hat{\mathcal{R}}$-matrix in the local braid relations is determined as part of the specification of the quantum symmetry. It comes from an element $R \in \mathcal{G}^* \otimes \mathcal{G}^*$ which has certain properties. A ((weak) quasi) Hopf algebra equipped with such an R is called quasitriangular. Given the above mentioned "reassociator" $\varphi = \sum_\sigma \varphi^1_\sigma \otimes \varphi^2_\sigma \otimes \varphi^3_\sigma$ one introduces $\varphi_{213} = \sum_\sigma \varphi^2_\sigma \otimes \varphi^1_\sigma \otimes \varphi^3_\sigma$. In this notation, the braid matrices are given by

$$\hat{\mathcal{R}}^{IJ>}_{kl,ij} = (\tau^I_{ki} \otimes \tau^J_{lj} \otimes \mathcal{U})(\varphi_{213}(R \otimes e)\varphi^{-1}) \in \mathcal{U}(\mathcal{G}^*) \quad \text{for all } ijkl \; . \tag{1.20}$$

The most important property of R is as follows. Given the coproduct Δ there exists another one, Δ'. If $\Delta(\xi) = \sum_\sigma \xi^1_\sigma \otimes \xi^2_\sigma$ then $\Delta'(\xi) = \sum_\sigma \xi^2_\sigma \otimes \xi^1_\sigma$. R is an intertwiner between these coproducts, that is

$$\Delta'(\xi)R = R\Delta(\xi) \text{ for all } \xi \in \mathcal{G}^* \; . \tag{1.21}$$

The transformation law of adjoint field operators Ψ^{J*}_j can be read of from eq.(1. 13) (assuming the properties of the antipode which are necessary for its validity). The result is of the form of a field transformation law (1. 8), if Δ is a *-homomorphism of algebras with the choice (1. 16) of the *-operation on $\mathcal{G}^* \otimes \mathcal{G}^*$. But with the choice (1. 17) of the *-operation this is not true. Instead the role of ξ^2 and ξ^1 in (1. 8) get interchanged. That is, Δ' appears in place of Δ. But one can use R to define a "covariant adjoint" $\bar{\Psi}^J_j$ which transforms like Ψ^J_j, except that the representation τ^J gets replaced by its contragredient $\tilde{\tau}^J$. The covariant adjoint is defined as follows.

$$\bar{\Psi}^J_j = \Psi^{J*}_k(\tilde{\tau}^J_{kj} \otimes \mathcal{U})(R) \; . \tag{1.22}$$

This is a complex linear combination of field components Ψ^{J*}_j times representation operators $\mathcal{U}(\xi)$ of elements $\xi \in \mathcal{G}^*$. It is similar to the Dirac adjoint $\bar{\psi} = \psi^*\gamma^0$ of a Dirac spinor.

2. GROUP ALGEBRAS

(Quasi) Quantum group algebras are generalizations of group algebras rather than of the groups themselves. Therefore we start with a review of the properties of traditional symmetries in quantum mechanics, which are groups. All the properties will be reformulated in such a way that they make no reference to the group elements themselves, but only to the group algebra.

If G is a finite group, the elements of its group algebra $\mathcal{G}^*$ are formal sums of group elements a with complex coefficients $\Xi(a)$,

$$\xi = \sum_{a \in G} \Xi(a) a \; . \tag{2.1}$$

Multiplication is performed with the help of the product in G,

$$\begin{aligned} \xi\phi &= \sum_{a\in G}\sum_{b\in G} \Xi(a)\Phi(b)ab & (2.\ 2)\\ &= \sum_{c\in G}(\Xi * \Phi)(c)c \; ; & (2.\ 3)\\ (\Xi * \Phi)(c) &= \sum_{a\in G} \Xi(a)\Phi(a^{-1}c) \; . & (2.\ 4)\end{aligned}$$

This generalizes to Lie groups G in the obvious fashion. If da is left invariant Haar measure then

$$\xi = \int_{a\in G} da \Xi(a) a \; , \tag{2.5}$$

where $\Xi(\cdot)$ is a generalized function on G [11]. $\mathcal{G}^*$ may be regarded as the dual of the space $\mathcal{G}$ of "good" functions f on G. The corresponding map

$$\xi : \mathcal{G} \mapsto \mathbf{C} \tag{2.6}$$

is defined by $\xi[f] = \int da \Xi(a) f(a)$. We assume that the constant function $1 \in \mathcal{G}$, and that $\xi \in \mathcal{G}^*$ only if $\xi[f]$ is defined for all $f \in \mathcal{G}$, including $f = 1$. In the Lie group case, the group algebra contains both the group and the universal envelopping algebra of the Lie algebra of G. The corresponding generalized functions Ξ are δ-functions and their derivatives.

We work in the context of second quantized quantum mechanics. There will be a Hilbert space of states $\mathcal{H}$, a Hamiltonian H which is a self adjoint operator on $\mathcal{H}$, a ground state $|0> \in \mathcal{H}$, and there will be field operators $\Psi(\mathbf{r}, t)$ which create particles from the ground state $|0>$.

Let us now suppose that a group G with unit element e is an internal symmetry of such a quantum mechanical system. This means the following:

- The Hilbert space $\mathcal{H}$ carries a unitary representation of G by operators $\mathcal{U}(a)$, $a \in G$, with $\mathcal{U}(e) = 1$. Unitarity means that

$$\mathcal{U}(a^{-1}) = \mathcal{U}(a)^* \; .$$

- The ground state and the Hamiltonian are invariant,

$$\mathcal{U}(a)|0> = |0> \; , \quad [\mathcal{U}(a), H] = 0 \;\; \text{for all } a \in G \; . \tag{2.7}$$

- The fields transform covariantly. This means the following. The fields come in multiplets (Ψ_j^J) where J distinguishes different multiplets, and j distinguishes the members of one multiplets. There are finite dimensional representations τ^J of G such that

$$\mathcal{U}(a)\Psi_j^J(\mathbf{r}, t) = \Psi_k^J(\mathbf{r}, t)\tau_{kj}^J(a)\mathcal{U}(a) \; . \tag{2.8}$$

Internal symmetries do not act on space time coordinates of fields. Summation over repeated indices is understood here and throughout.

From the properties stated above, one may work out the transformation properties of many particle states,

$$|I_N i_N, ... I_1 i_1 >= \Psi^{I_N}_{i_N}[f_N] ... \Psi^{I_1}_{i_1}[f_1]|0> , \tag{2.9}$$

where $\Psi^I_i[f] = \int d\mathbf{r}\Psi^I_i(\mathbf{r},t)f(\mathbf{r})$. The time t is thought to be fixed.

To compute $\mathcal{U}(a)|I_N i_N, ... I_1 i_1 >$ one shifts $\mathcal{U}(a)$ through field operators from left to right, using the field transformation law (2. 8), until it acts on the ground state. Then one uses invariance of the ground state, eq.(2. 7). As a result one finds that the many particle states transform according to the tensor product of the representations $\tau^I_N ... \tau^I_1$. The tensor product $\bigotimes$ of two group representations τ and τ' is defined by

$$(\tau \bigotimes \tau')_{ik,jl}(a) = \tau_{ij}(a)\tau'_{kl}(a) . \tag{2.10}$$

The tensor product of group representations is associative. Therefore multiple tensor products are uniquely defined. The transformation law of states comes out as

$$\begin{aligned} \mathcal{U}(a)|I, i > &= |I, k > \tau^I_{ki}(a) & (2.\ 11)\\ \mathcal{U}(a)|I_N i_N, ... I_1 i_1 > &= |I_N k_N, ... I_1 k_1 > (\tau^{I_N} \bigotimes ... \bigotimes \tau^{I_1})_{k_N ... k_1, i_N ... i_1}(a) . \end{aligned}$$

Summation over $k_N ... k_1$ is understood.

Now we wish to restate these properties and their consequences in terms of the group algebra.

Representations $\mathcal{U}$, τ^J etc. may be extended to the group algebra in an obvious way,

$$\mathcal{U}(\xi) = \int_{a\in G} da \Xi(a)\mathcal{U}(a) \text{ etc. .} \tag{2.12}$$

In particular, the trivial 1-dimensional representation is extended to a representation ϵ of $\mathcal{G}^*$,

$$\epsilon(\xi) = \int_{a\in G} da\Xi(a) = \xi[1] . \tag{2.13}$$

Being a representation, ϵ is a homomorphism of algebras $\epsilon : \mathcal{G}^* \mapsto \mathbf{C}$.

Invariance of the Hamiltonian and invariance of the vacuum, eqs.(2. 7) take the form

$$\mathcal{U}(\xi)|0 >= |0 > \epsilon(\xi) , \quad [\mathcal{U}(\xi), H] = 0 \text{ for all } \xi \in \mathcal{G}^* . \tag{2.14}$$

To state the transformation law of field operators, we introduce

$$\Delta(a) = a \otimes a \text{ for } a \in G . \tag{2.15}$$

This extends to a homomorphism of algebras

$$\begin{aligned} \Delta : \mathcal{G}^* &\mapsto \mathcal{G}^* \otimes \mathcal{G}^* , & (2.\ 16)\\ \Delta(\xi) &= \int_{a\in G} da\Xi(a) a \otimes a . & (2.\ 17) \end{aligned}$$

If τ and τ' are representations of an algebra $\mathcal{G}^*$ then the outer tensor product $\tau \otimes \tau'$ is a representation of $\mathcal{G}^* \otimes \mathcal{G}^*$. It is defined by

$$(\tau \otimes \tau')(\xi \otimes \eta) = \tau(\xi) \otimes \tau'(\eta) , \tag{2.18}$$

etc., where the right hand side involves the standard tensor product of matrices (or linear operators). Writing all or some of the matrix indices we have

$$\begin{aligned} (\tau \otimes \tau')_{ik,jl}(\xi \otimes \eta) &= \tau(\xi)_{ij}\tau'_{kl}(\eta) . & (2.\ 19)\\ (\tau_{ij} \otimes \mathcal{U})(\xi \otimes \eta) &= \tau_{ij}(\xi)\mathcal{U}(\eta) \text{ etc. .} & (2.\ 20) \end{aligned}$$

In this notation, the transformation law (2. 8) of field operators under the group extends to the group algebra as follows

$$\mathcal{U}(\xi)\Psi_j^J(\mathbf{r},t) = \Psi_k^J(\mathbf{r},t)\int_{a\in G} da\tau_{kj}(a)\mathcal{U}(\Xi(a)a) \quad (2.\ 21)$$

$$= \Psi_k^J(\mathbf{r},t)(\tau_{kj}^J \otimes \mathcal{U})(\Delta(\xi)) \ . \quad (2.\ 22)$$

The field transformation law (2. 21) permits to push representation operators $U(\cdot)$ of elements of $\mathcal{G}^*$ through field operators Ψ from left to right.

Using this fact and the invariance of the ground state (2. 14), one can work out the transformation law of many particle states under the group algebra in the same manner as before in the group situation. The result can again be stated by saying that the multiparticle states transform according to tensor products of representations. Let us explain this in more detail.

Using the definition (2. 15) of the coproduct Δ for the group G the tensor product (2. 10) of group representations can be written as

$$(\tau \bigotimes \tau')(\xi) = (\tau \otimes \tau')(\Delta(\xi)) \ , \quad (2.23)$$

for group elements $\xi \in G$. The right hand side involves the outer tensor product of representations as described before. Like every group representation, this representation can be extended to a representation of the group algebra $\mathcal{G}^*$. It is given by the same formula, where Δ is the extension (2. 17) of the coproduct (2. 15) to the group algebra. As a result, the field transformation law (2. 21) reduces to (2. 6).

The transformation law of many particle states comes out as

$$U(\xi)|I,i> = |I,k> \tau_{ki}^I(\xi) \quad (2.\ 24)$$

$$\mathcal{U}(\xi)|I_N i_N,...I_1 i_1> = |I_N k_N,...I_1 k_1> (..(\tau^{I_N} \bigotimes ... \bigotimes(\tau^{I_2} \bigotimes \tau^{I_1})..)_{k_N...k_1,i_N...i_1}(\xi)$$

for all $\xi \in \mathcal{G}^*$. The brackets in the tensor product are actually unnecessary, because the tensor product of group algebra representations is associative.

Consider the tensor product of some representation τ with the trivial 1-dimensional representation ϵ. This should reproduce τ. From the definitions and **C**-linearity of the representations we deduce

$$(\tau \bigotimes \epsilon)(\xi) = (\tau \otimes \epsilon)(\Delta(\xi)) = \tau((id \otimes \epsilon)\Delta(\xi)) \ , \quad (2.25)$$

and similarly for $\epsilon \bigotimes \tau$. id is the identity map. The result equals $\tau(\xi)$ if

$$(id \otimes \epsilon)\Delta = id = (\epsilon \otimes id)\Delta \ . \quad (2.26)$$

Since $\Delta(a) = a \otimes a$ and $\epsilon(a) = 1$ for group elements a, eq.(2. 26) is true when applied to group elements a. By linearity it is therefore also true on group algebra elements ξ.

Consider next triple tensor products of representations. From the definition (2. 23) we deduce

$$(\tau \bigotimes (\tau' \bigotimes \tau''))(\xi) = (\tau \otimes \tau' \otimes \tau'')((id \otimes \Delta)\Delta(\xi)) \ , \quad (2.\ 27)$$

$$((\tau \bigotimes \tau') \bigotimes \tau'')(\xi) = (\tau \otimes \tau' \otimes \tau'')((\Delta \otimes id)\Delta(\xi)) \ . \quad (2.\ 28)$$

Both sides are equal, i.e. the tensor product of representations is associative. This is true because

$$(id \otimes \Delta)\Delta = (\Delta \otimes id)\Delta \ . \quad (2.29)$$

Both sides of this equation give $a \otimes a \otimes a$ when applied to a group element a. Because group elements span $\mathcal{G}^*$, the equality holds on $\mathcal{G}^*$. A coproduct with the property (2.29) is said to be *coassociative* .

Let P act on $\mathcal{G}^* \otimes \mathcal{G}^*$ by interchange of factors, $P(a \otimes b) = b \otimes a$. Given one coproduct Δ, there exists another one,

$$\Delta'(\xi) = P\Delta(\xi) . \tag{2.30}$$

For group elements $\Delta(a) = a \otimes a = \Delta'(a)$. Therefore

$$\Delta = \Delta' \text{ for group algebras.} \tag{2.31}$$

A coproduct with this property is called *cocommutative* . Cocommutativity can be expressed in a fancier way,

$$\Delta'(\xi)R = R\Delta(\xi) \text{ for all } \xi \in \mathcal{G}^* , \tag{2.32}$$

with $\qquad R = e \otimes e.$

3. WEAK QUASITRIANGULAR QUASI HOPF ALGEBRAS AND THEIR INTERPRETATION AS A SYMMETRY

A weak quasitriangular quasi Hopf algebra $\mathcal{G}^*$ is a generalization of a group algebra. It is an algebra equipped with a counit ϵ, a coproduct Δ, a reassociator φ , an R-element $R \in \mathcal{G}^* \otimes \mathcal{G}^*$ and an antipode $\mathcal{S}$ with properties as follows.

$\mathcal{G}^*$ is an algebra with unit e. The counit

$$\epsilon : \mathcal{G}^* \to \mathbf{C} , \tag{3.1}$$

and the coproduct

$$\Delta : \mathcal{G}^* \to \mathcal{G}^* \otimes \mathcal{G}^* \tag{3.2}$$

are both homomorphisms of algebras, but Δ need not be unit preserving. The homomorphism property requires that $\Delta(\xi\eta) = \Delta(\xi)\Delta(\eta)$. Given the product Δ, there exists another one called Δ'. If $\Delta(\xi) = \sum \xi_p^1 \otimes \xi_p^2$, then Δ' is defined by

$$\Delta'(\xi) = \sum \xi_p^2 \otimes \xi_p^1 . \tag{3.3}$$

It is demanded that

$$(id \otimes \epsilon)\Delta = (\epsilon \otimes id)\Delta = id , \tag{3.4}$$

(id = identity map). Furthermore one demands that an element $\varphi \in \mathcal{G}^* \otimes \mathcal{G}^* \otimes \mathcal{G}^*$ is given which implements (weak) quasi–coassociativity of the coproduct, eq.(3. 10) below. We admit the possibility that $\Delta(e) \neq e \otimes e$, and we do not demand invertibility of φ, but only existence of a quasiinverse, still denoted by φ^{-1}, such that

$$\varphi\varphi^{-1} = (id \otimes \Delta)\Delta(e) \quad , \quad \varphi^{-1}\varphi = (\Delta \otimes id)\Delta(e), \tag{3. 5}$$

$$\begin{aligned}
(id \otimes id \otimes \epsilon)(\varphi) &= \Delta(e) , \\
(id \otimes \epsilon \otimes id)(\varphi) &= \Delta(e) , \\
(\epsilon \otimes id \otimes id)(\varphi) &= \Delta(e) .
\end{aligned} \tag{3. 6}$$

The statement that φ^{-1} is a quasiinverse of φ means that $\varphi\varphi^{-1}\varphi = \varphi$.

Finally there should exist $R \in \mathcal{G}^* \otimes \mathcal{G}^*$ such that

$$\Delta'(\eta)R = R\Delta(\eta) \quad \text{for all} \quad \eta \in \mathcal{G}^*. \tag{3.7}$$

We do not demand that R be invertible, instead it should have a quasiinverse R^{-1} such that

$$RR^{-1} = \Delta'(e) \quad , \quad R^{-1}R = \Delta(e) \ . \tag{3.8}$$

It is natural to impose also the following requirement.

$$(id \otimes \epsilon)(R) = e \quad , \quad (\epsilon \otimes id)(R) = e \ . \tag{3.9}$$

By definition, weak quasi–coassociativity demands that

$$\varphi(\Delta \otimes id)\Delta(\xi) = (id \otimes \Delta)\Delta(\xi)\varphi \quad \text{for all} \quad \xi \in \mathcal{G}^*. \tag{3.10}$$

Following Drinfeld the following relations between Δ,R, and φ are postulated.

$$(id \otimes id \otimes \Delta)(\varphi)(\Delta \otimes id \otimes id)(\varphi) = (e \otimes \varphi)(id \otimes \Delta \otimes id)(\varphi)(\varphi \otimes e) \ , \tag{3. 11}$$
$$(id \otimes \Delta)(R) = \varphi_{231}^{-1} R_{13} \varphi_{213} R_{12} \varphi^{-1} \ , \tag{3. 12}$$
$$(\Delta \otimes id)(R) = \varphi_{312} R_{13} \varphi_{132}^{-1} R_{23} \varphi \ . \tag{3. 13}$$

We used the standard notation. If $R = \sum r_a^1 \otimes r_a^2$ then

$$R_{13} = \sum r_a^1 \otimes e \otimes r_a^2 \ , \quad R_{12} = \sum r_a^1 \otimes r_a^2 \otimes e \ , \tag{3.14}$$

etc. If s is any permutation of 123 and $\varphi = \sum \varphi_\sigma^1 \otimes \varphi_\sigma^2 \otimes \varphi_\sigma^3$ then

$$\varphi_{s(1)s(2)s(3)} = \sum_\sigma \varphi_\sigma^{s^{-1}(1)} \otimes \varphi_\sigma^{s^{-1}(2)} \otimes \varphi_\sigma^{s^{-1}(3)} \ . \tag{3.15}$$

Eq. (3. 12), or (3. 13), implies validity of quasi Yang Baxter equations,

$$R_{12}\varphi_{312}R_{13}\varphi_{132}^{-1}R_{23}\varphi = \varphi_{321}R_{23}\varphi_{231}^{-1}R_{13}\varphi_{213}R_{12} \ , \tag{3.16}$$

and this guaranties that R together with φ determines a representation of the braid group as will be explained below. There should also exist an antipode

$$\mathcal{S} : \ \mathcal{G}^* \mapsto \mathcal{G}^* \tag{3.17}$$

which is an antihomomorphism of algebras. The coproduct is used to define the *tensor product of representations of* $\mathcal{G}^*$ as in eq.(3. 2). The antipode is used to define the *contragredient* $\tilde{\tau}$ to a representation τ by eq.(3. 12). One may want to impose it as an additional requirement on the antipode that $\tilde{\tau} \otimes \tau$ should contain the trivial representation ϵ for any representation τ. Drinfeld proposed conditions of this type for quasi Hopf algebras [4]. We shall not use them here.

In quantum mechanics it is important to have a *-operation. We say that $\mathcal{G}^*$ is a *weak quasitriangular quasi Hopf *-algebra* if it has the following properties in addition to those listed above.

Both ϵ and Δ are *-homomorphisms of algebras, with the *-operation on $\mathcal{G}^*$ given either by eq.(3. 16) or by eq.(3. 17). In addition we require that the antipode satisfies

$$m_r\left((\mathcal{S}^{-1} \otimes id \otimes id)(id \otimes \Delta)\Delta(\xi)\right) = e \otimes \xi \ , \tag{3. 18}$$
$$m_r'\left((id \otimes \mathcal{S}^{-1} \otimes id)(id \otimes \Delta)\Delta(\xi)\right) = \xi \otimes e \ , \tag{3. 19}$$

for all $\xi \in \mathcal{G}^*$, with multiplication operators defined by

$$m_r(\xi \otimes \eta \otimes \zeta) = \eta\xi \otimes \zeta \quad , \quad m'_r(\xi \otimes \eta \otimes \zeta) = \zeta \otimes \eta\xi \ .$$

This condition is used to obtain (3. 13) and the transformation law of adjoint field operators.

A weak quasitriangular quasi Hopf *-algebra$\mathcal{G}^*$ is called a *symmetry* of a quantum mechanical system if a unitary representation $\mathcal{U}$ of $\mathcal{G}^*$ in the Hilbert space of physical state is given, the Hamiltonian and the ground state are invariant, the field operators transform covariantly, and $\mathcal{U}(e) = 1$. All this was explained in the introduction.

Let us now discuss the implications of the assumptions. ¿From the transformation law of field operators and invariance of the ground state one deduces the transformation law of 1-particle states and multiparticle states (3. 9), eqs.(3. 24) in the same manner as for group algebras. The first identity (3. 26) is used in this. The brackets in the result are now important in general, because the tensor product need not be associative, unless the coproduct is coassociative. The identity (3. 4) guarantees that tensoring with the trivial 1-dimensional representation ϵ amounts to doing nothing, as in the case of group algebras.

The existence of the reassociator φ implies that the tensor product of representations is still quasi-associative in the sense that the equivalence (3. 9) of triple tensor products of representations holds true. If one of the three representations is the 1-dimensional representation ϵ, the reassociation of the triple tensor product is trivial. This is encoded in the relations (3. 6).

All this is a formal use of φ. Its physical use is to enable us to define products of field operators which transform covariantly under $\mathcal{G}^*$, and to write down $\mathcal{G}^*$-covariant operator product expansions for products of field operators in local quantum field theories. In addition it enters into the $\mathcal{R}$-matrix (3. 20) in the $\mathcal{G}^*$-covariant local braid relations (3. 19).

The coproduct is coassociative if φ is trivial, $\varphi = e \otimes e \otimes e$.

If the coproduct is not coassociative, then the ordinary product $\Psi^J_j \Psi^K_k$ of field operators is not $\mathcal{G}^*$-covariant. But we may define a covariant product $\times$ as follows

$$(\Psi^J \times \Psi^K)_{jk} = \Psi^J_l \Psi^K_m (\tau^J_{lj} \otimes \tau^K_{mk} \otimes \mathcal{U})(\varphi) \ . \tag{3.20}$$

This transforms according to the representation $(\tau^J \otimes \tau^K)$ of $\mathcal{G}^*$ in the sense of (3. 8). Note that $(\Psi^J \times \Psi^K)_{jk}$ is a sum of terms $\Psi^J_l \Psi^K_m$ multiplied from the right with representation operators $\mathcal{U}(\phi)$ of elements of $\mathcal{G}^*$. The proof of covariance uses the intertwining property (3. 10) of φ.

In this covariant notation, the local braid relations (3. 19) of field operators read as follows (with the same phase factor ω)

$$(\Psi^I \times \Psi^J)_{ij} = \omega(\Psi^J \times \Psi^I)_{lk} \mathcal{R}^{IJ>}_{kl,ij} \qquad \text{with} \tag{3. 21}$$
$$\mathcal{R}^{IJ>}_{kl,ij} = (\tau^I_{ki} \otimes \tau^J_{lj})(R) \ . \tag{3. 22}$$

This holds true under the same conditions on the arguments of the field operators, which we neglected to write, as in the braid relations (3. 19). It follows from the intertwining property (3. 21) that both sides of equation (3. 21) transform according to the representation $\tau^I \otimes \tau^J$ of $\mathcal{G}^*$. This shows that the local braid relations which we propose are $\mathcal{G}^*$-covariant. In the coassociative case $\varphi = e \otimes e \otimes e$, the covariant product is equal to the ordinary one, and the braid relations in the form (3. 19) involve a numerical $\mathcal{R}$-matrix.

The $\times$-product is not associative (unless φ is trivial). But the relation (3. 11) ensures that it is quasi-associative in the following sense.

$$((\Psi^I \times \Psi^J) \times \Psi^K)_{ijk} = (\Psi^I \times (\Psi^J \times \Psi^K))_{i'j'k'}(\tau^I \otimes \tau^J \otimes \tau^K)_{i'j'k',ijk}(\varphi) . \quad (3.23)$$

This equation tells us that the triple product on the left hand side is a complex linear combination of the triple products of field operators on the right hand side. The inverse relation is also true. It involves φ^{-1}.

If there is truncation, $\Delta(e) \neq e \otimes e$, then our stated requirement that $U(e) = 1$ has some important consequences. In particular it implies, together with eq.(3. 6), that

$$\begin{aligned}(\Psi^{I_N} \times (\Psi^{I_{N-1}} \times \ldots (\Psi^{I_2} \times \Psi^{I_1})\ldots))_{i_N \ldots i_1}|0> &= \Psi^{I_N}_{i_N}\Psi^{I_{N-1}}_{i_{N-1}}\ldots\Psi^{I_2}_{i_2}\Psi^{I_1}_{i_1}|0> \quad (3.\ 24)\\ &= |I_N i_N, \ldots I_1 i_1> .\end{aligned}$$

In words: On the ground state, ordinary products of field operators and covariant products performed in a definite order, are the same.

Finally we turn to the implications of the relations which involve R. It was already mentioned that the fundamental intertwining property (3. 21) of R is crucial to ensure the $\mathcal{G}^*$-covariance of local braid relations (3. 21).

The element $R \in \mathcal{G}^* \otimes \mathcal{G}^*$, which satisfies the quasi Yang Baxter equations, furnishes a representation of the braid group. Let us recall that the braid group B_n on n threads is generated by elements σ_i and σ_i^{-1} $(i = 1 \ldots n-1)$ which obey the Artin relations

$$\begin{aligned}\sigma_i\sigma_k = \sigma_k\sigma_i \quad \text{if } |k-i| \geq 2 \quad , \quad \sigma_i\sigma_{i+1}\sigma_i = \sigma_{i+1}\sigma_i\sigma_{i+1} ,\\ \sigma_i\sigma_i^{-1} = \iota = \sigma_i^{-1}\sigma_i . \qquad (3.\ 25)\end{aligned}$$

The unit element of B_n is written as ι. We introduce some notations. Write

$$e^n = e \otimes \ldots \otimes e \qquad \text{(n factors)} \qquad (3.26)$$

and similarly for id^n. In addition we abbreviate $\mathcal{G}^{*\otimes n} = \mathcal{G}^* \otimes \ldots \otimes \mathcal{G}^*$ (n factors), and

$$\begin{aligned}\Delta^n &= (id^{n-1} \otimes \Delta) \cdots (id \otimes \Delta)\Delta \quad \text{for } n \geq 2 , \qquad (3.\ 27)\\ \Delta^1 &= \Delta , \ \Delta^0 = id , \ \Delta^{-1} = \epsilon . \qquad (3.\ 28)\end{aligned}$$

For simplicity we assume that $\mathcal{G}^*$ is semisimple. Then the permutation of factors in $\mathcal{G}^* \otimes \mathcal{G}^*$ can be implemented by a multiplication operator $\mathcal{P} \in \mathcal{G}^* \otimes \mathcal{G}^*$ with the properties

(i) $\mathcal{P}\, \xi \otimes \eta = \eta \otimes \xi\, \mathcal{P}$ for all $\xi, \eta \in \mathcal{G}^*$,

(ii) $\tau \otimes \tau'\ (\mathcal{P}) = 0$ if τ, τ' are inequivalent irreducible representations of $\mathcal{G}^*$.

Theorem (Artin relations) *Suppose that $\mathcal{G}^*$ is semisimple. Let $\hat{R}^+ = \mathcal{P}R$ and $\hat{R}^- = R^{-1}\mathcal{P}$, and $n = r + k + 1, r \geq 0$. Define $\sigma_k^{n\pm} \in \mathcal{G}^{*\otimes n}$ by*

$$\sigma_k^{n\pm} = \Delta^{n-1}(e)(id^{n-k+1} \otimes \Delta^{k-2})(e^{n-k-1} \otimes \varphi(\hat{R}^\pm \otimes e)\varphi^{-1}) . \qquad (3.29)$$

Then the $\sigma_k^{n\pm}$ obey Artin relations (3. 25) with $\iota = \Delta^{n-1}(e)$.

The proof is spelled out in [12]. [If semisimplicity is not assumed, there is still a representation of B_n by linear maps $\mathcal{G}^{*\otimes n} \mapsto \mathcal{G}^{*\otimes n}$, but these linear maps are not necessarily implemented by multiplication operators.]

We remark that this representation of the braid group can act on the Hilbert space of physical states $\mathcal{H}$. In the algebraic approach to quantum field theory, this representation of the braid group is determined by the algebra of observables. It is the starting point from which one tries to recover the symmetry algebra - see Fredenhagens lectures at this school [13]. We will not go into this now.

Drinfelds relations (3. 12) and (3. 13) imply the quasi Yang Baxter equation. They imply also that covariant products of field operators obey local braid relations, under appropriate conditions on their arguments, if the individual fields do.

A field operator $\Phi(\mathbf{r},t)$ is called *$\mathcal{G}^*$-invariant* if it transforms according to the trivial 1-dimensional representation ϵ, or, equivalently, if $\mathcal{U}(\xi)\Phi = \Phi\mathcal{U}(\xi)$ for all $\xi \in \mathcal{G}^*$. It follows from eq.(3. 9) that $\mathcal{G}^*$-invariant fields Φ, which one may construct as composites from $\mathcal{G}^*$-covariant fields, for instance, obey ordinary local commutation relations with all braid-local fields, except for the possible appearance of a phase factor ω_1 which depends on the fields

$$\Phi(\mathbf{r},t)\Psi_j^J(\mathbf{r}',t) = \omega_1\Psi_j^J(\mathbf{r}',t)\Phi(\mathbf{r},t) \quad \text{if} \quad \mathbf{r} \neq \mathbf{r}' \; . \tag{3.30}$$

In algebraic field theory one demands that the phase factor ω_1 is 1 if Φ is an observable.

4.TRUNCATED QUANTUM GROUP ALGEBRAS

The best known examples of quasitriangular Hopf algebras are the deformations $U_q(sl_2)$ of the universal envelopping algebra (or the group algebra) of $SU(2)$. They are indexed by a complex number q.

The quantum group algebra $U_q(sl_2)$ is generated by elements $q^{\pm H/2}$ and $S_\pm$ subject to the relations

$$q^{H/2}q^{-H/2} = q^{-H/2}q^{H/2} = 1 \; , \tag{4. 1}$$

$$q^{H/2}S_\pm = q^{\pm\frac{1}{2}}S_\pm q^{H/2} \; , \tag{4. 2}$$

$$[S_+, S_-] = \frac{q^H - q^{-H}}{q^{\frac{1}{2}} - q^{-\frac{1}{2}}} \; . \tag{4. 3}$$

Here and in the following we write 1 in place of e for the unit element in $\mathcal{G}^* = U_q(sl_2)$.

We denote the comultiplication in $U_q(sl_2)$ by Δ_q in order to distinguish it from the comultiplication Δ for the truncated quantum group algebra which we are going to introduce. It is given by

$$\Delta_q(q^{\pm H/2}) = q^{\pm H/2} \otimes q^{\pm H/2} \tag{4. 4}$$

$$\Delta_q(S_\pm) = S_\pm \otimes q^{H/2} + q^{-H/2} \otimes S_\pm \; . \tag{4. 5}$$

If q is real, a *-operation can be defined by

$$S_\pm^* = S_\mp \; , \quad (q^{\pm H/2})^* = q^{\pm H/2} \; .$$

If $|q| = 1$, a *-operation can be defined by

$$S_\pm^* = S_\mp \; , \quad (q^{\pm H/2})^* = q^{\mp H/2} \; .$$

In both cases this is an automorphism satisfying $\Delta_q(\xi^*) = \Delta_q(\xi)^*$ provided we adopt the appropriate convention for the adjoint on $\mathcal{G}^* \otimes \mathcal{G}^*$, viz. (4. 16) if q is real, and (4. 17) if $|q| = 1$.

The counit is

$$\epsilon(q^{\pm H/2}) = 1 \quad , \quad \epsilon(S_\pm) = 0 \; ,$$

and the antipode $\mathcal{S}$ acts as

$$q^{\mp H/2} = \mathcal{S}(q^{\pm H/2}) \quad , \quad -q^{\pm 1/2} S_\pm = \mathcal{S}(S_\pm) \; . \tag{4.6}$$

There exists a canonical R-element for $U_q(sl_2)$ which we denote by R_q. It is given by

$$\mathcal{R}_q = q^{H\otimes H} \sum_{n\geq 0} \frac{(1-q^{-1})^n}{[n]!} q^{-\frac{1}{4}n(n-1)} q^{nH/2} S_+^n \otimes q^{-nH/2} S_-^n \tag{4.7}$$

If $|q| = 1$ and if we adopt the choice (4. 17) of the *-operation, viz. $(a \otimes b)^* = b^* \otimes a^*$ we get that R_q is *unitary,*

$$R_q^* = R_q^{-1} = (\mathcal{S} \otimes id)(R_q) \; .$$

The "quantum dimension" d_τ of a representation τ is defined by

$$d_\tau = tr\tau(q^H) \; . \tag{4.8}$$

If q is a (primitive p-th) root of unity, then $U_q(sl_2)$ is not semisimple, and tensor products of its irreducible representations are in general not fully reducible. Its irreducible representations τ^J with nonzero quantum dimension are called "physical" representations. They are labelled by $J = 0, \frac{1}{2}, \ldots \frac{1}{2}(p-2)$. They have dimension $2J+1$.

We denote the tensor product of $U_q(sl_2)$-representations by $\bigotimes_q$. The tensor product of $\tau^I \bigotimes_q \tau^J$ of two physical representations decomposes in general into physical representations, plus unphysical representations with quantum dimension 0. If we multiply $(\tau^I \bigotimes_q \tau^J)(\xi)$ with a projection operator P_{IJ} which cuts away the contribution with zero quantum dimension, one obtains what is known as the truncated tensor product of physical representations of $U_q(sl_2)$.

A weak quasitriangular quasi Hopf algebra $\mathcal{G}^*$ is canonically associated with $U_q(sl_2)$ with q a root of unity. As an algebra $\mathcal{G}^* = U_q(sl_2)/\mathcal{J}$, where $\mathcal{J}$ is the ideal which is annihilated by all the physical representations $\tau^I, 2I = 0 \ldots p-2$, of $U_q(sl_2)$. $\mathcal{G}^*$ is semisimple, its representations are fully reducible, and the irreducible ones are precisely the physical representations of $U_q(sl_2)$. Let

$$u(I,J) = \min\{|I+J|, p-2-I-J\} \tag{4.9}$$

and let P_{IJ} be the projector on the physical subrepresentations K, $|I-J| \leq K \leq u(I,J)$ of the tensor product $\tau^I \bigotimes_q \tau^J$ of $U_q(sl_2)$ representations. There exists $P \in \mathcal{G}^* \otimes \mathcal{G}^*$ such that $P_{IJ} = (\tau^I \otimes \tau^J)(P)$. The coproduct in $\mathcal{G}^*$ is determined in terms of the coproduct Δ_q in $U_q(sl_2)$ as

$$\Delta(\xi) = P\Delta_q(\xi) \; , \tag{4.10}$$

hence $\Delta(e) = P \neq e \otimes e$. This coproduct specifies a tensor product $\bigotimes$ which is equal to the truncated tensor product of physical $U_q(sl_2)$ representations. Thus

$$\tau^I \bigotimes \tau^J = \bigoplus_{|I-J|\leq K \leq u(IJ)} \tau^K \; . \tag{4.11}$$

There exists an element $\varphi \in \mathcal{G}^* \otimes \mathcal{G}^* \otimes \mathcal{G}^*$ such that $\varphi_{IJK} = (\tau^I \otimes \tau^J \otimes \tau^K)(\varphi)$ implements the well known unitary equivalence of the truncated tensor products

$\tau^I \otimes(\tau^J \otimes \tau^K)$ and $(\tau^I \otimes \tau^J) \otimes \tau^K$. A truncated tensor product $\otimes$ is defined also for basis vectors $\hat{e}^I_i$ in the dual representation spaces $\hat{V}^I$ on which $\mathcal{G}^*$ acts from the right, viz. $\hat{e}^I_i \otimes \hat{e}^J_j = \hat{e}^I_i \otimes \hat{e}^J_j P_{IJ}$. The map φ_{IJK} can be specified by its action on triple truncated products of basis vectors, together with the condition $\varphi = (id \otimes \Delta)\Delta(e)\varphi$, viz.

$$\sum_{ijkp} \begin{bmatrix} I & P & L \\ i & p & l \end{bmatrix}_q \begin{bmatrix} J & K & P \\ j & k & p \end{bmatrix}_q \hat{e}^I_i \otimes \hat{e}^J_j \otimes \hat{e}^K_k \varphi = \sum_{Q,ijkq} F_{PQ}\begin{bmatrix} J & I \\ K & L \end{bmatrix} \begin{bmatrix} I & J & Q \\ i & j & q \end{bmatrix}_q \begin{bmatrix} Q & K & L \\ q & k & l \end{bmatrix}_q \hat{e}^I_i \otimes \hat{e}^J_j \otimes \hat{e}^K_k \tag{4.12}$$

with fusion matrices given by 6J–symbols,

$$F_{PQ}\begin{bmatrix} J & I \\ K & L \end{bmatrix} = \begin{Bmatrix} K & J & P \\ I & L & Q \end{Bmatrix}_q .$$

$[\because]_q$ are Clebsch Gordan coefficients for $U_q(sl_2)$ (and at the same time for $\mathcal{G}^*$).

The R–element of $\mathcal{G}^* \otimes \mathcal{G}^*$ is given in terms of the R–element R_q for $U_q(sl_2)$ by

$$R = R_q\Delta(e) = \Delta'(e)R_q \ , \tag{4.13}$$

while antipode and counit are the same as in $U_q(sl_2)$. It is shown in ref. [1] that the defining properties of a weak quasitriangular quasi Hopf algebra are satisfied.

Since $\Delta(e)^* = \Delta(e^*) = \Delta(e)$, the unitarity of R_q implies that

$$R^* = R^{-1} \ . \tag{4.14}$$

5. A MODEL WITH QUANTUM SYMMETRY AND LOCAL BRAIDRELATIONS

The chiral Ising model provides an example of a conformal field theory which admits the truncated quantum group algebra $\mathcal{G}^*$ canonically associated with $U_q(sl_2)$ as a symmetry ($q = \pm i$). It has field operators which act in a positive definite Hilbert space of physical states, and which satisfy $\mathcal{G}^*$-covariant local braid relations. This was demonstrated in [2].

The chiral Ising model lives on a circle S^1. Its chief observable field is the stress energy tensor $T(z)$. $z \in S^1$ is regarded as a complex number of modulus 1. The stress tensor furnishes a Virasoro algebra with generators $L_n = L^*_{-n}$, $n = 0, \pm 1, \pm 2, ...$, and central charge $c = \frac{1}{2}$.

$$T(z) = \sum_{n\in\mathbf{Z}} L_n z^{-n-2} \ . \tag{5.1}$$

It can be constructed from a "universal Majorana algebra" Maj with generators b_a, $a = 0, \pm\frac{1}{2}, \pm 1, ...$, $b^*_a = b_{-a}$. Maj contains a central element $Y = 4b_0^2 - 1$ which has eigenvalues ± 1. The generators are the Fourier modes of a Fermi field,

$$\psi(z) = \sum_{a\in\frac{1}{2}\mathbf{Z}} b_a z^{-a-\frac{1}{2}} \ . \tag{5.2}$$

It is a two-valued function on the circle and satisfies anticommutation relations

$$\{\psi(z), \psi(w)\} = \pi i[\delta(z - w) - Y\delta(z - e^{2\pi i}w)], \tag{5.3}$$

or, equivalently

$$\{b_a, b_c\} = \frac{1}{2}[1 + (-1)^{2a}Y]\delta_{a,-c} \ . \tag{5.4}$$

The stress energy tensor is obtained as

$$T(z) = -\frac{1}{2}\lim_{w\mapsto z}\{\frac{1}{2}\psi(w)\psi'(z)+\frac{1}{2}\psi(we^{2\pi i})\psi'(ze^{2\pi i})-(w-z)^{-2}\}\ , \quad (5.\ 5)$$
$$\psi'(z) = \frac{\partial}{\partial z}\psi(z)\ .$$

The Virasoro algebra with central charge $c=\frac{1}{2}$ admits 3 inequivalent representations π^J which we label by $J=0,\frac{1}{2},1$. They act in Hilbert spaces $\mathcal{H}^J$ with lowest weight vectors $|\lambda_J>$ of weight $\lambda_J=0,\frac{1}{16},\frac{1}{2}$.

All these irreducible Virasoro representation spaces are actually representation spaces for an algebra of observables $\mathcal{A}$ which is spanned by all those even products of Majorana fields which are invariant under rotations of the circle by 2π. Inequivalent positive energy representations of the observable algebra are known in quantum mechanics as *superselection sectors* . Thus we have 3 superselection sectors, including the vacuum sector $\mathcal{H}^0$ which contains the ground state $|0>$. π^0 is called the vacuum representation of $\mathcal{A}$.

All these inequivalent irreducible positive energy representations of the algebra of observables can be obtained from the vacuum representation π^0 by composition with a suitable *-morphism ρ_J of the algebra of observables $\mathcal{A}$.

$$\pi^J(A)=\pi^0(\rho_J(A))\ . \quad (5.6)$$

By definition, a *-morphism ρ is a **C**-linear map $\mathcal{A}\mapsto\mathcal{A}$ which obeys

$$\rho(A_1)\rho(A_2)=\rho(A_1A_2)\ ,\quad \rho(A^*)=\rho(A)^*\ . \quad (5.7)$$

In our model, suitable morphisms are known explicitly. They are given by the following morphisms of the universal Majorana algebra. [10]

$$\rho_0 = id \quad (5.\ 8)$$

$$\rho_{1/2}(b_a) = \begin{cases} ib_{a+1/2} & a\geq\frac{1}{2} \\ \frac{i}{\sqrt{2}}(b_{1/2}-b_{-1/2}) & a=0 \\ -ib_{a-1/2} & a\leq-\frac{1}{2}\end{cases}\ ,\quad \rho_{1/2}(Y)=-Y \quad (5.\ 9)$$

$$\rho_1(b_a) = \begin{cases} -b_a & a\neq 0\pm\frac{1}{2} \\ b_{-a} & a=0,\pm\frac{1}{2}\end{cases}\ ,\quad \rho_1(Y)=Y. \quad (5.\ 10)$$

ρ_1 is an automorphism, i.e. bijective, but $\rho_{1/2}$ is not surjective on $\mathcal{A}$ because $b_r(b_{\frac{1}{2}}+b_{-\frac{1}{2}})$, $(r\in\mathbf{Z}+\frac{1}{2})$ and b_nb_0, $(n\in\mathbf{Z})$ are not in its range. The Hilbert space of physical states of our model will be the sum of $2J+1$ copies $\mathcal{H}^J_m$, $m=-J...J$ of each of our inequivalent representation spaces $\mathcal{H}^J$ of $\mathcal{A}$,

$$\mathcal{H}=\bigoplus_{J=0,\frac{1}{2},1}\bigoplus_{m=-J}^{J}\mathcal{H}^J_m\ . \quad (5.11)$$

A representation π of $\mathcal{A}$ on all of $\mathcal{H}$ is defined by its restrictions to the subspaces $\mathcal{H}^J_m$,

$$\pi(A)=\pi^J(A) \text{ on } \mathcal{H}^J_m\ (m=-J...J)\ . \quad (5.12)$$

All infinite-dimensional Hilbert spaces of the world are the same. In view of the equivalence of representations (5. 6) we may introduce identification maps i_{Jm} which identify the representation spaces $(\pi^J,\mathcal{H}^J_m)$ and $(\pi^0\circ\rho_J,\mathcal{H}^0)$. Thus,

$$i^*_{Jm}:\mathcal{H}^0\mapsto\mathcal{H}^J_m \quad (5.13)$$

is bijective and has the intertwining property

$$\pi^J(A) i^*_{Jm} = i^*_{Jm} \pi^0(\rho_J(A)) . \tag{5.14}$$

Now we are ready to describe the action of the truncated quantum group algebra $\mathcal{G}^*$ on the Hilbert space $\mathcal{H}$ of physical states.

$$\mathcal{U}(\xi) i^*_{Jm} |\psi> = i^*_{Jk} |\psi> \tau^J_{km}(\xi) \tag{5.15}$$

for arbitrary $|\psi> \in \mathcal{H}^0$. The symmetry acts as a *gauge symmetry (of first kind)*, i.e. all observables are invariant

$$[U(\xi), A] = 0 \ \text{ for all } A \in \mathcal{A}, \ \xi \in \mathcal{G}^* . \tag{5.16}$$

Field operators which make transitions between the sectors $\mathcal{H}^J_m$ with different J are constructed from observables by use of the general bosonization formula of Doplicher Haag and Roberts.

$$\Psi^J_m(w) = \Gamma^J_m A_J(w) . \tag{5.17}$$

All the dependence on the space time point w is in the second factor $A_J(w)$ which is an (unbounded) observable, i.e. a limit of elements in the observable algebra. These factors commute therefore with $\mathcal{U}(\xi)$, $\xi \in \mathcal{G}^*$. Only the "constant fields" Γ^J_m transform nontrivially under the symmetry. We will first describe some of their properties. Below we will give an explicit formula which will show that operators with these properties exist.

(i) intertwining property for representations of $\mathcal{A}$,

$$\pi(A)\Gamma^J_m = \Gamma^J_m \pi(\rho_J(A)) . \tag{5.18}$$

(ii) $\mathcal{G}^*$-covariance,

$$\mathcal{U}(\xi)\Gamma^J_m = \Gamma^J_k (\tau^J_{km} \otimes \mathcal{U})(\Delta(\xi)) . \tag{5.19}$$

(iii) Action on lowest weight vectors

$$\Gamma^J_m |0> = |\lambda_J>_m \in \mathcal{H}^J_m \tag{5.20}$$

The subscript m on $|\lambda_J>_m$ distinguishes between lowest weight vectors in the copies $\mathcal{H}^J_m$ of $\mathcal{H}^J$. The $\mathcal{G}^*$-covariance of the field operators (5. 17) follows from $\mathcal{G}^*$-covariance of Γ^J_m and $\mathcal{G}^*$-invariance of the observable factor. The intertwining property (i) and the multiplication law in $\mathcal{A}$ permit to push observables $T \in \mathcal{A}$ through the field (5. 17). Therefore their commutators can be computed in principle. The factor $A_J(w)$ in the bosonization formula is implicitly determined by the requirement that the field is relatively local to the observables, in particular to the stress tensor.

$$[\Psi^J_m(w), T(z)] = 0 \ \text{ if } \ w \neq z \ . \tag{5.21}$$

This determines A_J up to a Klein transformation, i.e. up to multiplication with an element in the center of the observable algebra. This arbitrariness can be eliminiated by suitable conventions, after a "point at infinity" is chosen on the circle. An illustrative calculation is found in ([15]) which shows how A_J is actually computed in a simpler model.

The existence of Γ^J_m needs only a limited amount of dynamical information: It must be possible to put the list of sectors into correspondence with the list of irreducible representations of the symmetry algebra in such a way that the fusion rules of the quantum field theory match with the tensor product decomposition of the symmetry algebra. To obtain local braid relations for the field operators more dynamical information will be needed. This will be discussed later on.

We turn to a discussion of the fusion rules. A product of morphisms is again a morphism. It is defined in the obvious way,

$$(\rho \circ \rho')(A) = \rho(\rho'(A)). \tag{5.22}$$

In our model the Hamiltonian is

$$L_0 = \sum_{a>0} a b_{-a} b_a + \frac{1}{8} b_0^2 . \tag{5.23}$$

Summation runs over $a = \frac{1}{2}, 1, \frac{3}{2}, \ldots$. It can be verified that $\rho(L_0)$ is a positive operator when ρ is equal to ρ_J or equal to a product of such morphism. Therefore $\pi^0 \circ \rho$ is a positive energy representation for any such ρ. *We write* $[\rho]$ *for the equivalence class of this positive energy representation of* $\mathcal{A}$.

If ρ is not one of the irreducible morphisms ρ_J then $[\rho]$ is in general a reducible representation of $\mathcal{A}$. It can be decomposed into irreducible representations. The result is as follows.

$$[\rho_{1/2}^2] = [\rho_0] + [\rho_1] \tag{5. 24}$$

$$[\rho_{1/2}\rho_1] = [\rho_1\rho_{1/2}] = [\rho_{1/2}] \tag{5. 25}$$

$$[\rho_1^2] = [\rho_0] \tag{5. 26}$$

and $[\rho_0\rho_J] = [\rho_J]$ for $J = 0, \frac{1}{2}, 1$. We see that this matches with the tensor product decompostion for the irreducible representations of the truncated quantum group algebra for $U_q(sl_2)$ with $q = \pm i$. Let us write $[J]$ for the equivalence class of the representation τ^J of $\mathcal{G}^*$. Formula (5. 11) yields

$$[1/2] \bigotimes [1/2] = [0] + [1] \tag{5. 27}$$

$$[1/2] \bigotimes [1] = [1] \bigotimes [1/2] = [1/2] \tag{5. 28}$$

$$[1] \bigotimes [1] = [0] . \tag{5. 29}$$

Let us briefly explain how the fusion rules are derived in the algebraic frame work. To do this one uses intertwiners T and their associated projectors $\Pi = T^*T$. The intertwiners are elements T of the observable algebra, and $\pi^0(T)$ are bounded operators in $\mathcal{H}^0$. To reduce $\pi^0 \circ \rho_K\rho_J$ into irreducible subrepresentations $\cong \pi_0 \circ \rho_L$, one needs projection operators $\pi^0(\Pi({}_L{}^J{}_K))$ which commute with the representation operators $\pi^0 \circ \rho_K\rho_J(A)$ for all $A \in \mathcal{A}$

$$\rho_K\rho_J(A)\Pi({}_L{}^J{}_K) = \Pi({}_L{}^J{}_K)\rho_K\rho_J(A) , \tag{5.30}$$

One needs also intertwining operators $\pi^0(T({}_L{}^J{}_K))$ which establish the unitary equivalence of $\pi_0 \circ \rho_K\rho_J$ restricted to $\pi^0(\Pi_L)\mathcal{H}_0$ with $\pi_0 \circ \rho_L$,

$$\rho_L(A)T({}_L{}^J{}_K) = T({}_L{}^J{}_K)\rho_K\rho_J(A) , \tag{5. 31}$$

$$T({}_L{}^J{}_K)^*T({}_L{}^J{}_K) = \Pi({}_L{}^J{}_K). \tag{5. 32}$$

We say that $T({}_L{}^J{}_K))$ is an intertwiner *"from $\rho_K\rho_J$ to ρ_L"*

Given J, K the number of linearly independent intertwining operators $\pi^0(T({}_L{}^J{}_K))$ is equal to the number of times that $[\rho_L]$ appears in $[\rho_K\rho_J]$. In our model this multiplicity is at most one. The projectors satisfy the completeness relation

$$\sum_L \Pi({}_L{}^J{}_K) = 1 \ . \tag{5.33}$$

The intertwiners are known explicitly [10]. For instance

$$T({}_{\frac{1}{2}}{}^1{}_{\frac{1}{2}}) = \frac{1}{2}[1-Y](1-2b_{-1/2}b_{1/2}) + \frac{1}{2}[1+Y]\sqrt{2}b_0(b_{-1}-b_1) \ . \tag{5.34}$$

This particular intertwiner is unitary. It follows that $[\rho_{1/2}\rho_1]$ is irreducible and equivalent to $[\rho_{1/2}]$.

With the knowledge that fusion rules and tensor product decomposition match, we can write down an expression for the "constant fields"

$$\Gamma^J_m = \sum_{K,L}\sum_{k,l} \left[\begin{smallmatrix} J & L & K \\ m & l & k \end{smallmatrix}\right]_q i^*_{Kk}\pi_0(T({}_K{}^J{}_L))i_{Ll} \tag{5.35}$$

$[\begin{smallmatrix}\cdots\\\cdots\end{smallmatrix}]_q$ are Clebsch Gordan coefficients for $U_q(sl_2)$ and for $\mathcal{G}^*$ at the same time. One verifies that Γ^J_m has all the properties listed above.

Now we have an action of the symmetry algebra in the Hilbert space $\mathcal{H}$ of physical states, and field operators on this space which are relatively local to the observables and which transform covariantly under $\mathcal{G}^*$. But we have used only part of the dynamical information. More is needed to get also the local braid relations.

The morphisms ρ_J of the algebra of observables and their intertwiners determine the 6J–symbols of the symmetry algebra, and a representation of the braid group for every J, as follows.

Intertwiners from $\rho_I\rho_J\rho_K$ to ρ_L can be constructed from the elementary intertwiners in two different ways, decomposing either $\rho_I\rho_J$ first, or $\rho_J\rho_K$. In either way one obtains a complete set of intertwiners. Therefore there must be a linear relation between these two sets of intertwiners. It is known as the "fusion identity" and reads

$$T({}_L{}^N{}_I)\rho_I(T({}_N{}^K{}_J)) = \sum_M F_{NM}\left[\begin{smallmatrix} J & I \\ K & L \end{smallmatrix}\right] T({}_L{}^K{}_M)T({}_M{}^J{}_I) \ . \tag{5.36}$$

in the absence of multiplicities. The explicit form of the fusion matrix depends on phase conventions. When suitable phase conventions are imposed then F is a numerical matrix which is given by 6J–symbols.

$$F_{PQ}\left[\begin{smallmatrix} J & I \\ K & L \end{smallmatrix}\right] = \left\{\begin{smallmatrix} K & J & P \\ I & L & Q \end{smallmatrix}\right\}_q \ .$$

The algebra of observable $\mathcal{A}$ and its morphisms determine a unitary representation of the braid group by elements of $\mathcal{A}$ of the following form.

$$\sigma_1 = \epsilon_J \quad , \quad \sigma_n = \rho_J^{n-1}(\epsilon_J) \ (n \geq 2) \tag{5.37}$$

This is discussed in some detail in Fredenhagens lectures at this school. There is a formula for ϵ_J, but it is easier to determine it from its properties. ϵ_J is an intertwiner from ρ_J^2 to ρ_J^2. In other words, it is in the commutant of $\rho_J^2(\mathcal{A})$. A nontrivial commutant exists only if ρ_J is not an automorphism. Otherwise, ϵ_J is a phase factor. In our model, the interesting case is $J = \frac{1}{2}$. The commutant of $\rho_{1/2}^2$ is spanned by the projection operators

$$\Pi_L \equiv \Pi\binom{\frac{1}{2}}{L\ \frac{1}{2}} \ , \ L = 0, 1 \ .$$

Thus

$$\epsilon_{1/2} = \gamma_0 \Pi_0 + \gamma_1 \Pi_1 \ . \tag{5.38}$$

The phase factors γ_L could a priori be elements of the center of $\mathcal{A}$, i.e. be linear combinations of 1 and Y. They can be determined, up to an overall phase factor, from the requirement that (5. 37) is actually a representation of the braid group, i.e. satisfy the Artin relations. The remaining phase factor is determined by Fredenhagens general spin and statistics theorem. As a result one obtains for the chiral Ising model

$$\gamma_0 = i\gamma \ , \ \gamma_1 = -\gamma \ \text{ with } \ \gamma = -e^{i3\pi/8} \ . \tag{5.39}$$

Using the explicit definition and homotopy invariance of ϵ_J as discussed in Fredenhagens lectures, one can show that the local braid relations of field operators (5. 17) with the same J are implied by the following property of the "constant fields"

$$(\Gamma^J \times \Gamma^J)_{ij} \epsilon_J = (\Gamma^J \times \Gamma^J)_{j'i'} \mathcal{R}^{JJ>}_{i'j',ij} \ . \tag{5.40}$$

All the quantities in this identity have been defined. It can be verified that the identity is indeed satisfied for $J = \frac{1}{2}$ [2]. The proof uses the hexagon- and pentagon identities of 6J–symbols. Validity of covariant operator product expansions can also be proven. They are valid on all of $\mathcal{H}$, i.e. on all superselection sectors. Validity of local braid relations for arbitrary I, J follows from their validity for $I = J = \frac{1}{2}$, from the operator product expansions for $I = J = \frac{1}{2}$ and from the fact that local braid relations for covariant products of fields follow from those for individual fields. This last fact is a consequence of Drinfelds relations (5. 12), (5. 13).

6. AN ALGEBRAIC APPROACH TO THE CONSTRUCTION OF BOSE FIELDS IN TWO SPACE TIME DIMENSIONS

The central idea of algebraic field theory is to start from an algebra of observable fields and its positive energy representations. The corresponding representation spaces are called superselection sectors. One constructs then new field operators which are relatively local with respect to the observables and which make transitions between superselection sectors. These new fields are constructed by use of the general bosonization formula of Doplicher Haag and Roberts, eq.(6. 17).

For the purpose of mathematical physics, any algebra of local Bose fields may be declared to be the algebra of observables to start with. One may choose to start with a smaller algebra of observables than the full algebra of observables which one would like to have. After construction of the new fields one may select among them those that are Bose fields, and obtain in this way a larger algebra of observables which contains the desired one.

A simple example of such a construction was given by Buchholz, Mack and Todorov in [15]. In that example, the $U(1)$-current algebra was the starting point, and the possible maximal extensions of it to larger algebras of observables were classified.

Here we wish to consider conformal field theory in 2-dimensional space time, with Lorentzian metric, i.e. on a tube $S^1 \times \mathbf{R}$. [17]. In particular, we consider the Wess-Zumino-Novikov-Witten models. This will enable the reader to compare with Ludwig

Faddeevs lectures [14]. We wish to explain how the extra variables ("zero modes") which appear in his approach in addition to the chiral currents would be reconstructed by an algebraic field theory approach. The considerations of this section could be made explicit for $SO(N)$-WZNW-models at level 1 because the morphisms of the chiral algebras are known in this case [18].

In the WZNW-model for group G one wishes to construct a renormalized version of the field operators

$$g : (x,t) \mapsto g(x,t) \in G \tag{6.1}$$

where the space coordinate x serves to fix the position on the circle and t is a time coordinate. We write $\partial_\pm$ for differentiation with respect to $x^\pm = \frac{1}{2}(t \pm x)$.

The ultimate algebra of observables $\mathcal{A}$ is supposed to be generated by the fields $g(x,t)$. A truely smaller algebra is generated by the chiral currents,

$$J_\pm = \partial_\pm g g^{-1} \quad . \tag{6.2}$$

This smaller algebra factors into chiral algebras,

$$\mathcal{A}^+ \times \mathcal{A}^- \quad . \tag{6.3}$$

Already at a classical level, the desired fields $g(x,t)$ cannot be reconstructed from the chiral currents alone, because a constant of integration g(0,0) is missing. The problem is to reconstruct the quantum version of this constant of integration.

In the algebraic approach one will first reconstruct chiral fields, using the bosonization formula

$$\psi_m^{I\pm}(x^\pm) = \Gamma_m^{I\pm} A_I^\pm(x^\pm) \tag{6.4}$$

This involves "constant fields" $\Gamma^{I\pm}$ and factors A which are constructed from the chiral currents. The formula for A in terms of currents will depend on I. We indicate this dependence. The constant fields will transform covariantly under a quantum symmetry while the chiral observables $A_I^\pm(x^\pm)$ are invariants under the quantum symmetry. The chiral fields $\psi^{I\pm}$ will satisfy local braid relations by construction.

Two dimensional fields of the form

$$\Gamma_i^{I+} A_I^+(x^+) \Gamma_j^{J-} A_J^-(x^-) \tag{6.5}$$

will also satisfy local braid relations. (Left and right moving building blocks commute with each other) .

It remains to pick out those linear combinations for which the local braid relations reduce to Bose commutation relations. The guiding principle here is that a nontrivial transformation law under a true quantum symmetry is in general not compatible with Bose commutation relations. Considering fixed time, one finds that the Bose fields should be invariant under the diagonal quantum symmetry algebra which acts on the left and right movers alike. One has the freedom of a choice of basis in index space so that $\sum_i \Gamma_i^{I+} \Gamma_i^{I-}$ are invariants. This gives a list of candidates for Bose fields indexed by I,

$$\phi^{(I)}(x) = \sum_i \Gamma_i^{I+} \Gamma_i^{I-} A_I^+(x^+) A_I^-(x^-) \; .$$

Which is g would have to be found out from further requirements. ¿From an algebraic point of view, this question is actually not so interesting. One is happy once one has the algebra of Bose fields. Not all the candidates are actually Bose fields in general, though. Covariantly defined composites of braid local fields are braid local fields, as a consequence of Drinfelds relations. This implies that $\mathcal{G}^*$-invariant fields obey "anyonic" commutation relations which may differ from Bose commutation relations by phase factors. Therefore one must work out these phase factors. In the case of the conformal Ising model which was considered in the last section, and which is very similar to the WZNW-case, $I = 0, \frac{1}{2}, 1$. For $I = 1$, the new field turns out to be a Fermi-field. The case I=0 is always trivial because $\Gamma^{0\pm} = 1$; therefore one gets back the old observables. The only new Bose field comes from $I = \frac{1}{2}$ in the case of the Ising model.

7. THE COMODULE PICTURE

In order to have a conventional picture of a symmetry, we used the algebra $\mathcal{G}^*$. An element ξ of $\mathcal{G}^*$ will be called a "quasi-rotation" in the following text. The Hilbert space of physical states is a representation space (=module) for $\mathcal{G}^*$. Thus there is a map

$$U : \mathcal{G}^* \otimes \mathcal{H} \mapsto \mathcal{H} \tag{7.1}$$

It takes $\xi \otimes |\psi>$ into $\mathcal{U}(\xi)|\psi>$. Given any quasirotation ξ, its action $\mathcal{U}(\xi)$ on any element $|\psi> \in \mathcal{H}$ is defined and produces another element of $\mathcal{H}$.

One may instead consider the action $\hat{\mathcal{U}}$ of an unspecified element of $\mathcal{G}^*$ on $|\psi> \in \mathcal{H}$. It produces an object which becomes an element of $\mathcal{H}$ when a quasirotation ξ is supplied to it. Mathematically

$$\hat{\mathcal{U}} : \mathcal{H} \mapsto \mathcal{H} \otimes \mathcal{G} \ , \tag{7.2}$$

where $\mathcal{G}$ is the dual bialgebra to $\mathcal{G}^*$. Given such a linear map (7. 2), one says that $\mathcal{H}$ is a comodule for $\mathcal{G}$, and $\hat{\mathcal{U}}$ defines a coaction of $\mathcal{G}$ on $\mathcal{H}$. The algebra $\mathcal{G}$ is a generalization of the algebra of (good) functions on a group. Its elements $f \in \mathcal{G}$ are linear maps

$$f : \ \mathcal{G}^* \mapsto \mathbf{C} \ . \tag{7.3}$$

The product $\cdot$ in $\mathcal{G}$ is furnished by the coproduct in $\mathcal{G}^*$,

$$f_1 \cdot f_2(\xi) = (f_1 \otimes f_2)(\Delta(\xi)) \ . \tag{7.4}$$

The matrix elements τ_{ij} of any representation of $\mathcal{G}^*$ are linear functions on $\mathcal{G}^*$, and may therefore be regarded as elements of $\mathcal{G}$. The formula (7. 4) for the product implies that

$$\tau_{ij} \cdot \tau'_{kl}(\xi) = (\tau \bigotimes \tau')_{ik,jl}(\xi) \tag{7.5}$$

In this sense, tensor products of representations of $\mathcal{G}^*$ become ordinary products of elements of $\mathcal{G}$. There is also a coproduct in $\mathcal{G}$ which comes from the product in $\mathcal{G}^*$.

We write $|\psi> f$ in place of $|\psi> \otimes f$ for elements of $\mathcal{H} \otimes \mathcal{G}$ and regard them as linear functions on $\mathcal{G}^*$ with values in $\mathcal{H}$, viz.

$$(|\psi> f)(\xi) = |\psi> f(\xi) \ .$$

In this notation, the coaction $\hat{\mathcal{U}}$ of $\mathcal{G}$ on $\mathcal{H}$ is given by

$$\hat{\mathcal{U}}|\psi> (\xi) = U(\xi)|\psi> \quad . \tag{7.6}$$

Imagine for a moment that a basis $|\phi_\alpha>$ has been introduced in $\mathcal{H}$, and define matrix elements of $\hat{\mathcal{U}}$ by

$$\hat{\mathcal{U}}|\phi_\alpha> = |\phi_\beta> \hat{\mathcal{U}}_{\beta\alpha} \ \text{ with } \ \hat{\mathcal{U}}_{\alpha\beta} \in \mathcal{G} \ .$$

The matrix elements $\mathcal{U}_{\alpha\beta}(\cdot)$ of $\mathcal{U}(\cdot)$ are functions on $\mathcal{G}^*$, hence elements of $\mathcal{G}$, and so are $\hat{\mathcal{U}}_{\alpha\beta}$. Both are in fact equal, by definition (7. 6),

$$\hat{\mathcal{U}}_{\alpha\beta}(\xi) = \mathcal{U}_{\alpha\beta}(\xi) \ .$$

$\hat{\mathcal{U}}$ and $\mathcal{U}$ are therefore "basically the same".

Since $\hat{\mathcal{U}}_{\alpha\beta}$ is an element of $\mathcal{G}$, the product $\hat{\mathcal{U}}_{\alpha\beta} \cdot \tau^J_{ij}$ is defined as a product in $\mathcal{G}$, and equal to the tensor product of representations $(\hat{\mathcal{U}} \otimes \tau^J)_{\alpha i, \beta j}$ by (7. 5). We continue to use this when the reference to a basis in $\mathcal{H}$ is dropped.

The 1-dimensional representation ϵ of $\mathcal{G}^*$ is also a function on $\mathcal{G}^*$, hence an element of $\mathcal{G}$. It is the identity in $\mathcal{G}$, i.e.

$$\epsilon \cdot f = f \ \text{ for all } f \in \mathcal{G} \ .$$

The defining properties of a symmetry (cp. introduction) can now be rewritten as follows

$$\hat{\mathcal{U}}|0> \ = \ |0> \epsilon \, , \qquad (7.\ 7)$$
$$\hat{\mathcal{U}} H \ = \ H \hat{\mathcal{U}} \, , \qquad (7.\ 8)$$
$$\hat{\mathcal{U}} \Psi^J_j \ = \ \Psi^J_k (\tau^J_{kj} \cdot \hat{\mathcal{U}}) \, . \qquad (7.\ 9)$$

On the right hand side of eq.(7. 8), the Hamiltonian H is understood to act on the first factor $\mathcal{H}$ of $\mathcal{H} \otimes \mathcal{G}$; a more careful notation would be to write $H \otimes \epsilon$ in place of H. Similarly, the factor Ψ^J_k on the right hand side of eq. (7. 9) is understood to act on $\mathcal{H}$, and so on.

The module-picture, which is used in the body of this paper, was introduced for quantum groups by Buchholz, Mack and Todorov [9]. The comodule picture was introduced by Reshetikhin and Smirnov [19]. We see that both pictures are equivalent.

8. SUPERSYMMETRY

It is instructive to consider supersymmetry as a special case of a quantum symmetry. Supersymmetry is not a purely internal symmetry, since it includes the space time translation generators P_μ among its generators. But this can be accomodated.

We adjoin fermionic parity $(-)^F$ to the generators of the supersymmetry algebra. It becomes a quasitriangular Hopf algebra ("quantum group algebra") by defining coproduct, counit and antipode as described below, and an R-element which reproduces the standard commutation / anticommutation relations. The coproduct is coassociative, therefore $\varphi = e \otimes e \otimes e$ and $\Delta(e) = e \otimes e$.

Apart from generators of groups, the supersymmetry algebra contains fermionic generators Q^A_α which anticommute with $(-)^F$. These fermionic generators obey anticommutation relations with Fermi field operators and commutation relations with Bose fields. Both are reproduced by the coproduct defined below.

$$\Delta(Q^A_\alpha) \ = \ Q^A_\alpha \otimes e + (-)^F \otimes Q^A_\alpha \qquad (8.\ 1)$$
$$R \ = \ \frac{1}{2}[e \otimes e + (-)^F \otimes e + e \otimes (-)^F - (-)^F \otimes (-)^F] \, , \qquad (8.\ 2)$$
$$\mathcal{S}(Q^A_\alpha) \ = \ (-)^F Q^A_\alpha . \qquad (8.\ 3)$$

Obviously, the coproduct is not cocommutative. It may be called "graded cocommutative". The counit is the obvious one, $\epsilon(Q^A_\alpha) = 0$ etc. All of Drinfelds relations are satisfied.

Let us also consider the comodule picture. We remember that representation functions τ^J_{ij} of $\mathcal{G}^*$ are elements of $\mathcal{G}$ whose product is given by the tensor product of representations of $\mathcal{G}^*$. Let us choose bases in the representation spaces such that fermionic parity $(-)^F$ is diagonal,

$$\tau^J_{jk}((-)^F) = s^J_j \delta_{jk} \ ,$$

and set

$$s^{JI}_{ji} \ = \ -1 \ \text{ if } s^J_j = -1 = s^I_i \qquad (8.\ 4)$$
$$= \ +1 \ \text{ otherwise} \qquad (8.\ 5)$$

The relation $\Delta'(\eta)R = R\Delta(\eta)$ translates into commutation relations in $\mathcal{G}$,

$$\tau^I_{ik} \cdot \tau^J_{jl} = \pm \tau^J_{jl} \cdot \tau^I_{ik} \ \text{ with } \ \pm = s^{IJ}_{ij} s^{IJ}_{kl} \ . \qquad (8.6)$$

One may express this result by saying that $\mathcal{G}$ is graded commutative.

The considerations of this section illustrate the point that the introduction of quantum symmetries continues a process of generalization which began with the introduction of supersymmetry.

References

[1] G. Mack, V. Schomerus, *Quasi Hopf quantum symmetry in quantum theory*, DESY 91–037, to appear in Nucl. Phys. B.

[2] G. Mack, V. Schomerus, *Quasi quantum group symmetry and local braid relations in the conformal Ising model* Phys. Letters **B267** (1991) 207

[3] T. Curtright et al (eds) *Quantum groups* World Scientific, Singapore 1991, and references therein

[4] V.G. Drinfel'd, *Quasi Hopf algebras and Knizhnik Zamolodchikov equations*, in: Problems of modern quantum field theory, Proceedings Alushta 1989, Research reports in physics, Springer Verlag Heidelberg 1989

[5] J.Fröhlich,*Statistics of fields, the Yang- Baxter equation and the theory of knots and links*, in: Nonperturbative quantum field theory, G.t'Hooft et al.(eds.), Plenum Press 1988

[6] R. Dijkgraaf, V. Pasquier, P. Roche, *Quasi-quantum groups related to orbifold models*, in: Proc. Intl. Colloquium on modern quantum field theory, TATA Institute of Fundamental Research, Bombay, January 1990

[7] L.Alvarez–Gaume, C.Gomez, G. Sierra,*Hidden quantum symmetry in rational conformal field theories*, Nucl.Phys. **B310** (1989)
L.Alvarez–Gaumé, C.Gomez, G. Sierra, *Quantum group interpretation of some conformal field theories*, Phys. Lett. **220B** (1989) 142
L.Alvarez–Gaumé, C.Gomez, G.Sierra, *Duality and quantum groups*, Nucl.Phys. **B330**, 347 (1990)

[8] V. Pasquier, H. Saleur,*Common structures between finite systems and conformal field theories through quantum groups*, Nucl.Phys.**B330**,523 (1990)

[9] D.Buchholz,G.Mack,I.T.Todorov, as reported in
I. Todorov, in: Proceedings of Conf. on Quantum Groups, Clausthal Zellerfeld, July 1989 (reviewed in ref.[10]).

[10] G. Mack, V. Schomerus, *Conformal field algebras with quantum symmetry from the theory of superselection sectors*, Commun. Math. Phys. **134**, 139 (1990)

[11] I.M. Gelfand, G.E. Shilov, *Generalized functions*, vol 1, Academic press, New York 1964

[12] G.Mack, V.Schomerus, *Action of truncated quantum group algebras on quasi quantum planes and a quasi associative differential calculus* to be submitted to Commun. Math. Phys.

[13] K. Fredenhagen, (contribution to these proceedings)

[14] L. D. Faddeev, (contribution to these proceedings)

[15] D.Buchholz,G.Mack,I.T.Todorov, *The current algebra on the circle as a germ of local field theories* Nucl. Phys B (Proc. Suppl.) **5B** (1988) 20 and
D. Buchholz, G. Mack, I.T.Todorov, *Localized automorphisms of the U(1) current algebra on the circle* in: D. Kastler (ed), Algebraic theory of superselection sectors and field theory, World Scientific, Singapore 1990

[16] M.Lüscher, G.Mack, *Global conformal invariance in quantum field theory* Commun.Math.Phys. **41** (1975) 203

[17] G.Mack, *Introduction to conformal invariant quantum field theories in two dimensions* in: *Nonperturbative quantum field theory*, G.t'Hooft et al. (eds.), Plenum Press New York 1988

[18] J. Fuchs, A. Ganchev, P. Vecsernyes, *Level 1 WZW superselection sectors* CERN-TH-6166-91

[19] N.Yu Reshetikhin and F. Smirnov, *Hidden quantum symmetry and integrable perturbations of conformal field theory*, Commun. Math. Phys. **131**, 157 (1990)

QUANTUM GROUPS IN LATTICE MODELS

V. Pasquier

Service de Physique Théorique*
Centre d'Etudes de Saclay F-91191 Gif-sur-Yvette cedex

Abstract:

The aim of these lecture notes is to give a pedestrian introduction to Yangians and affine quantum groups starting from integrable lattice models.

The first chapter presents the Bethe ansatz technique applied to the XXZ chain. The second is an introduction to the technique of diagonalisation of transfer matrices in the six vertex model case. The third is devoted to the study of some quadratic algebras. More specifically we consider the examples of Yangians and loop-algebras deformations.

Introduction:

I have tried to write these lecture notes for physicists who want to learn about developments in the theory of integrable systems, in particular affine quantum groups, from a basic knowledge of statistical mechanics.

The starting point in the first chapter is an elementary introduction to the Bethe-Ansatz technique, which we illustrate with the example of the XXZ chain. It is mainly a translation into English of parts of chapter 1 and 4 of M. Gaudin's book "La fonction d'onde de Bethe" [1]. The second chapter is devoted to the transfer matrix of the six-vertex model. Starting from Lieb's observation that the eigenvectors of the transfer matrix of the six-vertex model coincide with those of the XXZ chain, we derive the conditions under which such transfer matrices commute which naturally lead to the Yang-Baxter equation. (This overlaps with chapter III of Baxter's book). In order to prepare for the next chapter, we introduce Sklyanin's quadratic algebra and, as an application, we diagonalise the transfer matrix using the so-called "algebraic Bethe ansatz". The material of this chapter is covered

* Laboratoire de la Direction des Sciences de la Matière du Commissariat à l'Energie Atomique

New Symmetry Principles in Quantum Field Theory, Edited by
J. Fröhlich et al., Plenum Press, New York, 1992

in L. Faddeev "Les Houches" [3] lecture notes; in presenting it, I have adopted a statistical mechanic's approach close to the one followed in Baxter's book [2].

In the last chapter, I use the material of the first two chapters to introduce quantum groups, more specifically Yangians and affine algebras [4]. From this point of view, the quantum group is an algebraisation of the notion of transfer matrix in integrable models and many results can be intuitively understood this way. I have covered in relative detail the example of $SL(n)$ and $O(n)$ Yangians which can be understood without background of the theory of Lie-Algebras. I have also derived Drinfeld's second presentation of Yangians [5] from the quadratic algebra relations in the $SL(2)$ case. This presentation is quite useful when we look for vertex operator representations of affine algebras. When we replace the rational solution of the Yang-Baxter equations by its trigonometric counterpart, all the Yangian relations become the q-deformed loop algebra relations. I have used the vertex-operator representation [6] to introduce the affine algebra with a non zero central charge in the $SL(2)$ case. In the $SL(n)$ case, I show how to modify the braiding relations so as to obtain the correct affine algebra relations. Unfortunately, I have not been able to make the connection between affine algebras with a non zero central charge and integrable models, leaving this an open problem. I have included only a few references, a lot more can be found in the book [9].

1. Coordinate Bethe Ansatz

We consider the Heisenberg Hamiltonian

$$H_\Delta = \sum_{n=1}^{N} \sigma_i^+ \sigma_{i+1}^- + \sigma_i^- \sigma_{i+1}^+ + \frac{\Delta}{2}(\sigma_i^Z \sigma_{i+1}^Z - 1) \tag{1}$$

$\sigma_i^\pm, \sigma_i^Z$ are the 2×2 Pauli matrices acting on the ith site of a chain:

$$\begin{aligned} \sigma_i^+ &= 1 \otimes 1 \otimes ... \otimes \begin{pmatrix} 0 & 1 \\ 0 & 0 \end{pmatrix}_i \otimes 1 ... \otimes 1 \\ \sigma_i^- &= 1 \otimes 1 \otimes ... \otimes \begin{pmatrix} 0 & 0 \\ 1 & 0 \end{pmatrix}_i \otimes 1 ... \otimes 1 \\ \sigma_i^Z &= 1 \otimes 1 \otimes ... \otimes \begin{pmatrix} 1 & 0 \\ 0 & -1 \end{pmatrix}_i \otimes 1 ... \otimes 1 \end{aligned} \tag{2}$$

The chain is supposed to be cyclic so that

$$\vec{\sigma}_{N+1} = \vec{\sigma}_1. \tag{3}$$

A constant has been added to the Hamiltonian so that the energy of the reference ferromagnetic state $|S^Z = \frac{N}{2}\rangle$ is equal to zero. This Hamiltonian commutes with the total spin operator $S^Z = \frac{1}{2}\sum_{i=1}^{N} \sigma_i^Z$ and can therefore be diagonalised simultaneously with S^Z. An eigenstate is written:

$$
\begin{aligned}
|M\rangle &= \sum_{\{n\}} a(n_1, n_2, ..., n_M)\,|n_1, n_2 ..., n_M\rangle \\
&\text{with } |n_1 ... n_M\rangle = \sigma^-_{n_1} ... \sigma^-_{n_M}|0\rangle
\end{aligned} \tag{4}
$$

The summation is on all sets of integers $\{n\}$ such that:

$$1 \leq n_1 < n_2 ... < n_M \leq N. \tag{5}$$

Let us write the eigenvalue equation:

$$H|M\rangle = E|M\rangle \tag{6}$$

It yields:

$$\sum_{\{n'\}} (a(n') - \Delta\ a(n)) = E\ a(n) \tag{7}$$

where $\{n'\}$ are all the spin configurations satisfying condition (5) which can be obtained from $\{n\}$ by a permutation of neighbouring antiparallel spins:

$$n'_1 = n_1, ..., n'_k = n_{k\pm 1}, ..., n'_M = n_M \tag{8}$$

In order to properly take into account the cyclic boudary condition, we extend the definition of $\{n\}$ to:

$$n_1 < n_2 ... < n_M, \qquad n_M - n_1 < N \tag{9}$$

and impose the cyclicity condition:

$$a(n_1, ... n_2, ..., n_M) = a(n_2, n_3, ..., n_1 + N) \tag{10}$$

Let us solve equation (6) for the first values of M :

1) *case* $M = 1$:

Eq.(7) becomes $a(n+1) + a(n-1) - 2\Delta a(n) = E\ a(n)$ which is solved for $a(n) = z^n$, $E = z + z^{-1} - 2\Delta$. The cyclic condition (10) yields $z^N = 1$ so that $z = \exp\left(\frac{2\pi k i}{N}\right)$

2) *case* $M = 2$:

Equation (7) now reads

$$\sum_{\{n'\}} a(n'_1, n'_2) - \Delta\ a(n_1, n_2) = E\ a(n_1, n_2) \tag{11}$$

with the restriction that $n'_1 < n'_2$. It is convenient to extend the definition of $a(n_1, n_2)$ to the case where $n_1 = n_2$. We determine the coefficients so introduced by requiring that (11) is satisfied without the restriction $n'_1 \neq n'_2$. Call (11*) the new set of equations. It differs from (11) only in the case where $n_2 = n_1 + 1$; since both sets of equations are supposed to be satisfied, we can replace (11) by the equations obtained by subtracting them:

$$a(n, n) + a(n+1, n+1) - 2\Delta a(n, n+1) = 0 \tag{12}$$

(11*) is now solved by each term of the following (Bethe) Ansatz:

$$a(n_1, n_2) = A\ z_1^{n_1}\ z_2^{n_2} + B\ z_1^{n_2}\ z_2^{n_1} \tag{13}$$

with:

$$E = \sum_{i=1,2} (z_i + z_i^{-1} - 2\Delta) \tag{14}$$

Substituting the ansatz (13) in (12) gives:

$$A(1 + z_1 z_2 - 2\Delta z_2) = -B(1 + z_1 z_2 - 2\Delta z_1) \tag{15}$$

Finally, the cyclic condition (10) yields

$$\begin{aligned} A &= B\ z_1^N \\ B &= A\ z_2^N \end{aligned} \tag{16}$$

(15),(16) completely determine the wave function.

3) *general case:*

As for the $M = 2$ case, we extend the definition of $a(n_1, n_2, ..., n_M)$ to $n_1 \leq n_2 \leq ... \leq n_M \leq n_1 + N$ and require that (7) is also satisfied without the restriction $n_i \neq n_j$. We call (7*) the set of equations so obtained. When (7) differs from (7*), we replace it by (7)-(7*) to obtain:

$$\begin{aligned} &a(n_1, ...n_i, n_i, ..., n_M) + a(n_1, ...n_i + 1, n_i + 1, ...n_M) \\ &-2\Delta\ a(n_1, ..., n_i, n_i + 1, ..., n_M) = 0 \end{aligned} \tag{17}$$

Eq.(7*) is solved by each term of the following ansatz for $a(n_1, n_2, ...n_M)$:

$$a(n_1, n_2, ..., n_M) = \sum_{P \in \pi_M} A(P)\ z_{p_1}^{n_1}\ z_{p_2}^{n_2} ... z_{p_M}^{n_M} \tag{18}$$

where the sum is taken over the set of permutations of M elements π_M. The energy is given by

$$E = \sum_{i=1}^{M} (z_i + z_i^{-1} - 2\Delta) \tag{19}$$

Equation (17) yields for $i = 1$:

$$A(P)(z_{p_1} z_{p_2} + 1 - 2\Delta z_{p_2}) + A(P(12))(z_{p_1} z_{p_2} + 1 - 2\Delta z_{p_1}) = 0 \tag{20}$$

and the cyclic boundary condition equation (10) gives:

$$A(P) = A(PC) z_{p_1}^N \tag{21}$$

with C the cyclic permutation:

$$C = \begin{pmatrix} 12...M \\ 23...1 \end{pmatrix} = (12)(23)...(M-1, M) \tag{22}$$

We can use (20) repeatedly to express $A(PC)$ in terms of $A(P)$ in (20). Assumming that none of the denominators vanish, we obtain the consistency condition:

$$z_{p_1}^{-N} = \prod_{i \neq 1} - \frac{1 + z_{p_1} z_{p_i} - 2\Delta\ z_{p_i}}{1 + z_{p_1} z_{p_i} - 2\Delta z_{p_1}} \tag{23}$$

Since (23) has to be satisfied for any permutation, it can be rewriten as:

$$z_i^{-N} = \prod_{j \neq i} - \frac{1 + z_i z_j - 2\Delta z_j}{1 + z_i z_j - 2\Delta z_i} \tag{24}$$

This set of algebraic equations determine the z_i and hence the wave function completely.

Let us briefly discuss another problem where these equations also arise naturally. We consider the following Hamiltonian:

$$H = - \sum_{j=1}^{N} \frac{\partial^2}{\partial x_j^2} + 2c \sum_{i<j} \delta(x_i - x_j) \tag{25}$$

We want to find its eigenfunctions symmetric in $x_1 x_2 ... x_N$ (bosons) and continuous in $\mathbb{R}^N$. The eigenvalue equation decomposes into a wave equation:

$$(\Delta_N + E)\psi = 0 \tag{26}$$

in the fundamental domain:

$$D : x_1 < x_2 ... < x_N. \tag{27}$$

and boundary conditions at the frontier of D

$$\frac{\partial \psi}{\partial x_{i+1}} - \frac{\partial \psi}{\partial x_i} |_{x_i - x_{i+1} \longrightarrow 0} = c\psi|_{x_i = x_{i+1}} \tag{28}$$

Eq.(26) is analogous to (5*) and can be solved by considering a superposition of plane waves:

$$\psi(x_1, ... x_N) = \sum_{p \in \pi_N} A(P)\ e^{i(k_{p_1} x_1 + ... k_{p_N} x_N)} \tag{29}$$

Substituting (29) in (28) gives

$$(i(k_{p_{j+1}} - k_{p_j}) - c) A(P) = (i(k_{p_{j+1}} - k_{p_j}) + c) A(P(j, j+1)) \tag{30}$$

which is analogous to (20).

If we put the system on a circle of length L, it is natural to require the periodicity condition

$$\psi(x_1, x_2, ..., x_N) = \psi(x_2, x_3, ..., x_1 + L) \tag{31}$$

We obtain the analogous equation to (21) (24) which determines the k_i :

$$A(P) = A(PC)\ e^{ik_{p_1} L} \tag{32}$$

$$e^{-ik_i L} = \prod_{j \neq i} \frac{k_j - k_i + ic}{k_j - k_i - ic} \tag{33}$$

It is clear that (33) can be obtained as a limit of (24) if we set:

$$\begin{aligned} L &= N\epsilon \\ \Delta &= 1 - \epsilon\frac{c}{2} \\ Z_j &= e^{ik_j\epsilon} \end{aligned} \tag{34}$$

and $\epsilon \longrightarrow 0$, $N \longrightarrow \infty$ keeping L fixed. The direct equivalence can in fact be shown [1].

The simplest solution of (30) describes the ground state in the attractive case $c < 0$. If for k_j we take:

$$ik_j = k_0 + cj, \tag{35}$$

we observe that the only non zero amplitude is $A(1)$, the wave function:

$$\psi(x_1, ..., x_N) = e^{\frac{c}{2}\sum_{i<j}|x_i - x_j|} e^{ik_0(x_1+x_2+...x_N)}, \tag{36}$$

which describes a bound state of N particles propagating with momentum Nk_0 along the x direction.

In the repulsive case, $c > 0$, we take the logarithm of equation (33) to obtain

$$Lk_i = 2\pi\ I_i - 2\sum_{j\neq i} \mathrm{tg}^{-1}\frac{k_i - k_j}{c} \tag{37}$$

where I_i are integers or half integers depending on weather N is odd or even. Using a variational argument, one can show [1], that the ground state corresponds to the sequence $I_i = \{-\frac{N-1}{2}, -\frac{N-3}{2}, ..., \frac{N-1}{2}\}$.

In the thermodynamic limit, $N \longrightarrow \infty$, $\frac{N}{L} \longrightarrow \rho$, we can introduce a continuous variable $x = i/L$; the momentum k reaches a distribution $\rho(k)$ concentrated on the interval $[-k_0, k_0]$. In this limit, (37) becomes:

$$k = 2\pi\left(\int_{-k_0}^{k} \rho(k)dk - \frac{\rho}{2}\right) - 2\int_{-k_0}^{k_0} \mathrm{tg}^{-1}\left(\frac{k-k'}{c}\right)\rho(k')dk', \tag{38}$$

with the constraint

$$\rho = \int_{-k_0}^{k_0} \rho(k)\ \mathrm{d}\ k \tag{39}$$

Differentiating (38) with respect to k gives:

$$\frac{1}{2\pi} = \rho(k) - \frac{c}{\pi}\int_{-k_0}^{k_0} \frac{\rho(k')}{(k-k')^2 + c^2} \mathrm{d}\ k' \tag{40}$$

This equation can be solved explicitly in the limit $c \ll k_0$, $|k \pm k_0|$.

Using:

$$\frac{1}{\pi}\frac{c}{k^2 + c^2} = \delta(k) + \frac{c}{\pi}\frac{P}{k^2} + O(c^2) \tag{41}$$

we obtain to first order in c/k :

$$\int_{-k_0}^{k_0} \frac{P}{(k-k')}\rho(k')\mathrm{d}k' = \frac{k}{2c} \tag{42}$$

$$\int_{-k_0}^{k_0} \rho(k')\mathrm{d}k' = \rho \tag{43}$$

Setting $k/k_0 = \text{th}\,u$, (42) can easily be inverted by Fourier transform yielding [7]:

$$\rho(k) = \frac{1}{2\pi c}(2c\rho - k^2)^{1/2} \tag{44}$$

2. Transfer matrix

We have seen that the eigenvectors of the XXZ hamiltonian can be obtained by the Bethe-ansatz technique. The next step is to introduce a statistical mechanical model defined by its transfer matrix such that the eigenvectors of this matrix coincide with the eigenvectors of the XXZ hamiltonian. This model is known as the six vertex model [8].

Consider a square lattice with periodic boundary conditions. Configurations are obtained by putting arrows on each link in such a way that at each vertex there are as many arrows coming in that there are going out (Fig.1). For the 6 possible vertex configurations, we assign a Boltzmann weight as in Figure 1.

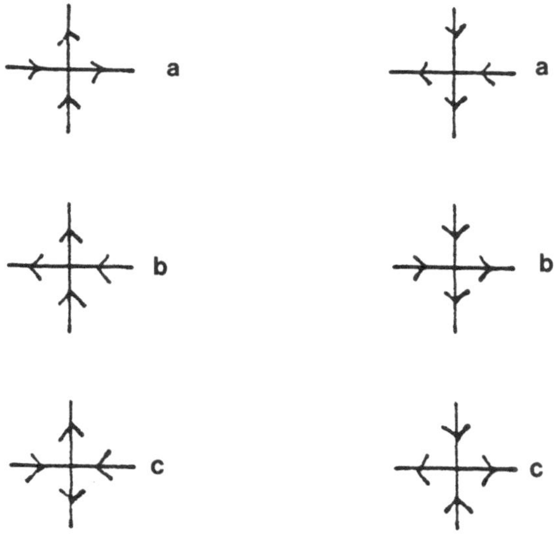

Figure 1

Here, a, b, c are positive numbers to be defined later. Define the partition function Z as the sum on all possible arrow configurations of the product of the Boltzmann weights at each vertex. Let us assign the value 1 to an arrow going up or to the right and -1 to an arrow going down or to the left and call $\omega(\alpha\beta|\mu\nu)$ the following Boltzmann weight

$$\omega(\alpha\beta|\mu\nu) \;= \tag{45}$$

$$\alpha + \mu \;= \beta + \nu$$

so that

$$\mathcal{Z} = \sum_{\text{conf}} \prod_{\text{vertices}} \omega(\alpha\beta|\mu\nu) \tag{46}$$

$\mathcal{Z}$ can be expressed as the trace of a transfer matrix $\bar{T}$ raised to the power M where M is the number of columns of the lattice. $\bar{T}$ is the sum of the product of the Boltzmann weights along a column of the lattice keeping the spins on the horizontal links fixed and summing over the spin configurations of the vertical links:

$$\bar{T}_{\alpha_1\alpha_2\ldots\alpha_N,\beta_1\ldots\beta_N} = \sum_{\substack{\{\mu_i\} \\ \mu_{N+1}=\mu_1}} \prod_{i=1}^{N} \omega\left(\mu_i, \mu_{i+1} | \, \alpha_i, \beta_i\right) \tag{47}$$

It is therefore a $2^N \times 2^N$ matrix with N the number of rows of the lattice.

$$\mathcal{Z} = \operatorname{tr} \bar{T}^M \tag{48}$$

L matrix:

Let us define a 2×2 matrix L with coefficients being themselves 2×2 matrices as follow:

$$L = \begin{pmatrix} L_{++} & L_{+-} \\ L_{-+} & L_{--} \end{pmatrix} = \begin{pmatrix} L_{11} & L_{12} \\ L_{21} & L_{22} \end{pmatrix} \tag{49}$$

and $(L_{\mu\nu})_{\alpha\beta} = \omega(\mu\nu|\alpha\beta)$

$$\begin{aligned} L_{11} &= \begin{pmatrix} a & 0 \\ 0 & b \end{pmatrix} \quad L_{12} = \begin{pmatrix} 0 & 0 \\ c & 0 \end{pmatrix} \\ L_{21} &= \begin{pmatrix} 0 & c \\ 0 & 0 \end{pmatrix} \quad L_{22} = \begin{pmatrix} b & 0 \\ 0 & a \end{pmatrix} \end{aligned} \tag{50 a}$$

The horizontal indices $(\alpha\beta)$ are conventionally called the quantum indices and the vertical indices $(\mu\nu)$ the matrix indices. It is clear from (47) that

$$\bar{T} = \sum_{\{\mu_i\}} L_{\mu_1\mu_2} \otimes L_{\mu_2\mu_3} \otimes \ldots \otimes L_{\mu_N\mu_1} \tag{50 b}$$

where the tensor product is taken in the quantum space. Introducing the notation $\cdot$ for taking the product in the matrix space we rewrite (50 b) as:

$$\bar{T} = \operatorname{tr} L \dot{\otimes} L \dot{\otimes} \ldots \dot{\otimes} L \tag{51}$$

The method to compute $\mathcal{Z}$ is to diagonalise $\bar{T}$. It is explained in chapter II of [2] and generalizes the coordinate Bethe-Ansatz described in the first part of this lecture. It turns out that provided the Bolzmann weights obey the following relations

$$\frac{a^2 + b^2 - c^2}{2\,ab} = \Delta, \tag{52}$$

the equations determining the eigenvectors of T coincide with those determining the eigenvectors of H_Δ. This strongly suggests that these matrices commute together and with H_Δ. Up to a homogenous transformation, they are defined by 2 parameters and there is one relation between them. We therefore expect to find a 1 parameter family of commuting transfer matrices. Consider 2 matrices $\bar{T}$ and $\bar{T}'$ (L and L') respectively defined in (50a,b) from two set of numbers abc, $a'b'c'$. We want to find under what condition

$$[\bar{T}, \bar{T}'] = 0 \tag{53}$$

Because (53) has to be satisfied for any number of rows N, which corresponds to the number of tensor products in (51), we expect to show it by recursion on N. Indeed, suppose that we can find an invertible matrix $R \in$ End $(\mathbf{C}^2 \otimes \mathbf{C}^2)$ such that

$$\sum_{\lambda' \rho'} R_{\mu\nu,\lambda'\rho'} L_{\lambda'\lambda} L'_{\rho'\rho} = \sum_{\lambda'\rho'} L'_{\nu\rho'} L_{\mu\lambda'} R_{\lambda'\rho',\lambda\rho} \tag{54}$$

In compact notations:

$$R \overset{1}{L} \overset{2}{L}{}' = \overset{2}{L}{}' \overset{1}{L} R \tag{55}$$

where $\overset{1}{L}$, $\overset{2}{L}{}'$ are respectively the matrices $L \otimes 1$ and $1 \otimes L'$. (Note that if L, L' were c number matrices, (55) would be trivially satisfied with $R = \mathrm{Id}$).
Multiplying (55) to the right by R^{-1} and taking the trace gives:

$$[\mathrm{tr}\ L,\ \mathrm{tr}\ L'] = 0 \tag{56}$$

which is relation (53) for $N = 1$.
Let us now define the matrix $T_{\mu\nu}$, the trace of which is $\bar{T}$ in (53).

$$T_{\mu\nu} = (L \dot{\otimes} L \dot{\otimes} \cdots \dot{\otimes} L)_{\mu\nu} \tag{57}$$

It is a consequence of (55) that

$$R \overset{1}{T} \overset{2}{T}{}' = \overset{2}{T}{}' \overset{1}{T} R \tag{58}$$

In the case $N = 2$, for example, we have the following chain of identities:

$$\begin{aligned} R(\overset{1}{L} \dot{\otimes} \overset{1}{L})(\overset{2}{L}{}' \dot{\otimes} \overset{2}{L}{}') &= R(\overset{1}{L}\overset{2}{L}{}') \dot{\otimes} (\overset{1}{L}\overset{2}{L}{}') \\ = (\overset{2}{L}{}' \overset{1}{L} R) \dot{\otimes} (\overset{1}{L}\overset{2}{L}{}') &= (\overset{2}{L}{}' \overset{1}{L}) \dot{\otimes} (R \overset{1}{L}\overset{2}{L}{}') \\ = (\overset{2}{L}{}' \overset{1}{L}) \dot{\otimes} (\overset{2}{L}{}' \overset{1}{L}) R &= (\overset{2}{L}{}' \dot{\otimes} \overset{2}{L}{}')(\overset{1}{L} \dot{\otimes} \overset{1}{L}) R \end{aligned} \tag{59}$$

Consequently, repeating the arguments following (55) we deduce the commutation relation (53). It remains to find a matrix R satisfying (55). Baxter [2] takes R in the same form as L:

$$R_{\mu\nu,\rho\sigma} = \omega''(\mu\rho, \nu\sigma), \tag{60}$$

so that equation (55) becomes:

$$R''_{12}\ R'_{13}\ R_{23} = R_{23}\ R'_{13}\ R''_{12} \tag{61}$$

In this form it is called the Yang-Baxter equation. In matrix form, (55) becomes:

$$\begin{pmatrix} a'' & \cdot & \cdot & \cdot \\ \cdot & b'' & c'' & \cdot \\ \cdot & c'' & b'' & \cdot \\ \cdot & \cdot & \cdot & a'' \end{pmatrix} \begin{pmatrix} L_{11}L'_{11} & L_{11}L'_{12} & L_{12}L'_{11} & L_{12}L'_{12} \\ L_{11}L'_{21} & L_{11}L'_{22} & L_{12}L'_{21} & L_{12}L'_{22} \\ L_{21}L'_{11} & L_{21}L'_{12} & L_{22}L'_{11} & L_{22}L'_{12} \\ L_{21}L'_{21} & L_{21}L'_{22} & L_{22}L'_{21} & L_{22}L'_{22} \end{pmatrix}$$

$$= \begin{pmatrix} L'_{11}L_{11} & L'_{12}L_{11} & L'_{11}L_{12} & L'_{12}L_{12} \\ L'_{21}L_{11} & L'_{22}L_{11} & L'_{21}L_{12} & L'_{22}L_{12} \\ L'_{11}L_{21} & L'_{12}L_{21} & L'_{11}L_{22} & L'_{12}L_{22} \\ L'_{21}L_{21} & L'_{22}L_{21} & L'_{21}L_{22} & L'_{22}L_{22} \end{pmatrix} \begin{pmatrix} a'' & \cdot & \cdot & \cdot \\ \cdot & b'' & c'' & \cdot \\ \cdot & c'' & b'' & \cdot \\ \cdot & \cdot & \cdot & a'' \end{pmatrix} \tag{62}$$

Substituting (50 a) in (62), one obtains the following equations: [2]

$$\begin{aligned} ac'a'' &= bc'b'' + ca'c'' \\ ab'c'' &= ba'c'' + cc'b'' \\ cb'a'' &= ca'b'' + bc'c'' \end{aligned} \tag{63}$$

Eliminating $a''b''c''$ leaves the single equation:

$$\frac{a^2+b^2-c^2}{2ab} = \frac{a'^2+b'^2-c'^2}{2a'b'} = \Delta \tag{64}$$

This is indeed what we wanted: provided abc, $a'b'c'$ define the same value of Δ, the 2 transfer matrices $\bar{T}$ and $\bar{T}'$ commute. It is convenient to use the following parametrisation for abc :

$$\begin{aligned} \Delta &= (q^{1/2}+q^{-1/2})/2 \\ a &= (q\lambda)^{1/2} - (q\lambda)^{-1/2} \\ b &= \lambda^{1/2} - \lambda^{-1/2} \\ c &= q^{1/2} - q^{-1/2} \end{aligned} \tag{65}$$

where a', b', c' are defined with λ' substituted λ in (65) and a'', b'', c'' solving (63) are obtained by substituting $\lambda'' = \lambda/\lambda'$ in (65).

As an application of the algebra (58), let us compute the eigenvectors of $\bar{T}$ using the so called "algebraic Bethe-ansatz" [3]. We introduce the notation $(x)_q = q^{x/2} - q^{-x/2}$, so that (65) reads:

$$\begin{aligned} a &= (u+1)_q \\ b &= (u)_q \\ c &= (1)_q \end{aligned} \tag{66}$$

We denote the matrix elements of T :

$$\begin{pmatrix} A & B \\ C & D \end{pmatrix} \tag{67}$$

For the method to be applicable, one needs an eigenstate $|\Omega\rangle$ of A and D anihilated by C.

$$\begin{aligned} C|\Omega\rangle &= 0 \\ A|\Omega\rangle &= \Delta_+(u)|\Omega\rangle \\ D|\Omega\rangle &= \Delta_-(u)|\Omega\rangle \end{aligned} \tag{68}$$

In the six-vertex model this is realised with: $|\Omega\rangle = \bigotimes_1^N |+\rangle, \Delta_+ = a^N, \Delta_- = b^N$.

The state $|\Omega\rangle$ is an eigenstate of $\bar{T} = (A+D)(u)$ with eigenvalue $\Delta_+(u) + \Delta_-(u)$.

We then try to find the other eigenstates in the form: $B(u_1)\cdots B(u_n)\,|\Omega\rangle$ where $u_1, u_2 \cdots, u_n$

are parameters to be determined. The eigenvalue equation is

$$(A+D)(u)B(u_1)\cdots B(u_n)|\Omega\rangle = \Lambda(u)B(u_1)\cdots B(u_n)|\Omega\rangle \tag{69}$$

To pass A and D through $B(u_i)$, we need the following relations among (58):

$$[B(u), B(u')] = 0 \tag{70}$$

$$\begin{aligned} a''BA' &= c''B'A + b''A'B \\ c''BD' + b''DB' &= a''B'D \end{aligned} \tag{71}$$

We rewrite (71):

$$\begin{aligned} (u-u_1)_q\, A(u)B(u_1) &= (u-u_1-1)_q\, B(u_1)A(u) + (1)_q\underline{B(u)A(u_1)} \\ (u-u_1)_q D(u)B(u_1) &= (u-u_1+1)_q\, B(u_1)D(u) - (1)_q\underline{B(u)D(u_1)} \end{aligned} \tag{72}$$

Were the underlined terms not present, the eigenvalue equation would automatically be satisfied and the eigenvalue would be equal to:

$$\Lambda(u) = \Delta_+(u)\prod_1^n \frac{(u-u_i-1)_q}{(u-u_i)_q} + \Delta_-(u)\prod_1^n \frac{(u-u_i+1)_q}{(u-u_i)_q} \tag{73}$$

The eliminations of the underlined terms will yield equations that determine the values of u_i. A typical unwanted term occurs when the spectral parameters u and u_i are exchanged between $A(u)$ and $B(u_i)$ when A passes through B so we expect that

$$\bar{T}(u)B(u_1)\cdots B(u_n)|\Omega\rangle = \Lambda(u)|\Omega\rangle + \sum_{i=1}^n C_i B(u)\prod_{\substack{k=1\\k\neq i}}^n B(u_k)|\Omega\rangle \tag{74}$$

There are only n unwanted terms due to the commutation of the B_s. The only way to obtain C_1 is to keep the unwanted term when $A(u)$ or $D(u)$ passes through $B(u_1)$ and to keep the wanted terms when $A(u_1)$ or $D(u_1)$ passes through $B(u_i)$, $i \neq 1$. This gives

$$C_1 \;= \frac{(1)_q}{(u-u_1)_q}\Big(\prod_2^n \frac{(u_1-u_i-1)_q}{(u_1-u_i)_q}\Delta_+(u_1) - \prod_2^n \frac{(u_1-u_i+1)_q}{(u_1-u_i)_q}\Delta_-(u_1)\Big) \tag{75}$$

Using the fact that the B commute between themselves, similar expressions are obtained for C_i. The Bethe equations result from setting $C_i = 0$:

$$\left(\frac{(u_i+1)_q}{(u_i)_q}\right)^N = \prod_{j\neq i}\frac{(u_i-u_j+1)_q}{(u_i-u_j-1)_q} \tag{76}$$

which are the same equations as (24) if we set

$$z_i^{-1} = \frac{(u_i+1)_q}{(u_i)_q} \tag{77}$$

We finally note that (76) is equivalent to the fact that (73) has no pole at $u = u_i$.

3. Quadratic algebras

In a more general way, we can view equation (55) as defining an associative algebra A (the Sklyanin algebra [8]). The matrix elements $L(u)$ are the generators and obey quadratic relations defined by (55) or (62). Given the product of $3L : \overset{1}{L}(u_1)\overset{2}{L}(u_2)\overset{3}{L}(u_3)$ using (55), we can reorder them into $\overset{3}{L}(u_3)\overset{2}{L}(u_2)\overset{1}{L}(u_1)$ by conjugating by $R_{23}(u_2-u_3)R_{13}(u_1-u_3)R_{12}(u_1-u_2)$ or $R_{12}(u_1-u_2)\,R_{13}\,(u_1-u_3)\,R_{23}(u_2-u_3)$ depending on the order in which we permute the terms. It is consistent to require that the two conjugacy matrices are proportional; in this way, we recover a weak form of the Yang-Baxter equation (61):

$$R_{23}(u_2-u_3)R_{13}(u_1-u_3)R_{12}(u_1-u_2) = c\; R_{12}(u_1-u_2)R_{13}(u_1-u_3)R_{23}(u_2-u_3) \tag{78}$$

It is also natural to look for the possible representations of the quadratic algebra (55). For a, b, c given by (65), (50 a) corresponds to the two dimensionnal representation. From this point of view, (50 b) defines a way to tensor product two representations. More precisely, there exists a homorphism Δ from A into its tensor product $A \otimes A$ (called a coproduct), defined by:

$$\Delta L(u) = L(u)\dot{\otimes}L(u) \tag{79}$$

The fact that ΔL obeys the same relations (55) as L follows from the chain of identities (59). Conversely, we can use any solution of the Yang-Baxter equation to define a quadratic algebra, we shall use this to define the Yangians later.

- *Universal R matrix* [4]:

Suppose now that instead of considering the column to column transfer matrix as in (47), we consider the line to line transfer matrix:

$$\bar{T}_{\mu_1\mu_2\cdots\mu_n,\nu_1\nu_2\cdots\nu_n} = \sum_{a_1,a_2\cdots a_n} (L_{\mu_1,\nu_1})_{a_1a_2}(L_{\mu_2\nu_2})_{a_2a_3}\cdots(L_{\mu_n\nu_n})_{a_na_1} \tag{80}$$

We can write a similar equation to (55) expressing the commutativity of two such matrices depending on spectral parameters u and v.
In components, the equation is:

$$\begin{aligned}&\sum_{\nu}\sum_{a''b''}\mathcal{R}_{ab,a''b''}(u-v)(L(u)_{\mu\nu})_{a''a'}(L(v)_{\nu\rho})_{b''b'}\\ &=\sum_{\nu}\sum_{a''b''}(L_{\nu\rho}(u))_{aa''}(L(v)_{\mu\nu})_{bb''}\mathcal{R}_{a''b'',ab}(u-v)\end{aligned} \tag{81}$$

We now view $L_{\mu\nu}(u)$ more generally as elements of an algebra A. So $\mathcal{R}$ itself is in $A\otimes A$ and the above equation becomes:

$$\mathcal{R}(u-v)(L(u)\dot{\otimes}L(v)) = \sigma(L(v)\dot{\otimes}L(u))\mathcal{R}(u-v) \tag{82}$$

with $\sigma(a\otimes b) = b\otimes a$. As before, the tensor product is in the quantum space and $\cdot$ means that the product is taken in the matrix space. This equation has a different interpretation from (55) because R carries no matrix index but is instead an element of $A\otimes A$. We denote by T_x the operation that shifts the spectral parameter by an amount x :

$$T_x(L)(u) = L(u+x) \tag{83}$$

(T_x is obviously an algebra automorphism since in (55), R depends only on the difference $u-u'$), and we rewrite (82) as follows:

$$\mathcal{R}(u-v)(T_u \otimes T_v)(\Delta L)(w) = (T_u \otimes T_v)(\sigma \circ \Delta)L(w)\mathcal{R}(u-v) \tag{84}$$

The point of rewriting the equation in this form is that it is not only true for the generators of the algebra $A : L_{\mu\nu}(w)$ but also for any element of A which can be substituted into $L(w)$. We shall see later that there can sometimes be more convenient generators to use than $L_{\mu\nu}$.

Another equation is obtained by applying $1 \otimes \Delta$ to (82) obtaining:

$$(1 \otimes \Delta)\mathcal{R}(u-v)(L(u)\dot{\otimes}L(v)\dot{\otimes}L(v)) = \sigma_{(1,2,3)}(L(v)\dot{\otimes}L(v)\dot{\otimes}L(u))(1 \otimes \Delta)\mathcal{R}(u-v)) \tag{85}$$

with $\sigma_{(1,2,3)}(a \otimes b \otimes c) = b \otimes c \otimes a$. We also have the equality which results from applying (82) twice:

$$\mathcal{R}_{13}(u-v)\mathcal{R}_{12}(u-v)(L(u)\dot{\otimes}L(v)\dot{\otimes}L(v)) = \sigma_{(1,2,3)}(L(v)\dot{\otimes}L(v)\dot{\otimes}L(u))\mathcal{R}_{13}(u-v)\mathcal{R}_{12}(u-v) \tag{86}$$

We therefore require the 2 intertwiners to be proportional; the proportionality coefficient can be absorbed in the normalization of $\mathcal{R}$ and set equal to 1. Finally we obtain:

$$(1 \otimes \Delta)(\mathcal{R}(u)) = \mathcal{R}_{13}(u)\mathcal{R}_{12}(u). \tag{87}$$

4. Yangians and loop algebras

– $SL(n)$ case:

Let us give a concrete example of the preceeding formalism in the simplest case of an R matrix obeying the Yang-Baxter equation. R is given by:

$$R_{12}(u) = u + \hbar P_{12} \tag{88}$$

where P_{12} is the permutation operator acting in a tensor space as:

$$P_{12}(x_1 \otimes x_2) = x_2 \otimes x_1 \tag{89}$$

and the constant $\hbar$ is introduced for later convenience. The Yang-Baxter equation:

$$R_{23}(u_2-u_3)R_{13}(u_1-u_3)R_{12}(u_1-u_2) = R_{12}(u_1-u_2)R_{13}(u_1-u_3)R_{23}(u_2-u_3) \tag{90}$$

relies only on the permutation properties

$$P_{ij}^2 = 1, (P_{12}P_{23})^3 = 1. \tag{91}$$

Let us look for $n \times n$ L matrices satisfying the quadratic relations

$$R(u_1-u_2) \overset{1}{L}(u_1) \overset{2}{L}(u_2) = \overset{2}{L}(u_2) \overset{1}{L}(u_1) R(u_1-u_2) \tag{92}$$

In components, (92) reads:

$$(u-u')\,[L_{ij}(u),L_{kl}(u')] = \hbar(L_{kj}(u')L_{il}(u) - L_{kj}(u)L_{il}(u')) \tag{93}$$

If we choose L in the form

$$L_{ij}(u) = \delta_{ij} + \frac{\hbar}{u}E_{ij} \tag{94}$$

and substitute it into (93), we obtain the $GL(n)$ relations:

$$[E_{ij},E_{kl}] = E_{kj}\delta_{il} - E_{il}\delta_{kj} \tag{95}$$

A possible motivation to introduce Yangians at this stage is to observe that $GL(n)$ is endowed with a natural coproduct preserving the relations (95):

$$\Delta E_{ij} = E_{ij}\otimes 1 + 1\otimes E_{ij}$$

However, the coproduct introduced for L_{ij} in (79) is not compatible with this one in the sense that

$$\Delta L_{ij}(u) = \sum_k L_{ik}(u)\otimes L_{kj}(u) \neq \delta_{ij}1\otimes 1 + \frac{\hbar}{u}\Delta(E_{ij}) \tag{96}$$

The left hand side contains the additional term $\frac{\hbar^2}{u^2}\sum_k E_{ik}\otimes E_{kj}$. The way out is to take $L(u)$ in the form of a serie:

$$L_{ij}(u) = \delta_{ij} + \sum_{n=0}^{+\infty}\hbar^{(n+1)}\frac{E^n_{ij}}{u^{n+1}} \tag{97}$$

which, when substituted into (93) gives the relations:

$$[E^{m+1}_{ij},E^n_{kl}] - [E^m_{ij},E^{n+1}_{kl}] = E^n_{kj}E^m_{il} - E^m_{kj}E^n_{il} \tag{98}$$

for $m,n \geq 0$ and for $m \geq 0$

$$[E^0_{ij},E^m_{kl}] = \delta_{il}E^m_{kj} - \delta_{kj}E^m_{il} \tag{99}$$

Now, we define the natural coproduct of E^n_{ij} inherited from the coproduct of $L(u)$:

$$\Delta L_{ij}(u) = \sum_k L_{ik}(u)\otimes L_{kj}(u) = \delta_{ij}1\otimes 1 + \sum_{n=0}^{\infty}\frac{\hbar^{n+1}}{u^{n+1}}\Delta(E^n_{ij}) \tag{100}$$

to obtain:

$$\Delta E^m_{ij} = E^m_{ij}\otimes 1 + 1\otimes E^m_{ij} + \sum_{p=0}^{m-1}E^p_{ik}\otimes E^{m-1-p}_{kj} \tag{101}$$

These relations define $YGL(n)$ (Y stands for Yangian [4]). It is generated by: $E_{ij} = E^0_{ij}$ and

$$\bar{E}_{ij} = E^1_{ij} - \frac{1}{2}\sum_k E^0_{ik}E^0_{kj} \tag{102}$$

obeying the relations:

$$[E_{ij}, E_{kl}] = \delta_{il}E_{kj} - \delta_{kj}E_{il} \tag{103}$$

$$\left[E_{ij}, \bar{E}_{kl}\right] = \bar{E}_{kj}\delta_{il} - \bar{E}_{il}\delta_{kj} \tag{104}$$

$$\begin{aligned} &\left[E_{ij}, \left[\bar{E}_{kl}, \bar{E}_{mn}\right]\right] - \left[\bar{E}_{ij}, \left[E_{kl}, \bar{E}_{mn}\right]\right] \\ &= \frac{1}{4}\sum_{p,q}\{[E_{ij}, [E_{kp}E_{pl}, E_{mq}E_{qn}]] - [E_{ip}E_{pj}, [E_{kl}, E_{mq}E_{qn}]]\} \end{aligned} \tag{105}$$

$$\Delta\ E_{ij} = E_{ij} \otimes 1 + 1 \otimes E_{ij} \tag{106}$$

$$\Delta\bar{E}_{ij} = \bar{E}_{ij} \otimes 1 + 1 \otimes \bar{E}_{ij} + \frac{1}{2}\sum_{k}(E_{ik} \otimes E_{kj} - E_{kj} \otimes E_{ik}) \tag{107}$$

The additional term in the definition of $\bar{E}_{ij}$ is introduced for the right hand side of (105) to contain only terms in E_{ij} (not $\bar{E}_{ij}$). If, instead of $L_{\mu\nu}$, we use this set of generators to rewrite (84). Using the fact that

$$\begin{aligned} T_{\hbar x}E_{ij} &= E_{ij} \\ T_{\hbar x}\bar{E}_{ij} &= \bar{E}_{ij} - xE_{ij} \end{aligned} \tag{108}$$

We obtain the following equations for $\mathcal{R}$:

$$\mathcal{R}(\hbar x)(E_{ij} \otimes 1 + 1 \otimes E_{ij}) = (E_{ij} \otimes 1 + 1 \otimes E_{ij})\mathcal{R}(\hbar x) \tag{109}$$

$$\begin{aligned} &\mathcal{R}(\hbar(x-y))((\bar{E}_{ij} - xE_{ij}) \otimes 1 + 1 \otimes (\bar{E}_{ij} - yE_{ij}) + \frac{1}{2}(\sum_{k}(E_{ik} \otimes E_{kj} - E_{kj} \otimes E_{ik})) \\ &= ((\bar{E}_{ij} - x\ E_{ij}) \otimes 1 + 1 \otimes (\bar{E}_{ij} - y\ E_{ij}) - \frac{1}{2}(\sum_{k}(E_{ik} \otimes E_{kj} - E_{kj} \otimes E_{ik}))\mathcal{R}(\hbar(x-y)) \end{aligned} \tag{110}$$

which are solved by

$$\mathcal{R}(x) = 1 + \frac{\hbar}{x}\sum_{i,j} E_{ij} \otimes E_{ji} + O(\hbar^2) \tag{111}$$

To define $YSL(n)$, we further require that the so called quantum determinant be equal to 1:

$$\mathrm{Det}(u) = \sum_{\sigma \in \pi_N} \varepsilon(\sigma)L_{\sigma_n,n}(u)L_{\sigma_{n-1},n-1}(u+\hbar)...L_{\sigma_1,1}(u+n\hbar) \tag{112}$$

This is consistent because (as will be clarified later) $\mathrm{Det}(u)$ commutes with all the algebra and

$$\Delta\ \mathrm{Det}(u) = \mathrm{Det}(u) \otimes \mathrm{Det}(u) \tag{113}$$

In terms of the generators $E_{ij}, \bar{E}_{ij}$, this amounts to impose the constraint:

$$\sum_i E_{ii} = \sum_i \bar{E}_{ii} = 0 \tag{114}$$

There is a simple mapping of $YSL(n)$ into $SL(n)$ which consists in taking L of the form

$$L_{ij} = \tilde{L}_{ij}/(\mathrm{Det}\tilde{L}(u)) \tag{115}$$

with $\tilde{L}$ given by (94). In particular, this shows that any representation of $SL(n)$ can be lifted to $YSL(n)$. (This is very specific to $SL(n)$).

This also suggests that a better presentation of the algebra can be obtained if we take appropriate ratio of matrix elements so as to get rid of the determinant constraint. Let us here obtain this other presentation in the $SL(2)$ case: $YGL(2)$ is defined by the following matrix

$$L(u) \quad = \begin{pmatrix} A(u) & B(u) \\ C(u) & D(u) \end{pmatrix} \tag{116}$$

Obeying the commutation relations specializing (93) which correspond to $\Delta = 1$ in (64):

$$\begin{array}{rl} (u-u')[A,B'] & = \hbar(A'B - AB') \\ (u-u')[A,C'] & = \hbar(C'A - CA') \\ (u-u')[A,D'] & = \hbar(C'B - CB') \\ (u-u')[B,C'] & = \hbar(D'A - DA') \\ (u-u')[B,D'] & = \hbar(D'B - DB') \\ (u-u')[C,D'] & = \hbar(C'D - CD') \\ [A,A'] = [B,B'] & = [C,C'] = [D,D'] = 0 \end{array} \tag{117}$$

Let us introduce the operators

$$\begin{array}{rl} S^+(u) & = CD^{-1}(u) \\ S^-(u) & = D^{-1}B(u) \end{array} \tag{118}$$

We then have:

$$\begin{array}{rl} (u-u')\left[S^\pm(u), S^\pm(u')\right] & = \pm\hbar(S^\pm(u)S^\pm(u') + S^\pm(u')S^\pm(u)) \mp \hbar(S^\pm(u)S^\pm(u) \\ & +S^\pm(u')S^\pm(u')) \end{array} \tag{119}$$

$$(u-u')\left[S^+(u), S^-(u')\right] = \hbar \ \mathrm{Det}(u')D^{-1}(u')D^{-1}(u'+\hbar) - \hbar \ \mathrm{Det}(u)D^{-1}(u)D^{-1}(u+\hbar) \tag{120}$$

with

$$\mathrm{Det}(u) = D(u)A(u+\hbar) - B(u)C(u+\hbar) \tag{121}$$

Let us set:

$$S(u) = \mathrm{Det}(u)D^{-1}(u)D^{-1}(u+\hbar) \tag{122}$$

then

$$[S(u), S(u')] = 0 \tag{123}$$

$$(u-u')\left[S(u), S^\pm(u')\right] = \pm\hbar(S^\pm(u')S(u) + S(u)S^\pm(u')) \mp \hbar(S^\pm(u)S(u) + S(u)S^\pm(u')) \tag{124}$$

When expanded in modes, these relations correspond to the second presentation of Yangians in [5]. They are straightforward to derive in the classical limit ($\hbar \longrightarrow 0$). In this limit, the variables A, B, C, D are commuting variables and the commutators divided by $\hbar$ are replaced by Poisson brackets. Relations (119–124) are recovered with $S^\pm, S$ commuting variables and $\hbar$ set equal to zero. Let us derive (124) in the quantum case: we set $\hbar = 1$ in this calculation:

$$\begin{array}{l} (u-u')\left[S^+(u), S^-(u')\right] = (u-u')\left[CD^{-1}(u), D^{-1}B(u')\right] \\ = (u-u')\left\{-CD^{-1}D'^{-1}[D,B']D^{-1} + D'^{-1}[C,B']D^{-1} - D'^{-1}[C,D']D'^{-1}B'D^{-1}\right\} \end{array} \tag{125}$$

Using (117) this becomes:

$$\begin{aligned}&-CD^{-1}D'^{-1}(DB'-D'B)D^{-1}+D'^{-1}(DA'-D'A)D^{-1}\\&-D'^{-1}(DC'-D'C)D'^{-1}B'D^{-1}\\&=(CD^{-1}BD^{-1}-AD^{-1})+D(-D'^{-1}C'D'^{-1}B'+D'^{-1}A')D^{-1}\end{aligned}\tag{126}$$

Using the relations obtained by putting $u' = u + \hbar$ in (117) we can reorder the terms in parentheses so as to make the determinant appear:

$$\begin{aligned}(126)\ &= D^{-1}(u+1)\,\mathrm{Det}(u)D^{-1}(u)\\&+D(u)D^{-1}(u')\,\mathrm{Det}(u')D^{-1}(u'+1)D^{-1}(u)\end{aligned}\tag{127}$$

Finally, using the fact that $\mathrm{Det}(u)$ is in the center, we obtain the result.

– *$O(N)$ case:*

It is instructive to carry out a similar exercise for the orthogonal group. We look for a solution of the Yang-Baxter equation in the form:

$$R_{12}(u) = 1_{12} + a\,P_{12} + b\,Q_{12}\tag{128}$$

P_{12} is the permutation operator defined in (89) and Q is defined by

$$Q(x_1 \otimes x_2) = (x_1 \cdot x_2)\sum_{i=1}^{n} \mathrm{e}_i \otimes \mathrm{e}_i\tag{129}$$

where e_i is an orthonormal bases in a n dimensionnal vector space. It is convenient to represent $1, P$ and Q graphically as follows:

$1 \;=$ 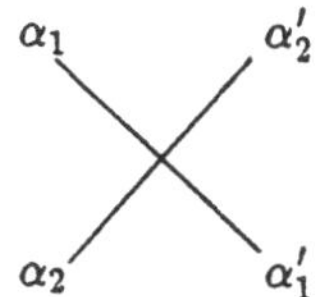 $= \delta_{\alpha_1\alpha_1'}\delta_{\alpha_2\alpha_2'}$

$P_{12} \;=$ 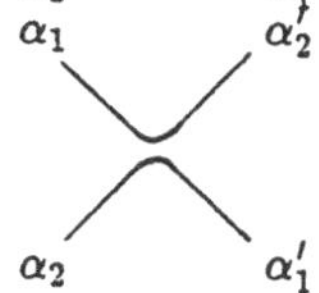 $= \delta_{\alpha_1\alpha_2'}\delta_{\alpha_2\alpha_1'}$ (130)

$Q_{12} \;=$ 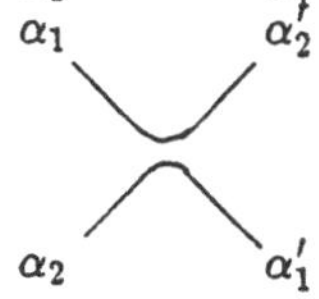 $= \delta_{\alpha_1\alpha_2}\delta_{\alpha_1'\alpha_2'}$

The Yang-Baxter equation is:

$$\begin{aligned}&(1+a''+P_{12}+b''Q_{12})(1+b\,P_{13}+a\,Q_{13})(1+a'\,P_{23}+b'\,Q_{23})\\&=(1+a'\,P_{23}+b'\,Q_{23})(1+b\,P_{13}+a\,Q_{13})(1+a''\,P_{12}+b''\,Q_{12})\end{aligned}\tag{131}$$

where, to obtain symmetric equations in a, b, we have inverted the order of a, b in the middle term. We represent the Yang-Baxter equation graphically:

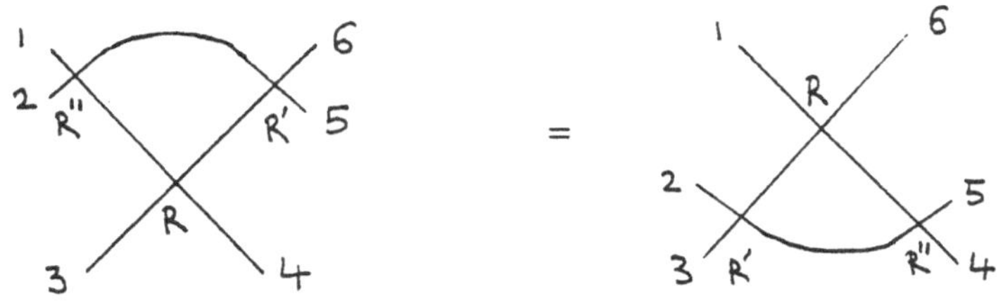

(132)

In (132), each R matrix is replaced by its expression given by (131). For each term of the expanded product, we replace $1, P, Q$ by its graphical expression, thus obtaining a figure with the external legs (numbered from 1 to 6) connected in pairs. We identify the figures with the same set of connected pairs. The equations are obtained by adding up the coefficients of identified figures multiplied by a factor n when a closed loop occurs. Non trivial equations arise for the following pairs:

$$(12)(34)(56) \quad : nb''b' + a''b'b + b''a'b + b''b'a + b''b' + b''b + b'b = a''a'a \tag{133}$$

$$\begin{array}{lll} (13)(26)(45) & aa' & = b''(a+a') \\ (16)(24)(35) & a'a'' & = b(a'+a'') \\ (15)(23)(46) & a''a & = b'(a''+a) \end{array} \tag{134}$$

The solution is given by:

$$\begin{array}{llll} a & = \dfrac{\hbar}{\hbar g - u_{13}} & b & = \dfrac{\hbar}{u_{13}} \\ a' & = \dfrac{\hbar}{u_{23}} & b' & = \dfrac{1}{\hbar g - u_{23}} \\ a'' & = \dfrac{\hbar}{u_{12}} & b'' & = \dfrac{1}{\hbar g - u_{12}} \end{array} \tag{135-136}$$

with $u_{ij} = u_i - u_j$ and $g = (1 - \frac{n}{2})$ so that

$$R(u) = 1 + \frac{\hbar P}{u} + \frac{\hbar Q}{(\hbar g - u)} \tag{137}$$

According to the general strategy, we define an algebra by:

$$R(u_1 - u_2) \overset{1}{L}(u_1) \overset{2}{L}(u_2) = \overset{2}{L}(u_2) \overset{1}{L}(u_1) R(u_1 - u_2) \tag{138}$$

with:

$$L_{\mu\nu}(u) = \delta_{\mu\nu} + \sum_{n=0}^{\infty} (\frac{\hbar}{u})^n X_{\mu\nu}^{(n)} \tag{139}$$

The relations obeyed by $X_{\mu\nu}^{(n)}$ follow naturally from (138) and their coproduct from (79). A natural representation for L inherited from (137) is given by

$$(L_{\mu\nu})_{\alpha\alpha'} = R_{\alpha\mu,\alpha'\nu} = \delta_{\mu\nu}\delta_{\alpha\alpha'} + \hbar\frac{\delta_{\alpha\nu}\delta_{\alpha'\mu}}{u} + \frac{\hbar\delta_{\alpha\mu}\delta_{\alpha'\nu}}{\hbar g - u} \tag{140}$$

Unlike in the $SL(n)$ case, there is no expression similar to (94) in terms of the generators of $O(n)$. In order to obtain $YO(n)$, we impose an additionnal constraint similar to (112). It relies on the idea of fusion which we explain here (it has already appeared in the derivation of (87)).

Suppose there exists a value of the spectral parameter $u = \mu$ for which the R matrix is proportional to a projector:

$$R(\mu) = \lambda \bar{P}, \bar{P}^2 = 1 \tag{141}$$

We rewrite the Yang-Baxter equation for $u_1 - u_2 = \mu$.

$$\bar{P} \overset{1}{L}(u+\mu) \overset{2}{L}(u) = \overset{2}{L}(u) \overset{1}{L}(u+\mu) \bar{P} \tag{142}$$

Let us denote by $\alpha(u)$ the matrix:

$$\alpha(u) = \bar{P} \overset{1}{L}(u+\mu) \overset{2}{L}(u) = \bar{P} \overset{1}{L}(u+\mu) \overset{2}{L}(u)\bar{P} \tag{143}$$

It is clear from above that the matrix elements of α are closed under coproduct:

$$\Delta\alpha(u) = \alpha(u)\dot{\otimes}\alpha(u) \tag{144}$$

Using similar arguments as those leading to (87), it is easy to show that they form a closed algebra and to find the R matrix for it. For example, the commutation relations of α with L are given by:

$$\tilde{R}(u-v)\alpha(u)L(v) = L(v)\alpha(u)\tilde{R}(u-v) \tag{145}$$

with:

$$\tilde{R}(u-v) = \bar{P}_{12}R_{13}(u+\mu)R_{23}(u) = R_{23}(u)R_{13}(u+\mu)\bar{P}_{12} \tag{146}$$

So, the matrix elements of $\alpha(u)$ can be taken as the generators of a Hopf algebra. If $\bar{P}$ is a rank one projector, $\tilde{R}$ is a scalar, so α is in the center and from (144) $\Delta\alpha = \alpha \otimes \alpha$. It is therefore consistent to impose $\alpha(u) = 1$.
In the case of the R matrix (137), $R(\hbar g)$ is proportional to $\frac{Q}{n}$ which is a rank one projector. We can therefore require:

$$\sum_i L_{ij}(u+\hbar g)L_{ik}(u) = \delta_{jk} \tag{147}$$

To first order in $\hbar$, (147) identifies X^0_{ij} with $-X^0_{ji}$ and it then follows from (138–139) that X^0_{ij} obey the $O(n)$ Lie algebra relations. Similar arguments can be used to obtain the quantum determinant (112).

– *Loop Algebras*:

It is natural to extend the relations (119–124) to the algebra defined in (62) to the case where $\Delta \neq 1$. We assume the following expression for $L(u)$ $(z = q^u)$:

$$\begin{aligned} A(z) &= K + \frac{a_0}{z} + \frac{a_1}{z^2} + \cdots \\ D(z) &= K^{-1} + \frac{d_0}{z} + \frac{d_1}{z^2} + \\ B(z) &= z^{-1/2}(b_{1/2} + \frac{b_{3/2}}{z} + \cdots) \\ C(z) &= z^{-1/2}(c_{1/2} + \frac{c_{3/2}}{z} + \cdots) \end{aligned} \tag{148}$$

Setting as before:

$$\begin{aligned} S^+ &= CD^{-1}(z) &= z^{-1/2}(S_1^+ + \frac{S_2^+}{z} + \cdots) \\ S^- &= D^{-1}B(z) &= z^{-1/2}(S_0^- + \frac{S_1^-}{z} + \cdots) \end{aligned} \tag{149}$$

we obtain the following relation for S^+, S^- :
(recall that $(x)_q = q^{x/2} - q^{-x/2}$)

$$(u-u'\mp 1)_q S^{\pm}(u)S^{\pm}(u')=(u-u'\pm 1)_q S^{\pm}(u')S^{\pm}(u)\mp(1)_q(S^{\pm}(u)S^{\pm}(u)+S^{\pm}(u')S^{\pm}(u')) \tag{150}$$

$$(u-u')_q\left[S^+(u),S^-(u')\right]=(1)_q(S(u)-S(u')) \tag{151}$$

with:

$$S(u)=\mathrm{Det}(u)D^{-1}(u)D^{-1}(u+1) \tag{152}$$

and:

$$\mathrm{Det}(u)=D(u)A(u+1)-B(u)C(u+1) \tag{153 a}$$

$$[S(u),S(u')]=0 \tag{153 b}$$

$$(u-u'\mp 1)_q S(u)S^{\pm}(u') \;=(u-u'\pm 1)_q S^{\pm}(u')S(u)\mp(1)_q(S^{\pm}(u)S(u)+S(u)S^{\pm}(u)) \tag{154}$$

Again, these relations can be expanded in modes leading to the q analog of the Yangian-relations.

This is only half of the loop algebra. To obtain the complete loop algebra, let us assume L has the following expansion in the vicinity of 0.

$$\begin{aligned} A(z) &= K+z\,a_{-1}+\cdots \\ D(z) &= K^{-1}+z\,d_{-1}+\cdots \\ B(z) &= z^{1/2}(b_{-1/2}+z\,b_{-3/2}+\cdots) \\ C(z) &= z^{1/2}(c_{-1/2}+z\,c_{-3/2}+\cdots) \end{aligned} \tag{155}$$

Denote by $X_>$, $X_<$ the expansion of an operator X around $z=\infty$ and $z=0$ respectively. If $|q|=1$, we can think of A,B,C,D as operators analytic for $|z|\gtrless 1$ but not for $|z|=1$. We define

$$\begin{aligned} S^+(z) &= z^{-1/2}(C_>D_>^{-1}(z)-C_<D_<(z))/(1)_q &= \sum_{-\infty}^{+\infty} S_n^+ z^{-n} \\ S^-(z) &= z^{1/2}(D_>^{-1}B_>(z)-D_<^{-1}B_<(z))/(1)_q &= \sum_{-\infty}^{+\infty} S_n^- z^{-n} \end{aligned} \tag{156}$$

and obtain the following relations:

$$(z-qz')S^+(z)S^+(z')=(qz-z')S^+(z')S^+(z) \tag{157}$$

$$(qz-z')S^-(z)S^-(z')=(z-qz')S^-(z')S^-(z) \tag{158}$$

$$\left[S^+(z),S^-(z')\right]=\frac{1}{(1)_q}\delta\left(\frac{z}{z'}\right)(S_>(z)-S_<(z)) \tag{159}$$

$$S_{\gtrless}(z)S^+(z')=\frac{qz-z'}{z-qz'}S^+(z')S_{\gtrless}(z) \tag{160}$$

$$S_{\gtrless}(z)S^-(z')=\frac{z-qz'}{qz-z'}S^-(z')S_{\gtrless}(z) \tag{161 a}$$

$$\left[S_{\gtrless}(z), S_{\gtrless}(z')\right] = [S_>(z), S_<(z')] = 0 \tag{161 b}$$

where $\delta(\frac{z}{z'}) = \sum_{n\in\mathbf{Z}}(\frac{z}{z'})^n$ and in (159–160), the ratio has to be expanded in $\frac{z'}{z}$ or $\frac{z}{z'}$ according to whether we consider $S_>$ or $S_<$. These relations can be taken as the defining relations for the q-analogue of the loop algebra of $SU(2)$. Indeed, if we consider the limit (not the classical limit):

$$\begin{aligned} q &= 1+\hbar \\ S_{\gtrless} &\longrightarrow 1+\hbar S_{\gtrless}(z) \\ S^{\pm}(z) &\longrightarrow S^{\pm}(z) \end{aligned} \tag{162}$$

and keep only the first non-zero term in the relations (157–161b), it gives:

$$[S^{\pm}(z), S^{\pm}(z')] = 0 \tag{163}$$

$$[S^+(z), S^-(z')] = \delta(\frac{z}{z'})(S_>(z) - S_<(z')) \tag{164}$$

$$[S_{\gtrless}(z), S^{\pm}(z')] = \pm\frac{z+z'}{z-z'}S^{\pm}(z') \tag{165 a}$$

$$\left[S_{\gtrless}(z), S_{\gtrless}(z')\right] = [S_>(z), S_<(z')] = 0 \tag{165 b}$$

which when expanded in modes give the usual loop algebra relations.

– *Affine algebras:*

It is interesting to try to modify these relations so as to recover the analog of a Kac-Moody algebra. From the definition of a Kac-Moody algebra, (165 a) and (165 b) are modified to:

$$[S^+(z), S^-(z')] = \delta(\frac{z}{z'})(S_>(z) - S_<(z')) + k\delta'(\frac{z}{z'}) \tag{166}$$

$$[S_>(z), S_<(z')] = 2k\frac{z'}{z}\frac{1}{(1-\frac{z'}{z})^2} \tag{167}$$

All the other relations remain unchanged. If we set $\gamma = q^{\hbar k/2}$ it is very natural to try to modify (159) to:

$$[S^+(z), S^-(z')] = \frac{1}{(1)_q}(\delta(\gamma\frac{z}{z'})S_>(z) - \delta(\gamma^{-1}\frac{z}{z'})S_<(z)) \tag{168}$$

which gives (166) in the limit $\hbar \longrightarrow 0$. Let us find a representation [6] of relations (157), (158) and (168) with the Hermiticity requirement that for $|q| = 1$:

$$\begin{aligned} (S^+)^+(z) &= S^-(z^{-1}) \\ (S^+_>)(z) &= S_<(z^{-1}) \end{aligned} \tag{169}$$

We look for S^+ and S^- in the following form:

$$\begin{aligned} S^+(z) &= e^{\sum_{n>0} x_{-n}a_n^+z^{-n}}e^{\sum_{n>0} x_{+n}a_nz^n}z^{q_0}q^{p_0} \\ S^-(z) &= e^{\sum_{n>0} \bar{x}_na_n^+z^{-n}}e^{\sum_{n>0} \bar{x}_{-n}a_nz^n}z^{-q_0}q^{-p_0} \end{aligned} \tag{170}$$

where a_n, a_n^+ obey the commutation relations:

$$\begin{aligned} [a_n, a_m^+] &= \delta_{n,m} \\ [a_n, a_m] &= [a_n^+, a_m^+] = 0. \end{aligned} \tag{171}$$

and

$$[p_0, q_0] = 2 \tag{172}$$

Using the standard rule for normal ordering;

$$\mathrm{e}^{\lambda a}\mathrm{e}^{\mu a^+} = \mathrm{e}^{\lambda\mu} : \mathrm{e}^{\lambda a + \mu a^+} : \tag{173}$$

We can satisfy (157) (158) if for $|z| < |z'|$ we have

$$\begin{aligned} S^+(z)S^+(z') &= (qz - z')(z - z') &: S^+(z)S^+(z') : \\ S^-(z)S^-(z') &= (z - qz')(z - z') &: S^-(z)S^-(z') : \end{aligned} \tag{174}$$

which implies:

$$\begin{aligned} x_n x_{-n} &= -\frac{1}{n}(1 + q^n) \\ \bar{x}_n \bar{x}_{-n} &= -\frac{1}{n}(1 + q^{-n}) \end{aligned} \tag{175}$$

Similarly, (168) is satisfied if for $|z| \le |z'|$ we have:

$$\begin{aligned} S^+(z)S^-(z') = & \frac{1}{(z' - \gamma z)(z' - \gamma^{-1}z)} \qquad : S^+(z)S^-(z') : \\ = \frac{1}{(\gamma - \gamma^{-1})z} & \left\{ \frac{1}{(z' - \gamma z)} : S^+(z)S^-(\gamma z) \quad : - \frac{1}{(z' - \gamma^{-1}z)} : S^+(z)S^-(\gamma^{-1}z) \right\} \\ & + \text{analytic terms around } |z/z'| = 1 \end{aligned} \tag{176}$$

The first equality implies that

$$x_n \bar{x}_n = \frac{1}{n}(\gamma^n + \gamma^{-n}) \tag{177}$$

and the similar one with the rôle of S^+, S^- exchanged gives:

$$x_{-n} \bar{x}_{-n} = \frac{1}{n}(\gamma^n + \gamma^{-n}) \tag{178}$$

For (176) to be compatible with (168), we must have

$$\begin{aligned} : S^+(z)S^-(\gamma z) : &= S_>(z) \\ : S^+(z)S^-(\gamma^{-1}z) : &= S_<(z) \end{aligned} \tag{179}$$

which imply that

$$\begin{aligned} x_n + \gamma^n \bar{x}_{-n} = 0 \\ x_{-n} + \gamma^n \bar{x}_n = 0 \end{aligned} \tag{180}$$

Equations (175),(177),(178) and (180) can be satisfied only if $\gamma = q^{1/2}$ which corresponds to $k = 1$ in (166) and:

$$x_n = x_{-n} = i(\frac{1 + q^n}{n})^{1/2} \tag{181}$$

$$\bar{x}_n = \bar{x}_{-n} = -i(\frac{1 + q^{-n}}{n})^{1/2} \tag{182}$$

Finally, if we set $\gamma = q^{1/2}$ and:

$$\begin{aligned} S_+(z) &= z\mathrm{e}^{i\sum_{n>0}(\frac{1+q^n}{n})^{1/2}a_n^+ z^{-n}} \mathrm{e}^{i\sum_{n>0}(\frac{1+q^n}{2})^{1/2}a_n z^n} z^{q_0} q^{p_0} \\ S_-(z) &= z\mathrm{e}^{-i\sum_{n>0}(\frac{1+q^{-n}}{n})^{1/2}a_n^+ z^{-n}} \mathrm{e}^{-i\sum_{n>0}(\frac{1+q^{-n}}{n})^{1/2}a_n z^n} z^{-q_0} q^{-p_0} \\ S_>(z) &= \mathrm{e}^{i\sum_{n>0}(\frac{q^{n/2}+q^{-n/2}}{n})^{1/2}(q^{n/2}-q^{-n/2})a_n^+ z^{-n}} z^{\frac{q_0}{2}} \\ S_<(z) &= \mathrm{e}^{i\sum_{n>0}(\frac{q^{n/2}+q^{-n/2}}{n})^{1/2}(q^{n/2}-q^{-n/2})a_n z^n} z^{-\frac{q_0}{2}}, \end{aligned} \tag{183}$$

it requires a little of algebra to obtain the relations similar to (157)–(161 b):

$$
\begin{aligned}
(z-qz')S^+(z)S^+(z') &= (qz-z')S^+(z')S^+(z)\\
(qz-z')S^-(z)S^-(z') &= (z-qz')S^-(z')S^-(z)\\
[S^+(z),S^-(z')] &= \frac{1}{(q^{1/2}-q^{-1/2})}\left\{\delta(q^{1/2}\frac{z}{z'})S_>(zq^{1/4})-\delta(q^{-1/2}\frac{z}{z'})S_<(zq^{-1/4})\right\}\\
S_>(z)S^+(z') &= \frac{q^{3/4}z-z'}{q^{-1/4}z-qz'}\quad S^+(z')S_>(z)\\
S_<(z)S^+(z') &= \frac{q^{5/4}z-z'}{q^{1/4}z-qz'}\quad S^+(z')S_<(z)\\
S_>(z)S^-(z') &= \frac{q^{1/4}z-qz'}{q^{5/4}z-z'}\quad S^-(z')S_>(z)\\
S_<(z)S^-(z') &= \frac{q^{-1/4}z-qz'}{q^{3/4}z-z'}\quad S^-(z')S_<(z)\\
[S_>(z),S_>(z')] &= [S_<(z),S_<(z')]=0.\\
S_<(z)S_>(z') &= \frac{(1-q^{3/4}\frac{z}{z'})(1-q^{-3/4}\frac{z}{z'})}{(1-q^{5/4}\frac{z}{z'})(1-q^{-5/4}\frac{z}{z'})}S_>(z')S_<(z)
\end{aligned}
\tag{184}
$$

The commutation relation (168) between S^+ and S^- has been slightly modified to account for the hermiticity condition (169). The more general relations involving γ are not difficult to establish, they are given in [5]. Let us now derive the affine algebra relations from the L matrix point of view [10]. We do it in the $SL(n)$ case and, as for Yangians, we start from the defining relation:

$$
R(u_1-u_2)\overset{1}{L}(u_1)\overset{2}{L}(u_2)=\overset{2}{L}(u_2)\overset{1}{L}(u_1)R(u-u_2)
\tag{185}
$$

where L is a $n\times n$ matrix and R is the q deformation of (88):

$$
\begin{aligned}
R_{ii,ii}(u) &= (u+1)_q\\
R_{ij,ij}(u) &= (u)_q \qquad \text{if}\quad i\neq j\\
R_{ij,ji}(u) &= (1)_q\ q^{1/2u} \quad \text{if}\quad i<j\\
&= (1)_q\ q^{-1/2u} \quad \text{if}\quad i>j,
\end{aligned}
\tag{186}
$$

all other matrix elements being equal to zero. It is convenient to multiply $R(u_1-u_2)$ by $q^{1/2(-u_1+u_2)}$ so that, if we set $z=q^u$, $z'=q^{u'}$, the R matrix becomes:

$$
R(z/z')=R^+-\frac{z'}{z}R^-
\tag{187}
$$

with:

$$
\begin{aligned}
R^+_{ii,ii} &= q^{1/2}\\
R^+_{ij,ij} &= 1 \qquad \text{if}\quad i\neq j\\
R^+_{ij,ji} &= (q^{1/2}-q^{-1/2})\quad \text{if}\quad i<j\\
R^-_{ii,ii} &= q^{-1/2}\\
R^-_{ij,ij} &= 1 \qquad \text{if}\quad i\neq j\\
R^-_{ij,ji} &= -(q^{1/2}-q^{-1/2})\quad \text{if}\quad i>j,
\end{aligned}
\tag{188}
$$

all other matrix elements being equal to zero.

Assuming L has a mode expansion: $L=\sum_{n\in\Gamma}L_nz^n$, the coefficient of $z^nz'^m$ of (185) becomes:

$$
R^+\overset{1}{L}_n\overset{2}{L}_m-R^-\overset{1}{L}_{n+1}\overset{2}{L}_{m-1}=\overset{2}{L}_m\overset{1}{L}_n\ R^+-\overset{2}{L}_{m-1}\overset{1}{L}_{n+1}\ R^-
\tag{189}
$$

As before, we assume L has two analytic expressions for $|z| > 1$ and $|z| < 1$ which we denote $L_>$ and $L_<$:

$$\begin{aligned} L_>(z) &= L_0 + \sum_{n=1}^{\infty} \frac{L_{-n}}{z^n} \\ L_<(z) &= \bar{L}_0 + \sum_{n=1}^{\infty} L_n z^n \end{aligned} \tag{190}$$

where L_0 and $\bar{L}_0$ are respectively upper and lower triangular matrices. (189) gives three types of relations:

$$\begin{aligned} (R^+ + \frac{z'}{z} R^-) L_>(z) L_>(z') &= L_>(z') L_>(z) (R^+ + \frac{z'}{z} R^-) \\ (R^+ + \frac{z'}{z} R^-) L_>(z) L_<(z') &= L_<(z') L_>(z) (R^+ + \frac{z'}{z} R^-) \\ (R^+ + \frac{z'}{z} R^-) L_<(z) L_<(z') &= L_<(z') L_<(z) (R^+ + \frac{z'}{z} R^-) \end{aligned} \tag{191}$$

$$\begin{aligned} L_0 &= \begin{pmatrix} K_1 & (1)_q K_1 f_1 & & * & \\ & K_2 & & (1)_q K_2 f_2 & \\ & 0 & & \ddots & \\ & & & & K_n \end{pmatrix} \\ \bar{L}_0 &= \begin{pmatrix} K_1^{-1} & & & \\ -(1)_q e_1 K_1^{-1} & K_2^{-1} & & 0 \\ & & * & \ddots & \\ & & & -(1)_q e_{n-1} K_n^{-1} & K_n^{-1} \end{pmatrix} \end{aligned} \tag{192}$$

We set $q^{H_i/2} = K_i K_{i+1}^{-1}$. The following relations are a consequence of (189) for $n = m = 0$

$$\begin{aligned} [H_i, H_j] &= 0 \\ q^{H_i/2}\ e_j &= q^{1/2 A_{ij}} e_j q^{H_i/2} \\ q^{H_i/2}\ f_j &= q^{-1/2 A_{ij}} e_j\ q^{H_i/2} \\ [e_i, f_j] &= \delta_{ij} \frac{q^{H_i/2} - q^{-H_i/2}}{q^{1/2} - q^{-1/2}} \\ e_i^2 e_j - (q^{1/2} + q^{-1/2})\ \ e_i e_j e_i + e_j e_i^2 &= 0 \qquad \text{if} \quad |i - j| = 1 \end{aligned} \tag{193}$$

where A_{ij} is the Cartan matrix of $SL(n) : 1 \leq i,j \leq n-1$

$$A = \begin{pmatrix} 2 & -1 & & 0 \\ -1 & 2 & \ddots & \\ & \ddots & \ddots & \ddots \\ & 0 & \ddots & -1 \\ & & -1 & 2 \end{pmatrix} \tag{194}$$

For the loop algebra, we add the two generators:

$$\begin{aligned} f^0 &= K_n^{-1} (L_{n1})_{-1} / (1)_q \\ e^0 &= -(L_{1n})_1 K_n / (1)_q \end{aligned} \tag{195}$$

If we set $q^{H_0/2} = K_n K_1^{-1}$, the relations obeyed by $H_i, e_i, f_i, 0 \leq i \leq n$ are the same as (193) with the matrix (194) replaced by the $n \times n$ matrix $A'_{ij}, 0 \leq i,j \leq n-1$:

$$A' = \begin{pmatrix} 2 & -1 & & & -1 \\ -1 & 2 & \ddots & & \\ & \ddots & \ddots & \ddots & \\ & & \ddots & \ddots & -1 \\ -1 & & & -1 & 2 \end{pmatrix} \tag{196}$$

It is still the analogue of a loop algebra because $q^{H_0/2}$ is not an independent generator:

$$q^{H_0/2} = \prod_{i=1}^{n-1} q^{-H_i/2} \tag{197}$$

To obtain the affine algebra, let us modify (191) as follows:

$$\begin{aligned} (R^+ - \frac{z'}{z}R^-)L_>(z)L_>(z') &= L_>(z')L_>(z)(R^+ - \frac{z'}{z}R^-) \\ (R^+ - \gamma^{-1}\frac{z'}{z}R^-)L_>(z)L_<(z') &= L_<(z')L_>(z)(R^+ - \gamma\frac{z'}{z}R^-) \\ (R^+ - \frac{z'}{z}R^-)L_<(z)L_<(z') &= L_<(z')L_<(z)(R^+ - \frac{z'}{z}R^-) \end{aligned} \tag{198}$$

Only the second relation is modified, with z'/z replaced by $\gamma^{-1}z'/z(\gamma z'/z)$ on the left (right) hand side of the equality. γ is assumed to commute with everything. Among the generators, the only modified relation is then:

$$[e^0, f^0] = \frac{\gamma K_n K_1^{-1} - \gamma^{-1} K_1 K_n^{-1}}{q^{1/2} - q^{-1/2}} \tag{199}$$

So, if we redefine the generators $q^{H_0/2}$ as follows:

$$q^{H_0/2} = \gamma K_n K_1^{-1}, \tag{200}$$

the new generators obey the same relations as the old one but now $q^{H_0/2}$ is independent of $q^{H_i/2}$ and (198) can be taken as defining the affine algebra $SL(n)^{(1)}$. The same arguments as those preceding (78) enable it to be shown that it is an associative algebra. It remains to find the coproduct. Let us define:

$$\Delta\gamma = \gamma \otimes \gamma \tag{201}$$

and set:

$$\begin{aligned} \Delta L_>(z) &= L_>(1 \otimes \gamma z)\dot{\otimes} L_>(z) \\ \Delta L_<(z) &= L_<(z)\dot{\otimes} L_<(\gamma\, z \otimes 1) \end{aligned} \tag{202}$$

It is immediate that the relations involving only $L_>$ or $L_<$ are preserved by this coproduct, so let us verify that the other one is also preserved:

$$\begin{aligned} &R(z\Delta\gamma/z')(\Delta \overset{1}{L}_> (z))(\Delta \overset{2}{L}_< (z')) \\ &= R(\gamma \otimes \gamma\frac{z}{z'}) \overset{1}{L}_> (1 \otimes \gamma z) \overset{2}{L}_< (z')\dot{\otimes} \overset{1}{L}_> (z) \overset{2}{L}_< (\gamma z' \otimes 1) \\ &= \overset{2}{L}_< (z') \overset{1}{L}_> (1 \otimes \gamma z)\dot{\otimes} R(\gamma^{-1}\frac{z}{z'} \otimes \gamma) \overset{1}{L}_> (z) \overset{2}{L}_< (\gamma z' \otimes 1) \\ &= \overset{2}{L}_< (z') \overset{1}{L}_> (1 \otimes \gamma z)\dot{\otimes} \overset{2}{L}_< (\gamma z' \otimes 1) \overset{1}{L}_> (z) R(\gamma^{-1}\frac{z}{z'} \otimes \gamma^{-1}) \\ &= (\Delta \overset{2}{L}_< (z'))(\Delta \overset{1}{L}_> (z)) R(\Delta\gamma^{-1}\frac{z}{z'}) \end{aligned} \tag{203}$$

Evaluated on the generators, the coproduct gives:

$$\begin{aligned} \Delta q^{H_i/2} &= q^{H_i/2} \otimes q^{H_i/2} & 0 \le i \le n \\ \Delta e_i &= e_i \otimes 1 + q^{H_i/2} \otimes e_i & 0 \le i \le n \\ \Delta f_i &= f_i \otimes q^{-H_i/2} + 1 \otimes f_i & 0 \le i \le n \end{aligned} \tag{204}$$

It finally remains to compare this presentation of the algebra with the one given in (184) in terms of S^+, S^-, S (for $\gamma = q^{1/2}$) in the $SL(2)$ case. Although the determinant cannot be set equal to 1 because it does not commute with L any more, it is easy to see that it commutes with the ratios of the matrix elements of L, in particular with $S^+ = CD^{-1}$ and $S^- = D^{-1}B$. It can therefore be eliminated by considering the algebra generated by S^+, S^- which coincides with (184). Note that expressing L in terms of $S^\pm, S$ should in principle give a vertex operator representation of relations (198). Unfortunately, the interpretation of such a L matrix as the monodromy matrix of a field theory is missing. It was the purpose of these lecture to motivate interest in affine algebras. So, we think it is a good point to stop, leaving the possible physical applications an open problem.

Acknowledgements

I would like to thank the organisers of the school for giving me the opportunity to lecture in the stimulating atmosphere of Cargese. I am particularly endebted to M. Gaudin and D. Bernard for valuable discussions, Philippe Roche for a careful reading of the manuscript and Anne Désalos for typing it.

References

[1] M. Gaudin, "La fonction d'onde de Bethe" collection scientifique du C.E.A. Masson (1983).

[2] R.J. Baxter "Exactly solved models in Statistical Mechanics" Academic, London 1982.

[3] L.D. Faddeev "Integrable models in (1 + 1) dimensional quantum field theory". Les Houches Lectures XXXIX Elsevier, Amsterdam (1982).

[4] V.G. Drinfeld translations Soviet. Math. Dokl. vol.32 (1985), n°1 p.254.

[5] V.G. Drinfeld translations Soviet. Math. Dokl. vol.36 (1988), n°2 p.212.

[6] I.B. Frenkel and N. Jing, *Proc. Nat. Acad. Sci. USA* **85** (1988), 9373.

[7] E. Lieb and W. Liniger, *Phys. Rev.* **130**, 4 (1963) p.1605 and 1616.

[8] E.K. Sklyanin, *Funkt. Anal. Appl.* **16** (1983) 263.

[9] M. Jimbo, "Yang Baxter Equation in Integrable systems". Advanced series in Mathematical Physics. Vol.10 (1990), World Scientific editors.

[10] N.Y. Reshetikin and M.A. Semenov-Tian-Shansky, Letters in Mathematical Physics **19** (1990) 133,142.

LAGRANGIAN CONFORMAL MODELS

Raymond Stora

Physique Théorique, B.P. 909
F 74019 Annecy Levieux Cedex

Free fields are studied from the point of view of resuming properly renormalized Feynman perturbative series as functional of complex structures. Quantization in $\mathbb{C}$ was summarized in the first lecture. The corresponding material can be found in *Progr. Theor. Phys.* **102**, (1990).

Quantization on a compact Riemann surface without boundary was the topic of the other lecture where the holomorphic factorization theorem of the vacuum amplitude at fixed non vanishing central charge was reinvestigated using the local families index formulae of J.M. Bismut, H. Gillet, C. Soulé. The result is that all the metrics which make the resummation possible, thanks to the positivity thereby introduced, can be eliminated.

The first part is published in: M. Knecht, S. Lazzarinie, R.S. *P.L.B* **262**, 1991, 25–31. The second part, which deals with free fields with values in an arbitrary holomorphic bundle is being prepared for publication in *P.L.B.* This is joint work with M. Knecht, S. Lazzarini, F. Thuillier.

New Symmetry Principles in Quantum Field Theory, Edited by
J. Fröhlich et al., Plenum Press, New York, 1992

SEMI-CLASSICAL LIOUVILLE THEORY, COMPLEX GEOMETRY OF MODULI SPACES, AND UNIFORMIZATION OF RIEMANN SURFACES

Leon Takhtajan

Program in Applied Mathematics
University of Colorado, Boulder, CO 80309-0526 U.S.A.

ABSTRACT

An introduction to the semi-classical theory for the Polyakov's functional integral approach to the quantum Liouville theory is presented with the emphasis on geometrical setup and interrelations between the Fuchsian uniformization of Riemann Surfaces and the complex geometry of Teichmüller, Schottky and moduli spaces.

LECTURE 1

1. Introduction

These lectures originated from an attempt to understand Polyakov's geometrical approach to non-critical, i.e. $D \neq 26$, bosonic strings [1]. In this formulation the famous Liouville action (with a coupling constant $\gamma = \frac{26-D}{48\pi}$) appears as a conformal anomaly. Since then there were many attempts to understand corresponding quantum Liouville theory. In the conventional approach Liouville theory is treated as a model of Conformal Field Theory (CFT). The result is that for $D < 1$ (or $D > 25$) it is equivalent (for rational values of $2\pi\gamma$) to minimal models of BPZ [2, 3, 4]. This approach could be interpreted as Euclidean version of the well-defined quantum field theory in Minkowsky space-time. As a result in the semi-classical limit Liouville field $\exp \phi(w, \bar{w})$ is not necessarily real valued (not to say positive), i.e. it does not correspond to Riemannian metric on a world sheet - a two-dimensional surface. On the contrary, Polyakov's functional integral formulation of Liouville theory (see Section 3) looks very elegant and is intrinsically geometric. This indicates that conventional approach (at least in its present form) does not serve the original purpose of understanding non-critical strings. It seems that realization of Polyakov's approach to quantum Liouville theory is one of the most

New Symmetry Principles in Quantum Field Theory, Edited by
J. Fröhlich et al., Plenum Press, New York, 1992

important and difficult problems in $2-d$ quantum field theory. (It should be noted that a seminal paper [5] emerges as an attempt to develop this approach). It should be also emphasized that Polyakov's formulation is equivalent to a certain phase of $2-d$ quantum gravity, which is one of the fundamental problems by itself. Therefore I am afraid that I will not contribute much to this subject as a whole. Concentrating on a less ambitious goal instead I will present here a semi-classical analysis of Polyakov's functional integral formulation of quantum Liouville theory and will show how it leads to novel results on complex geometry of moduli spaces of world sheets – Riemann Surfaces. Some of these results were conjectured quite a long time ago by Polyakov and Zamolodchikov [6, 7] and were proved and generalized in mathematical papers [8, 9, 10, 11]. In these lectures for the first time I will present "two sides of the coin": I will show heuristic derivation of these results based on the ideas of CFT and will give their proofs based on deformation theory of Riemann Surfaces. Consequently we will see interesting connections between semi-classical Liouville theory, uniformization theory of Riemann Surfaces and complex geometry of their moduli spaces. I hope that this presentation will partly compensate the incompleteness of results on quantum Liouville theory. In addition I will explicitly indicate subtle points of this approach by numerous remarks intended to clarify and not to confuse. I believe that these results indicate that Polyakov's functional integral treated non-perturbatively should be considered as the main object in the future "quantum geometry", a kind of a "modular geometry" of Friedan and Shenker [12]. *Acknowledgments*: I would like to thank the Organizing Committee and the Staff of Cargese Summer School "New Symmetry Principles in Quantum Field Theory" for the pleasant and stimulating atmosphere during the School. Also it was that occasion (and determination of the Organizing Committee!) which finally led to these lectures.

2. Uniformization of Riemann Surfaces

We will consider here Riemann Surfaces (RS) of type (g,n), i.e. of genus g with n punctures. If a RS X is of this type then its closure $\bar{X}$ is a compact RS of genus g – *algebraic curve* – and $X = \bar{X} \setminus \{x_1, \ldots, x_n\}$, where $x_1, \ldots, x_n$ are n distinct points in $\bar{X}$ called *punctures*. The uniformization theorem, in nowadays standard form, states that if $g + \frac{1}{2}n > 1$ then X can be represented as a quotient of the upper half-plane $H = \{z \in \mathbf{C} \mid \operatorname{Im} z > 0\}$ – called *hyperbolic* or *Lobachevsky plane* – by the action of a torsion-free finitely generated *Fuchsian group* Γ:

$$X \cong H/\Gamma. \tag{1}$$

It means that there exists a holomorphic covering $J : H \mapsto X$ with Γ acting as a group of automorphisms; as an abstract group Γ is isomorphic to a fundamental group $\pi_1(X)$ of a surface X.

Remaining case $g + \frac{1}{2}n \leq 1$ corresponds to RS's of the types: (1,0), (0,1), (0,2).

They can be uniformized either by elliptic functions for the type (1,0) or by elementary functions for the two latter types.

Remark 1 Let $G = PSL(2,\mathbb{R})$ be the group of motions of the hyperbolic plane H with a standard action:

$$\text{for } \gamma = \begin{pmatrix} a & b \\ c & d \end{pmatrix} \in G \text{ and } z \in H \quad \gamma z = \frac{az+b}{cz+d}.$$

A subgroup Γ of G is called Fuchsian if it acts *discontinuously* on H. (For subgroups of G discontinuity is equivalent to discreteness).

Though today the uniformization theorem can be found in any textbook on RS's (see, e.g. [13]) its proof was a high priority hundred years ago at the time of Klein and Poincaré. Namely, they made several ingenious (though incomplete) attempts to prove it until it was finally proved in 1907 by Poincaré and Koebe using different methods. It turns out that original (and almost forgotten) approaches of Klein and Poincaré are actually related to the semi-classical limit of the Polyakov's functional integral!

I will start with the approach which dates back to 1883 and is based on ordinary differential equations (cf. famous memoirs [14, 15]). To present it in the most transparent form let us consider the case of RS's of the type $(0,n)$, i.e. of genus zero with n punctures, which can be represented by a Riemann sphere $\mathbf{P}^1$ with n distinct points $w_1, \ldots, w_n$ removed. Two RS's of the same type are isomorphic if and only if they are related by an element of a group $G_{\mathbf{C}} = PSL(2,\mathbf{C})$ – a group of all automorphisms of $\mathbf{P}^1$ (obviously RS's of different types are non-isomorphic). Using this $G_{\mathbf{C}}$-freedom we can normalize RS's by setting $w_{n-2} = 0$, $w_{n-1} = 1$, $w_n = \infty$ so that they have a form $X = \mathbf{C} \setminus \{w_1, \ldots, w_{n-3}, 0, 1\}$.

Remark 2 Define a *space of punctures* as the following domain in $\mathbf{C}^{n-3}$:

$$\mathbf{W}_n = \{(w_1, \ldots, w_{n-3}) \in \mathbf{C}^{n-3} | w_i \neq w_j, \text{ for } i \neq j;\ w_i \neq 0, 1\}$$

so that points in $\mathbf{W}_n$ represent all RS's of type $(0,n)$. However different points in $\mathbf{W}_n$ could correspond to the same RS. This is the case if they belong to the same orbit of the action of symmetric group $Symm(n)$ on $\mathbf{W}_n$. This action consists in applying all permutations of n elements to the set $(w_1, \ldots, w_{n-3}, 0, 1, \infty)$ and then normalizing it back using $G_{\mathbf{C}}$-action so that the last three coordinates should be $0, 1$ and ∞ again. Namely, if σ_i, $i = 1, \ldots, n-1$, is an elementary transposition, then for $i \leq n-4$ its action on $(w_1, \ldots, w_{n-3})$ consists of a simple interchanging w_i and w_{i+1}, whereas for $n-3 \leq i \leq n-1$ we have explicitly $\sigma_i(w_1, \ldots, w_{n-3}) = (\tilde{w}_1, \ldots, \tilde{w}_{n-3})$, where

$$\tilde{w}_k = \begin{cases} (w_k - w_{n-3})/(1 - w_{n-3}) & k = 1, \ldots, n-4, \quad i = n-3, \\ w_{n-3}/(w_{n-3} - 1) & k = n-3, \quad i = n-3, \\ 1 - w_k & k = 1, \ldots, n-3, \quad i = n-2, \\ w_k/(w_k - 1) & k = 1, \ldots, n-3, \quad i = n-1. \end{cases} \tag{2}$$

Corresponding normalized RS's X and $\tilde{X}$ are isomorphic and are related by simple transformations $w \mapsto \tilde{w} = (w - w_{n-3})/(1 - w_{n-3})$, $i = n-3$, $w \mapsto \tilde{w} = 1 - w$, $i = n-2$

and $w \mapsto \tilde{w} = w/(w-1)$ for $i = n-1$. Quotient space $\mathbf{M}_{0,n} = \mathbf{W}_n/Symm(n)$ is isomorphic to the *moduli space* of RS's of the type $(0,n)$. When $n > 4$ a symmetric group $Symm(n)$ acts on $\mathbf{W}_n$ effectively so that $\mathbf{M}_{0,n}$ is a $n-3$-dimensional complex manifold. For $n = 4$ all RS's of the type $(0,4)$ have a non-trivial group of automorphisms isomorphic to $\mathbf{Z}/2\mathbf{Z} \oplus \mathbf{Z}/2\mathbf{Z}$ so that $Symm(4)$-action on $\mathbf{W}_4$ is not free and $\mathbf{M}_{0,4}$ is only an orbifold.

Now, together with Klein and Poincaré [14, 15], consider the following second-order linear differential equation on X:

$$\frac{d^2y}{dw^2} + \frac{1}{2}Q_X(w)y = 0, \quad w \in X, \tag{3}$$

where

$$Q_X(w) = \sum_{i=1}^{n-1}\left(\frac{1}{2(w-w_i)^2} + \frac{c_i}{w-w_i}\right). \tag{4}$$

The conditions

$$\sum_{i=1}^{n-1} c_i = 0, \quad \sum_{i=1}^{n-1} c_i w_i = 1 - \frac{1}{2}n \tag{5}$$

imply that differential equation (3) is of *Fuchsian type*, i.e. it has *regular singular points* $w_1, \ldots, w_{n-3}, 0, 1$ and ∞. Analytic continuation of the ratio y_1/y_2 of its two linearly independent solutions y_1 and y_2 along all noncontractible loops in X leads to a *monodromy representation* $Mon : \pi_1(X) \mapsto G_{\mathbf{C}}$ which is defined up to a conjugation. Its image in $G_{\mathbf{C}}$ is called a *monodromy group*; in our case it is generated by a parabolic transformations M_i, $i = 1, \ldots, n$, satisfying the relation $M_1 \ldots M_n = I$. The uniformization theorem is equivalent to the existence of such coefficients $c_1, \ldots, c_{n-1}$ that monodromy group of the differential equation (3) is a Fuchsian group Γ (up to a conjugation) which uniformizes X (i.e. $X \cong H/\Gamma$). Coefficients $c_1, \ldots, c_{n-3}$ (c_{n-2} and c_{n-1} can be expressed in terms of the others by means of (5)) are called *accessory parameters* of Fuchsian uniformization and they are single-valued functions of the singular points $w_1, \ldots, w_{n-3}$. Equivalence between these formulations of the uniformization theorem follows from the formulas $y_1/y_2 = J^{-1}$: $X \mapsto H$ (for suitably chosen y_1, y_2) and $Q_X = S(J^{-1})$, where

$$S(f) = \frac{f'''}{f'} - \frac{3}{2}\left(\frac{f''}{f'}\right)^2$$

stands for the *Schwarzian derivative* of a function f.

Remark 3 Though function J^{-1} on X is multi-valued, i.e. it has multiple branches related by transformations from Γ (such functions are called *linearly polymorphic*), due to the fundamental property of the Schwarzian derivative

$$S(A \circ f) = S(f) \quad \text{for all} \quad A \in G_{\mathbf{C}},$$

$S(J^{-1})$ is correctly defined and yields Q_X. Particular form (4) of Q_X and conditions (5) follow from the fact that J is an *automorphic function* for the Fuchsian group Γ and it is regular at all *cusps* of Γ (i.e. at fixed points of transformations M_i, $i = 1, \ldots, n$) except for the cusp ∞ where it has a simple pole (see, e.g. [9]). Sometimes function J is called *Klein's Hauptmodule.*

Remark 4 Differential equation (3) looks similar to the *null vector decoupling equation* in CFT for the field $\psi_{2,1}$ (see [5]). However it should be noted that monodromy groups of decoupling equations for minimal models of CFT are subgroups of $PSU(2)$, whereas monodromy group of equation (3) is a subgroup of $G = PSL(2, \mathbb{R})$ (cf. with [16]).

In 1898 Poincaré made another attempt to prove uniformization theorem [17] by proving that on X there exists a unique complete conformal metric of constant negative curvature -1. This metric (so-called *Poincaré* or *hyperbolic metric*) has a form $e^{\phi(w,\bar{w})}|dw|^2$ (recall that X has genus zero so that w is a global coordinate on it), satisfies *Liouville equation*

$$\frac{\partial^2 \phi}{\partial w \partial \bar{w}} = \frac{1}{2} \exp \phi \tag{6}$$

and has the following asymptotics near the punctures

$$\phi(w, \bar{w}) = \begin{cases} -2 \log |w - w_i| - 2 \log |\log |w - w_i|| + o(1) & \text{as } w \to w_i, \ i \neq n, \\ -2 \log |w| - 2 \log \log |w| + o(1) & \text{as } w \to w_n = \infty. \end{cases} \tag{7}$$

These asymptotics imply that punctures w_i are really points at infinity with respect to the Poincaré metric, i.e. a geodesic distance between points x_1 and x_2 in X goes to infinity as $x_2 \to w_i$, $i = 1, \ldots, n$. Moreover, the function

$$\exp \varphi(w, \bar{w}) = \frac{1}{r^2 \log^2 r},$$

where $r = \log |w|$, enjoys Liouville equation for $r \neq 0, 1$. With the help of the formula

$$\frac{\partial}{\partial \bar{w}} \left(\frac{1}{w}\right) = \pi \delta(w),$$

we can interpret singularities of the field $\exp \phi(w, \bar{w})$ near the punctures as sources needed to turn a punctured sphere X into the surface of constant negative curvature and rewrite (6) as the following equation on $\mathbf{P}^1$

$$\frac{\partial^2 \phi}{\partial w \partial \bar{w}} = \frac{1}{2} \exp \phi - \pi \sum_{i=1}^{n-1} \delta(w - w_i) + \pi \delta(w - w_n), \tag{8}$$

where near $w_n = \infty$ in $\mathbf{P}^1$ one should use a local coordinate $\tilde{w} = 1/w$. From this equation it follows, in agreement with Gauss-Bonnet, that

$$\int_X e^{\phi} d^2 w = 2\pi(n - 2), \quad \text{where } d^2 w = \left|\frac{dw \wedge d\bar{w}}{2}\right|.$$

Remark 5 It follows from (6) that the functional

$$T_\phi = \phi_{ww} - \tfrac{1}{2} \phi_w^2 \tag{9}$$

is *conserved* on solutions of classical equations of motion, i.e.

$$\frac{\partial}{\partial \bar{w}} T_{cl} = 0, \quad w \in X, \tag{10}$$

where T_{cl} is the value of T_ϕ on the solution of the Liouville equation. Moreover, from asymptotics (7) it follows that T_{cl} can be meromorphically continued to a rational function on $\mathbf{P}^1$ of the form (4). Functional T_ϕ can be interpreted as $(2,0)$-component of the modified ***stress-energy tensor*** for the classical Liouville theory. Note that a standard Noether stress-energy tensor for the formal (and naive) action

$$\tilde{S}(\phi) = \int_X (|\phi_w|^2 + e^\phi) d^2 w$$

is not traceless and does not "behave properly" under conformal change of coordinates. Modification of the Noether stress-energy tensor – term ϕ_{ww} – "cures" these properties. Namely, (10) means that modified stress-energy tensor is *traceless*. Moreover, since under the conformal change of coordinates $w \mapsto \tilde{w} = f(w)$ density e^ϕ of the conformal metric transforms like $(1,1)$-tensor:

$$e^\phi \mapsto e^{\tilde{\phi}} = (e^\phi \circ f)|f'|^2, \tag{11}$$

it follows from (9) that $T = T_\phi$ has the following "proper" transformation law

$$T \mapsto \tilde{T} = (T \circ f)(f')^2 + S(f). \tag{12}$$

In terms of the covering map J function e^ϕ has the following simple expression:

$$\exp \phi(w, \bar{w}) = \frac{|(J^{-1})'(w)|^2}{(\text{Im } J^{-1}(w))^2} \tag{13}$$

so that the metric $e^{\phi(w,\bar{w})}|dw|^2$ is just the projection on X of the Poincaré metric $(\text{Im } z)^{-2}|dz|^2$ on H. Since Poincaré metric on H is G-invariant Liouville field e^ϕ is correctly defined in terms of the linearly polymorphic function J^{-1} (cf. Remarks 1 and 3).

Remark 6 In Minkowsky space-time Liouville equation

$$\frac{\partial^2 \phi}{\partial \xi \partial \eta} = \frac{1}{2} \exp \phi, \quad \xi = x^1 + x^0, \quad \eta = x^1 - x^0$$

has a general solution parametrized by two arbitrary real-valued (up to an overall $G_{\mathbf{C}}$-action) functions u and v:

$$\exp \phi(\xi, \eta) = -4 \frac{u'(\xi) v'(\eta)}{(u(\xi) - v(\eta))^2}.$$

Therefore after the Wick rotation $\xi \mapsto w = x^1 + \sqrt{-1} x^0$, $\eta \mapsto \bar{w} = x^1 - \sqrt{-1} x^0$ this formula yields, in general, a complex solution of equation (6) and vice versa. This partially clarifies (at least at the classical level) why treatment of the Liouville equation by methods of CFT and Quantum Groups [2, 3, 4] corresponds rather to Minkowskian theory than to Euclidean one.

From representation (13) it follows that actually $T_{cl} = S(J^{-1}) = Q_X$. This formula connects approaches of Klein and Poincaré with the standard approach to the uniformization theory.

Remark 7 Transformation law (12) for the T_{cl} also follows from (13) and the *cocycle condition*

$$S(f \circ g) = (S(f) \circ g)(g')^2 + S(g)$$

for the Schwarzian derivative.

Remark 8 It follows from definition (9) that the real and single-valued function $\exp(-\frac{1}{2}\phi(w; w_1, \ldots, w_n))$ satisfies the differential equation (3):

$$\left(\frac{d^2}{dw^2} + \frac{1}{2}T_\phi\right)e^{-\frac{1}{2}\phi} = 0$$

so that it is tempting to treat it as a semi-classical limit of *n-point correlation function* in CFT. However, monodromy of (3) is real rather than unitary (cf. Remark 4) so that this interpretation is not at all clear. Moreover, $\exp(-\frac{1}{2}\phi(w; w_1, \ldots, w_n))$ does not correspond to the semi-classical limit of standard correlation functions for solvable models of CFT. Finally let us discuss Lagrangian formulation of the Liouville theory. Denote by $M(X)$ a class of (smooth) real-valued functions ϕ on X (recall that X has genus zero) satisfying asymptotics (7), i.e.

$$\exp\phi \cong \frac{1}{r_i^2 \log^2 r_i}, \quad i = 1, \ldots, n,$$

where $r_i = |w - w_i|$, $i = 1, \ldots, n-1$, and $r = |w|$ for $i = n$, and note that formal action $\tilde{S}$ (see Remark 5) diverges for $\phi \in M(X)$!

However, one can define in a natural way a *regularized action*:

$$S(\phi) = \lim_{\epsilon \to 0}\left\{\int_{X_\epsilon} (|\phi_w|^2 + e^\phi) d^2w + 2\pi n \log\epsilon + 4\pi(n-2)\log|\log\epsilon|\right\}, \tag{14}$$

where $X_\epsilon = X \setminus \bigcup_{i=1}^{n-1}\{|w - w_i| < \epsilon\} \bigcup \{|w| > 1/\epsilon\}$. It is correctly defined and yields (6) through Euler-Lagrange equation $\delta S = 0$. Let S_{cl} be a *classical action*, i.e. the value of S on the solution of classical equations of motion (a *critical value* of S). Evaluated for all (normalized) RS's of the given type $(0, n)$ it gives rise to a function $S_{cl} : \mathbf{W}_n \mapsto \mathbb{R}$.

Remark 9 Classical action S_{cl} is not invariant under the $Symm(n)$-action on $\mathbf{W}_n$ and therefore does not descend to a single-valued function on the moduli space $\mathbf{M}_{0,n}$. Indeed, although Poincaré metric is defined canonically, i.e. is independent of a particular normalization of a given RS, regularization procedure in (14) is not $Symm(n)$-invariant and leads to a *modular anomaly*. More precisely, as it follows from Remark 2, isomorphic RS's in the same $Symm(n)$-orbit are related by a transformation from $G_\mathbb{C}$; this transformation, in general, distorts radii of "small circles" in (14) and leads to the anomaly. One has explicitly (cf. [18])

$$S_{cl}(w_1, \ldots, w_{n-3})) \quad - \quad S_{cl}(\sigma_i(w_1, \ldots, w_{n-3})) = \tag{15}$$

$$= \begin{cases} 0 & i = 1, \dots, n-4, n-2, \\ 2\pi(n-2)\log|w_{n-3}-1| & i = n-3, \\ 4\pi\sum_{l=1}^{n-3}\log|w_l - 1| & i = n-1. \end{cases}$$

Remark 10 Formal action $\tilde{S}$, if it was convergent for all $\phi \in M(X)$, would lead to a $Symm(n)$-invariant classical action; in the next Lecture we will see that this is "too good to be true", i.e it would contradict certain global properties of the moduli space $\mathbf{M}_{0,n}$.

Remark 11 Formulas (15) admit the following geometrical interpretation (cf [18]). Let $\{f_\sigma\}_{\sigma \in Symm(n)}$ be a *1-cocycle* of the $Symm(n)$-action on $\mathbf{W}_n$ defined by

$$f_{\sigma_i}(w_1, \dots, w_{n-3}) = \begin{cases} 1 & i = 1, \dots, n-2, n-4, \\ (w_{n-3}-1)^{n-2} & i = n-3, \\ \prod_{l=1}^{n-3}(w_l - 1)^2 & i = n-1 \end{cases} \qquad (16)$$

and extended to all group $Symm(n)$ in accordance with the group cocycle property $f_{s_1 s_2} = (f_{s_1} \circ s_2) f_{s_2}$, $s_1, s_2 \in Symm(n)$. Cocycle f defines a *holomorphic line bundle*

$$\lambda_{0,n} = \mathbf{W}_n \times \mathbf{C}/Symm(n)$$

over the moduli space $\mathbf{M}_{0,n} = \mathbf{W}_n/Symm(n)$ as a quotient of a trivial line bundle $\mathbf{W}_n \times \mathbf{C} \mapsto \mathbf{W}_n$ by the following $Symm(n)$-action: $(\vec{w}, z) \mapsto (s(\vec{w}), f_s(\vec{w})z), \vec{w} \in \mathbf{W}_n$, $z \in \mathbf{C}$, $s \in Symm(n)$. The function $\exp(\frac{1}{\pi}S_{cl})$ has a transformation law:

$$(\exp(\frac{1}{\pi}S_{cl}) \circ s)|f_s|^2 = \exp(\frac{1}{\pi}S_{cl}), \quad s \in Symm(n),$$

and can be interpreted as a *Hermitian metric* in the line bundle $\lambda_{0,n} \mapsto \mathbf{M}_{0,n}$.

3. Polyakov's functional integral

Since the discovery of conformal anomaly in non-critical strings [1] quantum Liouville theory has became extremely popular. Although there are several different formulations of this theory the most profound one is Polyakov's original approach [6] via the following functional integral:

$$\langle X \rangle = \int_{M(X)} D\phi \exp(-\frac{1}{2\pi h}S(\phi)), \qquad (17)$$

where $h = 24/(26-D)$, $S(\phi)$ is the Liouville action and integration goes over functional space $M(X)$ of all complete conformal Riemannian metrics on a given normalized RS X of the type $(0,n)$. This beautiful formulation is intrinsically geometric and can be considered as a proper approach to $2-d$ quantum gravity without matter fields. It is tempting to say that the functional integral (17) should play a fundamental role in the future "quantum geometry of RS's"! However it turns out to be a very difficult problem and since Polyakov's 1982 Saint - Petersburg's seminar [6] (at that time it was called Leningrad!) until recently there were practically no published results (interesting Seiberg's paper [19] seems to be a rare exception).

The first question to ask is how one should understand the functional integral (17): does it represent a correlation function or it should be treated as a kind of partition function?

Remark 12 At first glance it seems that differential equation (8) supports the former interpretation. Namely, one can formally treat singularities of the Liouville field as sources at the punctures and argue that (17) can be rewritten in the form

$$\langle \Phi(w_1,\ldots,w_n)\rangle = \int D\varphi\ \Phi(w_1,\ldots,w_n)\exp(-\frac{1}{2\pi h}S(\varphi)), \tag{18}$$

where

$$\Phi(w_1,\ldots,w_n) = \exp(\frac{1}{h}\varphi(w_1))\ldots\exp(\frac{1}{h}\varphi(w_{n-1}))\exp(-\frac{1}{h}\varphi(w_n)),$$

and integration goes over all smooth conformal metrics on $\bar{X} = \mathbf{P}^1$. In other words it means that (17) should be treated as a correlation function of *puncture operators* $e^{\frac{1}{h}\varphi(w)}$ (in a sense of [19]) in the presence of a "charge" at $w_n = \infty$. However, although (17) and (18) yield the same classical equations of motion, "trade-off" singularities for sources is not so "harmless" as it looks! In particular, 2-point correlation functions of puncture operators, as well as the bare partition function, in this approach vanish in the semi-classical limit (since RS's of the types $(0,2)$ and $(0,0)$ do not carry hyperbolic metrics). In addition there are certain regularization problems in connection with (18).

Remark 13 Another natural question is whether one can use the Coulomb gas approach of Dotsenko-Fateev [20, 21] to get a closed expression for $\langle\Phi(w_1,\ldots,w_n)\rangle$. However, puncture operators, contrary to usual vertex operators, are not related to degenerate Verma modules, and do not fit naturally into the Coulomb gas picture. Nevertheless it could be possible [22] to realize $\langle\Phi\rangle$ as a certain large n limit of Dotsenko-Fateev's n-point correlation functions: a kind of "collective phenomena" effect in the Coulomb gas. Similar situation occurs in a simpler case of Ising model [23].

These remarks indicate (and we will adopt this point of view) that it is more "safe" to treat $\langle X\rangle$ as a partition function for a given normalized RS X. Evaluated for all X of the same type it yields a certain function on $\mathbf{W}_n$. Presumably it could be interpreted as a Hermitian metric in a certain "quantum line bundle" (i.e. depending on the Planck constant!) over the moduli space $\mathbf{M}_{0,n}$ in the spirit of Friedan-Shenker's modular geometry.

LECTURE 2

4. Ward Identities, Semi-Classical Limit and Symmetry Properties of Accessory Parameters

It would be extremely desirable to develop a non-perturbative approach to the partition function $\langle X\rangle$, i.e. to find a way to treat it for all values of h without "spoiling" its "beauty" by $h \to 0$ analysis. To achieve this one should use conformal invariance of the

Liouville field, i.e. the fact that e^{ϕ} transforms like a density of a Riemannian metric under conformal change of variables (see (11)). As it was discovered in [5], conformal invariance in $2-d$-quantum field theories leads to the infinite series of constraints for correlation functions – so-called *Conformal Ward Identities* (CWI). Their formal derivation is based on the conformal invariance of the action (in the functional integral formulation) and on the "fact" that conformal group in 2-dimensions is infinite-dimensional [5]. Needless to say that only Lie algebra of infinitesimal conformal transformations is well defined and is infinite-dimensional (it is a Virasoro algebra without central extension, i.e. a Witt algebra); however in the formal "proof" of CWI one should use global conformal transformations as well (as a change of variables in the action). Still there is absolutely no doubt that CWI play the major role in CFT; in particular, they yield complete solution of minimal models [5, 24], $\mathbf{Z}_N \times \mathbf{Z}_N$-parafermions [25, 26] and other solvable models of CFT. Lagrangian formulation for these theories is not so relevant so one can express their conformal invariance just by postulating CWI.

However in the Lagrangian formulation of the Liouville theory one should seriously consider that a true conformal group in 2-dimensions is not infinite-dimensional but rather is $G_{\mathbf{C}}$ and the main object $\langle X\rangle$ is rather a partition function than n-point correlator! Nevertheless Polyakov argued in [6] that CWI for the quantum Liouville theory still have the standard form

$$\frac{1}{h}\langle T_{\phi}(w)X\rangle=\sum_{i=1}^{n-1}\Big(\frac{\Delta(h)}{(w-w_i)^2}+\frac{1}{w-w_i}\cdot\frac{\partial}{\partial w_i}\Big)\langle X\rangle, \tag{19}$$

where $\langle X\rangle$ is considered as a function on the space of punctures $\mathbf{W}_n$ and $\Delta(h)$ is interpreted as a conformal weight of puncture operators. These CWI incorporate vector fields in "moduli directions" and look very suggestive. However they do not immediately contribute to the solution of the theory because one can not trivially convert (19) into differential equations for $\langle X\rangle$ (see Remark 13). Therefore it is instructive (and the only thing we actually can do!) to analyze the semi-classical limit of (19).

When $h\to 0$ in the leading order ("tree level") we get:

$$\langle X\rangle\sim\exp\Big(-\frac{1}{2\pi h}S_{cl}\Big),\quad \langle T_{\phi}(w)X\rangle\sim T_{cl}(w)\exp\Big(-\frac{1}{2\pi h}S_{cl}\Big),\quad \Delta(h)\sim\frac{1}{2h}$$

so that

$$\frac{1}{h}T_{cl}\langle X\rangle=\sum_{i=1}^{n-1}\Big(\frac{1}{2h(w-w_i)^2}+\frac{1}{w-w_i}\cdot\frac{\partial}{\partial w_i}\Big)\langle X\rangle,$$

or

$$T_{cl}=\sum_{i=1}^{n-1}\Big(\frac{1}{2(w-w_i)^2}-\frac{1}{2\pi}\frac{\partial S_{cl}}{\partial w_i}\frac{1}{(w-w_i)}\Big).$$

Comparing this expression with "triality" formula $T_{cl}=Q_X=S(J^{-1})$, where Q_X is given by (4), we get explicitly

$$c_i=-\frac{1}{2\pi}\frac{\partial S_{cl}}{\partial w_i}. \tag{20}$$

This formula looks striking: it states that classical action for the Liouville theory is the generating function (common anti-derivative) for the accessory parameters of Fuchsian uniformization! Since the time of Klein and Poincaré accessory parameters were considered to be "mysterious" objects: not only to compute them "explicitly", but even to prove that they exist and are single-valued functions of the punctures was a difficult problem. So formula (20) is rather dramatic: it gives additional information about the behaviour of accessory parameters as functions of the punctures, in particular, it implies the following symmetry properties

$$\frac{\partial c_i}{\partial w_j} = \frac{\partial c_j}{\partial w_i},$$

etc. Neither Klein nor Poincaré were aware of these results (though they might have appreciated them!). Formulas (20) appear as a conjecture in Polyakov's 82-seminar [6], their "derivation" which I presented here should be close to the original of Polyakov and Zamolodchikov [7]. However at that time (around 1983-1984) general attitude towards these hypothetical properties of accessory parameters was quite sceptical. First, if these formulas were true they should have been discovered by classics! Second, their derivation looks suspicious (not to speak about the rigour): proper treatment of the quantum field theory requires a regularization scheme to "throw away" infinities so that resulting finite values could, in principle, depend on a chosen scheme. And third, these formulas are "too nice to be true".

Nevertheless at that time Zograf and I started to work on a project to understand the true status of these conjectures. Using methods of Teichmüller theory, i.e deformation theory of RS's, we were able not only to prove (20) but to establish a deep and unexpected connection between accessory parameters, classical action and complex geometry of Teichmüller and moduli spaces (see [8, 9, 10, 11, 18]). These results naturally unify Fuchsian uniformization of RS's, complex geometry of their moduli spaces and classical Liouville theory and could be viewed as a "justification" of CWI at the semi-classical level. In order to present these results in accessible and transparent form I will first remind you of some necessary facts of Teichmüller theory.

5. Teichmüller Theory and Geometry of the Space of Punctures

Let X be a RS of the type $(0,n)$; it is called *marked* if one specifies a particular canonical system of generators (up to an inner automorphism) of its fundamental group $\pi_1(X)$; i.e. a marked RS $[X]$ is a pair consisting of X and a given canonical system of generators. Two marked RS's are isomorphic if they are isomorphic as RS's and corresponding isomorphism maps one set of generators into the other (up to a conjugation). This notion is more restrictive than the usual one for unmarked RS's so that corresponding quotient space – called *Teichmüller space* $\mathbf{T}_{0,n}$ of RS's of type $(0,n)$ – covers moduli space $\mathbf{M}_{0,n}$ (as well as the space $\mathbf{W}_n$) and is much easier to describe. Namely, points in $\mathbf{T}_{0,n}$ – marked RS's – can be obtained from a given marked RS $[X]$ (a marked point in $\mathbf{T}_{0,n}$) by deformations of its *complex structure* via *quasi-conformal mappings*. The easiest way to describe this procedure is to use uniformization theorem and work with the universal covering $J : H \mapsto X \cong H/\Gamma$. If $(1,0)$-form dz on H corresponds to the usual complex structure on H, then all complex structures correspond to $dz + \mu d\bar{z}$, where $\mu \in L^\infty(H)$ is a $(-1,1)$-form on H with the property $\|\mu\|_\infty < 1$. Such μ are called *Beltrami differentials.* A fundamental theorem from the theory of quasi-conformal mappings (a "heavy-duty" theorem from "elliptical engineering") states that all complex structures on H are isomorphic, i.e. there exists a diffeomorphism $f : H \mapsto H$ satisfying

$$\frac{\partial f}{\partial \bar{z}} = \mu \frac{\partial f}{\partial z}$$

–so-called *Beltrami equation* on H. These complex structures on H can not, in general, be projected on a RS $X \cong H/\Gamma$ unless μ has the following transformation law

$$\mu(\gamma z)\frac{\bar{\gamma}'(z)}{\gamma'(z)} = \mu(z), \quad \gamma \in \Gamma,$$

i.e. is a Beltrami differential on X. Even in this case isomorphism f in general will not be "compatible" with the Γ-action (i.e. $\Gamma^\mu = f \circ \Gamma \circ f^{-1}$ will not be conjugate to Γ) yielding RS $[X^\mu] = H/\Gamma^\mu$ as a deformation of a given RS $[X]$. Infinitesimal deformations can be described by the *cohomology group* $H^1(X, TX)$ (this is the main result of the *Kodaira-Spencer deformation theory*). Using the *Hodge theory* (for the choice of Poincaré metric on X) one can identify this cohomology group with the linear space $H^{-1,1}(X)$ of *harmonic* Beltrami differentials on X. This naturally introduces a complex structure on the Teichmüller space $\mathbf{T}_{0,n}$ with the space $H^{-1,1}(X)$ as the holomorphic tangent space $T_X\mathbf{T}_{0,n}$ to $\mathbf{T}_{0,n}$ at the point $[X]$. The inner product in $H^{-1,1}(X)$, induced by the Hodge $*$-operator, gives rise to a Hermitian metric on $\mathbf{T}_{0,n}$, called *Weil-Petersson metric*, which turns out to be Kähler. The corresponding holomorphic cotangent space $T^*_X\mathbf{T}_{0,n}$ to $\mathbf{T}_{0,n}$ at the point $[X]$ is isomorphic to a certain subspace $H^{2,0}(X)$ of *holomorphic quadratic differentials* on X (i.e. holomorphic tensors of the type $(2,0)$). Thus Teichmüller spaces (for general RS's of the type (g,n)) are complex manifolds of dimension $3g-3+n$ with a natural Weil-Petersson Kähler structure. These basic results on complex geometry of Teichmüller spaces are due to Ahlfors and Bers (see [9, 10] and references therein).

In the case of RS's of the type $(0,n)$ one could considerably simplify this picture by working with the space of punctures $\mathbf{W}_n$ which can be considered as an "intermediate" moduli space: it is covered by Teichmüller space $\mathbf{T}_{0,n}$ and covers true moduli space $\mathbf{M}_{0,n}$ (see [8]). Namely, in this case linear space $H^{2,0}(X)$ consists of all meromorphic functions on $\bar{X} = \mathbf{P}^1$ with at most simple poles at the punctures $w_1, \ldots, w_{n-1}$, having the order $O(|w|^{-3})$ as $w \to w_n = \infty$. This space is $n-3$-dimensional and is spanned by the functions

$$P_i(w) = -\frac{1}{\pi}\left(\frac{1}{w-w_i} + \frac{w_i - 1}{w} - \frac{w_i}{w-1}\right), \quad i = 1, \ldots, n-3.$$

Corresponding Petersson inner product is defined by

$$\langle P, Q\rangle = \int_{\mathbf{C}} P(w)\bar{Q}(w)e^{-\phi(w,\bar{w})}d^2w, \quad P, Q \in H^{2,0}(X).$$

If $\{Q_i\}_{i=1}^{n-3}$ is the bi-orthogonal basis to $\{P_i\}_{i=1}^{n-3}$ with respect to this inner product, i.e. $\langle P_i, Q_j\rangle = \delta_{ij}$, then harmonic Beltrami differentials $M_i = e^{-\phi(w,\bar{w})}\bar{Q}_i(w)$ determine (via Beltrami equation on $\mathbf{C}$) quasi-conformal mappings $F[\varepsilon\mu_i]$ which infinitesimally "move" only one puncture w_i, $i = 1, \ldots, n-3$, while keeping other punctures fixed. Mappings $F(\varepsilon) = F[\varepsilon_1\mu_1 + \ldots + \varepsilon_{n-3}\mu_{n-3}]$ (for sufficiently small $\varepsilon \in \mathbf{C}^{n-3}$) produce all RS's $X^{\varepsilon\mu} = \mathbf{C} \setminus \{F[\varepsilon](w_1), \ldots, F[\varepsilon](w_{n-3}), 0, 1\}$ from the neighborhood of $[X] \in \mathbf{T}_{0,n}$ so that vector fields $\partial/\partial w_i$ at the point $[X] \in \mathbf{T}_{0,n}$ are represented by infinitesimal deformations

$$\dot{F}_i = (\frac{\partial}{\partial\varepsilon}F[\varepsilon\mu_i])|_{\varepsilon=0}.$$

They satisfy $\bar{\partial}$-problem

$$\frac{\partial}{\partial\bar{w}}\dot{F}_i = M_i \qquad (21)$$

and have the property $\dot{F}_i(w_j) = \delta_{ij}$ of moving only the given puncture w_i, $i = 1, \ldots, n-3$. This is how one can describe the "motion" of punctures (singular points of the Fuchsian equation (3)) in terms of infinitesimal quasi-conformal deformations. Finally the Weil-Petersson metric on $\mathbf{T}_{0,n}$ is defined by the following inner product in $T_X\mathbf{T}_{0,n}$

$$\langle\frac{\partial}{\partial w_i}, \frac{\partial}{\partial w_j}\rangle_{WP} = \int_{\mathbb{C}} M_i\bar{M}_j e^{\phi(w,\bar{w})} d^2w, \quad i,j = 1, \ldots, n-3. \qquad (22)$$

Remark 14 Quasi-conformal deformations $F[\varepsilon\mu_i]$ should be used in a proper derivation of CWI (19) from the functional integral (17). In fact, they move only a given puncture w_i and should be considered as a replacement of the transformations from the "non-existing" infinite-dimensional conformal group. Actually they were already present implicitly in the standard CFT derivation of CWI when one used conformal transformations which were regular inside a domain containing w_i and singular outside. This means that in fact these transformations are quasi-conformal mappings with Beltrami differential supported at the set of singularities! However quasi-conformal mappings with harmonic Beltrami differentials seem to be the most natural ones since they correspond to the vector fields $\partial/\partial w_i$.

Using these results we can now describe dependence of the classical Liouville field $\exp\phi(w; w_1, \ldots, w_{n-3})$ on the punctures. Namely, we have the following conservation law

$$\frac{\partial}{\partial w_i}(e^{\phi}) = \frac{\partial}{\partial w}(e^{\phi}\dot{F}_i), \qquad (23)$$

or $\phi_{w_i} + \phi_w\dot{F}_i + \dot{F}_{iw} = 0$, $i = 1, \ldots, n-3$.

The idea of the proof is very simple. Note that the function ϕ_{w_i} satisfies linearized Liouville equation

$$\frac{\partial^2\phi_{w_i}}{\partial w\partial\bar{w}} = \frac{1}{2}e^{\phi}\phi_{w_i}.$$

Using (6), (21) and relation $M_i = e^{-\phi}\bar{Q}_i$ with holomorphic $Q_i(w), w \in X$, it is easy to check that the function $-\phi_w\dot{F}_i - \dot{F}_{iw}$ satisfies the same linearized equation. From (7) it follows that these functions have the same behaviour near the punctures so that they actually coincide.

Using (23) one gets the following result (see [8, 9]).

Theorem 1 *Classical action S_{cl} is a smooth function on $\mathbf{W}_n$ and*

$$-\frac{1}{2\pi}\frac{\partial S}{\partial w_i} = c_i, \quad i = 1, \ldots, n-3. \tag{24}$$

This proves the formula for accessory parameters conjectured by Polyakov and Zamolodchikov.

Remark 15 The proof of (24) is rather straightforward and consists in using (7), (23), "triality" relation and the harmonic property of Beltrami differentials. Together with the Stokes' formula they yield

$$\frac{\partial S_{cl}}{\partial w_i} = \lim_{\epsilon\to 0}\int_{X_\epsilon} 2T_{cl}\ M_i\ d^2w + \pi\sum_{j=1}^{n-1}\dot{F}_{iw}(w_j) - \pi\dot{F}_{iw}(\infty) = -2\pi c_i.$$

This derivation should be compared with those of CWI (cf. Remark 14 as well).

It is possible to proceed further and relate accessory parameters with Weil-Petersson geometry. Namely, using the deformation technique indicated above one can prove the following (see [8, 9])

Theorem 2 *Accessory parameters $c_1, \ldots, c_{n-3}$ are smooth functions on $\mathbf{W}_n$ and*

$$\frac{\partial c_i}{\partial \bar{w}_j} = \frac{1}{2\pi}\langle\frac{\partial}{\partial w_i}, \frac{\partial}{\partial w_j}\rangle_{WP}, \quad i,j = 1, \ldots, n-3. \tag{25}$$

Remark 16 It follows from Theorems 1 and 2 that the function $-S_{cl}$ is the *potential* for the Weil-Petersson metric on $\mathbf{W}_n$ (and, therefore, on $\mathbf{T}_{0,n}$) which immediately proves that it is a Kähler metric. This establishes a rather unexpecting connection between the Fuchsian uniformization, semi-classical Liouville theory and Weil-Petersson geometry of Teichmüller spaces.

Remark 17 Since the space of punctures $\mathbf{W}_n$ is a Kähler manifold it carries Weil-Petersson symplectic structure with the symplectic form $\omega_{WP} = \frac{\sqrt{-1}}{2}\langle,\rangle_{WP}$. Let $\{,\}_{WP}$ be the corresponding Poisson bracket. Then Theorems 1 and 2 imply that functions c_i on $\mathbf{W}_n$ are in involution i.e. $\{c_i, c_j\}_{WP} = 0$ and $\{c_i, w_j\}_{WP} = \frac{\sqrt{-1}}{\pi}\delta_{ij}, \quad i,j = 1, \ldots, n-3$. Needless to say that these formulas together with (24) look like canonical transformations in Hamiltonian mechanics.

Remark 18 Theorems 1 and 2 admit nice geometric interpretation in terms of the line bundle $\lambda_{0,n} \mapsto \mathbf{M}_{0,n}$ (cf. [18]). Namely, since $\exp(\frac{1}{\pi}S_{cl})$ is a Hermitian metric in $\lambda_{0,n}$ (see Remark 11) it defines a canonical metric connection compatible with the holomorphic structure in $\lambda_{0,n}$. The $(1,0)$-form θ of this connection is given by $\theta = \frac{1}{\pi}\partial S_{cl}$, where ∂ stands for $(1,0)$-component of exterior differential on $\mathbf{M}_{0,n}$ so that (24) is equivalent to $\theta = -2\sum_{i=1}^{n-3} c_i\ dw_i$. Curvature form of this connection is given by $\Theta = \bar{\partial}\theta$ so that (25) implies $\Theta = -\frac{2\sqrt{-1}}{\pi}\omega_{WP}$.

Remark 19 An interesting and non-trivial analytic problem is to investigate the behaviour of the classical action and accessory parameters at the boundary of $\mathbf{W}_n$, i.e. when punctures w_i coincide. This problem was solved by Zograf in [18] (partial results were also obtained in [27]). He proved the following results:

$$S_{cl}(w_1,\dots,w_{n-3}) = \begin{cases} 2\pi\log|w-w_i| + O(1) & \text{as } w_i \to w_k, \quad k \neq n, \\ 2\pi\log|w_i| + O(1) & \text{as } w_i \to w_n = \infty. \end{cases} \tag{26}$$

for the asymptotics of classical action and similar (but slightly more involved) formulas for accessory parameters (see [18]). In particular they mean that the "classical conformal dimension" of $\exp(-\frac{1}{2\pi}S_{cl})$ is $\frac{1}{2}$ in perfect agreement with Fuchsian differential equation (3). Asymptotics for accessory parameters contain logarithmic terms and imply that the semi-classical excitations spectrum of the Liouville theory is continuous. These asymptotics are also important for evaluation of Weil-Petersson volumes of moduli spaces $\mathbf{M}_{0,n}$ (see [28]).

Remark 20 It is well-known that the Teichmüller space $\mathbf{T}_{0,n}$ is topologically trivial (it is a cell in $\mathbf{C}^{n-3}$) whereas the modular space $\mathbf{M}_{0,n}$ has a non-trivial homology. In particular the Weil-Petersson symplectic form ω_{WP} realizes a non-trivial cohomology class $[\omega_{WP}] \in H^2(\mathbf{M}_{0,n}, \mathbf{R})$. This explains why classical action S_{cl} does not descend to a single-valued function on $\mathbf{M}_{0,n}$ (see Remark 10): otherwise it would have given rise to a single-valued potential for the form ω_{WP} on the moduli space so that it would be zero in cohomology. Therefore a naive action for the Liouville theory should be divergent!

LECTURE 3

6. Liouville Theory on Compact Riemann Surfaces

Let X be a RS of the type $(g,0)$, i.e. a compact RS of genus $g > 1$; it can be uniformized by a strictly hyperbolic Fuchsian group Γ generated by $2g$ hyperbolic transformations $A_1,\dots,A_g,B_1,\dots,B_g$ satisfying a single relation

$$\prod_{i=1}^{g} A_i^{-1}B_i^{-1}A_iB_i = 1.$$

Corresponding covering $J : H \mapsto X$ gives rise to a linearly-polymorphic function J^{-1} on X. In order to associate it with a Fuchsian differential equation on X we need to introduce an important notion of a *projective connection*. Namely, let $\{U_\alpha, u_\alpha\}$ be a complex-analytic atlas on X with transition functions $u_\alpha = f_{\alpha\beta}(u_\beta)$ on $U_\alpha \bigcap U_\beta$. A collection $R = \{r_\alpha\}$, where r_α is a holomorphic function on U_α, is called a (holomorphic) projective connection on X if on each intersection $U_\alpha \bigcap U_\beta$ we have

$$r_\beta = (r_\alpha \circ f_{\alpha\beta})(f'_{\alpha\beta})^2 + S(f_{\alpha\beta}). \tag{27}$$

The difference of two projective connections is a (holomorphic) quadratic differential so that the set $P(X)$ of all projective connections on X is the *affine space* over the vector space $H^{2,0}(X)$ of quadratic differentials. Fuchsian uniformization of X yields a canonical projective connection $R^F \in P(X)$ defined by the formula $R^F = S(J^{-1}) = \{S_{u_\alpha}(J^{-1})\}$, where S_{u_α} is the Schwarzian derivative with respect to the local coordinate u_α in U_α. Transformation law (27) follows from the cocycle condition (see Remark 7).

With any projective connection $R \in P(X)$ one can associate a second-order linear differential equation on X

$$\frac{d^2 y_\alpha}{d{u_\alpha}^2} + \frac{1}{2} r_\alpha y_\alpha = 0, \tag{28}$$

where $y = \{y_\alpha\}$ is a (multi-valued) differential of order $-\frac{1}{2}$ on X, i.e. it has a transformation law

$$y_\beta = (y_\alpha \circ f_{\alpha\beta})(f'_{\alpha\beta})^{-\frac{1}{2}} \tag{29}$$

on $U_\alpha \bigcap U_\beta$. Differential equation (28) is covariant under the transformations (27) and (29) and gives rise to a monodromy representation $Mon : \pi_1(X) \mapsto G_{\mathbf{C}}$. Fuchsian projective connection is characterized uniquely by the property that its monodromy group is the Fuchsian group Γ (up to a conjugation) which uniformizes X.

Let $e^\phi |du|^2 = \{e^{\phi_\alpha}|du_\alpha|^2\}$ be the density of Poincaré metric on X; locally it satisfies Liouville equation

$$\frac{\partial^2 \phi_\alpha}{\partial u_\alpha \partial \bar{u}_\alpha} = \frac{1}{2} \exp \phi_\alpha,$$

which obviously has a covariant form, and yields a classical stress-energy tensor $T_{cl} = \phi_{uu} - \frac{1}{2}\phi_u^2 = \{\phi_{u_\alpha u_\alpha} - \frac{1}{2}\phi_{u_\alpha}^2\}$, which is holomorphic (due to Liouville equation) and represents a projective connection (due to the transformation law (12)). Moreover, we have the formula $T_{cl} = R^F$, which relates these two approaches to the uniformization theorem.

In addition to Fuchsian uniformization of compact RS's there exists a uniformization by Schottky groups. (This uniformization was used in the "old days" of the string theory). Namely, a *Schottky group* is a free finitely generated *strictly loxodromic Kleinian group* and it has a connected *domain of discontinuity* which *limit set* is a Cantor set. Let Σ be a Schottky group of rank $g > 1$ and let $\Omega \subset \mathbf{P}^1$ be its domain of discontinuity; corresponding quotient space Ω/Σ is a compact RS of genus g. The Schottky uniformization theorem states that any compact RS X can be represented as such quotient. Moreover, if in addition X is a marked RS (i.e. there is a distinguished canonical system of generators $\alpha_1, \ldots, \alpha_g, \beta_1, \ldots, \beta_g$ of the fundamental group $\pi_1(X)$) then one can choose Σ such that the group of the covering $J_\Sigma : \Omega \mapsto X$ coincides with the smallest normal subgroup N in $\pi_1(X)$ which contains the elements $\alpha_1, \ldots \alpha_g$. The group Σ is then isomorphic to N and is defined uniquely up to a conjugation in $G_{\mathbf{C}}$. This indicates the importance of *marked* Schottky groups, i.e. groups with a distinguished system of relation-free generators. If $[\Sigma] = (\Sigma; L_1, \ldots, L_g)$ is a marked Schottky group then one can construct its *fundamental domain* $D \subset \Omega$ which is bounded by $2g$ disjoint Jordan curves $C_1, \ldots, C_g, C'_1, \ldots, C'_g$ such that $C'_i = -L_i(C_i), i = 1, \ldots, g$ (where the minus sign means reversal of orientation). Therefore Schottky uniformization of RS X can be considered as its "dissection" along the "A-cycles" so that X is realized by the plane domain D bounded by $2g$ curves (circles for many cases) which are identified pair-wise.

Since each element L_i, $i = 1, \ldots, g$, can be represented in the normal form

$$\frac{L_i w - a_i}{L_i w - b_i} = \lambda_i \frac{w - a_i}{w - b_i}, \quad w \in \mathbf{P}^1,$$

where a_i and b_i are respectively *attracting* and *repelling* fixed points of the transformation L_i, $0 < |\lambda_i| < 1$, we can always assume that $a_1 = 0$, $b_1 = \infty$ and $a_2 = 1$. Such Schottky groups are called *normalized*; the mapping

$$(\Sigma; L_1, \ldots, L_g) \mapsto (a_3, \ldots, a_g, b_2, \ldots, b_g, \lambda_1, \ldots, \lambda_g) \subset \mathbf{C}^{3g-3}$$

establishes a one-to-one correspondence between the set of normalized marked Schottky groups and a certain set $\mathbf{S}_g$ in $\mathbf{C}^{3g-3}$. Actually $\mathbf{S}_g$ is a domain in $\mathbf{C}^{3g-3}$ and is called a *Schottky space.* The Schottky space $\mathbf{S}_g$ is complex-analytically covered by the Teichmüller space $\mathbf{T}_g$ which is defined as a space of all normalized marked Fuchsian groups of the type $(g, 0)$.

The Schottky uniformization $J_\Sigma : \Omega \mapsto X$ gives rise to a Schottky projective connection $R^S = \{S(J_\Sigma^{-1})\}$ on X. Fuchsian and Schottky uniformizations are related by the following commutative diagram

$$\begin{array}{ccccc} & & J_\Omega & & \\ H & & \longmapsto & & \Omega \\ J & \searrow & & \swarrow & J_\Sigma \\ & & X & & \end{array}$$

i.e. $J = J_\Sigma \circ J_\Omega$. Here J_Ω is a meromorphic function on H with the property $J \circ A_i = J$, $J \circ B_i = L_i \circ J$, $i = 1, \ldots, g$, i.e. it is the covering of the domain Ω by the universal cover H. In its terms the difference between Fuchsian and Schottky projective connections has the form $S(J^{-1}) - S(J_\Sigma^{-1}) = (S(J_\Omega^{-1}) \circ J_\Sigma^{-1})((J_\Sigma^{-1})')^2$ and represents the quadratic differential $R^F - R^S \in H^{2,0}(X)$.

Now let us discuss Lagrangian formulation of the Liouville theory. First of all note that the naive action $\tilde{S}(\phi)$ (see Remark 5) makes no sense again since the term $|\phi_u|^2$ does not represent a $(1,1)$-form on X (it is a logarithm of the density of a conformal metric on X instead) and therefore can not be integrated against X. There are two possibilities to avoid this difficulty. The first one is to choose a *background metric* $e^{\phi_0}|du|^2$ and write Liouville equation for the *conformal factor* $\sigma = \phi - \phi_0$; since σ is a function on X one can easily present a corresponding Lagrangian. However, while treating all compact RS's of genus g one needs to make a canonical choice of such a background metric. The best choice is provided by the Poincaré metric itself but in this case classical solution for the conformal factor is trivial: $\sigma = 0$ so that the classical action vanishes. Certainly this is not what we need when considering the Polyakov's functional integral. The second possibility is to realize a given RS X as a plane domain and use the advantage of the existence of a global coordinate. Again one can trivially choose the universal cover H to be such a domain. However in this case corresponding classical action will depend on the parameters of the Fuchsian group Γ and not on RS X directly. Therefore we will choose the "intermediate" case which is provided by Schottky uniformization: X is represented by a plane domain D and still there is "half of the geometry" left. We will see that this choice yields interesting results which a posteriori justify it.

Namely, let us consider a Poincaré metric on X as a Σ-invariant hyperbolic metric on Ω. It has a form $e^{\phi}|dw|^2$ and in terms of the map J_Ω can be written explicitly

$$\exp\phi(w,\bar{w}) = \frac{|(J_\Omega^{-1})'(w)|^2}{(\operatorname{Im} J_\Omega^{-1}(w))^2}, \quad w \in \Omega \subset \mathbf{P}^1. \tag{30}$$

The field ϕ is a real-valued function on Ω with the following transformation law

$$\phi(Lw) = \phi(w) - \log|L'(w)|^2, \quad L \in \Sigma, \quad w \in \Omega$$

and it satisfies

$$T_{cl} = \phi_{ww} - \tfrac{1}{2}\phi_w^2 = S(J_\Omega^{-1}).$$

Corresponding action functional for the Liouville equation on the domain Ω (for a chosen normalized marked Schottky group Σ) has the form

$$\begin{aligned} S(\phi) &= \int_D (|\phi_w|^2 + e^{\phi})d^2w - \frac{\sqrt{-1}}{2}\sum_{i=2}^{g}\int_{C_i}\phi\left(\frac{\bar{L}_i''}{\bar{L}_i'}d\bar{w} - \frac{L_i''}{L_i'}dw\right) \\ &+ \frac{\sqrt{-1}}{2}\sum_{i=2}^{g}\int_{C_i}\log(|L_i'|^2)\frac{\bar{L}_i''}{\bar{L}_i'}d\bar{w} + 4\pi\sum_{i=2}^{g}\log|l_i|^2. \end{aligned}$$

Here D is the fundamental domain for Σ associated with a chosen set of generators $L_1,\ldots,L_g$ in Σ, $\partial D = \bigcup_{i=1}^{g}(C_i \cup C_i')$ and l_i is the bottom-left matrix element of the transformation L_i, $i = 1,\ldots,g$. The boundary terms in the formula for $S(\phi)$ ensure that it is a correctly defined functional, i.e. it is independent of a particular choice of the fundamental domain D for a normalized marked Schottky group Σ. The functional $S(\phi)$ yields Liouville equation through the Euler-Lagrange equation $\delta S = 0$ and corresponding classical action gives rise to a single valued function $S_{cl} : \mathbf{S}_g \mapsto \mathbb{R}$ on the Schottky space.

Remark 21 There is another possibility for the Lagrangian formulation of the Liouville theory. Namely, one can rewrite Liouville equation in the form $R(\gamma) = -1$, where $R(\gamma)$ is a Gaussian curvature of a Riemannian metric $\gamma_{ab}dx^a dx^b$ on X ($u = x^1 + \sqrt{-1}x^0, \bar{u} = x^1 - \sqrt{-1}x^0$) so that it can be considered as a two-dimensional Einstein equation with the cosmological term. Though the standard Einstein action in two dimensions is dynamically trivial and leads (according to Gauss-Bonnet) to a topological invariant

$$\int_X R(\gamma)\sqrt{\gamma}d^2x = 4\pi(1-g)$$

the action

$$\int_X (R(\gamma)+1)^2\sqrt{\gamma}d^2x$$

gives rise to the Liouville equation. Though this functional is obviously generally covariant it yeilds to a vanishing classical action so that it is not useful for our approach.

Now we can formulate the analogs of the main results from Lecture 2. Namely, since for all marked RS's X the difference $R^F - R^S$ is a quadratic differential, i.e. a cotangent vector to the Teichmüller space $\mathbf{T}_g$ (as well as to the Schottky space $\mathbf{S}_g$) at the point $[X]$, we can interpret $R^F - R^S$ as a $(1,0)$-form on $\mathbf{T}_g$ (and on $\mathbf{S}_g$). The analog of Theorem 1 has the form

Theorem 3 *Let ∂ be the $(1,0)$-component of the exterior differential on the Schottky space $\mathbf{S}_g$. Then*

$$R^F - R^S = \frac{1}{2}\partial S_{cl}.$$

whereas the analog of Theorem 2 is the following

Theorem 4

$$\bar{\partial}(R^F - R^S) = -\sqrt{-1}\omega_{WP},$$

where $\bar{\partial}$ stands for the $(0,1)$-component of the exterior differential on Schottky space $\mathbf{S}_g$.

As a corollary we get that the function $-S_{cl}$ is a potential for the Weil-Petersson metric on $\mathbf{S}_g$ which gives another proof that this metric is Kähler. Detailed proofs of these results as well as references on Schottky spaces can be found in [10].

Remark 22 It can be shown that families of Fuchsian and Schottky projective connections (for all RS's of genus g) define the sections of the *affine bundle* $\mathbf{P}_g$ of projective structures over the Teichmüller space $\mathbf{T}_g$ (the fiber of this bundle over the point $[X] \in \mathbf{T}_g$ is the affine space $P(X)$). Using for these sections the same notations, i.e. $R^F, R^S : \mathbf{T}_g \mapsto \mathbf{P}_g$ and using the fact that the Schottky projective connection R^S depends holomorphically "on moduli", we can reformulate Theorem 4 as $\bar{\partial}R^F = -\sqrt{-1}\omega_{WP}$.

Remark 23 It should be instructive to compare action functionals for the Liouville theory in punctured and compact cases. In these cases naive action $\tilde{S}(\phi)$ makes no sense: either it diverges or its density can not be integrated over a RS X. Therefore in the former case one should regularize divergencies due to the sources and in the latter case one should introduce additional line integrals with possible interpretation as tension terms. It would be interesting to understand the physical meaning of these modifications of the naive action.

Remark 24 It is clear that potential $-S_{cl}$ does not descend to a single-valued function on the true moduli space $\mathbf{M}_g$ of compact RS's of genus g. However using standard arguments in sheaf theory one can associate with S_{cl} a certain line bundle $\lambda_S \mapsto \mathbf{M}_g$ over $\mathbf{M}_g$ so that $\exp(\frac{1}{12\pi}S_{cl})$ is a Hermitian metric in λ_S [18]. Then it is possible to prove that the line bundle λ_S endowed with this metric is isometrically isomorphic to the Hodge line bundle $\lambda_H \mapsto \mathbf{M}_g$ over $\mathbf{M}_g$ endowed with a Quillen's metric (see [18] for details).

Remark 25 Let Im τ be the imaginary part of the period matrix τ for the marked RS $[X]$ and let $\det\Delta$ be the determinant of the Laplace operator for the Poincaré metric on X. Using the "Belavin-Knizhnik theorem", i.e. a local index theorem for the families of $\bar{\partial}$ -operators on RS's (see [29]; for the proof using Teichmüller theory see [30, 31]), one can deduce ([18]) that on $\mathbf{S}_g$ there exists a holomorphic function F (which is defined up to a constant factor of modulus 1) such that

$$|F|^2 = \frac{\det \Delta}{\det \operatorname{Im} \tau} \exp(\frac{1}{12\pi} S_{cl}).$$

The function F can be interpreted as a holomorphic determinant $\det \bar{\partial}$ of the family of $\bar{\partial}$-operators on RS parametrized by the Schottky space $\mathbf{S}_g$ (see [18] and [32]).

Remark 25 Needless to say that one can easily generalize these results for the case of RS's of the type (g, n) which requires both modifications of the naive action: regularization of the action integral together with a plane domain provided by the Schottky uniformization.

Remark 26 Using methods of deformation theory and the basic facts from the spectral theory of automorphic functions one can generalize "Belavin-Knizhnik theorem" to the case of RS's of the type (g, n) (see [33, 34]). A novel feature which appears when $n > 0$ is that in the local index theorem there is a cuspidal defect which yields a new Kähler metric on the moduli space $\mathbf{M}_{g,n}$.

This closes our discussion of the Liouville theory and Polyakov's functional integral. I hope that the material presented should convince a reader that this approach is worthwhile and is justified at the semi-classical level. Contrary to other quantum field theories where semi-classical limit was quite well understood, the semi-classical limit of the quantum Liouville theory in Polyakov's formulation turns out to be quite non-trivial and novel. Certainly one can go further and analize higher-order expansions in $\hbar$, i.e. to consider "higher loops corrections". Though mathematically this analysis is rather interesting and yields an infinite family of holomorphic quadratic differentials defined in terms of the Poincaré metric (proper status of this family has not been understood yet), it becomes quite involved and it seems that this is not the best way to treat the functional integral (17) and the right approach is still to come.

Though this paragraph could serve as a conclusion to these lectures I will still add the final remark.

Remark 27 Following the well-known similarity between the gravity and the Yang-Mills theory (the former can be interpreted as the latter with a gauge group being a diffeomorphism group) it is instructive to compare the Liouville theory, where one deals with RS's, with the Wess-Zumino-Witten (WZW) theory, where one deals with flat vector bundles over a given RS. Results of this comparison are summarized in the table below (in order to avoid unnecessary technicalities we will consider the case of compact RS's only). The reader can find all necessary details and references in [35]; interrogation mark "?" indicates that corresponding results are not known. In particular, the analog of the Schottky uniformization for vector bundles seems to be quite a hard problem in algebraic geometry. On the other hand it is clear that the proper analog of the classical Liouville action for vector bundles is the classical WZW-action, i.e. the critical value of the famous Novikov-Witten's functional.

Table 1

"Liouville Theory"	"WZW-Theory"
Compact RS's X of genus $g > 1$	Stable holomorphic vector bundles E of rank k and degree 0 over a compact RS X
Projective connections $R = \{r_\alpha\} \in P(X)$, second-order Fuchsian differential equations on X: $\frac{d^2 y_\alpha}{du_\alpha^2} + \frac{1}{2} r_\alpha y_\alpha = 0, \quad y = \{y_\alpha\}$	Flat connections $A = \{A_\alpha\} \in A(E)$ compatible with the holomorphic structure in E, first-order Fuchsian differential equations in E: $\frac{dY_\alpha}{du_\alpha} + A_\alpha Y_\alpha = 0, \quad Y = \{Y_\alpha\}$
Monodromy representation: $Mon : \pi_1(X) \mapsto PSL(2,\mathbf{C})$	Holonomy representation: $\rho : \pi_1(X) \mapsto GL(k,\mathbf{C})$
Fuchsian uniformization: $X \cong H/\Gamma$, $\Gamma \in PSL(2,\mathbf{R})$, Γ is Fuchsian	Narasimhan-Seshadri theorem: $E \cong H \times \mathbf{C}^k/\rho$, $\rho(\Gamma) \subset U(k)$, ρ is irreducible
Fuchsian projective connection $R^F \in P(X)$, linear-polymorphic function J^{-1} $= y_1/y_2 : X \mapsto H$ with the property $l_\gamma(J^{-1}(x)) = \gamma(J^{-1}(x))$, where l_γ stands for the analytic continuation along the loop $[\gamma]$, $\gamma \in \Gamma$, around $x \in X$	Narasimhan-Seshadri connection $d + A_{NS} \in A(E)$, multi-valued section Y of the bundle End$E$ $Y : X \mapsto \mathrm{End}E = E \otimes E^*$ with the property $l_\gamma(Y(x)) = Y(x)\rho(\gamma)$, where l_γ stands for the analytic continuation along the loop $[\gamma]$, $\gamma \in \Gamma$, around $x \in X$
Liouville equation: $\phi_{u\bar{u}} = \frac{1}{2}e^\phi$ for the conformal metric $e^\phi \lvert du\rvert^2$ on X of constant negative curvature -1; its solution $\exp\phi = \lvert(J^{-1})'\rvert^2/(\mathrm{Im}\, J^{-1})^2$ in terms of linearly-polymorphic function J^{-1} (Fuchsian condition is important)	WZW-equation: $(G_u G^{-1})_{\bar{u}} = 0$ for the Hermitian metric $G\lvert du\rvert^2$ in E of zero curvature; its solution $G = YY^\dagger$ in terms of multi-valued section Y (unitarity condition is important)
Schottky uniformization: $X \cong \Omega/\Sigma$, Schottky projective connection $R^S \in P(X)$	? ?
Deformations of complex structure on X: Beltrami differentials $\mu \in H^{-1,1}(X)$, operator $\bar\partial + \mu\partial$, Beltrami equation $f_{\bar z} = \mu f_z$, $\Gamma^\mu = f \circ \Gamma \circ f^{-1}$, $X^\mu = H/\Gamma^\mu$	Deformations of complex structure in E: End$E$-valued $(0,1)$ -forms $M \in H^{0,1}(X, \mathrm{End}E)$, $\bar\partial$-operator $\bar\partial + M$, $\bar\partial$-problem $F_{\bar z} = FM$, $\rho^\mu = F\rho F^{-1}$, $E^\mu = H \times \mathbf{C}^k/\rho^\mu$
Teichmüller space of compact RS's of genus $g > 1$: $\mathbf{T}_g = \mathrm{Hom}(\pi_1(X), PSL(2,\mathbf{R}))^0/PSL(2,\mathbf{R})$, (all Fuchsian representations) Teichmüller moduli group Mod_g, Moduli space $\mathbf{M}_g = \mathbf{T}_g/Mod_g$	Moduli space of stable vector bundles of rank k and degree 0 over a compact RS X: $\mathbf{N}(0,k) = \mathrm{Hom}(\pi_1(X), U(k))^0/U(k)$, (all irreducible representations) - - - - - -
Complex structure: $T_{[X]}\mathbf{T}_g = H^{-1,1}(X)$, $T^*_{[X]}\mathbf{T}_g = H^{2,0}(X)$	Complex structure: $T_E\mathbf{N}(0,k) = H^{0,1}(X, \mathrm{End}E)$, $T^*_E\mathbf{N}(0,k) = H^{1,0}(X, \mathrm{End}E)$

"Liouville Theory"	"WZW-Theory"
Kähler structure: Weil-Petersson metric $\langle\mu_1,\mu_2\rangle_{WP}=\int_X \mu_1\bar{\mu}_2 e^{\phi}d^2u$, $\mu_1,\mu_2\in H^{-1,1}(X)$; - - - - - - Weil-Petersson symplectic form: $\omega_{WP}=\frac{\sqrt{-1}}{2}\langle,\rangle_{WP},\quad d\omega_{WP}=0$	Kähler structure: Natural metric (Narasimhan-Atiyah-Bott) $\langle M_1,M_2\rangle=\int_X M_1\wedge *M_2$, $M_1,M_2\in H^{0,1}(X,\mathrm{End}E)$; Hodge $*$-operator is determined by metrics G in E and e^{ϕ} on X; natural symplectic form: $\omega=\frac{\sqrt{-1}}{2}\langle,\rangle,\quad d\omega=0$
Fuchsian projective connection R^F as a section of the affine bundle $\mathbf{P}_g\mapsto\mathbf{T}_g$ of projective structures on X and as a "$\bar\partial$-potential" for ω_{WP} : $\bar\partial R^F=-\sqrt{-1}\omega_{WP}$	Narasimhan-Seshadri connection A_{NS} as a section of the affine bundle $\mathbf{A}(0,k)\mapsto\mathbf{N}(0,k)$ of affine connections in E and as a "$\bar\partial$-potential" for ω : $\bar\partial A_{NS}=-2\omega$
Schottky projective connection R^S as a holomorphic section of $\mathbf{P}_g\mapsto\mathbf{T}_g$: $\bar\partial R^S=0$	? ? ?
Classical Liouville action S_{cl} as a potential for the Weil-Petersson metric: $\bar\partial\partial S_{cl}=-2\sqrt{-1}\omega_{WP}$	? ? ?
Hermitian metric $\exp(\frac{1}{12\pi}S_{cl})$ in the line bundle $\lambda_S\mapsto\mathbf{M}_g$, its connection form $\theta=\frac{1}{6\pi}(R^F-R^S)$ and curvature form $\Theta=-\frac{\sqrt{-1}}{6\pi}\omega_{WP}$	? ? ? ?
Local index theorem for families of $\bar\partial$-operators on X : curvature form of a Quillen's metric is proportional to the Weil-Petersson form ω_{WP}; family $\{X\}$ is endowed with the Poincaré metric for all $[X]\in\mathbf{T}_g$	Local index theorem for families of $\bar\partial$-operators in $\mathrm{End}E$: curvature form of a Quillen's metric is proportional to the natural form ω; family $\{\mathrm{End}E\}$ is endowed with the metric induced by G for all $[E]\in\mathbf{N}(0,k)$

References

[1] A.Polyakov. *Quantum geometry of bosonic strings*, Phys.Lett. B **103** (1981), 207-210.

[2] F.Smirnoff, L.Takhtajan. *Towards a quantum Liouville theory with* $c>1$, University of Colorado at Boulder PAM preprint no. **12**, March 1990.

[3] J.-L.Gervais. *The quantum group structure of 2d gravity and minimal models. 1*, Commun. Math. Phys. **130** (1990), 257-284.

[4] J.-L.Gervais. *Solving the strongly coupled 2d gravity: 1.Unitary truncation and quantum group structure*, Commun. Math. Phys. **138** (1991), 301-338.

[5] A.Belavin, A.Polyakov and A.Zamolodchikov. *Infinite conformal symmetry group in two-dimensional quantum field theory*, Nucl.Phys. B **241** (1984), 333-380.

[6] A.Polyakov. *Lecture at Steklov Institute in Leningrad*, 1982, unpublished.

[7] A.Polyakov, A.Zamolodchikov, *1982-1984 Seminars*, unpublished.

[8] P.Zograf, L.Takhtajan. *The Action of Liouville equation is a generating function for the accessory parameters and a potential of the Weil-Petersson metric on Teichmüller space*, Funktional Anal. i Prilozhen. **19** (1985), no.3, 67-68 (in Russian); English transl. in Funktional Anal. Appl. **19**, (1985).

[9] P.Zograf, L.Takhtajan. *On Liouville equation, accessory parameters, and the geometry of Teichmüller space for Riemann surfaces of genus 0*, Mat. Sb. **132** (1987), no.2, 147-166 (in Russian); English transl. in Math. USSR Sb. **60** (1988).

[10] P.Zograf, L.Takhtajan. *On uniformization of Riemann surfaces and the Weil-Petersson metric on Teichmüller and Schottky spaces*, Mat. Sb. **132** (1987), no.3, 304-327 (in Russian); English transl. in Math. USSR Sb. **60** (1988).

[11] L.Takhtajan. *Uniformization, local index theorem, and geometry of the moduli spaces of Riemann surfaces and vector bundles*, Proc. of Symposia in Pure Math. **49** (1989) 581-596.

[12] D.Friedan, S.Shenker. *The analytic geometry of two-dimensional conformal field theory*, Nucl. Phys. B **281** (1987), 509-545.

[13] H.Farkas, I.Kra. *Riemann Surfaces*, Springer-Verlag, 1980.

[14] F.Klein. *Neue beiträge zur Riemann'schen functiontheorie*, Math. Ann. **21** (1883), 201-312.

[15] A.Poincaré. *Sur les groupes des équations linéaries*, Acta Math. **4** (1884), 201-312.

[16] C.Gómez, G.Sierra. *Integrability and uniformization in Liouville theory: the geometrical origin of quantized symmetries*, CERN preprint, October 1990.

[17] A.Poincaré. *Les fonctions Fuchsiennes et l'equations* $\Delta u = e^u$, J. Math. Pures Appl. (5) **4**, (1898), 157-230.

[18] P.Zograf. *The Liouville action on moduli spaces and uniformization of degenerating Riemann surfaces*, Algebra and Analysis **1** (1989), no.4, 136-160 (in Russian); English transl. in Leningrad Math. J. **1** (1990).

[19] N.Seiberg. *Notes on quantum Liouville theory and quantum gravity*, Rudgers University preprint RU-90-29, June 1990.

[20] Vl.Dotsenko, V.Fateev. *Conformal algebra and multipoint correlation functions in 2d statistical models*, Nucl. Phys. B **240** (1985), 312-348.

[21] Vl.Dotsenko, V.Fateev. *Four-point correlation functions and the operator algebra in 2d conformal invariant theories with central charge* $c < 1$, Nucl. Phys. B **251** (1985), 691-734.

[22] V.Fateev, *Private communication*, 1991.

[23] A.Polyakov. *Private communication*, 1991.

[24] D.Friedan, Z.Qui and S.Shenker. *Conformal invariance, unitarity, and critical exponents in two dimensions*, Phys. Rev. Lett. **52** (1984), 1575-1578.

[25] A.Zamolodchikov, V.Fateev. *Nonlocal (parafermion) currents in two-dimensional conformal quantum field theory and self-dual critical points in* $\mathbf{Z}_N$*-symmetric statistical systems*, Zh. Eksp. Teor. Fiz. **89** (1985), 380-399 (in Russian); English transl. in Soviet Phys. JETP **62** (1985), 215-225.

[26] A.Zamolodchikov, V.Fateev. *Disorder fields in two-dimensional conformal quantum field theory and* $N = 2$ *extended sypersymmetry*, Zh. Eksp. Teor. Fiz. **90** (1986), 1553-1566 (in Russian); English transl. in Soviet Phys. JETP **63** (1986), 913-919.

[27] I.Kra. *accessory parameters for punctured spheres*, Trans. of the Amer. Math. Soc. **313** (1989), 589-617.

[28] P. Zograf. *The Weil-Petersson volume of the moduli space of punctured spheres*, (1991), to appear.

[29] A.Belavin, V.Knizhnik. *Algebraic geometry and the geometry of quantum strings*, Phys. Lett. B **168** (1986), 201-206.

[30] P.Zograf, L.Takhtajan. *Local version of the index theorem for the families of* $\bar{\partial}$*-operators on Riemann surfaces*, Uspechi Mat. Nauk **42** (1987), no.6, 133-150 (in Russian); English transl. in Russian Math. Surveys **42**:6 (1987).

[31] P.Zograf, L.Takhtajan. *A potential for the Weil-Petersson metric on Torelli space*, Zap. Nauch. Semin. LOMI **160** (1987), 110-120 (in Russian); English transl. in J. Soviet Math.

[32] V.Knizhnik. *Multiloop amplitudes in the theory of quantum strings and complex geometry. Riemann surfaces as branched covers*, preprint **ITF-87-62R**, Kiev, 1987 (in Russian).

[33] L.Takhtajan, P.Zograf. *The Selberg zeta function and a new Kähler metric on the moduli space of punctured Riemann surfaces*, J. Geom. Phys. **5**(4) (1988), 551-570.

[34] L.Takhtajan, P.Zograf. *A Local Index theorem for families of* $\bar{\partial}$*-operators on punctured Riemann surfaces and a new Kähler metric on their moduli spaces*, Commun. Math. Phys. **137** (1991), 399-426.

[35] P.Zograf, L.Takhtajan. *On the geometry of moduli spaces of vector bundles over a Riemann surface*, Izvestia Akad. Nauk SSSR, ser. matem. **53** (1989), no.4, 753-770 (in Russian) English transl. in Math. USSR Izvestiya, **35** (1990).

INTEGRABILITY PROPERTIES OF THE COLLECTIVE STRING FIELD THEORY

J. Avan

Brown University, Physics Department
Providence, RI, 02912, USA

1. Introduction

The collective coordinate formalism[1] was recently applied to the formulation of matrix models restricted to the singlet sector, i.e. the dynamics of the eigenvalues of the matrix variable. In particular, Das and Jevicki[2] considered the case of a 1-dimensional matrix model, introduced as a discretized version of a one-dimensional string theory[3]. The model is defined by the Lagrangian:

$$\mathcal{L}(M) = Tr\left[\frac{1}{2}\dot{M}(t)^2 - v(M)\right] \tag{1}$$

where M is a $N \times N$ hermitean matrix and v is a (polynomial) potential. The quantum theory is defined by the corresponding Hamiltonian:

$$\mathcal{H} = Tr\left(-1/2\frac{\partial^2}{\partial M^2} + v(M)\right) \tag{2}$$

where $\partial^2/\partial M^2$ is the hermitean-matrix Laplacian. Restricting the study to the singlet sector, one introduces a basis of vertex operators $\phi_m = Tr\ M^m$ to describe the dynamics of this sector. Das and Jevicki defined a collective field variable $\phi_k = Tr(e^{ikm})$, highly redundant in terms of the original matrix singlet variables, but relevant in the continuum, double scaling limit when one recovers the string field theory. In terms of eigenvalues:

$$\phi_k = \sum_{i=1}^{N} e^{ik\lambda_i} \iff \phi(x) = \int e^{-ikx}\,\phi_k\,dK = \sum_{i=1}^{N}\delta(x - \lambda_i(t)) \tag{3}$$

The dynamics of $\phi(x)$ is described by the hermitean field-theoretical Hamiltonian:

New Symmetry Principles in Quantum Field Theory, Edited by
J. Fröhlich et al., Plenum Press, New York, 1992

$$H = \int dx \left\{ \frac{1}{2} \partial_x \Pi \cdot \phi \cdot \partial_x \Pi + v(x)\, \phi(x) - \mu_F \left(\phi(x) - \frac{N}{v} \right) + \frac{1}{2} \dot{\phi}(x) \left[\int dy \, \frac{\phi(y)}{x-y} \right]^2 \right\}. \tag{4}$$

Here $\Pi(x)$ is the canonical conjugate momentum to $\phi(x)$; μ_F is the Fermi momentum implementing the constraint $\int \phi \, dx = N$; integration limits are taken to be the turning points of the potential $v(x) - \mu_F$. The last, non-local term in H is then expanded in powers of N^{-1} after re-establishing suitable dimensionalities. The leading order gives a local potential of the form $V_{\text{eff}}(\phi) = \int \frac{\pi^2}{6} \phi^3(x)\, dx$, plus non-local subleading terms. We consider here the local, leading order of the Hamiltonian; the effects of non-local terms were studied in Ref. 4.

This leading-order Hamiltonian was recently thoroughly studied in perturbation theory[5], and several results on correlation functions lead to believe that some form of integrability be present at the quantum level. Such a conjecture is backed by the equivalence between matrix models and free fermions in an external potential[6], and also the relation with Toda-type lattice models (before the double scaling limit)[7].

2. Classical Integrability[10]

There exists a far-reaching relationship between the string field model and the well-known integrable N-body Calogero model[8]. Reintroducing the field ϕ as (3) inside the leading-order local Hamiltonian:

$$H_e = \int \left(\frac{1}{2} \Pi_{,x} \cdot \phi \cdot \Pi_{,x} + \frac{\pi^2}{6} \phi^3 + (v(x) - \mu_F)\phi \right) dx \tag{5}$$

leads indeed to the Calogero hamiltonian:

$$H_{\text{cal}} = \sum_{i=1}^{N} \frac{p_i^2}{2m} - \sum_{i \neq j} \frac{1}{(x_i - x_j)^2} + v(x_i) \quad ; \quad p_i \simeq \dot{x}_i \tag{6}$$

when $v(x)$ is taken to be $-x^2$. This hamiltonian was shown to be classically[9] and quantum-mechanically[8b] integrable. Moreover the potential $v(x) \simeq x^2$ exhibits a remarkable feature, i.e. background independence. Defining canonically transformed fields:

$$\varphi(q, \tau) = \phi(x, t)\, ch\, t \qquad \Pi(q, \tau) = \Pi(x, t) \tag{7a}$$

$$q = x \cdot ch^{-1} t \quad ; \quad \tau = th\, t \tag{7b}$$

leads to a transformed Hamiltonian equivalent to (5) but without the $v(x) = -x^2$ term. This allows in particular a construction of classical solutions for the full, $-x^2$-potential case, starting from the simpler 0-potential case[10,11].

Finally the Hamiltonian (5) exhibits an infinite number of Poisson-commuting conserved charges, for any potential $v(x)$, making it (formally) Liouville-integrable[10]. This arises from considering a fundamental algebraic structure of the theory, obtained as follows.

Defining the two dynamical fields:

$$\alpha_{\pm}(x,t) = \Pi,x \pm \pi\, \phi(x,t) \tag{8}$$

with the Poisson structure of a classical $U(1) \times U(1)$ current algebra:

$$\{\alpha_{\pm}(x),\ \alpha_{\pm}(y)\} = \pm\, i\pi\, \delta'(x-y) \quad ; \quad \{\alpha_{+}, \alpha_{-}\} = 0 \tag{9}$$

and assuming that $\alpha_{\pm}(x)$ vanish fast enough at the boundaries which we implicitly consider, we introduce densities $h^{n}_{m}(x,t) = x^{m-1}\frac{\alpha_{\pm}^{m-n}}{m-n}$ and corresponding integrated charges:

$$H^{n(\pm)}_{m} = \int dx\, x^{m-1}\, \frac{\alpha_{\pm}^{m-n}}{m-n} \tag{10}$$

The Poisson brackets of $H^{(\pm)}$, up to boundary terms which we assume to vanish, given the above boundary conditions, describe a classical w_{∞}-algebra[12]:

$$\{H^{n_1}_{m_1},\, H^{n_2}_{m_2}\} = (m_2-1)n_1 - (m_1-1)n_2 \cdot H^{n_1+n_2}_{m_1+m_2-2} \tag{11}$$

This algebra was identified in a number of problems related to integrable systems and KP hierarchy[13], matrix models[14], and more recently in the Wilson-loop formulation of 2-d gravity[15], confirming our previous result for the classical case[10]. Note that the infinitesimal canonical transformations generated by the densities h^{n}_{m} build an exact w_{∞}-algebra unspoilt by any boundary effect.

The Liouville integrability of H_e follows from this algebraic structure. For a vanishing potential $v(x)$, $H_e \equiv H^{-2}_{1} + H^{0}_{1}$ hence it commutes with all $\{H^{n}_{1}\}$ generators. For a general polynomial potential $v(x)$, one shows similarly that the generators:

$$Q^{(m)}_{\pm} \equiv \int dx \int d\alpha\, \left(\alpha^{2}_{\pm} + 2v(x)\right)^{m} \tag{12}$$

commute with each other, generating thereby a hierarchy of (classically integrable) Hamiltonians. Since α_{+} and α_{-} commute, this construction actually gives rise to two commuting w_{∞}-algebras and hamiltonians can be taken as any combination inside the two Cartan subalgebras ($H_e \equiv Q^{+(1)} - Q^{-(1)}$, for instance). Let us finally note that besides this Cartan subalgebra, one also constructs a Virasoro subalgebra $\{L_n \sim H^{n}_{2}\}$ with 0-central charge, the first three generators of which play an important role in solving the Calogero model[9]. Although the classical Liouville integrability holds therefore for any potential $v(x)$ and not only the Calogero potential, we shall see that $v = -x^2$ nevertheless plays a privileged role when one studies the effective integrability of the quantum theory.

3. Quantum Integrability[17]

The classical w_∞-algebra structure smoothly converts into a quantum w_∞-algebra as can be explicitly checked from the commutation relations of $\alpha_\pm(x)$ and the symmetry properties of the polynomial expressions obtained when computing commutators of quantum operators:

$$H^n_m \equiv \int dx\, x^{m-1} \frac{\alpha^{m-n}}{m-n} \quad ; \quad \left[H^{n_1}_{m_1}\, ,\, H^{n_2}_{m_2}\right] = \Big((m_2-1)n_1-(m_1-1)n_2\Big)\, H^{n_1+n_2}_{m_1+m_2-2} \tag{13}$$

In this way one gets two copies of a w_∞-Lie algebra, and the classical conserved Poisson-commuting quantities (H^n_1) have natural quantum commuting counterparts. Furthermore, one is able to construct exact eigen-operators of the Hamiltonian $H_e = H^{-2}_1 + H^2_3$ with the potential $v(x) = \pm x^2$. This particular potential is indeed the only one which allows solutions of the diagonalizing equation

$$[H_e, \vartheta] = E(\vartheta)\cdot\vartheta \quad , \quad \vartheta \in w_\infty \tag{14}$$

such that ϑ is represented by a one-index sum $\sum_{n\in N} c(n)\, H^{q+2n}_{q+n}$ instead of the generic 2-indices sum $\sum_{n,m} c(n,m)\, H^n_m$. This degeneracy of the diagonalizing equation (13) allows to solve it for all finite series of the above type. (However, it is possible to construct explicit eigenoperators for more complicated potentials such as $x^2 + \gamma/x^2$, [20] the presence of the x^2 term being crucial anyway.)

One first constructs an algebraic basis $\{B^{(n)}\}$, generalizing the Perelomov-Calogero B-operators,[8b] defined as

$$B^{(n)}_\pm = \int \left(\alpha_\pm \pm \sqrt{2}x\right)^n dx \tag{15}$$

The commutator of $B^{(n)}_\pm$ with the Hamiltonian follows from shifting α to $\alpha_\pm\sqrt{2}x$ in the commutation relations (13) and in the Hamiltonian itself, transforming it from $H^{-2}_1 + H^2_3$ to $\tilde{H}^{-2}_1 + \tilde{H}^0_2$. One gets:

$$\left[H_e,\; B^{(n)}_\pm\right] = \pm\, i\sqrt{2}n\, B^{(n)}_\pm \tag{16}$$

The main result is that all finite-length diagonalizing operators ϑ are obtained from commutators of $B^{(n)}$-type operators:

$$\left[B^{(n)}_+\, ,\, B^{(m)}_-\right] \equiv B^{(n,m)} \qquad [B_+\, ,\, B_+] = [B_-\, ,\, B_-] = 0 \tag{17}$$

We note in particular that the $B^{(n)}$ themselves are such operators:

$$B^{(n,2)} \equiv B^{(n-1)}_+\, , \quad B^{(2,n)} \equiv B^{(n-1)}_- \tag{18}$$

These operators generate imaginary eigenvalued, zero-norm states as follows from their commutation relation (16) and the Jacobi identity:

$$\left[B^{(n,m)}\, ,\, H\right] = i\sqrt{2}\,(n-m)\, B^{(n,m)} \tag{19}$$

Such states indeed appear as imaginary poles in the vertex functions computed perturbatively in Ref. 5. They may be identified with zero-norm states described in Ref. 16, but this identification needs further clarification.

Another way to understand these states is as dual to the real-eigenvalued, positive-norm tachyon states. They can indeed be written as infinite combinations of such states, once the original dynamical fields $\alpha_{\pm}(x)$ are expanded in an oscillator basis[5].

$$\alpha_{+}(x) = \sum_{q=1}^{\infty} \frac{2\pi}{\sqrt{2}L} \sqrt{q} \left(e^{\mp\left(iq\frac{\pi}{L}\tau + iq\pi\right)} a_q + \text{complex conjugate} \right) \tag{20}$$

This "duality" relation also needs further clarification; it was established only at a formal level and one should get beyond that to use this duality to compute exact correlation functions and confront the result with perturbative answers. In some sense, this relation looks like a simplified version of the inverse scattering formalism, using back-and-forth "non-linear Fourier transformations" to go from free to interacting exact oscillators.

The $B^{(n,m)}$ operators build an alternative structure[12a] of w_∞-Lie algebra with different structure constants compared to (12):

$$\left[\tilde{B}^{(r_1,\, s_1)},\, \tilde{B}^{(r_2,\, s_2)}\right] = (r_2 s_1 - r_1 s_2)\ \tilde{B}^{(r_1+r_2,\, s_1+s_2-2)} \tag{21a}$$

$$\tilde{B}^{(r,s)} \equiv B\left(n = \frac{r+s}{2}\,,\ m = \frac{s-r}{2}\right) \tag{21b}$$

Note that, again, one has two commuting w_∞-algebras, corresponding to the choice of α_+ or α_- in $B^{(n)}$. This precise alternative w_∞-structure arises in the operator product expansion of chiral vertex operators for the first-quantized $d = 1$ string theory[18,19]. This indicates a very interesting and close relation between this first-quantized and our second-quantized formulation. Moreover the w_∞-algebra appears now as the major underlying structure of the $c = 1$, $d = 1$ string field theory.

Acknowledgements

This work was done in collaboration with A. Jevicki at Brown University; I wish to thank G. Mack and P. Mitter for their kind hospitality at Cargese Institute of Scientific Studies, and K. Demeterfi and S. Shenker for fruitful discussions.

References

1. A. Jevicki and B. Sakita, *Nucl. Phys.* **B165** (1980), 511.
2. S. Das and A. Jevicki, *Mod. Phys. Lett.* **A5** (1990), 1639.
3. S. Das, A. Dhar and S. Wadia, *Mod. Phys. Lett.* **A5** (1990), 799.
4. A. Jevicki, Brown Preprint HET-**807** (1991).
5. K. Demeterfi, A. Jevicki and J. Rodrigues, *Nucl. Phys.* **B362** (1991), 125 and Brown Preprint HET-**803** (1991).

6. a. D. Gross and I. Klebanov, *Nucl. Phys.* **B352** (1991), 671;
   b. I. Klebanov, Trieste Lectures 1991, Preprint PUPT 1271.

7. a. L. Alvarez-Gaumé, C. Gomez and J. Lacki, CERN-Th 5875/90 (1990);
   b. C. Ahn and K. Shigemoto, *Phys. Lett.* **B263** (1991), 44;
   c. E. Martinec, *Comm. Mat. Phys.* **138** (1991), 437.

8. a. F. Calogero, *Journ. Mat. Phys.* **12** (1971), 419;
   b. A.M. Perelomov, *Sov. Journ. Part. Nucl.* **10** (1979), 336.

9. G. Barrucchi, T. Regge, *Journ. Mat. Phys* **18** (1977), 1149 and references therein.

10. J. Avan and A. Jevicki, *Phys. Lett.* **B266** (1991), 35.

11. a. J. Polchinski, *Nucl. Phys* **B362** (1991), 125;
    b. D. Minic, J. Polchinski and Z. Yang, Texas UTTG **16/91** (1991).

12. a. C.N. Pope, L. J. Romans and X. Shen, *Phys. Lett.* **B236** (1990), 173;
    b. I. Bakas, *Phys. Lett.* **B228** (1989), 57.

13. a. M. Awada and S.J. Sin, Florida HEP **91-3** (1991)/**90-33** (1990)
    b. K. Yamagishi, *Phys. Lett.* **B259** (1991), 436.

14. a. M. Fukuma, A. Kawai and R. Nakayama, KEK Preprint TH-**251** (1990)
    b. R. Dijkgraaf, H. and E. Verlinde, *Nucl. Phys.* **B348** (1991), 435.

15. G. Moore and N. Seiberg, Rutgers Preprint RU **91-29** (1991).

16. D. Gross and I. Klebanov, *Nucl. Phys.* **B359** (1991), 3.

17. J. Avan and A. Jevicki, Brown HET-**824** (1991).

18. E. Witten, IASSNS-HEP **91-51** (Aug. 1991).

19. I. Klebanov and A.M. Polyakov, (to be published).

20. J. Avan and A. Jevicki, in preparation.

BRST ANALYSIS OF PHYSICAL STATES FOR $2D$ (SUPER) GRAVITY COUPLED TO (SUPER) CONFORMAL MATTER

Peter Bouwknegt[1], Jim McCarthy[2] and Krzysztof Pilch[3]

[1] CERN-TH, CH-1211 Geneva 23, Switzerland

[2] Dept. of Phys., Brandeis University, Waltham, MA 02254, USA

[3] Dept. of Phys., U.S.C., Los Angeles, CA 90089-0484, USA

ABSTRACT

We summarize some recent results on the BRST analysis of physical states of $2D$ gravity coupled to $c \leq 1$ conformal matter and the supersymmetric generalization.

1. INTRODUCTION

If the continuum theory of $2D$ gravity can be developed to the level achieved by matrix models, the combined approaches are expected to provide great insights into generally covariant quantum systems – one of particular interest being the critical "$2D$ string." A further benefit would be progress in $2D$ supergravity models, where matrix models seem difficult to apply. The continuum theory in the conformal gauge [1,2] has already had some success, in particular the computation of the spectrum of physical states employing a BRST analysis [3], as we will review here along the lines of [4]. One of the new results presented in this paper is the extension of these calculations to $N = 1$ supergravity models. Another is the result for representations on the boundary of the Kac table in a given Virasoro minimal model coupled to gravity. There is a large literature discussing BRST cohomology and/or $2D$ gravity which has greatly influenced the material presented here. Due to space limitation we must refer the reader to [4] for a more complete account of these references.

Mathematically, the problem is that of computing the cohomology of the BRST operator

$$d = \oint \frac{dz}{2\pi i} :(T^M(z) + T^L(z) + \tfrac{1}{2}T^G(z))c(z):\,, \tag{1.1}$$

on the tensor product $\mathcal{L}^M \otimes \mathcal{F}^L \otimes \mathcal{F}^G$ of modules of the Virasoro algebra (see *e.g.* [3]). Here, representing the matter sector, $\mathcal{L}^M$ is an irreducible highest weight module

New Symmetry Principles in Quantum Field Theory, Edited by
J. Fröhlich et al., Plenum Press, New York, 1992

$\mathcal{L}(\Delta^M, c^M)$ for $c^M < 1$, or a free scalar field Fock space $\mathcal{F}(p^M)$ for the $c^M = 1$ string. The Fock space $\mathcal{F}^L \equiv \mathcal{F}(p^L, Q^L)$ of a free scalar field with a background charge Q^L (*i.e.* a Feigin-Fuchs module) represents the Liouville mode, and $\mathcal{F}^G$ is the Fock space of the spin $(2,-1)$ bc-ghosts. Finally, $T^M(z) + T^L(z)$, is the stress energy tensor of the matter and the Liouville fields, while $T^G(z)$ is that of the ghosts. We will refer to the cohomology of d as the BRST cohomology of $\mathcal{L}^M \otimes \mathcal{F}^L$.

Any irreducible highest weight module of the Virasoro algebra admits a resolution in terms of Feigin-Fuchs modules [5,6]. Thus we may first compute the BRST cohomology of a product of two Feigin-Fuchs modules and then use a suitable resolution to project from the free field Fock space onto the irreducible representation in the matter sector. For the $c^M = 1$ string the second step is of course unnecessary.

It is particularly interesting, as discussed above, to study this problem in the supersymmetric category. One then considers the BRST cohomology of a super-Virasoro matter module coupled to a super-Liouville system.

This paper is organized as follows. In Section 2 we briefly summarize the results of [7,4] on the BRST cohomology of a product of two Feigin-Fuchs modules, and discuss explicit representatives of this cohomology in the case $c^M = 1$. Then, in Section 3, we use recently constructed resolutions to determine the cohomology for all matter modules with $c^M \leq 1$. Finally, in Section 4 we give several new results for $2D$ supergravity coupled to superconformal matter.

2. THE COHOMOLOGY OF TWO FEIGIN-FUCHS MODULES

The Feigin-Fuchs module $\mathcal{F}(p, Q)$ is the Fock space of a scalar field, $\phi(z)$, with a background charge Q and momentum p. We take (see [4]) $\langle \phi(z)\phi(w) \rangle = -\ln(z-w)$, $i\partial\phi(z) = \sum_{n\in\mathbb{Z}} \alpha_n z^{-n-1}$ and $T(z) = -\frac{1}{2} : \partial\phi(z)\partial\phi(z) : +iQ\partial^2\phi(z)$. The Fock space $\mathcal{F}(p, Q)$ has vacuum $|p\rangle$, $\alpha_0|p\rangle = p|p\rangle$. As a Virasoro module, $\mathcal{F}(p, Q)$ has central charge $c = 1 - 12Q^2$ and conformal dimension $\Delta(p) = \frac{1}{2}p(p - 2Q)$. We will distinguish between the Liouville and matter fields by writing superscripts L and M respectively.

We will study the cohomology of the BRST operator (1.1) on $\mathcal{F}(p^M, p^L) \equiv \mathcal{F}(p^M, Q^M) \otimes \mathcal{F}(p^L, Q^L) \otimes \mathcal{F}^G$. The ghost number (gh) is normalized such that the physical vacuum $|p^M, p^L\rangle \equiv |p^M\rangle \otimes |p^L\rangle \otimes |0\rangle_G$ has (gh) = 0, where the ghost vacuum $|0\rangle_G$ is annihilated by the modes c_n, b_n, $n \geq 1$ and by b_0. The nilpotency of d requires that the total central charge is zero, and thus $(Q^M)^2 + (Q^L)^2 = -2$. On defining $\alpha_\pm = (Q^M \pm iQ^L)/\sqrt{2}$ we then have $\alpha_+\alpha_- = -1$. Clearly, at this point the free parameters in the problem are α_+ (or, equivalently, α_-) determined by the background charges, and the momenta p^M and p^L. Let us parametrize the latter in terms of $r, s \in \mathbb{C}$ as follows

$$p^M - Q^M = \sqrt{\tfrac{1}{2}}(r\alpha_+ + s\alpha_-), \qquad i(p^L - Q^L) = \sqrt{\tfrac{1}{2}}(r\alpha_+ - s\alpha_-). \tag{2.1}$$

The BRST operator can be expanded into the ghost zero modes, $d = c_0 L_0 - b_0 M + \widehat{d}$, where $L_0 = \{d, b_0\}$ is the energy operator. By a standard argument the cohomology of d must be contained in the zero eigenspace of L_0. Moreover, the subspace of $\mathcal{F}(p^M, p^L)$ annihilated by L_0 *and* b_0 is invariant under d, which reduces there to $\hat{d}$. The relative cohomology of $\mathcal{F}(p^M, p^L)$ is defined as the cohomology of d on $\mathcal{F}(p^M, p^L) \cap \operatorname{Ker} L_0 \cap \operatorname{Ker} b_0$, as opposed to the absolute cohomology which refers to the entire space. We will denote them by $H^{(*)}_{rel}(\cdot, d)$ and $H^{(*)}_{abs}(\cdot, d)$, respectively.

In [4] (see also [7]) we proved the following theorem which summarizes all possible cases with nontrivial relative cohomology.

Theorem 2.1. *We can distinguish three different cases listed below in which the relative cohomology $H^{(*)}_{rel}(\mathcal{F}(p^M, p^L), d)$ is nontrivial. For given Q^M and Q^L, or, equivalently, α_+ and α_- satisfying $\alpha_+\alpha_- = -1$, they depend on discrete values of the momenta p^M and p^L parametrized by r and s as in (2.1).*

i) If either $r = 0$ or $s = 0$ then

$$H^{(n)}_{rel}(\mathcal{F}(p^M, p^L), d) = \begin{cases} \mathbb{C} & \text{for } n = 0\,, \\ 0 & \text{otherwise}\,. \end{cases}$$

ii) If $r, s \in \mathbb{Z}_+$ then

$$H^{(n)}_{rel}(\mathcal{F}(p^M, p^L), d) = \begin{cases} \mathbb{C} & \text{for } n = 0, 1\,, \\ 0 & \text{otherwise}\,. \end{cases}$$

iii) If $r, s \in \mathbb{Z}_-$ then

$$H^{(n)}_{rel}(\mathcal{F}(p^M, p^L), d) = \begin{cases} \mathbb{C} & \text{for } n = 0, -1\,, \\ 0 & \text{otherwise}\,. \end{cases}$$

In case *i)* the cohomology state is at level zero, *i.e.* it is just the vacuum, and the conformal dimensions of two Fock spaces satisfy $\Delta(p^L) = 1 - \Delta(p^M)$ [1,2]. The states corresponding to cases *ii)* and *iii)* are called discrete [8,9], and their level is equal to rs. Finally, we observe that in these two cases both $\mathcal{F}^M$ and $\mathcal{F}^L$ are reducible Virasoro modules [10,7].

The absolute cohomology is given by

$$H^{(*)}_{abs}(\mathcal{F}(p^M, p^L), d) \simeq H^{(*)}_{rel}(\mathcal{F}(p^M, p^L), d) \oplus c_0 H^{(*-1)}_{rel}(\mathcal{F}(p^M, p^L), d)\,. \tag{2.2}$$

More explicitly, each representative ψ of the relative cohomology gives rise to two states ψ and $c_0\psi - \chi$ in the absolute cohomology, with χ satisfying $M\psi = \hat{d}\chi$.

To reveal the physical consequences of the above cohomology it is important to construct explicit representatives, which is a nontrivial problem for cases *ii)* and *iii)*. A particularly interesting case, which we consider here, is $c^M = 1$ (*i.e.* $\alpha_+ = -\alpha_- = 1$), where the operator cohomology with $(r, s) = (-1, -2)$ and $(-2, -1)$ in $H^{(-1)}_{rel}(\mathcal{F}(p^M, p^L), d)$ generate a characteristic ring for the theory [11]. Further, the states in $H^{(0)}_{rel}(\mathcal{F}(p^M, p^L), d)$ with $r, s \in \mathbb{Z}_-$ are associated to symmetry currents which preserve this ring.

The matter Fock space for $c^M = 1$ decomposes into a direct sum of irreducible Virasoro modules, $\mathcal{F}(\frac{1}{\sqrt{2}}(r - s)) = \bigoplus_{\ell=0}^{\infty} \mathcal{L}(\frac{1}{4}(|r - s| + 2\ell)^2, 1)$. Moreover, the submodules obtained by restricting this sum to $\ell \geq l_0 \geq 0$ are isomorphic with $\mathcal{F}(\frac{1}{\sqrt{2}}(r_k - s_k))$, where $k = \max(0, s-r) + \ell_0$, $r_k = r + k$, $s_k = s - k$. Introduce a vertex operator $V^M = (1/2\pi i) \oint dz :\exp(-i\sqrt{2}\phi^M(z)):$. This isomorphism is then realized by the embedding $(V^M)^k : \mathcal{F}(\frac{1}{\sqrt{2}}(r_k - s_k)) \to \mathcal{F}(\frac{1}{\sqrt{2}}(r-s))$. It is also clear, as one can verify using the direct sum decomposition of both matter Fock spaces, that the extension of $(V^M)^k$ to an embedding of $\mathcal{F}(\frac{1}{\sqrt{2}}(r_k - s_k), p^L)$ into $\mathcal{F}(\frac{1}{\sqrt{2}}(r-s), p^L)$ will map a nontrivial cohomology class in $\mathcal{F}(\frac{1}{\sqrt{2}}(r_k - s_k), p^L)$ into a nontrivial cohomology class in $\mathcal{F}(\frac{1}{\sqrt{2}}(r-s), p^L)$.

This observation can be put to use as follows. To obtain (gh) $= 0$ representatives for case *ii)* say, we may take $k = s$ so that $s_k = 0$. By Theorem 2.1 *i)* the cohomology of $\mathcal{F}(\frac{1}{\sqrt{2}}(r_k - s_k), p^L)$ is given by the vacuum state $|\frac{1}{\sqrt{2}}(r_k - s_k), p^L\rangle$, which is then mapped into a singular vector in $\mathcal{F}(\frac{1}{\sqrt{2}}(r-s), p^L)$. For (gh) $= 0$ case *iii)* we take $k = -r$ so that $r_k = 0$ and proceed in the same way. Finally, the construction of the other ghost number representatives for cases *ii)* and *iii)* is similarly reduced to finding them for $s_k = 1$, $r_k \in \mathbb{Z}_+$, and $r_k = -1$, $s_k \in \mathbb{Z}_-$, respectively. This is quite easy to accomplish.

Before we state the result, let us recall that the elementary Schur polynomials $S_k(x)$, $x = (x_1, x_2, \dots)$, are defined through their generating function

$$\sum_{k \geq 0} S_k(x) z^k = \exp\left(\sum_{k \geq 1} x_k z^k\right), \tag{2.3}$$

and for convenience we put $S_k(x) = 0$ for $k < 0$. To any partition (Young tableau) $\lambda = \{\lambda_1 \geq \lambda_2 \geq \dots\}$ we then associate a Schur polynomial

$$S_{\lambda_1, \lambda_2, \dots}(x) = \det\left(S_{\lambda_i + j - i}(x)\right)_{i,j} . \tag{2.4}$$

Theorem 2.2. *The following are representatives of the discrete states for $c^M = 1$*

ii) For $r, s \in \mathbb{Z}_+$:

$$\psi^{(0)}_{r,s} = (V^M)^s |\sqrt{\tfrac{1}{2}}(r+s), p^L\rangle = (-1)^{\frac{1}{2}s(s-1)} s!\, S_{\underbrace{r, \dots, r}_{s}}(x) |p^M, p^L\rangle ,$$

$$\psi^{(1)}_{r,s} = (V^M)^{s-1} \sum_{q=1}^{r+s-1} S_{r+s-1-q}(-\sqrt{2}\alpha^M_{-j}/j) c_{-q} |\sqrt{\tfrac{1}{2}}(r+s-2), p^L\rangle .$$

iii) For $r, s \in \mathbb{Z}_-$:

$$\psi^{(0)}_{r,s} = (V^M)^{-r} | -\sqrt{\tfrac{1}{2}}(r+s), p^L\rangle = (-1)^{\frac{1}{2}r(r-1)} (-r)!\, S_{\underbrace{-s, \dots, -s}_{-r}}(x) |p^M, p^L\rangle ,$$

$$\psi^{(-1)}_{r,s} = (V^M)^{-r-1} \sum_{q=1}^{-(r+s+1)} S_{-(r+s+1)-q}(-\alpha^-_{-j}/j) b_{-q} | -\sqrt{\tfrac{1}{2}}(r+s+2), p^L\rangle ,$$

where $p^M = \sqrt{\frac{1}{2}}(r-s)$, $ip^L = \sqrt{\frac{1}{2}}(r+s+2)$, $\alpha_n^{\pm} = \sqrt{\frac{1}{2}}(\alpha_n^M \pm i\alpha_n^L)$, *and* $S_{\lambda_1,\lambda_2,\ldots}(x)$ *are the Schur polynomials with argument* $x_j = -\sqrt{2}\alpha_{-j}^M/j$.

For ghost number zero these explicit formulae for the physical states, or the simplest examples, were known to many people (see *e.g.* [8,12,13]) and can be traced back to [14,15]. For (gh) $= \pm 1$ with low values of r and s examples are given in [11]. In case *ii)* one can also write down explicit representatives in terms of Schur polynomials of the light-cone oscillators α_n^+ only [4], but we do not know of a comparably succinct representation in case *iii)*.

Representatives of the cohomology $H_{rel}^{(n)}(\mathcal{F}(p^M,p^L),d)$ for $c^M < 1$ are easily obtained from the $c^M = 1$ representatives by making use of the $SO(2,\mathbb{C})$ symmetry of [16,7].

It is interesting to observe that the pairing of discrete states for $c^M = 1$, which causes their cancellation in the character [17], can be understood from the existence of a "supersymmetry-like" operator $[d, i\phi^M(z)] = c(z)i\partial\phi^M(z)$ (discussed in [18]). Clearly, this operator maps between the (absolute) cohomolgies of d, and it can be shown that its zero mode indeed acts nontrivially.

3. BRST COHOMOLOGY OF HIGHEST WEIGHT IRREDUCIBLE MODULES

The projection onto $\mathcal{L}(\Delta^M, c^M)$ in the matter sector can be implemented by a suitable resolution. Using the results of Feigin and Fuchs on the submodule structure of the Fock space modules [10], one can show that for any irreducible highest weight module $\mathcal{L}(\Delta, c)$ there exists a complex of Feigin-Fuchs modules $(\mathcal{F}^{(n)}, \delta^{(n)})$ with cohomology $H^{(n)}(\mathcal{F},\delta) \simeq \delta^{n,0}\mathcal{L}(\Delta,c)$. Depending on the type of the irreducible module such a resolution is given by either a finite or an infinite (double sided) complex. Spaces $\mathcal{F}^{(n)}$ in the complex differ by their momentum, $p^{(n)}$, but have the same background charge Q. The differentials $\delta^{(n)}$ commute with the action of the Virasoro algebra, and can be constructed explicitly in terms of screening currents. For an explicit construction of resolutions for various types of modules with $c \leq 1$ (labeled as in the Feigin-Fuchs classification [10]) we refer the reader to: [5] for type III_- in the interior of the Kac table; [19] for type II; [6] for III_-^0 (the boundary of the Kac table), III_- in the exterior of the Kac table, and III_-^{00}.

Let $\mathcal{E}(\Delta, c) = \{\Delta(p^{(n)})\}$ be the set of conformal dimensions of Fock spaces labelled by the degree n in the resolution. For modules of type I, II and III_- it coincides with $E(\Delta, c)$, the set of conformal weights of singular vectors (including the highest weight vector) in the Verma module $\mathcal{V}(\Delta, c)$ (see [10] for explicit formulae). For modules of type III_-^0 and III_-^{00} it consists of two highest elements of $E(\Delta, c)$. Denote $\widetilde{\mathcal{E}}(\Delta, c) \equiv 1 - \mathcal{E}(\Delta, c)$, and for $\Delta = 1 - \Delta(p^{(n)})$ define $d(\Delta) = |n|$. One should note that $d(\cdot)$ does not depend on the choice of the Fock space resolution.

The BRST cohomology of $\mathcal{L}(\Delta^M, p^L) \equiv \mathcal{L}(\Delta^M, c^M) \otimes \mathcal{F}(p^L, Q^L)$ can be computed by considering a double complex $(\mathcal{F}^{(*)} \otimes \mathcal{F}(p^L, Q^L) \otimes \mathcal{F}^G, \delta, d)$, with two commuting differentials δ and d, and the corresponding grading by the degree in the resolution and the ghost number. The crucial observation of [4] is firstly that, since $\mathcal{L}^*(\Delta^M, c^M) \simeq \mathcal{L}(\Delta^M, c^M)$, one has a choice between a given resolution or its dual. Then, by explicit examination one always finds that, given p^L, there is a choice of resolution such that, when p^M ranges over the momenta of all Fock spaces in that complex, (2.1) has no solution for integral r and s with $rs > 0$. Then it follows that the BRST cohomology of $\mathcal{F}^{(n)} \otimes \mathcal{F}(p^L, Q^L)$ is nontrivial for at at most one degree n, say $n = n_0$, where it is given by case $i)$ of Theorem 2.1.[1] A standard argument for double complexes [20] then gives

$$\begin{aligned} H^{(n)}_{rel}(\mathcal{L}(\Delta^M, p^L), d) &= H^{(n)}_{rel}(H^{(0)}(\mathcal{F} \otimes \mathcal{F}(p^L, Q^L) \otimes \mathcal{F}^G, \delta), d) \\ &= H^{(n)}(H^{(0)}_{rel}(\mathcal{F} \otimes \mathcal{F}(p^L, Q^L) \otimes \mathcal{F}^G, d), \delta) \simeq \delta^{n,n_0} \mathbb{C} . \end{aligned} \tag{3.1}$$

We can summarize the complete result as follows.

Theorem 3.1. *For any highest weight irreducible module, the relative BRST cohomology of $\mathcal{L}(\Delta^M, c^M) \otimes \mathcal{F}(p^L, Q^L)$ is nontrivial if and only if $\Delta(p^L) \in \widetilde{\mathcal{E}}(\Delta^M, c^M)$, in which case $H^{(n)}_{rel}(\mathcal{L}(\Delta^M, c^M) \otimes \mathcal{F}(p^L, Q^L)) = \delta_{n,\eta(p^L)d(\Delta(p^L))}\mathbb{C}$, where $\eta(p^L) = \operatorname{sign} i(p^L - Q^L)$.*

For the BPZ minimal models (*i.e.* modules in the interior of the Kac table) this theorem has first been proven by a somewhat different method in [3], and then rederived as above in [4].

Let us illustrate our method in the simple example of a representation corresponding to the boundary of the Kac table. In this case $\alpha_+ = \sqrt{p'/p}$, $\alpha_- = -\sqrt{p/p'}$, where $p' \geq p \geq 1$ are relatively prime integers. The central charge is $c^M(p, p') = 1 - 6(p - p')^2/pp'$. Let $p^M(m, m')$ denotes a matter momentum as in (2.1) with $r = -m$ and $s = -m'$, $\mathcal{F}^M(m, m')$ the corresponding Fock space, and $\Delta^M(m, m')$ the conformal dimension. We have the following two (dual to each other) resolutions of the highest weight module $\mathcal{L}(\Delta^M(m, 0), c^M(p, p'))$, $1 \leq m \leq p - 1$ [6]

$$0 \xrightarrow{\delta^{(-2)}} \mathcal{F}^M(m - 2p, 0) \xrightarrow{\delta^{(-1)}} \mathcal{F}^M(m, 0) \xrightarrow{\delta^{(0)}} 0 \tag{3.2}$$

$$0 \xrightarrow{\delta^{(-1)}} \mathcal{F}^M(-m, 0) \xrightarrow{\delta^{(0)}} \mathcal{F}^M(-m + 2p, 0) \xrightarrow{\delta^{(1)}} 0 \tag{3.3}$$

Given p^L, integers r and s in (2.1) are only determined up to $r \to r + tp$ and $s \to s - tp'$, $t \in \mathbb{Z}$. However, using $p^M(m + p, m' + p') = p^M(m, m')$, one checks by inspection that – even with this ambiguity – in one of the resolutions (3.2) or (3.3) the p^M obtained from (2.1) for $rs > 0$ never appear, just as we have claimed above. Finally, one should note that whereas the corresponding Verma module has infinitely many singular vectors, by Theorem 3.1 the discrete states arise in only three ghost numbers. Strikingly different than for the modules in the interior of the Kac table!

---

[1] By the remark after Theorem 2.1 this is clearly true when $\mathcal{F}(p^M, Q^M) \simeq \mathcal{L}(\Delta^M, c^M)$ is irreducible.

4. THE SUPER EXTENSION

The results of the previous sections can be generalized to $2D$ supergravity coupled to $\hat{c} \leq 1$ superconformal matter (for the critical superstring see [21,22]). In this case we deal with a Fock space $\mathcal{F}^s(p,Q)$ of a scalar field $\phi(z)$ and a fermionic field $\psi(z)$, labelled by the scalar zero mode p and background charge Q as in Section 2. The generators of the $N=1$ superconformal algebra are given by [23]

$$T(z) = -\tfrac{1}{2} : \partial\phi\partial\phi : + iQ\partial^2\phi - \tfrac{1}{2}\psi\partial\psi\,, \qquad G(z) = i\partial\phi\psi + 2Q\partial\psi\,, \tag{4.1}$$

and the central charge equals $\hat{c} = \frac{2}{3}c = 1 - 8Q^2$. Physical states correspond to cohomology classes of the BRST operator

$$d = \oint \frac{dz}{2\pi i} : ((T^M + T^L + \tfrac{1}{2}T^G)(z)c(z) + (G^M + G^L + \tfrac{1}{2}G^G)(z)\gamma(z)) : . \tag{4.2}$$

In this case, nilpotency requires $\hat{c}^M + \hat{c}^L = 10$, *i.e.* $(Q^M)^2 + (Q^L)^2 = -1$. First we give the result for the relative cohomology of d on the Fock space $\mathcal{F}^s(p^M, p^L) \equiv \mathcal{F}^s(p^M, Q^M) \otimes \mathcal{F}^s(p^L, Q^L) \otimes \mathcal{F}^{sG}$ where $\mathcal{F}^{sG}$ is the Fock space of the bc-ghosts and their superpartners $\beta\gamma$. As in (2.1) we parametrize the momenta (p^M, p^L) in terms of $\alpha_\pm = (Q^M \pm iQ^L)/\sqrt{2}$ and $r, s \in \mathbb{C}$. Note, however, that now $\alpha_+\alpha_- = -\frac{1}{2}$.

In the following, relative cohomology means the cohomology relative to $\{L_0, b_0\}$ in the Neveu-Schwarz sector ($\kappa = \frac{1}{2}$), and the cohomology relative to $\{L_0, G_0, b_0, \beta_0\}$ in the Ramond sector ($\kappa = 0$).

Theorem 4.1. *Let (p^M, p^L) be parametrized by (r,s) as in (2.1). The relative cohomology of d on the Fock space $\mathcal{F}^s(p^M, p^L)$ is nontrivial only in the following three cases*

(i) If either $r = 0$ or $s = 0$ (i.e. $\Delta(p^M) + \Delta(p^L) = \frac{1}{2}$), then

$$H^{(n)}_{rel}(\mathcal{F}^s(p^M, p^L), d) = \begin{cases} \mathbb{C} & \text{if } n = 0\,, \\ 0 & \text{otherwise}\,. \end{cases}$$

(ii) If $r, s \in \mathbb{Z}_+ , r - s \in 2\mathbb{Z} + (1 - 2\kappa)$, then

$$H^{(n)}_{rel}(\mathcal{F}^s(p^M, p^L), d) = \begin{cases} \mathbb{C} & \text{if } n = 0, 1\,, \\ 0 & \text{otherwise}\,. \end{cases}$$

(iii) If $r, s \in \mathbb{Z}_- , r - s \in 2\mathbb{Z} + (1 - 2\kappa)$, then

$$H^{(n)}_{rel}(\mathcal{F}^s(p^M, p^L), d) = \begin{cases} \mathbb{C} & \text{if } n = 0, -1\,, \\ 0 & \text{otherwise}\,. \end{cases}$$

Note that "discrete states" occur only if both $\mathcal{F}^{sM}$ and $\mathcal{F}^{sL}$ are reducible. They occur at the same level, namely $\frac{rs}{2}$, as where the null vector in these modules occurs.

As in the bosonic case [4] the proof can be given by examining the spectral sequence associated to the filtered complex obtained by assigning the degree equal

+1 to the oscillators α_n^+, c_n, ψ_r^+, γ_r and -1 to α_n^-, b_n, ψ_r^-, β_r, respectively. In the NS sector, contrary to the bosonic case, the spectral sequence does not always collapse after the first term (*i.e.* the cohomology of the lowest degree differential $\widehat{d}_0$ is not always concentrated in a single degree), and one has to calculate also the second term, which then yields the final result. In the R sector there is an additional complication due to the fact that G_0 does not act reducibly on $\mathcal{F}^s(p^M, p^L)$. This is resolved by introducing "rotated" oscillators – exactly as in [24] for the critical string but without the exponential rescaling – which effectively diagonalize G_0 in a subspace of $\mathcal{F}^s(p^M, p^L)$. One can show that this subspace contains all states annihilated by G_0, and in terms of these new oscillators the cohomology computation is parallel to that in the NS sector.

Explicit representatives of the cohomology can be given in terms of "super Schur polynomials", and the physical states are paired by the action of the zero mode of the operator $[d, i\phi^M(z)] = c(z)i\partial\phi^M(z) + \gamma(z)\psi^M(z)$. Details will appear elsewhere.

Finally, $\hat{c} < 1$ super minimal models are parametrized by [25,23,26]

$$\hat{c}(p,p') = 1 - 2\frac{(p-p')^2}{pp'}, \quad \Delta(m,m') = \frac{(mp'-m'p)^2 - (p-p')^2}{8pp'} + \tfrac{1}{16}(1-2\kappa), \tag{4.3}$$

where $p, p' \in 2\mathbb{N}-1, \gcd(p,p') = 1$ or $p,p' \in 2\mathbb{N}, \frac{1}{2}(p-p') \in 2\mathbb{N}-1, \gcd(\frac{p}{2},\frac{p'}{2}) = 1$, and $1 \le m \le p-1, 1 \le m' \le p'-1, m-m' \in 2\mathbb{Z} + (1-2\kappa)$.

Let $\alpha_+ = \sqrt{p'/(2p)}$ and $\alpha_- = -\sqrt{p/(2p')}$. Define

$$p_\pm^{(n)}(\ell,\ell') = \begin{cases} \pm\sqrt{\tfrac{1}{2}}\,((\ell \pm 2np)\alpha_+ + \ell'\alpha_-) & \text{for } n \text{ even}, \\ \pm\sqrt{\tfrac{1}{2}}\,((-\ell \pm 2(n\pm1)p)\alpha_+ + \ell'\alpha_-) & \text{for } n \text{ odd}, \end{cases} \tag{4.4}$$

$$\Delta_\pm^{(n)}(\ell,\ell') = \tfrac{1}{2}\left(p_\pm^{(n)}(\ell,\ell') - Q\right)^2 - \tfrac{1}{2}Q^2 + \tfrac{1}{16}(1-2\kappa),$$

and $\widetilde{E}_{m,m'}(p,p') = \{\frac{1}{2} - \Delta_+^{(n)}(m,m'), n \in \mathbb{Z}\}$. Let $d(\Delta) = |n|$ for $\Delta = \frac{1}{2} - \Delta_+^{(n)}(m,m')$ and $\eta(p^L) = \mathrm{sign}(i(p^L - Q^L))$. Under the plausible assumption that the Felder resolution generalizes to the super-case we have (this result was also announced in [3])

Theorem 4.2.
(i) $H_{rel}^{(*)}(L(\Delta(m,m')) \otimes \mathcal{F}^s(p^L,Q^L) \otimes \mathcal{F}^{sG}) \neq 0$ iff $\Delta(p^L) \in \widetilde{E}_{m,m'}(p,p')$,
(ii) $\dim H_{rel}^{(n)}(L(\Delta(m,m')) \otimes \mathcal{F}^s(p^L,Q^L) \otimes \mathcal{F}^{sG}) = \delta_{n,\eta(p^L)d(\Delta(p^L))}$.

Acknowledgements

P.B. would like to thank the organizers for an enjoyable and stimulating conference, and V. Dotsenko, S. Mukhi, J. Sidenius and E. Verlinde for useful discussions. The work of J.M. was supported by the NSF Grant #PHY-88-04561 and the work of K.P. was supported in part by funds provided by the DOE Contract #DE-FG03-84ER-40168 and by the USC Faculty Research and Innovation Fund.

Note added: While preparing this paper we received [27] in which Theorem 4.1 is also presented. The proof given there for the NS sector is exactly as we discussed above. However, since $\widehat{d}^2 \neq 0$ on the $L_0 = 0$ subspace of the R sector ($\widehat{d} = d_R$ in [27]), the discussion for this case seems misleading.

References

[1] F. David, *Conformal field theories coupled to 2-D gravity in the conformal gauge*, Mod. Phys. Lett. **A3**, 1651 (1988).

[2] J. Distler and H. Kawai, *Conformal field theory and 2-D quantum gravity*, Nucl. Phys. **B321**, 509 (1989).

[3] B.H. Lian and G.J. Zuckerman, *New selection rules and physical states in 2D gravity; conformal gauge*, Phys. Lett. **254B**, 417 (1991).

[4] P. Bouwknegt, J. McCarthy and K. Pilch, *BRST analysis of physical states for 2D gravity coupled to $c < 1$ matter*, CERN-TH.6162/91.

[5] G. Felder, *BRST approach to minimal models*, Nucl. Phys. **B317**, 215 (1989); erratum, *ibid.* **B324**, 548 (1989).

[6] P. Bouwknegt, J. McCarthy and K. Pilch, *Fock space resolutions of the Virasoro highest weight modules with $c \leq 1$*, CERN-TH.6196/91.

[7] B.L. Lian and G.J. Zuckerman, *2D gravity with $c = 1$ matter*, Phys. Lett. **266B**, 21 (1991).

[8] A.M. Polyakov, *Selftuning fields and resonant correlations in 2D gravity*, Mod. Phys. Lett. **A6**, 635 (1991).

[9] D.J. Gross, I.R. Klebanov and M.J. Newman, *The two point correlation function of the one-dimensional matrix model*, Nucl. Phys. **B350**, 621 (1991).

[10] B.L. Feigin and D.B. Fuchs, *Representations of the Virasoro algebra*, in *Representations of infinite-dimensional Lie groups and Lie algebras*, Gordon and Breach, New York (1989).

[11] E. Witten, *Ground ring of two dimensional string theory*, preprint IASSNS-HEP-91/51.

[12] S. Mukherji, S. Mukhi and A. Sen, *Null vectors and extra states in $c = 1$ string theory*, TIFR/TH/91-25, May '91.

[13] I. Klebanov and A. Polyakov, *Interaction of discrete states in two-dimensional string theory*, preprint PUPT-1281.

[14] J. Goldstone, unpublished.

[15] V.G. Kac, *Contravariant form for infinite dimensional Lie algebras and superalgebras*, Lect. Notes Math. **94**, 441 (1979).

[16] K. Itoh, *$SL(2,\mathbb{R})$ current algebra and spectrum in two dimensional gravity*, CTP-TAMU-42/91.

[17] M. Bershadsky and I. Klebanov, *Genus-one path integral in two-dimensional quantum gravity*, Phys. Rev. Lett. **65**, 3088 (1990).

[18] R. Brooks, *Residual symmetries in $c = 1$ noncritical string theory*, MIT-CTP #1957.

[19] G. Felder, J. Fröhlich and G. Keller, *Braid matrices and structure constants for minimal conformal models*, Comm. Math. Phys. **124**, 647 (1989).

[20] R. Bott and L.W. Tu, *Differential forms in algebraic topology*, Springer-Verlag, New York, 1982.

[21] J. Figueroa-O'Farrill and T. Kimura, *The BRST cohomology of the NSR string: vanishing and "no ghost" theorem*, Comm. Math. Phys. **124**, 105 (1989).

[22] B.H. Lian and G.J. Zuckerman, *BRST cohomology of the super-Virasoro algebras*, Comm. Math. Phys. **125**, 301 (1989).

[23] M. Bershadsky, V. Knizhnik and M. Teitelman, *Superconformal symmetry in two dimensions*, Phys. Lett. **151B**, 31 (1985).

[24] M. Ito, T. Morozumi, S. Nojiri and S. Uehara, *Covariant quantization of Neveu-Schwarz-Ramond model*, Prog. Theor. Phys. **75**, 934 (1986).

[25] H. Eichenherr, *Minimal operator algebras in superconformal quantum field theory*, Phys. Lett. **151B**, 26 (1985).

[26] D. Friedan, Z. Qiu and S. Shenker, *Superconformal invariance in two-dimensions and the tricritical Ising model*, Phys. Lett. **151B**, 37 (1985).

[27] K. Itoh and N. Ohta, *BRST cohomology and physical states in 2D supergravity coupled to $\hat{c} \leq 1$ matter*, FERMILAB-PUB-91/228-T.

CORRELATION FUNCTIONS OF LOCAL OPERATORS IN 2D GRAVITY COUPLED TO MINIMAL MATTER

Vl.S. Dotsenko[1]

Current address: Université Pierre et Marie Curie, Paris VI
75252 Paris Cedex 05, France

ABSTRACT

Recent advances are being discussed on the calculation, within the conformal field theory approach, of the correlation functions for local operators in the theory of 2D gravity coupled to the minimal models of matter.

Here I would like to discuss the results available on the calculation of the correlation functions of local operators in the theory of 2D gravity coupled to the minimal matter, in the approach of conformal field theory [1, 2, 3, 4].

In the representation of David, Distler and Kawai [4] the local operators take the form:

$$\Phi_{s'.s}(z) = \phi^M_{s'.s}(z)\phi^L_{s'.s}(z) \sim V_{s'.s}(z)U_{-s'.s}(z) \equiv \exp(i\alpha_{s'.s}\varphi^M(z))\exp(\beta_{-s'.s}\varphi^L(z)) \quad (1)$$

where

$$\alpha_{s'.s} = \frac{1-s'}{2}\alpha_- + \frac{1-s}{2}\alpha_+, \quad \beta_{-s'.s} = \frac{1+s'}{2}\beta_- + \frac{1-s}{2}\beta_+ \quad (2)$$

$$\alpha_\pm = \alpha_0 \pm \sqrt{\alpha_0^2+2}, \quad \beta_\pm = \beta_0 \pm \sqrt{\beta_0^2-2} \quad (3)$$

Here the free field representation is assumed both for the matter and for the gravity (Liouville) factors of the operator (1). More specifically:

$$\langle\varphi^M(z)\varphi^M(z')\rangle = \langle\varphi^L(z)\varphi^L(z')\rangle = \log\frac{1}{z-z'} \quad (4)$$

$$T^M = -\frac{1}{2}\partial\varphi^M\partial\varphi^M + i\alpha_0\partial^2\varphi^M, \quad T^L = -\frac{1}{2}\partial\varphi^L\partial\varphi^L + \beta_0\partial^2\varphi^L \quad (5)$$

$$C_M = 1-12\alpha_0^2, \quad C_L = 1+12\beta_0^2 \quad (6)$$

[1]Permanent address: Landau Institute for Theoretical Physics, Moscow.

New Symmetry Principles in Quantum Field Theory, Edited by
J. Fröhlich et al., Plenum Press, New York, 1992

$$V_\alpha = \exp(i\alpha\varphi^M), \quad U_\beta = \exp(\beta\varphi^L) \tag{7}$$

$$\Delta_\alpha^M = \frac{1}{2}(\alpha^2 - 2\alpha\alpha_0), \quad \Delta_\beta^L = -\frac{1}{2}(\beta^2 - 2\beta\beta_0) \tag{8}$$

The constraint which couples the two theories is

$$C_M + C_L = 26 \tag{9}$$

and the constraint which couples the two representations of the corresponding Virasoro algebras is

$$\Delta^M + \Delta^L = 1 \tag{10}$$

Eq.(9) leads to

$$\beta_0^2 = \alpha_0^2 + 2, \quad \beta_\pm = \pm\alpha_\pm \tag{11}$$

while the expressions in (2) for α, β solve for (10), on account of eqs.(8) for Δ^M, Δ^L.

The notations for the Liouville part, made to be symmetric to the matter sector, are those used in [5].

The three-point functions, or amplitudes, take the form

$$\begin{aligned} A^{(3)} &= \langle \Phi_{s'.s}(0)\Phi_{n'.n}(1)\Phi_{m'.m}(\infty)\rangle \\ &= \langle \phi^M_{s'.s}(0)\phi^M_{n'.n}(1)\phi^M_{m'.m}(\infty)\rangle\langle \phi^L_{s'.s}(0)\phi^L_{n'.n}(1)\phi^L_{m'.m}(\infty)\rangle = A_M^{(3)}A_L^{(3)} \end{aligned} \tag{12}$$

The three-point functions of the matter sector had been calculated in [6], and can be given in the following form:

$$\begin{aligned} A_M^{(3)} &\propto (\pi\gamma(\rho'))^l(\pi\gamma(\rho))^k P(l,k)P(l-s',k-s)P(l-n',k-n)P(l-m',k-m) \\ &\times P^{-1}(l-s',k-s)P^{-1}(l-n',k-n)P^{-1}(l-m',k-m) \end{aligned} \tag{13}$$

Here

$$P(l,k) = \prod_{i=1}^{l}\frac{\Gamma(i\rho')}{\Gamma(1-i\rho')}\prod_{j=1}^{k}\frac{\Gamma(j\rho)}{\Gamma(1-j\rho)}\prod_{i=1}^{l}\prod_{j=1}^{k}\frac{(-1)}{(i\rho'-j)^2} \tag{14}$$

$$\rho = \frac{\alpha_+^2}{2}, \quad \rho' = \frac{\alpha_-^2}{2} = \frac{1}{\rho}, \quad \gamma(\rho) = \frac{\Gamma(1-\rho)}{\Gamma(\rho)} \tag{15}$$

$$l = \frac{s'+n'+m'-1}{2}, \quad k = \frac{s+n+m-1}{2} \tag{16}$$

The expression in (15) is different from that in [6] by normalization factors of individual operators, and is the same as in [5]. It corresponds to the Coulomb gas operators without extra normalization, apart from possible sign and $\rho^{(\cdots)}$ factors.

The expression for $A_L^{(3)}$ in (12) can be obtained by an analytic continuation of the result for $A_M^{(3)}$ (13) on using the following analytic continuation for the finite products with the negative integer upper bounds [5] :

$$\text{if} \quad q(l) = \prod_{i=1}^{l} f(i), \quad \text{then} \quad q(l = -|l|) = \prod_{i=0}^{|l|-1}\frac{1}{f(-i)} \tag{17}$$

Multiplying $A_M^{(3)}$ and $A_L^{(3)}$, eq.(12), one gets [5] :

$$A^{(3)} \propto (\mu)^S N_{s'.s} N_{n'.n} N_{m'.m} \tag{18}$$

Here

$$N_{s'.s} = \frac{\Gamma(1+s'\rho'-s)}{\Gamma(-s'\rho'+s)} = \frac{\Gamma(\frac{\beta^2-\alpha^2}{2})}{\Gamma(1-\frac{\beta^2-\alpha^2}{2})}, \quad \alpha = \alpha_{s'.s}, \quad \beta = \beta_{-s'.s} \tag{19}$$

$$S = \frac{1}{\beta_-}(2\beta_0 - \beta_{-s'.s} - \beta_{-n'.n} - \beta_{-m'.m}) \tag{20}$$

The three-point functions of the minimal model coupled to gravity first had been calculated in [7], using the KdV technique of [8]. In [9] they had been obtained in the Liouville theory approach, by integrating out first the Liouville field zero mode, the trick introduced in [10], and than by using the analytic continuation technique, though different from the one sketched on above.

Both in [7] and [9] the calculation had been restricted to the order operators, i.e. to the operators with

$$s' = s, \quad n' = n, \quad m' = m \tag{21}$$

The three-point functions for the general operators had been calculated in [5], with result (18). We remark that general operators is known to be quite a problem for the KdV technique, and may be this is the first instance when the field theory happens to be more powerful with respect to the otherwise quite successful matrix model approach [11] and the related KdV technique [8, 7].

We shall now discuss some of the properties of the three-point functions.

1. Invariance to switching the Fock space representations.

Each operator factor in (1) has two possible representations in the free field, or the Coulomb gas technique:

$$\phi^M: \quad V_{s'.s}, \quad V_{-s'.-s} \tag{22}$$

$$\phi^L: \quad U_{-s'.s}, \quad U_{s'.-s} \tag{23}$$

So we can use four representations:

$$\Phi_{s'.s}^{--} \sim V_{s'.s} U_{-s'.s} \tag{24}$$

$$\Phi_{s'.s}^{-+} \sim V_{s'.s} U_{s'.-s} \tag{25}$$

and two more, obtained with $V_{s'.s} \to V_{-s'.-s}$. The tree-point functions can be calculated for these different representations, with the analytic continuation technique sketched on above, to be used also for the matter sector, with the result as in (18) but with the normalization factors replaced according to:

$$\Phi_{s'.s}^{--} \sim N_{s'.s}^{--} = \frac{\Gamma(1+s'\rho'-s)}{\Gamma(-s'\rho'+s)} \tag{26}$$

$$\Phi_{s'.s}^{-+} \sim N_{s'.s}^{-+} = \frac{\Gamma(1-s'+s\rho)}{\Gamma(s'-s\rho)} \tag{27}$$

with two more obtained by switching signs. (See also the calculations in[12]). The form

$$N = \frac{\Gamma(\frac{\beta^2-\alpha^2}{2})}{\Gamma(1-\frac{\beta^2-\alpha^2}{2})} \tag{28}$$

for $\Phi = V_\alpha U_\beta$ always holds. In this sense the three-point functions are invariant with respect to switching the representations. This is the case also for the matter theory, i.e. the minimal model itself. The functions

$$\langle\phi_i\phi_j\phi_k\rangle, \quad \langle\phi_i\phi_j\tilde{\phi}_k\rangle, \quad \langle\phi_i\tilde{\phi}_j\tilde{\phi}_k\rangle, \quad \langle\tilde{\phi}_i\tilde{\phi}_j\tilde{\phi}_k\rangle, \tag{29}$$

($\phi_i \sim V_{s'.s}$, $\tilde{\phi}_i \sim V_{-s'.-s}$) differ by normalization factors of individual operators.

It appears reasonable to expect that for more-point functions there should exist some form of equivalence of the representations, if proper formulation is found. Since, whatever the representation, we are dealing with two coupled conformal theories, not with particular Fock spaces. Again, this is the case for the matter theory. E.g. for the four-point functions one usually uses the representation $\langle\phi\phi\phi\tilde{\phi}\rangle$ [6]. In fact, the representation $\langle\tilde{\phi}\tilde{\phi}\tilde{\phi}\phi\rangle$ could be used equally well, by employing antiscreening operators (or, a negative number of screening operators), as intertwiners of the Fock spaces. This extension of the technique would restore the direct - conjugate representation symmetry. By the way, the antiscreening operators could be useful objects from the point of view of the quantum group representation.

Returning to gravity, we remark still that the direct - conjugate representation symmetry had already been assumed in the form of the effective action for the quantum theory of Liouville [5]:

$$A[\phi^L] \propto \int [\partial\phi\bar{\partial}\phi + \mu\exp(\beta_-\phi^L) + a(\mu)^\rho \exp(\beta_+\phi^L)] \tag{30}$$

which is the free field representation with two screening operators, more proper to say. (Some arguments to such a representation had been given also in [13]).

In (30)

$$\mu\exp(\beta_-\phi^L) = \mu\Phi^{--}_{1.1} = \mu V_{1.1}U_{-1.1} \tag{31}$$

is the puncture operator, and

$$(\mu)^\rho \exp(\beta_+\phi^L) = (\mu)^\rho \Phi^{-+}_{1.1} = (\mu)^\rho V_{1.1}U_{1.-1} \tag{32}$$

is the L-sector conjugate representation operator. Then it should be natural to assume the possibility of using, on equal footing, the alternative representation for other operators as well, being multiplied by the μ-scaling compensating factors.

2. The result (18) assumes that the usual fusion rules of the minimal model are cancelled by gravity. In particular, the operators outside the basic conformal grid couple to the states inside, and so they become physical. This result is obtained also by the analytic continuation technique of [9], extended to more general set of operators in [14]. This disagrees with the KdV results of [7], for the order operators,

which support instead the usual fusion rules of the minimal model. On the other hand, coupling of operators outside the basic conformal grid may be in agreement with the BRST analysis of the paper [15], for the physical states (or operators) spectrum. This has been further investigated recently in [16, 17]. Although the extra states found in [15] (for the case of minimal $C < 1$ models coupled to gravity) involve ghosts, further analysis of [17] assumes that they could equivalently be represented by the states (operators) outside the basic conformal grid, without ghosts, with their ghost number grading replaced by the Felder BRST grading [18] (the number of block steps away from the basic conformal grid).

The conflicting evidence stated above, due to different techniques employed, shows that the problem with the fusion rules is still open. See also the discussions in [12, 19]. We expect that, as in the case of the minimal conformal theory itself, to fix the three-point functions we have to understand the four-point ones. They involve a lot more dynamics. We shall describe next some advances made in the calculation of four-point functions, in the field theory approach.

The Coulomb gas representation for the four-point functions has the form (we drop the μ scaling factor throughout):

$$\begin{aligned} A^{(4)} &= \int d^2z \langle \Phi(0)\Phi(z)\Phi(1)\Phi(\infty)\rangle \\ &= \int d^2z [(\prod_{i=1}^{l} \int d^2x_i)(\prod_{j=1}^{k} \int d^2y_j)\langle V_{\alpha 1}(0)V_{\alpha 2}(z)V_{\alpha 3}(1)V_{\alpha 4}(\infty)(\prod_{i=1}^{l} V_-(x_i))(\prod_{j=1}^{k} V_+(y_j))\rangle \\ &\times (\prod_{i=1}^{\tilde{l}} \int d^2u_i)(\prod_{j=1}^{\tilde{k}} \int d^2v_j)\langle U_{\beta 1}(0)U_{\beta 2}(z)U_{\beta 3}(1)U_{\beta 4}(\infty)(\prod_{i=1}^{\tilde{l}} U_-(u_i))(\prod_{j=1}^{\tilde{k}} U_+(v_j))\rangle] \end{aligned} \quad (33)$$

The matter and the Liouville correlation functions in (33) have separate expansions over the z singularities, $(z)^{-2(\Delta_1^M+\Delta_2^M+\Delta_{int}^M)}$, $(z)^{-2(\Delta_1^L+\Delta_2^L+\Delta_{int}^L)}$, which correspond to the intermediate primary operators (states). When multiplied, they produce the double expansion of the form:

$$\sum_{\tilde{p}',\tilde{p}} \sum_{p',p} \int d^2z |z|^{-2(\Delta_1+\Delta_2+\Delta_{int})} \times (\ldots) \quad (34)$$

where

$$\Delta_1 = \Delta_1^M + \Delta_1^L = 1, \quad \Delta_2 = \Delta_2^M + \Delta_2^L = 1, \quad \Delta_{int} = \Delta_{p'.p}^M + \Delta_{\tilde{p}'.\tilde{p}}^L \quad (35)$$

The terms in (34) with $\tilde{p'} = -p', \tilde{p} = p$ (the 'diagonal' ones), such that $\Delta_{int} = \Delta_{p'.p}^M + \Delta_{-p'.p}^L = 1$, they correspond to physical intermediate states. The integral (33) then gets log singularities, $\int d^2|z|^{-2}(\ldots)$.

To better define the integral (33) let us shift the values of the parameters (Coulomb gas charges, or states momenta) of the external states, slightly off their discrete values, but so that the physical states condition is kept, $\Delta_\alpha + \Delta_\beta = 1$, which requires $\beta_i = \alpha_i - \alpha_-$ (or $\beta_i = \alpha_+ - \alpha_i$, if conjugate representation is used):

$$\alpha_i = \alpha_i^{(0)} + \frac{\epsilon_i}{2}\alpha_-, \quad \beta_i = \beta_i^{(0)} - \frac{\epsilon_i}{2}\beta_-, \quad \sum \epsilon_i = 0 \quad (36)$$

Here

$$\begin{aligned}\alpha_1^{(0)} &= \alpha_{s'.s}, \quad \alpha_2^{(0)} = \alpha_{n'.n}, \quad \alpha_3^{(0)} = \alpha_{m'.m}, \quad \alpha_4^{(0)} = \alpha_{t'.t} \\ \beta_1^{(0)} &= \beta_{-s'.s}, \quad \text{so on} \end{aligned} \tag{37}$$

Then one checks that the integral (33) has the following expansion in z singularities:

$$A^{(4)} \sim \sum_{k_1,l_1}\sum_{\tilde{k}_1,\tilde{l}_1}\int d^2z|z|^{-2\gamma_{12}} \times Res \tag{38}$$

$$\begin{aligned}\gamma_{12} &= 1 + [(\tilde{l}_1 + l_1 + 1) + (\tilde{k}_1 - k_1)\rho] \\ &\times [(s' + n' + \tilde{l}_1 - l_1)\rho' + (1 - s - n + \tilde{k}_1 + k_1) - (\epsilon_1 + \epsilon_2)\rho'] \end{aligned} \tag{39}$$

Res in (38) stands for the residue amplitude; $\tilde{l}_1, \tilde{k}_1, l_1, k_1$ are numbers of screening operators integraded over the region close to 0 and z, and it is presumed that z is close to 0. The rest of screenings, $\tilde{l}_2 = \tilde{l} - \tilde{l}_1$, $\tilde{k}_2 = \tilde{k} - \tilde{k}_1$, $l_2 = l - l_1$, $k_2 = k - k_1$ are being away from $z \sim 0$. (The corresponding analysis in case of the minimal conformal theory of selecting the z-singularities of four-point functions by the configurations of the screening operators see in [6], and also in [20]). By summing over the configurations if screening operators in (38) we are summing over the intermediate states in (34). The terms in (38) with

$$\tilde{l}_1 = -s' - n' + l_1, \quad \tilde{k}_1 = s + n - 1 - k_1 \tag{40}$$

correspond to physical states. We get in this case

$$\gamma_{12} = 1 + (\epsilon_1 + \epsilon_2)\rho'(p' - p\rho) \tag{41}$$

$$p' = s' + n' - 1 - 2l_1, \quad p = s + n - 1 - 2k_1 \tag{42}$$

The subsum over the physical intermediate states, the 'diagonal' terms in (38), which produce $1/\epsilon$ singularities, takes the form:

$$(A^{(4)})_{sing} \sim \sum_{k_1,l_1(p',p)} \int d^2z|z|^{-2-2\rho'(\epsilon_1+\epsilon_2)(p'-p\rho)} \times Res \tag{43}$$

It is not difficult to realize that the sum for $(A^{(4)})_{sing}$ can be given in the form:

$$(A^{(4)})_{sing} \sim \sum_{p',p} \frac{1}{(\epsilon_1 + \epsilon_2)(p' - p\rho)} \langle \Phi^{--}_{s'.s} \Phi^{--}_{n'.n} \Phi^{-+}_{p'.p} \rangle \langle \Phi^{+-}_{p'.p} \Phi^{--}_{m'.m} \Phi^{--}_{t'.t} \rangle \tag{44}$$

Using the expressions for the three-point functions we obtain, for the residue in (44),

$$\sum_{p',p} \frac{1}{(p' - p\rho)} N_1 N_2 N^{-+}_{p'.p} N^{+-}_{p'.p} N_3 N_4 \tag{45}$$

$N_1 = N^{--}_{s'.s}$, so on. As

$$N^{-+}_{p'.p} N^{+-}_{p'.p} = \frac{\Gamma(1 - p' + p\rho)}{\Gamma(p' - p\rho)} \frac{\Gamma(1 + p' - p\rho)}{\Gamma(-p' + p\rho)} = (-1)(p' - p\rho)^2 \tag{46}$$

we obtain finally

$$(A^{(4)})_{sing} \sim N_1 N_2 N_3 N_4 \sum_{p',p} (p' - p\rho) \tag{47}$$

We have calculated the contribution of physical intermediate states only, and also just the singular piece, the residue at $1/\epsilon$ in $A^{(4)}$, coming from the region $z \sim 0$. Also, different representations for the external states (operators) can be used, but the general form of $(A^{(4)})_{sing}$ remains as in (47).

We notice that although we have chosen the representation $(--)$ for the external states, see (37),(44) and the difinitions (24),(25), the opposite representation (or the opposite Liouville dressing) states appear anyway, as intermediate states in the four-point functions, see(44). (Same is true of course for the often prefered representation, $\langle(--)(--)(--)(--)(+-)\rangle$, with one of the external operators chosen to be in the conjugate matter representation). This is another piece to the arguments given above on should be equal footing of the representations.

Several further comments are in order. We have calculated, in the ϵ - regularization, the residue at ϵ - singularity coming from the region $z \sim 0$ (s channel, in terminology of dual models) of the integral (33). Similar contributions should be considered also coming from $z \sim 1$ and $z \sim \infty$ (t and u channels). And they have to be summed over, with the coefficients which are not defined within the ϵ - regularization used above. In fact, the regularization fixes the calculation just for one particular channel, and leaves the relative coefficients of different channels not defined.

Related flaw is that the intermediate state violates the relation $\Delta^M + \Delta^L = 1$, by ϵ amount, which we kept for the external states. In this respect the ϵ - regularization is not fully consistent.

Other questions have been left open in the above calculation. We assumed that the correlation functions are given by the residues of the physical state singularities. What happens to the nonphysical intermediate states in the full sum (38)? One suggestion is that these problems can be resolved and a fully consistent regularization can be achieved via the analytic decomposition of the integral in (33) into a sum of products of contour integrals. This is the alternative way in which the four-point (and multipoint) functions can be defined, and the operator algebra calculated, in the usual conformal theory, see [6, 20]. The expectation is that proceeding in this way the contribution of nonphysical channels will get cancelled, presumably because there are pairs of intermediate states, with exponents

$$\Delta^M + \Delta^L, \quad \text{and} \quad (1 - \Delta^M) + (1 - \Delta^L) = 2 - \Delta^M - \Delta^L \tag{48}$$

which enter under $\sin \pi(...)$ in the coefficients at products of the corresponding conformal blocks. The two terms corresponding to (48) would cancel each other.

Also, in case of terms corresponding to physical states, it is likely that poles will contribute only (z going around 0, 1, ∞), as exponents are integer in this case.

Proceeding in this way it might be possible to actually check what form of factorization of the four-point functions is taking place.

One more technical question is the sum in (38), and in (47), over the numbers of the screening operators. By their origin, they have to be summed in the ranges:

$$0 \leq \tilde{l}_1 \leq \tilde{l}, \quad 0 \leq \tilde{k}_1 \leq \tilde{k}, \quad 0 \leq l_1 \leq l, \quad 0 \leq k_1 \leq k \tag{49}$$

where, by the charge conservation, $\sum \alpha_i = 2\alpha_0$, $\sum \beta_i = 2\beta_0$, the total numbers of the screenings are given by:

$$l = \frac{s' + n' + m' + t' - 2}{2}, \quad \tilde{l} = \frac{-s' - n' - m' - t' - 2}{2} = -l - 2 \tag{50}$$

$$k = \frac{s + n + m + t - 2}{2} = \tilde{k} \tag{51}$$

So we have to define a sum over a negative number of screenings - over l_1, in this particular representation of the external operators. A simple conjecture would be, by extending the analytic continuation trick for finite products, eq.(17), that the sums are to be defined as:

$$\text{if} \quad s(l) = \sum_{i=0}^{l} f(i), \quad \text{then} \quad s(l = -|l|) = -\sum_{i=-1}^{|l|-1} f(-i) \tag{52}$$

(l - integer).

For the moment the above remarks are only conjectures and further work is needed.

We mention that special multipoint functions, which involve no matter screenings, have been calculated in [21], in the Liuoville field theory approach, using the representation of Goulian and Li. Other special four-point functions could be obtained by just differentiating the general three-point functions (18) with respect to μ. These are the four-point functions which involve one puncture operator. Those are special isolated cases by which the general technique could in particular be verified.

Acknowledgements

I am grateful to the colleagues at LPTHE for their kind hospitality. The efforts and the hospitality of the organizers of the Cargèse Summer School, and of the French - Russian Workshop are also gratefully acknowledged. The discussions with L. Alvarez-Gaumé, P. Bouwknegt, G. Felder, C. Gómes, S. Mukhi, and especially with V. A. Fateev, were both useful and stimulating.

References

[1] A. M. Polyakov, *Phys.Lett.* **B103** (1981) 207

[2] T. Curtright and C. Thorn, *Phys.Rev.Lett.* **48** (1982) 1309; E. Braaten, T. Curtright, and C. Thorn, *Phys.Lett.* **B118** (1982) 115; *Ann.Phys.* **147** (1983) 365

J.- L. Gervais and A. Neveu, *Nucl.Phys.* **B238** (1984) 125; **B238** (1984) 396; **B257[FS14]** (1985) 59
E. D'Hocker and R. Jackiw, *Phys.Rev.* **D26** (1982) 3517

[3] A. M. Polyakov, *Mod.Phys.Lett.* **A2** (1987) 893; V. G. Knizhnik, A. M. Polyakov, and A. B. Zamolodchikov, *Mod.Phys.Lett.* **A3** (1988) 819

[4] F. David, *Mod.Phys.Lett.* **A3** (1988) 1651
J. Distler and H. Kawai, *Nucl.Phys.* **B321** (1989) 509

[5] Vl. S. Dotsenko, Preprint PAR-LPTHE 91-18, February 1991

[6] Vl. S. Dotsenko and V. A. Fateev, *Nucl.Phys.* **B251** (1985) 691; *Phys.Lett.* **154B** (1985) 291

[7] P. Di Francesco and D. Kutasov, *Nucl.Phys.* **B342** (1990) 589

[8] M. R. Douglas, *Phys.Lett.* **B238** (1990) 176

[9] M. Goulian and M. Li, *Phys.Rev.Lett.* **66** (1991) 2051

[10] A. Gupta, S. Trivedi, and M. Wise, *Nucl.Phys.* **B340** (1990) 475

[11] E. Brezin and V. A. Kazakov, *Phys.Lett.* **B236** (1990) 144
M. R. Douglas and S. H. Shenker, *Nucl.Phys.* **B335** (1990) 635
D. J. Gross and A. A. Migdal, *Phys.Rev.Lett.* **64** (1990) 127

[12] L. Alvarez-Gaumé, J. L. F. Barbón, and C. Gómez, Preprint SLAC-PUB-CERN-TH.6142/91, June 1991

[13] E. D'Hocker, *Mod.Phys.Lett.* **A6** (1991) 745

[14] Y. Kitazawa, Preprint HUPT-91/A013, March 1991

[15] B. Lian and G. Zuckerman, *Phys.Lett.* **254B** (1990) 417

[16] C. Imbimbo, S. Mahapatra, and S. Mukhi, Preprint GEF-TH 8/91, TIFR/TH/91-27, May 1991

[17] P. Bouwknegt, J. McCarthy, and K. Pilch, Preprint CERN-TH.6162/91, July 1991

[18] G. Felder, *Nucl.Phys.* **B317** (1989) 215; erratum, *Nucl.Phys.* **B324** (1989) 548

[19] L. Alvarez-Gaumé and C. Gómez, Preprint SLAC-PUB-CERN-TH.6175/91, July 1991

[20] Vl. S. Dotsenko, *Adv.Stud. in Pure Math.* **16** (1988)123

[21] P. Di Francesco and D. Kutasov, *Phys.Lett.* **B261** (1991) 385

$\mathcal{W}$-ALGEBRAS AND LANGLANDS-DRINFELD CORRESPONDENCE

Edward Frenkel

Department of Mathematics, Harvard University
Cambridge, MA 02138 USA

ABSTRACT

The $\mathcal{W}$-algebras, associated to arbitrary simple Lie algebras, are defined as the cohomologies of certain BRST complexes. This allows to prove many important facts about them, such as determinant formulas, duality and free field resolutions for generic values of the central charge. A classical limit of a $\mathcal{W}$-algebra can be identified with the center of the universal enveloping algebra of the corresponding affine Kac-Moody algebra. This gives some information on the geometric Langlands-Drinfeld correspondence for complex algebraic curves.

1. INTRODUCTION

The $\mathcal{W}$-symmetry is one of the most interesting among the new symmetries, recently discovered in conformal field theory (CFT). Unlike other symmetries of CFT, it is not generated by a Lie algebra: the relations in the corresponding conformal algebra, which is called the $\mathcal{W}$-algebra, are highly non-linear. The appropriate mathematical language for $\mathcal{W}$-algebras is the theory of vertex operator algebras (VOA) [1]. The VOA of the $\mathcal{W}$-algebra is its vacuum representation, in which every vector defines a linear operator, depending on a complex parameter: a local field in the holomorphic sector of the corresponding CFT. The operator product expansion (OPE) expresses the composition of two local fields in terms of other local fields.

In this work I will use the definition of the $\mathcal{W}$-algebra $\mathcal{W}(g)$, associated to a simple Lie algebra g, by means of the quantum Drinfeld-Sokolov reduction [2]. Namely, $\mathcal{W}(g)$ appears as the cohomology of a certain BRST complex, which involves the affine Kac-Moody algebra $\hat{g}$ of g and fermionic ghosts, associated to the currents to the maximal nilpotent subalgebra n of g. This approach has many advantages.

First of all, it works for any simple Lie algebra g and any level. Second, the OPE of the $\mathcal{W}$-algebra closes automatically. Third, we can use standard homological technique to prove important facts about the $\mathcal{W}$-algebra. In particular, we can compute the $\mathcal{W}$-algebra, using a spectral sequence. For generic central charge this allows us to identify the VOA of $\mathcal{W}(g)$ with a subalgebra of the VOA of free fields, associated to the Cartan

New Symmetry Principles in Quantum Field Theory, Edited by
J. Fröhlich et al., Plenum Press, New York, 1992

subalgebra of g, which lies in the kernel of certain vertex operators. Moreover, $\mathcal{W}(g)$ is the 0th cohomology of a certain complex, which looks like the Bernstein-Gelfand-Gelfand (BGG) resolution of the trivial representation of g. Higher cohomologies of this complex vanish for generic central charge. When the central charge tends to infinity, the $\mathcal{W}$-algebra degenerates into the Gelfand-Dikii algebra (or classical $\mathcal{W}$-algebra), which is obtained by the Hamiltonian reduction (due to Drinfeld and Sokolov [3]) of the Poisson algebra of local functionals on the dual space to $\hat{g}$. We are then able to prove that the $\mathcal{W}$-algebra is a quantization of the Gelfand-Dikii algebra in the usual sense: the Fourier components of the local fields of the $\mathcal{W}$-algebra are quantum deformations of elements of the Gelfand-Dikii algebra, and their Lie bracket is a quantum deformation of the Poisson bracket in the Gelfand-Dikii algebra.

It means that the $\mathcal{W}$-algebra is essentially finitely generated: there is a finite number of fields: $W_1(z), \ldots, W_r(z)$, where r is the rank of g, such that the vacuum representation is linearly generated by polynomials in their Fourier components, applied to the vacuum vector. In other words, the OPE of $W_i(z)$ and $W_j(w)$ can be expressed in terms of local fields, which are polynomials in $W_k(w), 1 \leq k \leq r$ and their derivatives. The conformal dimension of $W_i(z)$ is equal to $d_i + 1$, where d_i is the ith exponent of g.

For example, $\mathcal{W}(sl_2)$ is the Virasoro algebra, and $\mathcal{W}(sl_3)$ is the Zamolodchikov algebra [4]. For simply-laced Lie algebras $\mathcal{W}(g)$ have been defined by Fateev and Lukyanov [5] by means of the direct quantization of the Miura transformation (which is only possible in the simply-laced case). This gives explicit formulas for the embedding of the $\mathcal{W}$-algebra into the free fields, or, in other words, for the 0th cohomology of our free field resolution. The difference is that we do not need to prove the closure of the OPE, it is insured automatically from the very beginning.

Our definition of the $\mathcal{W}$-algebra also gives a functor from the category of positive-energy representations of $\hat{g}$ to a category of representations of $\mathcal{W}(g)$, which is studied in detail in [6]. It is possible to define minimal representations of $\mathcal{W}(g)$, compute their characters (as residues of the affine characters) and fusion coefficients.

There are two classical limits of $\mathcal{W}(g)$: when the level of $\hat{g}$ tends infinity, it degenerates into the Gelfand-Dikii algebra, associated to g, and when the level of $\hat{g}$ tends to minus dual Coxeter number, which is called the critical level, it degenerates into the Gelfand-Dikii algebra, associated to the Langlands dual Lie algebra g' [2, 7]. It has been proved by Feigin and myself [2, 7], that this limit can be also identified with the center of a local completion $U(\hat{g})_{\text{loc}}$ of the universal enveloping algebra of $\hat{g}$.

This result plays an important role in the recent development of the geometric Langlands correspondence, which is due to Drinfeld. The concept of Langlands correspondence (for an introduction to the Langlands Program see, e.g. [8], see also [9], where its relation to CFT was first suggested) is one of the most important and universal in mathematics. It relates local and global properties of a field k. Namely, there is a one-to-one correspondence between the automorphic representations of an algebraic group $G_{\mathbf{A}}$ over the ring $\mathbf{A}$ of adeles of k and homomorphisms from the Galois group of k to the Langlands dual group G'. A representation is automorphic, if it occurs in the decomposition of the space of functions on the coset space $G_{\mathbf{A}}/G_k$. An unramified automorphic representation, which contains an invariant vector with respect to the compact subgroup K of $G_{\mathbf{A}}$, defines a function on the double coset space $K \backslash G_{\mathbf{A}}/G_k$.

In our case k is the field of rational functions on a complex curve $\mathcal{E}$, and the role of the Galois group is played by the fundamental group of $\mathcal{E}$. The adele group $G_{\mathbf{A}}$ can be described as the product of the loop groups $LG_p, p \in \mathcal{E}$ of G over all points of the curve $\mathcal{E}$. Here LG_p is the group of maps from a small circle S_p around the point p to G. The compact subgroup is the product of the subgroups LG_p^+ of LG_p, where LG_p^+ consists of the holomorphic maps from the interior of the circle S_p to G. The double coset space is then the moduli space $\mathcal{M}_G(\mathcal{E})$ of stable G–bundles over $\mathcal{E}$.

In the geometric version of the Langlands correspondence, proposed by Drinfeld [10], the unramified representation of the adele group are replaced by certain holonomic systems of differential equations (or, more generally, by $\mathcal{D}$–modules) on $\mathcal{M}_G(\mathcal{E})$.

There is a surjective map from the center of $U(\hat{g})_{\text{loc}}$ at the critical level to the algebra of differential operators on the bundle of half-forms on $\mathcal{M}_G(\mathcal{E})$ [11]. Our description of the structure of the center implies "half" of the Langlands-Drinfeld correspondence, which looks very much like a suitably defined modular functor at the critical level.

We see that the $\mathcal{W}$-symmetry helps to understand a very deep algebro-geometric concept. This is yet another example of the impact that two-dimensional conformal field theory has had on modern mathematics in the last few years.

The results of this paper were obtained in collaboration with Boris Feigin.

2. DEFINITION OF THE $\mathcal{W}$-ALGEBRA

Let g be a complex simple Lie algebra, and $\hat{g}$ – its affine Kac-Moody algebra, which is the extension of $g \otimes \mathbf{C}[t, t^{-1}]$ by the central element K. It is linearly spanned by K and $A(m) = A \otimes t^m, A \in g, m \in \mathbf{Z}$. Let k be a complex number. We denote by $U_k(\hat{g})$ the quotient of the universal enveloping algebra of $\hat{g}$ modulo the relation $K = k$. We call k the level of $\hat{g}$.

Let V_k be the vacuum representation of $\hat{g}$, generated by the vacuum vector, annihilated by $g \otimes \mathbf{C}[t]$, on which K acts by multiplication by k. The space V_k carries the structure of vertex operator algebra (VOA) [1]: every vector of V_k defines a local field, which is a formal series in z and z^{-1}, whose coefficients are linear operators, acting on V_k. There is an operator product expansion (OPE) for the product of two local fields (c.f. [2] for details).

The Fourier coefficients of local fields form a Lie algebra, which we denote by $U_k(\hat{g})_{\text{loc}}$. It lies in a completion of $U_k(\hat{g})$ and is linearly spanned by Fourier coefficients of the normally ordered polynomials in the basic local fields $A(z) = \sum A(m) z^{-m-1}, A \in g$ and their derivatives [2]. The Lie bracket in $U_k(\hat{g})_{\text{loc}}$ is completely determined by the singular part of the OPE. Denote by n the upper-nilpotent subalgebra of g. It is spanned by the root vectors $e_\alpha, \alpha \in \Delta_+$, where Δ_+ is the set of positive roots of g. Put $\hat{n} = n \otimes \mathbf{C}[t, t^{-1}]$. Let us introduce the Clifford algebra, which is generated by $\psi_\alpha(m), \psi^*_\alpha(m), m \in \mathbf{Z}, \alpha \in \Delta_+$ with the standard anti-commutation relations:

$$[\psi_\alpha(n), \psi^*_\beta(m)] = \delta_{\alpha,\beta}\delta_{n,-m}.$$

Denote by Λ its irreducible representation, generated by the vacuum vector, annihilated by $\psi(m), m \geq 0, \psi^*(m), m > 0$. It is $\mathbf{Z}$–graded: $\deg\psi(m) = -1$, $\deg\psi^*(m) = 1$.

The space $V_k \otimes \Lambda$ has a structure of complex. The differential is given by $d = d_{st} + p$. Here

$$d_{st} = \int \Big(\sum_{\alpha \in \Delta_+} e_\alpha(z)\psi^*_\alpha(z) - \frac{1}{2} \sum_{\alpha,\beta,\gamma \in \Delta_+} c^\gamma_{\alpha\beta} \psi^*_\alpha(z)\psi^*_\beta(z)\psi_\gamma(z) \Big) dz,$$

where $c^\gamma_{\alpha\beta}$ are the structural constants of $\hat{n}$, is the standard differential of Lie algebra cohomology of $\hat{n}$, and

$$p = \int \sum_{1 \leq j \leq l} \psi^*_{\alpha_j}(z) dz.$$

We have: $d^2 = 0$. This complex is the quantum BRST complex, associated to the Drinfeld-Sokolov reduction [3, 12, 2].

The space $V_k \otimes \Lambda$ carries the structure of VOA, which is preserved by the differential d. Therefore the cohomology $\oplus_j H_k^j$ of this complex is again a VOA. H_k^0 is called the $\mathcal{W}$-algebra and is denoted by $\mathcal{W}_k(g)$. It is closed with respect to OPE. Feigin and I conjectured that all other cohomologies $H_k^j, j \neq 0$ vanish [2]. In the next Section I will prove it for generic k (i.e. k is not a rational number, greater or equal to $-h^\vee$, where $h^\vee$ is the dual Coxeter number of g).

We can also define another complex $C_k(g)$, which consists of all Fourier coefficients of the local fields in $V_k \otimes \Lambda$, that is normally ordered polynomials in $A(z), A \in g, \psi_\alpha(z) = \sum_{m \in \mathbf{Z}} \psi_\alpha(m) z^{-m-1}, \psi_\alpha^*(z) = \sum_{m \in \mathbf{Z}} \psi_\alpha^*(m) z^{-m}, \alpha \in \Delta_+$, and their derivatives. The differential d acts on $C_k(g)$ and preserves the structure of Lie algebra. Therefore the cohomology $\oplus_j H_{k,\mathrm{loc}}^j$ is a Lie algebra. It was proved in [2], that $H_{k,\mathrm{loc}}^j$ it is spanned by Fourier coefficients of local fields from H_k^j. The Lie bracket in $W_k(g)_{\mathrm{loc}} = H_{k,\mathrm{loc}}^0$ is defined by the singular part of the OPE in $\mathcal{W}_k(g)$.

Note that for any positive-energy module M over $\hat{g}$ we may consider the complex $M \otimes \Lambda$ with respect to the differenstial d. The cohomology $\oplus H_k^j(M)$ of this complex is a module over $W_k(g)_{\mathrm{loc}}$. This defines a functor from the category of positive-energy representations of $\hat{g}$ to a category of $W_k(g)_{\mathrm{loc}}-$modules, which is studied in detail in [6]. Note that $H_k^j = H_k^j(V_k)$.

In the conclusion of this Section I want to explain a finite-dimensional counterpart of the definition of $\mathcal{W}$-algebra. Let a be a Lie algebra, and b – its Lie subalgebra. Let $\chi : b \rightarrow \mathbf{C}$ be a character of b, which defines its one-dimensional representation $\mathbf{C}_\chi$. We can ask the following question: what is the (maximal) algebra, which acts on the space $(M \otimes \mathbf{C}_\chi)^b$ for any $g-$module M? Here $(M \otimes \mathbf{C}_\chi)^b$ denotes the space of invariants of the $b-$module $(M \otimes \mathbf{C}_\chi)$, which is the subspace of all vectors in M, such that any element $x \in b$ acts on them by multiplication by $-\chi(x)$. This algebra is called the Hecke algebra, associated to the triple (a, b, χ). Note that the algebra $\mathcal{H}_q$, which is usually called the Hecke algebra, appears in a similar context: when a is replaced by a finite group of Lie type over the finite field of q elements, b – by its Borel subgroup.

One possibility to find this algebra is the following. Recall that the space of $b-$invariants of a module N coincides with the 0th cohomology of b with coefficients in N. This cohomology is defined as the cohomology of the complex $N \otimes \Lambda(b)$ with respect to the standard (BRST) differential d_{st}. More precisely: let $\{e_i, i \in I\}$ be a basis in b. Introduce the Clifford algebra Cl, generated by $\psi_i, \psi_i^*, i \in I$ with the standard anti-commutation relations: $[\psi_i, \psi_j^*] = \delta_{i,j}$. Denote by $\Lambda(b)$ its representation, generated by the vacuum vector, annihilated by all ψ_i: this is the exterior algebra of the dual space to b. We have $d_{st} = \sum e_i \psi_i^* - 1/2 \sum c_{ij}^k \psi_k \psi_i^* \psi_j^*$, where c_{ij}^k are the structure constants of b.

If $N = M \otimes \mathbf{C}_\chi$, then this complex is isomorphic to the complex $M \otimes \Lambda(b)$ with respect to the differential $d = d_{st} + \sum \chi(e_i) \psi_i^*$. The algebra of all endomorphisms of the complex is $U(g) \otimes Cl$. The Hecke algebra consists of such endomorphisms of this complex, which commute with the differential d. The operator $D = [d, \cdot]$ – the supercommutator with d, defines a differential in $U(g) \otimes Cl$. All elements of $U(g) \otimes Cl$, which commute with d, act naturally on the cohomologies of $M \otimes \Lambda(b)$ for any $a-$module M, and those of them, which are of the form $[d, x]$ for some x, act by 0. Therefore the cohomologies of this complex with respect to D, i.e. the quotient of the kernel of D modulo the image of D, act on the cohomologies of $M \otimes \Lambda(b)$. The cohomologies of $U(g) \otimes Cl$ form an algebra. The 0th cohomology is the Hecke algebra we were looking for.

Here is an example. Let a be a simple Lie algebra g, b – its upper-nilpotent subalgebra n, and χ be such that $\chi(e_\alpha) = 1$, if α is a simple root, and 0, otherwise. Clearly, one always has a map from the center of $U(g)$ to the Hecke algebra. In general, this map may well have both a kernel and a cokernel: there may be other elements in the Hecke algebra, and some of the central elements may act by zero on the space $(M \otimes \mathbf{C}_\chi)^b$ for any g-module M. Kostant showed, however, [13] that in this example the Hecke algebra is isomorphic to the center of $U(g)$. He also showed that all other cohomologies of the complex $U(g) \otimes Cl$ vanish.

Our definition of the $\mathcal{W}$-algebra is a generalization of this construction. Moreover, as we have shown in [2, 7], there is an affine analogue of the Kostant's result on the cohomologies of our complex, when $k = -h^\vee$ (cf. Sect. 4).

3. FREE FIELD RESOLUTIONS

In this section we will construct free field resolutions of $\mathcal{W}$-algebras for generic values of the central charge. Our strategy is the following: first we will construct a resolution of the vacuum representation of the affine algebra $\hat{g}$ in terms of the Wakimoto modules for generic level. Then we will convert it into a resolution of $\mathcal{W}(g)$ in terms of Fock representations, using the vanishing of the cohomologies of the Wakimoto modules. The vanishing of higher cohomologies of this resolution for generic central charge c of $\mathcal{W}(g)$ will follow from the acyclicity of its classical limit, when c tends to infinity.

Let us denote by $W_{\lambda,k}$ the Wakimoto module over $\hat{g}$ of level k with highest weight λ [14, 15]. There exists an explicit realization of this module in terms of r free scalar fields and $|\Delta_+|$ $\beta\gamma$-systems. We know [16] that $H^j_k(W_{\lambda,k}) = 0$, if $j \neq 0$, and $H^0_k(W_{\lambda,k}) = \pi_{\lambda,k+h^\vee}$, where $\pi_{\lambda,k+h^\vee}$ is the Fock representation of the Heisenberg algebra of level $k+h^\vee$ with highest weight λ. This Heisenberg algebra is generated by $u(m), m \in \mathbf{Z}, u \in h$ – the Cartan subalgebra of g with the standard commutation relations:

$$[u(n), u(m)] = (k + h^\vee)n(u|v)\delta_{n,-m},$$

where $(|)$ denotes the invariant scalar product on h. Throughout this Section we will assume that $k \neq -h^\vee$. Therefore we can normalize the Heisenberg algebra by introducing new generators $\bar{u}(m) = u(m)/\nu$, where $\nu = \sqrt{k + h^\vee}$. Then $\pi_{\lambda,k+h^\vee}$ becomes $\pi_{\lambda/\nu,1} = \pi_{\lambda/\nu}$. Denote by $\alpha_i, 1 \leq i \leq r$, the set of simple roots of g.

We have the following complex: $C^*_k(g) = \oplus C^j_k(g)$, where

$$C^j_k(g) = \oplus_{l(s)=j} W_{s(\rho)-\rho,k},$$

where s runs over the Weyl group S of g and ρ is the half-sum of the positive roots of g.

Let me give explicit formulas for the differentials of this complex. Introduce the screening operators $s_i = \int s_i(z)dz$,

$$s_i(z) = \bar{e}_i(z)V_{-\alpha_i/\nu}(z),$$

where $V_{\gamma/\nu}(z)$ denotes the bosonic vertex operator, acting from $\pi_{\lambda/\nu}$ to $\pi_{(\lambda+\gamma)/\nu}$ (and therefore from $W_{\lambda,k}$ to $W_{\lambda+\gamma,k}$) and $\bar{e}_i(z)$ is a certain polynomial in $\beta\gamma$-fields, which comes from the right action of the Lie algebra $\hat{n}$ on $W_{\lambda,k}$ [17, 18]. These operators commute with the action of $\hat{g}$ on the Wakimoto modules.

Let me introduce, following [18], the operators

$$[s_{i_1} \ldots s_{i_n}] : W_{\lambda,k} \rightarrow W_{\lambda+\gamma,k},$$

where $1 \leq i_l \leq r$, and γ is equal to $\sum \alpha_{i_l}$.

$$[s_{i_1} \ldots s_{i_n}] = \int_\Gamma dz_1 \ldots dz_n s_{i_1}(z_1) \ldots s_{i_n}(z_n), \tag{1}$$

where Γ is a set of contours taken counterclockwise from the basepoint $z = 1$ to itself around 0 and nested according to $|z_1| > \ldots > |z_n|$, as in [18]. These operators obey the Serre relations of the quantum group $U_q(g)$, where $q = \exp \pi i/(k + h^\vee)$ [18, 19]. It means that there is a well-defined linear map ϵ_q from the nilpotent quantum group $U_q(n)$ to $\mathrm{End}(\oplus W_{\gamma,k})$, where γ runs over the root lattice of g, which sends the standard monomial $e_{i_1} \ldots e_{i_n}$ from $U_q(n)$ to $[s_{i_1} \ldots s_{i_n}]$.

Let M_λ be the Verma module over $U_q(g)$, free over $U_q(n)$, with lowest weight λ. It is known [20] that for any q M_0 contains singular vectors $w_s, s \in S$ and that the singular vector $w_{s'}$ is contained in the Verma submodule, generated by the singular vector w_s, if and only if $l(s') = l(s)+1$, and $s' > s$ [21]. For such pairs s, s' we have $w_{s'} = P_{s,s'} w_s$, where $P_{s,s'} \in U_q(n)$. It defines an embedding $M_{s'(\rho)-\rho} \rightarrow M_{s(\rho)-\rho}$ of the Verma modules over $U_q(g)$. We can then define a complex $B^q_*(g) = \oplus B^q_j(g)$, where $B^q_j(g) = \oplus_{l(s)=j} M_{s(\rho)-\rho}$, with the differential given by an alternating sum of such embeddings, as in [21]. For $q = 1$ this gives the standard Bernstein-Gelfand-Gelfand (BGG) resolution of the trivial representation of g. For generic q it gives a resolution of the trivial representation of $U_q(g)$ in terms of the Verma modules: the 0th cohomology is one-dimensional, and all higher cohomologies vanish.

It follows from [18], Theorem 3.5, that $\epsilon_q(P_{s,s'})$ is an intertwining operator between $W_{s(\rho)-\rho,k}$ and $W_{s'(\rho)-\rho,k}$, if $q = \exp \pi i/(k + h^\vee)$. The differential $d^j : C^j_k(g) \rightarrow C^{j+1}_k(g)$ is now given by an alternating sum of the maps $\epsilon_q(P_{s,s'}) : W_{s(\rho)-\rho,k} \rightarrow W_{s'(\rho)-\rho,k}$, where $l(s) = j, l(s') = j + 1, s' > s$. The signs are chosen in such a way that the product of them over every "fundamental square" of the Weyl group is equal to minus [21]. Then in the same way as in the case of the BGG resolution one proves that the differential is nilpotent: $d^{j+1} d^j = 0$.

It is easy to prove, c.f. [2], using the information on the structure of the Wakimoto modules, that for generic k ($k + h^\vee$ is not a positive rational number) the 0th cohomology of the complex $C^*_k(g)$ is isomorphic to the vacuum representation V_k of $\hat{g}$ and all higher cohomologies vanish.

We now want to apply the functor $H^j_k(\cdot)$ to this resolution. Since $H^j_k(W_{\lambda,k}) = 0$, if $j \neq 0$, and $H^0(W_{\lambda,k}) = \pi_{\lambda/\nu}$, $H^j_k(V_k)$ coincides with the jth cohomology of the complex $F^*_k(g) = \oplus F^j_k(g)$, where

$$F^j_k(g) = \oplus_{l(s)=j} \pi_{(s(\rho)-\rho)/\nu}.$$

The differential $D^j_k : F^j_k(g) \rightarrow F^j_k(g)$ of this complex is given by the same formula as the differential d_j of the complex $C^*_k(g)$, using the operators $[i_1 \ldots i_n]$, which are given by formula (1), where $s_i(z)$ is to be replaced by $V^{(i)}(z) = V_{-\alpha_i/\nu}(z)$. This two operators are equivalent, because they differ by a commutator with the differential of our BRST complex. The monodromy properties of the operators $V^{(i)}(z)$ are the same as the monodromy properties of the operators $s_i(z)$, therefore again $D^{j+1} D^j = 0$.

This complex is well-defined for any k. It is possible to extend this complex to the value $k = \infty$. In order to do that we should pass to a new basis in $\pi_{\lambda/\nu}$, which consists of all monomials in $u'(m) = u(m)/\nu^2, m < 0$. We should also rescale the differentials of the complex $F^*_k(g)$, passing from $[i_1 \ldots i_n]$ to $\nu^{2n}[i_1 \ldots i_n]$. In this limit $\mathcal{W}_k(g)_{\mathrm{loc}}$ degenerates into the Gelfand-Dikii algebra, in which the Poisson structure is given by the $(k + h^\vee)^{-1}-$linear term of the Lie bracket in $\mathcal{W}_k(g)_{\mathrm{loc}}$.

Now I want to explain, why higher cohomologies of this complex vanish for generic k.

We have a family of complexes $F^*_k(g)$. Each of them decomposes into a direct sum of finite-dimensional subcomplexes $F^*_k(g)_n, n \geq 0$, where $F^*_k(g)_n$ is the eigenspace of the operator $L_0 \in \mathcal{W}_k(g)_{\mathrm{loc}}$ with the eigenvalue n. The differential D^j_k decomposes into a sum of differentials $D^j_k(n) : F^j_k(g)_n \rightarrow F^{j+1}_k(g)_n$. As linear spaces $F^j_k(g)_n$ do not depend on k. The differentials $D_k^j(n)$ depend algebraically on $\nu^{-2} = (k + h^\vee)^{-1}$. Therefore the dimension of the kernel of $D_k^j(n)$ is the same for all but finitely many values of k, where it may jump *up*. The dimension of the image of $D_k^{j-1}(n)$ is the same for all but finitely many values of k, where it may jump *down*. Thus the dimension of the jth cohomology of the complex $F_k^*(g)_n$ is the same for all but finitely many of values of k, and in special points it may only jump *up*. In particular, if for some value of k the jth cohomology of the complex $F_k^*(g)$ vanishes, then it also vanishes for generic value of k (i.e. for all but countably many values of k).

So now in order to prove the vanishing of higher cohomologies for generic k it is enough to prove it for some value of k. I will do it for $k = \infty$. In fact, the vanishing of the higher cohomologies for $k = \infty$ follows from [3], Proposition 6.1. But I will give another proof, which will lead to a better understanding of the meaning of the complex $F_k^*(g)$.

When $\nu^{-1} \to 0$, the modules $\pi_{\lambda/\nu}$ get identified with π_0. The limit $\bar{V}_i$ of the operator $\nu^2 \int V^{(i)}(z)dz$ is equal to

$$\sum_{m<0} \partial/\partial\alpha_i'(m) \cdot \bar{V}_i(m+1),$$

where $[\partial/\partial u'(m), v'(n)] = (u|v)\delta_{n,m}, m, n < 0$, and $\bar{V}_i(l)$ are defined by the following generating function

$$\sum_{l\leq 0} \bar{V}_i(l)z^l = \exp(\sum_{m<0} \alpha_i'(m)z^m/m).$$

The composition operators $[i_1 \dots i_n]$ become the products $\bar{V}_{i_1} \dots \bar{V}_{i_n}$.

The operators $\bar{V}_i, 1 \leq i \leq r$ generate the nilpotent subalgebra n of g. It is a useful exercise to check this directly.

Therefore the complex $F_\infty^*(g)$ is isomorphic to $\mathrm{Hom}_n(B_*^1(g), \pi_0)$, where $B_*^1(g)$ is the BGG resolution of the trivial representation of g. Indeed, for any λ the module M_λ is a free $n-$module M. Therefore the module of $n-$homomorphisms $\mathrm{Hom}_n(M, \pi_0)$ is canonically isomorphic to π_0, namely, any non-zero homomorphism $x \in \mathrm{Hom}_n(M, \pi_0)$ defines a non-zero element in N: the image of the lowest weight vector of M. The embedding of M into itself, which is given by the singular vector $P_{s,s'}$, induces the homomorphism from π_0 to π_0, which sends $y \in \pi_0$ to $P_{s,s'} \cdot y$. Hence the differentials of the BGG resolution $B_*^1(g)$ go to the differentials of the complex $F_\infty^*(g)$.

It means that the cohomologies of the complex $F_\infty^*(g)$ coincide with the cohomologies of the Lie algebra n with coefficients in π_0 (with respect to the action by the operators $\bar{V}_i$). This action is "co-free" (i.e. the action on the dual module is free). In other words, we have $\pi_0 = \mathcal{W}_\infty(g) \otimes M^*$, where M^* is the dual module to the free $n-$module. It implies that the 0th cohomology is the space $\mathcal{W}_\infty(g)$ of invariants of π_0 under the action of n, and that all higher cohomologies vanish. Therefore higher cohomologies also vanish for generic k.

Recall that for any $\hat{g}-$module M the cohomolgy $H_k^j(M)$ is a module over $\mathcal{W}_k(g)_{\mathrm{loc}}$. In particular, $H_k^j(V_k)$ is the vacuum representation of $\mathcal{W}_k(g)_{\mathrm{loc}}$ (which is also the VOA $\mathcal{W}_k(g)$). So we may view the complex $F_k^*(g)$ as a BGG resolution of the vacuum representation of $\mathcal{W}_k(g)_{\mathrm{loc}}$ in terms of Fock representations. We have proved the following result.

Theorem 1 *For generic k there is a resolution $F_k^*(g) = \oplus F_k^j(g)$, where*

$$F_k^j(g) = \oplus_{l(s)=j} \pi_{(s(\rho)-\rho)/\nu},$$

of the vacuum representation of the $\mathcal{W}$-algebra $\mathcal{W}_k(g)_{\mathrm{loc}}$ in terms of the Fock representations, such that the 0th cohomology is the vacuum representation, and all higher cohomologies vanish.

For $g = sl_2$ this result follows from [22]. For $g = sl_n$ this result was obtained in [23] by other means.

It is natural to conjecture that there is a (highly non-local) action of the (nilpotent) quantum group $U_q(n)$ on π_0 for any k and $q = \exp \pi i/(k + h^\vee)$, such that it commutes with the action of $\mathcal{W}_k(g)_{\text{loc}}$ and with respect to their action the module π_0 decomposes as $\mathcal{W}_k(g) \otimes M^*$ for generic k. Theorem 1 would follow from this conjecture, because if it was true, then for any k the complex $F_k^*(g)$ would be isomorphic to $\text{Hom}_{U_q(n)}(B_*^q, \pi_0)$, and so its cohomologies would coincide with the cohomologies of $U_q(n)$ with coefficients in π_0. Therefore higher cohomologies would vanish for generic k, because the action of $U_q(n)$ on π_0 would be "co-free".

It follows from Theorem 1 that for generic k the $\mathcal{W}$-algebra $\mathcal{W}_k(g)$ is isomorphic to the intersection of the kernels of the operators $\int V^{(i)}(z) : \pi_0 \to \pi_{-\alpha_i/\nu}$. (In [2] we proved this result in a slightly different way.) Clearly, the elements of $\mathcal{W}_k(g)$ depend algebraically on k, therefore they define a VOA for any k. Note that this VOA is contained in $\mathcal{W}_k(g)$ (defined as the 0th cohomology of a BRST complex, cf. Sect. 2) even for non-generic k, so that we have a family of algebras, depending on a complex parameter k. It follows from our conjecture on the vanishing of cohomologies of the BRST complex, that it is isomorphic to $\mathcal{W}_k(g)$ for any k.

Note that the complex $F_k^*(g)$ is well-defined for any k, but for special values of k its cohomologies jump up, so that it is no longer a resolution of $H_k^0(V_k)$ (in other words, the action of $U_q(n)$ is no longer "co-free"). It happens exactly when $H_k^0(V_k)$ is not irreducible (as it is for generic k), but contains an extra singular vector. We then should take the quotient modulo this vector to obtain the irreducible vacuum representation in this case. The free field resolution of such module should be two-sided (as it is in the case of $\mathcal{W}(sl_2)$, which is the Virasoro algebra [24]). Such resolution was conjectured in [6, 19].

We can calculate the character (or partition function) of the $\mathcal{W}$-algebra $\mathcal{W}_k(g)$ for generic k. The simplest way to do that is to calculate it for $k = \infty$ (obviously, it will be the same for generic k). Using the decomposition $\pi_0 = \mathcal{W}_\infty(g) \otimes M^*$, we find:

$$\text{ch}\mathcal{W}(g) = \text{ch}\pi_0 \cdot (\text{ch}M^*)^{-1},$$

where $\text{ch}X = \sum \dim X_m t^m$. We have: $\text{ch}\pi_0 = \prod_{m>0}(1-t^m)^{-1}$. Since the operators $\bar{V}_i$ – the generators of the Lie algebra n – have degree 1, $\text{ch}M^* = \prod_{1\leq i\leq r}\prod_{1\leq m_i\leq d_i}(1-t^{m_i})^{-1}$, where $d_i, 1 \leq i \leq r$, are the exponents of g. Therefore,

$$\text{ch}\mathcal{W}(g) = \prod_{1\leq i\leq r}\prod_{m_i > d_i}(1-t^{m_i})^{-1}.$$

This implies that the $\mathcal{W}$-algebra is essentially finitely-generated: there are fundamental fields $W_i(z), 1 \leq i \leq r$, of conformal dimensions d_i+1, such that the ordered polynomials in their Fourier complonents $W_i(m_i), m_i < -d_i$ constitute the (Poincare-Birkhoff-Witt) basis of $\mathcal{W}_k(g)$ (or, in other words, the VOA is spanned by normally ordered polynomials in $W_i(z)$ and their derivatives. The formula for $W_1(z)$, whose Fourier components generate the Virasoro subalgebra of the $\mathcal{W}$-algebra, is given in [2] for any g.

Explicit formulas for other generators of the $\mathcal{W}$-algebra were given by Fateev and Zamolodchikov [25] for $g = sl_3$, and by Fateev and Lukyanov [5] for sl_n and $so_{2n}, n > 3$. Their formulas were obtained by direct quantization of the Miura transformation. It is easy to check that the corresponding fields are indeed in the kernel of the vertex operators $\int V^{(i)}(z)dz$. Since they have the correct conformal dimensions, they generate the whole $\mathcal{W}$-algebra. It is important that we do not need to prove that their operator product closes: it automatically follows from our definition of the $\mathcal{W}$-algebra. For non simply laced algebras one can not directly quantize the Miura transformation, and I do not know any anzatz for the fields $W_i(z)$ in this case.

From the description of the $\mathcal{W}$-algebra as the kernel of vertex operators and the existence of the Poincare-Birkhoff-Witt basis the determinant formulas [26] follow, using the explicit formulas for the intertwining operators as contour integrals of vertex operators $V^{(i)}(z)$. Further results on the representatation theory of $\mathcal{W}$-algebras can be found in [5, 6].

The Lie algebra $\mathcal{W}_k(g)_{\mathrm{loc}}$ of the Fourier components of the local fields from $\mathcal{W}_k(g)$ is a quantum deformation of the Gelfand-Dikii algebra $W_\infty(g)_{\mathrm{loc}}$, associated to g.

Let us pass to the basis in $\mathcal{W}_k(g)_{\mathrm{loc}}$, constituted by polynomials in $u'(m), u \in h, m \leq 0$. The 0th differential D_k^0 of the complex $F_k^*(g)$ (the sum of the vertex operators $\int V^{(i)}(z)dz$) is equal to $\sum_{j\geq 0} d_j^0(k+h^\vee)^{-j}$, where the constant term d_0^0 equals D_∞^0. The cohomology vanishing theorem guarantees that every $A_0 \in \mathcal{W}_\infty(g)$ can be extended to an element A of $\mathcal{W}_k(g)$, which a polynomial in $(k+h^\vee)^{-1}$ (this is the parameter of deformation), with constant term A_0. We have: $A = \sum_{i\geq 0} A_i(k+h^\vee)^{-i}$. The condition $D_k^0 A = 0$ for any k is then equivalent to the system of equations $\sum_{i+j=m} d_j^0 A_i = 0, m \geq 0$. I want to use these equations to determine $A_i, i > 0$.

The first equation $d_0^0 A_0 = 0$ is satisfied, since A_0 belongs to $\mathcal{W}_\infty(g)$. The second equation reads $d_0^0 A_1 = -d_1^0 A_0$. There exists A_1, which satisfies this equation. Indeed, we can also decompose the 1st diferential D_k^1 of the complex $F_k^*(g)$ as $\sum_{j\geq 0} d_j^1(k+h^\vee)^{-j}$, where d_0^1 is equal to D_∞^1. Since $D_k^1 D_k^0 = 0$, we have $\sum_{i+j=0} d_i^1 d_j^0 = 0, m \geq 0$. In particular, $d_0^1 d_1^0 = -d_1^1 d_0^0$. Therefore $d_0^1(d_0^0 A_1) = -d_0^1(d_1^0 A_0) = d_1^1(d_0^0 A_0) = 0$. But the first cohomology of the complex $F_\infty^*(g)$ is trivial. Therefore, if $d_0^1 B = 0$ for some $B \in F_\infty^1(g)$, then there exists $B' \in F_\infty^0(g)$, such that $B = d_0^0 B'$. In our case $B = -d_1^0 A_0$, and so there exists A_1, such that $d_0^0 A_1 = -d_1^0 A_0$.

We can proceed further in this way. On the nth step we will know $A_i, 0 \leq i \leq n-1$, and we will have to find A_n, which satisfies

$$d_0^0 A_n = -\Big(\sum_{1\leq j\leq n} d_j^0 A_{n-j}\Big). \tag{2}$$

If we apply d_0^1 to the right hand side of (2), we obtain

$$d_0^1\Big(-\sum_{1\leq j\leq n} d_j^0 A_{n-j}\Big) = \sum_{0\leq j\leq n-1} d_{n-j}^1 \cdot \sum_{0\leq i\leq j} d_i^0 A_{j-i} = 0.$$

Therefore $d_0^1(d_0^0 A_n) = 0$, and there exists A_n, which satisfies (2).

Thus, every element of $W_\infty(g)$ can be deformed to an element of $\mathcal{W}_k(g)$ for any k. Therefore any element a_0 of the Gelfand-Dikii algebra $W_\infty(g)_{\mathrm{loc}}$ can also be deformed to an element $a = a_0 + (k+h^\vee)^{-1} + \ldots$ of $W_k(g)_{\mathrm{loc}}$ for any k. The Lie bracket of a and b from $W_k(g)_{\mathrm{loc}}$ is given by $[a,b] = (k+h^\vee)^{-1}\{a_0, b_0\} + (k+h^\vee)^{-2}(\ldots)$, where $\{a_0, b_0\}$ is the Poisson bracket of a_0 and b_0 in the Gelfand-Dikii algebra. It is known [3] that $W_\infty(g)_{\mathrm{loc}}$ is the Poisson algebra of local functionals on the space of gauge classes of certain $g-$connections on the circle, which are intergrals of polynomials in $\bar{W}_i(z), 1 \leq i \leq r$ and their derivatives. For example, if $g = sl_n$, then the Gelfand-Dikii algebra is the space of local functionals on the differential operators on the cercle of the form $\partial^n + \bar{W}_1(z)\partial^{n-2} + \ldots + \bar{W}_{n-1}(z)$. In this case the local field $W_i(z)$ is precisely the quantum deformation of $\bar{W}_i(z)$.

In the conclusion of this Section, I would like to mention that the complex $F_k^*(g)$ can be defined for an arbitrary Kac-Moody algebra g. As I have just explained, if g is a simple Lie algebra, then the cohomology of this complex gives the $\mathcal{W}$-algebra, associated to g. In [27] Feigin and I have shown that if g is an affine algebra, then the cohomology of this complex gives integrals of motion in the quantum affine Toda field theory, associated to g. It is interesting, what is the meaning of this complex for non-affine Kac-Moody algebras.

4. CRITICAL LEVEL AND LANGLANDS-DRINFELD CORRESPONDENCE

The $\mathcal{W}$-algebra also has another classical limit: when k tends to $-h^\vee$. It was proved in [2, 7], that in this limit $W_k(g)_{\text{loc}}$ degenerates into a Poisson algebra, which is isomorphic to the Gelfand-Dikii algebra, associated to the Langlands dual Lie algebra g' (the Cartan matrix of g' is the transpose of the Cartan matrix of g). This result is the limit of the following duality of $\mathcal{W}$-algebra: $\mathcal{W}_k(g)$ is isomorphic to $\mathcal{W}_{k'}(g')$, if $r^\vee(k + h^\vee(g)) = (k' + h^\vee(g'))^{-1}$, where $r^\vee$ is the maximal number of edges, connecting two vertices of the Dynkin diagram of g. It follows from the description of the $\mathcal{W}$-algebra as the kernel of the vertex operators [2].

Let $U_k(\hat{g})_{\text{loc}}$ be the local completion of the universal enveloping algebra of $\hat{g}$ (cf. Sect. 2). This is a Lie algebra. Denote its center by $Z_k(g)$. Let $x \in V_k$ be a singular vector of *imaginary* degree. Then the operator product of the basic local fields of $\hat{g}$ with the local field $x(z)$, corresponding to x, has no singularities. Therefore all Fourier components of $x(z)$ belong to $Z_k(g)$. It is also possible to prove the converse statement [2]. It is known [28] that there is no singular vectors of imaginary degrees in V_k, if $k \neq -h^\vee$. Therefore $Z_k(g) = \mathbf{C}$, if $k \neq -h^\vee$.

For $k = -h^\vee$ there is a precise analogue of the Kostant's theorem (cf. Sect. 2): the center $Z_{-h^\vee}(g)$ is isomorphic to $W_{-h^\vee}(g)_{\text{loc}}$ [2, 7]. Combined with the duality theorem, it gives the following result.

Theorem 2 *The center of the local completion of the universal enveloping algebra of an affine Kac-Moody algebra $\hat{g}$ is isomorphic to the Gelfand-Dikii algebra, associated to g' – the Langlands dual Lie algebra to g.*

In fact, there is a natural Poisson structure on $Z_{-h^\vee}(g)$. We can identify $U_k(\hat{g})_{\text{loc}}$ as linear spaces for all k. If we have two central elements in $U_{-h^\vee}(g)_{\text{loc}}$, then their Lie bracket is a polynomial in $(k + h^\vee)$ with zero constant term. The $(k + h^\vee)$–linear term defines a Posson bracket on $Z_{-h^\vee}(g)$. This Poisson structure coincides with the Poisson structure on the Gelfand-Dikii algebra.

Let me give you an example. Define the Sugawara field:

$$S(z) = \sum_{n \in \mathbf{Z}} S(n) z^{-n-2} = \frac{1}{2} \sum u_i^2(z),$$

where u_i are the elements of an orthonormal basis of g with respect to the standard inner product. We have:

$$[S(n), A(m)] = -m(k + h^\vee) A(n+m),$$

for any $A \in g$. Therefore $S(n), n \in \mathbf{Z}$ belong to $Z_{-h^\vee}(g)$.

If $g = sl_2$, then the center is generated by $S(z)$: every central element is a Fourier component of a polynomial in $S(z)$ and its derivatives. We also have:

$$[S(n), S(m)] = (k+2)((n-m)S(n+m) + k/4 \cdot (n^3 - n)\delta_{n,-m}).$$

Therefore the Poisson bracket of $S(n)$ and $S(m)$ is equal to $(n-m)S(n+m)-1/2\cdot(n^3-n)\delta_{n,-m}$. Hence as a Poisson algebra $Z_{-2}(sl_2)$ is isomorphic to the Gelfand-Dikii algebra of sl_2, which is the space of local functionals on the dual space to the Virasoro algebra.

The operators $S(n)$ define a Poisson subalgebra of $Z_{-h^\vee}(g)$ for any g. Some explicit formulas for other elements of the center can be found in [29]. They are unknown in general, although it follows from Theorem 2 that their symbols give all $\hat{g}$−invariant local functionals on the dual space to $\hat{g}$.

Theorem 2 gives a possibility to assign $\hat{g}$−modules at the critical level to certain geometric data. Indeed, the Gelfand-Dikii algebra, associated to g is the algebra of functions (more precisely, local functionals) on the space $\mathcal{C}(g)$ of gauge classes of connections on the circle of special kind [3]. Let c be an element of $\mathcal{C}(g')$. It defines a map μ_c from the Gelfand-Dikii algebra of g' to $\mathbf{C}$: we take the value of each local functional at the point c. We can then define the module $V_{-h^\vee,c}$ over $\hat{g}$, taking the quotient of $V_{-h^\vee}$ modulo the relations $u=\mu_c(u)$ for any $u\in Z_{-h^\vee}(g)$. The module $V_{-h^\vee,c}\neq 0$, if and only if c can be holomorphically extended inside the circle, embedded into $\mathbf{C}$.

For example, $\mathcal{C}(sl_2)$ is the space of second order differential operators of the form $\partial_t^2+q(t)$ on the circle. An operator of this type can be extended inside the circle, if $q(t)$ is a polynomial in t: $q(t)=\sum_{m\leq-2}q(m)t^{-m-2}$. The module, corresponding to such operator, is the quotient of V_{-2} modulo its submodule, generated by the vectors $(S(m)-q(m))v_{-2}, m\leq-2$, where v_{-2} is the vacuum vector of V_{-2}.

Now I want to describe briefly a "globalization" of the Theorem 2, which was recently proposed by Drinfeld. It relates holonomic systems of differential equations on the moduli space $\mathcal{M}_G(\mathcal{E})$ of stable principal G−bundles on a smooth projective algebraic curve $\mathcal{E}$ over $\mathbf{C}$ to holomorphic G'−connections on $\mathcal{E}$.

This construction may be viewed as a quantization of a remarkable completely integrable hamiltonian system on the cotangent bundle to $\mathcal{M}_G(\mathcal{E})$, which is due to Hitchin [30]. Let me recall his construction. Let $T^*\mathcal{M}_G(\mathcal{E})$ be the cotangent bundle to $\mathcal{M}_G(\mathcal{E})$. A point of $T^*\mathcal{M}_G(\mathcal{E})$ is a pair (P,ξ), where P is a principal G−bundle over $\mathcal{E}$ and ξ is a cotangent vector to $\mathcal{M}_G(\mathcal{E})$ at the point $P\in\mathcal{M}_G(\mathcal{E})$ (sometimes it is called a Higgs field). The tangent bundle to P is isomorphic to $H^1(\mathcal{E},g_p)$, where $g_p=P\times_G g$ is the vector bundle, associated to P and the adjoint representation of G. So $\xi\in H^1(\mathcal{E},g_p)^*=H^0(\mathcal{E},g_p^*\otimes\Omega^1)$, where Ω^1 is the bundle of 1-differentials on E, by the Serre duality. We can define a map $\varphi: T^*\mathcal{M}_G(\mathcal{E})\rightarrow\mathcal{H}=\oplus H^0(\mathcal{E},\Omega^{d_i+1})$, which sends $(P,\xi)\in T^*\mathcal{M}_G(\mathcal{E})$ to $(\Delta_1(\xi),\ldots,\Delta_r(\xi))$, where Δ_i is a primitive generator of the algebra of G−invariant polynomials on the dual space g^* to g; Δ_i is of degree d_i+1, and it sends ξ to $H^0(\mathcal{E},\Omega^{d_i+1})$.

The miraculous fact is that the dimension of $\mathcal{H}$ is equal to the dimension of the moduli space, that is to one half of the dimension of $T^*\mathcal{M}_G(\mathcal{E})$. Indeed, the dimension of $\mathcal{M}_G(\mathcal{E})$ is equal to $(\mathbf{g}-1)\dim G$, where $\mathbf{g}$ is the genus of the curve $\mathcal{E}$, and the dimension of $\mathcal{H}$ is equal to $(\mathbf{g}-1)\sum(2d_i+1)=(\mathbf{g}-1)\dim G$.

The manifold $T^*\mathcal{M}_G(\mathcal{E})$ is equipped with the canonical holomorphic symplectic structure. Hitchin proved that the map φ is surjective and that it provides a completely integrable hamiltonian system on $T^*\mathcal{M}_G(\mathcal{E})$: any vector $v\in\mathcal{H}^*$ defines a function on $T^*\mathcal{M}_G(\mathcal{E})$, which is equal to $v\cdot\varphi$, and the Poisson bracket of two such functions is always equal to 0. In particular, the generic fiber of this map is a torus, and therefore the space of algebraic functions on $T^*\mathcal{M}_G(\mathcal{E})$ is isomorphic to the space of polynomials $\mathbf{C}[\mathcal{H}]$ on the space $\mathcal{H}$, which is a commutative Poisson algebra.

Now let us try to quantize this system. Quantization means that we assign to every classical hamiltonian h on $T^*\mathcal{M}_G(\mathcal{E})$ a differential operator D_h, acting on a certain line bundle over $\mathcal{M}_G(\mathcal{E})$. We also expect that the differential operators D_{h_1} and D_{h_2}, corresponding to two commuting hamiltonians h_1 and h_2, will also commute. It is known that every line bundle over $\mathcal{M}_G(\mathcal{E})$ is a power of the so-called determinant bundle ζ. In particular, the line bundle $\zeta^{-2h^\vee}$ is isomorphic to the canonical bundle (i.e the bundle of the volume forms) K over $\mathcal{M}_G(\mathcal{E})$. The bundle $K^{1/2} = \zeta^{-h^\vee}$ is called the bundle of half-forms.

It is natural to expect that the differential operators, which are the quantizations of the Hitchin's hamiltonians, would act on the bundle of half-forms. Indeed, if they acted on another bundle $L \neq K^{1/2}$, then the dual operators would act on the Serre dual bundle $KL^{-1} \neq L$, and we would have two different quantizations. This would be too much to hope for!

In fact, there is a deep relation between the center $Z_k(g)$ for integer k and the algebra of global differential operators, acting on the line bundle ζ^k.

The moduli space $\mathcal{M}_G(\mathcal{E})$ is an open set in a double coset space of the loop group of G [32]. Indeed, let p be a point of $\mathcal{E}$ and $\mathcal{D}$ - a small disc around this point. One can trivialize any bundle over both $\mathcal{D}$ and $\mathcal{E} - \mathcal{D}$. Therefore any bundle over $\mathcal{E}$ is defined by the transition function, which is an element of the loop group LG (maps from the boundary $\partial\mathcal{D}$ of $\mathcal{D}$ to G). The isomorphism classes of bundles are then identified with the double cosets $G_{\mathbf{out}}\backslash LG/G_{\mathbf{in}}$, where $G_{\mathbf{in}}(G_{\mathbf{out}})$ is the subgroup of LG, which consists of those maps, which can be extended inside (outside) of the disc $\mathcal{D}$.

The space $\mathcal{M} = LG/G_{\mathbf{in}}$ is a homogeneous space of the Lie group LG. Therefore the Lie algebra Lg embeds into the Lie algebra of vector fields on it. For any $k \in \mathbf{Z}$ one can construct a line bundle $\tilde{\zeta}^k$ over $\mathcal{M}$, such that the Kac-Moody algebra $\hat{g}$ of level k embeds into the Lie algebra of infinitesimal automorphisms of $\tilde{\zeta}^k$ [32]. This bundle is $G_{\mathbf{out}}$–equivariant, and it descends down to the bundle ζ^k over $\mathcal{M}_G(\mathcal{E})$.

The local completion $U_k(\hat{g})_{\mathbf{loc}}$ of the universal enveloping algebra of $\hat{g}$ at level k will map to the algebra of differential operators, acting on the bundle $\tilde{\zeta}^k$ over $\mathcal{M}$. Clearly, the differential operators, acting on $\tilde{\zeta}^k$, which lie in the image of the center $Z_k(g)$, commute with the action of $G_{\mathbf{out}}$, and therefore they define differential operators on the bundle ζ^k over the moduli space $\mathcal{M}_G(\mathcal{E})$.

But we know that $Z_k(g)$ is trivial for $k \neq -h^\vee$, therefore it does not produce any differential operators on the bundle ζ^k. In fact, it was proved in [31], that the only global differential operators on the bundle ζ^k over $\mathcal{M}_G(\mathcal{E})$ for $k \neq -h^\vee$ are constants. On the other hand, $Z_{-h^\vee}(g)$ produces a big algebra $Diff$ of global differential operators on the bundle of half-forms over $\mathcal{M}_G(\mathcal{E})$.

More precisely, the following is true: (1) $Diff$ is a commutative algebra, which is isomorphic to the algebra of polynomials on the space $\mathcal{H}$; (2) the map $Z_{-h^\vee}(g) \to Diff$ is surjective [11]. These results follow from the existence of the injective symbol map from $Diff$ to the space of algebraic functions on $T^*\mathcal{M}_G(\mathcal{E})$, Hitchin's description of this space, and Theorem 2.

The first statement allows to assign to any point $x \in \mathcal{H}$ a system of differential equations on half-forms on $\mathcal{M}_G(\mathcal{E})$. Namely, the point x defines a map $Diff \to \mathbf{C}$, taking $D \in Diff = \mathbf{C}[\mathcal{H}]$ to the value of the polynomial, corresponding to D, at the point x. We then consider the system of differential equations on half-forms η on $\mathcal{M}_G(\mathcal{E})$: $\{D \cdot \eta = x(D)\eta, D \in Diff\}$. By (1), this system is compatible, and it is holonomic, because the dimension of $\mathcal{H}$ is equal to the dimension of the moduli space $\mathcal{M}_G(\mathcal{E})$.

On the other hand, the statement (2) shows that $\mathcal{H}$ is embedded into the space of connections $\mathcal{C}(g')$. Indeed, by Theorem 2, the center $Z_{-h^\vee}(g)$ is identified with the Gelfand-Dikii algebra of g', which is the space of functions on $\mathcal{C}(g')$ of certain kind. We have a surjective map from $Z_{-h^\vee}(g)$ to the algebra $Diff$, which is the algebra of polynomial functions on $\mathcal{H}$, therefore we get an embedding of the corresponding spectra: $\mathcal{H} \to \mathcal{C}(g')$.

The space $\mathcal{C}(g')$ is the space of certain G'–connections on the circle. The image of $\mathcal{H}$ in $\mathcal{C}(g')$ consists of such connections, which can be holomorphically extended from the circle $\partial\mathcal{D} \subset \mathcal{E}$ to the whole curve $\mathcal{E}$.

Thus, we see that each point of $\mathcal{H}$ defines a system of differential equations on the moduli space $\mathcal{M}_G(\mathcal{E})$, and a holomorphic G'–connection on $\mathcal{E}$. Such a connection gives rise to a local system on $\mathcal{E}$, which defines a homomorphism from the fundamental group π_1 of $\mathcal{E}$ to G'. We have therefore established a correspondence between systems of differential equations on $\mathcal{M}_G(\mathcal{E})$ and homomorphisms $\pi_1 \to G'$. This is exactly "half" of the Langlands-Drinfeld corerespondence on $\mathcal{E}$.

Drinfeld conjectured [10] that there should be a one-to-one correspondence between the isomorphism classes of holonomic $\mathcal{D}$–modules on $\mathcal{M}_G(\mathcal{E})$ with singular support in the "global nilpotent cone" $\mathcal{N}$ and the isomorphism classes of homomorphisms from π_1 to G'. Here $\mathcal{N} = \varphi^{-1}(0)$ is the lagrangian subvariety in $T^*\mathcal{M}_G(\mathcal{E})$ [33]. Theorem 2 and the Hitchin's results give a correspondence between $\mathcal{D}$–modules of special kind (which are defined by systems of differential equations on half-forms) and homomorphisms $\pi_1 \to G'$ of special kind (which are defined by some holomorphic connections on $\mathcal{E}$). One can show that the space $\mathcal{H}$ of such homomorphisms gives a lagrangian subspace in the symplectic space of the isomorphism classes of all homomorphisms $\pi_1 \to G'$.

There is another way to relate these objects. Any element $c \in \mathcal{C}(g')$ defines the module $V_{-h^\vee,c}$ over $\hat{g}$. We can then take its coinvariants with respect to the Lie algebra $g_{\mathbf{out}}$ of the Lie group $G_{\mathbf{out}}$. This defines a certain modular functor on the $\hat{g}$–modules at the critical level. It is known, that for positive integer level this functor, applied to V_k, gives the space of conformal blocks in the WZW model [34]. For the critical level this functor, applied to $V_{-h^\vee,c}$, gives 0, if c can not be holomorphically extended to $\mathcal{E}$, and it gives the space of solutions of the corresponding system of differential equations on $\mathcal{M}_G(\mathcal{E})$, if it can be extended to a holomorphic connection on $\mathcal{E}$.

This construction can also be generalized to the moduli spaces of bundles over the curve $\mathcal{E}$ with punctures. Hopefully, it will allow to establish the Landlands-Drinfeld correspondence in general.

Acknowledgements: I would like to thank the organizers of the Cargese Summer School for financial support, excellent weather conditions, and the opportunity to present this work.

References

[1] I.Frenkel, J.Lepowsky, A.Meurman, *Vertex Operator Algebras and the Monster*, Academic Press 1988

[2] B.Feigin, E.Frenkel, *Affine Kac-Moody algebras at the critical level and Gelfand-Dikii algebras*, Preprint MSRI 04029-91, RIMS-796, to appear in Proceedings of RIMS-91 Program

[3] V.Drinfeld, V.Sokolov, Sov. J. Math. 30 (1985) 1975

[4] A.Zamolodchikov, Teor. Math. Phys. 65 (1985) 1205

[5] V.Fateev, S.Lukyanov, Int. J. of Mod. Phys. A3 (1988) 507

[6] E.Frenkel, V.Kac, M.Wakimoto, *Characters and fusion rules for W-algebras via quantized Drinfeld-Sokolov reduction*, Preprint, 1991

[7] E.Frenkel, *Affine Kac-Moody algebras at the critical level and quantum Drinfeld-Sokolov reduction*, Ph.D. Thesis, Harvard University, 1991

[8] S.Gelbart, Bull. Amer. Math. Soc. 10 (1984) 177

[9] E.Witten, Comm. Math. Phys. 113 (1988) 529

[10] V.Drinfeld, private communications;

A.Beilinson, *Affine algebras on the critical level and geometric Langlands correspondence (after V.Drinfeld)*, handwritten manuscript, and talks given at RIMS and MIT, 1991

[11] A.Beilinson, V.Drinfeld, B.Feigin, V.Ginzburg, *A commutative ring of differential operators on the moduli of G-bundles*, to appear

[12] B.Kostant, S.Sternberg, Ann. Phys. 176 (1987) 49

[13] B.Kostant, Invent. Math. 48 (1978) 101

[14] M.Wakimoto, Comm. Math. Phys. 104 (1986) 604

[15] B.Feigin, E.Frenkel, Russ. Math. Surv. 39, N5 (1988) 221

[16] B.Feigin, E.Frenkel, Comm. Math. Phys. 128 (1990) 161

[17] B.Feigin, E.Frenkel, in *Physics and Mathematics of Strings*, V.G.Knizhnik Memorial Volume, eds. L.Brink, e.a., 271-316, World Scientific 1990

[18] P.Bouwknegt, J.McCarthy, K.Pilch, Progr. Theor. Phys. Suppl. 102 (1990) 67

[19] P.Bouwknegt, J.McCarthy, K.Pilch, Preprint CERN, to appear in Proceedings of Stony Brook Conference, 1991

[20] C.De Concini, V.Kac, Progress in Math. 92, eds. A.Connes e.a., 471-506 (1990)

[21] J.Bernstein, I.Gelfand, S.Gelfand, in *Representations of Lie groups*, ed. I.Gelfand, 21-64, Wiley 1975

[22] B.Feigin, D.Fuchs, in *Representations of Lie groups and related topics*, eds. A.M.Vershik and D.P.Zhelobenko, 465-554, Gordon and Breach 1990

[23] M.Niedermaier, *Irrational free field resolutions for $\mathcal{W}(sl(n))$ and extended Sugawara construction*, Preprint, 1991

[24] G.Felder, Nucl. Phys. B324 (1989) 548

[25] V.Fateev, A.Zamolodchikov, Nucl. Phys. B280 [FS18] (1987) 644

[26] P.Bouwknegt, in Adv. Ser. in Math. Phys. 7, ed. V.Kac, 527-555, World Scientific 1989

[27] B.Feigin, E.Frenkel, *Free field resolutions in affine Toda field theories*, Preprint RIMS-827, 1991

[28] V.Kac, D.Kazhdan, Adv. Math. 34 (1979) 97

[29] T.Hayashi, Invent. Math. 94 (1988) 13
R.Goodman, N.Wallach, Trans. Amer. Math. Soc. 315 (1989) 1

[30] N.Hitchin, Comm. Math. Phys. 131 (1990) 347

[31] A.Beilinson, D.Kazhdan, *Projectively flat connections*, Harvard Preprint, 1990

[32] A.Pressley, G.Segal, *Loop Groups*, Cambridge University Press 1988

[33] G.Laumon, Duke Math. Journ. 57 (1988) 647

[34] A.Tsuchiya, K.Ueno, Y.Yamada, Adv. Stud. in Pure Math. 19, 459-565, World Scientific 1989

NON-TANNAKIAN CATEGORIES IN QUANTUM FIELD THEORY

Thomas Kerler

Institut für Theoretische Physik
ETH–Hönggerberg
CH-8093 Zürich, Switzerland

ABSTRACT*

We review the definitions of braided tensor-categories and relate them in the semisimple case to the structural data given by braid- and fusion-matrices. A number of duality-relations involving semisimple, Tannakian and weakly-Tannakian categories are summarized. We introduce a GNS-type construction to define the quotient of a rigid, abelian tensor-category onto a semisimple category. This yields a consistent calculus of truncated 6-j-symbols derived from non-semisimple Hopfalgebras. We illustrate how non-Tannakian categories are obtained in the examples of $U_q(sl_2)$, q a root of unity, and SU(2)-WZW-models. We show that the two categories are equal for $q = exp(\frac{i\pi}{k+2})$ and that $U_q(sl_2)$ is unique as a dual Hopf algebra if the fundamental representation is required to be two-dimensional. This leads us to formulate a duality-problem for non-Tannakian categories.

I. INTRODUCTION

As a mathematical discipline categories were invented by Eilenberg and MacLane [1] in the early fourties, as a means to organize families of maps between the abelian groups arising in homological algebra. The development proceeded very quietly, for the next decade, and was dominated by S. MacLane, who contributed the theory of abelian categories, the coherence theorem for monoids and the interpretation of the Tannaka-Kreĭn duality in categorial language. The subject experienced a great stimulus with the work of A. Grothendieck [2], where

* Notes on a talk presented at the Cargèse Summer Institute on "New Symmetry Principles in Quantum Field Theory", Cargèse, July 15-27, 1991.

New Symmetry Principles in Quantum Field Theory, Edited by
J. Fröhlich et al., Plenum Press, New York, 1992

the concepts of diagrams and schemes and the so-called A.B.5-categories were introduced. The prevailing concern of the research that followed was applications in algebraic geometry, like motives or Hodge structures. However, also results of general interest for the theory of categories were obtained, as for instance for universals and limits, adjoint functors and sheaves. One of particular interest to us has been given, recently, by P. Deligne [3], proving that general symmetric tensor categories fulfill the prerequisitives of the Tannaka-Kreĭn duality if certain dimensions are integer-valued. This will be discussed in some more detail in section III.1.). In mathematical physics, braided categories emerged from the study of conformal quantum field theories in two dimensions as the essential type of data characterizing a specific model. This insight has its foundations in the bootstrap approach proposed in the works of Polyakov and Belavin, Polyakov and Zamolodchikov [4], in which the infinite conformal symmetry is used to reduce the description of the abstract operator algebra to a finite set of structure constants, obtained from the conformal weights and the operator product expansions of the primary fields. Once the combinatorial selection structure is determined, the only constraints that remain to be solved, in this formulation, originate from the natural requirement of associativity of the field-algebra.

The subsequent study of vertex operators in particular models [5, 6] revealed that the conformal blocks appearing in the bootstrap equations are more appropriately replaced by transition matrices among them, called hereafter braid- and fusion matrices or also duality-matrices. The investigations focussed mainly on one special categorial relation, namely the Yang-Baxter-equation for the braid-matrices, which results from the manifest geometrical interpretation and permits the reconstruction of the conformal blocks as the solutions of a Riemann-Hilbert problem. A schematic type of categories, given in terms of sewing-compositions of punctured surfaces, appears in the axiomatic setting of conformal field theories, due to Segal [7] and to Friedan and Shenker [8]. In the latter approach, the consistency relations for conformal blocks are derived from the existence of a single-valued modular invariant partition function on the universal moduli-space, which contains the entire information of the conformal theory. Combining these points of view the entire set of categorial relations among the duality matrices has been established in [9] (including relations with the modular group which turn out to be redundant). In extending the Friedan-Shenker conjecture, one is thus led to believe that to every consistent set of duality-matrices, one cannot only reconstruct the conformal blocks but can also find the operator-algebra represented on Hilbert spaces for a "dual" conformal field theory.

In higher dimensional physics with permutation statistics and simple Poincaré-covariance, the concept of categories cristallized much earlier in the framework of algebraic field theory as the proper tool to describe the superselection structure of elementary particle physics. The basic ingredients in the formulation of Doplicher, Haag and Roberts [10] are localizable endomorphisms of the observable algebra, which are associated to each sector and intertwining operators reflecting the space-time properties. The interpretation of these elements as the constituents of a net of monoidal W^*-categories proved to be essential in an attempt of J.E. Roberts to classify general local field theories in terms of non-abelian local cohomology [11].

In cases where the observable algebra is found as the gauge-invariant part of a complete, unobservable field-algebra, with an adjoint action of a compact global gauge group, the sectors (e.g., charges, isospin ...) are given by the representations of the gauge group, and the W^*-category of the observable algebra is isomorphic to the representation-category of the group. A very natural and deep question – which is also of principal importance in the modelling of conventional quantum field theories from observable data – is whether every observable-algebra is given in this way. For theories with abelian statistics, i.e., invertible localized endomorphisms, the construction of the corresponding Haag-Araki-fields has already been performed in [12] and relies mainly on simple Pontrjagin-duality.

It was soon realized by Doplicher and Roberts [13, 14] that, in the presence of nonabelian- or para-statistics, the mentioned isomorphism of categories is not only a necessary, but also a sufficient condition for the construction of a field-algebra. The associated triviality problem of non-abelian cohomology is thus translated into a problem of generalized duality in which the gauge group has to be reconstructed from the abstract, a priori non-Tannakian, W^*-category. Still, it took more than a decade until Doplicher and Roberts [15] established their proof of duality, which differs strongly from that of Deligne in that it does not refer to the classical Tannaka-Kreĭn-theorem.

If the space-time dimension of a local field theory is two or if charges with localization in spacelike cones in three dimensions occur, permutation statistics is no longer stringent and often has to be replaced by braid statistics. The algebraic formalism of Doplicher, Haag and Roberts can be generalized to this situation and many applications of recent results for operator algebras were found [16]. Similarly, these theories define braided W^*-categories. However, the symmetry assumption in the duality theory of Doplicher and Roberts is of such central importance that braided categories are inaccessible to their methods, even in the most basic conclusions, like the integrality of statistical dimensions.

Indeed, since symmetry of a representation category is equivalent to cocommutativity of the represented algebraic object one has to rule out groups as possible candidates, but still can hope to find quasi-cocommutative, quasitriangular Hopf algebras. A large family of conformal models with a duality of this kind is given by those which are symmetric with respect to an affine Kac-Moody algebra at irrational level and have as generators chiral vertex operators on which the classical Lie group already acts as a global gauge group. The peculiarity of this example is that the Green functions are precisely the solutions to the Knizhnik-Zamolodchikov systems of differential equations.

Relating, by this observation, different types of short distance behavior, Drinfel'd [17] was able to define directly from the Green functions a Tannakian category and hence a quasi Hopf algebra which, as a bialgebra, is just the familiar gauge group. The uniqueness theorem on one-parameter deformations of Lie algebras yields a one-to-one correspondence between solutions to the Knizhnik-Zamolodchikov equations and invariant quantum group tensors, thus proving the Kohno theorem: that the two natural braid group representations are equivalent [18].

The difficulty with these models is that they are neither unitary nor minimal and are also not related to any simple Lagrangian, so that one does not expect them to have any physical relevance. This is different for the Wess-Zumino-Witten models, which are roughly speaking the limiting points of the previous ones at an integer level, k. However, the reconstruction procedure for the Hopf algebra breaks down for many reasons. The phenomena which appear here are that the admissible representations form only a finite subset of the ordinary group representations and vertex operators exist only for truncated fusion rules, so that the Green functions span a monodromy invariant, proper subspace of the solutions of the Knizhnik-Zamolodchikov equations. Since the algebra of chiral vertex operators, subject to these restrictions, is closed under braiding and fusion, it defines a category, in a similar manner, which is easily shown to be non-Tannakian, i.e., not the representation category of any bialgebra. Still, it turns out, as suggested by some formal continuity argument, that, for low ranks and for the fundamental representations, the braid matrices coincide with the 6-j-symbols of the respective quantumgroup whose deformation parameter is a root of unity. From this one infers the validity of the same truncation prescription as for the vertex operators. The naïve attempt to define an intertwiner calculus (as, e.g., in [19] for generic q) with analogously restricted 3-j-symbols, in order to explain a truncation phenomena for Hopf algebras, fails under most general assumptions; (for an example see [20] or section III.2. below). This subtlety caused many confusions, especially when one was looking at generalizations of field operator constructions with explicit quantum group symmetry.

In the Wess-Zumino-Witten model, one can circumvent this problem simply by viewing the chiral vertex operators as the field operators with a classical gauge group symmetry and apply the Drinfel'd procedure to the restricted set of Green functions. Since the truncated fusion rules may eliminate some group-invariant tensors in the short distance expansions, the associativity- and braid endomorphisms are no longer invertible, and one obtains the prototype of what Mack and Schomerus introduced as a weak quasi Hopf algebra [20], together with the respective local braid- and fusion relations for the vertex operators.

Assuming the equivalence of the braid- and fusion data in chiral SU(2)-Wess-Zumino-Witten models and the truncated 6-j symbols of $U_q(sl_2)$, one can also adjoin the vertex operators with a twist transformation, in such a way that the resulting field algebra has symmetry with respect to the truncated $U_q(sl_2)$ of [20]. As one can see from this example, the weak formalism of generalized symmetry leaves a considerable freedom in choosing a symmetry algebra and there is, a priori, no canonical choice or interpretation of the linear spaces on which the braid- and fusion morphisms are represented. Thus, the only truely invariant information is already contained in the braid- and fusion matrices, i.e., in the category of the field theory!

Nevertheless, for the particular model, the special choice in which the truncated quantumgroup expresses the symmetry has one advantage over others, namely that the associativity-morphism is a projector which is the unit on representations well below the level,k, so that we recover on these the original braid relations of [21]. The existence of an algebra resulting in almost local braid relations is intimately related to the fact that the categories defined by local quantum field theories are *strict* monoidal, a feature which is crucial in the analysis of Doplicher and Roberts. In a sense apparent in the Drinfel'd-Kohno theorem for generic, chiral models with current algebra symmetries and yet to be made precise for unitary Wess-Zumino-Witten models, the ordinary, coassociative quantumgroup can be regarded as the solution to the non-abelian cohomological problem that is associated to the requirement of strictness. Thus, despite truncation phenomena and non-integer dimensions, we expect a deeper connection between local quantum field theories with braid statistics, especially WZW-models, and usually non-semisimple, coassociative, quasitriangular Hopf algebras, like the Jimbo quantum groups at a root of unity. In an attempt to establish a generalized duality theory for field theories with strict monoidal, non-Tannakian categories in this sense, we are confronted with the following questions:

1.) How is it possible to find from a non-semisimple quantumgroup a semisimple category describing a local field theory with braid statistics? What exactly is the definition and the origin of the truncation prescription for quantum 6-j-symbols?

2.) Can one find to every local field theory a unique quasitriangular Hopf algebra, which defines the same category? Why is this true for SU(2)-WZW-models?

3.) In which sense does a quantum group act as a symmetry algebra, how should the concept of a global gauge group be generalized?

We shall give a complete answer to the first question by assigning to the representation category of a Hopf algebra a unique, semisimple category. The important ingredient here is a natural quotient of a general abelian category onto a semisimple category,to which we associate a projecting functor. For ordinary algebras, this is just the quotient by the radical of the algebra, and, for Hecke- and Birman-Wenzl-algebras (at special values), being subalgebras of certain C^*-category, this specializes to the definition of irreducible representations given in [22]. This procedure to define truncated 6-j-symbols and braid group representations has important applications in the construction of integrable RSOS-models [23] and invariants of links and 3-manifolds [24], where it is necessary to have a finite number of objects, in order to obtain finite state sums.

We will not be able to attack questions 2.) and 3.) in full generality, but rather propose them here as the natural extension of the duality problem solved by Deligne and by Doplicher and Roberts. However we will show that for a certain class of field theories which includes the SU(2)-WZW-model, $U_q(sl_2)$ is the unique dual Hopf algebra. The concept of charged fields in the ordinary sense does not occur naturally in this framework. Still, we shall argue that in WZW-models a modified concept of quantum group symmetry exists which views global gauge-algebras as a constructive tool, beyond the reproduction of the categories.

Survey of Contents

In chapter II.) we discuss the general definitions of braided tensor categories and the features implied by them. We shall lay special emphasis on the discussion of semisimplicity of abelian categories. More specifically, we review in section II.1.) the notions of abelian, relaxed or strict monoidal, braided, statistical, semisimple and finite categories. We discuss the peculiar rigidity structure and a set of special elements that are due to the braidedness of the category. In section II.2.) we introduce a procedure to quotient a non-semisimple, rigid, abelian category onto a semisimple category. This construction uses natural traces on braided categories and lies at the heart of the truncated 6-j-symbol calculus. The connection of the sometimes overly natural, mathematical formalism to the language more familiar to theoretical physicists working on conformal field theory is established in section II.3.), where we derive the structure constants of quantum categories and discuss in this context various notions of equivalence. We also address the problem of strictness in terms of structure constants and identify the special elements as functions on the set of irreducible objects.

In chapter III.) we investigate, how quantum categories can be realized by either Hopf algebras or conformal field theories. These are two versions of a duality problem for which we provide examples. In detail we explain in section III.1.) the role of Tannakian categories in the reconstruction of groups and Hopf algebras. In particular we give a series of duality relations for semisimple categories, which are accessible to blockmatrix-type proofs. We devote a larger part of this section to show how weak quasi-Hopf algebras, as introduced by G. Mack and V. Schomerus in [20] or these proceedings, fit into this frame. Finally we identify the special elements of a braided category with well known elements from quasi-triangular Hopf algebras. In section III.2.) we give a more detailed description of the semisimple quotient applied to the representation category of $U_q(sl_2)$. This yields a proof for the truncated 6-j-symbol calculus and at the same time explains why the correspondingly truncated relations involving 3-j-symbols do not hold. In section II.3.) we show how a non-Tannakian quantum category is derived from an SU(2)-WZW-model. We show that this is the same one as the semisimple quotient obtained from $\mathcal{R}ep(U_q(sl_2))$.

II. THE STRUCTURE OF QUANTUM-CATEGORIES

(1) General Definitions and Special Elements

We begin by briefly recalling the basic definitions of different types of categories, that are relevant for our purpose. Although important in field theoretic applications, we shall not consider *-structures in our exposition, but, instead, emphasize the more general aspects of semi-simplicity. More details and further implications related to the first part of our discussion may be found in the standard literature [25, 26].

The basic constituents of an *abelian* category over $\mathbb{C}$ are a set, $\mathcal{O}bj$, whose elements are called *objects*, and associated to each ordered pair, (X,Y), of objects a complex vectorspace, $\mathcal{M}or(X,Y)$, consisting of *morphisms* (or *arrows*) from X to Y. There shall be a bilinear and associative composition: $\mathcal{M}or(Y,Z)\otimes\mathcal{M}or(X,Y)\to\mathcal{M}or(X,Z)$ of morphisms and the so defined complex, associative algebra, $\mathcal{E}nd(X):=\mathcal{M}or(X,X)$, shall contain a unit, denoted $\mathbb{1}_X$. An object,X, is *finite* if $\dim(\mathcal{E}nd(X))<\infty$ and the category is called *locally finite* if $\dim(\mathcal{M}or(X,Y))<\infty$ for all $X,Y\in\mathcal{O}bj$. We say that two objects, $X,Y\in\mathcal{O}bj$, are *equivalent*,i.e., $X\approx Y$, if there exist $f\in\mathcal{M}or(X,Y)$ and $g\in\mathcal{M}or(Y,X)$, such that $f\circ g=\mathbb{1}_Y$ and $g\circ f=\mathbb{1}_X$.

We do not attempt here to discuss all the axioms of kernels and cokernels [25] that characterize an abelian category, but confine ourselves to the weaker subobject property, which will be sufficient for semisimple categories. It states that for any projector, $\Pi \in \mathcal{E}nd(X)$, there is an object, U, and morphisms, $P_U \in \mathcal{M}or(X, U)$ and $I_U \in \mathcal{M}or(U, X)$, such that $\Pi = I_U P_U$ and $P_U I_U = \mathbb{1}_U$. The diagram

$$U \underset{I_U}{\overset{P_U}{\rightleftarrows}} X \underset{I_V}{\overset{P_V}{\rightleftarrows}} V, \qquad (1)$$

where P_V, I_V and V are the elements associated to the projector, $1 - \Pi$, is called a *biproduct* or *direct sum*, denoted $U \oplus V$. Hence, in an abelian category, every finite object, X, can be decomposed into a direct sum of *indecomposable* objects, X_α, characterized by the property that $\mathcal{E}nd(X_\alpha) = \mathbb{C}\mathbb{1}_{X_\alpha} \oplus \mathcal{E}nd(X_\alpha)^0$, where $\mathcal{E}nd(X_\alpha)^0$ is a subalgebra consisting of nilpotent elements. An indecomposable object, X_α, is *irreducible* if $\mathcal{E}nd(X_\alpha)^0 = 0$. We denote the set of indecomposable objects by $\mathcal{D} \subset \mathcal{O}bj$ and the set of irreducible objects by $\mathcal{I} \subset \mathcal{O}bj$. An abelian category is also assumed to be *additive*, which means that to every pair of objects, $U, V \in \mathcal{O}bj$ there exists a direct sum, $U \oplus V$, with an object denoted by the same symbol. Note that the definition of a direct sum entails the isomorphisms $\mathcal{M}or(U \oplus V, X) \to \mathcal{M}or(U, X) \oplus \mathcal{M}or(V, X)$: $A \to (AI_U, AI_V)$ and $\mathcal{M}or(X, U \oplus V) \to \mathcal{M}or(X, U) \oplus \mathcal{M}or(X, V)$: $A \to (P_U A, P_V A)$. The definition of a (relaxed) monoidal category requires a map $\circ : \mathcal{O}bj \times \mathcal{O}bj \to \mathcal{O}bj : (X, Y) \to X \circ Y$, the *tensor product* of objects, and, parallel to this, a bilinear product of morphisms

$$\circ : \mathcal{M}or(X, X') \otimes \mathcal{M}or(Y, Y') \to \mathcal{M}or(X \circ Y, X' \circ Y') : I \otimes J \to I \circ J,$$

which is supposed to respect the composition in both arguments, i.e., $(I \circ J)(I' \circ J') = (II') \circ (JJ')$ whenever defined. Also, there shall be an element, $1 \in \mathcal{O}bj$, so that there are isomorphisms , $1 \circ X \approx X \approx X \circ 1$, for which rather obvious conditions are imposed. For a relaxed monoidal structure the tensor-product is assumed to be associative up to isomorphism, meaning that we have for each triple, $\{X, Y, Z\} \subset \mathcal{O}bj$, a specific, invertible morphism $\alpha(X, Y, Z) \in \mathcal{M}or(X \circ (Y \circ Z), (X \circ Y) \circ Z)$, providing $X \circ (Y \circ Z) \approx (X \circ Y) \circ Z$. It shall obey the *isotropy-equation*

$$\alpha(X', Y', Z')(I \circ (J \circ K)) = ((I \circ J) \circ K)\,\alpha(X, Y, Z) \qquad (2)$$

for all tensor-products of morphisms $I \in \mathcal{M}or(X, X')$, $J \in \mathcal{M}or(Y, Y')$ and $K \in \mathcal{M}or(Z, Z')$. Coherence [25] with respect to natural associativity is assured by only one constraint on the α's, namely the *pentagon equation*

$$\alpha(W \circ X, Y, Z)\,\alpha(W, X, Y \circ Z) = (\alpha(W, X, Y) \circ \mathbb{1}_Z)\,\alpha(W, X \circ Y, Z)\,(\mathbb{1}_W \circ \alpha(X, Y, Z)). \qquad (3)$$

"Commutativity up to equivalence" is imposed on the tensor-product by a specific, invertible morphism, $\varepsilon(X, Y) \in \mathcal{M}or(X \circ Y, Y \circ X)$, so that $X \circ Y \approx Y \circ X$ and for arbitrary $I \in \mathcal{M}or(X, X')$ and $J \in \mathcal{M}or(Y, Y')$ we have isotropy :

$$\varepsilon(X', Y')(I \circ J) = (J \circ I)\,\varepsilon(X, Y). \qquad (4)$$

A basic diagram involving α and ε is given by the hexagon, for which commutativity is expressed by

$$\alpha(Z,X,Y)\,\varepsilon(X\circ Y,Z)\,\alpha(X,Y,Z) \;=\; (\varepsilon(X,Z)\circ 1\!\!1_Y)\,\alpha(X,Z,Y)\,(1\!\!1_X\circ\varepsilon(Y,Z)). \quad (5)$$

Since we do not assume symmetry, $\varepsilon(X,Y)=\varepsilon(Y,X)^{-1}$, coherence with repect to natural associativity and commutativity holds only if for the admissible diagrams the effective braid group elements are trivial. Again we have canonical identities if one of the objects is 1. The final ingredient of the categories we wish to consider here is the existence of conjugate elements: To any object, X, there shall exist another object, $X^\vee$, and morphisms, $\vartheta_X \in \mathcal{M}or(1, X^\vee\circ X)$ and $\vartheta_X^\dagger \in \mathcal{M}or(X\circ X^\vee, 1)$, such that

$$(\vartheta_X^\dagger\circ 1\!\!1_X)\,\alpha(X,X^\vee,X)\,(1\!\!1_X\circ\vartheta_X) \;=\; 1\!\!1_X$$

and
$$(1\!\!1_{X^\vee}\circ\vartheta_X^\dagger)\,\alpha(X^\vee,X,X^\vee)^{-1}\,(\vartheta_X\circ 1\!\!1_{X^\vee}) \;=\; 1\!\!1_{X^\vee}. \quad (6)$$

There is a canonical element $q_X \in \mathcal{M}or(X, X^{\vee\vee})$, given by

$$q_X \;=\; (1\!\!1_{X^{\vee\vee}}\circ\vartheta_X^\dagger)\,\alpha(X^{\vee\vee},X,X^\vee)^{-1}\,(\varepsilon(X,X^{\vee\vee})\circ 1\!\!1_{X^\vee})\,\alpha(X,X^{\vee\vee},X^\vee)\,(1\!\!1_X\circ\vartheta_{X^\vee}) \quad (7)$$

with inverse

$$q_X^{-1} \;=\; (\vartheta_{X^\vee}^\dagger\circ 1\!\!1_X)\,\alpha(X^\vee,X^{\vee\vee},X)\,(1\!\!1_{X^\vee}\circ\varepsilon(X^{\vee\vee},X)^{-1})\,\alpha(X^\vee,X,X^{\vee\vee})^{-1}\,(\vartheta_X\circ 1\!\!1_{X^{\vee\vee}})$$

showing that braided categories with conjugate objects are *rigid* in the sense of [26].

The way the elements, q_X, depend on the choice of conjugates, $X \to (X^\vee, \vartheta_X)$, can be read from the relation $(q_X^{-1}\circ 1\!\!1_X)\vartheta_{X^\vee} = \varepsilon(X^\vee,X)\vartheta_X$. If for two objects, $X,Y \in \mathcal{O}bj$, the choice is such that $(X\circ Y)^\vee = Y^\vee\circ X^\vee$ and $\vartheta_{X\circ Y}$ is the canonical morphism obtained from ϑ_X and ϑ_Y (and similar for $Y^\vee$ and $X^\vee$) we find the factorization

$$q_{X\circ Y} \;=\; (q_X\circ q_Y)\,\mu(X,Y), \quad (8)$$

where $\mu(X,Y) := \varepsilon(Y,X)\,\varepsilon(X,Y) \in \mathcal{E}nd(X\circ Y)$ is the *monodromy*. Notice that we have a natural transposition, ${}^t : \mathcal{M}or(X,Y)\to\mathcal{M}or(Y^\vee,X^\vee)$, defined by

$$(1\!\!1_X\circ J)\vartheta_X \;=\; (J^t\circ 1\!\!1_Y)\,\vartheta_Y, \quad (9)$$

which obeys $J^{tt} = q_Y\,J\,q_X^{-1}$ and $(I\,J)^t = J^t\,I^t$ for compatible morphisms, I and J. As opposed to q_X, the isomorphism $g_X := (q_{X^\vee}^t)^{-1}q_X \in \mathcal{M}or(X, X^{\vee\vee\vee\vee})$, with $J^{tttt} = g_Y\,J g_X^{-1}$, factorizes completely, i.e., we have $g_{X\circ Y} = g_X\circ g_Y$ for a similar choice of conjugates. A set of endomorphisms, $\tau(X)\in\mathcal{E}nd(X)$, closely related to the asymmetry of the category is defined by $(1\!\!1_{X^\vee}\circ\tau(X))\vartheta_X = \mu(X^\vee,X)\vartheta_X$

or also $\tau(X^\vee)^{-1} := q_X^t q_{X^\vee}$, so that it is independent of the choice of conjugates and moreover behaves isotropically, i.e. $\tau(Y)J = J\tau(X)$ for any $J \in \mathcal{M}or(X,Y)$. It is related to the monoidal operations by $\tau(X) \circ \tau(Y) = \mu(X,Y)^2 \tau(X \circ Y)$ and $\tau(X^\vee) = \tau(X)^t$.

This observation naturally leads us to the notion of a *statistical* (or *balanced*) category, for which we require in addition the existence of an isotropic set of endomorphisms $\sigma(X) \in \mathcal{E}nd(X)$, with $\sigma(X)^2 = \tau(X)$, $\sigma(X) \circ \sigma(Y) = \mu(X,Y)\sigma(X \circ Y)$ and $\sigma(X^\vee) = \sigma(X)^t$. One easily proves this to be equivalent to the existence of isomorphisms $\xi_X \in \mathcal{M}or\ (X^{\vee\vee}, X)$, with relations $J^{tt} = \xi_Y\, J\, \xi_X^{-1}$, $\xi_{X \circ Y} = \xi_X \circ \xi_Y$, $\xi_X \xi_{X^{\vee\vee}} = g_X^{-1}$ and $\xi_X^t = \xi_{X^\vee}^{-1}$, by setting $\sigma(X)^{-1} = \xi_X q_X$.

It is a fact that every category of the kind described so far, which is also a C^*-category with unitary α and ε, is already statistical. The endomorphism $\sigma(X)$ can be chosen as the unitary part of the polar decomposition of λ_X, defined by $\vartheta_X^*(\mathbb{I}_{X^\vee} \circ \lambda_X) = \vartheta_X^\dagger\, \varepsilon(X^\vee, X)$. This completes the definition of an abelian, locally-finite, relaxed monoidal, braided (and occasionally statistical) category over $\mathbb{C}$ with conjugates, which we henceforth call a *quantum category*.

(2) A Semisimple Quotient

The following discussion is devoted to the study of certain natural expectations and traces on quantum categories. For this purpose consider the linear maps $E_X^{U,V} : \mathcal{M}or(U \circ X, V \circ X) \to \mathcal{M}or(U,V)$ given (independently of the particular conjugates) by

$$E_X^{U,V}(L) := (\mathbb{I}_V \circ \vartheta_X^\dagger)\, \alpha(V,X,X^\vee)^{-1}\, (L \circ \mathbb{I}_{X^\vee})\, \alpha(U,X,X^\vee)\, (\mathbb{I}_U \circ \varepsilon(X^\vee,X)\, \vartheta_X). \tag{10}$$

The functional, $Tr_X := E_X^{1,1} \in \mathcal{E}nd(X)^*$, proves to be cyclic with respect to the entire category in the sense that

$$Tr_X(IJ) = Tr_Y(JI), \quad \forall X,Y \in \mathcal{O}bj,\ I \in \mathcal{M}or(X,Y), J \in \mathcal{M}or(Y,X). \tag{11}$$

It is related to the monoidal product by

$$Tr_{X \circ Y}\left((A \circ B)\,\mu(X,Y)\right) = Tr_X(A)\, Tr_Y(B). \tag{12}$$

Moreover it is related to the expectations, $E_X^{U,V}$, by

$$Tr_{U \circ X}\left(M\mu(U,X)\right) = Tr_U(E_X^{U,U}(M))$$

for any $M \in \mathcal{E}nd(U \circ X)$. Using the relation

$$E_X^{U,W}\left((K \circ \mathbb{I}_X)\, L\right) = K\, E_X^{U,V}(L),$$

where $K \in \mathcal{M}or(V,W)$ and $L \in \mathcal{M}or(U \circ X, V \circ X)$, we find, with $U = W$,

$$Tr_{U\circ X}\left((K \circ 1\!\!1_X)\, L\right) \;=\; Tr_U\left(K\, E_X^{U,V}\left(L\mu(U,X)^{-1}\right)\right). \qquad (13)$$

The trace also obeys $Tr_{X^\vee}(I^t) = Tr_X(I)$, for $I \in \mathcal{E}nd(X)$. If the quantum category is statistical, a more convenient trace is given by $tr_X(I) := Tr_X(I\,\sigma(X)^{-1})$, which is also cyclic and has the monoidal properties $tr_{X^\vee}(I^t) = tr_X(I)$ and $tr_{X\circ Y}(I \circ J) = tr_X(I)\,tr_Y(J)$ for $I \in \mathcal{E}nd(X)$ and $J \in \mathcal{E}nd(Y)$. It provides us with the *statistical dimension*

$$d : \mathcal{O}bj \to \mathbb{C} : X \to d([X]) := tr_X(1_X),$$

depending only on the equivalence classes of objects. We find immediately

$$d(U \oplus V) = d(U) + d(V)\;,\; d(X \circ Y) = d(X)\,d(Y) \quad \text{and} \quad d(X^\vee) = d(X). \quad (14)$$

Clearly, in the C^*-case, tr_X is a positive state, so $d(X) \geq 0$. For symmetric C^*-categories one even has $d(X) \in \mathbb{N}$, [15].

Using the decomposition of morphisms of abelian categories into epi's and monics (see [25]) we can show in general that Tr_X vanishes on any nilpotent element, N. To this end we iteratively define for $N \in \mathcal{E}nd(X)$ a sequence of objects, X_i, epi's, $P_i \in \mathcal{M}or(X_{i-1}, X_i)$, and monic's $I_i \in \mathcal{M}or(X_i, X_{i-1})$ by $N = I_1 P_1$ and $P_i I_i = I_{i+1} P_{i+1}$. Nilpotency, $N^l = 0$, then implies $I_l = P_l = 0$ and thus by cyclicity, $Tr_X(N) = Tr_{X_{i-1}}(I_i P_i) = 0$.

In the remainder of this section we are primarily concerned with the existence and structure of *semisimple* categories. These are locally-finite, abelian categories for which $\mathcal{E}nd(X)$ is semisimple (hence a multimatrix-algebra) for any $X \in \mathcal{O}bj$. In addition a semisimple category is supposed to have no zero-divisors in the sense that the morphism spaces

$$\mathcal{M}or\,(X,Y)^\infty := \{I \in \mathcal{M}or(X,Y) : \quad IJ = JI = 0 \quad \forall\, J \in \mathcal{M}or(Y,X)\}$$

vanish for all $X, Y \in \mathcal{O}bj$.

We present a procedure that constructs semisimple categories as quotients of arbitrary quantum categories, using the traces introduced above. For this purpose we define the nullspaces of the trace-forms,

$$\mathcal{M}or(X,Y)^0 := \{I \in \mathcal{M}or(X,Y) : Tr_X(JI) = 0, \quad \forall\, J \in \mathcal{M}or(Y,X)\}. \quad (15)$$

The cyclicity of Tr_X implies that these are two-sided ideals for the composition, i.e.,

$$\mathcal{M}or(Y,Z)\,\mathcal{M}or(X,Y)^0 + \mathcal{M}or(Y,Z)^0\,\mathcal{M}or(X,Y) \subset \mathcal{M}or(X,Z)^0. \qquad (16)$$

Also the relation (13) shows $(\mathcal{M}or(V,U)^0 \circ \mathbb{I}_X) \subset \mathcal{M}or(V \circ X, U \circ X)^0$, from which we eventually conclude with (16) that the nullspaces are also tensor-ideals, i.e.,

$$\mathcal{M}or(V,U)^0 \circ \mathcal{M}or(V',U') + \mathcal{M}or(V,U) \circ \mathcal{M}or(V',U')^0 \subset \mathcal{M}or(V \circ V', U \circ U')^0. \tag{17}$$

We call $X \in \mathcal{O}bj$ a *ghost object* if $Tr_X \equiv 0$ or, equivalently, if $\vartheta_X \in \mathcal{M}or(1, X^\vee \circ X)^0$, and a *real object* otherwise. One easily finds that subobjects of ghosts are again ghosts and that the product $X \circ Y$ is real if and only if both factors, X and Y, are real. For a statistical category we see that $d(X) = 0$ if X is a ghost, and if X is indecomposable the converse holds, too. The objects of the quotient category are the real ones of the original category, with the restricted product, and the morphism spaces are defined by

$$\overline{\mathcal{M}or}(X,Y) := \mathcal{M}or(X,Y) \,/\, \mathcal{M}or(X,Y)^0,$$

with the induced composition- and tensor-products. As the monoidal isomorphisms $\bar{\alpha}(X,Y,Z) \in \overline{\mathcal{M}or}(X \circ (Y \circ Z), (X \circ Y) \circ Z)$ and $\bar{\varepsilon}(X,Y) \in \overline{\mathcal{M}or}(X \circ Y, Y \circ X)$, we use the respective images of α and ε so that the isotropy-, pentagonal- and hexagonal equations follow from the functoriality of the projection onto the quotients. Note that the existence of conjugates is immediate from the definition of real objects. We recall that for a finite-dimensional, associative algebra $\mathcal{A}$, with a trace, $tr \in \mathcal{A}^*$, the quotient $\pi : \mathcal{A} \twoheadrightarrow \bar{\mathcal{A}} = \mathcal{A}/\mathcal{A}_0$, where $\mathcal{A}_0$ is the nullspace of the trace-form, is semisimple, if tr vanishes on nilpotent elements. From this and the previous observation regarding Tr_X it follows that $\overline{\mathcal{E}nd}(X)$ is semisimple. Clearly we have $\mathcal{M}or(X,Y)^\infty \subset \mathcal{M}or(X,Y)^0$ so that the category is semisimple, if abelieness is preserved. In proving this (or at least the subobject property) it is convenient to use the existence of a *splitting* homomorphism $\psi : \bar{\mathcal{A}} \to \mathcal{A}$, with $\pi \circ \psi = \mathbb{I}$, in order to lift endomorphisms from $\overline{\mathcal{E}nd}(X)$ to $\mathcal{E}nd(X)$ and also the fact that $\mathcal{M}or(X,Y) = \mathcal{M}or(X,Y)^0$, if X or Y is a ghost.

Evidently, there is a natural functor onto the quotient category, once we add to it a dummy object, $\emptyset$, to which all ghosts are mapped, and which obeys $\mathbb{I}_\emptyset = 0$, $\mathcal{M}or(\emptyset, X) = \mathcal{M}or(X, \emptyset) = 0$ and $X \circ \emptyset = \emptyset \circ X = \emptyset$ so that $\emptyset$ has no conjugate.

(3) Structure Constants of Quantum categories

Before entering the matrix-description of semisimple categories we wish to introduce a notion of equivalence among general quantum categories, which requires the definition of functors respecting the monoidal properties:A *compatible tensor functor*, [26], between quantum categories, $\mathcal{C}_1 \to \mathcal{C}_2$, with monoidal structures $\{\circ_i, \alpha_i, \varepsilon_i\}$, $i = 1,2$, consists of a map $f : \mathcal{O}bj_1 \to \mathcal{O}bj_2$ with $f(1) = 1$ and $f(X \circ_1 Y) \approx f(X) \circ_2 f(Y)$, and linear maps $\mathcal{F} : \mathcal{M}or(X,Y) \to \mathcal{M}or(f(X),\, f(Y))$, with $\mathcal{F}(\mathbb{I}_X) = \mathbb{I}_{f(X)}$, such that $(f, \mathcal{F})$ is a functor of the mere abelian categories. Furthermore the definition includes a set of isomorphisms, $C(X,Y) \in \mathcal{M}or(f(X) \circ_2 f(Y), f(X \circ_1 Y))$, which intertwine the morphism products, ie.,

$$\mathcal{F}(I \circ_1 J) C(X,Y) = C(X',Y')(\mathcal{F}(I) \circ_2 \mathcal{F}(J)) \tag{18}$$

for all $I \in \mathcal{M}or(X,X')$ and $J \in \mathcal{M}or(Y,Y')$, and relate the monoidal isomorphisms as follows:

$$\alpha_2(f(X), f(Y), f(Z))$$
$$= (C(X,Y)^{-1} \circ_2 \mathbb{I}) C(X \circ_1 Y, Z)^{-1} \mathcal{F}(\alpha_1(X,Y,Z)) C(X, Y \circ_1 Z)(\mathbb{I} \circ_2 C(Y,Z))$$

and $$\varepsilon_2(f(X), f(Y)) = C(Y,X)^{-1} \mathcal{F}(\varepsilon_1(X,Y)) C(X,Y). \tag{19}$$

We call a compatible tensor functor $(f, \mathcal{F}, C)$ an *injection* if both f and $\mathcal{F}$ are injections, an *embedding* if f is an injection and $\mathcal{F}$ an isomorphism (whenever defined) and an *isomorphism of quantum categories* if f is also a bijection. Finally let us define a *reduction* of a quantum category, $\mathcal{C}$, to be a quantum category, whose objects are the equivalence classes, $\bar{X} = [X] \in \overline{\mathcal{O}bj}$, of objects, X, of $\mathcal{C}$, such that there exists an embedding $(f, \mathcal{F}, C)$ into $\mathcal{C}$, with $[f(\bar{X})] = \bar{X}$. Clearly, the tensor-product for the objects is associative and given by $\bar{X} \circ \bar{Y} = [X \circ Y]$, for any $X \in \bar{X}$ and $Y \in \bar{Y}$, independent of the particular reduction. It is straightforwardly verified that to any admissible injection of objects, $f\colon \overline{\mathcal{O}bj} \to \mathcal{O}bj$, and any choice of isomorphisms, $C(X,Y) \in \mathcal{M}or(f(\bar{X}) \circ f(\bar{Y}), f(\bar{X} \circ \bar{Y}))$, there exists a unique reduction with morphisms, $\overline{\mathcal{M}or}(\bar{X}, \bar{Y}) := \mathcal{M}or(f(\bar{X}), f(\bar{Y}))$, such that $(f, \mathcal{F} = id, C)$ is the embedding. Also, one easily shows that all reductions of $\mathcal{C}$ are isomorphic and that a reduction together with the "cardinalities" of the sets $\bar{X} \subset \mathcal{O}bj$ determines $\mathcal{C}$ completely up to isomorphisms. For these reasons, we content ourselves with the discussion of *irredundant* categories, meaning that $X \approx Y$ shall imply $X = Y$ or, equivalently, that $\mathcal{C}$ is isomorphic to its reduction.

A semisimple, irredundant quantum category can be described entirely by a finite set of linear maps, called R- and F-matrices, which obey relations analogous to (3) and (5). If the considered category is semisimple it follows immediately that all indecomposable objects are also irreducible and for reduced categories we have by $\mathcal{M}or(k,l)^{\infty} = 0$ and semisimplicity of $\mathcal{E}nd(k)$ that $\mathcal{M}or(k,l) = \delta_{kl}\, \mathbb{C}$ for all $k, l \in \mathcal{I} = \mathcal{D}$. Clearly this implies for finite categories, that every object, X, can be decomposed into irreducible objects. More concisely, we can express complete reducibility by the fact that the following is a sequence of isomorphisms:

$$\bigoplus_{k \in \mathcal{I}} \mathcal{M}or(k,X) \otimes \mathcal{M}or(X,k) \xrightarrow{\ i\ } \mathcal{E}nd(X) \xrightarrow{\ \mathcal{L}\ } \bigoplus_{k \in \mathcal{I}} \mathcal{E}nd(\mathcal{M}or(k,X)), \tag{20}$$

where i is the composition and $\mathcal{L}(a)$, with $a \in \mathcal{E}nd(X)$, acts on each $\mathcal{M}or(k,X)$ by left multiplication. In particular this implies the nondegeneracy of the multiplication $\mathcal{M}or(X,k) \otimes \mathcal{M}or(k,X) \to \mathbb{C}$, which identifies $\mathcal{M}or(X,k)$ and $\mathcal{M}or(k,X)$

as duals, with dimensions $N_{X,k}$. More generally, the left multiplication provides for semisimple categories the isomorphism

$$\mathcal{M}or(X,Y) \xrightarrow[\cong]{\mathcal{L}} \bigoplus_{k \in \mathcal{I}} Hom(\mathcal{M}or(k,X)\,,\,\mathcal{M}or(k,Y)), \tag{21}$$

which respects the composition of arrows in the obvious sense. Hence two objects, $X, Y \in \mathcal{O}bj$, are equivalent iff $N_{X,k} = N_{Y,k}$, $\forall k \in \mathcal{I}$, yielding a bijection $\overline{\mathcal{O}bj} \to \mathbb{N}^{\mathcal{I}} : X \to N_{X,\cdot}$, which is additive in the sense that $N_{X\oplus Y,k} = N_{X,k} + N_{Y,k}$, $\forall k \in \mathcal{I}, \forall X, Y \in \mathcal{O}bj$. The mere, reduced, abelian category, without any further structure, is thus completely reconstructed (up to isomorphisms) from the set $\mathcal{I}$, by setting $\mathcal{O}bj = \mathbb{N}^{\mathcal{I}}$, introducing $N_{X,k}$-dimensional vector spaces, $\mathcal{M}or(k,X)$, as building blocks and defining the morphisms and their compositions by (21). In order to describe how a monoidal structure of a reduced category is incorporated into this picture of the objects and morphisms, we consider the natural maps among the spaces built from $\mathcal{M}or(k,X)$'s:

$$\Gamma^k_{X,Y} : \bigoplus_{i,j \in \mathcal{I}} \mathcal{M}or(i,X) \otimes \mathcal{M}or(j,Y) \otimes \mathcal{M}or(k, i \circ j) \to \mathcal{M}or(k, X \circ Y), \tag{22}$$

which take $I \otimes J \otimes P$ to $(I \circ J) \cdot P$. Semisimplicity of the category implies that $\Gamma^k_{X,Y}$ is an isomorphism for any $k \in \mathcal{I}$ and $X, Y \in \mathcal{O}bj$ and thereby expresses the distributiveness of the tensor-product. Indeed, counting dimensions on both sides of (22) it follows that the composition $\mathbb{N}^{\mathcal{I}} \times \mathbb{N}^{\mathcal{I}} \to \mathbb{N}^{\mathcal{I}}$ induced by $\circ$ is not only associative but also distributive and is therefore given by the structure constants, $N_{ij,k} := N_{i\circ j,k}$ for $i, j, k \in \mathcal{I}$. Since the maps, $\Gamma^k_{X,Y}$, contain all the information on the tensor-product of the blocks $\mathcal{M}or(k,X)$, it can be used to extend the tensor-product to general morphisms with the help of the isomorphism $\mathcal{L}$. The defining formulae for $I \in \mathcal{M}or(X,X')$ and $J \in \mathcal{M}or(Y,Y')$ are

$$\mathcal{L}(I \circ J)\Gamma^k_{X,Y} = \Gamma^k_{X',Y'}(\mathcal{L}(I) \otimes \mathcal{L}(J) \otimes \mathbb{1}), \qquad \forall k \in \mathcal{I}. \tag{23}$$

The $\Gamma^k_{X,Y}$'s shall always be chosen such that they are the canonical maps for $X, Y \in \mathcal{I}$,$X = 1$ or $Y = 1$. There are two natural ways to construct from the isomorphisms (22) further distributiveness isomorphisms for three-fold tensor-products, denoted

$$\begin{aligned} P^t_{X(Y,Z)}, P^t_{(X,Y)Z} &: \\ &\bigoplus_{ijk \in \mathcal{I}} \mathcal{M}or(i,X) \otimes \mathcal{M}or(j,Y) \otimes \mathcal{M}or(k,Z) \otimes \mathcal{M}or(t, i \circ j \circ k) \\ &\qquad\qquad \longrightarrow \mathcal{M}or(t, X \circ Y \circ Z) \end{aligned} \tag{24}$$

such that for any $I \in \mathcal{M}or(X,X')$, $J \in \mathcal{M}or(Y,Y')$ and $K \in \mathcal{M}or(Z,Z')$ the relation

$$\mathcal{L}(I \circ (J \circ K))\, P^t_{X(Y,Z)} = P^t_{X'(Y',Z')}(\mathcal{L}(I) \otimes \mathcal{L}(J) \otimes \mathcal{L}(K) \otimes \mathbb{1}) \tag{25}$$

and the analogous one for $P^t_{(X,Y)Z}$ hold. The maps (24) allow us to express the isotropy axiom (2) equivalently as

$$\mathcal{L}(\alpha(X,Y,Z))\, P^t_{X(Y,Z)} = P^t_{(X,Y)Z}\big(\bigoplus_{ijk\in\mathcal{I}} \mathbb{I}^{\otimes 3} \otimes \mathcal{L}(\alpha(i,j,k))\big). \qquad (26)$$

Hence for general objects, $X,Y,Z \in \mathcal{O}bj$, $\alpha(X,Y,Z)$ can be computed from the maps (24) and the isomorphisms $\alpha(i,j,k)$, with $i,j,k \in \mathcal{I}$. It is convenient to express the latter in terms of the collection of invertible maps

$$F(i,j,k,t) : \bigoplus_{l\in\mathcal{I}} \mathcal{M}or(l,j\circ k)\otimes\mathcal{M}or(t,i\circ l) \longrightarrow \bigoplus_{l\in\mathcal{I}} \mathcal{M}or(l,i\circ j)\otimes\mathcal{M}or(t,l\circ k)$$

defined by the property

$$\mathcal{L}(\alpha(i,j,k))\,\Gamma^t_{i,j\circ k} = \Gamma^t_{i\circ j,k}\,F(i,j,k,t). \qquad (27)$$

For this presentation the pentagon equation for general $\alpha(X,Y,Z)$ can be reduced to an analogous equation for the F-matrices. More precisely, (3) holds for all $W,X,Y,Z \in \mathcal{O}bj$ if and only if

$$\begin{aligned}(\bigoplus_s F(i,j,k,s)\otimes\mathbb{I}_{N_{sl,t}})\,(\bigoplus_s \mathbb{I}_{N_{jk,s}}\otimes F(i,s,l,t))\,(\bigoplus_s F(j,k,l,s)\otimes\mathbb{I}_{N_{is,t}})\\ = (\bigoplus_s \mathbb{I}_{N_{ij,s}}\otimes F(s,k,l,t))\,T_{12}\,(\bigoplus_s \mathbb{I}_{N_{kl,s}}\otimes F(i,j,s,t)),\end{aligned} \qquad (28)$$

defined on $\bigoplus_{sr}\mathcal{M}or(s,k\circ l)\otimes\mathcal{M}or(r,j\circ s)\otimes\mathcal{M}or(t,i\circ r)$, is true for all $i,j,k,l,t\in\mathcal{I}$, where $T_{12}(I\otimes J\otimes K) := J\otimes I\otimes K$. Similarly, the isotropy of $\varepsilon(X,Y)$ is equivalent to the defining equation

$$\mathcal{L}(\varepsilon(X,Y))\,\Gamma^t_{X,Y} = \Gamma^t_{Y,X}\,T_{12}\,(\bigoplus_{ij}\mathbb{I}^{\otimes 2}\otimes r(i,j,t)), \qquad (29)$$

where $r(i,j,t) : \mathcal{M}or(t,i\circ j) \to \mathcal{M}or(t,j\circ i)$ is given by the restriction of $\mathcal{L}(\varepsilon(i,j))$. Note that, since $r(i,j,t)$ are isomorphisms, we have $N_{ij,k} = N_{ji,k}$ so that the fusion rule algebra is commutative. The hexagonal equation, (5), is, with (29), equivalent to

$$\begin{aligned}(\bigoplus_l r(i,k,l)\otimes\mathbb{I}_{N_{lj,t}})\;F(i,k,j,t)\;(\bigoplus_l r(j,k,l)\otimes\mathbb{I}_{N_{il,t}})\\ = F(k,i,j,t)\;(\bigoplus_l \mathbb{I}_{N_{ij,l}}\otimes r(l,k,t))\;F(i,j,k,t)\end{aligned} \qquad (30)$$

defined on $\bigoplus_l \mathcal{M}or(l,j\circ k)\otimes\mathcal{M}or(t,i\circ l)$.

Finally the conjugation is additive on $\mathcal{O}bj = \mathbb{N}^{\mathcal{I}}$ since we have a natural isomorphism $\varphi : \mathcal{M}or(i,X)^* \to \mathcal{M}or(i^\vee,X^\vee)$ describing the transposition for a fixed choice of conjugates. Irredundance and uniqueness of conjugates then implies $N_{ji^\vee,1} = \delta_{ji}$ for $j,i\in\mathcal{I}$, which entails invariance, $N_{ij,k} = N_{i^\vee j^\vee,k^\vee}$. For the basis vectors $\theta_i := (\Gamma^1_{i^\vee,i})^{-1}(\vartheta_i) \in \mathcal{M}or(1, i^\vee \circ i)$ and dual vectors $\theta_i^\dagger := (\Gamma^1_{i,i^\vee})^*(\vartheta_i^\dagger) \in \mathcal{M}or(1, i \circ i^\vee)^*$ we find that

$$(\theta_i^\dagger \otimes 1)\, F(i, i^\vee, i, i)\, (\theta_i \otimes 1) \;=\; (\theta_i^\dagger \otimes 1)\, F(i^\vee, i, i^\vee, i^\vee)^{-1}\, (\theta_i \otimes 1) \;=\; 1 \qquad (31)$$

holds on the respective blocks of the F-matrices. Conversely, vectors, $\theta_i, \theta_i^\dagger$, with (31) and any set of isomorphisms, $\varphi_{i,X} : \mathcal{M}or(i, X)^* \to \mathcal{M}or(i^\vee, X^\vee)$, implement the conjugation by

$$\vartheta_X \;=\; \Gamma^1_{X^\vee,X}\Big(\sum_{i\nu} \varphi_{i,X}(\xi_i^\nu) \otimes \xi_{i\nu} \otimes \theta_i\Big) \qquad \in \mathcal{M}or(1, X^\vee \circ X)$$

$$\text{and} \qquad \vartheta_X^\dagger \;=\; (\Gamma^1_{X,X^\vee}{}^*)^{-1}\Big(\sum_{i\nu} \xi_i^\nu \otimes (\varphi_{i,X}^{-1})^*(\xi_{i\nu}) \otimes \theta_i^\dagger\Big) \in \mathcal{M}or(1, X \circ X^\vee)^*, \qquad (32)$$

where $\xi_{i\nu} \in \mathcal{M}or(i, X)$ and $\xi_i^\nu \in \mathcal{M}or(i, X)^*$ are dual bases. In the analysis presented so far we have extracted the following pieces of information from a semisimple, irredundant quantum category: On the one hand we have a fusion rule algebra, i.e., a distributive, commutative and associative product on $\mathbb{N}^{\mathcal{I}}$, with structure constants $N_{ij,k}$, and a conjugation, $^\vee$, on $\mathcal{I}$, with $N_{ij^\vee,1} = \delta_{ij}$. On the other hand we find invertible matrices

$$r(i,j,k) : \mathbb{C}^{N_{ij,k}} \to \mathbb{C}^{N_{ji,k}}$$

$$\text{and} \qquad F(i,j,k,t) : \bigoplus_l \mathbb{C}^{N_{jk,l}} \otimes \mathbb{C}^{N_{il,t}} \to \bigoplus_l \mathbb{C}^{N_{ij,l}} \otimes \mathbb{C}^{N_{lk,t}}$$

acting on the spaces, $\mathbb{C}^{N_{ij,k}}$, modelled from the fusion rules, and satisfying the equations (28) and (30). Reversing the logic of the above derivation, we can in fact reconstruct the quantum category, up to isomorphisms, from this data alone. The first step in the procedure is to introduce a collection of vector spaces, $\{\mathcal{M}or(k, X)\}$, of dimensions $X(k) = N_{X,k}$, where $k \in \mathcal{I}$ and $X \in \mathcal{O}bj := \mathbb{N}^{\mathcal{I}}$, with given monoidal product. From these we construct the morphism spaces, $\mathcal{M}or(X, Y)$, as the homomorphisms spaces in (21) with the obvious composition. Next we choose *any* set of isomorphisms,

$$\widetilde{\Gamma}^k_{X,Y} \;:\; \bigoplus_{ij} \mathcal{M}or(i, X) \otimes \mathcal{M}or(j, Y) \otimes \mathbb{C}^{N_{ij,k}} \longrightarrow \mathcal{M}or(k, X \circ Y),$$

with $\widetilde{\Gamma}^k_{1,X} = \widetilde{\Gamma}^k_{X,1} = \mathbb{I}$. The tensor-product of morphisms is then defined by the intertwining relation (23) and the monoidal isomorphisms are obtained from the r- and F-matrices, using equations (26), (27) and (29). From the pentagon equation we infer that the F-matrix elements

$$F(i, i^\vee, i, i)^1_1,\; (F(i^\vee, i, i^\vee, i^\vee)^{-1})^1_1 \;:\; \mathcal{M}or(1, i^\vee \circ i) \longrightarrow \mathcal{M}or(1, i \circ i^\vee),$$

coincide for all $i \in \mathcal{I}$, giving rise to vectors , θ_i and $\theta_i^\dagger$, with (31). From these we obtain for any set of isomorphisms, $\varphi_{i,X}$, by (32) the morphisms, $\vartheta_X \in \mathcal{M}or(1, X^\vee \circ X)$ and $\vartheta_X^\dagger \in \mathcal{M}or(1, X \circ X^\vee)^*$.

Observe that in this setting we have additional isomorphisms, denoted $\widetilde{\Gamma}^k_{i,j} : \mathbb{C}^{N_{ij,k}} \to \mathcal{M}or(k, i \circ j)$, yielding $\Gamma^k_{X,Y} = \widetilde{\Gamma}^k_{X,Y}(\bigoplus_{ij} \mathbb{I}^{\otimes 2} \otimes (\widetilde{\Gamma}^k_{i,j})^{-1})$

for the isomorphisms previously defined. Here, the concept of "uniqueness up to equivalence" is specified in the remark that two quantum categories, constructed from datas $(r, F, \mathcal{M}or, \widetilde{\Gamma})$ and $(r', F', \mathcal{M}or', \widetilde{\Gamma}')$ are isomorphic with isomorphism $(f = id, \mathcal{F}, C)$, iff there exist isomorphisms, $F_X = \bigoplus_i F_{X,i}$, mapping $F_{X,i} : \mathcal{M}or(i, X) \xrightarrow{\cong} \mathcal{M}or(i, X)'$, $\forall\, i \in \mathcal{I}$, $X \in \mathcal{O}bj$, and $T^k_{ij} \in Gl(\mathbb{C}^{N_{ij,k}})$, $\forall\, i, j, k \in \mathcal{I}$, such that the following relations hold:

$$\begin{aligned} \mathcal{L}(\mathcal{F}(I)) &= F_Y\, \mathcal{L}(I)\, F_X^{-1}\,, \qquad \forall \ \ I \in \mathcal{M}or(X,Y) \\ \widetilde{\Gamma}^k_{X,Y}{}'\,(\bigoplus_{ij} F_{Y,i} \otimes F_{Y,j} \otimes T^k_{ij}) &= C(X,Y)^{-1}\, F_{X\circ Y,k}\, \widetilde{\Gamma}^k_{X,Y} \end{aligned} \tag{33}$$

and further

$$\begin{aligned} F'(i,j,k,t)\,(\bigoplus_l T^l_{jk} \otimes T^t_{il}) &= (\bigoplus_l T^l_{ij} \otimes T^t_{lk})\, F(i,j,k,t) \\ r'(i,j,t) \;\; T^t_{ij} &= T^t_{ji} \;\; r(i,j,t). \end{aligned} \tag{34}$$

It is obvious that to any sets of isomorphisms, $\{\widetilde{\Gamma}\}$, $\{\widetilde{\Gamma}'\}$, on vector spaces, $\{\mathcal{M}or\}$, $\{\mathcal{M}or'\}$, and any collection of T's, consistent with (34), the defining ingredients, $C(X,Y)$ and F_X, of an isomorphism of categories can be determined such that (33) holds. Thus, the equivalence class of a category only depends on the r- and F-matrices and two sets of matrices, $(r, F), (r', F')$, define the same class iff they are cohomologous by a natural transformation of the form (34). In most field-theoretic applications (e.g. [14]) the arrow sets are realized as the intertwiners of representations of local algebras so that the tensor-product is properly associative not only for the objects but also for the morphisms. Indeed, we find the stronger condition: $\alpha(X,Y,Z) = \mathbb{1}_{X\circ Y\circ Z}$. We take this as a justification to study *strict* monoidal quantum categories, i.e., categories for which there are isomorphisms $C(X,Y) \in \mathcal{E}nd(X \circ Y)$, which solve the equation $\alpha(X,Y,Z) = (C(X,Y)^{-1} \circ \mathbb{1}_Z) C(X \circ Y, Z)^{-1} C(X, Y \circ Z) (\mathbb{1}_X \circ C(Y,Z))$, where we accounted for the extended notion of equivalence. On the level of the r-F-matrix data strictness is detected iff there exist isomorphisms, $\widetilde{\Gamma}^k_{X,Y}$, such that for all $X, Y, Z \in \mathcal{O}bj$

$$\begin{aligned} &\bigoplus_{ijk} \mathbb{1}^{\otimes 3} \otimes F(i,j,k,t) = \\ &T_{34}\,(\bigoplus_{lk} (\widetilde{\Gamma}^l_{X,Y})^{-1} \otimes \mathbb{1}_{N_{Z,k}} \otimes \mathbb{1}_{N_{lk,t}})\,(\widetilde{\Gamma}^t_{X\circ Y,Z})^{-1} \widetilde{\Gamma}^t_{X,Y\circ Z}\,(\oplus_{il}\, \mathbb{1}_{N_{X,i}} \otimes \widetilde{\Gamma}^l_{Y,Z} \otimes \mathbb{1}_{N_{il,t}}) \end{aligned} \tag{35}$$

holds or, equivalently, if the F-matrix is determined by (27) for the specialization $X = 1, Y = j, Z = k \in \mathcal{I}$ and further if there exist $\Gamma^t_{X,Y}$, (with $\Gamma^t_{i,k} = id$), such that the maps from (24) are equal. Of course, every F-matrix of the form (35) is a solution of the pentagon equation, (28). A family of nonstrict quantum categories are provided by θ-categories, defined by the property that $k \circ k^\vee = 1$,

$\forall\ k \in \mathcal{I}$, so that $\mathcal{I}$ is a discrete group. The inequivalent monoidal structures for given $\mathcal{I}$ can be set in one-to-one correspondence with the Eilenberg-MacLane space $H^4(\mathcal{I},2,U(1))$ and the strict ones are those annihilated by the suspension map $S^* : H^4(\mathcal{I},2,U(1)) \to H^3(\mathcal{I},1,U(1))$, see [27].

Since we have $X = X^{\vee\vee}$ for an irredundant category any choice of conjugates, $(\theta_j, \varphi_{j,X})$, is naturally characterized by the quantities $s_j \in \mathbb{C}$, given by $r(j^\vee, j, 1)\theta_j = s_j \theta_{j^\vee}$, and $A_{j,X} := \varphi_{j^\vee, X^\vee}(\varphi_{j,X}^{-1})^* \in End(\mathcal{M}or(j,X)) \subset \mathcal{E}nd(X)$, $\forall\ j \in \mathcal{I},\ X \in \mathcal{O}bj$. These allow us to express the elements, associated to the braided rigidity of the category, as endomorphisms, $q_X = \bigoplus_j s_j^{-1} A_{j,X}$ and $g_X = \bigoplus_j s_{j^\vee} s_j^{-1} A_{j,X}^2$. Observe that every isotropic set of endomorphisms, $f(X) \in \mathcal{E}nd(X)$, $X \in \mathcal{O}bj$, is given analogous to $\alpha(X,Y,Z)$ and $\varepsilon(X,Y)$, by $f(X) = \bigoplus_j f(j)\, \mathbb{1}_{\mathcal{M}or\,(j,X)}$, so that for example the elements $\tau(X)$ are already determined by $\tau(j) = s_j s_{j^\vee} = \mu(j, j^\vee, 1)$, where $\mu(i,j,k) := r(j,i,k) r(i,j,k)$. Now it is always possible to choose $\varphi_{j,X}$ such that $I^{tt} = I$ for all morphisms, I, or, equivalently that $q_X = q(X)$ is isotropic, in which case we have $A_{j,X} = 1$. Further we can find θ_j's, for which $s_j = s_{j^\vee}$ holds, so that $q(X)^{-1} = q(X^\vee)^t$ and $g_X = 1$ follow. In this frame the definition of a statistical category is reduced to the existence of numbers, $\sigma(j) \in \mathbb{C}, j \in \mathcal{I}$, with $\mu(i,j,k) = \sigma(i)\sigma(j)\sigma(k)^{-1}\, \mathbb{1}_{N_{ij,k}}$, $\sigma(j) = \sigma(j^\vee)$ and $s_j^2 = \sigma(j)^2$. Evidently, the $\sigma(j)'$s can always be multiplied by a $\mathbb{Z}_2$-grading, $\mathcal{I} \to \{\pm 1\} : j \to \varepsilon_j$, of the fusion rule algebra, but are otherwise unique. By a suitable rescaling of the θ_j's we can arrange that $\xi_j := \sigma_j s_j^{-1} = 1, \forall\ j \in \mathcal{I}$ with $j \neq j^\vee$. We call a selfconjugate object $j \in \mathcal{I}$ real if $\xi_j = 1$ and pseudoreal for $\xi_j = -1$, depending on the choice of σ but not on the choice of the conjugates. Hence, if reality is a $\mathbb{Z}_2$-grading, as it is the case for C^*-categories, we can set $\sigma(X)' = q(X)^{-1}$. In this situation there exists $\varphi_{j^\vee, X^\vee} = \varphi_{j,X}^*$, such that $\vartheta_{X \circ Y}$ is the canonical morphism obtained from ϑ_X and ϑ_Y, and the statistical dimension is given by $d(i) = \theta_{i^\vee}^t \theta_i$. From (14) we infer at once that they diagonalize the fusion rules, i.e., $\Sigma_k N_{ij,k} d(k) = d(i)d(j)$.

III. THE DUALITY PROBLEM AND EXAMPLES

(1) Tannakian and weakly Tannakian Categories

In order to provide examples for quantum categories we discuss next *Tannakian* categories, a property which is very powerful in duality theory, but too restrictive for the interesting applications in mathematical physics. The frame of a Tannakian category is the category $\mathcal{V}ec_{\mathbb{C}}$, whose objects are the finite-dimensional $\mathbb{C}$-vectorspaces and whose morphism spaces are given by $Hom_{\mathbb{C}}(V,W)$ with the usual composition. $\mathcal{V}ec_{\mathbb{C}}$ carries a monoidal product which is given for the objects as well as for the morphisms by the ordinary tensor-product for vectorspaces with $Hom_{\mathbb{C}}(V \otimes W, V' \otimes W')$

$=Hom_{\mathbb{C}}(V,V')\otimes Hom_{\mathbb{C}}(W,W')$. A monoidal category is thus called *Tannakian* if it possesses an injection into $\mathcal{V}ec_{\mathbb{C}}$. More specifically, this implies an assignment $X \to V_X$ of vector spaces to the objects and a faithful representation of the category, $\mathcal{F} : \mathcal{M}or(X,Y) \to Hom_{\mathbb{C}}(V_X, V_Y)$.

Moreover we have isomorphisms $C(X,Y) \in Hom_{\mathbb{C}}(V_X \otimes V_Y, V_{X\circ Y})$ such that

$$C(X',Y')\,(\mathcal{F}(I)\otimes\mathcal{F}(J))\,C(X,Y)^{-1} = \mathcal{F}(I\circ J) \tag{36}$$

for $I \in \mathcal{M}or(X,Y')$ and $J \in \mathcal{M}or(Y,Y')$. Further there are monoidal isomorphisms, $\alpha(U,V,W) \in End_{\mathbb{C}}(U\otimes V\otimes W)$ and $\varepsilon(V,W) \in Hom_{\mathbb{C}}(V\otimes W, W\otimes V)$, induced by (19), which also obey the hexagonal- and pentagonal-equations. The cardinal example of a Tannakian quantum category is the representation category, $\mathcal{R}ep(\mathcal{A})$, of a quasitriangular, quasicoassociative Hopf algebra, $\mathcal{A}$, with elements $\Delta : \mathcal{A} \to \mathcal{A}^{\otimes 2}$, $\epsilon : \mathcal{A} \to \mathbb{C}$, $S : \mathcal{A} \to \mathcal{A}$, $R \in \mathcal{A}^{\otimes 2}$ and $\phi \in \mathcal{A}^{\otimes 3}$ as in [28]. The objects of $\mathcal{R}ep(\mathcal{A})$ are representations, (ρ, V), of $\mathcal{A}$, where $\rho : \mathcal{A} \to End(V)$, with product $(\rho,V)\circ(\rho',V') = (\rho\otimes\rho'\Delta, V\otimes V')$ and conjugation $(\rho,V)^\vee = (\rho^\vee, V^*)$ with $\rho^\vee(a) = \rho(S(a))^t$. The morphisms are the intertwinerspaces, $Hom_{\mathcal{A}}(V,W)$, with ordinary tensor-product and we set

$$\varepsilon(V,W) := T_{12}\rho_V\otimes\rho_W(R) \qquad \text{and} \qquad \alpha(U,V,W) := \rho_U\otimes\rho_V\otimes\rho_W(\phi^{-1}).$$

The functor to $\mathcal{V}ec_{\mathbb{C}}$ simply "forgets" the interpretation of the vectorspaces as modules.

A map $\varphi : \mathcal{A}_1 \to \mathcal{A}_2$, which is a homomorphism of quasitriangular quasi Hopf algebras for some *twist equivalent* structure of $\mathcal{A}_2$,see [17], yields a compatible tensor functor $\varphi^* : \mathcal{R}ep(\mathcal{A}_2) \to \mathcal{R}ep(\mathcal{A}_1)$ .The properties triangularity, coassociativity (up to twists) and semisimplicity of $\mathcal{A}$ translate into symmetry, strictness and semisimplicity of $\mathcal{R}ep(\mathcal{A})$ respectively. The simplest type of semisimple Tannakian quantum categories are the *classical* Tannakian categories, characterized by $\alpha(U,V,W) = \mathbb{1}$ and $\varepsilon(V,W) = T_{12}$. To a general Tannakian category, $\mathcal{T}$, with conjugation we associate the group $G(\mathcal{T})$, given by the selections, $\tilde{g}$, of endomorphisms, $X \to \tilde{g}(X) \in End(V_X)$, which are isotropic with respect to the subsets $\mathcal{M}or(X,Y) \subset Hom(V_X, V_Y)$ and obey $C(X,Y)\tilde{g}(X\circ Y) = \tilde{g}(X)\otimes\tilde{g}(Y)C(X,Y)$. For the special case $\mathcal{T} = \mathcal{R}ep(\mathbb{C}[G])$, where G is a compact group with $\Delta(g) = g\otimes g$ and $R = \phi = 1$, Tannaka's theorem asserts that any such selection is of the form $\tilde{g}(X) = \rho_X(g)$, for some $g \in G$, so that we have $G \cong G(\mathcal{R}ep(\mathbb{C}[G]))$. This is complemented by Kreĭns theorem, showing $\mathcal{T} \cong \mathcal{R}ep(\mathbb{C}[G(\mathcal{T})])$ for any classical Tannakian $\mathcal{T}$ (see [29]). Imitating this construction, we find from every Tannakian quantum category, $\mathcal{T}$, a quasitriangular quasi Hopf algebra, $\mathcal{A}(\mathcal{T})$, for which elements, $\xi \in \mathcal{A}(\mathcal{T})^{\otimes n}$, are given as assignments,

$(X_1, \dots, X_n) \to \xi(X_1, \dots, X_n) \in End(V_{X_1} \otimes \dots \otimes V_{X_n})$, obeying

$$\xi(Y_1, \dots, Y_n) \quad I_1 \otimes \dots \otimes I_n = I_1 \otimes \dots \otimes I_n \quad \xi(X_1, \dots, X_n), \tag{37}$$

for any sequence $I_k \in \mathcal{M}or(X_k, Y_k)$, $k = 1, \dots, n$. For $a \in \mathcal{A}(\mathcal{T})$ the coproduct is defined by

$$\triangle(a)(X,Y) := C(X,Y)\, a(X \circ Y)\, C(X,Y)^{-1}, \tag{38}$$

the co-unit by $\epsilon(a) := a(1)$ and the antipode by $S(a)(X^\vee) := \varphi_X\ a(X)^t\ \varphi_X^{-1}$, where $\varphi_X \in Hom(V_X^*, V_{X^\vee})$ is canonically given by $C(X^\vee, X)\, \mathcal{F}(\vartheta_X) \in V_{X^\vee} \otimes V_X$ and is invertible by (6). The selections of endomorphisms

$$\phi^{-1} : (X,Y,Z) \to \alpha(V_X, V_Y, V_Z) \qquad \text{and} \qquad R : (X,Y) \to T_{12}\, \varepsilon(V_X, V_Y)$$

are by isotropy of α and ε in $\mathcal{T}$ elements of $\mathcal{A}(\mathcal{T})^{\otimes 3}$ and $\mathcal{A}(\mathcal{T})^{\otimes 2}$ and thus provide us with the remaining constituents. In analogy to the situation of the Tannaka-Kreĭn-theorem, we have a Hopf algebra monomorphism $\mathcal{A} \hookrightarrow \mathcal{A}(\mathcal{R}ep(\mathcal{A}))$ and an injection of quantum categories $\mathcal{T} \hookrightarrow \mathcal{R}ep(\mathcal{A}(\mathcal{T}))$, which one easily proves to be isomorphisms if $\mathcal{A}$ and $\mathcal{T}$ are semisimple, using $\mathcal{A} \cong \bigoplus_{j \in \mathcal{I}} End_{\mathbb{C}}(V_j)$.

Note that the last remark only proves the easier part of the Tannaka-Kreĭn duality, which in addition implies that any semisimple, cosemisimple, coassociative and cocommutative Hopf algebra is given by $\mathcal{A} = \mathbb{C}[G(\mathcal{A})]$, where the group is defined by $G(\mathcal{A}) := \{g \in \mathcal{A} : \triangle(g) = g \otimes g\}$.

In essence, however, duality theory reduces to the question, which categories are Tannakian? A necessary condition is, of course, that there are integers, $d_X \geq 0$, for all $X \in \mathcal{O}bj$, given by $d_X = dim\, V_X$, which have the same properties with respect to products and direct sums as the statistical dimensions, $d(X)$, in (14).

Recently, Deligne showed that this is also sufficient for symmetric categories (actually assuming $d(X) = d_X$). We want to show that the prerequisite of symmetry is in fact not a very crucial one in this result, by giving a pedestrian construction for *braided* quantum categories, which are *semisimple* . In this case it is by $d_X = \Sigma_{j \in \mathcal{I}} N_{X,j}\, d_j$ enough to assume integers, $d_j \geq 0$, $j \in \mathcal{I}$, with $d_1 = 1$, $d_j = d_{j^\vee}$ and $d_i\, d_j = \Sigma_k N_{ij,k}\, d_k$. The basic components are a set of vectorspaces, $\{V_j\}_{j \in \mathcal{I}}$, with $dim\, V_j = d_j$, and any set of isomorphisms, $C_{ij} : \bigoplus_{k \in \mathcal{I}} \mathcal{M}or(k, i \circ j) \otimes V_k \longrightarrow V_i \otimes V_j$, between the spaces of equal dimensions. For any $X \in \mathcal{O}bj$ the respective object in $\mathcal{V}ec_{\mathbb{C}}$ is given by $V_X := \bigoplus_{j \in \mathcal{I}} \mathcal{M}or(j, X) \otimes V_j$ and for morphisms, $I \in \mathcal{M}or(X,Y)$, we set $\mathcal{F}(I) = \bigoplus_{j \in \mathcal{I}} \mathcal{L}(I) \otimes 1\!\!1_{d_j} \in Hom(V_X, V_Y)$. The natural isomorphisms of the tensor functor to $\mathcal{V}ec_{\mathbb{C}}$ are

$$C(X,Y) = T_{23}\,(\bigoplus_{i,j} 1\!\!1_{N_{X,i}} \otimes 1\!\!1_{N_{Y,j}} \otimes C_{ij})(\bigoplus_k (\Gamma^k_{X,Y})^{-1} \otimes 1\!\!1_{d_k}), \tag{39}$$

with $C(i,j) = C_{ij}$.

For semisimple quantum categories with $|\mathcal{I}| < \infty$ the dimension criterion becomes even more rigid: We recall, that the fusion rules define by $(\mathbb{N}_j)_{ik} = N_{ij,k}$ commuting, positive and normal matrices, which have no proper common invariant blocks and are closed under transposition, so that by Perron-Frobenius theory there exists a *unique* vector of positive numbers, $(d_j) \in (\mathbb{R}^+)^{\mathcal{I}}$, with $d_1 = 1$, and $d_i\, d_j = \Sigma_k\, N_{ij,k}\, d_k$, implying $d_j = d_{j^\vee}$ by $\mathbb{N}_{j^\vee} = \mathbb{N}_j^t$. Thus a category is decided to be Tannakian iff this particular vector consists only of integers, which fails to hold for the categories associated to, e.g., minimal conformal models or quantum groups at roots of unity!

A much larger class of categories – let us call them *weakly-Tannakian* – is obtained if we relax the conditions on the functor to $\mathcal{V}ec_{\mathbb{C}}$, requiring $C(X,Y)$ only to be an epimorphism with right inverse $C(X,Y)^{-1}$, such that equation (36) still holds but $P_{X,Y} := C(X,Y)^{-1}\, C(X,Y)$ can be a proper projector on $V_X \otimes V_Y$. Again $\varepsilon^{\pm 1}$ and $\alpha^{\pm 1}$ of a quantum category define by (19) respective homomorphisms in $Vec_{\mathbb{C}}$, which are now invertible only between the subspaces given by the projectors

$$
\begin{aligned}
P_{X,Y} &= \varepsilon(V_X, V_Y)^{-1}\, \varepsilon(V_X, V_Y) = \varepsilon(V_Y, V_X)\varepsilon(V_Y, V_X)^{-1}, \\
\tilde{P}_{(X,Y)Z} &:= (C(X,Y)^{-1} \otimes \mathbb{I})\, P_{X\circ Y,Z}(C(X,Y) \otimes \mathbb{I}) \\
&= \alpha(V_X, V_Y, V_Z)\, \alpha(V_X, V_Y, V_Z)^{-1}, \\
\tilde{P}_{X(Y,Z)} &:= (\mathbb{I} \otimes C(Y,Z)^{-1})\, P_{X,Y\circ Z}\, (\mathbb{I} \otimes C(Y,Z)) \\
&= \alpha(V_X, V_Y, V_Z)^{-1} \alpha(V_X, V_Y, V_Z),
\end{aligned}
$$

but still satisfy the pentagon- and hexagon-equations. Thus for a weakly-Tannakian quantum category, $\mathcal{T}$, the algebra $\mathcal{A}(\mathcal{T})$ is obtained in the same way as in the usual Tannakian case, equipped with elements $\Delta, \phi^{\pm 1}$ and $R^{\pm 1}$. For these the axioms of a *weak-quasi-Hopf algebra*, as proposed by Mack and Schomerus (see [20] or contribution to this volume), are immediately verified and we have $\Delta(1_{\mathcal{A}})(X,Y) = P_{X,Y}$. It is obvious that a weak Hopf algebra, $\mathcal{A}$, defines a weakly-Tannakian quantum category, $\mathcal{R}ep(\mathcal{A})$, and if either is semisimple, a duality correspondence as for ordinary Hopf algebras is in existence.

Addressing the question which categories are weakly-Tannakian, we observe that we still need additive, positive integer dimensions, d_X, for all objects X, but multiplicativity with respect to $\circ$ is replaced by the inequality $d_{X\circ Y} \leq d_X d_Y$. If semisimplicity is given, the respective condition, $d_i\, d_j \geq \Sigma\, N_{ij,k}\, d_k$ for some $d_j \in \mathbb{N}$, with $d_1 = 1$ and $d_j = d_{j^\vee}$, is indeed sufficient and the construction of the functor to $\mathcal{V}ec_{\mathbb{C}}$ proceeds in a way analogous to the reinterpretation of C_{ij} as an epimorphism with right inverse C_{ij}^{-1}. Summarily, we can view weak-quasi-Hopf algebras as a reframing of quantum categories, obeying the rather mild dimensionality condition, with an arbitrary direct sum decompositions $V_X \otimes V_Y = V_{X\circ Y} \oplus V_{X,Y}^{\text{unphys.}}$, with $V_{X,Y}^{\text{unphys.}} := ker\, C(X,Y)$, $\forall\ X,Y \in \mathcal{O}bj$, chosen in the intermediating weakly-Tannakian category. A consequence of this procedure is that automorphisms, $(id, \mathbb{I}, C)$, of the monoidal structure on a category $\mathcal{T}$ induce on $\mathcal{A}(\mathcal{T})$ only those twist transformations which preserve the direct sums, $V_{X \circ Y} \oplus V_{X,Y}^{\text{unphys.}}$, so that there is no natural notion of coassociativity on $\mathcal{A}(\mathcal{T})$ which parallels that of strictness in $\mathcal{T}$.

In the course of the classification of quantum field theories with braid statistics and for the construction of link-invariants , integrable lattice models and eventually weak-quasi-Hopf algebras one wishes to extract categorial data ,e.g., in terms of solutions to (28) and (30), from ordinary, mostly coassociative and typically non-semisimple Hopf algebras, of which plenty are known. Although, a priori ,they only yield the (non-semisimple) Tannakian categories, $\mathcal{R}ep(\mathcal{A})$, their semisimple quotients, denoted $\overline{\mathcal{R}ep(\mathcal{A})}$, can in general be non-Tannakian and for $\mathcal{A} = U_q(\mathfrak{g})$, with $q^{2(k+2)} = 1$, they turn out to be also rational, i.e., $\mid \mathcal{I} \mid < \infty$. In order to exemplify the construction of quotients, as described in the previous section, for the case of representation categories it is useful to establish relations between the special morphisms in $\mathcal{R}ep(\mathcal{A})$ and the special elements in $\mathcal{A}$ (or $\mathcal{A}(\mathcal{R}ep(\mathcal{A}))$). For simplicity we only treat the coassociative case ($\phi = 1$) so that we can choose $\vartheta_X = \vartheta_X^\dagger \in V_X^* \otimes V_X$, to be the canonical elements. The special elements for general ϕ, where ϑ_X and $\vartheta_X^\dagger$ involve further isomorphisms, are obtained in the same way and can be taken from [17].

With these assumptions the transposition on $\mathcal{M}or(X,Y)$ is the one inherited from $Hom(V_X, V_Y)$, where $J^{tt} K_X = K_Y J$ for $J \in \mathcal{M}or(X,Y)$ and $K_X : V_X \to V_{X^{\vee\vee}} = (V_X)^{**}$ is the canonical isomorphism. For twice-conjugates we find

$$\rho_{X^{\vee\vee}}(S^2(a))K_X = K_X \rho_X(a), \ \forall\, a \in \mathcal{A}$$

. It follows that $K_X^{-1} q_X$ is isotropic, so we can write $q_X = K_X \rho_X(u_3^{-1})$, where the respective element in the Hopf algebra is given by $u_3 = m(1 \otimes S^{-1})(\sigma R)$ and $u_3^{-1} = m(1 \otimes S)(\sigma R^{-1})$, with $m(a \otimes b) = ab$ and $\sigma(a \otimes b) = b \otimes a$. Since $q_X \in Hom_{\mathcal{A}}(V_X, V_{X^{\vee\vee}})$ we have $u_3\, a\, u_3^{-1} = S^2(a)$, $\forall a \in \mathcal{A}$. The factorization of q_X implies $M := (\sigma R)R = (u_3 \otimes u_3)\, \Delta(u_3)^{-1}$. By isotropy we can also write $2\ \tau(X) = \rho_X(z)$, where $z \in \mathcal{A}$ is found to be $z = u_3\, S(u_3)$, lies , by $\tau(X) \in End_{\mathcal{A}}(V_X)$, in the center of $\mathcal{A}$ and has the coproduct $\Delta(z) = (z \otimes z)M^{-2}$. Similarly, we have $g_X = K_{X^{\vee\vee}}\, K_X\, \rho_X(g^{-1})$, where $g := u_3\, S(u_3)^{-1}$ is the Lyubashenko-Radford-element with $\Delta(g) = g \otimes g$ and $g\, a\, g^{-1} = S^4(a)$. For more details see [30] and references therein.

In [31] *ribbon Hopf algebras* are defined as quasitriangular Hopf algebras possessing a *central* element $v \in \mathcal{A}$, which satisfies $v^2 = z$ and $M = (v \otimes v)\, \Delta(v)^{-1}$. Setting $\sigma(X) = \rho_X(v)$, we see immediately that $\mathcal{A}(\mathcal{T})$ is a ribbon Hopf algebra if $\mathcal{T}$ is statistical and conversely for any ribbon Hopf algebra, $\mathcal{A}$, the category $\mathcal{R}ep(\mathcal{A})$ is statistical. In many cases it is more practical to verify this property by finding the element $\kappa := u_3 v^{-1} \in \mathcal{A}$, related to $\xi_X = \rho_X(\kappa) K_X^{-1}$. For this the relations $\Delta(\kappa) = \kappa \otimes \kappa$, $\kappa^2 = g$ and $\kappa a \kappa^{-1} = S^2(a)$, $\forall a \in \mathcal{A}$, have to hold. Using the special elements, it is possible to express the traces entering the quotient construction in terms of the canonical trace, tr_V, on a module, V. We have for any $I \in \mathcal{E}nd(X) = End_{\mathcal{A}}(V_X)$ the indentities

$$Tr_X(I) = tr_{V_X}(\rho_X(u_3) I), tr_X(I) = tr_{V_X}(\rho_X(\kappa) I) \text{ and } d(X) = tr_{V_X}(\kappa).$$

In particular we can identify an indecomposable representation, X, as a ghost iff $\rho_X(u_3)$ or $\rho_X(\kappa)$ is traceless.

For the quantum groups, $U_q(\mathfrak{g})$, we can choose for κ the element $q^{2\delta}$ of the Cartan-algebra, with $\delta = \frac{1}{2}\Sigma_{\alpha>0} h_\alpha$, and the resulting $d(X)$ is often called the *q-dimension.*

(2) Quantumgroup Duality, an Example

Next, we shall illustrate how the $\mathcal{M}or(X,Y)^{\circ}$-spaces appear in the category $\mathcal{R}ep(U_q\ (sl_2)), q = exp\left(\frac{i\pi}{k_l+2}\right)$, $k_l \in \mathbb{N}$, and in which sense they can be discarded. Among the objects of $\mathcal{R}ep(U_q(sl_2))$ are irreducible representations, V_λ, of highest weights $\lambda = 0, 1, \cdots, k_l$, dimensions $dim(V_\lambda) = \lambda + 1$ and q-dimensions, $d(\lambda) = [\lambda + 1]_q > 0$. Further we have a set of indecomposable,projective modules, $W_\lambda, \lambda \in \mathbb{Z}$, with vanishing q-dimension, $dim(W_\lambda) = k_l + 2$ for $(\lambda + 1) \in (k_l + 2)\mathbb{Z}$ and $dim(W_\lambda) = 2(k_l + 2)$ elsewise. For $0 \leq \lambda \leq k_l$, $\mathcal{E}nd(W_\lambda)$ contains besides $\mathbb{1}$ a nilpotent element, N_λ (given for example by the nilpotent part of the quartic casimir) with a monic-epi-presentation $N_\lambda = \tilde{I}_\lambda \tilde{P}_\lambda$, where $\mathbb{C}\tilde{P}_\lambda = \mathcal{M}or(W_\lambda, V_\lambda)$, $\mathbb{C}\tilde{I}_\lambda = \mathcal{M}or(V_\lambda, W_\lambda)$ and $\tilde{P}_\lambda \tilde{I}_\lambda = 0$ so that $N_\lambda^2 = 0$.

The decomposition of tensor-products of irreducible representations into direct sums can be given in terms of injections and projections,

$$\begin{aligned} &I_{\lambda\mu,\eta} \in Hom_{\mathcal{A}}(V_\eta, V_\lambda \otimes V_\mu),\ \ P_{\eta,\lambda\mu} \in Hom_{\mathcal{A}}(V_\lambda \otimes V_\mu, V_\eta), \\ &\qquad \text{for}\quad |\lambda - \mu| \leq \eta \leq min(\lambda + \mu, 2k_l - \lambda - \mu),\ \ \eta \equiv \lambda + \mu \bmod 2 \end{aligned} \tag{40}$$

and

$$\begin{aligned} &\hat{I}_{\lambda\mu,\eta} \in Hom_{\mathcal{A}}(W_\eta, V_\lambda \otimes V_\mu),\ \ \hat{P}_{\lambda\mu,\eta} \in Hom_{\mathcal{A}}(V_\lambda \otimes V_\mu, W_\eta), \\ &\qquad \text{for}\quad 2(k_l + 1) - (\lambda + \mu) \leq \eta \leq k_l + 1,\ \ \eta \equiv \lambda + \mu \bmod 2, \end{aligned} \tag{41}$$

so that

$$\Pi_\eta^{\lambda\mu} := I_{\lambda\mu,\eta} P_{\eta,\lambda\mu} \qquad \text{and} \qquad \hat{\Pi}_\eta^{\lambda\mu} := \hat{I}_{\lambda\mu,\eta} \hat{P}_{\eta,\lambda\mu} \tag{42}$$

form a partition of one into projectors. Hence, if $k_l \geq (\lambda + \mu)$, the tensor-product is completely reducible into real objects. In this case we can use (27) to define the F-matrices. Inserting $I_{\mu\nu,\eta} \otimes I_{\lambda\eta,\gamma}$ we have

$$(\mathbb{I}_\lambda \otimes I_{\mu\nu,\eta})\, I_{\lambda\eta,\gamma} \;-\; \sum_{\eta'} F(\lambda,\mu,\nu,\gamma)_\eta^{\eta'} \,(I_{\lambda\mu,\eta'} \otimes \mathbb{I}_\nu)\, I_{\eta'\nu,\gamma} \;=\; 0, \tag{43}$$

thus the matrix elements of F are the q-analogue 6-j-symbols with

$$F(\lambda,\mu,\nu,\gamma)_\eta^{\eta'} \;=\; P_{\gamma,\eta'\nu}\,(P_{\eta',\lambda\mu} \otimes \mathbb{I}_\nu)(\mathbb{I}_\lambda \otimes I_{\mu\nu,\eta})\, I_{\lambda\eta,\gamma}. \tag{44}$$

If k_l is too small, we can still consider the expression in (43), where the summation is restricted to η', such that the F-matrix elements can be defined by (44), i.e., all triples of indices occurring in (43) shall obey the restriction from (40) so that the respective projections and injections exist. It is remarked in [20] that the resulting expression, $I^\circ_{k_l 11,k_l}$ is in general no longer zero, as we can see from the specialization $\lambda = \gamma = k_l \geq 2$, $\mu = \nu = 1$ and $\eta = 0$. In this case the only admissible summation index is $\eta' = k_l - 1$. For the computation of $I^\circ_{k_l 11,k_l}$ we conveniently assume $I_{11,0} = \vartheta_1$ and $(\mathbb{I}_{(k_l-1)} \otimes \vartheta_1^\dagger)\,(I_{(k_l-1)1,k_l} \otimes \mathbb{I}_1) = P_{(k_l-1),k_l 1}$ so that $F(k_l,1,1,k_l)_0^{(k_l-1)} = 1$ and $(P_{(k_l-1),k_l 1} \otimes \mathbb{I}_1)\,(\mathbb{I}_{k_l} \otimes \vartheta_1) = I_{(k_l-1)1,k_l}$. We find

$$I^\circ_{k_l 11,k_l} \;=\; (\mathbb{I} - (\Pi^{(k_l-1)}_{k_l 1} \otimes \mathbb{I}_1))(\mathbb{I}_{k_l} \otimes \vartheta_1) \;=\; (\hat{\Pi}^{(k_l+1)}_{k_l 1} \otimes \mathbb{I}_1)(\mathbb{I}_{k_l} \otimes \vartheta_1) \tag{45}$$

which is apparently not zero. However, we infer from

$$(\hat{P}_{(k_l+1),k_l 1} \otimes \mathbb{I}_1)(\mathbb{I}_{k_l} \otimes \vartheta_1) \;\in\; Hom_{\mathcal{A}}(V_{k_l}, W_{k_l+1} \otimes V_1) \;\cong\; Hom_{\mathcal{A}}(V_{k_l}, W_{k_l}),$$

that $I^\circ_{k_l 11,k_l} = R\,\tilde{I}_{k_l}$, with $\tilde{I}_{k_l} \in \mathcal{M}or(V_{k_l}, W_{k_l})$, where we have set for some normalization $R := (\hat{I}_{k_l 1,(k_l+1)} \otimes \mathbb{I}_1)\hat{I}_{(k_l+1)1,k_l}$. Hence for any morphism $L \in \mathcal{M}or(V_{k_l} \otimes V_1 \otimes V_1, V_{k_l})$, we have $L\, I^\circ_{k_l 11,k_l} = 0$, since $L\,R \in \mathcal{M}or(W_{k_l}, V_{k_l})$ has to be proportional to P_{k_l}. This proves

$$I^\circ_{k_l 11,k_l} \;\in\; \mathcal{M}or(V_{k_l}, V_{k_l} \otimes V_1 \otimes V_1)^\circ, \tag{46}$$

so that in particular (43) remains true if it is read modulo the nullspaces of the traces.

As remarked earlier we have for general $\mathcal{A}$ a splitting homomorphism, ψ, of

$$0 \;\rightarrow\; End_{\mathcal{A}}(V_i \otimes V_j)^\circ \;\hookrightarrow\; End_{\mathcal{A}}(V_i \otimes V_j) \;\underset{\pi}{\overset{\psi}{\rightleftarrows}}\; \overline{End_{\mathcal{A}}(V_i \otimes V_j)} \;\rightarrow\; 0 \tag{47}$$

such that for the minimal projectors $e^k_{\nu\nu} \in \overline{End_{\mathcal{A}}(V_i \otimes V_j)}$, $\nu = 1, \ldots, N_{ij,k}$, $k \in \mathcal{I}$, $\Pi^{k,\nu}_{ij} := \psi(e^k_{\nu\nu})$ is a projector on a unique, real and indecomposable module V_k. If $P^{(\nu)}_{k,ij}$ and $I^{(\nu)}_{ij,k}$ are epis and monics, with $\Pi^{k,\nu}_{ij} = I^{(\nu)}_{ij,k}\, P^{(\nu)}_{k,ij}$ and

$l_k^\nu \in \mathcal{M}or(k, i \circ j)^*$, $e_k^\nu \in \mathcal{M}or(k, i \circ j)$ are dual bases, we find for the linear maps $C_{ij} := \bigoplus_{k\nu} e_k^\nu \otimes P_{k,ij}^{(\nu)}$ and $C_{ij}^{-1} := \bigoplus_{k\nu} l_k^\nu \otimes I_{ij,k}^{(\nu)}$ that $C_{ij} C_{ij}^{-1} = \mathbb{I}$ and $P_{ij} = C_{ij}^{-1} C_{ij} = \psi(\mathbb{I})$. Thus any $\overline{\mathcal{R}ep(\mathcal{A})}$ is weakly-Tannakian with explicit $C(X,Y)$ and $C(X,Y)^{-1}$ given by (39). Moreover we recover for this choice of $C(X,Y)$'s the relations

$$\phi = id \otimes \Delta(\Delta(1))\, \Delta \otimes id(\Delta(1)) \quad \text{and} \quad \phi^{-1} = \Delta \otimes id(\Delta(1))\, id \otimes \Delta(\Delta(1))$$

in the algebra $\mathcal{A}(\overline{\mathcal{R}ep(\mathcal{A})})$, expressing coassociativity of the original $\mathcal{A}$. (For an example see the truncated $U_q(sl_2)$ in [20].

A question that arises naturally in this context is, whether a quasitriangular Hopf algebra $\mathcal{A}$ can be reconstructed from the abstract category $\overline{\mathcal{R}ep(\mathcal{A})}$, in which the ghost-representations of $\mathcal{A}$ and all connections of $\mathcal{R}ep(\mathcal{A})$ to $\mathcal{V}ec_{\mathbb{C}}$ disappeared. As an example we can consider $\mathcal{A} = U_q(sl_2)$, q a root of unity, so that $\overline{\mathcal{R}ep(\mathcal{A})}$ is no longer Tannakian. Still it contains enough information to derive $U_q(sl_2)$ as the unique quasitriangular Hopf algebra, which is dual to $\overline{\mathcal{R}ep(\mathcal{A})}$ in the extended sense, assuming that its fundamental representation V_1 is irreducible and of dimension two and that $k_l > 2$. Since ghost-representations are never one-dimensional it follows from $(1) \circ (1) = (0) \oplus (2)$ that $dim(V_2) = 1 \, \text{or}\, 3$. However $dim(V_2) = 1$ and $dim(V_3) \geq 1$ contradict the dimension- inequality for $(2) \circ (1) = (1) \oplus (3)$ so that we have the direct sum decomposition into "real" $\mathcal{A}$-modules , $V_1 \otimes V_1 = 1 \oplus V_2$. The conditions (6) and $\vartheta_1^\dagger \vartheta_1 = d(1) \neq -2$ completely determine (up to isomorphisms of V_1) the elements $\mathcal{F}(\vartheta_1) \in V_1^{\otimes 2}$ and $\mathcal{F}(\vartheta_1^\dagger) \in (V_1^{\otimes 2})^*$ and thereby also the Temperley-Lieb projector, $\Pi_{11}^0 = \mathcal{F}(e) \in End(V_1 \otimes V_1)$, with $e := d(1)^{-1} \vartheta_1 \vartheta_1^\dagger$. The braid-isomorphism in $\overline{\mathcal{R}ep(\mathcal{A})}$ is of the form $\varepsilon((1),(1)) = z((q^2+1)e - 1) \in \overline{\mathcal{E}nd}((1) \circ (1))$ so that

$$\varepsilon(V_1, V_1) = z((q^2+1)\Pi_{11}^0 - 1) + N$$

with a possible, nilpotent contribution $N \in End_{\mathcal{A}}(V_1 \otimes V_1)^\circ$.

Using $\vartheta_1^\dagger N = N \vartheta_1 = 0$ and the fact that the expectations ,$E_{V_1}^{V_1,V_1}$ as in (10), vanish on N, we find $N = n \otimes n$ for some nilpotent $n \in End(V_1)$. The isotropy equation (4) applied to $X = X' = V_1 \otimes V_1$, $Y = Y' = V_1$, $J = 1$ and $I = N$ then yields , with the factorization of $\varepsilon((1) \circ (1), (1))$ according to (5), $n = 0$. Hence $\varepsilon(V_1, V_1)$ is unique. Once the matrix, $R_{11} := T\varepsilon(V_1, V_1) \in End(V_1 \otimes V_1)$, is known one can recontruct $U_q(sl_2)$ by the procedure of Faddeev, Reshetikhin and Takhtadzhyan, [32] . Assuming for $\mathcal{A}$ that $\bigcap_n ker\, \rho_1^{\otimes n} = 0$, the construction entails that $\mathcal{A}$ is a sub-Hopf-algebra of $U_q(sl_2)$. Also, $\mathcal{A}$ should be quasitringular with the same element $R \in \mathcal{A}^{\otimes 2}$, which implies $\mathcal{A} = U_q(sl_2)$.

Of course we could take this result as a motivation to conjecture on an extended duality theorem, asserting that to every semisimple, weakly Tannakian, strict monoidal quantum category $\mathcal{T}$, there exists a quasitriangular Hopf algebra $\mathcal{A}$

with $\mathcal{T} \cong \overline{\mathcal{R}ep(\mathcal{A})}$. Yet we have to keep in mind that in this example we made substantial use of two special properties of $\overline{\mathcal{R}ep(U_q(sl_2))}$. Namely, the fusion rule algebra is generated by a single fundamental representation, π, and that $C(\pi^n, \pi^m)$ are isomorphisms for small $n, m \in \mathbb{N}$, such that $\phi = 1$ on $V_\pi^{\otimes 3}$.

(3) Non-Tannakian Categories in Wess-Zumino-Witten models

A simple but nevertheless instructive example from quantum field theory, which yields a quantum category, is the SU(2)-WZW-model. In the following outline of the derivation of categorial data we shall profit from the very thorough analysis of this model, existent in the literature, especially [5]. The loop-group symmetry, manifest at classical level, is recovered as a current algebra symmetry in the quantized model of integer levels, k_l, and is related to the Virasoro-algebra of conformal invariance by Sugawara's construction. Each field has a unique decomposition into a holomorphic and an anti-holomorphic part so that the theory is entirely determined by a chiral field theory with braid statistics. The fact that the physical fields have to be single valued and the symmetry algebra is unitarily implemented on the same Hilbert spaces on which the chiral fields act leads to unitarity of the respective braid-group representations and eventually to a C^*-structure on the associated tensor-category. The generators of the chiral field algebra are the vertex operators $\phi^\lambda(z) : \mathcal{H} \to \overline{\mathcal{H}'} \otimes V_\lambda$, where $\mathcal{H}$ and $\mathcal{H}'$ are representation spaces of $\widehat{SU}(2)_{k_l}$ and $Vir(c = 3k_l/(k_l+2))$ and V_λ is a simple SU(2)-module of highest weight $\lambda \in \mathbb{N}$ and highest weight vector $v_\lambda \in V_\lambda$. The definition of a vertex operator further implies the usual commutation relations with the generators of the current algebra and the Virasoro algebra with conformal weight $\Delta(\lambda) = \lambda(\lambda+2)/4(k_l+2)$. The construction of vertex operators is most convenient if the representation spaces are Verma modules, $\mathcal{M}_\mu = U(\widehat{SU}(2)_{k_l}) \otimes_\sim V_\mu$, where $\sim$ means the positive part of the current algebra with trivial action and the ordinary action of $SU(2) \subset \widehat{SU}(2)_{k_l}$ on V_μ so that $v_\mu \in V_\mu \subset \mathcal{M}_\mu$ is also a highest-weight vector of $\mathcal{M}_\mu$. We denote by $\Pi_\mu : \mathcal{M}_\mu \twoheadrightarrow V_\mu$ the projection, whose kernel is given by the negative part of $U(\widehat{SU}(2)_{k_l})$. For any vertex operator $\phi^\lambda : \mathcal{M}_\nu \to \mathcal{M}_\mu$ one shows by rotational covariance that there exists some $P^\circ_{\mu\lambda,\nu} \in Hom_{SU(2)}(V_\mu \otimes V_\lambda, V_\nu)$ such that

$$\Pi_\mu \phi^\lambda(z) v \quad = \quad z^{(\Delta(\mu)-\Delta(\lambda)-\Delta(\nu))} \; P^\circ_{\mu\lambda,\nu} v \tag{48}$$

for all $v \in V_\nu$, and conversely, to any Clebsch Gordan matrix, $P^\circ_{\mu\lambda,\nu}$, there is a unique ϕ^λ for which (48) holds, see[5]. However these vertex operators cannot be used for the construction of WZW-models since none of the representations, $\mathcal{M}_\mu$, is unitarizable. In order to find the unitary ones we have to restrict the highest weights to $\mu = 0, \ldots, k_l$ and divide $\mathcal{M}_\mu$ by the Verma-submodule, $\mathcal{J}_\mu$, generated by $E_{-1}^{(k_l+1-\mu)} v_\mu$. (Here we use the standard notation for the generators $E \in SU(2)$, $E_{-1} \in \widehat{SU}(2)_{k_l}$ of the respective Borel-algebras).

The vertex operators, $\phi^\lambda_{\mu\nu}(z)$, acting on the Hilbert spaces $\mathcal{H}_\nu = \mathcal{M}_\nu \,/\, \mathcal{J}_\nu$, are obtained from those $\phi^\lambda(z)$ which map $\mathcal{J}_\nu$ to $\mathcal{J}_\mu$. In terms of initial data (48) this condition is equivalent to

$$(\mathbb{I}_\mu \otimes E^{(k_l+\nu-1)})\, P^\circ_{\mu\lambda,\nu}\, v_\nu = 0 \tag{49}$$

implying the fusion rules we found in the $U_q(sl_2)$-example:

$$\mu+\lambda+\nu \equiv 0 \ mod2 \qquad \text{and} \qquad 2max(\mu,\lambda,\nu) \le \mu+\lambda+\nu \le 2k_l.$$

Products, $\phi^\lambda_{\mu\varepsilon}(z_2)\,\phi^\eta_{\varepsilon\nu}(z_1) : \mathcal{H}_\nu \to \mathcal{H}_\mu \otimes V_\lambda \otimes V_\eta$, can be defined for $| z_2 | > | z_1 |$, with analytic continuations to the covering space, $\widetilde{M}_2$, of the configuration space $M_2 = \{(z_2, z_1) \in (\mathbb{C}^*)^2 : z_2 \neq z_1\}$. The four point functions are defined by

$$G_4(z_2,z_1)_\varepsilon := \Pi_\mu\, \phi^\lambda_{\mu\varepsilon}(z_2)\,\phi^\eta_{\varepsilon\nu}(z_1)\, P^\circ_{\nu\nu,0} \quad \in Inv_{SU(2)}(V_\mu \otimes V_\lambda \otimes V_\eta \otimes V_\nu).$$

They obey the Knizhnik-Zamolodchikov equations, [33] ,

$$\begin{aligned} &\Big((k_l+2)\frac{\partial}{\partial z_2} - \frac{t_{\lambda\nu}}{z_2} - \frac{t_{\lambda\eta}}{z_2 - z_1}\Big)\, G_4(z_2,z_1) = \\ &\qquad\qquad \Big((k_l+2)\frac{\partial}{\partial z_1} - \frac{t_{\eta\nu}}{z_1} - \frac{t_{\lambda\eta}}{z_1 - z_2}\Big)\, G_4(z_2,z_1) = 0, \end{aligned} \tag{50}$$

where $t := \frac{1}{2} H \otimes H + E \otimes F + F \otimes E$, and in addition

$$\begin{aligned} &\mathbb{I}_\mu \otimes \Big(\frac{1}{z_2 - z_1} E \otimes \mathbb{I}_\nu - \frac{1}{z_1}\, \mathbb{I}_\lambda \otimes E\Big)^{(k_l+1-\eta)} (\mathbb{I}_\mu \otimes \mathbb{I}_\lambda \otimes \zeta_\eta \otimes \mathbb{I}_\nu)\, G_4(z_2,z_1) = 0 \\ &\mathbb{I}_\mu \otimes \Big(\frac{1}{z_1 - z_2} E \otimes \mathbb{I}_\nu - \frac{1}{z_2}\, \mathbb{I}_\eta \otimes E\Big)^{(k_l+1-\lambda)} (\mathbb{I}_\mu \otimes \zeta_\lambda \otimes \mathbb{I}_\eta \otimes \mathbb{I}_\nu)\, G_4(z_2,z_1) = 0 \end{aligned} \tag{51}$$

where $\zeta_\eta \in V^*_\eta$ are such that $(\zeta_\eta \otimes \mathbb{I})\, P^\circ_{\eta\eta,0} = v_\eta$.

In [5] it is shown that the $G_4(z_2,z_1)_\varepsilon$, for admissible ε, are in fact a basis for the solutions of (50) and (51). This implies that any fieldoperator on $\widetilde{M}_2$, $\Phi(z_2,z_1) : \mathcal{H}_\nu \to \mathcal{H}_\mu \otimes V_\lambda \otimes V_\mu$, which transforms with respect to current- and Virasoro algebra in the same way as products, $\phi^\lambda(z_2)\,\phi^\mu(z_1)$, and for which the four point function $\Pi_\mu\, \Phi(z_2,z_1) \restriction V_\nu$ obeys (50) and (51), is a linear combination of products of vertex operators. A particular way of producing such an operator is to consider for some ε the analytic continuation of the product to $[(z_1,z_2),\sigma] \in \widetilde{M}_2$, where σ is a path in M_2 starting at (z_2,z_1), $| z_2 | > | z_1 |$, and exchanging the two radially ordered positions counter-clockwise. We denote the result by $(\phi^\lambda_{\mu\varepsilon}(z_1)\,\phi^\eta_{\varepsilon\nu}(z_2))_\sigma$. If we further apply the flip $T : V_\lambda \otimes V_\eta \to V_\eta \otimes V_\lambda$ to this, we can verify all conditions needed to write

$$T(\phi^\lambda_{\mu\varepsilon}(z_1)\,\phi^\eta_{\varepsilon\nu}(z_2))_\sigma = \sum_{\varepsilon'} R^+(\mu,\lambda,\eta,\nu)^{\varepsilon'}_\varepsilon\, \phi^\eta_{\mu\varepsilon'}(z_2)\,\phi^\lambda_{\varepsilon'\nu}(z_1) \tag{52}$$

where $R^+(\mu,\lambda,\eta,\nu)$ is clearly an invertible matrix. Repeating the continuation along σ^2 we obtain the monodromy of (50) in this path basis as $M(\mu,\lambda,\eta,\nu)^{\varepsilon}_{\varepsilon''} = \Sigma_{\varepsilon'} R(\mu,\lambda,\eta,\nu)^{\varepsilon}_{\varepsilon'} R(\mu,\eta,\lambda,\nu)^{\varepsilon'}_{\varepsilon''}$, which can be proven to be unitary from the properties of the Knizhnik-Zamolodchikov connection. Thus performing an eigenspace decomposition of M we can write

$$\phi^{\lambda}_{\mu\varepsilon}(z_1)\,\phi^{\eta}_{\varepsilon\nu}(z_2) \;=\; \sum_i \psi_i(z_1,z_2)_{\varepsilon},$$

where $(z_1-z_2)^{\delta_i}\,\psi_i(z_1,z_2)_\varepsilon$ are single valued functions in a simply connected domain without 0. Further we have $\delta_i \in [0,1[$ with $\{e^{-2\pi i\delta_i}\} = spec(M)$ and each ψ_i is again a combination of products of vertex operators. From the differential equations (50) we see that the divergence of matrix elements of an operator product can be at most a power divergence as $z_1 \to z_2$ of degree $\leq \|t_{\lambda\eta}\|$, so that the corresponding Laurent expansion for ψ_i has to terminate in $(z_1-z_2)^{-1}$. Hence there is a smallest number $n_1 \in \mathbb{Z}$ and operators $\widetilde{\psi}_i(z_1,z_2)_\varepsilon$ analytic in $(\mathbb{C}^*)^2$ such that $\psi_i(z_1,z_2) = (z_1-z_2)^{-(\delta_i+n_i)}\,\widetilde{\psi}_i(z_1,z_2)_\varepsilon$ for all ε. The operator $\Xi_i(z)_\varepsilon := \widetilde{\psi}_i(z,z)_\varepsilon$ has the transformation properties of a chiral vertex operator with conformal weight $\Delta = \Delta(\eta)+\Delta(\lambda)-(\delta_i+\mu_i)$ and SU(2)-components in $V_\lambda \otimes V_\mu$, so that there exists some $\varepsilon' = 0,\ldots,k$ with $\Delta \doteq \Delta(\varepsilon')$ and

$$\Xi_i(z)_\varepsilon \;=\; F(\mu,\lambda,\eta,\nu)^{\varepsilon'}_{\varepsilon}\,(1_\mu \otimes P^{\circ}_{\lambda\eta,\varepsilon'})\,\phi^{\varepsilon'}_{\mu\nu}(z), \tag{53}$$

for constants $F \neq 0$, if $\Xi_{i\varepsilon} \neq 0$ and n_i is chosen properly.

If we take the limit $z_1 \to z_2$ in the constraint (51) we reproduce (49) for $P^{\circ}_{\varepsilon'\lambda\eta}$, hence the triple $(\varepsilon',\lambda,\mu)$ has to obey the fusion rules, and we choose the normalization of the Clebsch-Gordon matrices such that $F(\mu,\lambda,\eta,0)=1$. Using that the operators $\psi_i(z,w)_\varepsilon$ are single valued we find from the short distance expansion of (52)

$$\begin{aligned}\sum_{\varepsilon''} R(\mu,\lambda,\eta,\nu)^{\varepsilon''}_{\varepsilon}\,F(\mu,\eta,\lambda,\nu)^{\varepsilon'}_{\varepsilon''} \;&=\; F(\mu,\lambda,\mu,\nu)^{\varepsilon'}_{\varepsilon}\,r(\eta,\lambda,\varepsilon')\\ \text{with}\qquad r(\eta,\lambda,\varepsilon) := e^{i\pi(\Delta(\eta)+\Delta(\lambda)-\Delta(\varepsilon))} \;&=\; R(\varepsilon,\lambda,\eta,0).\end{aligned} \tag{54}$$

The definition of the F-matrices thus far is of course insufficient since we may have $dim M_i > 1$ for the eigenspaces, $M_i \subset \bigoplus_\varepsilon \mathbb{C}^{N_{\mu\lambda,\varepsilon}} \otimes \mathbb{C}^{N_{\varepsilon\eta,\nu}}$, of the monodromy. For operators, ψ_i, lying in the kernel, M_i^1, of $F^{\varepsilon'} : M_i \to \mathbb{C}^{N_{\lambda\eta,\varepsilon'}} \otimes \mathbb{C}^{N_{\mu\varepsilon',\nu}}$, we can find $\varepsilon'' < \varepsilon'$, with $\Delta(\varepsilon'') \equiv \Delta(\varepsilon')\ mod 1$ so that the analogous linear form $F^{\varepsilon''} : M_i^1 \to \mathbb{C}^{N_{\lambda\eta,\varepsilon''}} \otimes \mathbb{C}^{N_{\mu\varepsilon'',\nu}}$ is well defined and nonzero.

Hence we find by iteration a natural flag $M_i \supset M_i^1 \supset \cdots \supset M_i^{k_i} = \mathbb{C}$ of operators with the same monodromy but different conformal weights so that F is triangular with respect to this flag, but only the diagonal elements are determined. However, there is a (up to isomorphism) unique choice such that F or r

satisfy the hexagonal equation (30). In order to see this we consider the operator $\phi^{\lambda}_{\mu\varepsilon}(z)\ \phi^{\eta}_{\varepsilon\nu}(w)\,\phi^{\nu}_{\nu 0}(\xi)$, for $|\,\xi\,|<|\,z\,|<|\,\xi'\,|$, where σ moves ξ to ξ' around a small common neighborhood of the pair w and z, which is kept fixed. Obviously this operation commutes with the monodromy in z and w, so that the result of the continuation expressed by vertex operators can again be decomposed into single valued contributions:

$$T_{12}T_{23}(\tilde{\psi}_{i\varepsilon}\cdot\phi^{\nu}_{\nu 0})[(z,w,\xi');\sigma] = \sum_{\varepsilon':\Delta(\varepsilon')=\Delta_i \, mod\, 1} (z-w)^{(\Delta(\varepsilon')-\Delta_i)}\, r(\nu,\mu\,\varepsilon)R(\mu,\lambda,\nu,\eta)^{\varepsilon'}_{\varepsilon}\,\phi^{\nu}_{\mu\varepsilon'}(\xi')\,\tilde{\psi}^{\circ}_{i\varepsilon'}(z,w) \quad (55)$$

where $\tilde{\psi}^{\circ}_{i}(z,w) := (z-w)^{(\Delta(\varepsilon')-\Delta(\lambda)-\Delta(\eta))}\,\phi^{\lambda}_{\varepsilon'\eta}(z)\,\phi^{\eta}_{\eta 0}(w) \rightarrow \mathbb{I}_{\varepsilon'}\otimes P^{\varepsilon'}_{\lambda\eta,\varepsilon'}(w)$ as $z\rightarrow w$. Comparing the asymptotic behaviour for $z\rightarrow w$ and continuing the limits back to $(w,\xi)\in M_2$ we find that

$$F(\mu,\lambda,\eta,\nu)^{\varepsilon'}_{\varepsilon} = r(\nu,\mu,\varepsilon)\,R(\mu,\lambda,\nu,\eta))^{\varepsilon'}_{\varepsilon}\,r(\varepsilon',\nu,\mu)^{-1} \quad (56)$$

is a definition of F-matrices consistent with the above interpretation as leading coefficients in short distance expansions. In particular (56) shows that F is invertible. Combining (54) with (56) we immediately find the hexagonal equation (30) (up to a reordering of indices).

In general we may consider n-point functions $W_{(\lambda)}:\widetilde{M}_n\rightarrow V_{\eta_n}\otimes\cdots\otimes V_{\eta_1}$,

$$W_{(\lambda)}(z_n,\ldots,z_1) := \langle\Omega,\,\phi^{\eta_n}_{\lambda_n\lambda_{n-1}}(z_n)\,\phi^{\eta_{n-1}}_{\lambda_{n-1}\lambda_{n-2}}(z_{n-1})\ldots\phi^{\eta_1}_{\lambda_1\lambda_0}(z_1)\,\Omega\rangle \quad (57)$$

for admissible paths, (λ), which form a basis of solutions to

$$\frac{\partial}{\partial z_i}W_{(\lambda)}(z_n,\ldots,z_1) = \frac{1}{k_l+2}\sum_{j\neq i}\frac{t_{ij}}{z_i-z_j}\,W_{(\lambda)}(z_n,\ldots,z_1) \quad (58)$$

and

$$\Big(\sum_{j\neq i}\frac{1}{z_i-z_j}E_j\Big)^{(k+1-\eta_i)}(1\otimes\cdots\otimes\zeta_{\eta_i}\otimes\cdots\otimes 1)\,W_{(\lambda)}(z_n,\ldots,z_1) = 0, \quad (59)$$

for $i=1,\ldots,n$ and $E_j=1\otimes\cdots\otimes\rho_{\eta_j}(E)\otimes\cdots\otimes 1$.

The F-matrices provide a one to one correspondence, coherent with the braiding operations, of the set of solutions to the joint system (50) and (51) and the types of short-distance behavior that are compatible Clebsch-Gordan matrices as in (49) yields associativity endomorphisms between different SU(2)-invariant subspaces of $V_{\eta_n}\otimes\cdots\otimes V_{\eta_1}$ which are no longer invertible but obey the weak relations of [20]. From this one proves in a fashion similar to [17] that the F-matrices also satisfy the pentagonal-equation (28). Thus we have found a quantum category with an explicit weakly-Tannakian realization.

The computation of structural data for the SU(2)-WZW-models yields precisely the same F- and r-matrices, as those obtained from $U_q(sl_2)$, with $q = exp(\pm \frac{i\pi}{k_l+2})$. This surprising fact has a rather simple explanation: In both cases the fusion rule algebra of the category has an irreducible, selfconjugate object, (1), which generates the entire algebra, has a decomposition $(1) \circ (1) = (0) \oplus (2)$, with $(0) = 1$ and (2) irreducible, and has a nonscalar monodromy $\mu(1,1)$. A specialization of results in [27] characterizes all C^*-quantum categories with these properties as subcategories of a product of the semisimple category obtained from $U_q(sl_2)$ for q being a primitive root of unity and one of the four θ-categories with $\mathbb{Z}_2$-fusion rules. For the $\mathbb{Z}_2$-graded cases the fusion rules are those of the category $\overline{\mathcal{R}ep(U_q(sl_2))}, q = exp(\pm \frac{i\pi}{k_l+2})$, with irreducible objects $(\lambda) = (0), (1), \cdots, (k)$, and the spins, $\sigma(\lambda) = e^{2\pi i \theta_\lambda}$, are given by

$$\theta_\lambda \;=\; \pm\frac{\lambda(\lambda+2)}{4(k+2)} \;+\; \omega\,\lambda^2 \quad mod\,1 \tag{60}$$

where $\omega \in \frac{1}{4}\mathbb{Z}/\mathbb{Z}$. The category is uniquely determined by the quantities k and ω and for $\omega = 0$ reduces to the ordinary quantum group category. The category in the ungraded case has as objects the representations of even weight $(0), (2), \cdots, (2k')$, for $k = 2k' + 1$, and the fundamental object is $(2k')$ with spin $\theta = \frac{k'(k'+1)}{1+2(k'+1)}$ and $(2k') \circ (2k') = (0) \oplus (2)$. In particular all categories can be distinguished by their fundamental spin, so that the isomorphy of quantum group category and WZW-category is immediately verified. Also we have learned from an earlier line of arguments that the quantum group, producing this category is unique, if we require the fundamental representation to have dimension two. Thus for SU(2)-WZW models and a large class of similar models, e.g., SU(2)-Chern-Simons- or $C = 1 - \frac{6}{(k+1)(k+2)}$, minimal models, $U_q(sl_2)$-duality can be considered stringent.

In the special case of WZW-model we have presumably not reached the end of the story. Remark that the total set of solutions to the differential equation (50) can be obtained as expectation values of the vertex operators acting on the Verma modules $\mathcal{M}_j$, thus we can also define a category obtained from a non-unitary, local field theory.

As a matter of fact close relation between this larger Tannakian-type category and the complete representation category of a corresponding quantum group, for negative integer values of the level, have been found in a recent proof of the Kazhdan-Lusztig conjecture [34], extending the generic results of [17,18]. The existence of some contractions between the categories of positive and negative integer level WZW-models suggests that the physical subspace of solutions to (50), which is characterized by (51), has to be the space orthogonal to the ghost ideal of the representation category of this quantum group. A closer investigation of these connections should provide further insight in the categorial structure of affine

Lie algebras, by means of translation to finite problems of quantum group representations. Finally from a constructive field theorist's point of view it might be more convenient to start from solutions of partial differential equations and analyze the associated quantum group structure in order to find unitary theories, then to explicitly construct the entire operator algebra.

ACKNOWLEDGMENTS

I wish to thank Jürg Fröhlich for proposing the problem of the Vertex-RSOS transformation to me, which initiated this work. I am also grateful for his constant support and encouragement.

REFERENCES

1. *Eilenberg, S., Mac Lane, S.*: Natural Isomorphisms in Group Theory. Proc. Nat. Acad. Sci. U.S. **28**, 537-543 (1942); General Theory of Natural Equivalences. Trans. Am. Math. Soc. **58**, 231-294 (1945).

2. *Grothendieck, A.*: Sur Quelques Points d'Algèbre Homologique. Tôhoku Math. J. **9**, 119-221 (1957).

3. *Deligne, P.*: Catégories Tannakiennes. The Grothendieck Festschrift, Vol. II, Progress in Mathematics, Birkhäuser, Boston (1990).

4. *Belavin, A., Polyakov, A., Zamolodchikov, A.B.*: Infinite Conformal Symmetry in Two-Dimensional Quantum Field Theory. Nucl. Phys. **B241**, 333-380 (1984).

5. *Tsuchiya, A.; Kanie, Y.*: Vertex Operators in Conformal Field Theory on $\mathbb{P}^1$ and Monodromy Representations of Braid Groups. Advanced Studies in Pure Mathematics **16**, 297-372 (1988).

6. *Felder, G., Fröhlich, J, Keller, G.*: Braid Matrices and Structure Constants for Minimal Conformal Models. Commun. Math. Phys. **124**, 647-664 (1989).

7. *Segal, G.*: unpublished, see [9] .

8. *Friedan, D., Shenker, S.*: The Analytic Geometry of Two-Dimensional Conformal Field Theory. Nucl. Phys. **B281**, 509-545 (1987).

9. *Moore, S., Seiberg, N.*: Polynomial Equations for Rational Conformal Field Theories, Phys. Lett. **B212**, 451-460 (1988); Classical and Quantum Conformal Field Theory, Commun. Math. Phys. **123**, 177-254 (1989).

10. *Doplicher, S. Haag, R., Roberts, J.E.*: Local Observables and Particle Statistics I/II, Commun. Math. Phys. **23**, 199-230 (1971) / **35**, 49-85 (1974).

11. *Roberts, J.E.*: New Light on the Mathematical Structure of Algebraic Field Theory. In: Proceedings of Symposia in Pure Mathematics, Vol. **38**, Pt. 2, 523-550 (1982).
Ghez, P., Lima, R., Roberts J.E.: W^*-Categories. Pacific Journal of Mathematics, Vol. **120**, No. 1, 79-109 (1985).

12. *Doplicher, S., Haag, R., Roberts, J.E.*: Fields, Observables and Gauge Transformations I/II, Commun. Math. Phys. **13**, 1-23 (1969) / **15**, 173-200 (1969).

13. *Doplicher, S., Roberts. J.E.*: Fields, Statistics and Non-Abelian Gauge Groups, Commun. Math. Phys. **28**, 331-348 (1972); *Roberts, J.E.*: Must there be a Gauge Group? Proc.Steklov Inst., 211-214(1978).

14. *Doplicher, S., Roberts, J.E.*: Why There is a Field Algebra with a Compact Gauge Group Describing the Superselection Structure in Particle Physics, Commun. Math. Phys. **131**, 51-107 (1990).

15. *Doplicher, S., Roberts, J.E.*: A new duality for compact groups. Invent. math. **98**, 157-218 (1989).

16. *Fröhlich, J., Gabbiani, F., Marchetti, P.A.*: Superselection Structure and Statistics in Three-dimensional Local Quantum Field Theory, Proc. 12th John Hopkins Workshop on "Current problems in High Energy Particle Theory "Florence 1989, G. Lusanna (ed.). Braid Statistics in Three-Dimensional Local Quantum Theory, Proceedings of the Banff Summer School in Theoretical Physics, "Physics Geometry and Topology", August 1989.
Fröhlich, J., Gabbiani, E.: Braid Statistics in Local Quantum Theory, Reviews in Mathematical Physics, **3**, No. 2 (1991).
Longo, R.: Index of Subfactors and Statistics on Quantum Fields I/II. Commun. Math. Phys. **126**, 217-247 (1989); **130**, 285-309 (1990).
Fredenhagen, K., Rehren, K.H., Schroer, B.: Superselection Sectors with Braid Group Statistics and Exchange algebras, Commun. Math. Phys. **125**, 201-226 (1989).

17. *Drinfel'd, V.G.*: Quasi-Hopf Algebras, Leningrad Math. J. **1**, No. 6, 1419-1457 (1990).

18. *Kohno, T.*: Monodromy Representations of Braid Groups and Yang-Baxter Equations, Ann. Inst. Fourier, Grenoble, **37.4**, 139-160 (1987);
Kohno, T.: Quantized Universal Enveloping Algebras and Monodromy of Braid Groups, Nagoya University preprint, Nagoya 1988.

19. *Kirillov, A.N., Reshetikhin, N.Yu.*: Representations of the Algebra $U_q(sl_2)$, q-Orthogonal Polynomials and Invariants of Links, in "New Developments in the Theory of Knots", T. Kohno (ed.), Advanced Series in Mathematical Physics Vol. **11**, World Scientific Singapore, 1989.

20. *Mack, G., Schomerus, V.*: Conformal Field Algebras with Quantum Symmetry from the Theory of Superselection Sectors, Commun. Math. Phys. **134**, 139 (1990).
Mack, G., Schomerus, V.: Quasi Hopf Quantum Symmetry in Quantum Theory, Desy preprint 91-037, May 1991.

21. *Fröhlich, J.*: Statistics of Fields, the Yang-Baxter Equation and the Theory of Knots and Links, in: Nonperturbative Quantum Field Theory, G. t' Hooft et al. (eds.), Plenum Press, 1988.

22. *Wenzl, H.*: Hecke Algebras of Type A_n and Subfactors, Invent. math., **92**, 349-383 (1988);
Wenzl, H.: Representations of Braid Groups and the Quantum Yang-Baxter equation, Pacific J. Math. 1990;
Wenzl, H.: Quantum Groups and Subfactors of Type B, C and D, preprint 1989.

23. *Pasquier, V.*: Etiology of IRF Models, Commun. Math. Phys. **118**, 355-364 (1988).

24. *Reshetikhin, N., Turaev, V.G.*: Invariants of 3-Manifolds via Link-Polynomials and Quantum Groups, Invent. math. **103**, 547-597 (1991);
Turaev, V.G., Viro, O.Y.: State Sum Invariants of 3-Manifolds and Quantum 6-j-symbols, Lomi-preprint, Leiningrad (1990).

25. *MacLane, S.*: Categories for the Working Mathematician, Graduate Texts in Mathematics 5, Springer-Verlag, New York, Heidelberg, Berlin, 1971.

26. *Rivano, N.S.*: Catégories Tannakiennes, Lecture Notes in Mathematics, 265, Springer-Verlag, Berlin, Heidelberg, New York, 1972.

27. *Fröhlich, J., Kerler, T.*: On the Role of Quantumgroups in Low Dimensional Quantum Field Theory, Lecture Notes in Mathematics, Springer-Verlag, to appear.

28. *Drinfel'd, V.G.*: Quantum Groups, Proc. Internat. Congr. Math., Berkeley, Ca., 1986, Am. Math. Soc., Providence, R.I., 798-820 (1987).

29. *Kirillov, A.A.*: Elements of the Theory of Representations, Grundlehren, 220, Springer-Verlag, Berlin, Heidelberg, New York, 1976.

30. *Drinfel'd, V.G.*: On Almost Cocommutative Hopf Algebras, Leningrad Math. J., **1**, 321-342 (1990), and references therein.

31. *Reshetikhin, N.Yu., Turaev, V.G.*: Ribbon Graphs and their Invariants Derived from Quantum Groups. Commun. Math. Phys. **127**, 1-26 (1990).

32. *Faddeev, L.D., Reshetikhin, N.Yu., Takhtajan, L.A.*: Quantization of Lie Groups and Lie Algebras, LOMI-preprint (1987).

33. *Knizhnik, V.G., Zamolodchikov, A.B.*: Current Algebra and Wess-Zumino Models in Two Dimensions, Nucl. Phys. **B247**, 83-103 (1984).

34. *Kazhdan, D., Lusztig, G.*: Affine Lie Algebras and Quantum Groups, MIT-preprint, 1991.

EXTRA STATES IN $c < 1$ STRING THEORY*

Sunil Mukhi

Tata Institute of Fundamental Research
Homi Bhabha Road, Bombay 400005, India

Abstract

A construction of elements of the BRS cohomology of ghost number ± 1 in $c < 1$ string theory is described, and their two-point function computed on the sphere. The construction makes precise the relation between these extra states and null vectors. The physical states of ghost number $+1$ are found to be exact forms with respect to a "conjugate" BRS operator.

1. Introduction

The no-ghost theorem[1] for critical string theory states that the BRS cohomology classes can be represented by

$$c(z)\bar{c}(\bar{z})\, V(z,\bar{z}) \tag{1}$$

where $V(z,\bar{z})$ is a dimension $(1,1)$ primary field of the $c = 26$ matter conformal field theory, $c(z)$ is the holomorphic spin -1 ghost and $\bar{c}(\bar{z})$ is its antiholomorphic counterpart. There is an interesting exception to this theorem: the identity field of the combined matter-ghost theory is certainly in the BRS cohomology, but is not of the form given in Eq.(1). Indeed, if the ghost number of the states in Eq.(1) is chosen by convention to be $(0,0)$, then that of the identity becomes $(-1,-1)$. In addition, one finds a state of ghost number $(1,1)$ and a few more of mixed ghost number $(1,0),(-1,0)$ and so on.

The chiral BRS operator for bosonic string theory in a given background is

$$\begin{aligned} Q_B &= \oint dz\ :c(z)\left(T^{(M)}(z)+\frac{1}{2}T^{(G)}(z)\right): \\ &= \sum_{n=-\infty}^{\infty} c_n (L_{-n}^{(M)} - \frac{1}{2}\sum_{m,n=-\infty}^{\infty}(m-n) :c_{-m}c_{-n}b_{m+n}:) \end{aligned} \tag{2}$$

* Based on work done in collaboration with C. Imbimbo and S. Mahapatra

where $T^{(M)}(z), T^{(G)}(z)$ are the holomorphic stress-energy tensors of the $c = 26$ matter theory and the $c = -26$ ghost system respectively , and $L_n^{(M)}$ are the modes of $T^{(M)}(z)$. In the closed string theory, we are interested in the cohomology of $Q_B + \bar{Q}_B$, the sum of the chiral and antichiral BRS charges. More precisely, we must restrict to the cohomology on the subspace annihilated by $b_0^- \equiv b_0 - \bar{b}_0$ where $b(z)$ is the chiral antighost field. This is an example of a "relative" cohomology.

It can be shown that the cohomology for the closed string can be reconstructed from a knowledge of that for the open string, in other words, the cohomology of the chiral BRS operator alone. Even more, one can restrict to the "relative" chiral cohomology, where we consider only the subspace annihilated by b_0. The no-ghost theorem for the relative chiral cohomology of the critical string says that the only physical states are of the form $c_1|V\rangle$ where $|V\rangle$ is a chiral primary state of dimension 1. The vacuum state $|0\rangle$, along with a finite number of other states, provides an exception to the theorem. It is important to note that all these states have zero 26-momentum. Such exceptional states are very few within the enormous classical phase space of the critical string, and do not seem to be associated with significant physical effects.

The situation is very different for non-critical string theories in conformal backgrounds with $c \leq 1$. In the present article I will concentrate only on the case $c < 1$, where the background matter theory is a minimal model[2]. In this case, the full Hilbert space is obtained by taking the direct product of the matter CFT, the Liouville theory (which is believed to describe the effect of quantized two-dimensional gravity when the matter is non-critical), and the ghost system. The BRS operator described above is generalized by adding the Liouville stress-energy tensor to the matter one, wherever the latter appears.

We parametrize the matter and Liouville central charges as

$$\begin{aligned} c_M &= 13 - 6/t - 6t \\ c_L &= 26 - c_M = 13 + 6/t + 6t \end{aligned} \tag{3}$$

with $t = q/p > 0$. Here, p and q are two positive, coprime integers. The Liouville stress-energy tensor is

$$T^{(L)}(z) = -\frac{1}{2}(\partial\Phi^{(L)}\partial\Phi^{(L)} + Q_L\partial^2\Phi^{(L)}) \tag{4}$$

where

$$\begin{aligned} Q_L &= \sqrt{\frac{25 - c^{(M)}}{3}} \\ &= \sqrt{2}(\sqrt{t} + 1/\sqrt{t}) \end{aligned} \tag{5}$$

The Virasoro generators following from this are

$$L_n^{(L)} = \frac{1}{2}\sum_{m=-\infty}^{\infty} :\alpha_{n-m}^{(L)}\alpha_m^{(L)}: + \frac{iQ}{2}(1+n)\alpha_n^{(L)} \tag{6}$$

We follow the convention that Liouville vertex operators are defined as $V_{k_L}^{(L)} = :e^{k_L \Phi^{(L)}}:$, with conformal dimension $\Delta_{k_L}^{(L)} = -\frac{1}{2}k_L(k_L - Q_L)$ and $\alpha_0^{(L)}$ eigenvalue $-ik_L$. Thus a Liouville primary of given conformal dimension can have two possible values of momentum, denoted $k_L^{\pm}$, where $k_L^+ > \frac{Q_L}{2}$, $k_L^- < \frac{Q_L}{2}$ and $k_L^+ + k_L^- = Q_L$. We will denote the Fock space above a momentum $k_L > \frac{Q_L}{2}$ as the "(+)-Fock space", and the other one as the "(−)-Fock space".

Then, a class of physical states in the (relative, chiral) cohomology is again given by states like $c_1|V\rangle$, where this time $|V\rangle$ is the direct product of a matter primary and a Liouville momentum state:

$$|V\rangle = |\Psi\rangle_M \otimes |k_L\rangle_L \tag{7}$$

Here, the Liouville momentum is adjusted such that the Liouville dimension Δ_L and the matter primary dimension Δ_M satisfy $\Delta_L + \Delta_M = 1$. Such states (two for every matter primary, because of the two possible values of the Liouville momenta) are known as "DDK states" [3].

Just like the critical string, the non-critical string also has exceptional states which are not of the above kind, among which one example is the vacuum state. However, it has been shown by Lian and Zuckerman[4] that in this case there are *infinitely many* exceptional physical states (we will call them "LZ states") in the relative chiral cohomology, at *all* positive and negative values of the ghost number. At the same time, the DDK states, analogous to the dimension 1 primaries of the critical string, are finite in number for minimal model backgrounds, since each minimal model has only finitely many primary fields. Thus in these theories, the number of exceptional states is a lot larger than that of the "normal" states, a situation quite different from the critical string. One may expect these states to play an important physical role in the analysis of non-critical string theories. (Various arguments have been put forward of late to support the idea that there are actually infinitely many DDK states in minimal backgrounds, because the decoupling of null vectors fails in the present of gravity due to "contact terms" at boundaries of moduli space. It remains true that the proportion of exceptional states of non-trivial ghost number to the states of ghost number 0 is significantly greater than for the critical string.)

The DDK states are often thought of as matter primaries "dressed" by a Liouville momentum state to have total dimension 1. The principal observation in Ref. 4 is that given a primary field of the matter CFT, exceptional physical states appear in the module whenever the Liouville momentum is one which "dresses" a *null vector* over the matter primary. In minimal models, each primary has an infinite chain of null vectors over it. The ghost number of the physical state turns out to be equal in magnitude to the distance of the associated null vector in this chain from the original matter primary. The sign of the ghost number is positive or negative if the Liouville momentum lies in the (+) or (−) Fock space respectively. The dimension of the relative chiral BRS cohomology is precisely 1 in every such case. Thus there is an infinite set of physical states for each matter primary, one for each null vector over it and for each sign of the ghost number.

In the rest of this article I will summarise an explicit method of construction for a large class of LZ states, in arbitrary minimal models coupled to gravity. The details, including proofs of a number of theorems, can be found in Ref.[5].

2. Null Vectors

Degenerate fields in a $c < 1$ CFT are those which have null vectors in the Verma module above them. Their dimensions are given by the Kac formula:

$$\Delta_{r,s}^{(M)} = \frac{(r^2-1)}{4}\frac{1}{t} + \frac{(s^2-1)}{4}t - \frac{(rs-1)}{2} \tag{8}$$

with $t = q/p$, $1 \leq r \leq q-1$, $1 \leq s \leq p-1$.

Each such field has infinitely many null vectors above it. We will refer to the lowest of these null vectors in a module as "primitive". The primitive (r,s) null vector over $\Psi_{r,s}^{(M)}$ has dimension

$$\begin{aligned} \Delta_{r,s}^{(M)(null)} &= \Delta_{r,s}^{(M)} + rs \\ &= \frac{(r^2-1)}{4}\frac{1}{t} + \frac{(s^2-1)}{4}t + \frac{(rs+1)}{2} \end{aligned} \tag{9}$$

Now, an LZ state will occur whenever we consider a Liouville momentum which "dresses" the above dimension:

$$\begin{aligned} \Delta_{r,s}^{(L)} &= 1 - \Delta_{r,s}^{(M)(null)} \\ &= \frac{(1-rs)}{2} - \frac{(r^2-1)}{4}\frac{1}{t} - \frac{(s^2-1)}{4}t \end{aligned} \tag{10}$$

From the dimension formula above for Liouville vertex operators, it follows that the two Liouville momenta are

$$k_L^{\pm} = \frac{1}{\sqrt{2}}\left(\frac{(1\pm r)}{\sqrt{t}} + (1\pm s)\sqrt{t}\right) \tag{11}$$

According to the Lian-Zuckerman theorem, a physical state of ghost number $+1$ occurs in the Fock module above k_L^+, while a state of ghost number -1 occurs above k_L^-.

There exists a rather neat formula, due to Benoit and Saint-Aubin[6] for the null vector in the Verma module of any CFT above a primary satisfying the Kac formula(8) for the special case $r = 1$, s arbitrary:

$$|\Psi_{1,s}^{(null)(Vir)}\rangle = \sum_j \sum_{p_1+p_2+\cdots+p_j=s,\ p_i\geq 1} t^{s-j} \frac{(s-1)!^2}{\prod_{i=1}^{j-1}(p_1+\cdots p_i)(r-p_1-\cdots-p_i)} L_{-p_1}L_{-p_2}\cdots L_{-p_j}|\Psi_{1,s}\rangle \tag{12}$$

where t parametrizes the central charge via the second equation in Eq.(3). We will ultimately apply this to Liouville theory with $c > 25$, hence positive t.

One can ask how this null vector descends to the Fock space, for theories with a Fock space description. It may in principle vanish identically when re-expressed in oscillators, in which case it is in the kernel of the projection map from the Verma module to Fock space. In this case, from well-known arguments, there must be a state in the Fock space which is not in the image of the projection. Alternatively, the projection to Fock space may have no kernel in this module, in which case every state in the Fock space lies in the image of the projection. This situation was analyzed some years ago by Kato and Matsuda[7]. For our purposes we need a stronger result than that of Ref.[7], namely:

Theorem 1: For $t > 0$,

$$|\Psi_{1,s}^{(null)(Vir)}\rangle \to 0, \quad k_L < \frac{Q_L}{2}$$

$$|\Psi_{1,s}^{(null)(Vir)}\rangle \to \prod_{k=1}^{s}(kt+1)\,|\Psi_{1,s}^{(null)(Fock)}\rangle, \quad k_L > \frac{Q_L}{2}$$

where the state $|\Psi_{1,s}^{(null)(Fock)}\rangle$ is a null state in the Fock module which is non-vanishing for all values of t, and the arrow indicates the projection map.

The proof of this theorem is given in Ref.[5]. It shows that in general, for $t > 0$, the projection to the $(-)$ Fock space has a kernel, while the projection to the $(+)$ Fock space does not. But in fact we learn something even for $t < 0$. In this case, the result of Theorem 1 holds with a possible interchange of $(+)$ and $(-)$ Fock spaces, so that for the values $t = -1/k$, $k = 1, \cdots, s$, the projection to *each* Fock space has a kernel. In other words, for these special values of t, null vectors vanish when expressed in oscillators, in both Fock spaces. This result will be crucial in the subsequent analysis. Note that this in particular holds for $t = -1$ at every value of s, confirming the well-known result that Virasoro null vectors in $c = 1$ CFT vanish identically in terms of oscillators.

3. Construction of LZ States

Although the LZ states occur in correspondence with matter null vectors, they are not themselves null in any sense. They are to be found in the module above a matter primary and a Liouville momentum which has the right *dimension* to dress a matter null vector. This means that the Liouville momentum k_L corresponds to a conformal dimension $\Delta^{(L)}$ which satisfies

$$\Delta^{(L)} + \Delta^{(M)(null)} = 1 \tag{13}$$

where $\Delta^{(M)(null)}$ is the dimension of some null vector above the chosen matter primary.

Because of the well-known relation[8] between null vectors for two theories of central charge c and $26 - c$, we can re-state the above result in a complementary way: whenever there is a Liouville null vector (in the Fock module above a given momentum state) of dimension $\Delta^{(L)(null)}$ satisfying

$$\Delta^{(L)(null)} + \Delta^{(M)} = 1 \tag{14}$$

for some matter primary of dimension $\Delta^{(M)}$, the module contains an LZ state.

To find these extra physical states, we start with a primitive $(1,s)$ Liouville null vector (in the Liouville Verma module) and the $(1,s)$ matter primary, combined into the state

$$|X^{(0)}\rangle = |\Psi^{(L)(null)}(k_L^{\pm})\rangle_L \otimes |\Psi_{1,s}^{(M)}\rangle_M \otimes c_1|0\rangle_G \tag{15}$$

The total dimension of this state is zero, by virtue of Eq.(14) and the fact that the ghost mode c_1 has dimension -1. It has the same ghost number as DDK states, which we have chosen to call 0 by convention.

We will construct physical states of ghost number ± 1 starting from the state defined above (this state is itself null, since its Liouville sector is null.) Let us first define a conjugation operator

$$\mathcal{C}: \quad (+) \text{ Fock space } \rightarrow (-) \text{ Fock space} \tag{16}$$

which, acting on a state, simply replaces k_L by $Q_L - k_L$. Clearly, $\mathcal{C}$ acts in either direction, and $\mathcal{C}^2 = 1$. Using this operator we define a "conjugate" BRS charge:

$$Q_B^* \equiv \mathcal{C}\, Q_B\, \mathcal{C} \tag{17}$$

which is nilpotent:

$$(Q_B^*)^2 = 0 \tag{18}$$

by virtue of the nilpotence of Q_B and the fact that $\mathcal{C}^2 = 1$.

Now, the state in Eq.(15) has the following property: because of Theorem 1, $|X^{(0)}\rangle$ is in the kernel of the projection to the $(-)$ Fock space, so it vanishes identically when expressed in oscillators. On the other hand, as long as $t \neq -1/k$ (in fact t is positive for Liouville theory), this state is not in the kernel of the projection to the $(+)$ Fock space. Hence it descends to a non-vanishing Fock space state, which is, however, both primary and secondary and hence null in the usual sense. In particular this means that it is Q_B-exact, and hence of course closed. Nevertheless, the operator Q_B^* that we have just defined has a non-trivial action on it. Indeed, define

$$|X^{(1)+}\rangle \equiv Q_B^* |X^{(0)+}\rangle \tag{19}$$

Here, the $+$-superscript indicates that we are dealing with states in the $(+)$ Fock space. The action of Q_B^* is to first conjugate the state to the $(-)$ Fock space, (where it becomes a non-primary, non-secondary state with respect to the Virasoro algebra!), then act with Q_B, which produces a nonzero result on such a state, and finally conjugate back to the $(+)$ Fock space. The result is a state of ghost number $+1$ in the $(+)$ Fock space, and we have:

Theorem 2: The state $|X^{(1)+}\rangle$ defined in Eq.(19) above is a LZ state of ghost number +1.

This is a remarkably simple result, and clarifies a conceptual point: although LZ states are not themselves null, they are Q_B^*-variations of null states. The fact that Q_B^* has very different properties from Q_B reflects a deep property of minimal matter coupled to gravity: the (+) and (−) Liouville Fock spaces are very different, a fact which has played an important role in several contexts[7][9][10][11][5].

Before discussing the proof of this theorem, we state the corresponding result for the conjugate LZ state, which according to Ref.[4] should lie in the (−) Fock space and have ghost number −1. Let us define the operator

$$\mathcal{K} \equiv \sum_{n=1}^{\infty} n c_{-n} c_n \tag{20}$$

This operator is produced by anticommuting Q_B and Q_B^*, as one can easily check by explicit computation:

$$\{Q_B, Q_B^*\} = (k_L^+ - k_L^-)^2 \mathcal{K} \tag{21}$$

A property of $\mathcal{K}$ that can be checked is that it has no kernel on states of ghost number −1, hence its inverse exists on ghost number +1, and is given by

$$\mathcal{K}^{-1} = \sum_{n=1}^{\infty} \frac{1}{n} b_{-n} b_n \tag{22}$$

Consider now the Fock space state

$$|X^{(-1)+}\rangle = \mathcal{K}^{-1} |X^{(1)+}\rangle \tag{23}$$

of ghost number −1. Since it is in the (+) Fock space, where the projection from the Verma module has no kernel, we can rewrite the left hand side of this equation in terms of (Liouville and matter) Virasoro secondaries acting on suitable primary states. Then, project this to the (−) Fock space. This procedure defines a state which we call $|X^{(-1)-}\rangle$, of ghost number −1. This brings us to

Theorem 3: The state $|X^{(-1)-}\rangle$ defined above is a LZ state of ghost number −1.

Indeed, it is clear that $|X^{(-1)-}\rangle$ and $|X^{(1)+}\rangle$ are built on the same Liouville primary, but the Liouville momenta are respectively k_L^- and k_L^+, whose sum is precisely Q_L. Thus this pair of states can have a non-vanishing inner product between them. The computation of this inner product proves Theorems 2 and 3, according to which these states are genuinely in the cohomology, for the following reason. The state $|X^{(+1)+}\rangle$ is closed, because of its definition Eq.(19), the anticommutation relation Eq.(21), and the fact that $|X^{(0)}\rangle$ is annihilated by both Q_B and $\mathcal{K}$. This in turn implies that $|X^{(-1)-}\rangle$ is closed, since Q_B commutes with $\mathcal{K}$ (an immediate consequence of Eq.(21)). Now if both states are closed, and if their inner product is nonvanishing, it follows that neither of them is exact, which proves Theorem 2 and 3 above. We find that in fact the inner product is:

$$\iota X^{(-1)-}|c_0|X^{(1)+}\rangle = s(s-1)!^2 \prod_{n=1}^{s-1} \left((nt)^2 - 1\right) \tag{24}$$

which is in fact nonvanishing for all t other than $\pm 1, \pm\frac{1}{2}, \ldots, \pm\frac{1}{s-1}$. Apart from the first pair of values, which correspond to $c = 1, 25$, the remaining are "unphysical" minimal models. Hence for all genuine $c < 1$ minimal models, the inner product above is nonvanishing, and our construction gives the LZ states associated to the matter primaries of type $(1, s)$, at ghost number ± 1. For $c = 1$ it does not work, a fact which may merit further investigation.

The detailed proof of Eq.(24), which is the principal result of this work, is given in Ref.[5], along with various other expressions for the physical states, and the interrelations among them. Here I will just sketch the proof. Starting from the Benoit-Saint-Aubin formula, Eq.(12), one can convince oneself that all the states that we have constructed above, and their inner products, are polynomials in t. (This is not true for null vectors of type (r, s) for $r \neq 1$, where one finds polynomials in t and $\frac{1}{t}$.) From the asymptotic behaviour of the formula, one can evaluate the asymptotic behaviour, for large t, of the inner product. This suffices to fix the degree of the polynomial and the leading coefficient. It remains only to determine the zeroes. Precisely half of them are obtained from Theorem 1 above, since the special values of t at which the null vector vanishes on projecting to *both* Fock spaces are clearly values for which the states we constructed above, and hence their inner product, vanish. The remaining zeroes follow from the fact that the transformation $t \to -t$ interchanges matter and Liouville sectors, and one can argue that the inner products have a definite parity under this transformation. This completes the derivation.

Although this derivation only works starting from the special null vectors of type $(1, s)$, this appears to be only a technical limitation. One can check in explicit examples that the same construction works also in situations where $r \neq 1$, but a general proof for this is not available at present. Another, more serious, limitation is the restriction to ghost numbers ± 1.

4. Conclusions

More than a year after the discovery of infinitely many extra physical states of every ghost number in the $c < 1$ string, their physical interpretation remains unclear. (Recently there has been very interesting progress in the corresponding problem for the $c = 1$ string[12][13].) An understanding of the role played by these states is likely to help clarify the situation regarding correlation functions in $c < 1$ strings, on which a lot of work has been done but a clear understanding reconciling all the different approaches remains to be achieved (see the lecture of V. Dotsenko in this volume, and references therein.)

The present understanding of these correlation functions from the continuum approach is based on analytic continuation and the eventual insertion of a fractional and/or negative number of screening charges in the Liouville sector. The fact that even vertex operators outside the minimal table acquire non-vanishing correlators in this framework, described by the heuristic expression that "null vectors do not decouple in the presence of gravity", tends to agree with the qualitative feature that there are infinitely many scaling fields in the matrix model and topological approaches. On the other hand, we now have infinitely many LZ physical states, which one might also be tempted to identify with this infinity of scaling fields. It needs to be understood whether these states are in some sense an alternate representation of the null vectors which "do not decouple", or should be interpreted in some different way. (Some aspects of the relation between matrix model and continuum CFT fields are discussed in Ref.[14].)

The present work is an attempt not at addressing this question directly, but rather at formulating and analysing the explicit form of LZ states, which are much more non-trivial to write down than DDK states. An explicit algorithm (quite distinct from the "brute-force" method that one can always use) was obtained to construct the extra LZ states. Explicit examples are worked out in Ref.[5]. The algebraic structure related to the conjugate BRS operator Q_B^* remains somewhat mysterious, and perhaps once this is clarified then the physical interpretation of LZ states for $c < 1$ will become more evident.

5. Acknowledgements

I wish to thank the organizers of the Cargese School for making it possible for me to participate and for their kind invitation to give this talk. I am grateful to C. Imbimbo, S. Mahapatra, S. Mukherji and A. Sen for their collaboration and for many useful discussions.

References

1. C. Thorn, Nucl. Phys. **B286** (1987) 61.

2. A.A. Belavin, A.M. Polyakov and A.B. Zamolodchikov, Nucl. Phys. **B241** (1984) 333.

3. F. David, Mod. Phys. Lett. **A3** (1988) 1651;
J. Distler and H. Kawai, Nucl. Phys. **B321** (1989) 509.

4. B.H. Lian and G.J. Zuckerman, Phys. Lett. **B254** (1991) 417.

5. C. Imbimbo, S. Mahapatra and S. Mukhi, Genova/Tata Institute Preprint GEF-TH 8/91, TIFR/TH/91-27, May 1991.

6. L. Benoit and Y. Saint-Aubin, Phys. Lett. **B215** (1988) 517.

7. M. Kato and S. Matsuda, in "Conformal Field Theory and Solvable Lattice Models", Advanced Studies in Pure Mathematics **16**, Ed. M. Jimbo, T. Miwa and A. Tsuchiya (Kinokuniya, 1988).

8. B.L. Feigin and D.B. Fuchs, Funct. Anal. and Appl. **16** (1982) 114.

9. J. Polchinski, Texas preprint UTTG-39-90.

10. N. Seiberg, Rutgers Preprint RU-90-29 (1990).

11. S. Mukherji, S. Mukhi and A. Sen, Phys. Lett. **B266** (1991) 337.

12. E. Witten, IAS Princeton preprint IASSNS-HEP-91/51, August 1991.

13. I.R. Klebanov and A.M. Polyakov, Princeton University preprint PUPT-1281, September 1991.

14. G. Moore, N. Seiberg and M. Staudacher, Rutgers-Yale preprint RU-91-11, YCTP-P11-91.

W(sl(n)) : EXISTENCE, CARTAN BASIS AND INFINITE ABELIAN SUBALGEBRAS

Max R. Niedermaier

II. Institute for Theoretical Physics, University of Hamburg
Luruper Chaussee 149, 2000 Hamburg 50, F.R.G.

INTRODUCTION

Classical integrable systems are amenable to a variety of different solution techniques. The two most prominent ones are the calculus of pseudo differential operators and τ-functions and the inverse scattering method centered around the classical Yang Baxter equation. The transition to the quantum regime, however, has so far only been successful for the case of the inverse scattering method. The classical field theories for which the resulting quantum inverse scattering method can be implemented have by now become paradigmatic examples of integrable quantum field theories. Supplemented by the S-matrix- and form factor bootstraps many of the physically relevant questions can be answered. There are, however, many classically integrable field theories for which even the first step, the construction of the quantum monodromy matrix solving a Yang Baxter equation, has not been achieved. This motivates the search for possible alternatives. In particular, there are indications that the quantum analogues of the first of the mentioned classical techniques should reveal some of the required mathematical structure. The seminal work of A.B. Zamolodchikov on the integrable deformations of 2-dimensional CFTs [?] suggests the existence of integrable structures in a large class of meromorphic CFTs. The existence of an infinite set of conserved charges in involution is then equivalent to the vanishing of a certain subset of the structure constants. This is a dynamical problem for which no general solution techniques are known. For the low members of the $\mathcal{W}(\mathrm{sl}(n))$ series it has been solved with reference to the classical case (involution) and a specific induction scheme visible in the free field realization (existence). For Casimir algebras $\mathcal{W}(g)$ in general one expects that for generic central charge infinite dimensonal abelian subalgebras are present with generators having weights equal to the exponents of the underlying simple Lie algebra g modulo the dual Coxeter number. The first period of (one set of) generators constitutes a basis of the Cartan subalgebra of $\mathcal{W}(g)$.

New Symmetry Principles in Quantum Field Theory, Edited by
J. Fröhlich et al., Plenum Press, New York, 1992

1. ABELIAN SUBALGEBRAS OF A MEROMORPHIC CFT

Basically a meromorphic CFT (mCFT) or vertex operator algebra[2, 3] is an infinite dimensional Lie algebra which contains the Virasoro algebra as a distinguished subalgebra and for which all fields have integer or halfinteger conformal weight. In more detail, a mCFT consists of a of a Hilbert space $\mathcal{H}$ and an assignment $|P\rangle \to P(z)$, which associates a unique field operator $P(z)$ to any state $|P\rangle$ in (a dense subspace of) $\mathcal{H}$. The bilinearform may be indefinite or degenerate. The space $\mathcal{H}$ is a vacuum representation space of the Virasoro algebra i.e. there exists a distinguished state $|L\rangle$ for which the modes L_n, $n \in \mathbb{Z}$ of the associated field $L(z)$ form a copy of the Virasoro algebra and which define a unique $su(1,1)$ invariant vacuum by $L_s|v\rangle = 0$, $s = 0, \pm 1$. The dense subspace $\underline{\mathcal{H}}$ is that of finite L_0 grade and an element $|P\rangle$, the associated field operator satisfies $P(z)|0\rangle = e^{zL_{-1}}|P\rangle$ $(*)$ as well as a number of additional conditions. The additional conditions force the spectrum of L_0 on $\mathcal{H}$ to be integer or halfinteger and in particular guarantee the injectivity of the assignment $(*)$. Let $\mathcal{H}$: $\mathcal{H}_\Delta = \{|P\rangle \in \mathcal{H} : L_0|P\rangle = \Delta|P\rangle\}$ denote the subspaces of fixed L_0-grade, where Δ is called the (conformal) weight of $|P\rangle$ or $P(z)$. The operators $P(z)$ are linear operators which map $\underline{\mathcal{H}}$ to infinite sums of elements in $\underline{\mathcal{H}}$: $P(z) : \underline{\mathcal{H}} \to \bigoplus_\Delta \mathcal{H}_\Delta$. They are completely determined by their matrix elements $(P_{\Delta\Delta'})_{\Delta,\Delta' \in \mathbb{Z}_+}$, where $P_{\Delta\Delta'} : \mathcal{H}_\Delta \to \mathcal{H}_{\Delta'}$. The product $P(z)Q(w)$ exists (for $|z| > |w|$) if the series $\sum_\Delta \langle P^i | P_{\Delta\Delta_i} Q_{\Delta\Delta_k} | P^k \rangle$ is absolutely convergent for all $|P^i\rangle \in \mathcal{H}_{\Delta_i}$, $|P^k\rangle \in \mathcal{H}_{\Delta_k}$, which is the last condition stipulated. The associativity of the product is then guaranteed by the absolute convergence. For the product of two fields $P(z)$, $Q(z)$ of weights Δ_P, Δ_Q one has the series expansion

$$P(z)Q(w) = \sum_{k=-\Delta_P-\Delta_Q}^{\infty} (z-w)^k (P_{-k-\Delta_P} Q_{-\Delta_Q})(w) , \quad |z| > |w| , \tag{1.1}$$

where $(P_{-k-\Delta_P} Q_{-\Delta_Q})(w)$ is the field corresponding to the state $P_{-k-\Delta_P} Q_{-\Delta_Q}|v\rangle$. In particular, $(P, Q)(z) := (P_{-\Delta_P} Q_{-\Delta_Q})(z)$ is a natural definition of the normal ordered product of both fields. The usual contour deformation argument then shows that (1.1) amounts to the specification of the Lie brackets $[P^i_m , P^j_n]$. The Jacobi identity is implied by the associativity of the operator product expansion and hence is guaranteed whenever the product is well defined on $\mathcal{H}$. A convenient basis for the Lie algebra is obtained by decomposing the Hilbert space $\mathcal{H}$ w.r.t. the action of the $su(1,1)$ subalgebra of the Virasoro algebra generated by $\{L_{\pm 1}, L_0\}$. (For notational simplicity we will from now on drop the distinction between $\underline{\mathcal{H}}$ and $\mathcal{H}$.) The $su(1,1)$ highest weight states satisfy $L_1|P\rangle = 0$ and such states (or the corresponding fields) are called quasiprimary. The subspace of quasiprimary states in $\mathcal{H}$ will be denoted by $\widehat{\mathcal{H}}$. The $su(1,1)$ descendences $L^n_{-1}|P\rangle$ of a basis in $\widehat{\mathcal{H}}$ make up a basis of $\mathcal{H}$. Further one has $\widehat{\mathcal{H}} \cong \mathcal{H}/L_{-1}\mathcal{H}$. In terms of the fields this amounts to considering equivalence classes modulo total derivatives, $n!(P_{-(n+\Delta)})(z) = (L^n_{-1} P_{-\Delta})(z) = \partial^n_z P(z)\, , n \geq 0$.

Let $|P^i\rangle$, $i \in I$ be a set of quasiprimary states of weight Δ_i that form a basis of

$\widehat{\mathcal{H}}$. Choose the P^i to be real for Δ_i even and purely imaginary for odd Δ_i. Put

$$\begin{aligned}
&\langle v|P^i_{\Delta_i}P^j_{-\Delta_j}|v\rangle =: D^{ij}\\
&\langle v|P^k_{\Delta_k}P^i_{-\Delta_j-\Delta_k}P^j_{-\Delta_j}|v\rangle =: C^{ijk}\,,
\end{aligned}\tag{1.2}$$

which in the above convention are real. The $su(1,1)$ covariance then fixes the commutator of the modes of any two quasiprimary fields with only the constants (1.2) as dynamical input[3]

$$[P^i_m\,,\,P^j_n] = \sum_k C^{ij}_k\, p^{\Delta_i\Delta_j}_{\Delta_k}(m,n)P^k_{m+n} + D^{ij}\binom{n+\Delta_i-1}{2\Delta_i-1}\delta_{m+n}\,,\tag{1.3}$$

where $C^{ij}_m D^{mk} = C^{ijk}$ and the sum is over fields of non-zero weight. The $p^{\Delta_i\Delta_j}_{\Delta_k}(m,n)$ are polynomials in m, n

$$p^{\Delta_i\Delta_j}_{\Delta_k}(m,n) = \sum_{r,s\geq 0}\delta_{r+s,\Delta_{ijk}-1}\,c_{rs}\binom{m+\Delta_i-1}{r}\binom{n+\Delta_j-1}{s}\tag{1.4}$$

The notations are:

$$c_{rs} = (-)^s\frac{(2\Delta_k-1)!}{(\Delta_i+\Delta_j+\Delta_k-2)!}(\Delta_i-\Delta_j+\Delta_k)_s(-\Delta_i+\Delta_j+\Delta_k)_r\,,\tag{1.5}$$

where $(x)_n = (x)(x+1)\ldots(x+n-1)$ and $\Delta_{ijk} = \Delta_i+\Delta_j-\Delta_k$.

Consider now in particular the subalgebra stabilizing the vacuum, $P^i_n|v\rangle = 0$, $-\Delta_i+1\leq n\leq\Delta_i-1$. The Jacobi identity implies

$$[L_l\,,[P^i_m\,,\,P^j_n]] + [[L_n\,,\,P^i_m]\,,\,P^j_n] + [P^i_m\,,[L_l\,,\,P^j_n]] = 0\tag{1.6}$$

i.e. L_l acts on the commutator as on a $su(2)$ tensor product. The polynomials $p^{\Delta_i\Delta_j}_{\Delta_k}$ play the role of $su(2)$ Clebsch-Gordan coefficients. It follows that

$$|\Delta_i-\Delta_j|\leq\Delta_k-1\leq\Delta_i+\Delta_j-2\,.\tag{1.7}$$

Of particular interest are the generators $P^i_{-\Delta_i+1}$, P^i_0, $P^i_{\Delta_i-1}$. One has $p^{\Delta_i\Delta_j}_{\Delta_i+\Delta_j-1} = 1$, $p^{\Delta_i\Delta_j}_{\Delta_k}(\pm(\Delta_i-1),\pm(\Delta_j-1)) = 0$ for $\Delta_{ijk}>1$ and

$$\begin{aligned}
&p^{\Delta_i\Delta_j}_{\Delta_k}(0,0) = 0 &&\text{for } \Delta_{ijk}\text{ even}\\
&p^{\Delta_i\Delta_j}_{\Delta_k}(0,0) = \frac{(2\Delta_k-1)!}{(2\Delta_k-2+\Delta_{ijk})!}\,\frac{(-)^s}{s!}\prod_{r=1}^{s}(\Delta_i-r)(\Delta_j-r)(\Delta_k+r-1)\\
& &&\text{for } \Delta_{ijk}\text{ odd}\,,
\end{aligned}\tag{1.8}$$

with $s = (\Delta_{ijk}-1)/2$. In particular $p^{\Delta_i\Delta_j}_{\Delta_k}(0,0)\neq 0$ iff $1\leq\Delta_{ijk}\leq 2\min(\Delta_i,\Delta_j)-1$. The first equation follows from $p^{\Delta_i\Delta_j}_{\Delta_k}(-m,-n) = (-)^{\Delta_{ijk}+1}p^{\Delta_i\Delta_j}_{\Delta_k}(m,n)$ and the second follows by induction on Δ_{ijk} from the general expression (1.4). Together

$$
\begin{aligned}
[P^i_{\pm(\Delta_i-1)}, P^j_{\pm(\Delta_j-1)}] &= \sum_{\{k:\Delta_{ijk}=1\}} C^{ij}_k P^k_{-(\Delta_i+\Delta_j-2)} \\
[P^i_0, P^j_0] &= \sum_{\{k:\Delta_{ijk}\, odd,\, 1\leq\Delta_{ijk}\leq 2\min(\Delta_i,\Delta_j)-1\}} p^{\Delta_i\Delta_j}_{\Delta_k}(0,0)\, C^{ij}_k\, P^k_0 \, .
\end{aligned} \tag{1.9}
$$

The existence of an abelian subalgebra therefore is equivalent to the vanishing of the respective structure constants in (1.9). Clearly the existence of the a zero-mode abelian subalgebra implies the commutativity of the $P^i_{\pm(\Delta_i-1)}$ modes. The correctness of the converse will depend on the dynamics of the theory considered but will in general cease to hold. Zero-mode abelian subalgebras are of particular interest because they are compatible with the L_0 graduation.

An abelian subalgebra $[P^i_{\pm(\Delta_i-1)}, P^j_{\pm(\Delta_j-1)}] = 0$ also gives rise to a zero-mode abelian subalgebra without additional input from the dynamics. Define a map $\gamma : \mathcal{H} \to \mathcal{H}$ by

$$
\gamma|P\rangle = \sum_{\{\nu:|\nu|\leq\Delta\}} \gamma_\nu\, L_\nu|P\rangle \, , \tag{1.10}
$$

where $L_\nu = L_{n_l} \ldots L_{n_1}$ and $\nu \in Par(1)$ (eqn. (2.2)), $|\nu| = n_1 + \ldots + n_l$, $l = l(\nu)$. The coefficients γ_ν are independent of the central charge and are recursively defined [3] by $\gamma_\emptyset = 1$ and

$$
|\nu|\gamma_\nu = \text{ coeff. of } L_\nu \text{ in } \qquad \sum_{m=1}^{\infty} \frac{(-)^{m+1}}{m(m+1)} L_m \sum_{\{l(\nu')\geq l(\nu)-1\}} \gamma_{\nu'}\, L_{\nu'} \tag{1.11}
$$

In particular $\gamma|L\rangle = |L\rangle - c/24$. The states $\gamma|P\rangle$ are no longer of homogenous L_0 grade. The fields $\gamma P(\sigma) = z^\Delta(\gamma|P\rangle)(z)$, $z = e^{i\sigma}$ can be shown to coincide with that obtained from the conformal transformation $z \to e^z$. The commutation relations take the form

$$
\begin{aligned}
[(\gamma P^i)_m, (\gamma P^j)_n] = \sum_{\{k:\Delta_{ijk}\geq 1\}} C^{ij}_k\, \tilde{p}^{\Delta_i\Delta_j}_{\Delta_k}(m,n)\,(\gamma P^k)_{m+n} + \\
D^{ij}\frac{m^{2\Delta_i-1}}{(2\Delta_i-1)!}\delta_{m+n,0}
\end{aligned} \tag{1.12}
$$

where the structure constants are the same as in (1.3). The universal polynomials are now given by

$$
\begin{aligned}
\tilde{p}^{\Delta_i\Delta_j}_{\Delta_k}(m,n) &= \frac{1}{(d-1)!}\frac{(2\Delta_k-1)!}{(2\Delta_k+d-2)!}\sum_{r=0}^{d-1}(-)^r\binom{d-r}{r}\, c_r\, m^{d-1-r} n^r \\
c_r &= \frac{(2\Delta_j-2-r)!}{(2\Delta_j-d-1)!}\frac{(2\Delta_i-d-1+r)!}{(2\Delta_i-d-1)!} \, ,
\end{aligned} \tag{1.13}
$$

where $d = \Delta_{ijk}$. In particular it follows

$$
[\, P^i_{\pm(\Delta_i-1)}, P^j_{\pm(\Delta_j-1)}\,] = 0 \quad \text{iff} \quad C^{ij}_k = 0,\ k:\Delta_{ijk}=1 \quad \text{iff} \quad [\,(\gamma P^i)_0, (\gamma P^j)_0\,] = 0. \tag{1.14}
$$

2. THE CASIMIR ALGEBRA $\mathcal{W}(\mathrm{sl}(n))$

W-algebras are special mCFTs. The basic point is that one does not take all quasiprimary fields as the generators of the algebra but allows the use of normalordered products to generate the algebra.

Definition: A *W-algebra of rank r* is an infinite dimensional Lie algebra whose Lie bracket is inherited from a meromorphic CFT. The meromorphic CFT is generated by the operations ∂ and $\mathcal{N}$ from r quasiprimary fields $W^1(z) = L(z)$, $W^2(z), \ldots, W^r(z)$. The bilinearform on $\bigoplus \widehat{\mathcal{H}}_{\Delta_i}$, $1 \leq i \leq r$ is non-degenenerate.

Here $\mathcal{N}(\,,\,)$ is a $su(1,1)$ covariant normalordering prescription. It differs from $(\,,\,)$ (induced by (1.1)) by a finite number of derivative terms. The choice of the normalordering is in principle irrelevant, but $\mathcal{N}(\,,\,)$ is a convenient one[3]. The basis $|W^i\rangle$ is unique up to linear transformations in the sector $\bigoplus \widehat{\mathcal{H}}_{\Delta_i}$, $1 \leq i \leq r$. A basis $\overline{W}^i$ is called a *Cartan basis* if its zero modes satisfy $[\overline{W}_0^i\,,\,\overline{W}_0^j] = 0$, $1 \leq i,j \leq r$. A drawback of this definition is that it does not specify the commutator of arbitrary monomials in the modes W_n^i. To study the representation theory commutators of the type $[\mathcal{N}(W^{i_1} \ldots W_r^i)_n\,,\, W_{n_1}^{j_1} \ldots W_{n_k}^{j_k}]$ are needed, which can not directly be traced back to the operator product expansion. The evaluation from the $[W_m^i\,,\,W_n^j]$ commutators on the other hand involves infinite sums of generators, whose convergence at intermediate stages is not guaranteed. This means that a regularization prescription is required. Clearly the detailed form of the regularization should be irrelevant and, if possible, direct reference to it should be avoided. In the case of the so-called Casimir algebras we shall therefore adopt the following, slightly stronger definition. To prepare this set

$$c \in C_r = \{0 \leq c \leq r \,|\, c = r - 12\rho^2(s_+^2 + s_+^{-2} - 2),\ s_+^2 \text{ irrational }\}\,, \tag{2.1}$$

where ρ is the Weyl vector of g and r the rank. Further take $\mathcal{H}(g)$ to be the (completion of the) span of the lexicographically ordered states of the form

$$\begin{aligned} &W^1_{-\nu_1} \ldots W^r_{-\nu_r}|v\rangle\,, \qquad \text{with} \qquad \nu_i \in Par(\Delta_i) := \\ &\{\nu = (n_1, \ldots, n_l) \in \mathbb{Z}^l : n_j \geq n_{j+1} \geq \Delta_i,\ 1 \leq j \leq l,\ l \geq 0\} \end{aligned} \tag{2.2}$$

Definition': For a complex Lie algebra g a *Casimir algebra* $W(g)$ is a W-algebra for which the weights of the generating fields coincide with the orders of the independent Casimirs of g. For $c \in C_r$ the algebra is of rank r and the L_0-graded highest weight module satisfying $W_n^i|v\rangle = 0$ iff $n > -\Delta_i$ is irreducible and coincides with $\mathcal{H}(g)$.

The additional condition guarantees that any regularization prescription employed to evaluate the missing commutators yields the same answer which is moreover compatible with the parts directly fixed by the operator product expansion. The basic representation theoretic concepts, highest weight vectors, Verma modules, characters etc. can now be defined in the usual way and the study of the representation theory is a well posed problem[5].

For a finite set of c values the rank of the algebra may actually be smaller than r. At these values one or more of the generating fields becomes composite i.e. some linear combination of generators decouples from all conformal blocks. The coset space obtained by dividing out the submodule generated by the associated state may form the Hilbert space of a W-algebra of lower rank. Consider the commutators $[P^i_m\,,\,P^j_n]$ of the quasiprimary fields of weight Δ. The coefficients (1.2) of the c-number term form a matrix D_Δ which yields the bilinearform on the corresponding vector space $\widehat{\mathcal{H}}_\Delta$. The vanishing of the determinant of this bilinearform gives a criterion for the decoupling. Call a basis regular if the structure constants are polynomial in c. If such bases exist the rank can be calculated from the

Lemma: For a Casimir algebra in a regular basis the rank is given by $r-s$, where s is the number of $1 \leq i \leq r$ for which $\det D_{\Delta_i}$ vanishes.

It should, however, be emphasized that the existence of Casimir algebras has not been established in general. For low rank cases one can explicitly solve the associativity constraint implicit in the definition of a mCFT. For generic rank the existence of Casimir algebras can be established by constructing a realization. The problem is then shifted to the proof that the proposed set of field generators closes. Of particular interest are free field realizations in that they also provide a powerful starting point for the study of the representation theory. Candidates for such such free field realizations have been proposed by Fateev and Lykyanov [4]. The basic fields are functionals in a set of $r = \text{rank}(g)$ bose fields and are defined in terms of a generating functional which takes the form of a generalized Miura transformation. The existence of the realization and, in particular, of the corresponding type of Casimir algebras has been established in [5] for the $sl(r+1)$ series. The extension at least to simply laced Lie algebras is unproblematic.

Introduce r scalar fields $\phi^a(z) = q^a - ip^a \ln z + i\sum_{n\neq 0}\frac{1}{n}a^a_n z^{-n}$ with modes having free oscillator commutation relations $[a^a_n\,,a^b_m] = m\delta^{ab}\delta_{n+m,0}$, $[p^a, q^b] = -i\delta^{ab}$. In the envelopping algebra of the oscillator algebra introduce r field operators $W^i(z)$ by means of a *symmetrized* Miura transformation

$$\begin{aligned}\tau\ [2s_0\partial_z + i\hat{h}_{r+1}\cdot\partial_z\phi][2s_0\partial_z + i\hat{h}_r\cdot\partial_z\phi]\ \ldots\ [2s_0\partial_z + i\hat{h}_1\cdot\partial_z\phi] = \\ = -\sum_{i=0}^{r+1} W^i(z)(2s_0\partial)^{r+1-i}\,,\end{aligned} \tag{2.3}$$

where α_j are the simple roots, $\alpha_j =: \hat{h}_{j+1} - \hat{h}_{j+2}$ $(\hat{h}_{r+2} = \hat{h}_1)$, $2s_0 = s_+ - s_+^{-1}$ and normalordering shall be implicit. τ projects onto the sector invariant under the automorphism $\tau : \alpha_i \to -\alpha_{r+1-i}$, $s_+ \to -s_+$ of the Dynkin diagram, which is implemented by the maximal element of the Weyl group (for simplicity we use the same symbol for the automorphism and the associated projection operator). This symmetrization is cruical for the following results. For the generators one finds, in particular, $W^0 = -1$, $W^1 = 0$

and $L(z) := W^2(z)$ generates a Virasoro algebra of central charge $c = r - 48{s_0}^2\rho^2$. The fields W^i , $2 \leq i \leq r+1$ are of L_0-weight i, but in general neither primary nor quasiprimary relative to $L(z)$. By adding suitable normalordered products of $W^{i-1}, \ldots, W^2$ to W^i one can try to promote W^i to a quasiprimary or primary field. Since $\tau L(z) = L(z)$ the invariance under τ is clearly a necessary condition for this to be possible. As there is no possible 'counterterm', W^3 is always primary. For the other generators the projection onto quasiprimary or primary fields is nontrivial. The projection onto quasiprimary fields turns out to be unproblematic and will in the following often implicitly assumed to be performed. The projection onto primary fields may fail for certain values of the central charge and will be discussed below.

Theorem:(Existence) The (quasiprimary projection of the) symmetrized Miura fields $W^i(z)$ generate a $\mathcal{W}(\mathrm{sl}(r+1))$ algebra in the sense of definition'. The structure constants are polynomials in the central charge.

The proof first shows that the Miura generators close for $c \in C_r$. This is done by showing that the (object which a posteriori becomes the) singlet representation space contains the correct number of independent states. This again is traced back to a finite dimensional quantum group problem, which is equivalent to giving a constructive proof of a q-deformation of the Bernstein-Gelfan'd-Gelfan'd resolution of $\mathcal{U}_q(sl(r+1))$-modules, for q not a root of unity. The latter has been done in [6], which completes the proof for $c \in C_r$. In addition the structure constants in the basis of fields $W^i(z)$ are polynomial in the central charge. Thus, after the OPEs or commutation relations have been reconstructed, the range of definition of the algebra can be extended to all values of the central charge.

The Miura fields $W^i(z)$ do, however, not form a Cartan basis. A Cartan basis can be obtained as follows: Let Ω be the generator of the cyclic group Z_{r+1} acting by $\Omega : (\alpha_1, \ldots \alpha_r, -\theta) \to (\alpha_2, \ldots, \alpha_r, -\theta, \alpha_1)$ on the root system; where θ is the highest root. In terms of the fundamental reflections r_i, $1 \leq i \leq r$ of the Weyl group, Ω is given by the Coxeter element $\Omega = r_1 r_2 \ldots r_r$. The Dynkin automorphism τ is implemented by the maximal element of the Weyl group. Together Ω and τ generate a Coxeter subgroup of the Weyl group with relations $\Omega^{r+1} = 1$, $\tau^2 = 1$, $(\Omega\tau)^2 = 1$. These are the defining relations of the dihedral group D_{r+1} i.e. the symmetry group of a regular polygon ($r+1$-gon). Let $P[s_0, \alpha_i \cdot \partial_z\phi]$ be a (normalordered) functional in s_0, $\alpha_i \cdot \phi$. By $\Omega P[s_0, \alpha_i \cdot \partial_z\phi] = P[s_0, (\Omega\alpha_i) \cdot \partial_z\phi]$ and $\tau P[s_0, \alpha_i \cdot \partial_z\phi] = P[-s_0, (\tau\alpha_i) \cdot \partial_z\phi]$ one has an induced action of the dihedral group. Let D_{r+1} denote the projector onto the dihedral invariant subsector and set

$$D_{r+1} \ [2s_0\partial_z + i\hat{h}_{r+1} \cdot \partial_z\phi][2s_0\partial_z + i\hat{h}_r \cdot \partial_z\phi] \ \ldots \ [2s_0\partial_z + i\hat{h}_1 \cdot \partial_z\phi] = = -\sum_{i=0}^{r+1} D^i(z)(2s_0\partial)^{r+1-i} \ , \tag{2.4}$$

In particular, $D^0 = -1$, $D^1 = 0$, $D^2 = -\frac{1}{2}\partial_z\phi \cdot \partial_z\phi$. The fields are not quasiprimary relative to $L(z) = W^2(z)$ in (2.3). Define

$$\overline{W}^i(z) = \mathcal{N} D^i(z) , \tag{2.5}$$

where $\mathcal{N}$ denotes the projection onto quasiprimary fields (i.e. $L_1|\overline{W}^i\rangle = 0$).

Proposition:($r \leq 4$) The fields $\overline{W}^i(z)$, $2 \leq i \leq r+1$ form a Cartan basis of $\mathcal{W}(\mathrm{sl}(r+1))$.

We expect this to be correct in general.

As a non-trivial example consider the $sl(4)$ case and set $L = W^2$, $W = W^3$, $V = W^4$. The Cartan basis is given by $\overline{L} = L$, $\overline{W} = W$, $\overline{V} = V + \frac{1}{8}\Lambda + \frac{c-3}{200}\partial^2 L$, where $\Lambda = \mathcal{N}(L,L) = (L,L) - \frac{3}{10}\partial^2 L$. In the Cartan basis the commutation relations have the following form:

$$[L_m\,,\,\overline{W}_n] = (2m-n)\overline{W}_{m+n} \tag{2.6.a}$$

$$\begin{aligned}[\overline{W}_m\,,\,\overline{W}_n] = {} & p_{334}(m,n)\left[4\overline{V}_{m+n} + \frac{1}{2}\Lambda_{m+n}\right] + \\ & \frac{1}{5}(c+7)p_{332}(m,n)\,L_{m+n} + \frac{1}{30}c(c+7)\binom{m+2}{5}\delta_{m+n,0}\end{aligned} \tag{2.6.b}$$

$$[L_m\,,\,\overline{V}_n] = (3m-n)\overline{V}_{m+n} + \frac{1}{1200}(7c+114)m(m^2-1)\,L_{m+n} \tag{2.6.c}$$

$$[\overline{W}_m\,,\,\overline{V}_n] = \frac{1}{50}(7c+114)p_{343}(m,n)\,\overline{W}_{m+n} + \frac{5}{4}p_{345}(m,n)\,\mathcal{N}(L,\overline{W})_{m+n} \tag{2.6.d}$$

$$\begin{aligned}[\overline{V}_m\,,\,\overline{V}_n] = {} & p_{446}(m,n)\left[\frac{3}{4}\mathcal{N}(\overline{W},\overline{W})_{m+n} + \right. \\ & \left.\frac{1}{8}\mathcal{N}(L,\Lambda)_{m+n} + \frac{1}{320}(c-6)\,\mathcal{N}(L,\partial^2 L)_{m+n}\right] - \\ & p_{444}(m,n)\left[\frac{50}{3}(c-23)\overline{V}_{m+n} - \frac{1}{160}(7c+114)\,\Lambda_{m+n}\right] + \\ & \frac{1}{6000}(7c+114)(8c+41)p_{442}(m,n)\,L_{m+n} + \\ & \frac{1}{48000}c\,(7c+114)(8c+41)\binom{m+3}{7}\delta_{m+n,0}\end{aligned} \tag{2.6.e}$$

The following quasiprimary normalordered products enter.

$$\begin{aligned}\Delta = 5: \quad & \mathcal{N}(L,\overline{W}) = (L,\overline{W}) - \frac{3}{14}\partial^2\overline{W} \\ \Delta = 6: \quad & \mathcal{N}(L,\Lambda) - \frac{1}{6}\partial^2\Lambda \\ & \mathcal{N}(L,\partial^2 L) = (L,\partial_z{}^2 L) - \partial(L,\partial_z L) + \frac{2}{9}\partial^2(L,L) - \frac{1}{42}\partial^4 L \\ & \mathcal{N}(L,\overline{V}) = (L,\overline{V}) - \frac{1}{6}\partial^2\overline{V}\end{aligned}$$

$$\mathcal{N}(\overline{W},\partial L) = (\overline{W},\partial L) - \frac{2}{5}\partial(\overline{W},L) + \frac{1}{20}\partial\overline{W}$$
$$\mathcal{N}(\overline{W},\overline{W}) = (\overline{W},\overline{W}) - \frac{5}{9}\partial^2\overline{V} - \frac{5}{72}\partial^2\Lambda + \frac{c+7}{840}\partial^4 L \tag{2.7}$$

We add a number of comments. Evidently, the algebra is well defined for all values of the central charge. To determine the rank according to the lemma we note that $\det D_\Delta = \frac{1}{2}c,\ \frac{1}{30}c(c+7),\ \frac{1}{1200}c(c+2)(c+7)(7c+114)$ for $\Delta = 2,\ 3,\ 4$, respectively. This means that the algebra is of rank 3 except for $c \in \{0,\ -2,\ -7,\ -114/7\}$, where it is of rank 0, 2, 1, 2, respectively. Notice that these points lie in the $\mathcal{W}(\mathrm{sl}(4))$-minimal spectrum.

A second source of exceptional c values is the projection onto primary fields. Let $\underline{W}^k$ denote the projections of the Cartan field generators onto primary fields, normalized s.t. $\langle \underline{W}^k, \underline{W}^k \rangle = c/k$. In the above example one finds

$$\underline{W} = \sqrt{\frac{10}{c+7}}\,\overline{W}\,,$$
$$\underline{V} = \sqrt{\frac{300(5c+22)}{(7c+114)(c+7)(c+2)}}\Big(\overline{V} - \frac{7c+114}{40(5c+22)}\Lambda\Big)\,. \tag{2.8}$$

The structure constants in this basis are more complicated roots of rational functions, which enter also in the definition of the quasiprimary fields. We have verified explicitly that the transformation $\overline{W}^k \rightsquigarrow \underline{W}^k$ leads to the commutation relations given in [7]. For the c values $\{-2,\ -\frac{22}{5},\ -7,\ -\frac{114}{7}\}$ the basis transformation is singular which introduces a corresponding set of singularities in the structure constants. Three of these singular points can (a posteriori) be removed by relaxing the normalization condition, but at $c = -22/5$ the projection onto primary fields fails. Notice that this point does not lie in the $\mathcal{W}(\mathrm{sl}(4))$-minimal spectrum. For general rank one can at least put constraints on the the set of c values where the projection onto primary fields might fail[5](In the preprint this and the previous exceptional class where mixed up). For the subset disjoint from the first set, the algebra is well defined and of maximal rank, but its representation spaces can not be decomposed into Virasoro-irreducible sectors.

3. INFINITE ABELIAN SUBALGEBRAS

The generators of the Cartan subalgebra do in fact constitute the lowest period of an *infinite* sequence of conserved charges in involution. Together they form an infinite dimensional abelian subalgebra of the mCFT. The naming 'conserved charges' comes from the significance of these generators in the context of affine Toda theories. In extension to the field generators of the Cartan basis these charges arise as the modes of field generators which, in the free field realization (2.3) are dihedral invariant modulo total derivatives. From here one can already conclude that, if they exist, these charges will be mutually commuting. The actual construction of the conserved charges amounts to a dynamical problem for which no general results are known. For low members of the $sl(r+1)$ series (r=1, 2) the existence has been established by means of a specific induction scheme, visible in the free field realization[8]. Set $\mathcal{H} = \mathcal{H}(sl(r+1))$ and consider the the quasiprimary states as equivalence classes modulo total derivatives $\widehat{\mathcal{H}} \cong \mathcal{H}/L_{-1}\mathcal{H}$. The principle can then be summarized as follows:

Proposition:(r=1, 2) There exists a basis $|P_s^i\rangle$, $s = 1, \ldots, \dim\widehat{\mathcal{H}}_i$ of $\widehat{\mathcal{H}}_i$ and an ordering relation '$\succeq$' on the space of Fock monomials such that the elimination process of the D_{r+1}-noninvariant Fock monomials in order of '$\succeq$' iterates (induction step). The first elimination step can be performed (induction basis) if and only if $i \neq r+2 \bmod (r+1)$.

The basis $|P_s^i\rangle$ is induced by the Fock space realization of the monomial Verma module basis of Cartan generators in $\widehat{\mathcal{H}}$. The ordering relation '$\succeq$' is given by

$$\begin{aligned} sl(2): &\quad [H^p] \succeq [H^{p'}] \quad \text{if } p \geq p' \\ sl(3): &\quad [H^p H^{*q}] \succeq [H^{p'} H^{*q'}] \quad \text{if } \begin{cases} p+q > p'+q' & \text{or} \\ p+q = p'+q' & \text{and } |p-q| \geq |p'-q'|, \end{cases} \end{aligned}$$

where the fields

$$H_j = \frac{1}{\sqrt{r+1}} \sum_{k=1}^{r} (1 - \omega^{k(r+1-j)}) i\alpha_k \cdot \partial\phi\,, \qquad \omega^{r+1} = 1\,, \qquad 1 \leq j \leq r\,,$$

diagonalize the Coxeter element Ω and $[\,H^p H^{*q}\,]$, $p \geq q$ is a symbolic shorthand for an equivalence class modulo total derivatives of real Fock monomials of power $p+q$ and Ω eigenvalue $\omega^{\pm(p-q)}$ for the $H^p H^{*q}$, $H^{*p} H^q$ pieces, respectively. Similar induction schemes should exist for higher rank.

An immediate consequence is that there exists states $|P^i\rangle \in \widehat{\mathcal{H}}$, $i \neq r+2 \bmod (r+1)$ for which $[\,P^i_{\pm(i-1)}\,,\, P^j_{\pm(j-1)}\,] = 0$ or equivalently $[\,(\gamma P^i)_0\,,\,(\gamma P^j)_0\,] = 0$. In contrast to the operators of the first period, constituing a basis of the Cartan subalgebra, the zero modes of the the homogenous fields $P^i(z)$ of the higher periods will in general not commute.

REFERENCES

[1] A.B. Zamolodchikov, in: Adv. Stud. Pure Math. **19**(1989) 641.

[2] P. Goddard, in: 'Infinite dimensional Lie algebras', C.I.R.M. Luminy, 1988.
I. Frenkel, J. Lepowsky and A. Meurman, *Vertex operator algebras and the monster*, Academic Press (1988).

[3] W. Nahm, in: 'Recent developments in conformal field theory', Trieste, 1989.
W. Nahm, in: 'Topological Methods in QFT', Trieste 1990, Int. J. Mod. Phys. **A6** (1991) 2837.

[4] V.A. Fateev and S.L. Lykyanov, Int.J.Mod.Phys.**A3** (1988) 507-520; Yad. Fiz. **49** (1989) 1491.
S.L Lykyanov, Funct. Anal. Appl. **22** (1990) 1.

[5] M.R. Niedermaier, *Irrational free field resolutions of* $\mathcal{W}(\mathrm{sl}(n))$ *and extended Sugawara construction*, Comm. Math. Phys., to appear.

[6] V.K. Dobrev, *Singular vectors of quantum group representations for straight Lie algebra roots*, preprint Göttingen, 1991.
V.K. Dobrev, in: 'II Symposium on Topological and Geometrical Methods in Field Theory', May/June 1991, Turku, Finland.

[7] R. Blumenhagen, M. Flohr, A. Kliem, W. Nahm, A. Recknagel and R. Varnhagen, Nucl. Phys. **B361**(1991) 255
H.G. Kausch and G.M.T. Watts, Nucl. Phys. **B354** (1991)740-768.

[8] M.R. Niedermaier, Nucl. Phys. **B 364** (1991) 165.

ASPECTS OF QUANTIZING LORENTZ SYMMETRY

Arne Schirrmacher

Max–Planck–Institut für Physik
München*

ABSTRACT

After a short reminder of quantum groups, some recent progress in understanding the q-deformation of the Lorentz symmetry from the research group of J. Wess is reported. A simple argument is given for extracting the six physical generators of the Lorentz algebra from an initial number of seven emerging from the deformation procedure.

1 INTRODUCTION

Nowadays, quantum groups occupy various spots in theoretical physics. They appeared, for example, as somewhat hidden symmetries of conformal field theories that had to be 'unravelled' or they are of interest since, in the case that the deformation parameter is a root of unity, the representation theory changes such that it allows for truncated tensor products [1]. In this talk quantum groups and quantum symmetries shall be considered directly as deformations of space-time coordinates. Since the deformation carries us *smoothly* away from well-known theories we can also learn much about the initial; degeneracies split and mathematical subtleties get regularized. As the analogy to other deformations in physics such as special relativity or quantum mechanics teaches us, a lot of work is needed to understand the generalized notions. According to Faddeev one can at least say that in particular the deformed Lorentz group now *allows* us to speak about what quantized space-time may be, an otherwise vague idea quoted from time to time.

The seminal work of Faddeev, Reshetikhin, and Takthajan [2] provides a comprehensive account of the "Quantization of Lie groups and Lie algebras". In particular for all simple groups deformations are worked out. There are, however, two directions in which new results have been added:
(1) R-matrices with more than one deformation parameter have been found revealing

*postal address: Föhringer Ring 6, W-8000 München 40, Germany; e-mail: ARS at DM0MPI11.

New Symmetry Principles in Quantum Field Theory, Edited by
J. Fröhlich et al., Plenum Press, New York, 1992

independent domains of different quantization parameters. We will briefly discuss this for the two-dimensional case.
(2) The entries of the quantum matrices are no more numbers and an equivalent to complex conjugation is not canonically given. For real forms of quantum groups with arbitrary signature no general theory is known yet. The Lorentz group is one example where this crucial problem has been solved that has important consequences also for Lie algebra deformations.

Addressing the first point, recall the basic structures of quantum groups from the Faddeev-Reshetikhin-Takthajan approach [2] for a two-dimensional example. Given a matrix $R \in M(4 \times 4, \mathbf{C})$ solving the Yang–Baxter equation(YBE)

$$R_{12}R_{13}R_{23} = R_{23}R_{13}R_{12} \ , \tag{1}$$

the deformations of the group, the coordinate space, and the algebra can be defined as follows:

	notation	*relations*
quantum group	$T = \left(\begin{smallmatrix} a & b \\ c & d \end{smallmatrix}\right)$	$R_{12}T_1T_2 = T_2T_1R_{12}$
quantum plane	$\mathbf{x} = \left(\begin{smallmatrix} x \\ y \end{smallmatrix}\right)$	$(\hat{R}-1)(\mathbf{x}\otimes\mathbf{x}) = 0$
quantum Lie algebra	$T^{\pm}, H, K$	$R_{12}L_2^{\pm}L_1^{\pm} = L_1^{\pm}L_2^{\pm}R_{12}$ $R_{12}L_2^{+}L_1^{-} = L_1^{-}L_2^{+}R_{12}$

(2)

where $\hat{R} = PR$ (P the permutation matrix) with the normalization $\hat{R}^{ii}{}_{ii} = 1$.

Adopting the point of view that the basic entities to be deformed are the space variables and the Lie algebra, we might require the space to be a q-plane

$$xy = q\, yx \tag{3}$$

and the Lie algebra being deformed in the Kulish-Reshetikhin manner [3]

$$[T^+,T^-] = [H]_r \equiv \frac{r^H - r^{-H}}{r - r^{-1}} \ , \qquad [T^\pm, H] = \pm T^\pm, \qquad [K, ...] = 0 \ . \tag{4}$$

It turns out that q and r are independent deformation parameters and that the group deformation of $GL(2)$ is a derived result, i.e. for

$$R = R_{r^2;q} = \begin{pmatrix} 1 & 0 & 0 & 0 \\ 0 & \frac{1}{q} & 0 & 0 \\ 0 & 1-\frac{1}{r^2} & \frac{q}{r^2} & 0 \\ 0 & 0 & 0 & 1 \end{pmatrix} \tag{5}$$

equation (3) results from definition (2). The entries of the matrices $L^\pm$ obey relations similar to those of the quantum group whereas with the following identification

$$L^+ = \begin{pmatrix} p^{\frac{H+K}{2}} & 0 \\ \lambda\,(\frac{q}{r})^{H/2}\,T^+ & q^{\frac{-H+K}{2}} \end{pmatrix} , \qquad L^- = \begin{pmatrix} q^{-\frac{H+K}{2}} & -\lambda\,(\frac{q}{r})^{H/2}\,T^- \\ 0 & p^{\frac{H-K}{2}} \end{pmatrix} , \tag{6}$$

where $\lambda = r - \frac{1}{r}$, relations (4) are reproduced. One can also find an equivalent algebra using deformed commutators $[A,B]_q = qAB - q^{-1}BA$:

$$[T^+, T^-]_r = T^3 , \qquad [T^\pm, T^3]_{r^{\pm 2}} = [2]_r T^\pm \qquad (T^3 \sim 1 - \lambda r^H) . \tag{7}$$

In this form the left hand side is linear and deformed structure constants can be defined. (This form of the algebra naturally comes out for the vector fields on the quantum group [4, 6].)

The quantum group relations for this R-matrix turn out to be:

$$\begin{aligned} ab &= \tfrac{r^2}{q}\, ba, & ac &= q\, ca, \\ cd &= \tfrac{r^2}{q}\, dc, & bd &= q\, db, \\ bc &= \tfrac{q^2}{r^2}\, cb, & ad - da &= \frac{r^2-1}{q}\, bc, \end{aligned} \tag{8}$$

which we call the quantum group $GL_{r^2;q}(2)$ (or $GL_{p,q}(2)$ where $p \equiv r^2/q$).

A second quantum plane of exterior variables ξ, η ($\xi^2 = \eta^2 = 0$) is also defined that can be understood as the differentials of x and y [5, 7]:

$$\xi\eta = -\frac{q}{r^2}\,\eta\xi . \tag{9}$$

Unless the $\hat{R}$-matrix for the two-parameter deformation is similar to that of the standard one-parameter deformation

$$\hat{R}_{r;q} = S^{-1}\hat{R}_r S , \tag{10}$$

$S = diag(1, \sqrt{r/q}, \sqrt{q/r}, 1)$, we have to remember that $\hat{R}$ (or R) acts in the tensor square of the quantum space (group) algebra. S, however, does not decompose into a tensor product, $S \neq S_0 \otimes S_0$, thus the quantum groups are inequivalent.

The quantum determinant $\mathcal{D} = ad - p\,bc$ is in general not central and hence before one can demand unimodularity one has to fix $r = q$ in order to arrive at $SL_q(2)$.[2]

Since the entries of the quantum matrix T are no more numbers we have to define a conjugation (antilinear involution) in order to get real forms. There are, however, different possibilities. In two dimensions we can have:

$$\begin{aligned} &(a) \quad \overline{x} = y, \quad \overline{y} = x , \\ &(b) \quad \overline{x} = x, \quad \overline{y} = y . \end{aligned} \tag{11}$$

conjugation of the plane relations yields:

$$xy = qxy \longrightarrow \begin{aligned} &(a) \quad xy = \overline{q} yx , \\ &(b) \quad xy = \tfrac{1}{\overline{q}} yx . \end{aligned} \tag{12}$$

For the group parameters conjugation has to be:

$$\begin{aligned} &(a) \quad \overline{T} = STS = \left(\begin{smallmatrix} d & c \\ b & a \end{smallmatrix}\right), \qquad S = \left(\begin{smallmatrix} 0 & 1 \\ 1 & 0 \end{smallmatrix}\right), \\ &(b) \quad \overline{T} = T . \end{aligned} \tag{13}$$

[2]Manin in [8] and Burdik and Hlavaty in [9] are obviously mistaken in talking of "$SL_{p,q}(2)$". For supergroups, however, there is a two parameter $SL_{p,q}(1|1)$ possible due to the nilpotencies [10].

Consistency with (6) gives

$$\begin{aligned} &(a) \quad q, r \in \mathbf{R} \setminus \{0\}, \\ &(b) \quad q = e^{i\vartheta},\ r^2 = e^{i\Theta}, \quad \vartheta, \Theta \in \mathbf{R}, \end{aligned} \tag{14}$$

i.e. both deformation parameters are either real or pure phases. These groups may be denoted as $GL_{r^2;q}(2,\mathbf{R})$ and $GL_{\Theta;\vartheta}(2,\mathbf{R})$, respectively. For $SL(2,\mathbf{R})$ we need again $r = q$.

For $U(2)$ we set

$$T^+ = T^{-1} \tag{15}$$

giving the group $U_{r^2;\vartheta}(2)$, where $q = r\,e^{i\vartheta}$, and $SU_r(2)$, where $q = r \in \mathbf{R}$, with two and one real deformation parameters, respectively. For $n > 2$ see [7, 11].

2 EUCLIDIAN AND LORENTZ SYMMETRY

In turning to the question of quantizing the Lorentz group we are, roughly speaking, confronted with the problem to find a R-matrix solution of the Yang-Baxter equation that also obeys relations from orthogonality and reality and elaborating the 'generator content' of the appropriate L-matrices. One can, however, try to exploit the classical isomorphy of the Lorentz group

$$SO(3,1) \sim SL(2) \times \overline{SL(2)}\,. \tag{16}$$

This idea has been worked out by Carow-Watamura, Schlieker, Scholl und Watmura [12], we will refer to.[3] To approach the problem pedagogically let us first look at the isomorphy

$$SO(4,\mathbf{R}) \sim SU(2) \times \widetilde{SU(2)} \tag{17}$$

where the tilde reminds us that in the quantized case one deals with two independent copies of $SU_q(2)$. The $SO_q(4,\mathbf{R})$ quantum space and group variables are given by

$$\mathbf{x}^{ii'} = x^i \otimes x^{i'}, \tag{18}$$

$$\mathbf{T}^{ii'}{}_{kk'} = T^i{}_k \otimes \tilde{T}^{i'}{}_{k'}, \tag{19}$$

and the corresponding R-matrix is

$$\mathcal{R}^{ii'jj'}{}_{kk'll'} = R^{ij}{}_{kl} \otimes R^{i'j'}{}_{k'l'}\,. \tag{20}$$

Note, the isomorphy used here does not hold in general, in the multiparametric case it breaks down, e.g. for $SO_{p,q}(4,\mathbf{R})$ [11]:

$$R(SO_{r^2;q}(4)) \neq R(SU_p(2)) \otimes R(SU_{p'}(2)) \tag{21}$$

for all p, p'.

[3] Implicitly, this idea appeared already in [13] where the discussed quantum double construction can be understood as product of two conjugate algebras leading to an equation similar to (28).

$\mathcal{R}$ obeys the YBE trivially since the tensor product decouples. Graphically, the braids corresponding to the unprimed indices do not "see" those of the primed ones [12]:

$$\hat{R}^{ij}{}_{kl} =$$

$$, \quad \hat{\mathcal{R}}^{ii'jj'}{}_{kk'll'} =$$

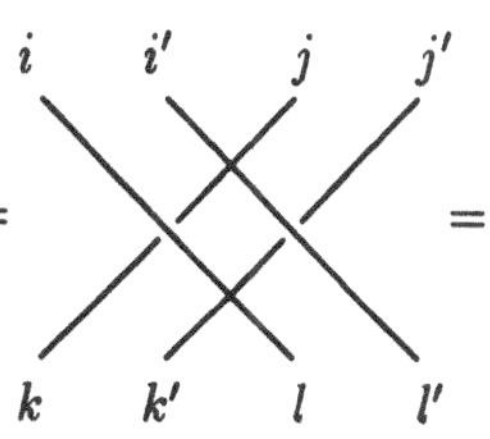

$$= \hat{R}^{ij}{}_{kl}\,\hat{R}^{i'j'}{}_{k'l'} \qquad (22)$$

The deformed Lie algebra can be constructed from duality [2]:

$$\langle L^{\pm}|T\rangle = R^{\pm}\,, \qquad R^{+} = PRP, \quad R^{-} = R^{-1}\,. \qquad (23)$$

Here, we simply find:

$$T = t \otimes \tilde{t}\,, \qquad L^{\pm} = \tilde{l}^{\pm} \otimes l^{\pm}\,, \qquad (24)$$

or

$$|T\rangle = |t\rangle|\tilde{t}\rangle\,, \qquad \langle L^{\pm}| = \langle\tilde{l}^{\pm}|\langle l^{\pm}|\,, \qquad (25)$$

then

$$\langle L^{\pm}|T\rangle = \langle\tilde{l}^{\pm}|\langle l^{\pm}|t\rangle|\tilde{t}\rangle = R^{\pm} \otimes R^{\pm} = \mathcal{R}^{\pm} \qquad (26)$$

where small letters refer to the $SU(2)$ groups and algebras.

Drinfeld and Jimbo showed that the algebra deformation (3) of Kulish and Reshetikhin generalizes for simple Lie algebras with generators $T_i^{\pm}, H_i (i = 1...rank(g))$ [14, 15]. This, however, does not apply anymore for the Lorentz group since the generators of the two $SU(2)$ subalgebras are related by conjugation and do not commute.[4] For the same reason the pairing duality (23) does not anymore decouple into a simple tensor product.

Commutation relations among x^i and $\overline{x}^j$ have to be introduced. This can be done with the help of the R-matrix again:

$$x^i \overline{x}^j = q\, R^{ij}{}_{kl} \overline{x}^k x^l\,. \qquad (27)$$

Now the crossings of primed and unprimed lines in the pictorial representation also give R-matrices [12]:

$$\hat{\mathcal{R}}^{ii'jj'}{}_{kk'll'} =$$

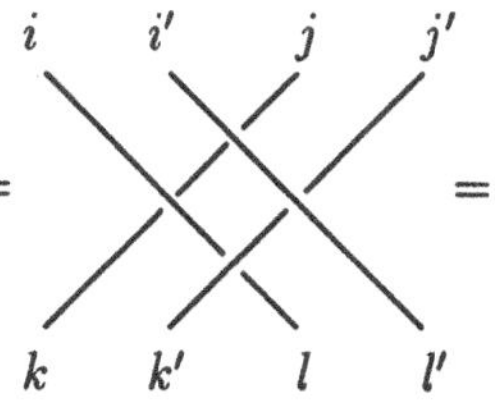

$$= \hat{R}^{i'j}{}_{I'J}\hat{R}^{iJ}{}_{kI}\hat{R}^{I'j'}{}_{J'l'}\hat{R}^{-1\,IJ'}{}_{k'l} \qquad (28)$$

Here, summation goes over capital letters. The YBE is satified since the primed braids lay above the unprimed ones.

If we try to exploit duality in the Lorentz case and attempt to reduce the relations of the generators to those for the $SL(2)$ algebras, we first set

$$T = t \otimes \bar{t}\,, \qquad \langle L^{\pm}|T\rangle = \mathcal{R}^{\pm}\,, \qquad (29)$$

[4]Celeghini *et.al.* recently demonstrated that Lorentz signature can also be archieved with the standard $SO(4)$ R-matrix [16]; here, however, no rotation subgroup is present.

and demand, using the $sl_q(2)$ generators l,

$$L^{\pm} = \mathbf{X} \cdot (\bar{l}^{\pm} \otimes l^{\pm}) \tag{30}$$

i.e.

$$L^{\pm ii'}{}_{ll'} = X^{\pm ii'}{}_{ll'}{}^{bb'}{}_{aa'} \, \bar{l}^{\pm a}{}_{b} \, l^{\pm a'}{}_{b'} \,. \tag{31}$$

Calculating the components of $\mathbf{X}$, s.t. (29) is obeyed, one finds:

$$X^{ii'}{}_{ll'}{}^{bb'}{}_{aa'} = \hat{R}^{-1\,kb}{}_{aj} \hat{R}^{-1\,k'b'}{}_{a'j'} \widehat{\mathcal{R}}^{ii'jj'}{}_{kk'll'} \,. \tag{32}$$

Unfortunately, $\mathbf{X}$ cannot be interpreted and simplified by pictorial means but gives a mess of lines!

Despite the fact that the Lorentz algebra has been found by a different method that we will refer to later, here a simple argument shall be given to illuminate what happens. The duality (23) can also be used differently: The pairing of L and T gives a representation of the Lie algebra generators L on the quantum group variables T:

$$(L^{I}{}_{L})^{J}{}_{K} = \widehat{\mathcal{R}}^{IJ}{}_{KL} \tag{33}$$

where $I = (ii')$ etc. In our case the representation is by 4×4 matrices that can easily be read off from $\mathcal{R}$. Some of the generators vanish and 18 linear independent matrices are present. Direct inspection shows that in the classical limit some of the matrices coincide or vanish, while in the deformed case there are obvious relations among them. Assuming that these matrices come from the square subspace of a more fundamental algebra as required in (30) one easily finds that a number of seven independent generators suffices. The main point in our argument is now to note that there is a decomposition of the $L^{\pm}$ matrices that replaces the tensor product structure of the $SO(4, \mathbf{R})$ case:

$$PL^{+}P = \begin{pmatrix} q^{H} & \lambda J^{+} \\ 0 & q^{-H} \end{pmatrix} \otimes \begin{pmatrix} q^{M} & \lambda K^{+} \\ \lambda K^{-} & q^{N} \end{pmatrix} \quad \text{or} \quad L^{+}_{21} = \mathcal{L}^{+}_{r} \otimes \mathcal{L}_{b} \,, \tag{34}$$

$$PL^{-}P = \begin{pmatrix} q^{M} & \lambda K^{+} \\ \lambda K^{-} & q^{N} \end{pmatrix} \otimes \begin{pmatrix} q^{-H} & 0 \\ \lambda J^{-} & q^{H} \end{pmatrix} \quad \text{or} \quad L^{-}_{21} = \mathcal{L}_{b} \otimes \mathcal{L}^{-}_{r} \,, \tag{35}$$

where $\otimes$ denotes the Kronecker product. This decomposition in particular reveals substructures of the comultiplication, i.e. it is canonical for $\mathcal{L} = \mathcal{L}^{\pm}_{r}\,, \mathcal{L}_{b}$:

$$\Delta\mathcal{L} = \mathcal{L}\dot{\otimes}\mathcal{L} \,, \tag{36}$$

where $\dot{\otimes}$ denotes a combination of matrix multiplication and tensoring within the resulting matrix. It can be verified that this decomposition is independent of the representation using the general seven parameter deformation of Schmidke, Wess and Zumino [17]: By consideration of the action of the generators on spinors with help of the ansatz

$$Tx^{i} = x^{i}T + a^{i}_{j}x^{j} \tag{37}$$

the $\mathcal{L}^{\pm}_{r}$ generators $J^{\pm}$ and H were found giving the rotation part, while the more general ansatz

$$T^{a}x^{i} = B^{ai}{}_{bj}x^{j}T^{b} + A^{ai}{}_{j}x^{j} \tag{38}$$

yields a four generator algebra isomorphic to that of $\mathcal{L}_{b}$ with generators $K^{\pm}$, M and N (a, A, B numbers). These generators mix boosts with rotations as can be seen from their action on spinors.

In this picture of the decompositions (34) and (35) the reduction of the seven generator algebra to the six physical generators is now becoming very transparent by the following observation. $\mathcal{L}_r^{\pm}$ and $\mathcal{L}_b$ can be understood as being quantum matrices themselves! The matrices of the rotation generators obey $SL_{1/q^2}(2)$ commutation relations as that of the boost part is a $GL_{q^2}(2)$ quantum matrix. While the determinant of the rotation part matrices is trivially unity, for the boost part matrix we learn immediately from

$$det_{q^2}\,\mathcal{L}_b = q^M q^N - q^2\lambda^2 K^+K^- \tag{39}$$

that this quadratic combination of generators is central in $\mathcal{L}_b$. Moreover it also commutes with the rotation generators and hence does not belong to the Lorentz algebra. Requiring

$$q^M q^N - q^2\lambda^2 K^+K^- = 1 \tag{40}$$

consistently eliminates this spurious generator. (It was a phase transformation on spinors not affecting the four-vector [17].)

The reason that one has to face seven generators first is now obvious. The constraint (40) is not a linear dependence of generators and eliminating one of the generators by this relation results in higher order terms as outcome of the Lie brackets, i.e. if we understand deformed Lie algebra relations as

$$[T^i, T^j]_{q_{ij}} = f^{ij}{}_k T^k \tag{41}$$

with the q-commutator $[A,B]_q = qAB - q^{-1}BA$ we have to deal with the relation (40) separately as with a constraint[5].

Clearly, the question how a quantum deformation is described best — with L-generators (2), q-brackets (4) or q-commutators (7) — is not answered yet and we also do not know whether there are other deformations of the Lorentz Symmetry meeting our understanding of a quantized space-time.

Acknowledgements

The author would like to thank J. Wess for suggesting the topic of this talk and is indebted to L. D. Faddeev for stimulating remarks and for references.

References

[1] G. Mack, V. Schomerus: *Quasi Hopf symmetry in quantum theory*, DESY preprint 91-037;
A. Alekseev, L. Faddeev: *The unravelling of the quantum group structure in the WZNW theory*, CERN preprint TH–5981/91.

[2] N. Yu. Reshetikhin, L. A. Takhtadzhyan, L. D. Faddeev,: *Quantization of Lie groups and Lie algebras*, Leningrad Math. J. 1, 193–225, (1990).

[5]In [18], however, a chiral decomposition of the six generator algebra has been found obeying q-commutator relations; but the spinor action is then non-linear and as in the presented case no simple relation by conjugation between the two (chiral) subalgebras is present.

[3] P. P. Kulish, N. Yu. Reshetikhin: *Quantum linear Problem for the Sine-Gordon equation and higher representations*, LOMI **101**, 101–110, (1981).

[4] A. Schirrmacher, J. Wess, B. Zumino: *The two-parameter deformation of GL(2) its differential calculus, and Lie algebra*, Z. Phys. C **49**, 317, (1991).

[5] J. Wess, B. Zumino: *Covariant differential calculus on the quantum hyperplane*, Nucl. Phys. B (Proc. Suppl.)**18B**, 302-312, (1990).

[6] S. L. Woronowicz: *Twisted SU(2) group. An example of noncommutative differential calculus*, Publ. Res. Inst. Math. Sci. **23**, 117–181, (1987).

[7] A. Schirrmacher: *The multiparameter deformation of GL(n) and the covariant differential calculus on the quantum vector space*, Z. Phys. C **50**, 321, (1991).

[8] Yu. I. Manin: *Notes on quantum groups and quantum de Rahm complexes*, Bonn preprint MPI/91–60 (1991).

[9] C. Burdik, L. Hlavaty : *A two parametric quantization of sl(2)*, J. Phys. A **24**, L165–168, (1991).

[10] L. Dabrowski, L. Wang: *Two-parameter quantum deformation of* $GL(1|1)$, Phys. Lett. B **266**, 51–54, (1991).

[11] A. Schirrmacher: *Multiparameter R-matrices and their quantum groups*, preprint MPI-Ph/91-24 to appear in J. Phys. A.

[12] U. Carow-Watamura, M. Schlieker, M. Scholl, S. Watamura: *A quantum Lorentz group*, Int. J. Mod. Phys. A **6**, 3081–3108, (1991).

[13] N. Yu. Reshetikhin, M. A. Semenov-Tian-Shansky: *Quantum R-matrices and factorization problems*, J. Geom. Phys. **5**, 533–550, (1988).

[14] V. G. Drinfel'd: *Quantum groups*, Proc. Int. Congr. Math. (Berkeley), **1**, Academic Press, New York, 798–820, (1986).

[15] M. Jimbo: *Quantum R-matrix for the generalized Toda system*, Commun. Math. Phys. **101**, 537–547, (1986).

[16] E. Celeghini, R. Giachetti, A. Reyman, E. Sorace, M. Tarlini: $SO_q(n+1, n-1)$ *as a real form of* $SO_q(2n, C)$, preprint Firenze DFF 138/6/91.

[17] W. B. Schmidke, J. Wess, B. Zumino; *A q-deformed Lorentz algebra*, preprint MPI-Ph/91-15.

[18] O. Ogievetsky, W. B. Schmidke, J. Wess, B. Zumino: *Six generator q-deformed Lorentz algebra*, preprint MPI-Ph/91-51.

FOCK SPACE REPRESENTATIONS OF $A_1^{(1)}$ AND TOPOLOGICAL REPRESENTATIONS OF $U_q(sl_2)$

G. Felder[1] and C. Wieczerkowski[2]

[1] Mathematik, ETH-Zentrum, CH-8092 Zürich
[2] Institut für Theoretische Physik 1, WWU, D-4400 Münster

ABSTRACT

We apply topological representations of $U_q(sl_2)$ to the Fock space representations of the untwisted affine Kac-Moody algebra $A_1^{(1)}$. We show how singular vectors in quantum group Verma modules determine Fock space representations of BRST operators, primary fields, and conformal blocks.

1. FOCK SPACE REPRESENTATIONS OF $A_1^{(1)}$

Let us recall the basic facts about Fock space representations of $A_1^{(1)}$. The algebra $A_1^{(1)}$ is generated by J_n^a, $a \in \{+,-,0\}$ and $n \in \mathbf{Z}$, and a central element K with relations

$$\begin{aligned} [J_n^0, J_m^\pm] &= \pm J_{n+m}^\pm, \\ [J_n^0, J_m^0] &= \frac{n}{2}\delta_{n,-m}K, \\ [J_n^+, J_m^-] &= 2J_{n+m}^0 + n\delta_{n,-m}K, \\ [K, J_n^a] &= 0, \end{aligned} \tag{1}$$

supplemented by a derivation d satisfying

$$\begin{aligned} [d, J_n^a] &= -nJ_n^a, \\ [d, K] &= 0. \end{aligned} \tag{2}$$

Following [6, 7, 8] we require a bosonic $\omega - \omega^+$ system together with a free bosonic field ϕ. Let Γ^1 be the algebra generated by ω_n and ω_n^+, $n \in \mathbf{Z}$, with relations

$$\begin{aligned} [\omega_n, \omega_m^+] &= \delta_{n,-m}, \\ [\omega_n, \omega_m] &= [\omega^+, \omega_m^+] = 0, \end{aligned} \tag{3}$$

and F^1 the Fock space with vacuum vector v^1 such that, for $0 \le n$,

$$\omega_{n+1}v^1 = \omega_n^+ v^1 = 0. \tag{4}$$

New Symmetry Principles in Quantum Field Theory, Edited by
J. Fröhlich et al., Plenum Press, New York, 1992

Let Γ^2 be the algebra generated by a_n, $n \in \mathbf{Z}$, with relations

$$[a_n, a_m] = n\delta_{n,-m}, \tag{5}$$

and $F^2_{J,k}$ the Fock space of charge $\frac{J}{\gamma}$, $\gamma = \sqrt{\frac{k+2}{2}}$, generated from a vacuum vector $v^2_{J,k}$ satisfying, for $0 \leq n$,

$$a_n v^2_{J,k} = \frac{J}{\gamma}\delta_{n,0} v^2_{J,k}. \tag{6}$$

Then define $\Gamma = \Gamma^1 \otimes \Gamma^2$ and $F_{J,k} = F^1 \otimes F^2_{J,k}$. This tensor product of Fock spaces is made a highest weight module over $A_1^{(1)}$ at level k through the following construction due to [6]. Let us introduce field operators

$$\begin{aligned} \omega(z) &= \sum_{n=-\infty}^{\infty} \omega_n z^{-n}, \\ \omega^+(z) &= \sum_{n=-\infty}^{\infty} \omega_n^+ z^{-n-1}, \\ j(z) &= \sum_{n=-\infty}^{\infty} a_n z^{-n-1}, \end{aligned} \tag{7}$$

and $\phi(z)$ such that $j(z) = i\partial\phi(z)$. In terms of these we can build currents

$$\begin{aligned} J^+(z) &= \omega^+(z) \otimes id, \\ J^0(z) &= :\omega(z)\omega^+(z): \otimes id + \gamma\, id \otimes j(z), \\ J^-(z) &= -\left(:\omega(z)^2\omega^+(z): + k\,\partial\omega(z)\right) \otimes id - 2\gamma\,\omega(z) \otimes j(z). \end{aligned} \tag{8}$$

Then it can be shown that the coefficients of the developments

$$J^a(z) = \sum_{-\infty}^{\infty} J^a_n z^{-n-1} \tag{9}$$

satisfy the relations (1). The free field stress energy tensor

$$T(z) = -:\partial\omega(z)\omega^+(z): \otimes id + \frac{1}{2} id \otimes \left(:j(z)^2: - \frac{1}{\gamma}\partial j(z)\right) \tag{10}$$

agrees with the Sugawara form. Expanding, as usual,

$$T(z) = \sum_{n=-\infty}^{\infty} L_n z^{-n-2} \tag{11}$$

$F_{J,k}$ also becomes a highest weight module over the Virasoro algebra *Vir* with central charge $c = \frac{3k}{k+2}$. The generators of $A_1^{(1)}$ have the explicit form

$$\begin{aligned} J_n^+ &= \omega_n^+ \otimes id, \\ J_n^0 &= \sum_{m=-\infty}^{\infty} :\omega_m \omega^+_{n-m}: \otimes id + \gamma \otimes a_n, \\ J_n^- &= -\left(\sum_{m,l=-\infty}^{\infty} :\omega_m \omega_n \omega^+_{n-m-l}: - kn\,\omega_n\right) \otimes id - 2\gamma \sum_{m=-\infty}^{\infty} \omega_m \otimes a_{n-m}. \end{aligned} \tag{12}$$

The generators with $n = 0$ generate an sl_2 subalgebra. $F_{J,k}$ is, in particular, a module over sl_2. The vacuum is also an sl_2 highest weight vector with

$$\begin{aligned} J_0^+ v_{J,k} &= 0, \\ J_0^0 v_{J,k} &= J v_{J,k}. \end{aligned} \tag{13}$$

As an sl_2-module, $F_{J,k}$ thus contains a highest weight representation with spin J generated from the vacuum. This representation is irreducible since $(J_0^-)^{2J+1} = 0$.

Basic objects in the following constructions are vertex operators, which we take to be defined as

$$V_{I,k}(z) = \exp\left(\frac{I}{\gamma} b_0\right) \exp\left(\frac{I}{\gamma} \ln(z) a_0\right) \exp\left(\frac{I}{\gamma} \sum_{n=1}^{\infty} a_{-n} \frac{z^n}{n}\right) \exp\left(-\frac{I}{\gamma} \sum_{n=1}^{\infty} a_n \frac{z^{-n}}{n}\right). \tag{14}$$

Here $[a_n, b_0] = \delta_{n,0}$ and we have an operator from $F^2_{J,k}$ to $F^2_{J+I,k}$. To be precise, $V_{J,k}(z) =: \exp(i\frac{I}{\gamma}\phi(z))$: defines a bilinear form on $(F^2_{J+K,k})^* \otimes F^2_{J,k}$ (restricted dual). Finally let

$$V_{I,k}(z) = \tilde{V}_{I,k}(z) z^{I\gamma^{-1} a_0}, \tag{15}$$

splitting off the factor which depends on the zero mode of $j(z)$.

2. INTERTWINERS

As a first application of the topological representations, we will construct operators mapping one Fock space to another which intertwine the representations of $A_1^{(1)}$ and, consequently, *Vir*. As a basic ingredience, introduce

$$V^-(z) = \omega^+(z) \otimes V_{-1,k}(z), \tag{16}$$

the screening operator. $V^-(z)$ maps $F_{J,k}$ to $F_{J-1,k}$. The operator product expansion

$$T(z)V^-(w) = \frac{1}{(z-w)^2} V^-(w) + \frac{1}{z-w} \frac{\partial}{\partial w} V^-(w) + O(1), \tag{17}$$

where $O(1)$ sums up regular terms, shows that V^- has conformal weight 1. The operator product expansions of V^- with the currents have the form

$$\begin{aligned} J^+(z)V^-(w) &= O(1) \\ J^-(z)V^-(w) &= 2\gamma^2 \frac{\partial}{\partial w} \{\frac{1}{z-w} id \otimes V_{-1,k}(w)\} \\ J^0(z)V^-(w) &= O(1) \end{aligned} \tag{18}$$

They follow from basic operator product expansions of the free field, diverging terms cancelling each other neatly. Due to the total derivative, $V^-(z)$ does not intertwine the action of $A_1^{(1)}$ on $F_{J,k}$ with that on $F_{J-1,k}$. To produce intertwiners we have to integrate products of screening operators. This gives

$$\begin{aligned} Q_C^R &= \int_C V^-(z_1) \dots V^-(z_R)\, dz_1 \dots dz_R \\ &= \int_C : \tilde{V}^-(z_1) \dots \tilde{V}^-(z_R) : \prod_{i=1}^{R} z_i^{-\gamma^{-1} a_0} \prod_{1 \le i < j \le R} (z_i - z_j)^{\gamma^{-2}}\, dz_1 \dots dz_R, \end{aligned} \tag{19}$$

a map from $F_{J,k}$ to $F_{J-R,k}$.

Taking matrix elements, these integrals give rise to topological representations of $U_q(sl_2)$ at $q = \exp(\frac{\pi i}{2\gamma^2})$, possibly a root of 1. See [10]. Let $X_R(0) = (D\backslash\{0\})^R \backslash \cup_{1\leq i\leq j\leq R} \{z_i = z_j\}/S_R$ be the configuration space of R indistinguishable screening charges in the disc $D = \{|z| \leq 1\}$, punctured at 0. Fix a point p_- on the boundary of D, e.g., $p_- = -1$. Let L_R be the local system determined by the monodromy of the multi valued form on X_R obtained by evaluating the integrand on Fock vacua. Note that different values of J yield different L_R.

Let $A_R(0)$ be the space of linear combinations of families Γ of non-intersecting loops in X_R based at p_- together with sections of $\Gamma^* L_R$ modulo equivalence relations reflecting the possibility to homotopically deform or reparametrize Γ.

Let $\hat{E} : A_R(0) \to A_{R-1}(0)$, $\hat{F} : A_R(0) \to A_{R+1}(0)$, and $\hat{K}^2 : A_R(0) \to A_R(0)$ be the topological operators introduced in [10]. They satisfy the $U_q(sl_2)$ relations. $\hat{E}$ is the combinatorial boundary operator composed with a map which identifies $(R-1)$-chains in X_R with $(R-1)$-chains in X_{R-1}. We find

$$\begin{aligned} \hat{K}^2\hat{E} &= q^2\hat{E}\hat{K}^2 \\ \hat{K}^2\hat{F} &= q^{-2}\hat{F}\hat{K}^2 \\ [\hat{E},\hat{F}] &= \hat{K}^2 - \hat{K}^{-2}, \end{aligned} \tag{20}$$

endowing $\bigoplus_{R=0}^{\infty} A_R(0)$ with the structure of a module over the quantum group algebra $U_q(sl_2)$.

In the applications we have in mind, $q^p = 1$. In this case, $\hat{E}^p = 0$ and $\hat{K}^{4p} = 1$. This is proved by an explicit computation using a basis to describe $A_R(0, w)$ as a space. Families of loops which contain p homotopic deformations of a single loop represent null homologous cycles. To prove this one retracts the p loops to p particles on the single loop path ordering the arguments. The observation is that the prefactor entering through the path ordering vanishes whenever one has p particles. This can be taken into account by adding another relation in the definition of the space $A_R(0)$. This relation puts families of loops which contain p homotopic loops equivalent to the null family. As a consequence, also $\hat{F}^p = 0$. Using a generalized version of Poincare duality this could possibly be understood in terms of the kernel of the topological intersection pairing.

For q a root of unity the conclusion is that we find a representation of the reduced quantum group algebra $U_q^{\rm red}(sl_2)$, the algebra obtained from $U_q(sl_2)$ dividing by the ideal generated by the central elements E^p, F^p, and $K^{4p} - 1$. Let us only consider this case in the following. For a detailed account on $U_q^{\rm red}(sl_2)$ we recommend [11].

As a quantum group module, $\bigoplus_{R=0}^{p-1} A_R(0)$ is isomorphic to the $U_q^{\rm red}(sl_2)$ Verma module $V(n)$, $n = 2J + 1$, generated from a singular vector $v_0(n)$ such that

$$\begin{aligned} Ev_0(n) &= 0 \\ K^2 v_0(n) &= q^{n-1} v_0(n). \end{aligned} \tag{21}$$

Let $\varphi : \bigoplus_{R=0}^{p-1} A_R(0) \to V(n)$ be the isomorphism. For $0 \leq R \leq p-1$, we define weight spaces $V(n)_R = \mathbf{C}F^R v_0(n)$. K^2 acts on $V(n)_R$ by multiplication with q^{n-1-2R}.

Restricting our attention to the most interesting case, let $2\gamma^2 = k+2 = p$ be a positive integer and let $n \in \mathbf{Z}$. The BRST construction of [7, 8] then produces unitary integrable irreducible highest weight representations of $A_1^{(1)}$ with $1 \leq n \leq p-1$, $n = 2J+1$, on the homology. Here we work directly with the representations on the Fock spaces.

Theorem 2.1 *(1) Consider $Q^{(R)}$ as a map from $V(n)$ to the space of linear operators*

$\mathrm{Hom}_{\mathbf{C}}(F_{J,k}, F_{J-R,k})$. *Then*

$$Q^{(R)} : \ker(E : V(n)_R \to V(n)_{R-1}) \to \mathrm{Hom}_{A_1^{(1)}}(F_{J,k}, F_{J-R}) \tag{22}$$

maps to the space of $A_1^{(1)}$ *intertwiners. (2) Let* $n = 2J+1$ *and* $\bar{n} = n\,mod$, $0 \leq \bar{n} \leq p-1$. *For* $\bar{n} = 0$,

$$\ker(E) = \mathbf{C}v_0(n). \tag{23}$$

For $1 \leq \bar{n} \leq p-1$,

$$\ker(E) = \mathbf{C}v_0(n) \oplus \mathbf{C}F^{\bar{n}}v_0(n). \tag{24}$$

(3) For $1 \leq \bar{n} \leq p-1$, *the only non-vanishing intertwiner besides the identity is* $Q_C^{(R)}$ *with* $R = \bar{n}$ *and* $C = \varphi^{-1}(F^{\bar{n}}v_0(n))$, *unique up to a normalization constant.*

Thus the structure of singular vectors in $U_q(sl_2)$-Verma modules, in our case a root of 1, provides all data needed to decide wether intertwiners exist and, furthermore, gives explicit formulas for them.

Corollary 2.2 *(1) For* $n = 2J+1$, *let* $F_n = F_{J,k}$. *For* $1 \leq \bar{n} \leq p-1$, *the non-vanishing intertwiners form an infinite sequence*

$$\ldots \to F_{2p-\bar{n}} \overset{Q^{(p-\bar{n})}}{\to} F_{\bar{n}} \overset{Q^{(\bar{n})}}{\to} F_{-\bar{n}} \to \ldots \tag{25}$$

(2) This sequence is a complex.

The homology of this complex is isomorphic to the irreducible $A_1^{(1)}$ highest weight module of weight J and level k. Further intertwiners can be constructed out of cycles with np path ordered arguments, $n \in \{1, 2, \ldots\}$. They correspond to powers of generators $F^p[p]!^{-1}$ in Lusztig's version of $U_q(sl_2)$. They can be used to construct complexes whose homology gives non-integrable $A_1^{(1)}$ modules [14]. Here $[n] = q^n - q^{-n}$ and $[n]! = [1]\ldots[n]$.

3. CHIRAL PRIMARY FIELDS

The second application of topological representations concerns the construction of chiral primary fields $\phi_{J,I}^K(w) : F_{J,k} \otimes \tilde{U}_I \to F_{K,k}$. $\tilde{U}_I$ is a representation of $A_1^{(1)}$ at level 0 to be defined below. Consider operators

$$\psi_{I,\mu}(z) =: \omega(z)^{I-\mu} : \otimes V_{I,k}(z) \tag{26}$$

mapping $F_{J,k}$ to $F_{J+I,k}$. The operator product expansion with $T(z)$

$$T(z)\psi_{I,\mu}(w) = \frac{\frac{I(I+1)}{2\gamma^2}}{(z-w)^2}\psi_{I,\mu}(w) + \frac{1}{z-w}\partial\psi_{I,\mu}(w) + O(1) \tag{27}$$

shows that $\psi_{I,\mu}(z)$ has conformal weight $h_I = \frac{I(I+1)}{2\gamma^2}$. The operator product expansion with the current $J^a(z)$ has the form

$$J^a(z)\psi_{I,k}(w) = \frac{1}{z-w}\sum_{\nu=-I}^{I} \psi_{I,\nu}(w) D_{\nu,\mu}^{(I)}(J^a) + O(1). \tag{28}$$

$D^{(I)}(J^a)$ is a spin I representation matrix of sl_2. For $u \in U_I$, the representation space of the representation $D^{(I)}$, define

$$\psi_I(u; w) = \sum_{\mu=-I}^{I} \psi_{I,\mu}(w) u_\mu. \tag{29}$$

The operator product expansion implies that

$$[J_n^a, \psi_I(u;w)] = \psi\left(D^{(I)}(J^a)u; w\right) w^n. \tag{30}$$

With these preparations in mind let us consider the operators

$$\phi_{I,\mu}(w)_C^{(R)} = \int_C \psi_{I,\mu}(w) V^-(z_1)\dots V^-(z_R) dz_1 \dots dz_R \tag{31}$$

from $F_{J,k}$ to $F_{K,k}$ with $K = J + I - R$. Using Wick's theorem

$$:\omega(w)^{I-\mu} : \prod_{i=1}^{R} \omega^+(z_i) = \sum_{\nu=0}^{min\{I-\mu,R\}} C(I-\mu,\nu) \sum_{\substack{\mathcal{I} \subset \{1,\dots,R\} \\ |\mathcal{I}| = \nu}} \; : \omega(w)^{I-R-\mu+\nu} \prod_{i\in\mathcal{I}} \omega^+(z_i) : \prod_{i\in\{1,\dots,R\}\setminus\mathcal{I}} (z-w)^{-1} \tag{32}$$

we normal order the integrand. $C(I-\mu,\nu)$ is a combinatorial factor counting the number of contractions. The result is

$$\begin{aligned}\phi_{I,\mu}(w)_C^{(R)} = & \int_C \sum_{\nu=0}^{min\{I-\mu,R\}} C(I-\mu,\nu) \sum_{\substack{\mathcal{I} \subset \{1,\dots,R\} \\ |\mathcal{I}| = \nu}} \\ & : \omega(w)^{I-R-\mu+\nu} \prod_{i\in\mathcal{I}} \omega^+(z_i) : \otimes : \tilde{V}_{I,k}(w)\tilde{V}_{-1,k}(z_1)\dots\tilde{V}_{-1,k}(z_R) : \\ & w^{I\gamma^{-1}a_0} \prod_{i=1}^{R} z^{-\gamma^{-1}a_0} \prod_{i\in\{1,\dots,R\}\setminus\mathcal{I}} (w-z_i)^{-1} \prod_{i=1}^{R} (w-z_i)^{-I\gamma^{-2}} \\ & \prod_{1\le i<j\le R} (z_i - z_j)^{\gamma^{-2}} dz_1 \dots dz_R. \end{aligned} \tag{33}$$

The chain C is taken to be an element of $A_R(0,w)$. The local system $L_R(0,w)$ involved in the definition of $A_R(0,w)$ is determined by the monodromy of the multi valued R-form obtained by evaluating the integrand on Fock vacua. As a $U_q(sl_2)$-module, the space $\bigoplus_{R=0}^{2p-2} A_R(0,w)$ turns out to be isomorphic to the tensor product of Verma modules $V(n) \otimes V(m)$. Let $(V(n)\otimes V(m))_R$ be the weight space on which K^2 acts by multiplication with $q^{n+m-2-2R}$.

Lemma 2.1 *Consider $\phi_I(w)^{(R)}$ as a map from $V(n)\otimes V(m)$ to the space of bilinear operators $\mathrm{Hom}_{\mathbf{C}}(F_{J,k}\otimes U_I, F_{K,k})$ with $K = J + I - R$. Here U_I is the sl_2 representation space. For $\varphi(C) \in \ker\left(E : (V(n)\otimes V(m))_R \to (V(n)\otimes V(m))_{R-1}\right)$, we find the commutation relation*

$$[J_n^a, \phi_I(u;w)_C^{(R)}] = \phi_I\left(D^{(I)}(J^a)u; w\right)_C^{(R)} w^n. \tag{34}$$

For C a chain, which represents an absolute cycle in homology, we obtain a chiral primary field. Its construction goes as follows. Expanding

$$\phi_I(u;w)_C^{(R)} = \sum_{n=-\infty}^{\infty} \phi_{I,n}(u)_C^{(R)} w^{-n+h_K-h_I-h_J} \tag{35}$$

it follows that

$$[J_m^a, \phi_{I,n}(u)_C^{(R)}] = \phi_{I,m+n}\left(D^{(I)}(J^a)u\right)_C^{(R)}. \tag{36}$$

Let

$$\tilde{U}_I = U_I \otimes w^{-h_K+h_I+h_J}\mathbf{C}[w, w^{-1}] \tag{37}$$

be the $A_1^{(1)}$-module with J_n^a acting as $D^{(I)}(J^a) \otimes w^n$, d as $w\frac{\partial}{\partial w}$, and K as zero. Then

$$\xi \otimes \left(u \otimes w^{n-h_K+h_I+h_J}\right) \rightarrow \phi_{I,n}(u)_C^{(R)}\xi \tag{38}$$

defines a map from $F_{J,k} \otimes \tilde{U}_I$ to $F_{K,k}$. To be precise, for $u \in U_I$,

$$\phi_I(u; w)_C^{(R)} \in \mathrm{Hom}_{\mathbf{C}}(F_{J,k}, F_{K,k}) \otimes w^{-h_K+h_J+h_I}\mathbf{C}(w, w^{-1}), \tag{39}$$

is a formal power series in w whose coefficients are linear operators from the Fock space $F_{J,k}$ to the Fock space $F_{K,k}$. For $u \otimes \epsilon \in \tilde{U}_I$, we obtain a well defined operator through

$$\phi_I(u \otimes \epsilon)_C^{(R)} = \frac{1}{2\pi}\oint \phi_I(u; w)_C^{(R)}\epsilon(w)dw. \tag{40}$$

Here C is moved with w through the Gauss-Manin connection on $\mathbf{C} \setminus \{0\}$. That is, we associate operators to Laurent polynomials ϵ. This map is an $A_1^{(1)}$ homomorphism.

Theorem 2.2 *Consider $\phi_I^{(R)}$ as a map from $(V(n) \otimes V(m))_R$ to the space of bilinear operators $\mathrm{Hom}_{\mathbf{C}}(F_{J,k} \otimes \tilde{U}_I, F_{K,k})$ with $K = J + I - R$. Then*

$$\phi_I^{(R)} : \ker\left(E : (V(n) \otimes V(m))_R \rightarrow (V(n) \otimes V(m))_{R-1}\right) \rightarrow \mathrm{Hom}_{A_1^{(1)}}(F_{J,k}\otimes\tilde{U}_I, F_{K,k}) \tag{41}$$

maps to the space of $A_1^{(1)}$-intertviners.

The general decomposition of the tensor product of Verma modules $V(n) \otimes V(m)$, which contains among other things the structure of singular vectors, has been carried out in [12]. For the sake of brevity, it cannot be reproduced here. Let us only mention the following partial results.

Proposition 2.3 *Let $n = 2J + 1$, $m = 2I + 1$, $1 \leq n, m \leq p - 1$, and $0 \leq R < n, m$. Then*

$$\ker\left(E : (V(n) \otimes V(m))_R \rightarrow (V(n) \otimes V(m))_{R-1}\right) = \mathbf{C}\left(\sum_{l=0}^{R} x_l F^{R-l}v_0(n) \otimes F^l v_0(m)\right) \tag{42}$$

with x_l, $0 \leq l \leq R$, determined by $x_0 = 1$ and the recursion relation

$$[R - l]_q[n - R + l]_q x_l + q^{n-1-2(R-l-1)}[l + 1]_q[n - l - 1]_q x_{l+1} = 0, \tag{43}$$

where $[x]_q = q^x - q^{-x}$.

The intertwiner corresponding to this cycle is denoted by $\phi_{J,I}^K(u \otimes \epsilon)$ with $K = J + I - R$. It is unique up to a normalization constant which can be fixed as follows using the three point function

$$\langle v_{K,k,\lambda}, \phi_{J,I}^K(e_\mu; w)v_{J,k,\nu}\rangle = w^{h_K-h_I-h_J}N_{J,I}^K\begin{bmatrix} K & J & I \\ \lambda & \nu & \mu \end{bmatrix}, \tag{44}$$

involving a (classical) sl_2 Clebsch-Gordan coefficient and the fusion coefficient $N_{J,I}^K$. The three point function is conveniently normalized such that the fusion coefficient takes the values 0 or 1. $\phi_{J,I}^K(u \otimes \epsilon)$ is the chiral primary field in the Fock space representation.

Proposition 2.4 *Let $n = 2J+1$, $m = 2I+1$, and $l = 2K+1$ be such that $1 \leq n, m, l \leq p-1$. Define a triple $\begin{pmatrix} K \\ JI \end{pmatrix}$ to be admissible if $|I-J| \leq K \leq I+J$, $I+J+K \in \mathbf{Z}$, and $I+J+K \leq p-2$. Then the chiral primary field has the following properties: (1) For $\begin{pmatrix} K \\ JI \end{pmatrix}$ not admissible, $N^K_{J,I} = 0$. (2) For $\begin{pmatrix} K \\ JI \end{pmatrix}$ admissible, $N^K_{J,I} = 1$.*

The quantum group data encoded in the fusion rules of $\phi^K_{J,I}(w)$ is here inherited from the representation $V(n) \otimes V(m)$. The most important property of $\phi^K_{J,I}(w)$ is BRST invariance. It is a necessary condition to proceed through the BRST construction. The BRST invariance also has a cohomological meaning on the quantum group level. We plan to return to this point in a forthcoming paper [15].

Conformal blocks are vacuum expectation values of products of chiral primary fields. A product

$$\phi^{J_{s+1}}_{J_s,I_s}(u_s \otimes \epsilon_s) \ldots \phi^{J_3}_{J_2,I_2}(u_2 \otimes \epsilon_2)\phi^{J_2}_{J_1,I_1}(u_1 \otimes \epsilon_1) \tag{45}$$

defines an element of

$$\mathrm{Hom}_{A^{(1)}_1}(F_{J_1,k} \otimes \tilde{U}_{I_1} \otimes \ldots \otimes \tilde{U}_{I_s}, F_{J_{s+1},k}). \tag{46}$$

Using (33), such operators can be expressed in terms of free fields. Products of expressions of the form (33) are well defined (as absolutely convergent matrix product) and give rise to elements of (46) when integrated as in (40), provided the time ordering condition is satisfied. This condition states that : $U_1(z_1) \ldots U_p(z_p)$ :: $U_{p+1}(z'_1) \ldots U_{p+q}(z'_q)$: is well defined whenever $|z_i| \geq |z'_j|$ for all i, j. Here U_i stands for ω, ω^+ or a vertex operator. This condition gives the cycle C of integration for the expression (45) as a product of nested cycles $C^{J_{i+1}}_{J_i,I_i}$ used to integrate the factors. The resulting cycle is identified with a singular vector of $V(m_0) \otimes \ldots \otimes V(m_s)$, with $m_i = 2J_i + 1$, $I_0 = J_1$, by [10]. Algebraically, the expression of C in terms of $C^{J_{i+1}}_{J_i,I_i}$ is understood as follows: Identify $C^{J_{i+1}}_{J_i,I_i}$ with an element of

$$\ker|_{(V(n_i)\otimes \ldots V(m_i))_{R_i}} \cong \mathrm{Hom}_{U_q(sl_2)}(V(n_{i+1}), V(n_i) \otimes V(m_i)) \tag{47}$$

where $m_i = 2J_i + 1$ and $R_i = I_i + J_i - J_{i+1}$. Then

$$\begin{aligned} C &= \left(C^{J_2}_{J_1,I_1} \otimes 1 \otimes \ldots \otimes 1\right) \ldots \left(C^{J_s}_{J_{s-1},I_{s-1}} \otimes 1\right) C^{J_{s+1}}_{J_s,I_s} \\ &\in \mathrm{Hom}_{U_q(sl_2)}(V(n_{s+1}), V(m_0) \otimes V(m_1) \otimes \ldots \otimes V(m_s)) \\ &\cong \ker E|_{(\bigotimes V(m_i))_{\sum R_i}}. \end{aligned} \tag{48}$$

These quantum group intertwiners and path spaces have been investigated in [11].

Acknowledgment

It is a pleasure to thank M. Crivelli, K. Gawedzki and T. Kerler for discussions.

References

[1] A. A. Belavin, A. M. Polyakov, A. B. Zamolodchikov, Nucl. Phys. B241 (1984) 333-380

[2] V. G. Knizchnik, A. B. Zamolodchikov, Nucl. Phys. B247 (1984) 83-103

[3] E. Witten, Comm. Math. Phys. 92, 455-472 (1984)

[4] A. Tsuchiya, Y. Kanie, Adv. Stud. Math. 16 (1988) 297-372

[5] G. Felder, R. Silvotti, Phys. Lett. B Vol. 231 No. 4 (1989), 411-416

[6] M. Wakimoto, Comm. Math. Phys. 104, 605-609 (1986)

[7] D. Bernard, G. Felder, Comm. Math. Phys. 127, 145-168 (1990)

[8] B. L. Feigin, E. V. Frenkel, preprint (1989)

[9] A. B. Zamolodchikov, V. A. Fateev, Sov. J. Nucl. Phys. 43 (4), 657-664 (1986)

[10] G. Felder, C. Wieczerkowski, Comm. Math. Phys. 138, 583-605 (1991)

[11] J. Fröhlich, T. Kerler, On the Role of Quantum Groups in Low Dimensional Quantum Field Theory, Springer Lecture Notes draft (1991)

[12] T. Kerler, private communication

[13] P. Bouwknegt, J. McCarthy, K. Pilch, 1990 Yukawa International Seminar on Common Trends in Mathematics and Quantum Field Theory

[14] P. Bouwknegt, J. McCarthy, K. Pilch, preprint CERN-TH 6196/91

[15] G. Felder, C. Wieczerkowski to appear

[16] V. A. Fateev, S. L. Lukyanov, preprint PAR-LPTHE 91-16

[17] G. Felder, R. Silvotti, Conformal blocks of minimal models on Riemann surfaces, preprint, November 1990

Index

6j-symbols, 49f, 164, 343

Abriskov vortex, 210
Accessory parameters, 386, 391
Aharonov-Bohm effect, 196, 209ff, 224, 228f
Aharonov-Casher effect, 196, 211f, 227ff
Airy functions, 118f, 127
Algebra,
  affine ∼, 375ff
  Borchers ∼, 300
  Casimir ∼, 493, 497f
  Clifford ∼, 61, 73, 435
  current ∼, 18, 126, 195, 234, 409
    chiral ∼∼, 197, 234
    quantum ∼∼, 22ff, 25
  de Rham ∼, 61f
  differential ∼, 74
  exchange ∼, 167
  Gelfand-Dikii ∼, 434, 442
  graded ∼, 58, 62
  group ∼, 333ff
  Hecke ∼, 436
  Hopf ∼, 7, 12, 330
    quasi ∼∼, 330
    weak quasi ∼∼, 330, 337, 451
  Kac-Moody ∼, 13, 22, 25, 161, 168f, 171ff, 197, 375, 433, 435, 513
  loop ∼, 11, 19, 367, 373ff
  observable ∼, 178ff, 187, 344, 348
  operator ∼, 146ff, 178ff, 290
  quantum group ∼, (*see* Quantum groups)
    truncated ∼∼, 341
  quantum universal enveloping ∼, 13
Algebra (continued)
  Sklyanin ∼, 366
  vertex-operator ∼, 433, 435, 494
  Virasoro ∼, 97, 125ff, 138, 152, 344, 434, 494, 514
  von-Neumann ∼, 180ff
  $\mathcal{W}$-∼, 125ff, 154, 433, 435
  $W(sl(n))$, 493ff
Anderson-Higgs mechanism, 196, 214
Anomaly, 229
  $U(1)$, 140, 233f
Antipode, 7, 331
Anyon, 209f, 212, 225, 227, 22
Artin relations, 340
Asymptotic expansion, 105, 118
Asymptotic particles, 2, 28
Asymptotic states, 26, 31
Atiyah's Axioms, 129f

Background independence, 408
Baker-Akhiezer function, 98, 112, 117, 119, 122, 309ff
Bethe Ansatz, 356ff, 364
Bimodule, 77ff, 91
Biproduct, 456
Black hole, 247, 259ff, 275ff, 283f
Boundary terms, 219, 228f, 231ff
Braid-commutativity, 331
Braid statistics, 195f, 332, 451, 470
  abelian ∼, 209, 227
  non-abelian ∼, 212, 227f
Braiding relation, 8, 26, 50, 332
BRS cohomology, 483
BRST, 134, 138, 148, 150f, 413ff, 513, 520
  ∼ cohomology, 413f, 417f, 433

Cartan matrix, 13
Category, 449ff
braided ~, 449f
rigid balanced ~, 49
quantum ~, 455, 458, 460
Tannakian ~, 466f
non-~~, 449, 474
Charge, 29
background ~, 140
BRST ~, 150
central ~, 138, 140, 146, 161, 254, 437, 483
~ conservation, 101, 106, 146
conseved ~, 493, 501
non-local ~~, 27
topological ~, 9
Chern-Simons term, 229f
Chevalley-Serre relations, 13
Chiral spin liquid, 212, 225
Class
Chern ~, 101f, 104–109, 121
fractional ~, 107
Mumford ~, 107, 121
Classical isotropic top, 162ff
Classical limit, 111, 171, 370
Cluster properties, 218
Co-free action, 439
Coassociative, 329, 337
quasi ~, 330
Cocommutative, 329, 451
graded ~, 351
Cocycle equation, 301f
Coloring, 47
Comodule, 350
Conformal block, 252, 450, 513, 520
Connection,
Grasmannian ~, 66
Lax ~, 14f, 19
projective ~, 397, 399
spin ~, 199
$U(1)$-Knizhnik-Zamolodchikov ~, 225, 227
Yang-Mills ~, 33
Contact term, 152f
Coproduct, 7, 29, 330, 335
Correlation functions, 5f, 101, 105, 107, 128f, 132f, 137, 148, 151, 423ff, 429
selection rules for ~, 140
Correlator, 137, 142, 152
Coset theory, 249ff, 259ff, 268ff
Coulomb gas, 427
Counit, 7
Covering, 40ff
~ transformation, 40
Critical level, 442
Critical phenomena, 197
Current,
~ algebra, 18
chiral ~, 197
conserved ~, 6, 16, 139, 215
non-local ~~, 12, 16, 20f, 22, 25f
symmetric ~, 415

Deformation, 164, 505
Deformed KZ equation, 31
Descent equation, 135
Differentials, 296
Dilation equation, 102
Dimensional reduction, 173
Dirac operator, 53ff, 65, 72f
Diximier trace, 53, 55f
Doplicher-Haag-Roberts criterion, 178
Dressing transformation, 14ff
Dual Coxeter number, 146
Duality,
Alexander ~, 43
Haag ~, 183, 185
Poincaré ~, 77
~ problem, 466ff
quantum group ~, 471
Tannaka-Kreĭn ~, 449

Effective action, 215, 217
Einstein-de Haas (-Barnett) effect, 196, 213f
Electromagnetism, 286ff
Entire cyclic Cohomology, 297, 299f, 301
Equation,
cocycle ~, 301f
deformed KZ ~, 31
descent ~, 135
dilation ~, 102
Kadomtsev-Petviashvili (KP) ~, 310, 313
Knizhnik-Zamolodchikov ~, 452
Korteweg-de Vries (KdV) ~, 96, 110
Lax ~s, 318ff
linear respose ~, 221

Gravity, 95, 130, 134, 177, 248, 413
Group,
braid, 189, 227, 340
deformed ~, 169
fundamental ~, 40, 51
gauge ~ (*see* Gauge group)
Lie ~,
ADE-classification, 145
loop ~, 19
deformed ~~, 173
quantum ~, 166f, 170, 227
quasi ~~, 329ff
renormalization ~, 144
Schottky ~, 398f
symmetry ~, 190

Haag-Kastler net, 181
Hamiltonian methods, 159
Hawking radiation, 20
Higgs
boson, 78, 80
field, 85, 89
potential, 70
Highest weight module, 417f
Horizon shift, 281f

Incompressible system, 196ff
Integrability
classical ~, 408
Liouville ~, 409
quantum ~, 410
Integrable hierarchies, 95–158
Gelfand-Dikii, 110
Korteweg-de Vries (KdV) ~, 95, 98, 101ff, 105, 107, 109, 118, 121, 147
dispersionless ~~, 111
generalized ~~, 110
tau-function of ~~, 96, 107, 112
Kadomtsev-Petviashvili (KP) ~, 113, 115, 122, 124, 310, 312
Integrable models, 1, 309, 470, 493ff
Intersection
~ number, 101, 153
~ theory, 95, 99–109
Intertwiner, 173, 186, 191, 515
Invariance,
gauge ~, 60, 195f, 208ff, 214, 216, 218, 230
modular ~, 145
Invariance (continued)
topological, 128, 133, 137
Weyl ~, 162
yangian ~, 11
Invariant, 297 (*see* also Polynomial)
combinatorial ~, 47
knot ~, 37–52,
link ~, 46
manifold ~, 37–52

JLO-Cocycle, 302f
Jones index, 188

K-cycle, 56, 77
K-theory, 48, 53, 77
Kac table, 417f
Kadomtsev-Petviashvili (KP) equation, 310, 313
Kähler potential, 144
Knizhnik-Zamolodchikov equation, 452
Knot, 37–52
Korteweg-de Vries (KdV) equation, 96, 110

Langlands-Drinfeld correspondence, 433ff, 443
Larmor's theorem, 208
Laughlin vortices, 197, 224f, 230
Lax,
~ connexion, 14f, 19
~ equations, 318ff
~ formulation, 109
~ operator, 118, 124, 148
~ pair, 115, 123
~ representation, 14
Linear respose equation, 221
Liouville
~ equation, 387
~ theory, 247ff, 383ff, 397, 403f, 426, 483
Local net, 181
Locality, 23, 56, 178, 332
Loop, 516
~ algebra, 11, 19, 367, 373ff
~ equations, 323ff
~ group, 19
deformed ~~, 173
~ operator, 97, 104f
Wilson ~, 33
Loss of information, 276

Equation (continued)
  Liouville ~, 387
  loop ~, 323ff
  ~ of motion, 14, 18
  Pauli ~, 202
  puncture ~, 102, 107
  string ~, 118f
  Whitham ~s, 315, 316
  Yang-Baxter ~, 11, 13, 17, 165, 191, 338, 363f, 450, 506
  Zakharov-Shabat ~s, 109
Exchange relations, 28
Expectation values, 300

Factorization, 129f, 132, 152ff
Famous factorial growth, 133
Feign-Fuchs module, 414f
Fermion,
  chiral ~, 62, 65, 78, 113
  spinless ~, 231
Field
  ~ multiplet,9ff, 29f
  ~ operator, 9ff
  primary ~, 107, 127, 485, 513, 517, 520
Field theory, (*see also* Model, Quantum field theory)
  classical ~, 159
  collective string ~, 407ff
  conformal ~, 2, 32, 47, 98, 159, 167, 383, 423, 433, 450
    meromorphic ~~, 494
    minimal ~~, 97
    non-rational ~~, 247
    topological ~~, 137–148, 309
  open string ~, 103f
  topological ~, 95f, 107, 121, 127f, 131, 136
Finite space, 53, 71, 81ff
Flux quantization, 210
Fock space, 113f, 116, 120, 200, 414f, 485
  ~ representation, 513f
Four-point function, 427
Free field realization, 493
Free field resolution, 437
Fuchsian, 384ff, 393
Function,
  Airy ~s, 118f, 127
  Baker-Akhiezer ~, 98, 112, 117, 119,
Function (continued)
  Baker-Akhiezer ~, (continued) 122, 309ff
  correlation ~s, 5f, 101, 105, 107, 128f, 132f, 137, 148, 151, 423ff, 429
    selection rules for ~~, 140
  partition ~s, 5, 128f, 132f, 136, 145, 253f, 260ff, 266f, 361f, 440
    string ~~, 95, 101
  tau-~, 96, 107, 112, 116, 125
Functional integral, 268
  Polyakov's ~, 383, 390f
Fusion rules, 346, 427, 520

Gap labelling theorems, 197
Gauge,
  ~ group, 64, 451
    $End_{\mathcal{A}}(\mathcal{E})$, 75
    $SU(2)$, 226, 230
    $U(1)_{em} \times SU(2)_{spin}$, 196, 209ff, 216, 218
    $U(1) \times SU(2) \times SU(3)$, 78ff
    $U(1) \times U(1)$, 69
    $U(1) \times U(2)$, 71
    $U(N)$, 65
  ~ field, 89, 216, 226
  ~ invariance, 60, 196, 208ff, 214, 216, 218, 230
  ~ potential, 54, 67ff, 215
  ~ theory, 54, 69, 83, 87, 134, 150, 177, 295
    noncommutative ~~, 76
  tidal ~ field, 196
  ~ transformation, 15, 60, 64, 168, 173, 177, 199f, 250
Gauged spinors, 64
Ghost, 150, 414
  ~ number, 483, 488
GNS-construction, 180
Goldstone
  boson, 221
  theorem, 197, 221
Graded commutator, 297
Graded tensor product, 74
Graded spacelike commutativity, 184
Grading, 67f, 88, 186, 297
Grassmannian, 66, 113f, 116
  Sato's ~, 126
Grassmann variables, 215

Marginal term, 217f
Matrix,
  Cartan ~, 13
  fusion ~, 343
  ~ integrals, 95, 105, 118, 121–125
  L-~, 362
  mixing ~, 85
  ~ models, 95f, 98, 101
  ~ potential, 98
  R-~, 164, 167, 227, 333, 366, 505f
  S-~, 1, 28, 275ff, 280, 281, 284
  transfer ~, 5, 361ff
  transition ~, 450
  transport ~, 168
'matter theory', 127, 130
Meissner effect, 210
Metric, 53ff
  dynamical ~, 149
  fluctuating ~, 134
Mini superspace, 254, 256, 266
Miura field, 499
Model, (*see also* field theory)
  black hole sigma ~, 259ff
  BPZ minimal ~, 383, 418
  Calogero, 408f
  Chern-Simons ~, 173, 195, 197, 251
  chiral Ising ~, 343ff
  Glashow-Weinberg-Salam (GWS) ~, 76
  Gross-Neveu ~, 28
  3D Ising ~, 33
  lattice ~, 4, 355
  matrix ~, 95f, 98, 101, 407
  minimal conformal ~, 32, 423, 483
  non-linear sigma-~, 144
  $(p, 1)$, 98, 119, 121, 126
  Penner's ~, 125
  Sine-Gordon ~, 3
  six vertex ~, 361
  standard ~, 54, 78–90, 289ff
  superconformal, 137, 140, 143, 145
  topological ~, 98
    ~ Landau-Ginzburg ~, 143ff, 148, 322
    ~ minimal ~, 118, 121, 145, 154, 322
    ~ sigma-~, 140, 149
  vertex ~, 4
  Wess-Zumino-Novikov-Witten (WZNW) ~, 159ff, 227, 348ff
Model (continued)
  Wess-Zumino-Witten WZW ~, 25, 403f, 452, 474
    non-compact ~~, 247ff
Moduli space, 95, 99–109, 121ff, 133, 309, 383ff
Monodromy, 162, 170, 386
Mysterious, 98, 167, 228f

Non-locality, 12, 16, 20f, 22, 25f, 135ff
Noncommutative geometry, 53–93, 169, 173, 295–308
Null vectors, 486f

Object, 455
  conjugate, 459
Observables, 149, 177ff
  non-local ~, 135
Odd denominator rule, 225, 241
Operator,
  antiscreening ~, 462
  Dirac ~, 53f
  disorder ~, 290
  field ~, 9ff
  grading ~, 56
  Lax ~, 118, 124, 148
  local ~, 128, 423ff
  loop ~, 97, 104f
  non-local, 136
  screening, 426, 430, 515
  vertex ~, 235, 407, 433, 452f, 475, 483
Operator product expansion (OPE), 6, 22, 24f, 339, 433, 435, 515
Osterwalder-Schrader construction, 306

Parafermion, 7, 33, 263
Particle, 268f
Partition functions, 5, 128f, 132f, 136, 145, 253f, 260ff, 266f, 361f, 440
  string ~, 95, 101
Path integral, 103f, 129
Pauli equation, 202
Pentagon equation, 456
Perturbation, 136ff, 146
  ~ theory, 315f
Plack mass, 277f
Poisson bracket, 17, 111, 160, 167f, 305, 370, 409

Polynomial,
Alexander ~, 37, 43
Conway ~, 37, 39
HOMFLY ~, 189
Jones ~, 44, 46, 189
Positive Energy representation, 179, 182ff, 348
Propagator, 103ff
Puncture, 385ff
~ equation, 102, 107

Q-cohomology, 133
q-deformation, 505
Quantum dimension, 342
Quantum field theory, 128, 133, 177, 295 (*see also* Field theory)
algebraic ~, 178ff, 348, 450
2D massive ~, 1–35
supersymmetric ~, 53
Quantum fluids, 195–246
Quantum group, 166f, 170, 297, 329, 355, 440, 505ff
$SU(2)_q$, 28, 172
$U_q(sl_2)$, 227, 332
~ symmetries, 1, 191, 329
Quantum Hair, 278f
Quantum Hall effect, 195–208, 224
Quantum invariance, 7
Quantum inverse scattering method, 169
Quantum space, 303ff
Quasi commutativity, 331
Quasi particle excitation, 224
Quasiinverse, 337
QUEA, 13

Reassociator, 331
Reidemeister movers, 39f, 44
regular ~, 45
Relevant term, 217f
Ribbon graph, 103
Riemann Hilbert problem, 15, 19, 450
Riemann-Roch, 99, 106

Scaling dimension, 23
Scaling limit, 215, 217f
Scattering theory, 50
Sector, 451
composition of ~s, 187
matter ~, 417
Neveu-Schwarz ~, 419f
Sector (continued)
Ramond ~, 419
singlet ~, 407
soliton ~, 178
superselection ~, 178, 191, 344, 348
vacuum ~, 188
Seifert surface, 41
Skein rules, 44
Small phase space, 121, 127
Soliton, 14, 309
Spectrum condition, 184
Spin manifold, 203
Spinon, 226ff, 230
Spinor,
~ bundle, 199
gauged ~, 64
Stability, 248, 269
State, 131
asymptotic ~, 28, 31
~of a black hole, 279, 288
~ of a $C*$-Algebra, 55, 179f
DDK ~, 485
density matrix ~, 180
incomming ~, 132, 281f
intermediate ~, 429
Kaufman's ~ algorithm, 37ff
~ of a knot, 44
LZ ~, 485, 487
normal ~, 181
outgoing ~, 132, 281f
physical ~, 413ff, 485, 488
quasiprimary ~, 494
~ space, 177, 182ff, 256f, 264, 270
tachyon ~, 411
vector ~, 180
Statistical dimension, 190, 459
Statistical phase, 190, 209
Statistics, 178, 188, 197, 200, 332
braid ~, 191, 195f, 332, 451, 470
abelian ~~, 209, 227
non-abelian ~~, 212, 227f
fractional ~, 212
~ operator, 187
para-~, 451
String equation, 118
String free energy, 118
String theory, 95, 287, 407ff, 483ff
topological ~, 97, 125, 127, 148–154
Sugawara field, 442
Super extension, 419f

Super-KMS-condition, 299f
Superconductor, 196, 210, 214
Supercurrent, 139, 214
Superfluid, 196, 209ff, 215–243
Supergravity, 413, 419
Superpotential, 144ff
Supersymmetry, 35, 134f, 138, 351
Symmetry, 239f, 278, 339, 451
  ∼ algebra, 2
  ∼ breaking, 67, 70, 85
  external ∼, 177
  ∼ group, 190
  internal ∼, 177, 184, 217, 329, 334
  Lie-Algebra ∼, 6
  Lorentz ∼, 505, 508
    deformed ∼∼, 505ff, 508ff
  non-abelian, 1
  permutational ∼, 142
  quantum ∼, 1, 159, 191f, 195f, 329
  slightly broken ∼, 177
  $sl_q(2)$, 3

Tau-function, 96, 107, 112, 116, 125
The unique theory, 278
Teichmüller deformation, 393f
Tensor product of representations, 330, 335f
Thermal equilibrium, 204
Thermodynamics, 276f
Three-point function, 424ff
Toeplitz quantization map, 305ff
Topological charge, 9
Topological gravity, 95f, 107, 148ff
Topological representation, 513
Transformation,
  BRST ∼, 148
  covering ∼, 40
  dressing ∼, 14ff
Transformation (continued)
  gauge ∼, 15, 20, 60, 168, 173, 177, 199f
  super-Möbius ∼, 141
Twist, 254
Twisting, 137, 140f

Ultralocal, 168f
Uniformization, 383, 384f, 393, 398
Unimodulatity, 91f

Vacuum representation, 184, 190, 344, 433, 435
Vandermonde determinant, 117, 122f
Virasoro constraints, 309
Vorticity, 210

Ward identities, 2, 30, 196, 208, 216f, 391ff
Whitham
  ∼ equations, 315, 316
  ∼ theory, 309
Wilson
  ∼ line, 251
  ∼ loop, 33

XXZ chain, 355

Yang-Baxter equation, 11, 13, 17, 165, 191, 338, 363f, 450, 506
Yang-Mills functional, 53–93
Yangian, 11ff, 22, 28, 355f, 367
  $SL(n)$, 367

Zakharov-Shabat equations, 109
Zeeman terms, 201f, 207
Zero-curvature relation, 109